THE TURTLES OF MEXICO

The Turtles of Mexico

LAND AND FRESHWATER FORMS

JOHN M. LEGLER and
RICHARD C. VOGT

UNIVERSITY OF CALIFORNIA PRESS
Berkeley Los Angeles London

University of California Press, one of the most distinguished university presses in the United States, enriches lives around the world by advancing scholarship in the humanities, social sciences, and natural sciences. Its activities are supported by the UC Press Foundation and by philanthropic contributions from individuals and institutions. For more information, visit www.ucpress.edu.

University of California Press
Berkeley and Los Angeles, California

University of California Press, Ltd.
London, England

Library of Congress Cataloging-in-Publication Data

Legler, John M.
The turtles of Mexico : land and freshwater forms / John M. Legler and Richard C. Vogt.
pages cm
Includes bibliographical references and index.
ISBN 978-0-520-26860-9 (cloth : alk. paper)
1. Turtles—Mexico—Classification. 2. Turtles—Mexico—Identification. I. Vogt, Richard Carl. II. Title.
QL666.C5L42 2013
597.920972—dc23 2012043367

Manufactured in China
19 18 17 16 15 14 13
10 9 8 7 6 5 4 3 2 1

The paper used in this publication meets the minimum requirements of ANSI/NISO Z39.48-1992 (R 2002) (*Permanence of Paper*). ♾

Cover images: Top, *Rhinoclemmys pulcherrima,* Mazunte, Oaxaca.
Bottom left, head of adult *Claudius angustatus.*
Bottom center, copulating pair of *Rhinoclemmys rubida perixantha,* near Chamela, central coastal Jalisco.
Bottom right, adult female *Rhinoclemmys r. rubida,* near Mazunte, Oaxaca.
All photos by Richard C. Vogt except bottom center, photo by E. Goode and M. Rodrigues, used by permission.

We dedicate this book to Hobart Muir Smith (b. 1912), who has achieved a greatness in the field of herpetology generally and was a 20th-century pioneer in studies of Mexican amphibians and reptiles. With Rozella B. Smith (1911–1987), he produced the first book relating only to Mexican turtles.

Hobart Muir Smith (1912–); photo ca. 1983. (Courtesy of Larry L. Miller, Kansas Heritage Photography, Wakarusa, Kansas.)

CONTENTS

ACKNOWLEDGMENTS

Because this book spans the bulk of our careers in chelonology, our acknowledgments are numerous and heartfelt. Although this book is based chiefly on our own work, we have received significant help from others. The following remarks are lamentably cursory.

The inception of Legler's research on turtles began at a small lake in southern Minnesota and was encouraged by his mentors at Gustavus Adolphus college—Professors C. Hamrum and A. Glass. This interest was expanded at the University of Kansas under the mentorship of Henry S. Fitch (1909–2009), Edward H. Taylor (1889–1978), and E. R. Hall (1902–1986). This interest became centered on Mesoamerican chelonians during an internship at the U.S. National Museum (1958), where there was access to exotic taxa (e.g., *Dermatemys*, *Staurotypus*, and *Claudius*) that were then known only to the few scientists who had traveled to Mexico. Doris M. Cochran (1891–1968) enhanced the internship and provided access to the research notes of Dr. L. H. Stejneger (1851–1943). The first trip to Mexico (August 1959) was sparked by the discovery of aquatic algae on the "rare" *Terrapene coahuila*. Legler and W. L. Minckley were first guided to the various aquatic habitats in the basin of Cuatro Ciénegas, Coahuila, by Daniel Rodriquez-Villareal (a local school teacher) in August 1958, resulting in the discovery of two new turtles and several new fishes.

William G. Reeder and E. Elizabeth Pillaert (University of Wisconsin Zoological Museum) arranged financing for Vogt's first expeditions to Mexico (Sonora in 1968 and Veracruz in 1970). Dr. C. J. McCoy sponsored a postdoctoral fellowship for Vogt at the Carnegie Museum of Natural History (1968), which constituted the inception of his interest in Mexican turtles. Without his support and mentorship, this book would be less rich in the parts dealing with neotropical species. McCoy introduced Vogt to Gustavo Casas Andreu who, in turn, introduced him to Oscar Flores-Villela, who became a graduate student and field companion. Flores introduced him to the Los Tuxtlas Tropical Research Station (1980), where Vogt and lived and worked for the next 18 years. Facilities at Los Tuxtlas were expanded and improved with the assistance of Gonzalo Perez Higareda, Miguel Martinez, and Rodolfo Dirzo. Don Miguel Alvarez del Toro of ZOOMAT (a zoo in Tuxtla Gutierrez, Chiapas) encouraged Vogt's early work and allowed access to his live and preserved collection of turtles and his philosophical insights.

John B. Iverson generously shared unpublished data and provided insights and remarks on several species of *Kinosternon*. Others who were generous with unpublished manuscripts, theses, observations, and illustrations were P. Medica, David Germano, R. M. Winokur, and James F. Berry. George M. Ferguson provided photos and detailed notes on habitat for *Terrapene nelsoni klauberi*. Ms. Shaun Copus facilitated Legler's observations of bolson tortoises at the Research Ranch in Elgin, Arizona.

George R. Zug always tolerated lengthy telephone calls concerning various frustrations with the progress of the book and reminding us of things we either did not know or had forgotten.

Prominent among those who shared knowledge was James R. Buskirk, who consistently provided photos, field notes, and firsthand information from his many field trips to most of Mexico. He also was of great help in accessing the substantial literature in European journals and unpublished observations from turtle enthusiasts. Buskirk also introduced us to Martha Harfush (Centro Mexicana de Tortugas, Mazunte, Oaxaca), who permitted use of unpublished morphometric and reproductive data, hundreds of photographs, and valuable observations on diet and reproduction from a captive colony of chelonians in Pacific coastal Oaxaca—especially *Rhinoclemmys* sp.

This book straddles the advent of personal computers and digital photography, and many of Legler's color photos are on film exposed more than half century ago. Permission to duplicate and use photographs from private collections was granted by Patty Scanlan (Gladys Porter Zoo), Phil Medica, Francis Rose, Cecil Schwalbe, C. W. Painter, Brian Horne, R. M. Winokur, James F. Berry, and P. P. van Dijk. Gerald Smith gave permission to modify and use the drainage map of Mexico made by Robert R. Miller.

The original line drawings were made by Elizabeth Lane, Kerry Matz, Lucy Remple, and Connie Spitz. Eric C. Felt made the dental castings used for jaw armature (Figure 2.7).

We thank W. L. Minckley (1935–2001) for sharing his vast knowledge of Mexican fishes and the history of drainage systems in Mexico, especially the basin of Cuatro Ciénegas. Suzanne McGaugh was generous with data, photographs, and advice on Cuatro Ciénegas, as were R. M. Winokur, J. Howeth, W. S. Brown, and D. A. Hendrickson.

All of the turtles Vogt collected in Mexico from 1968 to 1977 are at the University of Wisconsin Zoology Museum. Those from 1978 to 1980 and some from 1980 to 1992 are at Carnegie Museum of Natural History, Pittsburgh. Most of the turtles collected while at Los Tuxtlas from 1980 to 2000 were deposited in the Los Tuxtlas collection. However, that collection was disbanded (after Vogt's departure) and now resides at the Chelonian Research Institute in Oveido, Florida. Most of the turtles collected by Legler are in the "Legler turtle collection" at the Museum of Natural History, University of Utah (UU); a few (1953–1959) are housed in the Museum of Natural history, University of Kansas.

When we began our work in Mexico, most of the scientists interested in Mexican turtles were academics who did their fieldwork during academic recesses (chiefly in the northern temperate summer). Vogt became a resident of Mexico for 20 years (1980–2000) and was able to spend full time on studies of turtles in Mexico. Thereafter he moved his work to Brazil. Legler's fieldwork in Mexico was done mostly during year-long sabbatical leaves and other leaves of absence (chosen by season), with or without salary.

For assistance with fieldwork, Legler thanks Abbot Gaunt, W. Berg, Charles Long, Craig Nelson, E. O. Moll, Elias Alfaro, Frank V. Nabrotsky, James L. Christiansen, James S. Peebles, R. Wimmer, Nowlan K. Dean, Raymond Lee, Robert Bolland, Robert G. Webb, Roger Conant, S. Anderson, S. H. Bartz, R. Shine, W. L. Minckley, and the University of Kansas Museum of Natural History.

The Legler family (Avis J. Legler and children—Austin F., Edward P., Gretchen T., and Allison K.), beginning at ages as young as 5, enjoyed rather than suffered the vicissitudes of many Mesoamerican camps where their father went forth each day to rummage for turtle shells in middens or left with a pack full of traps and came back with lots of odd-looking turtles. Austin Legler learned to tell time in Yucatán the same year he learned to catch turtles. Avis coordinated the vital aspects of family life with the research.

Most of the fieldwork for this study involved living and working in a country where we were aliens. We were always treated with courtesy, generosity, and respect. Specifically, we thank the government agencies, institutions, and landowners for permission to travel, collect, and export specimens.

Vogt directed the research of many Mexican undergraduate and graduate students, some of whom continued their work with him for 20 years. Notable among these were Marco Lazcano Barrios, Claudia Lopez Carnevale, Marlet Dath, Veronica Ernestina Espejel Gonzales, Salvador Guzman Guzman, Nora Lopez Leon, Marco Antonio Lopez Luna, Mardocheo Felix Palma, Gracia Gonzales Porter, Chucho Ramirez Ramos, Mario Ramos, Francisco Soberon, Martha Gomez Soto, Adrian Morales Verdeja, Claudia Zenteno-Ruiz, and Miguel Angel de la Torre.

Gustavo Casas Andreu and Enrique Gonzalez (UNAM faculty) were always available for assistance and advice of all kinds. Oneide Ferreira da Cruz spent 8 years with Vogt in and out of the field, collecting turtles and maintaining the captive colony at Los Tuxtlas. Marcelo Paxtian Sinaca and Santos Mataperro Ventura participated significantly in lab and fieldwork in Veracruz and Chiapas. The following persons aided intermittently in various ways at the field stations in Los Tuxtlas, Veracruz, and Montes Azules, Chiapas: Gustavo Aguirre-León, M. A. Ewert, Carlos Guichard, Camilo Hernandez, Augustine Lara, Eretmechengi Perez, Michael Pappas, Basilio Sanches, and Bob Waide.

Preparation of this book took place chiefly at the University of Utah; on the campus of the Instituto Nacional de Pesquisas da Amazônia (INPA) in Manaus, Brazil; and in our homes. Facilities and secretarial help were provided by the Department of Biology, University of Utah, and included working space and stenographic and other support services of significant magnitude. Outstanding among these was the help of Ms. Karen Zundel with a plethora of tasks and in preparing illustrations using sophisticated graphic software and techniques far beyond our ken. Professor Mark T. Nielsen advised us on graphic technology. Other technical support for computer operations was supplied by Spencer Streeter, Larry Okun, Jon McGowan, and Jude Higgins.

For assistance with building Legler's extensive databases and digitizing field journals and catalogs, we thank Maurine Vaughan, Laura Eberhardt-Morehead, Gloria Cuellar, Cameron S. Denning, Karen Price, and Katrina Heiner-Childs. Special attention was given to Legler's library needs by Barbara Cox, Robin Reed, and April Love.

Numerous people and organizations were responsible for granting collecting permits, facilitating access to reserves, and obtaining specimens and data: Padre Kino (Catholic bishop of Villahermosa), Felipe Brizuela, Marcos Brizuela, Morena La Barriga, Adrian Cerda, Graciela de la Garza, Javier de la Masa, Ejido de Laguna Escondida, Estación de Biología Tropical Los Tuxtlas, Claudia Gonzales, Pedro Gonzales, Ramon Perez Gil, Tony Gamble, Victor Gonzales, the Instituto Nacional Ecología y Recursos Bioticos (INEREB-CHIAPAS), Instituto de Biología, Universidad Nacional Autonoma de Mexico, Maria la Bandida-Ramiro

Berron Lara, Pepe Lugo, Veronica Lopez, Horatio Merchant, Patricia Roman Perullero, Ejido de las Margaritas, S. G. Platt, Pedro Ramirez, Thomas Rainwater, SEDUE, PESCA (Secretario de Desarrollo Urbano y Ecología), and Bob Waide, Facultad de Ciencias, UNAM.

We wish to thank the following persons for permitting the examination or long-term loan of specimens in their care: W. E. Duellman and John Simmons, University of Kansas (KU); C. Holton, American Museum of Natural History (AMNH); J. A. Holman, personal collection (JAH); R. F. Inger and H. Marx, Field Museum of Natural History (FMNH); A. G. Kluge and Greg Schneider, University of Michigan (UMMZ); L G. Lowery, Louisiana State University (LSU); J. A. Peters, D. M. Cochran, and George Zug, U. S. National Museum (USNM); Jose Rosado, Harvard University (MCZ); W. W. Tanner, Brigham Young University (BYU); and the P. C. H. Pritchard collection (PCHP).

The following have granted favors and assistance in ways that are too numerous to categorize: Gary Adest, T. Akrea, Elias Alfaro, Otto Aquino Cruz; E. N. Arnold, Gustavo Aguire, Clyde Barbour, John Bickham, Karen Bjourndahl, C. M. Bogert, Dennis Bramble, Bryce C. Brown, Michael R. Bruno, Fred Caporaso, Joe Collins, Roger Conant, Justin Congdon, R. H. Dean, W. E. Duellman, M. A. Ewert, George M. Ferguson, Aaron Flesch, M. J. Foquette, M. Forstner, Wayne Frair, Michael Goode, A. G. C. Grandison, Harold Heringhi, D. Holland, J. A. Holman, Bill Hoy, R. F. Inger, Dale Jackson, Clyde Jones, R. T. Kazmaier, E. King, A. G. Kluge, Trip Lamb, J. Lee, Ernest Liner, C. H. Lowe, J. Lovich, R. Lovich, H. Marx, C. McArthy, Phil Medica, Peter Meylan, R. Minckley, Patrick Minx, D. Moll, E. O. Moll, Judy Moll, David Morafka, Carrie Morjan, Michael Pappas, David and Esther Pendergast, R. Peterson, John Polisar, Gregory Pregill, P. Pritchard, David Propst, A. Rentfro, J. Roberts, Francis Rose, Karl P. Schmidt, N. J. Scott, M. Seidel, C. H. Shaw, C. Sheil, Rick Shine, G. R. Smith, M. L. Smith, A. Stimson, J. N. Stuart, Father James Thompson (Newman Center, SLC), W. W. Tanner, Henry van der Schalle, T. Van Devender, Eric Wallace, J. Ward, R. G. Webb, E. E. Williams, Gordon R. Willey, Dawn Wilson, R. Wimmer, R. M. Winokur, and John W. Wright.

Both authors personally supported various periods of the fieldwork. Legler received grants from the American Philosophical Society (Johnson Fund), three grants from the National Science Foundation ([NSF]; G 17659, GB 2068, and GB 16249), plus numerous smaller grants from the University of Utah Research Committee, the Biomedical Research Committee, the Park Foundation, and the Department of Biology.

Vogt received support from the following agencies: the Oneil Fund of Carnegie Museum, expedition to Cozumel; Consejo Nacional de Biodiversidade (CONABIO) for studies of populations and distribution of turtles in Chiapas, Veracruz, and Tabasco; the Secretaría de Medio Ambiente Recursos Naturales y Pesca (SEMARNAP) for permissions to collect and for logistic support; the Consejo Nacional de Ciencia y Tecnología (CONACYT) for funding various turtle ecology projects, including temperature-dependent sex determination in neotropical turtles; and the Nacional de Ciencia y Tecnología Sector Industrial del Golfo (CONACYT-SIGOLFO) for funding of population studies of neotropical turtles in Mexico. Studies of *Dermatemys mawi* at Selva Lacandona (Montes Azules Biosphere Reserve) in Chiapas were funded by Conservation International and elsewhere by the Wildlife Conservation Society. The Turtle Survival Alliance provided funding for other studies of *Dermatemys*.

The following institutions, persons, and journals are thanked for permission to use published illustrations: *Journal of Morphology*; University of Kansas Science Bulletin; University of Kansas Publications, Museum of Natural History; Los Angeles County Natural History Museum, Contributions to Science; Bulletin Florida State Museum of Biology and Science; Catalogue of American Amphibians and Reptiles; American Museum of Natural History Bulletin and Novitates; J. F. Berry, PhD Thesis, 1978; J. W. Gibbons (Ed.), *Life History and Ecology of the Slider Turtle*.

Finally, our relationships with the University of California Press were enhanced by Chuck Crumly's knowledge of chelonian biology and the cheerful help of Kate Hoffman, Lynn Meinhardt, and Deepti Agarwal with the final stages of preparation.

TURTLES

A PARADIGM OF VARIABILITY,
VENERABILITY, AND VULNERABILITY

Introduction

Objectives

This book concerns the diverse chelonian fauna of Mexico: 15 families, 14 genera, 38 species, and 66 terminal taxa of freshwater and terrestrial chelonians occurring within the political boundaries of Mexico (Los Estados Unidos de Mexico). These represent all of the families of nonmarine cryptodires that occur in the Western Hemisphere.

The authors of this book (hereinafter JML and RCV) have a cumulative experience of nearly a century with chelonian studies in general and with Mexican turtles in particular.

Smith and Smith (1979) gave complete and accurate synonymies for all Mexican testudinates and we make no attempt to do so. The current book makes no significant taxonomic decisions; it attempts to present an account of the species occurring in Mexico with easily used keys and distribution maps, and to cover what is known of their natural history.

Our use of recent nomenclature at the genus and species level, where there is a choice, is conservative. We have tried to separate the identification of specimens from the diagnosis of taxa. Keys are based chiefly on external characters and are made to be used with whole specimens. They do not, as Ernest Williams once put it, require "X-ray vision." Diagnoses may, however, rely heavily on morphological states, often osteology, that cannot easily be perceived outside of a good museum or without dissection.

We hope this book will serve as a one-stop guide, a *vade mecum*, for those who will study and learn new things about Mexican turtles and turtles in general. We hope the book will also foster communication among persons who study turtles as a hobby and those who consider it a profession.

Most of the available information on Mexican turtles concerns taxonomy and geographic distribution, and it is at this point that most of the knowledge of Mexican chelonians rests. We seek to remedy this situation by organizing what is known about their natural histories and augmenting this information with inferences drawn from the same species or closely related species in countries other than Mexico (usually the United States).

We seek particularly to stimulate studies of turtles in Mexico by Mexican students of cheloniology. Our hope is that this book will encourage many substantial studies of natural history such as the few that have been done at this point (*Terrapene coahuila* by Brown, 1974; *Gopherus flavomarginatus* by Morafka and colleagues, 1980–1997; and studies of various neotropical species by Vogt and students, 1980–2011 and several unpublished theses in English and Spanish such as Dean [1980, on *Staurotypus salvini*] and Zenteno-Ruiz 1993 and 1999). Above all, we wish to chronicle what is currently known about Mexican turtles before their extinction, a possibility that is being hastened by human works.

Although we have consulted and cited an extensive body of literature, much of our information is gleaned from our own, previously unpublished sources—personal observations, field notes, and databases.

The History of Turtle Biology in Mexico

The progression of knowledge of the turtle fauna of Mexico has followed a familiar path, beginning with basic exploration during which a few taxa were named from an occasional specimen reaching one of the large overseas museums. The type specimens of *Dermatemys mawi* and *Trachemys scripta ornata* were obtained by British naval officers. Descriptions were made by scientists who knew little about the country of origin. Eventually, there were zoological explorations and floral and faunal surveys with paid collectors who sent their material to overseas institutions. Even then the collector or purveyor and the describing scientist were never the same. Localities often were vague, and natural history information was nonexistent or ignored in taxonomic descriptions, a circumstance that persisted into the early 1950s.

Turtles were typically no more than addenda to larger general collections, probably because they were difficult to obtain, bulky, and difficult to preserve. Special techniques for their study, capture, and preservation had not been developed. New taxa were commonly described from one or two specimens, the provenance of which was unknown or based on dubious information. Series of specimens were almost never available, sought, or valued, and information from dissections and viscera was not routinely harvested. More than half of the Mexican turtles described from 1827 to 1985 were from type localities outside Mexico and were subsequently shown to occur in Mexico.

From 1758 to 1895, European workers described 20 of the 31 new chelonian taxa and workers in the United States described 11. There was a hiatus in activity from 1895 to 1922 when no new taxa were described, including during World War I. From 1922 to 1997, 29 new taxa were described, all by workers in the United States. This change reflected the ascendancy of science in the New World, which, in turn, was correlated with discovery, colonization, and the establishment of academic institutions—the coming of age of North American science following a period of European hegemony.

Hobart M. Smith (HMS), Rosella B. Smith (RBS), and E. H. Taylor and associates ushered Mexican herpetology into the 20th century by making, studying, and reporting on huge general herpetological collections (somewhere in excess of 30,000 specimens gathered in Mexico from 1932 to 1941, with HMS and RBS concentrating the work from 1938 to 1941). This firmly established the course of Mexican herpetology but not the study of Mexican turtles. There were fewer than 100 turtles in the collection (pers. comm., H. M. Smith). An impressive total of 752 scientific papers resulted from these collections but included no substantive papers on turtles. Knowledge of Mexican turtles achieved a logical plateau in the annotated checklist of Smith and Taylor (1950). Virtually no information was available on natural history at the time.

In retrospect, the foregoing statistics clearly signaled that "general" herpetological collecting in Mexico (or anywhere) would not produce the wherewithal for erudite studies of a chelonian fauna. At the middle of the 20th century, all biological knowledge of chelonians lagged at least 50 years behind that for other groups of vertebrates. This lag was directly correlated with the lack of adequate collections and special techniques for study.

A significant change was affected in the last 40 years of the 20th century. Turtle specialists (mostly young, all zealous) at U.S. universities made expeditions specifically to study and collect turtles, developed their own techniques, and published their own accounts of the results. Among these were James F. Berry, John Iverson, John M. Legler, and Robert G. Webb, whose names were included in the authorship of 14 of the 16 new taxa of Mexican turtles in the years 1959 to 1997 (see Literature Cited). In this same period, Brown (1974) conducted a detailed autecological study of *Terrapene coahuila*, Morafka and McCoy (1988) spearheaded natural history and conservation studies of *Gopherus flavomarginatus* in the Bolsón de Mapimí, and Vogt and his students (e.g., Vogt and Flores-Villela, 1992a,b; and Vogt, 1990) began long-term studies of aquatic turtle populations in neotropical Mexico. This surge of interest and activity in a formerly moribund field has persisted to the present time.

Chelonian Studies by Mexican Biologists

Although Mexican biologists have not described new taxa of turtles, they have contributed significantly to the literature on Mexican chelonians.

Professor Eduardo Caballero y Caballero (Universidad Nacional Autónoma de México, or UNAM) published numerous papers on the intestinal parasites of freshwater turtles from various parts of Mexico over a period of about 30 years, one of the earliest (Caballero y Caballero and Sokoloff, 1935) being on *Dermatemys mawi*.

Don Miguel Álvarez del Toro (1917–1996) originally from Colima, studied the natural history of Chiapas beginning in 1942. Although he had no formal education, he was an outstanding naturalist with a predilection for freshwater turtles and held honorary doctorates from the Universities of Chiapas and Chapingo. He developed a modern zoo in Tuxtla Gutiérrez that includes his name (ZOOMAT). His notable contributions were his early observations of *Dermatemys* in the Rio Lacantún (Álvarez del Toro, 1960) and comments on economic use of turtles in Chiapas (Álvarez del Toro, et al., 1979). He published three editions of a book, *Los Reptiles de Chiapas* (1960, 1973, and 1982) that included natural history data on turtles. A fourth greatly expanded edition was prepared (edited by RCV) but never published. The zoo in Tuxtla Gutiérrez (ZOOMAT) maintains breeding populations of native Chiapan turtles, and the keepers there continue to record observations. Don Miguel was also well known for his studies of arachnids, birds, and mammals.

Studies of Mexican freshwater turtles by Gustavo Casas-Andreu (Instituto de Biología, UNAM) culminated in the publication (1967) of a useful work that included illustrations (19 plates), keys, descriptions, geographic distributions with maps, and notes on natural history. Subsequently, his students have undertaken various projects on Mexican turtles. One of his prominent students, Oscar Flores-Villela, wrote an undergraduate thesis that included Mexican freshwater turtles (Flores-Villela, 1980) and then began studies of the ecology of *Claudius angustatus,* under the supervision of RCV, which was extended and completed as a master's thesis by Veronica Gonzáles Espejel (2004)

(Instituto de Ecología, Jalapa). Oscar Flores-Villela wrote a doctoral thesis on biogeography of a regional Mexican herpetofauna and was an author on several publications concerning turtles in southern Mexico based on data collected from 1980 to 1988 (Flores-Villela and Vogt, 1984, 1992a; Vogt and Flores-Villela, 1992b). Flores is currently coordinating an atlas of the herpetofauna of Mexico and is professor and curator of the chief herpetological collection at UNAM—Muséo de Zoologia, Facultad de Ciencias.

A work on the herpetofauna of coastal Jalisco by Garcia Aguayo and Ceballos (1994) contained observations on freshwater turtles.

Dr. Gustavo Aguirre-León and associates (Instituto de Ecología A. C., Jalapa, Veracruz) have continued and extended the studies of *Gopherus flavomarginatus* in the Bolsón de Mapimí (see citations in account of *Gopherus flavomarginatus*). He and his students have also continued the long-term studies of several aquatic turtles that were begun by Vogt and students in 1980. Others continuing these projects are Claudia Zenteno and students (Universidad Juárez Autonoma de Tabasco) and Carlos Guichard (ZOOMAT, Tuxtla Gutiérrez , Chiapas).

According to Flores-Villela, Smith, and Chiszar (2004), a total of about 70,000 specimens of reptiles and amphibians exists in 20 Mexican collections. Only a few of these are turtles. The largest collection of turtles in Mexico, some 1200 specimens, was made and curated by Vogt during his tenure (1980–2000) at the Estación de Biología Tropical Los Tuxtlas, San Andres Tuxtla, Veracruz. After his departure for Brazil, there was little interest in the collection, and it was given to a private foundation in the United States.

Other Books on Turtles

Turtles form a small, unique group and have been a favorite subject for scientists and popular writers. The works of Walbaum (1782) and Schoepff (1792–1801) on turtles constitute the earliest monographs on any group of reptiles. The 19th century brought forth works such as those by Bell (1831) and Sowerby and Lear (1872) in which artfully crafted plates, often hand colored, were self-standing but text, if any, was simply an addendum. Most works on turtles of this period were attempts to survey the turtles of the world and were based on museum collections.

In an intermediate category was Louis Agassiz's classic *Contribution to the Natural History of the United States of America. First Monograph*, which comprised three parts bound in two volumes (1857a and 1857b). Parts II and III were on turtles and consisted of extensive text containing accounts of what was then known of form, function, growth, sexual dimorphism, fossils, and some life history, followed by the most complete account of chelonian embryology and organogenesis of the time. These parts of the book were only slightly clouded by part one in which his creationist views (following his mentor, G. Cuvier) were espoused. Parts II and III are best known for their 34 exquisite supporting plates (by artists A. Sonrel and H. J. Clark), seven of life-sized neonates and eggs, 18 on developmental stages, and two in color of *Trachemys* and *Pseudemys*. The accounts and illustrations included at least nine taxa that were later shown to occur in Mexico. The book is a compendium that stands as a landmark of utility and excellence for 19th-century cheloniology.

Books exclusively devoted to turtles and relying mostly on what appeared in the text were characteristic of the 19th century. In the years before the current work, books devoted to turtles have consisted chiefly of textual information and were illustrated by photographs and line drawings. These works initiated an era characterized by far less elegance of illustration but far more information in the text. Three of these (Pritchard 1967, 1979; Ernst and Barbour, 1989) were on turtles of the world, and four were on turtles of the United States and Canada (Pope, 1939; Carr, 1952; Ernst and Barbour, 1972; Ernst et al., 1994). Four annotated checklists of great utility have appeared: Iverson (1992a) contains maps and cladograms but no drawings; Wermuth and Mertens (1961) lacks maps but has useful line drawings of most taxa; Smith and Taylor (1950) presented keys, brief synonymies, and geographic distributions for the turtles of Mexico; Bickham et al. (2007) produced a summary of taxonomy and nomenclature that we have usually followed in this book.

It was Carr (1952) who set a modern trend in writing both knowledgeably and delightfully about turtles. His book has been an inspiration for the current book. The only prior work devoted solely to the turtles of Mexico (Smith and Smith, 1979) also stands as the most thorough account ever prepared on the literature, taxonomy, and geographic distribution of turtles in a major region of the world.

The Definition of a Turtle

Turtles are members of the Order Chelonia (also termed Testudines or Testudinata). The distinction of turtles as a unique group of the Class Reptilia and the monophyly of the group has never seriously been questioned. The geographic distributions of humans and turtles overlap broadly, and turtles are familiar to humans. There is a word for "turtle" in most languages, and many children can draw a crude (but diagnostic) picture of a turtle before they can spell the word. If turtles were known only from fossils, they would be regarded with substantially more awe than they are as familiar animals.

The most common English names for members of the order are *turtle*, *tortoise*, and *terrapin*. Each name may have a special significance in local areas, but all are vernacular. The terms *turtle*, *testudinate*, and *chelonian* are commonly used in reference to all members of the order. The word *tortoise* is used by most turtle biologists in reference to the completely terrestrial testudinids—but in Australia for any member of the Family Chelidae.

Extant members of the Order Chelonia can be succinctly diagnosed as follows: toothless, oviparous, quadrupedal, pentadactyl reptiles having a unique trochlear system for the common tendon of the jaw adductor muscles and a shell that encloses (lies peripheral to) the limb girdles. The principal and unequivocal characters of this suite are the shell (incorporating the rib cage) and changes in form and function dictated by a rigid thorax (Figures 2.1 and 2.2).

Origin and Phylogeny of Turtles

Until recently, the oldest known turtle was *Proganochelys quenstedti* from the Upper Triassic of Bavaria, dated conservatively at 200 million years ago (MYA). Baur (1887) first used the name in a footnote to an article on ichthyosaurs. Descriptions by Jaekel (1914, 1918) followed under two synonyms (*Stegochelys dux* and *Triassochelys dux*, respectively) (Gaffney, 1990). More recently Gaffney (loc. cit.) has prepared an excellent descriptive account of a series of six specimens, including some that are virtually complete skeletons. These restorations and illustrations suggest an animal with habits, habitat, and body form similar to a large North American snapping turtle (Chelydra serpentina) (Figure 1.1).

Proganochelys is therefore not only an early turtle fossil but one of the best known turtles in a skeletal sense. The taxon has some primitive characters that do not occur in living chelonians, but it is otherwise quite clearly a bona fide turtle. By late Triassic times, it had already achieved all or most of the adaptive modifications that are diagnostic of turtles. Most particularly, the shell had fully evolved and surrounded the limb girdles (Figures 2.1 and 2.2).

Despite our good knowledge of the *Proganochelys* skeleton, it provides little or no information about the evolutionary steps leading from the generalized tetrapod body plan (from which it logically must have evolved) to the unique turtle body plan.

At the middle of the 20th century, it was comfortably logical and convenient to place the ancestry of turtles somewhere among the anapsid "stem reptiles." Since 1991, turtle origins have been vigorously and eruditely contested. Clever writers of turtle lore simply avoided the issue. There are now four hypotheses that are taken seriously enough to have been published in prestigious scientific journals.

Gaffney and Meylan (1988) use the Captorhinidae, a group of anapsids that place turtle origins in the early Permian or even the late Carboniferous, 100 million years (MY) before *Proganochelys*. Reisz and Laurin (1991) invoke the Procolophonids, small tetrapod reptiles from the Permian of South Africa. Lee (1993a, 1994, 1996, 1997) uses the Pumiliopareia, a group of heavily armored dwarf Pareiasaurs in the late Permian.

Rieppel and deBraga (1996) and deBraga and Reippel (1997) make a morphological case for diapsid ancestry. Two recent papers (Hedges and Poling, 1999; Zardoya and Meyer, 1998) present molecular data said to support the diapsid affinities of turtles. The origin from (or membership in) diapsids is the greatest departure from conventional schemes and, by the authors' own statement, "heretical." Some of these analyses are cladistic (see Zug, Vitt, and Caldwell, 2001) and subject to the vicissitudes of number and kinds of characters studied and to the number of taxonomic groups included in the analysis.

Most of the hypotheses neither address the evolution of the turtle body plan nor affect our understanding of evolution in the Triassic. The Pareiasaur hypothesis is attractive in that Lee illustrates with fossils (Pumiliopareia) or hypothesizes the intermediate steps necessary to evolve a testudinate body plan from a generalized tetrapod plan and discusses the selective advantages of each step. In short, Lee hypothesizes a small (<1 m), broad, flat pareiasaur with reduced number of dorsal vertebrae, widened ribs and an armor of osteoderms, variably fused inter se and to the ribs and dorsal vertebrae. These constitute the structural requisites for a chelonian ancestor—regardless of the taxonomic group to which it belongs. Indeed, there have been other reptilian experiments with a "shell." *Eunotosaurus* (Permian, South Africa) had broadened ribs, and some placodont reptiles (e.g., *Henodus*, Upper Triassic, Europe) had dermal elements fused to the axial skeleton, but neither animal was a turtle nor a likely ancestor to turtles (Carroll, 1988).

Even the best explanations of turtle origins do not explain how the rib cage came to surround the limb girdles. It once seemed logical that the limb girdles had migrated to the inside of the rib cage, both ontogenetically and phylogenetically. Watson (1914) hypothesized a chelonian ancestor having a broad flat thorax with broadened, horizontally projecting (rather than ventrally curved) ribs overlain by heavy dermal armor (osteoderms); the pectoral girdle lay anterior to this prototypic "carapace" and eventually migrated to a position beneath its anterior edge. Ruckes (1929) thought that the anlagen of the limb girdles remained in place while the carapace primordium overgrew them. More recently, there is fossil (Lee, 1993b) and developmental evidence (Burke, 1989 a, b, c) that a posterior migration of the pectoral girdle primordium actually occurred. Lee regards Watson's hypothesis as the most plausible explanation to date.

Deraniyagala (1939) is seemingly the only author bold enough to illustrate a whole hypothetical turtle ancestor. His swamp dwelling "saurotestudinate" (Permian or Carboniferous), although fanciful, had some of the characters that modern turtle biologists would predict in an ancestral group.

Aside from an initial and radical modification of the skeleton, turtles have remained relatively generalized in most other aspects of anatomy, physiology, and behavior. There has been a modest adaptive radiation over the approximately 200 million years they have existed. The extant extremes in this radiation are the nearly complete terrestrialism in land tortoises (Testudinidae), the marine adaptations of sea turtles (Cheloniidae and Dermochelyidae), and the extreme shell modifications of trionychids.

FIGURE 1.1 Artist's conception of *Proganochelys* as a semiaquatic turtle in a late Triassic environment, showing prosauropod dinosaurs, a pterosaurid, hybodont sharks, and plants of that period. (From Gaffney, 1990: Figure 133, with permission.)

Turtles have not only survived but also prospered while witnessing the triumphs and failures of the major events in amniote evolution.

Proganochelys had most of the characteristics seen in modern turtles. The exceptions were small palatal teeth on the vomers, palatine, and and pterygoid bones; 12 pairs of supramarginal scutes and 16 or 17 pairs marginals; consistently four large central scutes; and a rodlike cleithral bone on each side, extending from the epiplastron to the anterior edge of the carapace. In any case, *Proganochelys* was a turtle in general appearance and would be recognized as such by anyone who had seen a modern turtle.

Proganochelys was a large animal (up to 1 meter CL). Gaffney's series of seven specimens had calculated carapace lengths of 450 to 670 mm; he regarded the smallest specimen to be immature on the basis of carapace fontanels.

At the time we started this book, *Proganochelys quenstedi* Baur stood more or less alone as a late Triassic enigma, representing the early fossil history (dated variously from 210 MYA to 204 MYA and traditionally rounded to 200 MYA. It had nearly all of the characters included in a diagnosis of modern chelonians. But it told us little or nothing about turtle ancestry or what we might expect in a primitive turtle.

Currently, the oldest turtle is *Odontochelys semitestacea* (Li et al., 2008), a complete skeleton from 220-million-year-old marine deposits in China. The authors allege it to be ancestral to all other turtles.

Odontochelys demonstrates a unique and puzzling set of characters as follows: small size (length of trunk 210 mm); probably marine; teeth in dentary and maxillary bones as well as palatal teeth; a well-develped plastron (including two pairs of mesoplastra) forming a well-developed bridge on each side; only known turtle of any period in which dorsal rib heads articulate with a facet at midcentrum; ribs broadened; neurals present; no evidence of costals or peripherals; cleithra as in *Proganochelys*.

In the same issue of *Nature*, Reisz and Head (2008) made their own evaluation of *Odontochelys*. They argue that the virtual lack of dermal armour (costals and peripherals) is a derived condition resulting from simple lack of ossification. This logic is reasonably acceptable because the bridge evolved to support the carapace, and the expanded ribs (present) and the costals and peripheral elements develop from the same anlagen. This scenario would produce a paedomorphic, leathery carapace, lacking surface topography and supported chiefly by ossified, expanded ribs and thoracic vertebrae and neural bones.

We agree with Reisz and Head (2008) that *Odontochelys* is already too specialized to be considered as the ancestor of all other turtles. Discoveries such as *Odontochelys* give us brief glimpses of a chelonian radiation that was in progress far more than 220 million years ago and from which the venerable turtle Bauplan emerged in the form of *Proganochelys*.

A Classification of Turtles with a List of the Species Occurring in Mexico

There are two subordinal groups of extant turtles: the Cryptodira and the Pleurodira. Cryptodires retract the neck in a vertical sigmoid curve (straight back, Dalrymple 1979), whereas pleurodires flex the neck laterally (in either direction) and tuck it under the anterior edge of the carapace without truly retracting it (Figure 2.8). The pelvis is fused to the plastron in pleurodires but free in crytodires. Pleurodires now occur only on southern continents. Cryptodires occur worldwide and are more diverse than Pleurodires.

Living turtles comprise 12 to 15 families (depending on subfamial versus familial rank in certain groups) (Iverson, 1992a; Bickham et al., 2007). Eight families of nonmarine turtles with 14 genera, 38 species, and 66 terminal taxa are considered for Mexico. Taxa marked with an asterisk (*) do not occur in Mexico. Subspecies are covered in the text of species accounts. The suffixes *-oidea* and *-idae* are used for superfamilial and familial groups, respectively. Suprafamilial categories are based broadly on Shaffer, Meylan, and McKnight (1997), Zug et al. (2001), and Gaffney and Meylan (1988).

Class Reptilia (Subclass Parareptilia)

Order Chelonia

Suborder Pleurodira*

Family Pelomedusidae*

Family Podocnemidae*

Family Chelidae*

Suborder Cryptodira

Superfamily Chelonoidea (sea turtles)

Family Cheloniidae

Caretta caretta

Chelonia agassizi

Chelonia mydas

Eretmochelys imbricata

Lepidochelys kempi

Lepidochelys olivacea

Family Dermochelyidae

Dermochelys coriacea

Superfamily Trionychoidea

Family Dermatemydidae (river turtles)

Dermatemys mawi

Family Kinosternidae (musk turtles)

Claudius angustatus

Staurotypus triporcatus

Staurotypus salvini

Kinosternon herrerai

Kinosternon sonoriense (2 subspecies)

Kinosternon hirtipes (6 subspecies)

Kinosternon flavescens

Kinosternon arizonense

Kinosternon durangoense

Kinosternon scorpioides (2 subspecies)

Kinosternon integrum

Kinosternon alamosae

Kinosternon chimalhuaca

Kinosternon oaxacae

Kinosternon acutum

Kinosternon creaseri

Kinosternon leucostomum

Sternotherus odoratus

Family Trionychidae (soft shelled turtles)

Apalone atra

Apalone spinifera

Family Carretochelyidae (pig nose turtles)*

Superfamily Testudinoidea

Family Geoemydidae (Batagurid pond turtles)

Rhinoclemmys areolata

Rhinoclemmys pulcherrima (3 subspecies)

Rhinoclemmys rubida (2 subspecies)

Family Emydidae (Emydid pond turtles)

Actinemys marmorata

Chrysemys picta

Pseudemys gorzugi

Trachemys scripta (11 subspecies)

Terrapene ornata (2 subspecies)

Terrapene nelsoni (2 subspecies)

Terrapene mexicana

Terrapene yucatána

Terrapene coahuila

Family Testudinidae (Tortoises)

Gopherus agassizi

Gopherus berlandieri

Gopherus flavomarginatus

Superfamily Chelydroidea (snapping turtles)

Family Chelydridae

Chelydra rossignoni

Chelydra serpentina

Family Platysternidae (big headed turtles)*

Geographic Distribution and Biogeography

Turtles have a cosmopolitan geographic distribution. Breeding populations occur in the temperate and tropical regions of the world on major continents, large islands, some oceanic islands, and in the seas around these land masses. There are epicenters of diversity in the southeastern United States and southern Asia. Cryptodires are now the only turtles in the Northern Hemisphere, but they occur also in the Southern Hemisphere.

In temperate zones with permanent water, temperature is a limiting factor; turtles are oviparous and must rely on a warm season long enough for nesting and at least partial embryonic development. Overwintering of hatchlings in the nest is known (Carr, 1952; Ernst and Barbour, 1972; Ewert, 1979, Ernst et al., 1994). Substrates that do not permit the digging of nests may limit distributions in various areas of otherwise favorable habitat.

Breeding populations of *Chrysemys* and *Chelydra* occur in Canada (ca. 52°N), at or near the northern limits for oviparous reptiles, and *Chelonoidis* (formerly *Geochelone*) ranges southward to 43°S in Argentina. Sea turtles may range widely north and south from their breeding ranges; *Dermochelys coriacea* has the greatest foraging range (66°N to 47°S) of any chelonian. *Trachemys scripta*, if truly a single polytypic species, has the greatest breeding range of any freshwater turtle (42°N to 36°S) (Legler, 1990).

A few of the U.S. species have distributions that stop abruptly in northern Mexico. *Rhinoclemmys* reaches subtropical western Mexico as far north as 27° but has its main distribution in Central America and northern South America. In southern Mexico, straddling the Isthmus of Tehuantepec, are the remnants of an old Mesoamerican component, shared with Guatemala, Belize and El Salvador, consisting of the genera *Claudius*, *Staurotypus*, and *Dermatemys*. The snapping turtles (*Chelydra*) and the box turtles (*Terrapene*) are widely distributed in the United States and Canada and then extend, as a series of species or as subspecies through most of Mexico in isolated populations—probable relics of former continuous distributions. Within Mexico there has been a radiation within the genus *Kinosternon* and to a lesser extent within the geographic races of *Trachemys*.

The large river systems of the southeastern United States are a chief epicenter for emydine diversity. *Trachemys scripta* demonstrates what little emydine diversity there is in Mexico. *Dermatemys* occupies a niche similar to that of basking aquatic emydines in southern Mexico. The paleotropical region is an epicenter of batagurine and trionychid diversity. Trionychids are poorly represented in Mexico, reaching the southern limits of their New World distribution in the extreme northeastern part of the country.

Topography and Climate of Mexico

Map 4 (Based on Raisz, 1959; Meek, 1904; Leopold, 1959; Tamayo, 1949 & 1962)

More than half of the border between Mexico and the United States is defined by the course of the Rio Grande or Rio Bravo del Norte. Westward from El Paso, the border consists chiefly of neat straight lines defined by surveyors that separate California, Arizona, and New Mexico from Baja California, Sonora, and Chihuahua.

The topography of mainland Mexico is dominated by a high (1000–2500 m) Mesa Central or central plateau, flanked east and west by mountain ranges and forming a "V" with its apex at the Isthmus of Tehuantepec. The mountains are the Sierra Madre Occidental and the Sierra Madre Oriental, both of which descend steeply to coastal plains or directly into the sea in some places. The plateau tilts slightly from the south (central Jalisco and western Hidalgo) to its lowest part along the United States border. For zoogeographic purposes, it is instructive to divide the Mesa Central into a basin-and-range

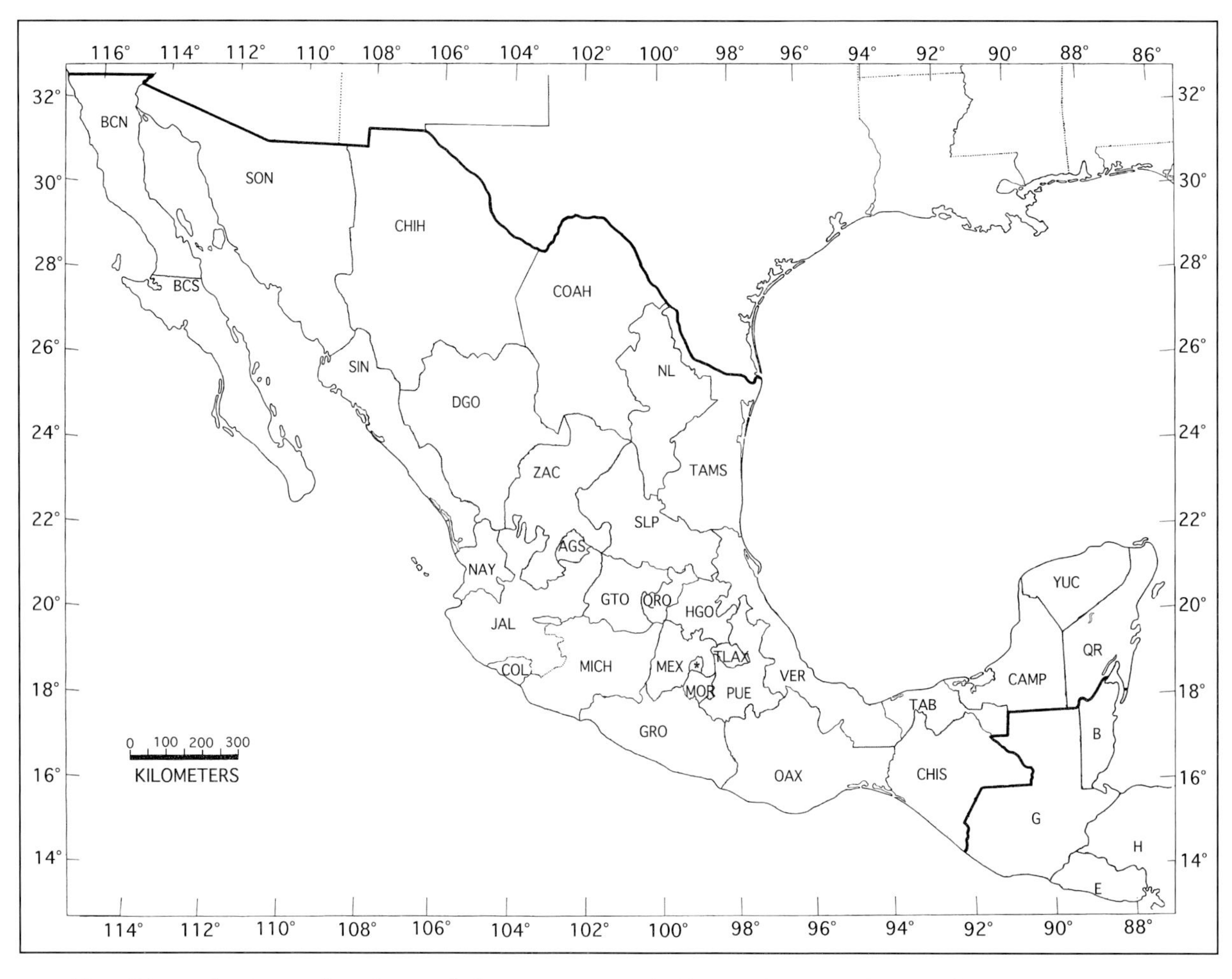

MAP 1 Map of Mexico showing state boundaries and adjacent countries with abbreviated names: AGS, Aguascalientes; B, Belize; BCN, Baja California Norte; BCS, Baja California Sur; CAMP, Campeche; CHIS, Chiapas; CHIH, Chihuahua; COAH, Coahuila; COL, Colima; DGO, Durango; E, El Salvador; G, Guatemala; GTO, Guanajuato; GRO, Guerrero; H, Honduras; HGO, Hidalgo; JAL, Jalisco; MEX, México; MICH, Michoacán; MOR, Morelos; NAY, Nayarit; NL, Nuevo León; OAX, Oaxaca; PUE, Puebla; QRO, Querétaro; QR, Quintana Roo; SLP, San Luís Potosí; SIN, Sinaloa; SON, Sonora; TAB, Tabasco; TAMS, Tamaulipas; TLAX, Tlaxcala; VER, Veracruz; YUC, Yucatán; ZAC, Zacatecas. The Distrito Federal (Mexico, D.F.) is indicated by an asterisk (*).

portion north of a transverse range at the latitude of Torreon (Sierra de Jimulco and Sierra de Parras) and the central plateau proper to the south. The basin-and-range portion is sparsely vegetated except along stream courses.

South of the central plateau is a shallow basin, the Bajío, that contains a rich agricultural area. The Bajío is bordered to the south by the Sierra Volcánico Transversal (neovolcanic cordillera), a transverse zone of recent volcanism that contains all of the large volcanos (e.g., Citlaltépetl, 5594 m, and Popocatépetl, 5400 m) of Mexico and many smaller ones (e.g., Ceboruco, 2164 m). Tamayo (1949) presents a spectacular cross section of the cordillera at latitude 19°N. The Valley of Mexico is one of the many closed basins within this region. South of the volcanic cordillera is the Rio Balsas basin, which drains to the Pacific and constitutes a deep penetration of tropical habitat into the interior. The Balsas basin is bounded to the south by the Sierra Madre del Sur and on the east by the Sierra Madre de Oaxaca. The convergence of the Sierra Madre Occidental and Sierra Madre Oriental with these shorter ranges forms the blunt southern apex of the aforementioned mountainous "V" just west of the Isthmus of Tehuantepec.

East of the isthmus and south of the state of Tabasco lies the Chiapas–Guatemala upland, which extends southeastward as a crescent of mountains to the Gulf of Honduras and forms the base of the Yucatán Peninsula.

The Sonoran Desert gives way to a coastal plain watered by perennial streams near Guaymas (ca. 27°N). South of Guaymas the plain becomes progressively narrower and is reduced to virtually nothing at Cabo Corrientes, Nayarit, occurring farther southeastward in isolated patches with coastal marshes, the largest of which is near Acapulco. The coastal plain is continuous again from Salina Cruz, Oaxaca, to El Salvador.

The Atlantic or gulf coastal plain is vast by comparison, narrowing from about 240 km (latitude of

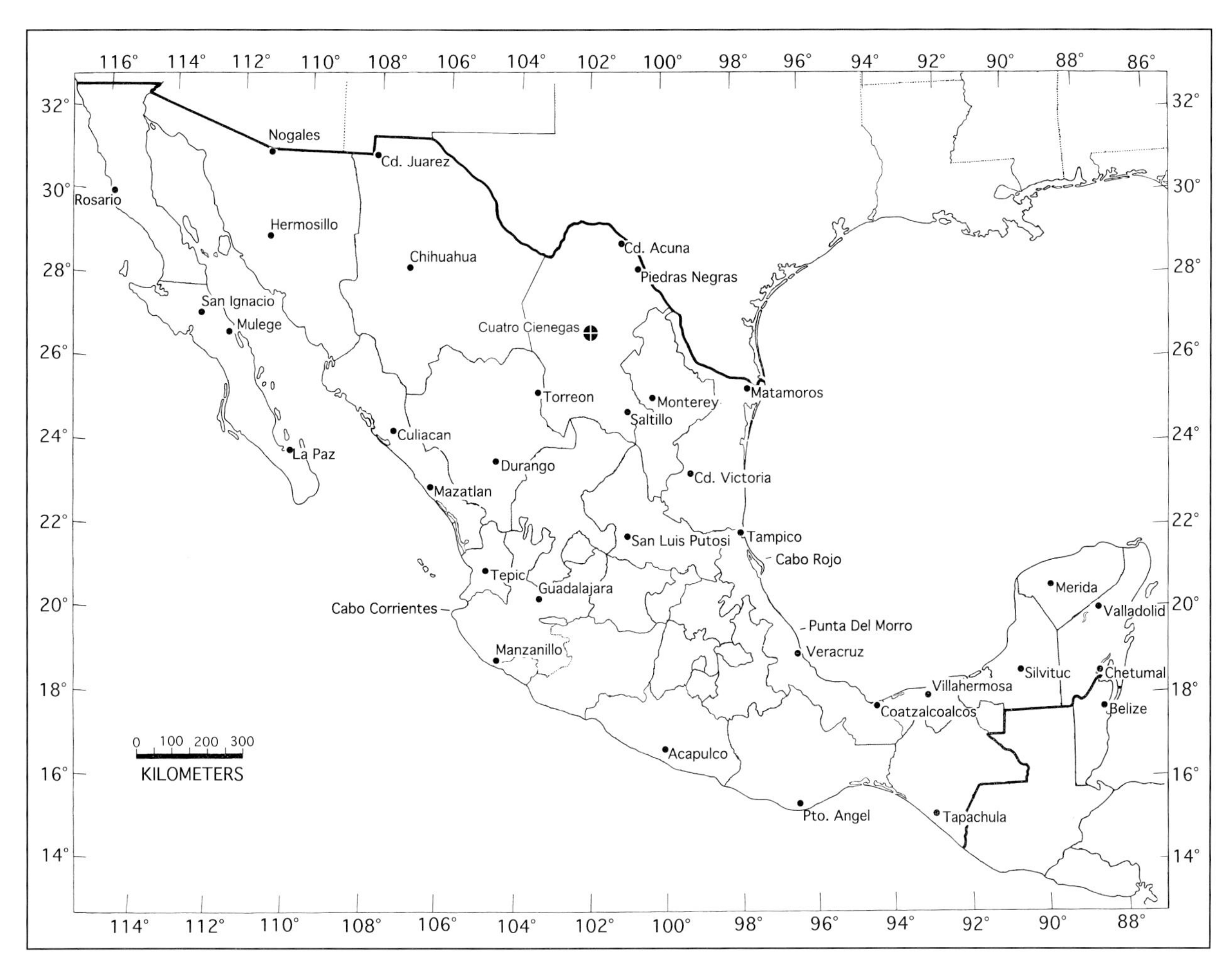

MAP 2 Map of Mexico and adjacent countries showing state boundaries and place names mentioned in text.

Matamoros–Brownsville) to virtually nothing at Punta del Morro (just north of Ciudad Veracruz) and widening again to the south and east to become continuous with the flat, low limestone country of the Yucatán Peninsula. In this region, it includes the drainages of the wettest regions of Mexico (e.g., Rios Papaloapam, Coatzacoalcos, and Usumacinta–Grijalva). The Isthmus of Tehuantepec (lowest point ca.100 m) narrowly separates the two coastal plains.

Temperature is the factor determining the separation of tropical and temperate zones and is governed by altitude and latitude. About 70% of Mexico is temperate and 30% tropical. Temperate climates occur above 1500 m and have a well-marked winter season despite their southern latitudes. The frost-free tropics of Mexico occur chiefly in the coastal strips and on the Yucatán Peninsula. In general, the Mesa Central has low rainfall and desert-type vegetation but is broken by smaller mountain ranges, many of which are high enough to have more luxuriant plant growth. Raisz (1959) summarizes these physiographic provinces with elegant simplicity on a single exquisite map.

In general, northern Mexico is dry; annual rainfall is less than 400 mm in the basin-and-range region. The region southeast of Veracruz receives more than 1600 mm annually, and the central sector is a mixture of wet and dry areas conforming to the vicissitudes of topography. The wettest areas mentioned are neotropical.

The rainfall patterns on the Pacific side of the Isthmus of Tehuantepec are distinctly different from those of the Gulf Coast. A distinct short dry season runs from March through May on the Gulf Coast followed by sporadic summer thundershowers from June to August. Torrential monsoon rains are common in September and October, and from November to February cold north storms (nortes) prevail, typified by cooler temperatures, light rains, and cloud cover lasting from 5–12 days. The nesting season of the neotropical endemics (*Dermatemys*, *Claudius*, *Staurotypus triporcatus*, *Kinosternon leucostomum*, and *Kinosternon acutum*) on the Gulf Coast commences in September and terminates in March; their eggs hatch with the beginning of the summer thunderstorms in June. Incubation time is controlled by temperature as well as embryonic diapause and embryonic aestivation. The temperate-zone invaders (*Chelydra* and *Trachemys*) begin nesting on the Gulf Coast after the winter nortes become warmer in January and terminate in May

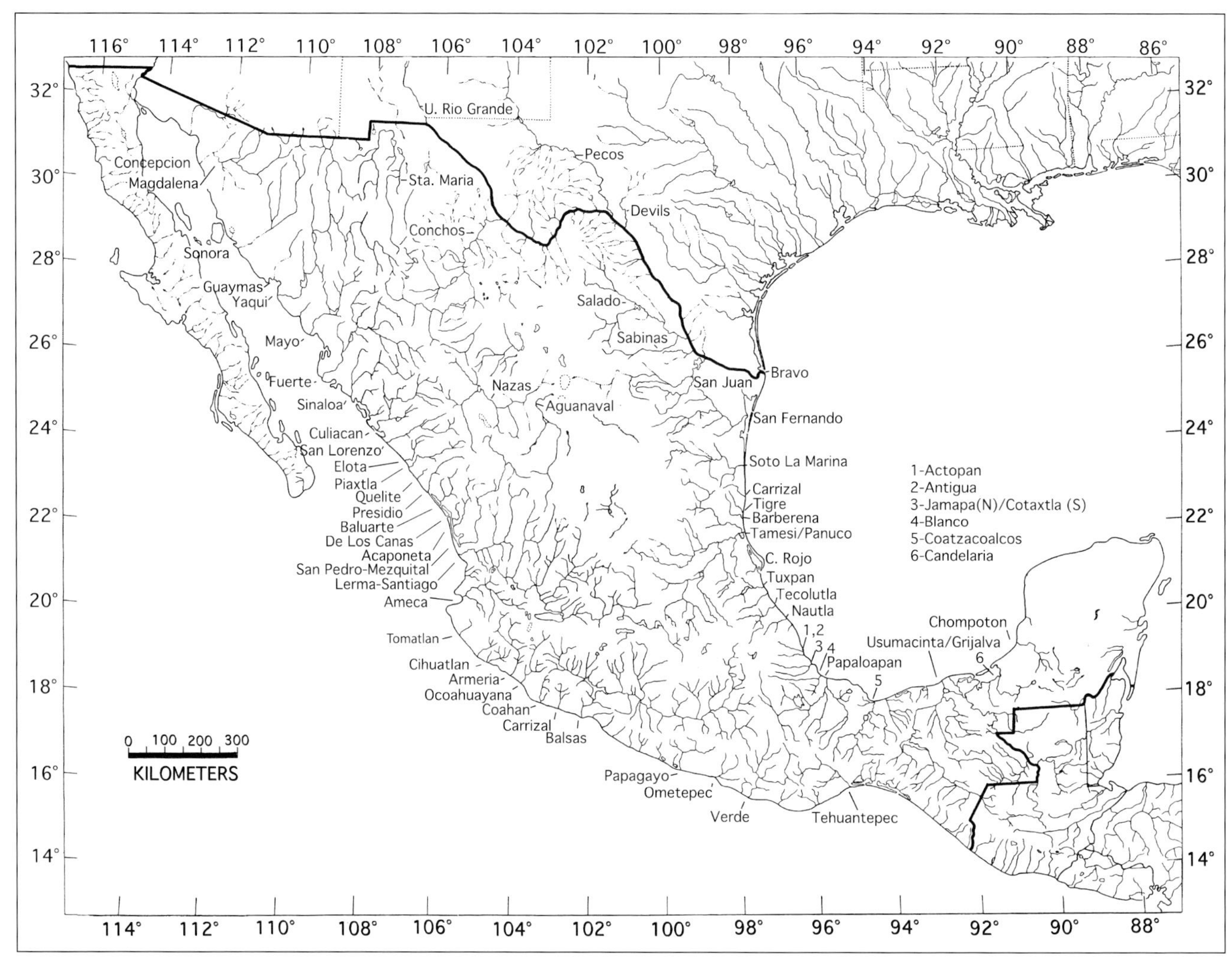

MAP 3 Map of Mexico and adjacent countries showing river drainages. Many minor streams have been omitted for clarity. (Modified from a drainage map of Mexico by Robert Rush Miller, Museum of Zoology, University of Michigan.)

before the summer rains begin. Their eggs have direct development and hatch in 60–90 days, depending on incubation temperature. The dry season on the Pacific coast of Chiapas is more pronounced, lasting from November to April, and the rainy season from May to October.

Between Tampico and Monterrey, rains begin in late May and continue until October. In Chihuahua, the rainy season commences in early June and extends into September; in the wet season, many isolated streams in northern Mexico form terminal lakes of variable size (e.g., Lago de Guzman, Lago de Patos, Lago de Santa Maria) and then cease to flow in drier periods. Northern Mexico was better watered in the past than now.

Cenozoic Paleogeography

By rough definition, most of Mexico is in North America, but there is a tendency to think of North America as beginning somewhere near the U.S.–Mexican border, where the continent begins to narrow dramatically. In this book, we use the term *North America* quite loosely to include most of Mexico and the terms *Mesoamerica*, *Middle America*, and *Central America* interchangeably in reference to an isthmus (78°–95°W) connecting North and South America, extending southeastward from the Isthmus of Tehuantepec (Tabasco and Chiapas), and through all of the countries of Central America to the South American continent. The isthmus is dramatically narrowed at Tehuantepec (ca. 200 km) and in Panama (ca. 50 km) and is as wide as 400–500 km across Nicaragua.

This intercontinental connection was interrupted at various times in the Cenozoic by open portals—gaps through which the oceans could communicate—thus separating parts of the bridge and isolating parts of its flora and fauna for various periods of time.

These portals had a significant effect on the evolution of many groups of vertebrates, including turtles. The portals

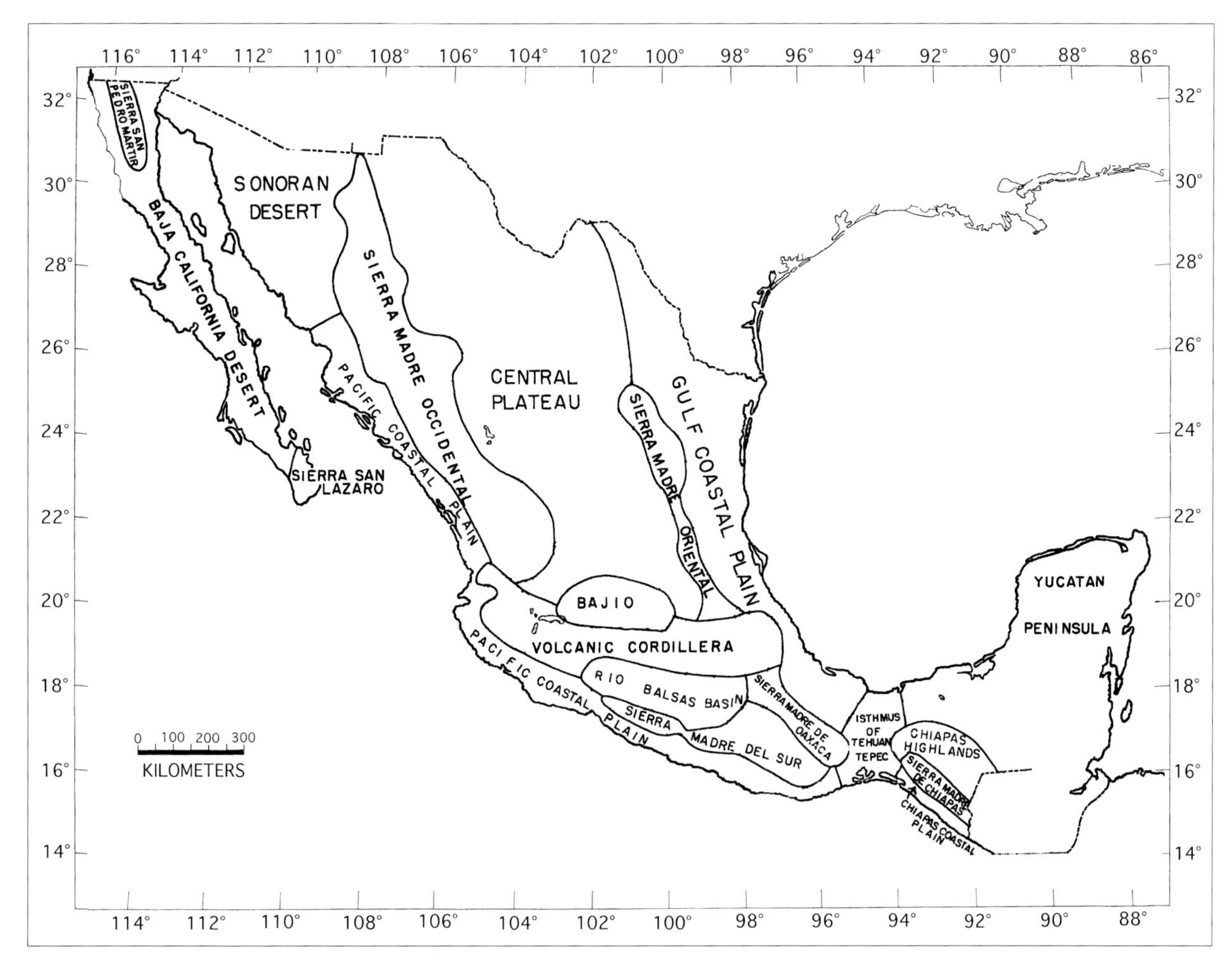

MAP 4 Map of Mexico and adjacent countries showing major geophysical features. Abbreviation: IDT, Isthmus of Tehuantepec. (Modified from Leopold, 1959.) Not shown is a short (ca. 85 km) volcanic range (Sierra del los Tuxtlas) closely paralleling the Gulf coast between 94°25′W and 95°25′W (between Lerdo de Tejada and Coatzacoalcos), containing Lago Catemaco and near the Los Tuxtlas Tropical Research Station—Vogt's headquarters from 1980 to 2000 (see Raisz, 1959).

were as follows (seriatim, southeastward from Mexico to Colombia) (Stuart, 1950: 71, 1951:28; Duellman, 1958:127):

1. The Tehuantepec portal, centered on the current Isthmus of Tehuantepec, was narrow and of relatively short duration—from the early Pleistocene to possibly the late Miocene.
2. The Nicaraguan portal, from southern Nicaragua to northern Costa Rica, lasted from the late Eocene to the late Miocene. In its later stages, it fluctuated in size and probably was an archipelago.
3. The Panamanian portal from the late Eocene to the mid-Miocene ran eastward from isthmian Panama to connect with the Colombian portal to form an archipelago that became eastern Panama.
4. The Colombian portal was an extension of the current Gulf of Uraba in northwestern Colombia through the Colombian Chocó (approximately the current Atrato drainage) that separated South America from Central America from the late Paleocene to the mid-Pliocene.

The substantial land mass isolated northwest of the Nicaraguan portal became "nuclear Central America" (Stuart, 1966) and has been exposed continuously since Cretaceous times. In its late stages, it was seemingly an archipelago rather than a stark oceanic barrier. In any case, the Nicaraguan portal remained a barrier of varying magnitude to movements between continents (in both directions) for its entire existence. Most of the Yucatán Peninsula was submerged for most of the Cenozoic period.

Nuclear Central America corresponds approximately with the ranges of several endemic neotropical taxa covered in this book (e.g., *Dermatemys*, *Staurotypus*, *Claudius*), and the entire history of the opening and closing of the portals is reflected in the phylogeny of other reptilian groups. Duellman (1958) provided a lucid review in his monograph on *Leptodira* and contributed substantially to a series of papers published in *Copeia* in 1966 (by Duellman, Miller, Myers, Savage, and Stuart).

Materials and Methods Used for This Book

How to Use This Book

This book is intended for use as a reference source. It is not a field guide. Few people will read it from cover to cover.

The first part of the book presents general information on turtles: origins, structure, function, natural history, techniques for study, and conservation measures. Much of this information is general, some of it is philosophical, and much of it is applicable to all chelonians.

The rest of the book contains accounts of taxonomic groups that we hope include all or most of existing knowledge about these taxa from the level of order to terminal taxon.

The taxonomic keys work. But no key is flawless. We have attempted to use external characters in keys. Morphological characters that require dissection or the use of skeletal material are used only in diagnoses of major groups.

To identify a specimen (preferably of known provenance), one would work seriatim through the dichotomous keys, which are appropriately placed in the text and listed in the table of contents and in the index.

Presentation

All literature cited is merged at the end of the book. The account of each terminal taxon includes most of what we cover for that taxon. The amount of detail in this book is directly related to our own familiarity with the subject or taxon, the amount of information we can glean from a sparse (but growing) literature, and the unpublished observations of generous colleagues and associates.

This book is written in English but, being about Mexico, is replete with Spanish place names. We have done our best to properly use diacritical marks on geographic and author names, with *Mexico* as the major exception.

We make free use of the detailed etymological information in Smith and Smith (1979) and various dictionaries, but our final sources of authority on etymology are Brown (1954) and Jaeger (1955). In most cases, we list two or three vernacular names, most often following those used in Iverson (1992a, English) and Liner (1994, Spanish) and the accounts in the *Catalogue of American Amphibians and Reptiles*. In areas where we have personally worked, we cite one or more locally used names. Vernacular names are given in the order of our preference.

In preparing species accounts from the work of others, we were frustrated by the lack of order and syntactical equivalency in published accounts. We made every attempt to organize diagnoses, general descriptions, and the major components of each species account in a manner by which the reader would have little difficulty when looking for a single element of information.

In surveys of literature, if a subject is summarized or reviewed in a scholarly way, we cite the review rather than each paper in it. We have attempted to use our own unpublished knowledge to augment what is known or, in some cases, to serve in lieu of nothing whatever.

The use of the double "i" ending (e.g., *bellii, salvinii, agassizii*) is legally optional according to the International Code of Zoological Nomenclature. We follow Smith and Smith (1979) in using the single "i" protocol (e.g., *belli, salvini, agassizi*).

Terminology

Terminology for chelonian shell structures varies among authors, often without justification (Zangerl, 1969). In Zangerl's words, "No effort has been made to standardize the terminology internationally, and authors have persisted in using their favorite terminologies." We unabashedly continue this custom and use a terminology based on Carr (1952). Our terminology—such as that in Figure 2.3—has been used (with a few exceptions) by Legler (1960c), Moll and Legler (1971), Legler and Cann (1980), Legler (1990: Fig. 7.1), Legler (1993b: Figs. 16.1, 16.2), Cann and Legler (1994: Fig. 1), and Smith and Smith (1979).

Abbreviations Used in This Book

Nearly all abbreviations are clearly defined in the map and figure legends or in the text with the following exceptions. CBL stands for *condylobasilar length* as measured from the posteriormost point of the occipital condyle to the anteriormost projection of the prefrontal bones (usually on the lower rim of the nasal apperture) (Figure 2.6). Abbreviations for Mexican state names are on Map 1. For the species of *Kinosternon*, "Basic Proportions" (based on our data) are followed by "Plastral Proportions" (based on data from John Iverson). Abbreviations unique to the latter were FL (length of forelobe), HL (length of hind lobe), BL (length of bony bridge), PS1 (length of plastral scute 1), and PS 6 (length of plastral scute 6).

Museum Acronyms

AMNH, American Museum of Natural History; BYU, Brigham Young University; CAS, California Academy of Sciences; CM, Carnegie Museum; CNHM, Chicago Natural History Museum; FMNH, Field Museum of Natural History; KU, University of Kansas Museum of Natural History; LACM, Los Angeles County Museum; MCZ, Museum of Comparative Zoology (Harvard); MSU, Michigan State University; PCHP, P. C. H. Pritchard Collection; UCM, University of Colorado Museum of Natural History; USNM, U.S. National Museum; UU, University of Utah.

Turtle Measurements and Basic Shell Proportions

All measurements, distances, and altitudes are stated in metric units—weights in grams or kilos and temperatures in degrees centigrade. In all cases, "size" is expressed as carapace length (CL). Size at sexual maturity was determined by gonadal examination or the development of secondary sexual characters. The overlap in CL of the largest immature and smallest mature individuals was useful in stating size ranges at puberty. Age of puberty is often unknown but could be hypothesized from growth rings in a few cases (e.g., *Staurotypus*).

For nearly all terminal taxa, "Basic Proportions" are given as part of the general description. Shell width (CW) and height (CH) are expressed as percentages of carapace length as are CH to CW ratios, and they constitute a simple expression of shell shape—the smallest rectilinear box into which the shell would fit.

Unless otherwise noted, CL and all other measurements were made in the manner shown in Figure 5.1 and have been used by us and our students for thousands of specimens. Carapace length is regarded as a standard for size, but note that a turtle can "grow" without increasing in length. Several authors have gone into some detail on how a turtle should be measured (Legler in Cann and Legler, 1994; Legler, 1990; Carr, 1952; Mosimann 1956). These accounts refer to *straight-line* measurements with calipers. *Curved-line* measurements, made along the curve of the shell, enjoy some popularity because they can be made easily with an inexpensive flexible tape or a piece of string and a ruler. Straight-line measurements require more work and special tools. There seems to be no accurate way to convert curved measurements to straight ones.

For anyone living in or near a rectilinear structure with reasonably plumb floor and walls, the following procedure will produce accurate straight-line measurements of the carapace. Place one end of the shell against a wall with the midline axis aligned at right angles to the wall; find an object that will form an approximate right angle with the floor (a brick, a thick book or a piece of scrap lumber), place it (parallel to wall) against the rear edge of the carapace and then remove the turtle and measure the distance between the wall and the object against the carapace. We have used this technique to measure very large chelonians that exceed the limits of available calipers.

Paired plastral scutes are measured along their median interlaminal seams. Because their transverse boundaries frequently are not in perfect alignment, the difference is judged by eye and the measurement made to the middle of the overlap (see Figure 5.1). *Plastral formula* is simply the order of midline interlaminal lengths from longest to shortest. Having discussed plastral scute homologies in some detail (see the account of family Kinosternidae), we decided to express all plastral formulae numerically rather than by abbreviations. These formulae are variable and expressed as modes. Two scutes expressed as "1 = 5" are of subequal length. Thus, the modal plastral formula of *Trachemys scripta gaigeae* would be $4 > 6 > 3 > (1 = 5) > 2$.

Mass or weight is often used as a standard for size. It is valuable and probably accurate in aquatic turtles after they have been handled and have voided water from the cloaca (in fact, Legler regarded weight, under controlled conditions, to be a meaningful datum for Australian chelid turtles). For terrestrial species—*Terrapene ornata* (Legler, 1960c) and *Gopherus* (Germano, 1993)—weight is less useful because of the variable amount of water stored in the accessory bladder. Also, a gravid female will weigh more before laying than after. Anyone fortunate enough to observe a female before and after oviposition should record body weight in both states as well as the weight (and number of) eggs in the clutch.

Most of the numeric data in this book were calculated from raw data in our own databases. The calculations are usually associated the following descriptive statistics: mean ± one standard deviation, extremes, and number of cases. Statistical expressions of size and frequently used ratios are given separately for adult males, adult females, and various classes of nonadult stages.

Geographic Range and Distribution Maps

Distributional ranges are stated and plotted using all available sources (but chiefly Iverson, 1992a); we sacrifice some precision by not showing dots on the maps but by using precision under the heading of "Geographic Distribution" in the text. The most frequently used reference maps were 1:250,000 maps (Anon, 1987), the 1:1,000,000 World Aeronautical Charts and Operational Navigation Charts (Defense Mapping Agency; Aerospace Center, St. Louis Air Force Center; Army Map Service, Corps of Engineers; U.S. Coast and Geodetic Survey), and many road maps. Tamayo (1946) was used as the authoritative source for river names. A large, composite aerial photo of Mexico (consisting of many conjoined prints) and the superb topographic relief map of Mexico by Irwin Raisz (1959) were immensely helpful in understanding the relationships of topography, drainage areas, and geographic ranges.

Our base map was modified from a large-scale drainage map of the New World, in three parts, purchased from the Museum of Zoology, University of Michigan (presumably based on Miller, 1968). Our various digitized versions of the map show topography and political boundaries (Maps 1, 2, and 4). A version showing virtually every major drainage (Map 3), labeled by name and some place names, is the basis for the distribution maps. Our method of graphic exposition of geographic distribution ranks somewhere between very simple maps that show only political boundaries (e.g., Carr, 1952; Ernst et al., 1994) and the detailed maps of Iverson (1992a), which show precisely placed dots for each locality and most rivers. None of the foregoing works employ marks of longitude and latitude or a scale of kilometers. Our maps are visual approximations, including the entire known range plus likely extensions of the range within the same drainage system. Our maps are not intended for use by workers who wish to publish notes on range extensions. Any exactitude in our expression of geographic distribution is to be found in the account of the taxon. Hard to find localities are stated in latitude and longitude to the nearest minute; localities on coastlines or on streams are given in terms of the intersection of longitude or latitude with the linear feature mentioned. An Internet source of distributional data is the "Emysystem World Turtle Database" (http://emys.geo.orst.edu/). Although the dots are often misplaced, each dot is associated with accurate coordinates of latitude and longitude.

Structure and Function

The Chelonian Shell

The chelonian shell (see Figure 2.3) consists of a dorsal carapace and a ventral plastron joined on each side by a bridge (to which both contribute). The limb girdles lie *inside the shell* and are *surrounded by the axial skeleton*—a bauplan that is unique among vertebrates (Figure 2.2).

The shell is a combination of discrete bony elements and epidermal elements (Figure 2.2). From outside to inside, it consists of three layers: skin, in the form of large keratinized scutes; a layer of usually planar dermal bones which articulate *inter se;* and the endochondral bones of the axial skeleton—vertebrae and ribs. The scutes are modifications of ordinary scales. The ribs and vertebrae are endochondral in origin, having cartilaginous developmental precursors (see Figure 3.5). Most parts of the shell can be homologized with skeletal elements in generalized tetrapods. There are 10 dorsal vertebrae and 10 pairs of ribs articulating with the vertebrae. Homologies of the ribs and vertebrae are clear. The plates of dermal bone probably evolved from amalgamations of smaller osteoderms and are broadly analogus to the osteoderms of other reptiles. (e.g., crocodilians).

Counting dermal and endochondral bones separately (even if fused), there are some 80 osseous elements in the carapace and 9 in the plastron of a generalized shell; these elements form well over 200 articulations (inter se and with the girdles). Most of these articulations are sutures between dermal plates.

The dermal plates are arranged in three series: a median series of as many as seven neural plates extending from nuchal to pygal bones; eight costal elements on each side; and a series of peripheral elements on each side. Most of the neurals and all of the costals are inextricably fused to endochondral elements—the neural arches and the ribs, respectively.

Neural plates may be lost at either or both ends of the neural series (e.g., *Terrapene, Kinosternon*). In these cases, the costal plates articulate mid-dorsally. Australian chelid turtles have lost the entire neural series (Legler, 1993b).

The dermal bones articulate by sutures. Each neural consists of a polygonal plate fused to the neural arch of a vertebra. The anterior nuchal element and the posterior pygal and suprapygals are not fused to vertebrae.

Each costal element clearly contains a rib. Ribs 1 and 2 are fused to the first costal element. Ribs 9 and 10 fuse with C8. Although the dermal and endochondral parts are fused (even in early embryos), a rib head clearly protrudes from the proximal end and a rib tip protrudes from the distal end, articulating with a pit on a peripheral bone or into a suture between two peripherals. The peripherals form the rim of the carapace and lack an endoskeletal component (see Figures 3.4 and 3.5).

The plastron consists of eight or nine dermal plates (Figure 2.3), an anterior unpaired entoplastron flanked by two epiplastra, and posteriorly paired hyoplastra, hypoplastra, and xiphiplastra. The entoplastron is fused to a vestige of the interclavical and the epiplastra (probably) to vestigial clavicles (see exceptions in accounts of kinosternids).

Most of the plastron is nearly flat. On each side, the hyo- and hypoplastra form more nearly vertical processes of bone termed *buttresses* (axillary and inguinal). The buttresses form a bridge that articulates suturally with the peripherals (in the range of P3 to P7).

The dermal elements of the carapace range in shape from triangular and rectangular to variously polygonal (chiefly hexagonal). Modifications of the number and shape of carapace bones influence the shape of the shell. For example, if one compares the shell profiles of a *Trachemys* (Figure 42.1) to a *Terrapene* (Figure 49.1) and a *Gopherus* (Figure 30.1),

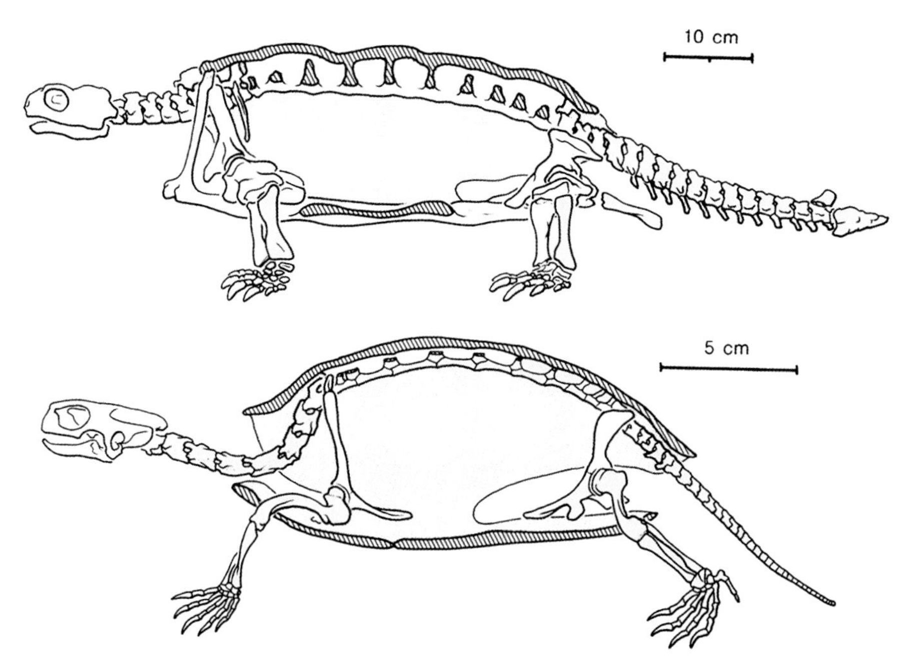

FIGURE 2.1 Comparison of the chelonian Bauplan in ancient and modern turtles. *Proganochelys* (top) and *Emys orbicularis* (bottom). Note the cleithrum passing dorsad between the anterior edges of the shell in *Proganochelys* and the general differential in neck length and shell overhang. (Modified from Gaffney, 1990).

it is clear that the latter two have a posterior carapace that descends rather sharply to the posterior edge and that this profile is facilitated by loss or shortening of the neural-pygal series (whereas in *Trachemys* the anterior and posterior ends of the carapace slope gradually to the rim and form a protective overhang). Likewise, the highly domed shell of *Gopherus* and other testudinids is facilitated in part by increasing the height of the peripheral bones of the bridge to about 30–60% of total shell height and narrowing one end of each costal element to form a long narrow triangle .

The Epidermal Elements

Superimposed on the bony shell is a series of well-defined epidermal scutes (large epidermal scales, also termed *laminae*) separated by interlaminal seams. The epidermal scutes are keratinous, thin, and separated from the underlying bony elements by a thin layer of soft and vascular germinative tissue. Most scutes overlap two or more bony elements and increase the strength of the shell. Carapace scutes are arranged as follows (Figure 2.3): a median row of 5 large central scutes flanked by 4 large lateral scutes on each side plus a circumferential series of 11–12 marginal scutes on each side. The precentral is a small, unpaired anterior-median scute. Axillary and Inguinal scutes usually are present on the anterior and posterior edges of the bridge. They are vestiges of an inframarginal series that is complete only in *Dermatemys*.

There are typically six pairs of large scutes on the plastron: the gulars, humerals, pectorals, abdominals, femorals, and anals (respectively from anterior to posterior) (Figure 2.3).

The shell illustrated in Figure 2.3 is regarded as "typical" only because it has a full complement of shell bones and scutes. Yet more than half of the turtles considered in this book depart significantly from this plan, usually by reducing the number of elements in the shell. These "exceptions" are discussed in the accounts of families Chelydridae, Dermatemydidae, Kinosternidae, and Trionychidae (see also Figure 7.1).

The lamination, fusion and overlap of three layers in the shell produces rigidity and strength. A dried small adult shell of *Trachemys scripta venusta* (193 mm) easily bears the weight of an 80 kg human.

The chelonian shell is versatile in having modified the shape and number of dermal elements. It is venerable in having evolved such changes without altering the basic and ancient chelonian bauplan (for 200 million years). The shell is best regarded as an extremely specialized body wall. Its initial, most important, and most enduring selective advantage is surely that of mechanical protection of soft parts that would otherwise be exposed to damage—essentially the limbs, head, and coelomic viscera.

Protection is meant to include that from predators of all kinds and sizes and the danger of being stepped on and crushed by larger animals. *Proganochelys* was probably unable to retract the head, neck, and limbs. The neck was short (despite having one more vertebra than modern turtles) and heavily armored with osteodermal spikes. The V-shaped aperture formed by the cleithral bones would have impeded head and neck retraction (see Gaffney, 1990, Figures 96 and 133) (see Figure 2.1).

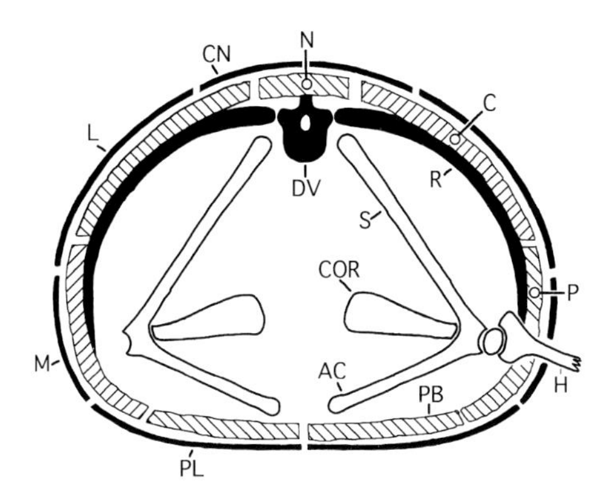

FIGURE 2.2 Ideogrammatic cross-section of the unique chelonian Bauplan, in which the limb girdles are surrounded by the axial skeleton. Principal parts, from exterior to interior, are epidermal scutes (black), dermal bone (diagonal hatching), and endochondral bone (black) in the form of trunk vertebrae and ribs that are fused to the dermal plates. Gaps between scutes are interlaminal seams; those between dermal bones are sutures. Note that the epidermal laminae overlap osseous sutures except along the mid-ventral line. In a fully retracted turtle, the head, neck, tail, and limbs and the entire contents of the coelomic cavity are protected to some extent by the shell. The rib tips may form gomphoses with the peripherals but do not usually co-ossify with them. The term "element" is used for a part consisting of a fusion of endochondral and dermal bone. Abbreviations: AC, acromion; C, costal element; CN, central scute; COR, coracoid; DV, dorsal vertebra; H, humerus; L, lateral scute; M, marginal scute; N, neural element; P, peripheral bone; PB, plastral bone; PL, plastral lamina; R, rib; S, scapula. (Modified from Legler, 1993a.)

The compromises for a rigid shell are constraint of locomotion and any other function normally associated with a flexible body wall (e.g., pulmonary function and the varying coelomic capacity dictated by ovarian follicles, eggs, and gut contents).

Shell Kinesis

The dermal elements of the shell articulate with each other by sutures. A suture is a simple (syndesmotic) joint consisting of a series of, usually shallow, osseous interdigitations separated by a layer of fibrous connective tissue. Sutures of the chelonian plastron are chiefly straight and relatively long. Any suture in a live turtle (but not in a dried shell) is at least slightly flexible. Flexibility in any suture is enhanced by a reduction of interdigitation and an increase in the amount and quality of the intervening softer fibrous connective tissue. Plastral hinges form in exactly this way. In the two cases actually studied by Legler (*Kinosternon flavescens* and *Terrapene ornata*), plastral hinges begin development at a late embryonic stage (see account of *K. flavescens*) or within the first postnatal year (Legler, 1960c). Among Mexican turtles, regularly occurring, functional plastral hinges occur in all kinosternids and all *Terrapene*.

Movable plastral lobes quite clearly enhance the protective function of the shell by closing its anterior or posterior orifices in the manner of hatches. This applies to most kinosternids and all *Terrapene*. In other cases (e.g., *Staurotypus* and *Claudius*), a movable plastral lobe or entire plastron just as clearly enhances the ability to more widely *open* the anterior shell orifice. Movable plastral lobes have evolved independently and for various purposes in a variety of other chelonians (see summary below). An example of tight plastral closure is seen in *Kinosternon leucostomum*, in which it is nearly impossible to pry either lobe open, even with tools. Tight closure could impede desiccation as well as enhancing protection. For taxa that display an intimidating wide gape, depression of an anterior lobe permits an anterior shell orifice to be filled with a gaping mouth (Figure 8.8).

Unmodified sutures of any kind permit growth of the articulating dermal bones. Co-ossification occurs when growth stops; conversely, growth stops if co-ossification occurs. If *Proganochelys* had been genetically programmed for small adult size and sutural ankylosis, then turtles would have become an interesting dead end in the fossil record. Plastral kinesis is an example of a simple modification of a good and enduring body plan.

Summary of Shell Kinesis in Modern Chelonians

Most hinges are plastral. *Kinixys* is the only reported example of carapacal kinesis. The long, straight sutures of the planar plastral bones lend themselves to hinge formation. Among pleurodires, plastral hinges have evolved only once (*Pelusios*). Fusion of the pelvis to the xiphiplastron constrains the development of a posterior hinge.

Plastral hinges have evolved, independently, a total of at least 11 times in chelonians and have various functions. A summary follows.

Pleurodira
- Pelomedusidae
  - *Pelusios*: an anterior hinge, hyoplastron-mesoplastron, pectoral-abdominal. Pleurodires have the pelvic skeleton firmly fused to both carapace and plastron, making a posterior hinge impossible. However, Küchling (1999, Figure 4.1) states that when the large eggs of *Pseudemydura unbrina* (a rare Australian pleurodire) are laid, the carapace and plastron "move apart."

Cryptodira
- Kinosternidae
  - *Kinosternon*: 2 lobes; 2 hinges (except in *K. herrerai*, which has only an anterior lobe and hinge); anterior hinge between scutes 3 and 4 (a huge divided humeral), epiplastron-hyoplastron; posterior hinge between scutes 4 and 5 (humeral-femoral).
  - *Sternotherus*: 1 anterior lobe; 1 hinge between scutes 2 and 3 or scutes 3 and 4 (in either case dividing a huge humeral as in *Kinosternon*); epiplastron-hyoplastron; no posterior hinge.

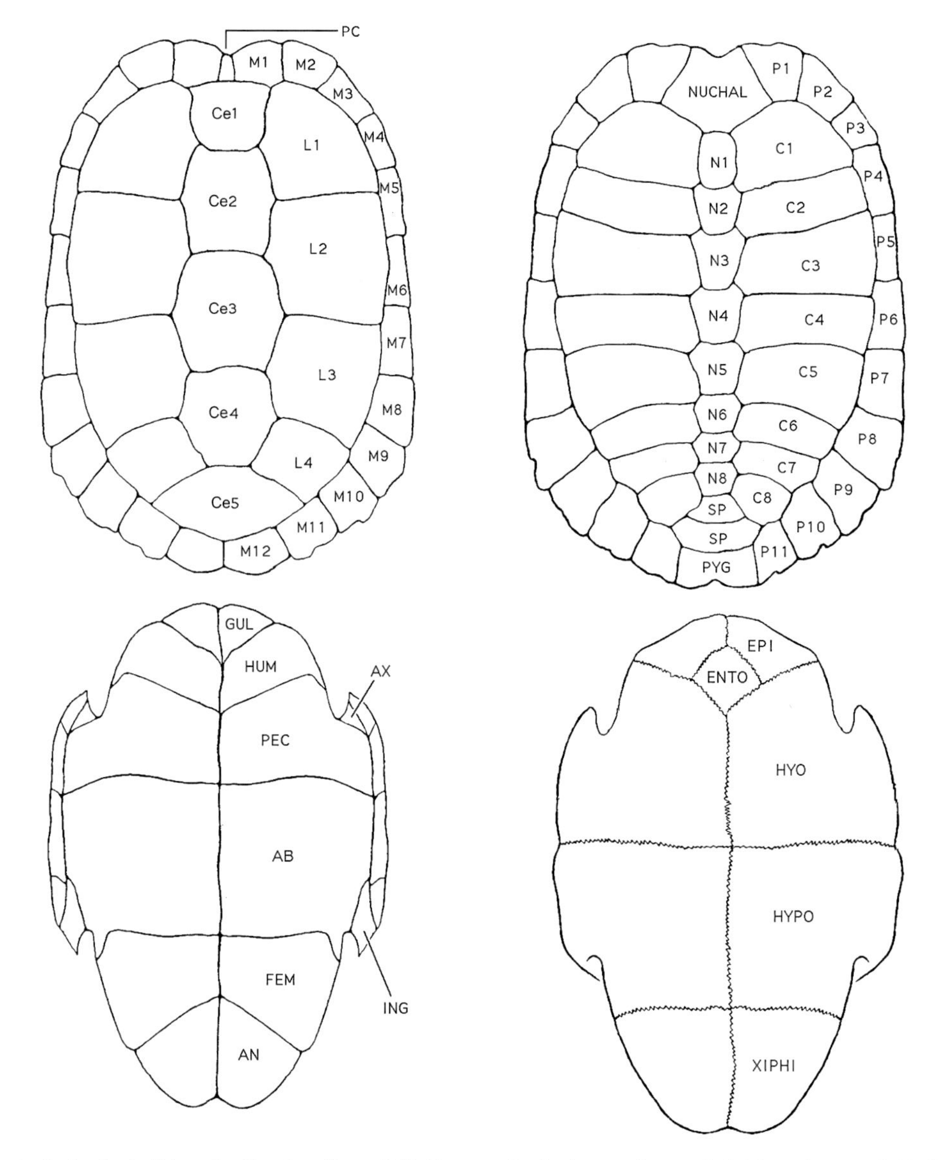

FIGURE 2.3 The parts of a turtle shell (see also Bauplan, Figure 2.2). Carapace (top), plastron (bottom). Epidermal scutes (left), dermal bone elements (right). Based on a generalized emydine turtle (*Trachemys scripta* ssp. ca. 180 mm CL). See also Figure 7.1 for derived features of the kinosternid shell. The axillary and inguinal scutes define limits of bridge (upon which some of the marginal scutes can also be seen). Abbreviations: AB, abdominal; AN, anal; AX, axillary; C, costal; Ce, central; ENTO, entoplastron; EPI, epiplastron; FEM, femoral; GUL, gular; HUM, humeral; HYO, hyoplastron; HYPO, hypoplastron; ING, inguinal; L, lateral; M, marginal; N, neural; P, peripheral; PC, precentral; PEC, pectoral; PYG: pygal; SP, suprapygal; XIPHI, xiphiplastron.

Staurotypus: 1 anterior lobe and hinge, epiplastron-hyoplastron between scutes 1 and 2; no posterior hinge.

Claudius: no transverse plastral hinges; a bridge without buttresses movably articulated to carapace.

Carettochelyidae

Carettochelys insculpta: one anterior lobe and hinge; hinge formed by epiplastra and entoplastron anteriorly, hyoplastra posteriorly.

Emydidae

Emys, Emydoidea, Terrapene: two lobes, one hinge; hyoplastron-hypoplastron; pectoral-abdominal.

Geoemydidae

Notochelys, Cyclemys, Cuora, Cistoclemmys, Pyxidea: two lobes, one hinge; hyoplastron-hypoplastron; pectoral-abdominal.

Testudinidae

Testudo (sensu stricto): 1 hinge, both sexes; posterior lobe; hypoplastron-xiphiplastron; abdominal-femoral.

Gopherus berlandieri: 1 hinge, females; posterior lobe; hypoplastron-xiphiplastron; abdominal-femoral.

Pyxis arachnoides: 1 hinge, both sexes; anterior lobe; epiplastron-hyoplastron; humeral-pectoral.

Kinixys: Carapacal hinge between costals 4 and 5 and peripherals 7 and 8, the intercostal suture aligned with posterior edge of inguinal buttress.

Periodic Plastral Kinesis and Large Eggs

Nearly all the foregoing examples of shell kinesis are of discrete plastron hinges that are derived from relatively straight plastral sutures (prenatally or during early post natal ontogeny); the sutures lose the interdigitating processes, and the fibrous connective tissue is replaced by a more elastic tissue usually referred to as *ligamentous*. These hinges remain for life. Their functions are reviewed elsewhere. In Mexican chelonians, typical movable plastral lobes occur in all kinosternines, *Staurotypus*, *Terrapene*, and *Gopherus berlandieri* and are the result of at least four independent evolutionary events.

Overgrowth of the posterior carapace margin (Figure 30.1) may enhance protection of the hind quarters, but it impedes oviposition and coital access. Among Mexican turtles, this circumstance obtains in *Gopherus berlandieri* and in the three species of *Rhinoclemmys* (probably all members of the genus).

Several kinds of Mexican turtles produce large hard-shelled eggs—singly or in small clutches. This is part of the "tropical" reproductive pattern proposed by Moll and Legler (1971). Passing a large egg is not a problem for any turtle with a hinged plastron. Nonhinged emydid plastra have a posterior carapacal overhang that allows adequate space between the posterior edges of plastron and carapace (Figure 42.1).

Large eggs are a problem for tortoises and for *Rhinoclemmys sp.*; the large brittle eggs are too large to pass through this posterior shell aperture (without breaking). The solution to this problem is periodic and usually brief bouts of shell kinesis. We refer to this as *periodic plastral kinesis*. The lowering of the plastron at oviposition has actually been observed only by Joel Friedman (Bell Research laboratories, pers. comm.) in *R. punctularia* and in *Gopherus berlandieri* by Judd and Rose (1989).

Thus far, a sparse body of data suggests that periodic shell kinesis occurs in all *Gopherus* and *Rhinoclemmys* and in any chelonian where the apposition of the posterior shell edges forms a passageway significantly smaller than the smallest dimension of a hard-shelled egg. The diameter of the pelvic ring is large enough to pass an egg in all the cryptodires examined by us.

The whole sequence of events in the successful laying of large eggs is probably finely coordinated.

M. Ewert (unpublished data cited by Clark, Ewert, and Nelson, 2001) stimulated a *Rhinoclemmys areolata* to lay a large egg with oxytocin; the egg broke during oviposition.

The following observations tell us a little about what might be happening.

In 1964, Legler and E. O. Moll dissected a series of *Rhinoclemmys funerea* in Nicaragua for future skeletal preparation. The adult females were gravid. It was our standard procedure to remove the entire visceral mass in a single step. The eggs would not pass through the effective posterior or the inguinal orifices and had to be removed through the anterior orifice. All eggs could pass easily through the pelvic ring. There was no evidence of plastral kinesis at that time (but the turtles had been stressed by 48 hours in wet bags).

In 2010, Legler and Mark Nielsen examined a dried shell of a *Rhinoclemmys areolata* (Belize, CL 174, UU 6505). The female contained two oviducal eggs with mean width of 27 mm. The effective posterior shell orifice (as accurately measured with a lab stopper) was 17mm. We noted a broad gap (in the dried shell) where dermestids had consumed the soft tissue of the weakly sutural joint between the bases of the hypoplastral buttresses and the adjacent peripheral bones (Figure 2.5, arrows). We soaked the shell overnight in tap water. The results were a general softening of connective tissues and the ability to widen the posterior orifice to 30mm (and beyond) without signs of bone breakage or any other damage to the shell. Slight movement at the hypo-hyoplastra sutures could be detected with force. In a live turtle, this kinesis would have stretched the soft tissue of the buttress-peripheral articulation to about twice its normal breadth. Rose and Judd (1989) conducted a similar experiment with *Gopherus berlandieri* with much the same results.

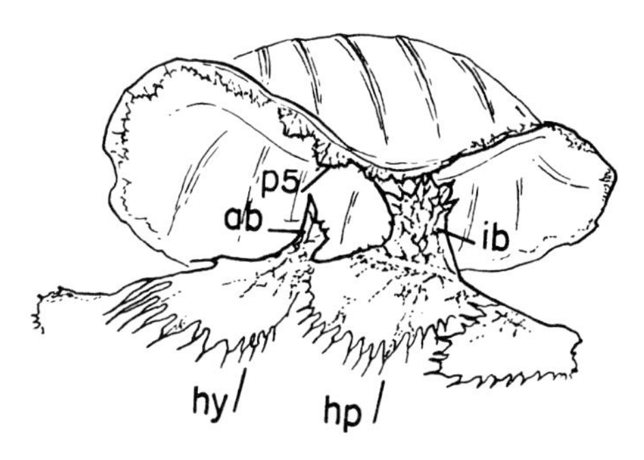

FIGURE 2.4 Neonate shell skeleton of *Terrapene ornata* (drawn from cleared and stained specimen) demonstrating early sutural articulations of plastron. A single hinge will form between the hyoplastron and hypoplastron well after hatching. Note the short axillary buttress. Abbreviations: ab, axillary buttress; hp, hypoplastron; hy, hyoplastron; ib, inguinal buttress; p5, peripheral 5. (Modified from Legler, 1960.)

From the preceding notes, we conclude that some physiological phenomenon softens the connective tissue in plastral sutures for the brief period of oviposition of large eggs. It is known that gravid females of most turtles lay their eggs where and when they please. This would explain the lack of kinesis in a gravid female that is about to be dissected.

We propose that the basic mechanisms of periodic shell kinesis are controlled by a hormone named *relaxin*.

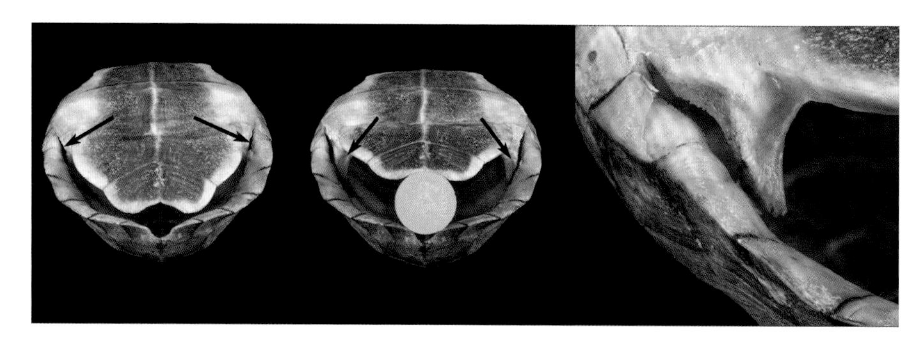

FIGURE 2.5 Plastral kinesis in *Rhinoclemmys areolata* (ventral views of UU 6505, CL 174, gravid female). Left: Dried shell from routine preparation. Note where dermestids have consumed fibrous connective tissue between hypoplastral bridge and peripherals (arrows). Effective diameter of posterior shell orifice, 17 mm. Middle: Same shell after soaking in water. Posterior aperture of shell easily enlarged to mean width of eggs (27 mm) by insertion of laboratory stopper. Note movement of posterior buttresses (arrows) and slight depression of hyo-hypoplastral suture (between scutes 4 and 5). Right: Close-up of right posterior buttress at maximal plastral depression. (Photos by Mark T. Nielsen.)

Relaxin is a peptide hormone of the insulin group that is produced by the female reproductive system (probably the corpus luteum). Relaxin was discovered in 1926 by Fredercik Hisaw. Hisaw (and others) had noted the "extraordinary separation" of pelvic components in pregnant guinea pigs. He reproduced this state in virgin "post oestrum" guinea pigs by injections of blood serum from pregnant guinea pigs and rabbits. Noticeable "relaxation" of pelvic structures appeared in 6–8 hours and persisted for 2–3 days. Serum from males and nonpregnant females produced no relaxation. Similar results (relaxation) were produced by injections of amnionic fluid and saline extracts from rabbit placentae. Hisaw concluded that the ovary was of some importance but not entirely the cause of these early experiments.

Since Hisaw's work, relaxin has been found in various vertebrates and its origin attributed to various parts of the female reproductive system, especially the ovary, its corpus luteum, and progesterone. These more recent findings concentrated on humans (Koob, 1998; Sherwood, 2004).

Collagen (a fibrous protein) is a cheif component of fibrous connective tissue. Relaxin inhibits collagen synthesis (Mookergee et al., 2006). In practical terms, this would loosen and soften the fibrous connective tissue component of a suture and permit a change from rigidity to flexibility. Koob (1998) allows that relaxin has not been identified in reptiles but predicts its presence.

Hofmeyr, Henen, and Loehr (2005) discuss the successful laying of large eggs by a small southern African tortoise (*Homopus signatus*) and attribute this to "pelvic kinesis." They suggest that "hormones may coordinate shell and pelvic kinesis."

The study of periodic plastral kinesis and relaxin is an open niche for further study.

Skull and Jaws

The turtle skull has historically been termed *anapsid* because it was considered to lack true fenestrae (Romer, 1956; Gaffney and Meylan, 1988). This seemed simple enough several decades ago when it was assumed that the anapsid condition in turtles was directly related to that of the stem reptiles. There are now some erudite alternative hypotheses about the nominal anapsid condition in modern turtles (Rieppel and deBraga, 1996; deBraga and Rieppel, 1997).

Roofing of the temporal region, be it primitive or derived, ranges from complete to nil in turtles. The edges of the temporal roof are variously emarginated from behind, from below, or both. Posterior emargination interrupts the parietosquamosal contact and, when coupled with at least slight ventral emargination, creates a lateral temporal arch (*zygomatic* arch) formed from various combinations of the quadratojugal, jugal, and postorbital. This partial reduction is typical of most cryptodires. In some cryptodires, the lateral arch has been lost entirely (Figure 49.2, *Terrapene* skulls). The two families of sea turtles show little or no emargination of any kind (Gaffney, 1979; Boulenger, 1889).

Turtles have relatively large heads, suggesting (to the uninitiated) a large brain and massive intelligence. A cross section of a turtle head, passing through the cranial cavity, reveals the space accommodating the brain to be minuscule in comparison to the total cross section, the mandibular adductor muscles occupying most of the space beneath the temporal roofing, or skull "roof." The quadrate is always firmly attached to the braincase and to the basicranial elements. The dermal roof serves no function in bracing the jaw articulation (Romer, 1956).

Skulls of various Mexican turtles are shown in Figures 6.8, 18.2, 20.3, 25.2, 27.2, 30.2, 48.1, 49.2, 56.1, and 58.7.

The jaw musculature of turtles is unique. In nonchelonian reptiles, the mandibular adductors pursue a fairly direct course to a common insertion on the coronoid process of the lower jaw. In turtles, the same muscles must pass over the enlarged otic capsule to the same insertion (Gaffney, 1975a; Legler, 1993b, Figure 16.5). The otic capsule acts as a pulley or trochlea around which the adductor tendon passes, smoothly translating force from a horizontal

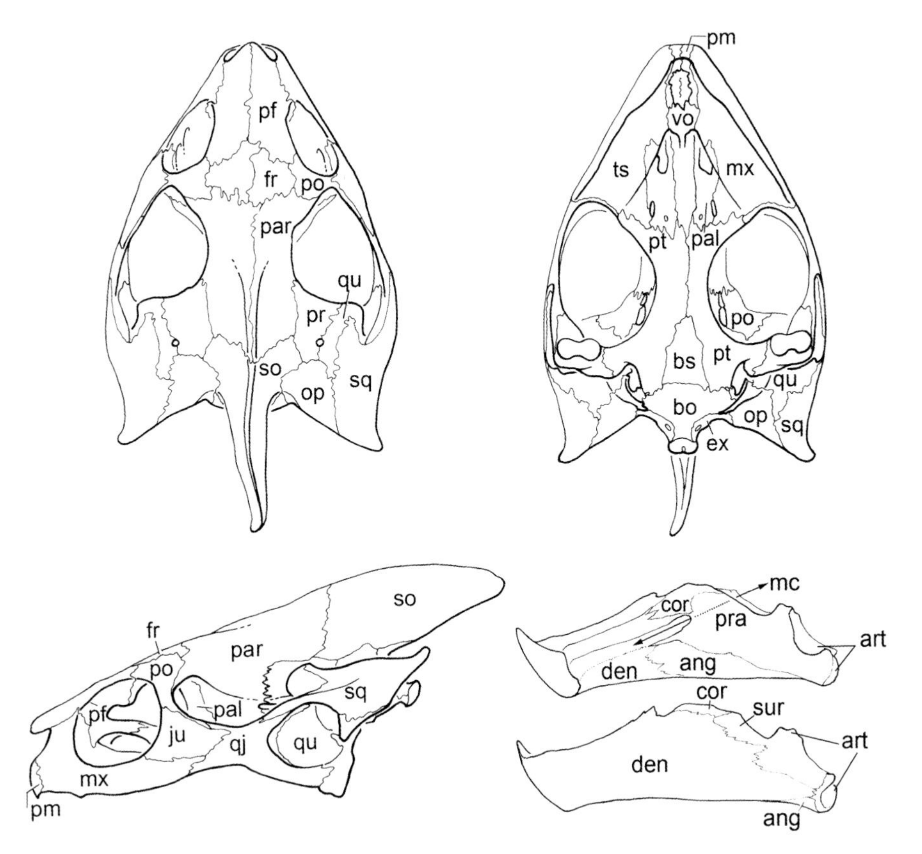

FIGURE 2.6 Bones of the chelonian cranium (Claudius angustatus, AMNH 65865, near actual size, upper row and left below); lowerright, bones of the chelonian mandible, medial and lateral views (*Chelydra serpentina*, AMNH 67015). Abbreviations: ang, angular; art, articular; bo, basioccipital; bs, basisphenoid; cor, coronoid; den, dentary; ex, exoccipital; fr, frontal; ju, jugal; mx, maxillary; mc, Meckelian canal; op, opisthotic; pra, prearticular; pal, palatine; par, parietal; pf, prefrontal; pm, premaxillary; po, postorbital; pr, prootic; pt, pterygoid; qj, quadratojugal; qu, quadrate; so, supraoccipital; sq, squamosal; sur, surangular; ts, triturating surface; vo, vomer. (All figures modified from Gaffney, 1979.)

to a vertical vector. The movement of tendon against bone is lubricated by a true synovial capsule in cryptodires (Gaffney, 1975a).

The upper jaw consists chiefly of the maxillary bones and is solidly braced to the cranium. The lower jaw consists of two rami fused at the mandibular symphysis and each consisting of six bones: the dentary, coronoid, articular, prearticular, angular, and surangular (Figure 2.6). A seventh bone, the splenial, has been lost in most cryptodires. The anlage of each mandible is an elongate mandibular (Meckel's) cartilage, the proximal end of which forms the articular bone. The other five bones are of dermal origin and form a bony sheath around the vestige of the mandibular cartilage (which is exposed along the medial surface of each mandibular ramus).

True teeth have been replaced by horny epidermal sheaths (ramphothecae) over the jaw bones in recent turtles. These sheaths have sharp, vertical tomial (cutting) edges, the lower sheath fitting inside the upper and creating an efficient shearing mechanism. Any reference to "teeth" in turtles alludes to serrations in the horny jaw sheaths and their underlying bone. Serration enhances shearing efficiency. Medial to the tomia there is an horizontal crushing or triturating ledge. The entire triturating surface is covered by a continuation of the jaw sheath. Upper and lower triturating surfaces occlude and, in some turtles, are modified for the crushing of seeds, mollusks, and other hard objects (durophagy).

The triturating surfaces may bear one or more ridges. Ridging has evolved independently several times in both suborders of turtles (e.g., *Dermatemys*, *Batagur*, *Elseya dentata* group) and is associated with herbivory. Both the tomial edge and the triturating surfaces may develop serrations or denticles analogous to teeth.

In Mexican turtles, the jaw armature varies from simple and generalized (e.g., *Chelydra* and most emydids to highly specialized in *Staurotypus* sp. and *Dermatemys*) (Figure 2.7 and individual species accounts).

In *Dermatemys*, we see the quintessence of a vegetation chopping occlusion, in *Staurotypus* a well-developed and braced crushing system, and in *Chelydra* a relatively simple occlusion used principally for shearing or for use as a holdfast device (Figure 2.7) (see also "Feeding Mechanisms" under "Natural History").

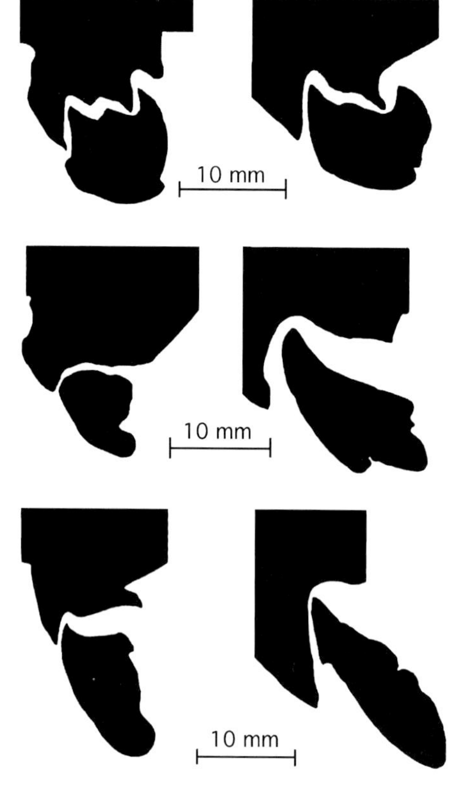

FIGURE 2.7 Sections of three grades of jaw occlusion made from plaster casts. In each row, the section on the left is a transverse, frontal view, midway between the coronoid process and dentary symphysis, lateral to left; on the right is a midsagittal section of the dentary symphysis and overlying cranial elements, anterior to left. From top: *Dermatemys mawi*, *Staurotypus triporcatus*, and *Chelydra acutirostris*. These represent, respectively, a sophisticated vegetation chopping mechanism, crushing plates, and a relatively simple shearing mechanism. (Castings made by Eric C. Felt.)

Other Skeletal Elements

The pectoral girdle is distinctive and triradiate (Figure 2.2). The scapula is L-shaped, its contribution to the glenoid cavity lying at the angle. The longer rodlike scapular limb extends dorsad to the inside of the carapace and the shorter arm, the acromion process, extends anteromediad. The coracoid bone joins the scapula at its angle and contributes to the glenoid cavity; it extends posteromediad in the same plane as and at right angles to the acromion. The arms of the pectoral girdle therefore relate to one another as do the three edges at the corner of a box, the glenoid cavity lying at the corner. One or more suprascapular elements, hinged to the tip of the scapular rod in *Terrapene,* facilitate closure of the anterior plastral lobe.

In cryptodires, the pelvis is movable and free of ankylosed shell attachments; it is fused to the xiphiplastron and the eighth costal element in pleurodires.

The usual and probably primitive phalangeal formula for turtles is 2, 3, 3, 3, 3 (or 2) for manus and pes. *Proganochelys* was already specialized in having a 2, 2, 2, 2, 2 formula. Hyperphalangy is known in the family Trionychidae. Modifications in the flippers of sea turtles result from phalangeal elongation rather than hyperphalangy.

The Cleithrum

The cleithrum is a bone of the old pectoral girdle that has been lost in all modern tetrapods but was present in *Proganochelys* and other turtles until the early Jurassic. It articulated between the anteriormost edges of carapace and plastron in *Proganochelys* (see The Chelonian Shell, Figure 2.1).

Jaekel (1918) correctly recognized the presence of well-developed cleithra in *Proganochelys*; Gaffney (1990) recognized, discussed, and illustrated these elements but referred to them as *dorsal processes* of the epiplastron or *clavicles*.

The cleithrum (Gaffney, 1990, Figure 133) could well have acted as an anterior carapacoplastral buttress in early turtles that had a minimum of anterior shell overhang (Figure 2.1). Loss of the cleithra probably enhanced the ability to retract the head and neck. *Proganochelys* had an extensive covering of osteodermal spines on its soft parts. The cervical vertebral column was comprised of nine short vertebrae. This, considered with the large head, its elaborate osteodermal armor, and the juxtapostion of these members to the V-shaped aperture formed by the cleithra, strongly suggest that the head and neck could not be retracted, nor could the animal achieve any sort of striking behavior with the head and neck (see useful comparisons in Gaffney, 1990, Figures 133 and 137).

In both cryptodires and pleurodires, the cervical vertebral column (eight elements) is modified for extreme flexion at the articulation of cervical vertebrae 5 and 6 and at the articulation of C8 with the first dorsal vertebra. In pleurodires, the C2–C3 joint is also modified (Figure 2.8).

Making room for retraction of head and limbs is a major drawback of a rigid thorax. Obese captives have trouble retracting these members.

Integument

The vertebrate epidermis consists of stratified squamous epithelium that forms an outer cornified layer of variable thickness. The cornified layer consists of dead cells, is protective, and is expendable; it can wear away gradually or be shed in pieces of variable size.

The "soft parts" of turtles consist of the limbs, head, neck and tail (i.e., parts exclusive of the shell or lying outside it). The limbs, tail, and head bear a variable number of distinctive scales, but most of the soft skin lacks discrete, typically reptilian scales, appearing smooth and divided only by shallow grooves or folds. The large scutes of the shell consist chiefly of keratinized epidermis (see the preceding accounts of chelonian shell and upcoming accounts of growth).

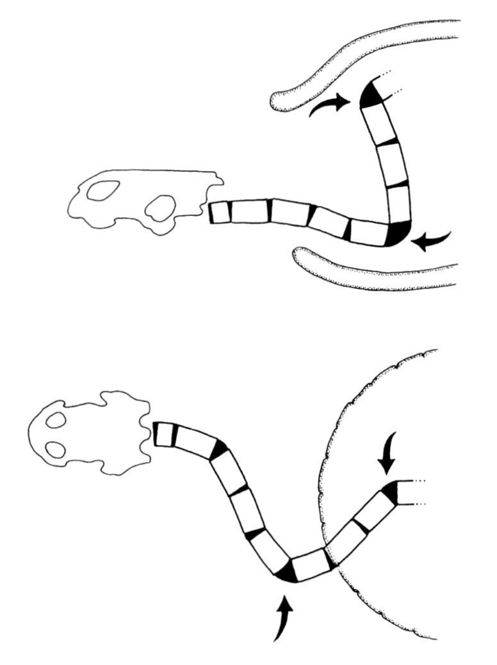

FIGURE 2.8 Neck retraction in the two living suborders of chelonians: Cryptodira (top), Pleurodira (bottom). Maximal flexure occurs at C5-C6 and between C8 and the first dorsal vertebra in both groups, but plane of flexion differs. See Williams (1950) for details of articular surfaces.

Recent studies of chelonian skin have revealed a variety of heretofore unknown integumentary organs (Winokur and Legler, 1974, 1975; Legler and Winokur, 1979; Winokur, 1982b).

Winokur (1982b) recognized three kinds of integumentary appendages in chelonian skin: tubercles, barbels, and fimbriae. Integumentary organs consist of an evagination or invagination of the epidermis associated with the dermis and subcutaneum. Tubercles are conical extensions of the integument that are usually wider than long, only slightly movable, but sometimes cornified and sharp-tipped. Tubercles occur chiefly on the neck and tail and are often in dorsolateral rows. Gaffney (1990) shows neck tubercles but not barbels in an hypothetical reconstruction of *Proganochelys*.

Gular barbels (also termed *mental barbels* or *chin barbels*) are single or paired, usually cylindrical or conical, somewhat longer than wide, blunt-ended projections of the skin just posterior to the mandibular symphysis. Barbels are at least somewhat flexible and can be moved easily by contact with a foreign object and probably by water flow.

Well-developed barbels occur in the Chelydridae, Kinosternidae, Pelomedusidae, and Chelidae and probably evolved independently in these four groups. Some *Dermatemys* have very small tubercular barbels, and a few batagurids have small, paired tubercular barbels associated with mental glands.

As projections of the body surface, integumentary appendages may produce tactile stimuli or a sense of fluid flow before the main surface of the body is contacted, mimicking some of the sensory functions of mammalian hair follicles. Murphy and Lamoreaux (1978) have described the use of barbels in the mating behavior of Australian chelids.

Rostral pores (Figure 2.9) are epidermal invaginations on the narrow isthmus of skin between the nostrils. They may be large and melanistic, or cryptic. They occur in all families of turtles except the two sea turtle families and *Carettochelys*. These epidermal invaginations range from simple to highly branched. Dead cells from the stratum corneum tend to fill the lumen and sometimes form a projecting plug of waxy to keratinous tissue. The dermal papillae associated with rostral pores are highly innervated. Rostral pores may facilitate deep mechanical stimulation of the dermal papillae, via the denser core of keratinized tissue, in a manner analogous to the peritrichal receptors of mammalian hair follicles.

Musk glands (known also as Rathke's glands—Seifert et al., 1994) occur in all turtle genera except those of the family Testudinidae and the *Chrysemys* complex of the family Emydidae (e.g., *Chrysemys*, *Pseudemys* and *Trachemys* in Mexico). The taxonomic occurrence of musk glands suggests that they are primitive structures and that their absence in testudinids and most emydines is derived (Waagen, 1972). Kinosternids are well known for the quantity and pungency of the musk they secrete.

Musk glands (Legler, 1993b) are located within the shell, between the buttresses, in the angle formed by the peripheral bones (the sternal cavity). They develop from ectodermal invaginations that migrate internally before the shell forms. In the course of development, they acquire a tunic of striated muscle and are innervated by the posterior rami of the inner intercostal nerves (Ogushi, 1913; Vallen, 1944, Stromsten, 1917). Their ducts go (usually through osseous foramina or grooves) to orifices near the union of the marginal scutes and the soft skin anterior and posterior to the bridge. Waagen (1972) recognized 10 morphological patterns of glands, ducts, and orifices.

Musk is most likely to be secreted when turtles are traumatized; secretion diminishes or stops in captives that are accustomed to handling. Eisner, Jones, Meinwald, and Legler (1978) isolated the following compounds from the musk of *Chelodina longicollis*: oleic acid, linoleic acid, palmitoleic acid, palmitic acid, stearic acid, citronellic acid, and beta-ionone. *Chelonia mydas* produces PAS-positive, protein-rich nonacidic musk. In *Sternotherus odoratus*, droplets of free lipid are present in the musk (Ehrenfeld and Ehrenfeld, 1973).

The function of musk is moot. Kool (1981) was unable to demonstrate experimentally that the musk of *Chelodina longicollis* deterred various native Australian predators. Dorrian and Ehmann (1988) presented anecdotal evidence that eels and crocodiles are repulsed by *Chelodina longicollis* musk. Musk might function as an advertisement to other turtles that a nearby turtle is being traumatized (JML, pers. obs.).

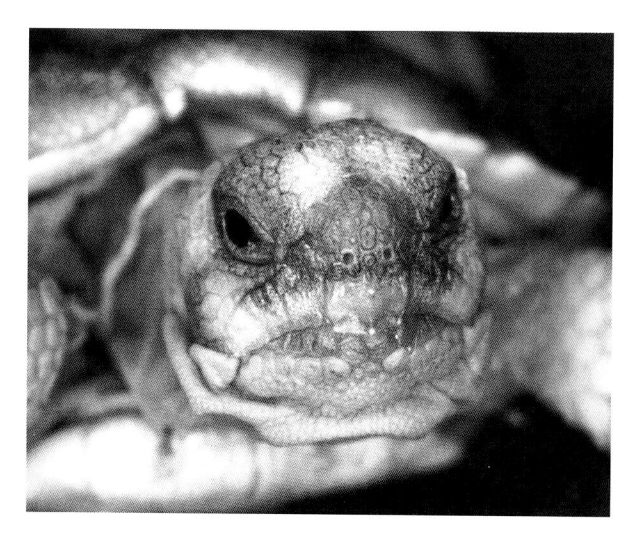

FIGURE 2.9 Rostral pores in a juvenile *Gopherus agassizi*. Three pores in vertical series beginning between nostrils. (Photo by J. M. Legler.)

FIGURE 2.10 Fully developed, secreting mental glands in a mature male *Gopherus agassizi*, ventral view. (Photo by R. M. Winokur). See also Figure 31.1.

Mental glands (Figure 2.10) are paired integumentary glands (or glandular vestiges) on the anterior throat skin. They occur in the families Emydidae (*Deirochelys, Actinemys, Glyptemys insculpta*), Geoemydidae (17 of 22 genera), Platysternidae, and Testudinidae. They are large, highly specialized pheromonal structures in *Gopherus* but are usually cryptic and only marginally secretory in the other taxa. *Gopherus* are the only Mexican chelonians known to have well-developed mental glands (see account of *Gopherus agassizi*). Mental glands were not found in any Mexican *Rhinoclemmys* (Winokur and Legler, 1975) but, in view of their spotty distribution in other members of the genus, they should be sought.

Rostral pores and mental glands may be vestiges of glandular structures in primitive amniotes that have lost their original function but have been re-exploited in groups where the primordia persist (Winokur, 1982b).

The Gut Tube

Parsons and Cameron (1977) review the general internal topography of the gut and Luppa (1977) its histology and histochemistry. Both works are well illustrated and present the turtles in the context of the Class Reptilia. Jacobshagen (1937) represents a substantial early study of the gut.

From mouth to cloaca, there is a continuous tube, most of it derived from the primitive gut and all of it devoted principally to digestion. Digestive processes can be divided into mechanical breakdown, chemical breakdown, absorption, and feces production. The internal topography of the gut is seldom simple and smooth. The various folds and ridges of the tunica mucosa increase the internal surface area (for secretion and/or absorption) and act as pleats for distension. The pattern of ridging and folding is usually longitudinal and less commonly (chiefly in trionychids) transverse or oblique. It would seem that longitudinal folds allow free passage of materials through the lumen whereas transverse or oblique folds impede it (Parsons and Cameron, 1977).

The esophagus is a distensible tube of transmission from pharynx to stomach. Food travels through the esophagus either whole or in large chunks. The tube has no known function in digestive breakdown. Its lining consists typically of longitudinal folds. *Lissemys* is unusual in having transverse or oblique folds (Parsons and Cameron, 1977).

In all sea turtles, there are long, sharp, keratinized, posteriorly directed papillae in the back of the pharynx and the esophagus. They are thought to facilitate swallowing without excessive imbibing of saltwater. Esophageal papillae have also been observed in podocnemid turtles in South America (Vogt, Sever, and Moreira, 1998). Uniquely developed esophageal glands occur in Australian chelids but have not been found (or sought) elsewhere (Legler and Georges, 1993b).

Circulatory System

The erythrocytes are oval and nucleated, and they make up 20–30% of blood volume and carry about the same amount of hemoglobin as mammals. Erythrocyte life span is 600–800 days in one cryptodire (*Terrapene*) (Bellairs, 1969; Dessauer, 1970).

Three main arterial vessels emanate from the heart. From left to right in ventral view these are the right aortic arch, the left aortic arch, and the pulmonary trunk, which almost immediately bifurcates into right and left pulmonary arteries. The right aortic arch gives rise to a massive brachiocephalic trunk that vascularizes the forelimbs, head, neck, and some anterior viscera. The left aortic arch vascularizes most of the coelomic viscera before anastomosing with the right arch to form the single dorsal aorta, a vessel giving rise to all other arteries in the posterior half of the body. Systemic venous drainage is to the right atrium via the sinus venosus; pulmonary drainage is to the left atrium via the pulmonary veins. There is a renal portal system.

The chelonian heart is "imperfectly" four-chambered because the interventricular septum is structurally incomplete (White, 1976). However, the heart performs a separation of blood flow more efficient than its anatomy suggests. The gap in the interventricular septum actually permits cardiopulmonary options in turtles that would be impossible in the completely divided avian or mammalian heart.

The ventricle is actually a single chamber with three nominally interconnected subdivisions. Through a system of internal baffling and pressure regulation, selective shunting of the blood (rather than indiscriminate admixture) is achieved. Adequate studies of flow through the heart have been made in *Trachemys scripta* and *Chelydra serpentina*. During normal air breathing, systemic venous blood is returned to the right atrium, flows through the right atrioventricular valve into the ventricle to the right of the septum and thence is pumped to the lungs via the pulmonary arteries for oxygenation. Oxygenated blood is returned to the left atrium via the pulmonary veins, flows through the left atrioventricular valve into the ventricle to the left of the septum and is pumped into the two aortic arches. The hiatus in the septum is termed the *interventicular canal*. Some of the blood passing through the left atrioventricular valve is shunted across the interventricular canal and back into the pulmonary circuit. Therefore, under aerobic conditions, blood going to the lungs is of higher quality than that entering the right atrium. During a dive, when the lungs are not in use, most of the systemic blood being returned to the right side of the heart circumvents the pulmonary circuit and is routed, via the interventricular canal, directly into the aortae. The details of this process are complex, and the reader is referred to White (1976).

The heart beats more slowly during a dive and, if a turtle remains inactive, energy requirements are less than in air. Glycolysis and other anaerobic pathways then meet vital needs. Turtles store more tissue glycogen than other reptiles. Tolerance of turtles to anoxia exceeds that of all other tetrapod vertebrates. Some *Chrysemys picta* have lived for 3 to 4 months submerged at 1.5–3.5°C (Dessauer, 1970). Anoxia tolerance for various turtles studied by Belkin (1963) ranged from 1.9–33.0 hours. *Chelonia mydas* spends the winter underwater at 15°C in the Gulf of California (Felger, Clifton, and Regal, 1976), and *Caretta caretta* hibernates under water off Florida (Carr et al., 1980).

Respiration

All turtles have lungs that function in typical pneumatic breathing. Many aquatic turtles have evolved alternative respiratory structures analogous to gills. No turtle is known to have reduced the size or efficiency of the lung.

The basic lung plan in turtles consists of eight distinct subdivisions or lobes; four of these are small and medial and four are lateral, larger and much wider on the transverse axis than the longitudinal axis. This plan is typical of generalized cryptodires (Gräper, 1931; Wolf, 1933; Legler, unpublished data). Specializations of the lung result from subdivision of the eight basic lobes. The intermediate septa are complete and the subordinate lobes individually ventilated in sea turtles and trionychids. In kinosternines and staurotypines, the basic lobes are incompletely divided. Tortoises (e.g., *Gopherus*) have a large, boxy, seemingly thinner-walled lung with the requisite eight lobes but with a minimum of intralobular subdivision.

Because the intercostal spaces are filled with bone, the intercostal musculature has become obsolete. There is some evidence that striated muscle lies directly on the lung and is homologous to the intercostal muscles (George and Shah, 1954). The homologues of the lateral belly muscles (obliquus abdominis and transversus abdominis) remain and are used in respiration.

The lungs are about half the length of the carapace when inflated. They extend roughly from the pectoral to the pelvic girdle, occupy the dorsal part of the peritoneal space and are vulnerable to dorsal puncture wounds such as spears, bullets, and crocodilian teeth.

The coelomic cavity has two divisions: a pericardial cavity containing the heart and a common peritoneal cavity for everything else. There is no diaphragm and no dedicated cavity for the lungs. In some testudinids, the common cavity is incompletely divided by a sheet of connective tissue beneath the lungs (Gans and Hughes, 1967).

Nonchelonian reptiles can exploit the flexibility of the thoracic wall to affect increase or decrease in coelomic volume and aid in lung ventilation. With a rigid thorax, turtles do not have this option and nowhere are the constraints of this rigidity more evident than in ventilatory mechanics.

The definitive modern works on lung ventilation are Gans and Hughes (1967) for a testudinid (*Testudo graeca*) and Gaunt and Gans (1969) for an aquatic turtle (*Chelydra*).

Turtles utilize the revised lateral belly muscles to affect ventilatory movements. The transverse abdominis muscle is a contractile sling running beneath the peritoneum and the posterior viscera and attaching to the inner surface of the carapace on each side. When it contracts, it lifts and compresses the visceral mass and indirectly exerts pressure on the lungs. The oblique abdominis muscle lies just beneath the subcutaneum of the inguinal pocket and attaches to the edges of the carapace above and the plastron below. When it contracts, it flattens the inguinal depression and decreases intracoelomic pressure. There is another muscle, the diaphragmaticus in *Chelydra*, that works synergistically with the transversus to compress the lungs (Gaunt and Gans, 1969). Testudinids achieve further coelomic compression by rotating the pectoral girdles inward (Gans and Hughes, 1967). Complete or partial retraction of the head and limbs can reduce the volume of the coelom in any turtle (often accompanied by an audible hiss).

Despite being equipped with an efficient heart and good lungs, aquatic turtles have evolved various alternative strategies to extract oxygen from the water. Ward (1970) has demonstrated that various unmodified parts of the gut can extract oxygen from water.

Cloacal bursae are dorsolateral diverticula of the cloaca and are unique to turtles. Cloacal bursae are better developed in aquatic than in terrestrial turtles. Cloacal bursae range from small, thin-walled sacs with a smooth simple lining to large bladderlike structures with an elaborate, vascular, papillose lining. They are absent in the Trionychoidea, in the two families of sea turtles, in all testudinids, and in some (usually terrestrial) emydids and geoemydids. They reach a quintessence of development in certain Australian chelids (e.g., *Rheodytes*) (Legler and Georges, 1993; Mark Nielsen and Legler, unpublished ms.).

All aquatic turtles practice some form of buccopharyngeal pumping, whereby the volume of the mouth and pharynx is increased by lowering the hyoid apparatus. Water is drawn into the cavity through the nostrils or the mouth or both and then forced out through either or both passages. The buccopharyngeal mucosa, especially the roof of the mouth, is vascular, and exchanges of oxygen and carbon dioxide can occur there (Ward, 1970). *Apalone* and *Dermatemys* have vascular, branched, buccopharyngeal papillae that function as gills (Winokur, 1988).

Seemingly, no turtle taxon has both cloacal bursae and specialized structures for buccopharyngeal respiration. The two families of sea turtles lack both. All alternative respiratory structures are in need of study.

Alternative modes of respiration may suffice to support a turtle at rest for hours, days, or months, depending on environmental conditions (e.g., water temperature). Under conditions of strenuous activity (e.g., being chased by a human diver), they must revert to pulmonary respiration. Legler and James F. Berry crudely tested this idea in northern Florida in May 1975. We swam together on the surface until we observed a large adult female *Apalone ferox* fleeing from a place of concealment. Thereafter we worked as a team, one man under water in chase and the other swimming at the top. The turtle became progressively slower and weaker and ultimately rose to the surface in complete exhaustion and bobbed up and down in a period of recovery using pulmonary respiration. It could be approached and handled with impunity. In the same period, we tried the same experiment with a large adult *Chelydra serpentina*. *Chelydra* has relatively small, simple cloacal bursae. The results were essentially the same. These two incidents are in sharp contrast to certain Australian chelids (e.g., *Rheodytes*, Legler and Cann, 1980) with huge cloacal bursae, that when, under chase, swim with the cloaca wide open (exposing a red interior lining of long vascular fimbriae) and can sustain a chase longer than a team of divers.

Thermoregulation

Turtles are ectothermic heliotherms, deriving heat chiefly by moving into the sun or into a warmer medium. Ectotherms achieve a comfortable operating temperature, conduct the important functions of life while maintaining this temperature, and retire to safety when this is no longer possible. There is thus a period of vulnerability at emergence. Ectothermy permits survival (including recovery from injury) for long periods with minimal sustenance or breathing. Reptiles lack the insulation imparted by the fur, feathers, and subcutaneous fat of mammals and birds, but the chelonian shell is probably better insulation than the dermis of other reptiles.

Turtles can acclimate to a wide variety of environmental temperatures. Preferred temperature range in most turtles is 25–33°C (Hutchison, 1979). Critical maxima range from 39–42°C (Bellairs, 1969; Moll and Legler, 1971).

Moll and Legler (1971) demonstrated that a large *Trachemys* warmed by basking could maintain a deep-core body temperature above that of the water for several hours after reentry. Basking turtles tolerated basking temperatures close to the lethal maximum and showed signs of distress before entering the water.

The leatherback sea turtle *Dermochelys* feeds as far north as the Arctic Circle and can maintain deep core temperatures as high as 18°C above cold seawater by virtue of large size, use of peripheral tissues as insulation, and countercurrent exchange in the limbs (Paladino, O'Connor, and Spotila, 1990; Frair, Ackman, and Mrosovsky, 1972; Greer, Lazell, and Wright, 1973; Standora, Spotila, Keinath, and Shoop, 1984; Spotila and Standora, 1985).

Urinary System

Fox (1977) presents a review of the urogenital system in reptiles and a few illustrations of urogenital structures in turtles. Adult turtles have a metanephric kidney. Turtles have a cloaca, with a common external orifice, for gut, urinary, and reproductive systems. The posteroventral ureter is short and has a two-layered tunica muscularis. A large urinary bladder joins the ventral wall of the cloaca. The ureters and bladder communicate separately with the cloaca. Renal arterial circulation penetrates the kidney dorsally as interlobar arteries. There are several thousand nephron units in both kidneys (13,000–16,000 in one cryptodire studied). Glomerular diameter varies from 50 to 110 micrometers in several cryptodires reported. Dantzler (1976) gives the relative lengths of renal tubule segments in one freshwater turtle (*Chrysemys*) as proximal tubule 1.5 mm, intermediate segment 0.3 mm, and distal tubule 1.0. The chelonian kidney lacks a Henle's loop, a renal pelvis, pyramids, and the tubular "sex segment" of squamates.

The amniote kidney has two important functions: elimination of nitrogenous waste and maintenance of water balance. Dantzler (1976) reviews the form of urinary nitrogen in chelonians. Values for aquatic turtles are ammonia 20–25%, urea 20–25%, and urates 50%. Data given by Dantzler and Schmidt-Nielsen (1966) contrast an aquatic emydid (*Trachemys scripta*) and a fully terrestrial desert tortoise (*Gopherus agassizi*) as (1) *Trachemys*, ammonia 4–44%, urea 45–95%, and urates 1–24%; and (2) *Gopherus*, ammonia 3–8%, urea 15–50%, and urates 20–50%. Seemingly little water is absorbed through the skin of aquatic turtles; if excess water enters the gut, it can be passed in dilute urine (Bellairs, 1969).

Chelonians can store water in the urinary bladder, in the cloacal bursae, and seemingly in coelomic fluid. Presumably these reserves can be reabsorbed if necessary. It is common for aquatic turtles and tortoises of all kinds to squirt varying amounts of liquid from the cloaca when handled.

Osmoregulation

Water comprises about 65% of body mass in turtles (Bellairs, 1969). Freshwater turtles live in a low-sodium environment, whereas the converse is true of sea turtles. Plasma osmolarity is seemingly variable. Range of variation in seven testudinids (no species given) was 278–400 mos/L. *Trachemys* and *Pseudemys* occur in water salty enough to support mangroves, and *P. concinna* enters saltwater regularly in the Gulf of Mexico (Moll and Legler, 1971; Carr, 1952).

The chief osmoregulatory strategy in freshwater turtles seems to be the production of dilute, low-sodium urine. *Apalone spinifera* has special cells in the pharynx that can absorb sodium from water containing as little as 5 mM/liter. If sea turtles swallow saltwater, they must absorb salts from the gut and then eliminate them. Sea turtles and some estuarine turtles (*Malaclemys*) have enlarged lacrimal glands that can secrete salty tears. The tears of *Caretta* have a concentration of 810–992 mM chloride (732–878 mM sodium) compared to 470 mM sodium in normal seawater (Dessauer, 1970; Bellairs, 1969; Dunson, 1976).

Eye and Vision

Turtles have a generalized eye that functions well under a variety of conditions. Details of eye structure appear in Walls (1942), Underwood (1970), Peterson (1992), and Granda and Sisson (1992).

The pupil is always round in turtles. The surrounding iris may be dark or pale (golden, yellowish, or greenish) and contrasts sharply with the pupil. There is often a bright metallic-looking ring at the rim of the pupil. A bright iris is very evident under water. The iris is often disrupted by patches of melanin; these may blend with the pupil to form a longitudinal stripe or be arranged in a stellate pattern. These patterns seem to camouflage the iris. It is of interest that a horizontal stripe is always aligned approximately perpendicular to Earth's gravitational force (Figure 39.1 and JML, pers. obs.).

The supportive layer of the eyeball contains a ring of 6–13 platelike bones, the scleral ossicles. The entire numeric range of sclerotic ossicles for the reptiles can be seen in the turtles (Underwood, 1970).

During immersion in water, light is focused on the retina by modifying the shape of the flexible lens. The lens is surrounded by a ring of muscle, the ciliary body. The lens is "set" at distant vision when the ciliary body is relaxed. When it contracts, it squeezes the lens and thickens it from anterior to posterior for near vision. The cornea does much of the work of focusing in aerial vision (Romer and Parsons, 1977). Many turtles are adapted to vision in both air and water. Underwood (1970) compared eye structure in *Emys, Testudo* (sensu lato), and *Caretta. Emys* (semiaquatic) has a strongly convex cornea and a thick lens; out of water, the cornea is the principal focusing surface. *Testudo* (terrestrial) has a cornea similar to that of *Emys,* but the lens is flatter and the ciliary body less well developed. In *Caretta* (marine), corneal curvature is reduced and the lens is significantly more curved than that in *Emys.*

The retina contains single and double cones and one kind of single rod. The double retinal cells may serve as detectors of polarized light. Nocturnal turtles have more cones than rods. Reported percentages of cones in the retina are *Chelydra serpentina* 60%, *Chrysemys picta* >90%, and *Emydoidea blandingi* 75% (Underwood, 1970).

Two kinds of glands are associated with the eyeball. Harderian glands lie anteromedial to the eyeball, and lacrimal glands lie posterolateral. Harderian ducts enter the ventral part of the space between the nictitating membrane and the eyeball. Lacrimal ducts enter this space from the inner surface of the lower eyelid. Lacrimal glands are of universal occurrence in chelonians; the secretion lubricates and cleans the conjunctival cavity. The lacrimal glands are large and secrete highly concentrated tears in sea turtles and some estuarine turtles (*Malaclemys*). Good descriptions of Chelonia and *Caretta* glands appear in Abel and Ellis (1966).

The eyelids are modifications of the body skin. With the eye closed, the edges of the eyelids tilt upward from front to back. This aligns the edges of the eyelids with the water surface when a turtle is floating diagonally.

Nictitating membranes are of common occurrence in reptiles and birds; they constitute a third, transparent inner eyelid that can be drawn from front to back across the eye to protect it without obscuring vision.

Nictitating membranes are absent in the families Kinosternidae, Dermatemydidae, and Carettochelyidae. They are present and well developed in all other families of cryptodires. See Legler (1993b) and Bruno (1983) for occurrence in the pleurodires. Taxa that lack or have rudimentary nictitating membranes also lack Harderian glands and ducts, and tend to have translucent to transparent lower eyelids (Legler, 1993b; see also account of *Kinosternon flavescens*).

The Chelonian Ear and Vocalization

The external ear is absent, the middle ear is unique and complex, and the inner ear is poorly known (Baird, 1970). The tympanic membrane lies flush with the side of the head, is attached to the circular rim of the quadrate bone, and is covered with ordinary body skin. The air-filled tympanic cavity lies immediately below the tympanic membrane. A distal expansion of the columella is attached to the inner surface of the tympanum. The eustachian tube passes from the floor of the tympanic cavity to the pharynx. The slitlike pharyngeal eustachian orifices are easily seen when the mouth gapes widely. In preserved turtles, pressure exerted on the tympanic membranes will eject fine streams of fluid from these orifices.

The stapes or columella is the only osseous element in the middle ear (Baird, 1970). Its distal part is the cartilaginous extrastapes, an expansion of which is attached to the inner surface of the tympanic membrane. The extrastapes joins the stapedial shaft, which passes through a foramen in the quadrate, traverses the recessus cavi tympani, and expands into a vertically oval footplate articulated to the vestibular window.

Part of the stapedial footplate abuts a fluid-filled sac, the paracapsular sinus. The shaft of the stapes lies on the sinus and indents it dorsally. The sinus occupies most of the recessus cavi tympani, and its posterior part lies against the lateral surface of the periotic sac. According to Baird (1970: 262), the paracapsular sinus is a "dampening mechanism" unique to turtles and has no homologue in other reptiles.

The range of audible frequencies in the human ear is from 20 to 20,000 Hz, usually decreasing in the higher ranges with age. This range extends over 11 octaves, of which 7 are used in musical instruments such as the piano. Ordinary conversation falls within the range of 300–3,000 Hz (Kiernan, 2004; Kandel, Schwartz, and Jessell, 2000). Speed of transmission through water is three times as fast in air and even faster in a solid medium.

Isolated notes have been published on airborne sounds emitted by chelonians: various testudinids (Bogert, 1960; Auffenberg, 1964, 1978; Campbell and Evans, 1967; Jackson and Awbrey, 1972; Sacchi, Galeotti, and Fasola, 2003; Sacchi, Galeotti, Fasola, and Gerzeli, 2004; Galeotti et al., 2005a; and Galeotti, Sacchi, Rosa, and Fasola, 2005b); a sea turtle, *Dermochelys* (Mrosovsky, 1972); and a few freshwater species—*Pseudemys floridana, Glyptemys insculpta, Platysternon megacephalum* (Allen, 1950; Kaufmann, 1992; Campbell and Evans, 1972). Carr (1952) commented that *Gopherus polyphemus* made short rasping calls and another sound like the mewing of a kitten.

None of these reports described complex acoustic displays, and none suggested vocalization as a means of communication. Most of the sounds produced were attributed to involuntary exhalations. However, Sacchi, Galeotti, and Fasola (2003, *Testudo marginata*) and Galeotti et al. (2005a, 2005b, *Testudo hermanni*) have shown that the sounds produced by these tortoises are important in various breeding activities, including mate selection.

Most tortoise vocalizations have been associated with breeding. The roaring or bellowing of *Chelonoidia nigra* are well known (Bogert, 1960; Jackson and Aubrey, 1972), as are the clucking sounds made by *Chelonoidia carbonaria* (Campbell and Evans, 1967).

The first published report of infrasound being used for communication by chelonians is that of Ashton and Ashton (2008) for *Gopherus polyphemus*. Males responded to sounds made by other males in their burrows. The sounds were of low frequency (ca. 200 Hz), near the lower limit of human perception. They were interpreted as territorial calls. Additional work has been done by Ray Ashton (www.ashtonbiodiversity.org) on low-frequency sound perception in *Testudo hermanni, Geochelone sulcata,* and *G. elegans.*

Allen (1950) first reported sounds made by freshwater turtles. Captive slider turtles (*Pseudemys floridana suwanniensis* in a large "pen") made short, low, deep-throated grunts while floating at the surface with the head held high and the mouth closed. The neck pulsated with each grunt. Allen thought this behavior plays some part in courtship.

The most recent scientific advance in our knowledge of vocalization in freshwater turtles is that of Giles (2005) as a doctoral student at Murdoch University, Western Australia. Her thesis on sound production (hereinafter "vocalization") was based on a long-necked chelid turtle (*Chelodina oblonga*). By recording with special hydrophones in aquaria, she discovered that the turtles produced a repertoire of sounds in the range of 120–3,800 Hz, with the dominant frequencies in the range of 160–1,800 Hz and a duration of 0.07–57.0 seconds. These sounds were made underwater with the mouth closed.

Giles identified 17 different sounds, all within the audible range of humans. Bouts of one or more vocalizations included rolling pulselike sounds of 2.7 seconds each that continued for 1–5 minutes. One pulse bout lasted 9.5 minutes. Vocalizations ranged from 100 Hz in percussive sounds to 3,500 Hz in complex calls. Some of the non-percussive sounds are reminiscent (to us) of sounds made by cetaceans.

Vocalizations were categorized subjectively as clacks, clicks, squawks, hoots, short chirps, high short chirps, medium short chirps, long chirps, high calls, cries or wails, cat whines, grunts, growls, blow bursts, staccatos, wild howls, and drum rolling. Chirp calls were the most common and were the only calls recorded for juveniles. Complex vocalizations included harmonically related elements and different rates of frequency modulation. Clicks extended above the 20,000 Hz upper limit of the recording equipment used. Giles's work is reported in Giles, Davis, McCauley, and Kuchling (2009).

Giles later worked with Vogt in Brazil (in 2006), where they successfully recorded underwater vocalizations of *Podocnemis expansa*. Since then, Vogt and students have recorded underwater vocalizations (in the field and laboratory) from other South American pleurodires (*Podocnemis erythrocephala* and *Peltocephalus dumerilianus*) and two North American cryptodires (*Chrysemys picta* and *Emydoidea blandingi*).

Vogt and students are now in the early stages of determining the circumstances under which different sounds are emitted and if and how other turtles respond to them, either by action or vocalization. Their studies are being conducted in the field with a very social species of turtle, *Podocnemis expansa*, that is known to migrate to the nesting beaches in groups, bask on the beaches in groups, nest in arribadas, and remain in front of the nesting beaches in deep holes in the river for as long as two months after nesting (Vogt, 2008).

Vogt and students have recently discovered that hatchlings are vocalizing in the nest and during their scampering to the water. Vocalization intensity increases when they enter the water. Females respond to this with their own vocalizations. Females and hatchlings were found surfacing together and subsequently moving downriver together. These data suggest that females and hatchlings migrate from the nesting beaches together and are using vocalizations to maintain the group. Vocalizations could also be used to initiate other kinds of grouping, migrations, emergence from the water to bask or nest, as well as in escape behavior—an entire group returning to the water when threatened during basking or nesting. Maternal behavior and vocal communication seem to be better developed in *Podocnemis expansa* than in any other species yet studied.

The overlap in sound frequencies produced and perceived by turtles and humans is of particular interest. We have both dived for turtles for many years. It would have been good to know what to listen for or to spend a night listening in the dark of an aquarium room. However, the amplitude of sounds produced by turtles underwater is low, and vocalizations are difficult to hear even when recorded with filtered background sounds. Thus far it has been necessary to make recordings in isolated environments (e.g., swimming pools or artificial ponds) where there were no sounds from fishes, invertebrates, dolphins, wind, or rain. Once the frequencies of turtle vocalizations are determined, sounds emanating from other sources ("noise") can be removed from the recording. Perhaps turtles do this naturally.

Exactly where and how turtles make these sounds is uncertain. Sacchi, Galeotti, Fasola, and Gerzeli (2004) described the skeletal framework of the larynx in tortoises to be a rather simple structure of three cartilages—the cricoid and two arytenoids. The opening and closing of the glottis is controlled by two pairs of muscles. They found two bands of elastic fibers on the anterior wall of the larynx. These fibers were capable of vibration, suggesting that they might be the equivalent of vocal cords.

Although it is clear that chelonians make sounds (other than simple aspirations) and that other turtles are capable of hearing them, there are currently few clear data that other turtles respond vocally to them, thus the term *communication* is used hypothetically. Giles (2005) did, however, report that when a 14-second recording of a vocalization was played back, turtles within range assumed an "alert" posture.

Vocalization by chelonians, whether on land or in the water, is a reality that urgently requires study. It is currently assumed that other turtles can perceive these sounds and respond to them. We suggest that a vocal signal could, at the very least, be an advertisement of presence (e.g., "I am here"). These signals and their perception by other turtles could be important in almost any aspect of life history that necessitated the locating of one turtle by another—whether to avoid it or seek it—especially in murky water that impedes vision (e.g., anything from mating to avoidance of injury when feeding in a group or group aggregation for nesting).

As this book goes to press, it is parsimonious to predict that all aquatic turtles and probably all chelonians can vocalize and can perceive such vocalizations by other chelonians. This matter is a fertile field for laboratory and field investigation.

Smell and Taste

The nose is a passageway from the external nares to the internal nares in the roof of the buccal cavity. The nose is used in normal aerial breathing and for olfaction and aquatic respiration. Detailed sections of the nasal region in several kinds of turtles can be found in Parsons (1970). He states that turtles have the most distinctive nasal anatomy of the reptiles, and he regards this pattern as primitive for reptiles.

Immediately inside the external naris is an expanded vestibule followed by a short, narrow passageway and then the expanded main nasal chamber (cavum nasi proprium). The roof of this chamber contains the olfactory epithelium (Romer and Parsons, 1977; Parsons, 1970). Olfaction and gustation rely on bringing molecules into close proximity with chemoreceptive nerve endings. Most turtles have widely scattered taste buds in the buccopharyngeal region (Winokur, 1988). In an aquatic medium, both olfactory and gustatory senses are probably used to sense nuances of chemical change. Experienced turtle collectors usually bait their traps with aromatic substances (e.g., sardines and bananas).

Locomotion

Most chelonians achieve forward motion by moving their limbs in various versions of what can be called a trot (Walker, 1973, 1979; Zug, 1971). The front and hind limbs on opposite sides provide propulsive force while the contralateral limbs recover. This pertains to swimming, bottom walking, and terrestrial locomotion. The major forces to be counteracted are resistance in water and gravity on land. All chelonians can move about to some extent on land and in water, and most are fairly efficient at both.

Our observations of turtles underwater result from chasing them and are heavily biased toward escape behavior. A turtle resting at the surface usually is oriented at an angle of about 45° to the surface; the hind limbs are extended, the toes and interdigital webbing are spread, and the foot is extended to a nearly horizontal position. When startled, the turtle quickly pulls itself backward and underwater with the dorsal surfaces of the hind feet and thereafter initiates a stereotypical forward locomotion using all four limbs. Typical escape behavior in deeper water is toward the bottom in a long, curving, banking path. At the bottom, the turtle usually pauses momentarily, turns its body by obtaining purchase on the substrate, and swims off again at speed, usually in a different direction. This has been termed *terminal reverse behavior* and has been observed in many kinds of freshwater turtles (Legler, 1993b). It is difficult for a good swimmer with ordinary snorkeling gear to equal the speed of an escaping turtle. Turtles are best caught by turning inside the turtle's trajectory and working in concert with another diver.

Unless chased, turtles seldom move rapidly in nature. Observations of undisturbed turtles moving in clear water show slow deliberate movements with frequent pauses for rest, circumspection, and feeding. Rapid movements are brief, quite obvious, and normally associated with feeding, aggression, or mating. Most freshwater and terrestrial chelonians probably have little reason to move long distances.

Natural History

Autecological Studies

Knowledge of Mexican turtles was chiefly at an alpha taxonomic level at mid-twentieth century. As collections were enlarged and field observations were recorded and analyzed, it became possible to form autecological profiles of many species. This phase is still ongoing and should ultimately lead to the same quality and magnitude of synecological studies that have produced massive bases of knowledge for species such as *Chrysemys picta* and *Chelydra serpentina*. If we do not understand the elements of an animal's life history—how it conducts its life from day to day, from season to season, and from natality to death (with an emphasis on diet and reproduction)—we cannot proceed with studies of its relationship to the ecosystem.

Simply observing, recording, and then collating what an animal does from day to day and season to season—especially what it eats and how it reproduces—is a worthwhile endeavor (albeit belittled in certain branches of ecology). It is only with such information that sound hypotheses can be formed on life-history strategies, and it is the workers who gather the data themselves who produce the best hypotheses. Observations need not depend on grants. The work of Legler's mentor, Henry S. Fitch (1909–2009) at the University of Kansas, bears this out and emphasizes the value of autecological studies.

Both authors of this book started their careers prior to the age of grants and before their association with graduate programs and mentors. Collectively, we and our students have personally studied nearly all of the taxa covered in this book and worked in most of the places mentioned.

Significant observations and gathering of data can begin with little more than average vision and hearing, literacy, a pocket notebook, and a writing instrument. Binoculars and a camera help but are not initially essential. We encourage young Mexican scientists to gather biological data in this manner and to enjoy the circumstance of discovery.

Habitat

Many chelonians have experimented with terrestriality (e.g., *Terrapene* in North America and *Cuora* in Asia). Some chelonians can be categorized as semiaquatic—spending a substantial part of their lives in water but predictably emerging from water for purposes other than nesting. Semiaquatic turtles include most of the *Clemmys* group (now placed in the genera *Clemmys, Actinemys,* and *Glyptemys*) in the New World and many geoemydids, including *Rhinoclemmys* in Mexico. Semiaquatic turtles have fewer specializations for terrestrial life than true tortoises. Indeed, the earliest turtles may have been semiaquatic or even terrestrial, and the selective factors related to the evolution of the unique shell may well have been a protective advantage in the terrestrial phase of this existence.

Diet and Feeding Behavior

Acquiring nutrition is the most important thing an animal can do. The terms *carnivory, herbivory,* and *omnivory* should reflect overall, long-term diet and any suite of modifications that has evolved with these dietary habits. The terms should not reflect momentary preferences or what captives are fed.

Most chelonians are opportunistic omnivores or retain the potential to be so. They survive and prosper by eating what is available. There are relatively few examples of complete, obligatory herbivory or carnivory in turtles. North American slider turtles *(Trachemys scripta)* and box turtles

(*Terrapene* sp.) are good examples of versatile omnivory. There is a general tendency for juveniles of omnivorous and herbivorous species to be more carnivorous than adults (Moll and Legler, 1971). Major exceptions to these generalities in Mexico are the nearly complete herbivory of *Gopherus* and *Dermatemys* and the carnivorous specializations of *Claudius*. Despite these generalizations, chelonians seldom ignore a windfall (e.g., figs) or an unexpected chunk of animal protein, including carrion. Flexible dietary habits have probably played a major role in chelonian venerability.

Foraging

The final solution to what an animal eats lies in its gut. Knowledge of natural diet has accelerated since the development of stomach flushing (Legler, 1977; Legler and Sullivan, 1979), a process whereby a canula is inserted through the esophagus into the stomach of a live animal and the stomach contents are displaced by water without harm to the animal. The neck of the turtle must be straightened to avoid puncture of the esophagus. The initial technique used a continuous pump syringe as a controlled source of pressurized water (Figure 5.8). For larger turtles, we have since used various pressurized water sources such as plant sprayers and direct connections to a plumbing system with a foot-operated control valve.

Stomach flushing is benign; we have collectively flushed several thousand turtle stomachs with only a few fatalities due to the procedure. Stomach flushing obviates the need to sacrifice animals to study diet. The technique provides incontestable evidence of what an animal is eating. The animal can be released unharmed after a brief period of recovery.

However, gut contents represent only a brief point in dietary time (Legler's students came to call this "the last supper"). A series of many flushings over a longer period would better define general dietary trends in a population. For example, the dietary profile of a population in a small stream may be significantly different before a flood scours the algae from rocks and the silt from the bottom, producing a rapid change from herbivory to carnivory (Legler, pers. obs.).

Feeding Mechanics

The mechanics of feeding are medium dependent (Shafland 1968; Bramble, 1973; Wake and Bramble, 1985). Suction feeding is employed in water. A turtle approaches a food item and simultaneously opens the mouth and lowers the hyoid apparatus, rapidly increasing volume and creating negative pressure in the buccal cavity and pharynx. The food item is sucked in with a substantial volume of water, the mouth closes, excess water is expelled through the nostrils and the narrowly opened mouth, and swallowing occurs. The initial "bite" may be a fast, well-coordinated strike (e.g., *Chelydra*) or a series of gentler bites in the course of unhurried foraging. Small food items may pass into the mouth without touching the jaws. More often the object ingested is too large to pass in completely and is shifted backward by rapid inertial feeding movements—forward movements of the head and neck and the repeated regrasping of the object, orienting it, and moving it gradually backward into the mouth. The forefeet are used to tear apart large objects held in the jaws. Typical aquatic feeding combines suction, inertial movements, and the tearing apart of large objects. Suction feeding reaches its quintessence in certain chelids (*Chelus* in South America and *Chelodina expansa* in Australia; Legler and Georges, 1993).

Suction feeding is clearly impossible out of the water. Most aquatic turtles are unable to swallow when out of the water, although they may pull food into the water and swallow it there.

The swallowing mechanism of tortoises (Testudinidae) depends heavily on the tongue and is called *lingual feeding*. It occurs in advanced testudinids and relies on tongue kinesis synchronized with protraction and retraction of the mandible. The food mass is bitten off, crushed somewhat by the jaws, and then captured between the upper surface of the tongue and the palate. The tongue slides rearward and pushes the bolus into the esophagus. Bramble and Wake (1985) show a radiographic sequence of swallowing in *Gopherus*.

Lingual feeders (e.g., *Gopherus*) have a rather small hyoid apparatus and a large papillose, glandular, and complexly muscular tongue. The palate is vaulted to accommodate the bulk of the tongue and to facilitate breathing through the nostrils during feeding. Aquatic suction feeders have a relatively large hyoid; a small, simple tongue; and a broad, flat palate.

Some mechanical breakdown is accomplished by the tomial edges of the jaw sheaths and any accessory alveolar ridging within the mouth (see Structure and Function). This usually has more of a crushing than a chopping effect and is of little use for fragmenting muscular tissue. Long blades of vegetation can be ingested whole by scoring and then folding them repeatedly, accordion style, into swallowable masses (Legler, pers. obs.).

Reproduction

Mating

Mating usually occurs early in the season of activity but may occur any time. Foreplay ranges from rapelike ramming and biting of the female by the male to complex and beautifully coordinated swimming behavior in male *Trachemys scripta* and related emydids. Ultimately, intromission and insemination are accomplished clumsily when the male mounts the rear of the female's carapace and holds on by one means or another for a variable period of time (Figure 57.5). Zug (1966) described and illustrated the penis of cryptodires. Details of mating, where known, are covered in the accounts of species. An engorged penis may fit so tightly in the female cloaca that a female can actually tow a male about by the penis on land or in water.

Precopulatory behavior commonly involves the touching of the male snout to the cloacal region of the female (often referred to as *trailing* and *sniffing*). This suggests some sort of olfactory cue emanating from the cloaca or perhaps musk gland secretions in the wake of the moving female. The phenomenon deserves study.

Gametogenesis

The gonads of both sexes lie immediately ventral to the kidneys. Testes are usually oval. They may become elongated during the cycle of enlargement. The efferent ducts consist of a long coiled epididymis and a short vas deferens that enters the dorsal wall of the cloaca just medial to the ureter. There is a seasonal testicular cycle of hypertrophy and regression. Stages of spermatogenesis can be approximated by degree of enlargement (Moll, 1979). Engorgement and pallidity of the epididymides are accurate indicators of motile, mature sperm.

Meylan, Schuler, and Moler (2002) describe three basic patterns in chelonians that have been studied: postnuptial, prenuptial, and acyclic. Most turtles that have been studied (and most of the turtles in Mexico) have the postnuptial pattern (*Chelydra, Apalone, Kinosternon, Rhinoclemmys,* and all emydids). Sperm mature late in the season of activity after ovulation and oviposition have occurred. Ova are therefore fertilized by sperm produced in the preceding year. Separate hormonal cues initiate spermatogenesis and mating. In Mexico, only *Dermatemys mawi* (Polisar, 1996), *Gopherus agassizi* (Rostal, Lance, Grumbles, and Alberts, 1994), and the cheloniid sea turtles show the prenuptial pattern (spermatogenesis completed in the same season as mating), and only *Claudius angustatus* (Flores-Villela and Zug, 1995) is acyclic (spermatogenesis continuous). For an erudite discussion, see Meylan, Schuler, and Moler (2002).

Sperm Storage

Ewing (1943) and Finneran (1948) recorded female box turtles *(Terrapene carolina)* that produced fertile eggs for up to 4 years in the absence of males. Oviducal sperm storage has been demonstrated for long periods in a wide variety of turtle taxa (Gist and Jones, 1989) and could reasonably be predicted for all extant turtles. Sperm storage facilitates multiple paternity (Galbraith, 1993; Harry and Briscoe, 1988; Pearse, Dastrup, Hernandez, and Sites, 2006; Refsider, 2009). Stored sperm are most likely to be used in the fertilization of second and subsequent clutches (Gist and Congdon, 1998).

The ovary is a sheetlike organ in contrast to the compact mass of tissue typical of mammals. Presumptive ovarian follicles are detectable as pale surface granules in juveniles. The oviducts lie ventral to the ovaries and enter the cloaca ventrolaterally. The part of the oviduct nearest the ostium is flattened and folded (accordion style), whereas the part communicating with the cloaca is rounded, has a thicker and more muscular wall, and is unfolded.

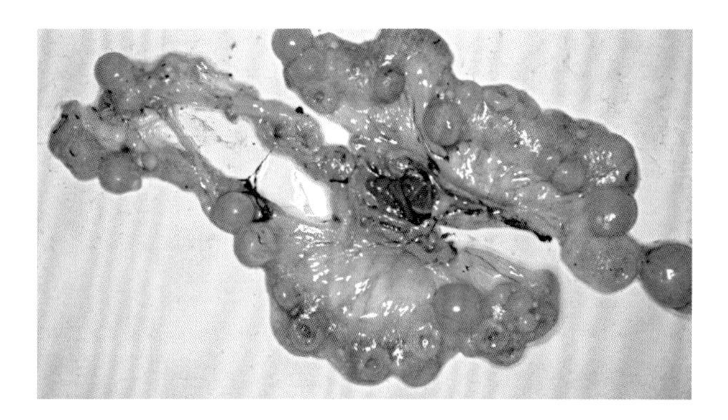

FIGURE 3.1 Ovaries of a *Dermatemys mawi*—typical for early to mid season of ovarian activity. Largest follicle (right, ca. 35 mm) is probably atretic. Corpora lutea (ca. 14 visible, clearest at bottom) appear as doughnutlike depressions with vascular centers. Developing follicles (bright yellow) are in two distinct sets, representing potential ovulations. Atretic follicles have a tan color. These ovaries show a reproductive potential of at least 30 eggs in three clutches for the season.

Anatomical drawings show a rather neat arrangement of these structures in the coelomic cavity, but when large follicles or oviducal eggs are present, they take up most of the "spare" coelomic space and impinge on other coelomic organs, including the lungs.

Presumptive ova in ovarian follicles begin to acquire yolk (vitellogenesis) at puberty and enlarge to ovulatory size at the beginning of each annual cycle. A follicle consists of the thinly stretched ovarian wall surrounding a yolk-laden ovum. At ovulation, the follicular wall ruptures and releases the ovum (Figure 3.2). Ova are free in the coelomic cavity (at least momentarily) before entering the oviducal ostium. Extra-uterine migration is common; ova from one ovary may go into either oviduct (Legler, 1958). This may involve coelomic migration of ova, movement of the oviducal ostia, or both. Ovulation of all follicles in a preovulatory size group seems usually to be simultaneous.

The empty follicle collapses and becomes a distinct, cup-shaped, glandular corpus luteum (Figure 3.1). A fresh corpus luteum has a bloody orifice and is about 15–25% the diameter of the mature follicle. Corpora lutea begin to regress almost immediately but remain visible throughout the breeding season (and sometimes into the next). They constitute absolute evidence of ovulation and can usually be accurately counted. Corpora lutea occur in different sized groups (i.e., stages of regression) if more than one set of ovulations occurs in one breeding season. Multiple ovulations can be predicted by counting enlarging follicles in different sized groups. Annual reproductive potential (ARP) can be calculated by simple ovarian examination: Total enlarged follicles + total corpora lutea of all stages = ARP. If oviducal eggs are present, they should equal the number

FIGURE 3.2 Ovary from a fresh dissection of *Kinosternon leucostomum*. The large follicle in foreground is in the process of ovulation (revealing part of ovum proper on left. The process is swift, and the remaining parts of follicle quickly become a corpus luteum.

of fresh corpora lutea. If not, one or more eggs have been laid or one shell has formed around two ova to produce dizygotic twins. It is common for turtles to produce multiple clutches in the same season.

Reproductive efficiency is often expressed by *relative clutch mass* (RCM), which can approximate the energy expended to produce a clutch of eggs. RCM is simply the weight of the clutch expressed as a percentage of the postpartum weight of the female.

Depending on environmental factors, all enlarged follicles may not be ovulated and the reproductive potential may not be realized. Follicles not ovulated in a given season become dark and flaccid (atretic) and are reabsorbed in most species studied. Atretic follicles are usually identifiable late in the season. The ovaries bear a lot of information about reproduction and can be harvested from road kills, preserved specimens, or any other dissection.

Eggs

All turtles are oviparous. Females lay shelled eggs on land. Most turtles dig a nest as deep as a hind leg can reach and bury their eggs in earth. Several kinds of neotropical turtles (e.g., *Rhinoclemmys* sp., *Kinosternon* sp.) may cover the eggs shallowly in ground litter. Nesting seems always to be a secretive affair and is seldom observed except when there is synchronous nesting of many females. Individual females searching for a place to nest are usually wary and easily discouraged by intruders; once they start the processes of excavation and oviposition, however, they enter an almost trancelike state during which little deters or distracts them until nesting is completed.

As amniotes, turtles produce eggs with a full set of extraembryonic membranes—an amnion, a chorion, an allantois, and a yolk sac. The yolk sac contains enough yolk to sustain the growing embryo to hatching and emergence. The amnion encloses the embryo proper in a fluid-filled space, and the allantois is a recipient for nitrogenous wastes. Soon after development begins, there is gaseous exchange through the porous shell. This is the "land egg" of textbooks. It circumvents the need for (and the risks of) a free-living embryo or larval stage.

Most turtle eggs are elliptical; those of *Chelydra, Apalone,* and *Gopherus* are spherical (as are all sea turtle eggs). Much of what one would want to know about turtle eggs and embryology can be found in two major studies by M. A. Ewert (1979 and 1985). Agassiz (1857b) contains seven exquisite lithographic plates depicting life-sized neonates and eggs of North American chelonians.

The remaining parts of a shelled egg are secreted around the ovum while it is in the oviduct. The width of the shelled egg is determined by the diameter of the ovum. Peripheral to the yolk are a clear albuminous layer, a shell membrane, and the hard part of the shell. These events proceed quickly (probably in an hour or so) to the stage of a thin shell. The shell then thickens more gradually over a period of days. Unshelled oviducal eggs are rarely found in dissections. The shell membrane has multiple layers, each with fibers running in different directions.

The mineralized shell layer begins with a single layer of shell units, each of which is an inverted polyhedron with its apex abutting on the shell membrane. These units begin as crystallization nuclei (Ewert, 1985) and then grow to final size by the addition of thousands of aragonite crystals. Growing shell units are at first conical; they become multihedral as they abut on one another. The crystallization nucleus becomes a hollow space near the base (apex) of each shell unit (Legler, 1993b, Figure 16.8).

The structure of eggshells ranges among taxa from thin and flexible (e.g., *Trachemys* and *Terrapene*) to thick, hard, and brittle (e.g., *Rhinoclemmys* and *Kinosternon*) (Legler, 1985; Ewert, 1985). The shell may continue to thicken in long-retained eggs. Eggs laid with thin, incompletely developed shells rarely develop but are not necessarily "infertile." Shell strength, rigidity, and thickness depend on the height of the shell units and the extent of their fusion to one another. Under normal circumstances, there is a single layer of shell units. In thick-shelled eggs, these are high; in thin-shelled eggs with flexible shells, they are lower.

Flexible-shelled eggs are actually fairly rigid at laying but swell and become turgid as they absorb water during incubation. To achieve this expansion, movement between the shell units must occur. Even hard-shelled eggs of medium shell thickness eventually expand but not without the cracking and flaking of part of the hard shell. The hardest and thickest-shelled eggs (e.g., *Rhinoclemmys* and *Kinosternon*) don't expand at all. Water absorption late in development causes cracks in the shell through which there may be an oozing of albumen and vitelline blood (Figure 3.8). These eggs remain viable. Thick-shelled eggs tend to have well-defined pores that are formed at points common to three or four shell units. Thin-shelled eggs usually lack well-defined pores.

Embryonic Development

Embryonic development advances to a gastrula stage in the oviduct and does not resume until laying. This developmental phenomenon is well known (Agassiz, 1857b; Ewert, 1985) but was only recently explained (Etchberger, Ewert, Phillips, Nelson, and Prange, 1992) in terms of higher oxygen concentrations outside the oviduct. Shelled eggs can be held in the oviduct until environmental conditions are suitable for laying. This is a "decision" of high selective value made by the female.

When development resumes after laying, a white chalky patch appears on the uppermost surface of the otherwise translucent shell. The patch enlarges to form a "dorsal" saddle and eventually spreads over the entire shell except for parts in direct contact with a moist medium. The process is called *chalking,* a term seemingly coined by Ewert (1985).

Thompson (1985) quantified chalking as follows. The chalky patch forms progressively by partial drying of the shell and its membranes where they are in close juxtaposition to the underlying chorio-allantoic membrane; this facilitates exchanges of water and respiratory gases through the shell. The conductance of O_2 and CO_2 is low before chalking but increases as chalking progresses, permitting gas exchange to keep pace with the increasing demands of the developing embryo.

At hatching, a yolk sac of variable size projects from an umbilical opening in the plastron. The sac eventually retracts and becomes part of the small intestinal wall. This yolk is a source of neonatal sustenance. Time of incubation to hatching is correlated with nest temperature in species with direct development (e.g., *Terrapene, Chelydra*).

Predation on fresh nests in the first 24 hours is high (often >90%) (Alford, 1980; Moll and Legler, 1971). Emerging hatchlings disperse quickly and secretively. Hatchlings of most species are rarely seen.

Hatching and Emergence

Hatching signals the end of life within the egg and the rapid transition from an embryo to an air-breathing, free-living turtle. Escape from the egg involves puncture of extra-embryonic membranes (amnion and chorio-allantois) and the breaching of the harder shell. Initial perforation of the shell is referred to as *pipping* and is the point at which incubation is regarded as complete.

Full-term embryos bear a caruncle—a chalky white, conical, and sharply pointed projection on the upper part of the maxillary sheath below the nostrils. The caruncle serves, at least, to puncture the extra-embryonic membranes and often to initiate the tearing or breaking of the shell. Once punctured, the shell is torn or broken open by forceful movements of the head and limbs.

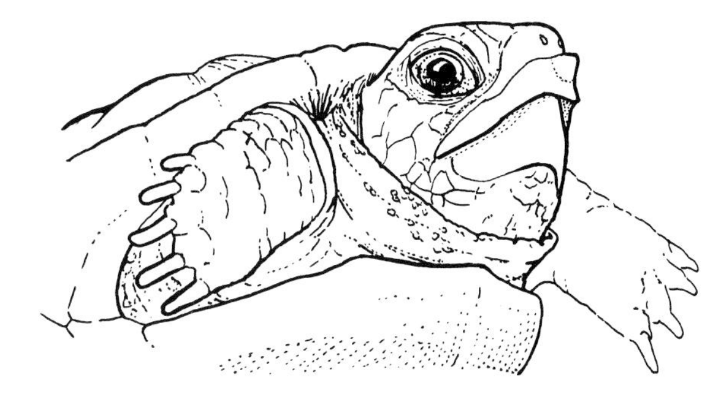

FIGURE 3.3 A typical caruncle (conical, below nostrils) in a hatchling of *Terrapene ornata.* (From Legler, 1960, with permission.)

The color of the caruncle has suggested to some workers (Gadow, 1909; Legler, 1960c) that it was calcareous. Actually, it forms by hyperplastic epithelial growth and consists chiefly of cornified epithelium underlain by a thin stratum intermedium. The latter probably accounts for the clean plane of fracture from the maxillary sheath. The caruncle is usually lost in the first two weeks of life after hatching. Its presence is reliable evidence of recent hatching. Caruncles have analogous function and are truly homologous in crocodilians, turtles, rhynchocephalians, and birds. True "egg teeth" are formed as part of the premaxillary dentition in lizards and snakes (De Beer, 1949; Fioroni, 1962; Edmund, 1969; Guibe, 1970).

Detailed accounts of hatching are few. Ewert (1985) shows the process of pipping in *Staurotypus triporcatus,* and Moll and Legler (1971) describe it in *Trachemys scripta venusta.* Testudinids and most emydids have a large external yolk sac at pipping and are not ready to leave the eggshell. At time of pipping in kinosternids, all extra-embryonic membranes (including the yolk sac) have been reduced to a small padlike structure at the umbilicus.

Sources of Eggs for Study

Presence of oviducal eggs and degree of eggshell development can be determined by palpation through the walls of the inguinal pocket (Legler, 1960c). Special tools and equipment are required for radiography (Gibbons and Green, 1979), ultrasonography, and laparoscopy (Küchling, 1999).

A gravid female can be kept for long periods without damage to the eggs. Oviposition can be stimulated and eggs harvested by injections of pituitary hormones (Ewert and Legler, 1978). This has been done with various kinds and grades of pituitary extract, the best of which is commercial or clinical Oxytocin conveniently supplied in 10 unit/mL ampoules. Dosages and techniques are covered in Ewert and Legler (1978). Egg damage can be minimized by placing the injected female in water and harvesting the eggs as laid. Injected females may dig a nest if a proper substrate is provided. Terrestrial chelonians can be suspended over sand or water in a sling.

Eggs are easily incubated under simple controlled laboratory conditions for studies of development (Agassiz, 1857b;

Ewert, 1985; Legler, 1985; Yntema, 1968). Early stages of development can be viewed by transmitted light. It is important to make the distinction between infertility and egg inviability (Legler, pers. obs.). Early development (and fertility) can be detected by observing vascularity with transmitted light. Various kinds of incubation medium have been promulgated by various workers. If eggs will be examined regularly during incubation (e.g., measured, weighed, and viewed by transmitted light), they must be free of debris, and this favors an incubation medium that does not adhere to the eggs. It must be possible to add water easily to the incubation medium without directly wetting the eggs. Legler (1985) favors large laboratory tissues (Kimwipes), first saturated and then wrung out to form a concave pad. Ewert (1985) and Bull and Vogt (1979) have used vermiculite with various amounts of water. Sealable plastic food containers can be used as incubation chambers and eggs identified by numbering with a graphite pencil (Figure 5.7).

The allure of keeping and examining eggs transcends the simple need for scientific knowledge. The uncounted eggs incubated by Rathke (1848), Agassiz (1857b), Legler (1960c, 1985), Bull and Vogt (1979, 1981), Ewert (1979, 1985), and many others bespeak also the mystique and charm of watching the early ontogeny of a vertebrate and then witnessing the fully formed neonate as it hatches (whether these experienced scientists would admit it or not).

Embryonic Diapause and Synchronous Emergence in the Neotropics

Most turtles lay multiple clutches in the same season. In the neotropics they generally lay more clutches of larger eggs over a longer period. Therefore, in a given area, eggs in nests may differ in age by as much as several months. Yet emergence of hatchlings from these nests tends to be synchronous and at a time when optimal conditions exist. In Mexico, this is the tropical rainy season. Extensive studies by Vogt and students in tropical Mexico (Chiapas, Tabasco, and Veracruz) form the basis of the following discussion (Morales-Verdeja and Vogt, 1997a).

The "normal" or basic sequence of reproductive events in turtles is vitellogenesis, ovulation, formation of egg coverings, embryonic development suspended at a gastrula stage in the oviduct, oviposition, embryonic development resumed and completed in the nest, hatching, and emergence from nest. Synchronization of emergence is achieved by delaying this basic cycle at one or more points.

Delayed emergence occurs in tropical *Trachemys s. venusta* and *Chelydra rossignoni* (Moll and Legler, 1971, Panama; Vogt and Flores-Villela, 1992a; Vogt, 1990, Veracruz), a simple strategy whereby eggs develop and hatch and the young (of different ages) lie dormant until rains drench or inundate the nests.

Embryonic diapause is a post-ovipositional continuation of the delayed development that occurs first in the oviduct. The following Mexican species display an embryonic diapause that may last for as long as 10 months: *Dermatemys mawi, Claudius angustatus, Staurotypus triporcatus, Kinosternon acutum, K. leucostomum,* and *K. scorpioides* (Vogt, 1990, 1997a; Vogt and Flores-Villela, 1992a, 1992b; Vogt et al., 1997; Ewert, 1991). All of these species occur in the tropical region of Mexico, and most of them are neotropical endemics. There was no evidence for embryonic diapause in *Kinosternon flavescens* in northern Mexico (Vogt, Bull, McCoy, and Houseal, 1982).

Embryonic aestivation (Ewert, 1985, 1991) is a strategy in which embryonic development is completed, but the eggs do not actually hatch until a stimulus for emergence occurs. Embryonic aestivation may serve in lieu of embryonic diapause but is often an addendum to diapause.

Details of diapause and aestivation are presented in the species accounts. The occurrence of diapause in *Kinosternon leucostomum* (southern Veracruz, Morales-Verdeja and Vogt, 1997a) is given here as an example. Several clutches are laid. Nesting begins at the height of the rainy season in late August and continues until May. Hatching is synchronous and coincides with the first heavy rains in June or July at the beginning of the "summer" rainy season. Controlled incubation in the laboratory ranged from 90 to 265 days at 26°C. Eggs incubated at constant high humidity remained in diapause and died. Eggs in diapause began to develop normally when subjected to lower humidity and exposed to 2 weeks of lower temperature (21–22°C) and then shifted back to 26°C. Eggs with fully developed young passed into aestivation, and hatching was induced by raising the humidity of the incubation medium to near 100%. Projecting these findings to natural nests, it is hypothesized that diapause is terminated at various times in dormant eggs and that development is completed in approximately 90 days, at which point the eggs enter a period of embryonic aestivation. Synchronous hatching and emergence are triggered by the first heavy rains of the next rainy season. The strategy combines embryonic diapause and embryonic aestivation.

Synchronous emergence in the wet season probably increases the survivorship of hatchlings by simplifying the transition from nest to water (or some other refugium) and by producing a glut of prey that satiates hatchling predators. Furthermore, sustenance of a dormant egg or fully developed but unhatched embryo is metabolically cheaper than sustenance of a free-living, air-breathing hatchling (Congdon and Gibbons, 1983). There is more yolk in an ovum than is necessary for embryonic development alone (Nagle, Burke, and Congdon, 1998). The remaining yolk, usually in a slightly protruding yolk sac, must also sustain the hatchling until it can enter the general habitat and begin feeding. If less yolk is used before emergence then more is available for postemergence sustenance.

Sex Determination in Turtles

Until the late 20th century, sex determination in turtles was thought to be genetically controlled. Heteromorphic sex chromosomes had been demonstrated in *Staurotypus* (Bull, Moon, and Legler, 1974; Moon, 1974). Pieau (1976) and Yntema (1976) first suggested that incubation temperature might determine sex, but their studies were clouded by high mortality of hatchlings.

Bull and Vogt (1979) first demonstrated unequivocally that sex was determined by incubation temperature in controlled lab experiments and natural nests of *Graptemys* sp. in Wisconsin. Temperatures of 25–28°C produced males, temperatures of 31°C or greater produced females, and both sexes resulted from intermediate temperatures. In other species, females, but no males, were produced also at 23°C (Vogt, Bull, McCoy, and Houseal, 1982). Following this pioneering work, many papers on the subject have appeared. This relatively new field of knowledge was summarized by Valenzuela and Lance (2004).

The intermediate temperatures that produce both sexes are "threshold" temperatures and may vary by 2–3°C from one taxon to another or in different populations of the same species. Sex determination is affected by exposure to a temperature above threshold for about 4 hours per day in the middle third of incubation. In the range of temperatures that produces males, the hormone precursor aromatase is inhibited, and the bipotential gonad becomes a testis; it becomes an ovary in the presence of aromatase (Valenzuela and Lance, 2004). The precise manner in which the indifferent gonad is induced to form a testis or an ovary is still not clear. Temperature-dependent sex determination (TSD) of some sort occurs in most chelonian species. TSD occurs also in some lizards and in all crocodilians.

The adaptive significance of TSD, although subject to a vast range of postulation, "remains an enigma in many ways" (Bull, 2004, in Valenzuela and Lance, 2004).

Genetic sex determination (GSD) occurs in the families Chelidae (Pleurodira) and Trionychidae and in the kinosternid genera *Claudius* and *Staurotypus* in which sex is not affected by incubation temperature. *Staurotypus* has sex chromosomes, but *Apalone* and *Claudius* lack them. In GSD, sex is determined at approximately the same stage of embryonic development as in TSD (stages 18–20). Sex differentiation begins immediately in GSD, and gender is henceforth immutable, whereas gender is reversible until about stage 22 in turtles with TSD. This could account for evidence of some sexual dimorphism at hatching in *Apalone* (see Greenbaum and Carr, 2001, for details).

Natural Predators and Parasites

Availability of eggs and hatchlings and the horrendous losses of these stages to a variety of predators is discussed in Reproduction. A turtle shell of any kind is ipso facto protective—relative to the nonarmored thorax of more generalized quadrupeds. There are few predators (other than humans) that can make a living on turtles. Many predators will eat a turtle or its eggs but, to our knowledge, no predator is specialized for turtle predation.

Consuming a whole turtle requires a large enough gape and sufficient jaw power to crush the turtle or swallow it whole. As a turtle becomes larger, the number of predators that can do this diminishes dramatically. Crocodilians of some sort have occurred with aquatic turtles for most of the time turtles have existed—and still do in some temperate and nearly all tropical regions of Earth. The same probably applies to sharks.

Aquatic turtles have the ability to escape crocodilian attacks and show evidence of this in missing limbs and tooth marks on the shell. Such wounds are directly correlated with turtle size—smaller turtles seldom or never escape from crocodilian attacks. Medium-sized turtles may be swallowed whole. But it is difficult to assess crocodilian predation except by observing survivors (see account of *Trachemys scripta venusta*).

Sharks are not often mentioned as predators of freshwater turtles. Bull sharks (Tiburon Sarda, *Carcharhinus leucas*) have a cosmopolitan distribution on the continental coasts of all tropical and subtropical seas, and it is well known that they travel far up warm rivers and into freshwater lakes. Mexican records exist for estuarine and freshwater habitats on the Gulf coast from the Rio Bravo to the Usumacinta–Grijalva drainage and, on the Pacific Coast, from the Rio Presidio in Sinaloa and the Mar Muerto in Chiapas. Bull sharks grow to 3.4 m, and gape is large enough to swallow turtles whole (Castro-Aguire, 1978). Legler saw and caught bull sharks in northern Australia and attributed severe lacerations of turtles in trammel nets and net damage to them. The only published record of predation on freshwater turtles is that for Lake Nicaragua (Compagno, 1984).

Larger terrestrial predators probably do not bother with a turtle. Domestic dogs and cats do a lot of barking, hissing, circling, and perhaps some gnawing but seem to lose interest at some point before actually killing and eating the turtle.

Smaller mammals (anywhere from rats to coyotes) can kill turtles that are out of water by waiting for a head or limb to be exposed, but this takes a maximum of patience for a minimum of food. Large birds (e.g., magpies, crows, vultures) may be able to kill turtles and can certainly and cleanly eviscerate a dead turtle. But all these things occur on land. Adult turtles that live in the water have relatively few predators other than crocodilians and otters. Few vertebrates (birds, reptiles, mammals) can pass up the opportunity to consume a nest of eggs.

Epizoic Algae

Basicladia is a genus of filamentous green alga that grows on the shells (chiefly the carapace) of most North American freshwater turtles. Although it is usually considered a unique turtle epizoon, it grows less commonly on other substrata. Other algae (blue and green) also grow on turtle shells. The combined algal flora on a turtle shell has a camouflaging effect that is lost if and when the scutes are shed (Proctor, 1958). *Basicladia* can survive long periods of drying (see account of *Terrapene coahuila*).

Shell Staining and Encrustation

The shells of many aquatic turtles bear a dark stain that could be mistaken for natural coloration. This ranges from simple reddish-brown discoloration of parts of the shell to a thick encrustation of the entire shell. In the latter case, the color is a dark brown with a purplish iridescence. This external stain can be scrubbed off (or prised away in flakes) to reveal the natural coloration of a scute. Areas of new growth are free of the stain, and all stain is lost with a shed scute. The staining seems to be characteristic of turtles living in the bottom sediment of dark, still, tropical waters and is much less common in fluviatile and temperate habitats. We have observed it in *Kinosternon* (Figure 3.7), *Trachemys,* and *Dermatemys* in Mexico and in *Podocnemis* in Brazil. Vogt (1981a) reported the condition for *Chrysemys picta* in Wisconsin but attributed it to tannin. Dr. James L. Christiansen (pers. comm., 1965) analyzed some of the thick flakes from a specimen of *Kinosternon leucostomum* (UU 8793) taken in a dark backwater in Panama. Particles of the crust examined under a microscope were semicrystalline and reddish amber with a luster. They consisted of iron oxide (82%) and iron phosphate (18%). Turtles could acquire the iron by frequenting microhabitats with high iron content. The phenomenon is probably related to iron content of the water, the bottom substrate, or both. It requires further study.

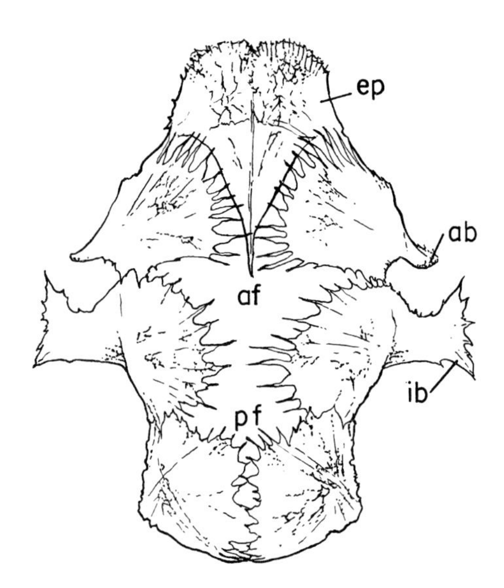

FIGURE 3.4 Plastron of hatchling *Terrapene ornata* (PL ca. 27 mm, Kansas, drawn from cleared and stained specimen) showing anterior (af) and posterior (pf) fontanelles—gaps that persist to varying extent in most immature turtles. Other abbreviations: ep, epiplastron; ab, axillary buttress; ib, inguinal buttress. Note the reduced axillary buttress, which will become part of a plastral hinge. (Modified from Legler, 1960.)

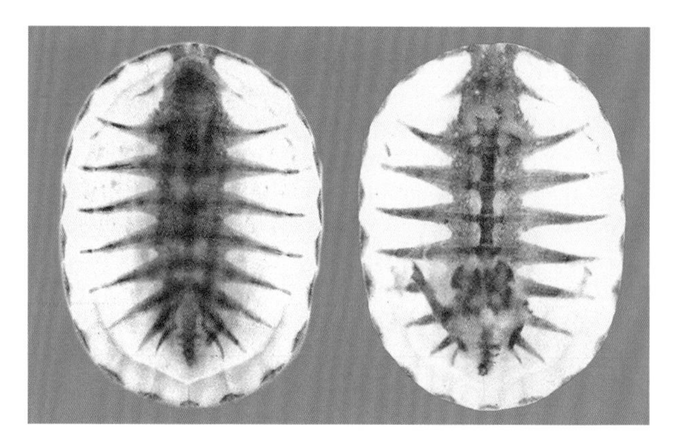

FIGURE 3.5 Dorsal (left) and ventral (right) views of *Kinosternon leucostomum* hatchling cleared and stained with alizarin red; CL 35 mm (plastron removed, pelvis and femur removed, on dorsal view). Anterior end upward. Note nearly complete formation of endochondral ribs and the early formation of dermal bone in the form of peripheral, costal, and neural plates. Anteriormost riblike elements are nuchal costiform processes. (Preparation and photos by JML.)

Ontogeny

Size

There is a vast range of size in turtles. The smallest adults (*Sternotherus* and *Glyptemys muhlenbergi* in the United States) mature at carapace lengths of less than 100 mm. *Dermochelys* is the largest living turtle (maximum length and weight: 219 cm, 900 kilos) and ranks with the largest reptiles. Cheloniid sea turtles reach a larger size than recent freshwater turtles. *Caretta* seems to be the largest with carapace lengths of about 1 m. Freshwater turtles are of modest size; most adults longer than 300 mm would be considered large. Lengths of more than 400 mm are not uncommon in Asiatic river geoemydids, trionychids, and chelydrids. Tortoises attain moderately large size on islands (e.g., Galapagos), but they never reach a size at which they become graviportal, and there is a reduced need to increase the strength of the limb bones. Terrestrial modifications usually include reduction of bone in the shell, especially in larger tortoises (as well as the relatively small *Terrapene ornata*).

Dermatemys mawi (650 mm CL, 22 kg) and *Gopherus flavomarginatus* (371 mm CL known; 1 m and 35–50 kg anecdotal), respectively, are the largest freshwater and terrestrial chelonians in Mexico.

Sexual Dimorphism

Males can usually be distinguished from females by sexually dimorphic body form. The cloaca accommodates the penis and accounts for a bulkier and longer tail base in males. In some taxa (e.g., *Gopherus*), this is slight; in some kinosternids, however, the male tail is long, thick, and prehensile and bears a clawlike tip. Females of most species tend to have a relatively higher, more commodious shell. The plastron of males is frequently concave and noticeably less extensive than in females. A pronounced sexual dichromatism (eye color and head color) occurs in *Terrapene*.

Berry and Shine (1980) demonstrated convincing trends in sexually dimorphic size differences. Males are larger in terrestrial, semiaquatic, and bottom-walking species

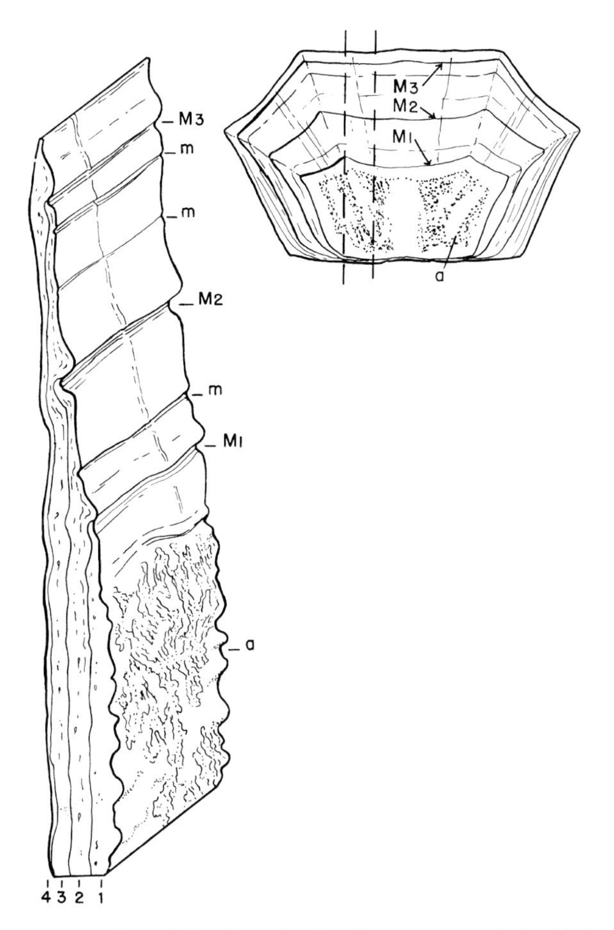

FIGURE 3.6 Growth-ring formation in a *Terrapene ornata* in its third full season of growth. The entire second central scute (upper right) was removed and longitudinal section cut from it by hand (dashed lines). Oblique view of section on left. Scute shows three major (M1–3) and three minor (m, unnumbered) growth rings. Thickness of section is exaggerated for clarity. Major rings form when growth stops (winter in temperate regions). A new layer is formed beneath the entire scute when growth resumes. Minor rings form when growth slows during the season and are not accompanied by a new layer of epidermis. Zone peripheral to most recent growth ring (with distinct color and texture) is a protrusion of the newest, deepest layer (4). The areola (a) is the natal scute, which has a distinctive granular texture and grows variably in the time between hatching and hibernation, ending at M1. *Terrapene* and other terrestrial chelonians do not shed the older layers of the scute. Most aquatic emydids do. (From Legler, 1960, with permission.)

(chelydrids, most kinosternids, and large testudinids) in which male combat and forced insemination occurred. Females are dramatically larger than males in fully aquatic taxa (most emydids, at least some trionychids, and sea turtles) in which a female could accept or reject the advance of a male. Sexual dimorphism is most dramatic in *Graptemys* and *Trachemys scripta* in the southeastern United States, where males are minuscule, have elongated foreclaws, and practice a complex liebespiel. The nature and degree of this dimorphism changes remarkably in populations of *Trachemys* south of the Rio Grande (Legler, 1990).

Growth Rings

Legler (1960c) demonstrated the morphological bases of growth-ring formation in *Terrapene o. ornata* in a northern temperate environment. Moll and Legler (1971) elaborated on this to explore these phenomena in a neotropical population of *Trachemys scripta*.

As turtles grow, the scutes must keep up with everything they cover; general growth is periodic or seasonal—not continuous. The germinal layer of a scute forms a new layer at the beginning of each new growth period. This is marked by a major growth ring where the newer, deeper layer projects past the older, more superficial layer. Rate of growth may vary during a season and produce minor growth rings without producing a new layer. This accounts for the qualitative differences between major growth rings and minor growth rings.

A major growth ring is therefore a new interlaminal seam defining the periphery of the most recent zone of growth and signaling the end of a growth period. Major growth rings are accurate indicators of age in temperate regions where turtles hibernate in winter and are active in summer. In older turtles that are still growing, growth zones narrow as growth slows and may descend into the interlaminal seam like a microscopic staircase. Growth rings can also be formed during periods of aestivation or other periods of inactivity (say, recovery from an injury) or any other factor that influences the ability to acquire and metabolize food.

Ergo, growth rings are real and can be measured to express growth increments. Their association with time lies in the skill with which an observer interprets them.

Aquatic turtles in the tropics may form several major growth rings in a year. Moll and Legler (1971) determined that each major flood in Panama produced a major growth ring in *Trachemys scripta venusta* and that minor growth rings could be produced by leaving turtles in a cool, wet bag for several days.

Growth rings can be useful in studies of growth rate but are not necessarily accurate indicators of age (Moll and Legler, 1971; Legler, 1960c). The areolae and growth rings quickly become obscure in turtles that shed scutes. Germano (1994b) gives a useful summary of growth as related to diet in various chelonians, including sea turtles. It will be of great interest to see studies of growth rings in turtles that aestivate during dry times and emerge only when there is an abundance of standing surface water (e.g., *Kinosternon creaseri*).

Many aquatic turtles (e.g., emydids) shed scutes periodically as growth occurs. Moll and Legler (1971) show the basic steps in the growth of bone and scute and of scute shedding. Near the end of a new cycle of growth, the new layer of epidermis forms a zone of fracture beneath the existing scute and the new, and the former is shed. Each molt reduces the distinctiveness and usefulness of growth rings. When the shell ceases to grow, the epidermal laminae tend to fuse inter se. At the other extreme,

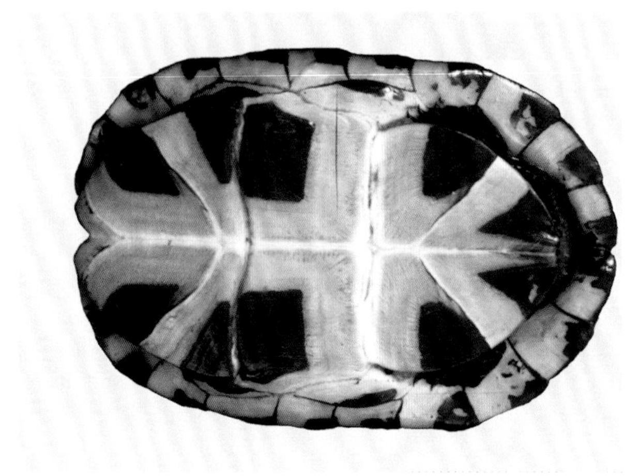

FIGURE 3.7 (Top) A heavily encrusted shell in a *Kinosternon leucostomum* from a black water swamp. (Bottom) Plastron of adult *Kinosternon leucostomum* revealing heavy staining/encrustation at earlier age and a normally pallid plastron in later zones of growth.

FIGURE 3.8 Eggs of *Kinosternon flavescens* in a late stage of development. The inflexible shell has cracked, and vitelline blood and albumen can be seen oozing from the cracks. All four eggs produced normal hatchlings with closed umbilici.

land tortoises of the family Testudinidae and terrestrial emydids (e.g., *Terrapene*) do not regularly shed scutes; scute growth occurs in the manner mentioned, but the old layer is retained and a low pyramid of layers builds up over time (Figure 3.6, and Moll and Legler, 1971).

The original scute of a neonate remains as the distinctively textured areola, the initial epicenter of laminal growth. In nonmolting taxa, it eventually wears away. Otherwise, it disappears with the first molt. The growth rings radiate from the areola (Figure 3.6). Growth of most plastral scutes is anteromediad (but see account of Kinosternidae).

At hatching, the bony shell is incompletely formed. The endochondral elements of the carapace have achieved their basic form, but only the anlagen of the dermal elements exist, leaving wide, unossified gaps between the ribs and between rib tips and peripherals. In some cases, the peripheral anlage does not appear until well after hatching. The chiefly dermal plastron forms a ring of dermal ossification containing two large gaps (fontanelles), one anterior to the interhypoplastral suture and one posterior to it. Gaps and fontanelles sometimes persist almost to maturity and constitute vulnerable soft areas comparable to the "soft spot" on a human infant's head (Figures 3.4 and 3.5).

Longevity

Chelonians in general and large testudinids in particular are often associated with long life. Among major classes of vertebrates, chelonians and crocodilians are the longest lived (Gibbons, 1976, 1987; Gibbons and Semlitsch, 1982). Members of both groups also attain large size. However, even relatively small turtles (e.g., *Terrapene*) attain old ages.

Some primates, several reptiles, and a few birds are known to have lived more than 50 years in captivity. Few other vertebrates live so long. Chelonians rank near the top in longevity with records of 54–57 years for *Testudo graeca,* 62–63 years for *Geochelone gigantea,* and more than 70 years for *Emys orbicularis*. The age of 70 years is stated for the genus *Terrapene* (Dodd, 2001). The record age is seemingly 150 years for a captive *Geochelone* sp. kept on Mauritius Island.

Although captive data are of interest, these animals are not subject to the vicissitudes of a natural environment and demonstrate only the potential for the given ages.

Long-term capture-and-release studies tell us a somewhat different but incomplete story. In most cases, estimated longevity is less than that in captives, and marked individuals may disappear, so cause or time of death is seldom known.

For *Trachemys scripta* in Panama (Moll and Legler, 1971), maximum age was estimated subjectively at 35 years. For a natural population of *Trachemys scripta* in Georgia (Gibbons and Semlitsch, 1982), only 1% of the individuals were expected to attain an age of 20 years, and predicted maximum longevity was ca. 30 years.

Long-term studies of *Emydoidea blandingi* (35 years, George Reserve, Michigan) show that some individuals survive and remain reproductive at ages of "greater than 75 years" (Congdon, Nagle, Kinney, and Van Loben Sels, 2001). *Terrapene* adults in various studies have been accurately aged at 20–33 years (Legler, 1960c; Blair, 1976; Metcalf and Metcalf, 1985). Legler (1960c) guessed that maximum age could be 50. Estimates of age in *T. carolina* range as high as 100 years (Graham and Hutchison, 1969; Oliver, 1955). The population of *Terrapene o. ornata* studied by Legler (1960c) may no longer exist (see account of species).

Gibbons (1987) attributes the venerability of chelonians to "low metabolic activity," an absence of physiological or anatomical senility, a large investment in a protective shell, and a life history with a long maturation period.

Chelonians may stop growing in terms of shell length, but increase in mass may continue with advanced age. Fusion of scutes (usually on plastron) may accompany cessation of shell growth. The largest female (345 mm CL) of *Trachemys scripta venusta* in the Panama study (Moll and Legler, 1971) had scute fusions but was still laying eggs. Dissections of both sexes showed no evidence of gonadal senility.

Turtles, Humans, and Research

Economic Use, Conservation, and the Law

Human Exploitation

Turtles have lived in balance among their natural predators for more than 200 million years, and it may be assumed that they will continue to do so. Their human predators tend not to live in balance with anything. Turtles are now most at risk when subject to anthropogenic habitat degradation (e.g., draining of wetlands—see account of *Terrapene coahuila*) and to commercial collecting for food and the pet trade.

Human primates have fancied turtles and their eggs as food, ornaments, and pets since the dawn of intelligence. The hunter-gatherer way of life probably had minimal impact on turtle populations (but see remarks on humans and Meiolanids on Pacific islands, Pregill and Steadman, 2004; Gaffney, Balouet, and de Broin, 1984).

There is widespread use of turtles and their eggs for food in Mexico. Although *Dermatemys mawi* is preferred and served in restaurants, freshwater turtles of all kinds and sizes are eaten in Tabasco, Veracruz, and Chiapas. Villahermosa, Tabasco, and Alvarado, Veracruz, are the epicenters for the market sale of turtles. Undetermined numbers are caught in Guatemala by hunters from Chiapas and sold in Villahermosa.

Turtle meat is a traditional food for rural people who live near rivers and lakes, and who supplement their diet by opportunistic hunting and fishing. Major decimation of known populations of turtles by market hunters took place from 1960 to 2000. The once marvelous abundance and diversity of freshwater turtles in southern Mexico attracted market hunters. This was an unoccupied niche. Hundreds of turtles could be caught and sent to market in a day or two. In 1966 illegal commercial collecting took place within sight and earshot of paved roads in Campeche and Tabasco (JML, pers. obs.). As turtle populations declined with heavy collecting, the hunters moved (usually upstream) to more productive waters. Boats were used when populations could not be reached by motor vehicle, and exploitation of turtles paralleled the construction of new roads.

Professional tortugueros are not impoverished peasants. Their expeditions are well funded, and they work with modern equipment—thousands of meters of deep nets that are hauled with large, motorized fiberglass boats. They are well-enough armed to intimidate interference—official or otherwise. At this point there begins an unfortunate but predictable cycle. Poaching on a large scale was lucrative only when turtles were abundant. As supply decreased, demand and price increased. In 2005 black market prices for a large *Dermatemys* in Villahermosa were the equivalent of US$200. This market was in turn fueled by the needs and wishes of upper-class urban people. Sales are especially high around Easter because of a loophole in religious dogma—turtle flesh is not regarded as being from an "air breathing animal or bird born on land" and is not subject to the usual Lenten restrictions on red meat (pers. comm., Father James Thompson, Newman Center, Salt Lake City).

The rural subsistence hunter can not afford to eat the one or two *Dermatemys* he might catch in a month. These animals could be taken by bus to a place like Villahermosa and sold on the street for enough money to feed a family for a month. There is nothing cunning about these rural people at all—they may not even know they are breaking the law. But they are more likely to be arrested and penalized than the commercial hunters. Furthermore, whereas their modest use of a resource may have been innocuous in the past, it may now be removing the last few reproductive adults from populations that are on the brink of extinction (Figure 4.1).

Despite more stringent rules and a greater likelihood of apprehension, professional hunters were still organizing

FIGURE 4.1 (Left) Mayan girl at roadside stand with stuffed *Dermatemys*. (Right) Individual turtle hunter at roadside with day's catch of *Trachemys scripta venusta* and *Kinosternon leucostomum*.

dry season expeditions to seek out concentrations of turtles in backwater lakes of the Rio Papaloapan in 2005 (pers. comm. to RCV by Miguel Angel de la Torre). Some of the effects of commercial hunting on turtle communities are covered in the account of *Trachemys scripta venusta*.

Rural people are aware of population fluctuations and especially of declines in the aquatic animals they hunt. They know little or nothing of endemism or antiquity in a unique neotropical ecosystem, but they are capable of understanding these things if they are explained properly. Most of them would probably take pride in participating in anti-poaching measures and in conserving species that occurred nowhere else on Earth. This sort of pride has been evoked in the local populace of the Bolsón de Mapimí to successfully conserve the Bolsón Tortoise (Morafka, Berry, and Spangenberg, 1997; see account of *Gopherus flavomarginatus*).

Until at least 1995 stuffed turtles, particularly *Kinosternon herrerai, K. leucostomum, K. acutum,* and *Trachemys scripta venusta,* were sold in markets, shops, and roadside stands along the Gulf coast of Mexico. The turtles were mummified, varnished, and mounted in grotesque upright positions playing musical instruments (Vogt, pers. obs.; Mittermeier, 1971; Figure 4.2).

Comparative Quality of Turtle Flesh

In the course of our studies and travels we have had the opportunity to sample the meat of many kinds of chelonians. Vogt (pers. obs.) regards *Dermatemys* as the best-tasting turtle meat and attributes this to its rapid growth and herbivorous diet. Legler and family ate home-cooked (grilled or pan broiled) meat of *Chelydra, Trachemys,* and *Dermatemys* and found it all good with little difference in quality. Turtle meat is seldom eaten this way in restaurants or upper-class homes. Rather, it is elaborately prepared and seasoned with pepper, lime, and other condiments. Eggs are eaten raw or cooked within the turtle.

Captive Propagation: Turtle Farms

As populations of popular market turtles (e.g., *Dermatemys* and *Trachemys*) declined, the idea of captive propagation grew. The basic idea was if turtles could be kept in captivity and at least some of them would lay eggs then this might be the answer to a serious conservation problem. Raising turtles for market would cost less than fixing an environmental problem, and the farmed turtles could be sold at a lower price. Turtle farms in the United States and China were said to be successful.

THE NACAJUCA TURTLE FARM

The state of Tabasco established an experimental turtle farm at Nacajuca, near Villahermosa, in 1978. The main objective of this farm was to explore the potential and biological efficacy of captive propagation. If successful, it would then advise and help to establish other turtle farms, some of which would be in the private sector.

The prime objectives of turtle farming would be to raise turtles for food and the pet trade, to provide breeding stock

FIGURE 4.2 Mummified and varnished turtles (*Kinosternon herrerai* in this photo) "singing" and playing musical instruments. These grotesque items were popular as souvenirs in roadside stands along the main highways of Gulf Coastal Mexico until the 1990s. (P. C. H. Pritchard collection, used with permission.)

for other turtle farms, and, most importantly, to replenish decimated populations by stocking hatchlings.

The Nacajuca farm has been a success in some respects. The breeding stock for *Dermatemys* (55 adults, both sexes) was confiscated from market hunters and was of known origin. RCV marked 20 of these at the time of confiscation in 1985 near the mouth of the Rio Lacantún, Selva Lacandona Chiapas; the turtles supposedly had been collected in the Rio Salinas, a tributary of the Rio Usumacinta that enters Guatamala. In 2006 the captive population of *Dermatemys* was approximately 1,000, at least 90% of which were produced at the farm from eggs laid by captives. The 20 marked specimens were still present. The population of *Trachemys* was near 8,000. *Rhinoclemmys areolata* has also been reared to reproductive size. Success has not been as good with *Chelydra rossignoni, Staurotypus triporcatus, Claudius angustatus,* and *Kinosternon leucostomum.*

Some of these turtles are now being transferred to private ownership. Vogt and students have advised this facility since 1990. A major problem at the Nacajuca facility is the lack of a natural water supply. The well goes dry in the dry season and water must be trucked in. Being a state government institution, Nacajuca suffers also from political problems.

OTHER FARMS

The Institute of Ecology at Jalapa, Veracruz, was initiating the construction of a prototype facility for raising *Dermatemys* in natural wetlands, about 25 km N of Veracruz near La Mancha, in 2006. However, the project was shut down by 2009 due to the facility being taken over for mangrove swamp regeneration. There will also be a community-based program in a wetland (Lerdo de Tejada) near Alvarado, Veracruz. Both of these were joint efforts with the local fishermen, and both are under the guidance of Gustavo Aguirre and Miguel Angel de la Torre. The Lerdo area was a proposed RAMSAR site (International Treaty for the Conservation and Wise Use of Wetlands). Due to pressure from cattle ranchers the govenador of Veracruz refused to sign the treaty and the project was terminated in 2008.

Basilio Sanchez, a landowner at La Florida, Veracruz (near Cabada), began breeding turtles in 1998 with legally obtained wild stock. He has inexpensively built earthen tanks that are watered by a permanent stream. He has been able successfully to produce hatchlings yearly for the following species: *Trachemys s. venusta, Dermatemys mawi, Kinosternon leucostomum, Kinosternon acutum, Staurotypus triporcatus,* and *Chelydra rossignoni. Claudius angustatus* has not adapted well to captivity. The chief aim of the La Florida farm is to produce juvenile turtles for the international pet trade.

The Centro Mexicana de Tortugas in Mazunte, Oaxaca, directed by Martha Harfush, is a facility dedicated to the education of local people with an emphasis on turtle conservation. It maintains an exhibit of live turtles, particularly the species indigenous to Oaxaca, and gives guided tours of the facility explaining to the public its ideas about turtle conservation and biology. It has excellent breeding stocks of *Trachemys scripta grayi, Rhinoclemmys pulcherrima,* and *Rhinoclemmys rubida.* It is producing hatchlings from these species annually with the idea of eventually repopulating decimated areas. As of 2011 the facility is still functioning but has been remodeled as a tourist attraction, with more emphasis on environmental education than breeding freshwater turtles.

Turtle farms in the northern temperate region (southeastern United States) have problems with the health of captive adults and must frequently replace this breeding stock with wild caught adults (Ernst, Lovich, and Barbour, 1994). The success of the Nacajuca turtle farm suggests that this problem does not exist in the neotropical environment.

In any case, the success or failure of captive propagation for conservation purposes in Mexico will probably not be clear until at least 2025. Mexican regulations require that introduction of hatchlings and other conservation experiments be subject to follow-up studies. To our knowledge, such studies have not been done.

Gopherus flavomarginatus once suffered severe decimation but is now subject to a successful program of conservation, repropagation, husbandry, and head-start programs. Morafka, Berry, and Spangenberg (1997) go into some detail on what has worked and what has not (see account of *G. flavomarginatus*).

It is our joint personal opinion that the time, money, and energy invested in turtle farms could be more efficaciously spent on basic conservation and enforcement measures with wild populations. Also, we eschew the bending of taxonomic conclusions to achieve conservation goals.

Laws and Enforcement in Mexico

Collecting Permits

The Mexican Constitution (Constitución Federal de Los Estados Unidos Mexicanos, Article 27) states that all natural resources belong to the nation. The native biota of Mexico,

TABLE 1

Conservation Status of Mexican Turtles

If data are insufficient to evaluate a taxon, it is "unlisted."

Taxon	Norma Oficial Mexicana[a]	CITES[b]	IUCN Red List[c]	USFWS
Apalone ater	Special Protection	I	Critically Endangered	Unlisted
Apalone spinifera	Special Protection	Unlisted	Unlisted	Unlisted
Chrysemys picta	Rare	Unlisted	Unlisted	Unlisted
Claudius angustatus	Endangered	Unlisted	Near Threatened	Unlisted
Chelydra serpentina	Special Protection	Unlisted	Unlisted	Unlisted
Clemmys marmorata	Unlisted	Unlisted	Vulnerable	Unlisted
Dermatemys mawi	Endangered	II	Critically Endangered	Unlisted
Gopherus agassizi	Threatened	II	Unlisted	Threatened
Gopherus berlandieri	Threatened	II	Unlisted	Unlisted
Gopherus flavomarginatus	Endangered, Endemic	II	Unlisted	Endangered
Kinosternon acutum	Special Protection	Unlisted	Near Threatened	Unlisted
Kinosternon alamosae	Special Protection, Endemic	Unlisted	Unlisted	Unlisted
Kinosternon creaseri	Unlisted	Unlisted	Near Threatened	Unlisted
Kinosternon herrerai	Special Protection, Endemic	Unlisted	Unlisted	Unlisted
Kinosternon hirtipes	Special Protection	Unlisted	Unlisted	Unlisted
Kinosternon integrum	Special Protection, Endemic	Unlisted	Unlisted	Unlisted
Kinosternon leucostomum	Special Protection		Unlisted	Unlisted
Kinosternon oaxacae	Rare	Unlisted	Near Threatened	Unlisted
Kinosternon scorpioides	Special Protection	Unlisted	Unlisted	Unlisted
Kinosternon sonoriense	Unlisted	Unlisted	Vulnerable	Unlisted
Pseudemys gorzugi	Rare	Unlisted	Near Threatened	Unlisted
Rhinoclemmys areolata	Threatened	Unlisted	Unlisted	Unlisted
Rhinoclemmys pulcherrima	Threatened	Unlisted	Unlisted	Unlisted
Rhinoclemmys rubida	Rare, Endemic	Unlisted	Vulnerable	Unlisted
Staurotypus salvini	Special Protection	Unlisted	Near Threatened	Unlisted
Staurotypus triporcatus	Special Protection	Unlisted	Near Threatened	Unlisted
Terrapene carolina	Special Protection	II	Near Threatened	Unlisted
Terrapene coahuila	Special Protection, Endemic	I	Endangered	Endangered
Terrapene nelsoni	Special Protection, Endemic	II	Data Deficient	Unlisted
Terrapene ornata	Special Protection	II	Near Threatened	Unlisted
Trachemys scripta	Special Protection	Unlisted	Near Threatened	Unlisted
Trachemys gaigeae	Unlisted	Unlisted	Vulnerable	Unlisted

a. The Norma Oficial Mexicana (2001) lists all protected species of Mexican flora and fauna that are in any risk category of extinction. Taxonomic names in this table follow the Norma Oficial.

b. CITES criteria define various degrees of risk in three appendixes as follows. Appendix I: Species that are considered most threatened with extinction; no collection or international trade is allowed except for scientific study with a special permit. Appendix II: Species that could be threatened with extinction in the near future; international trade is allowed only with export permits from the country of origin and import permits from the country of destination. Appendix III: This category is designed to monitor trading activities for a particular species in a country that suspects a high level of trading; a member nation can insist on reporting the numbers of a listed species being exported from that country; permits are not issued.

c. The IUCN RED LIST criteria used in this table are as follows. (1) Extinct: no reasonable doubt that the last individual has died. (2) Critically Endangered: facing extremely high risk of extinction in the immediate future. (3) Endangered: facing extinction in the near future. (4) Vulnerable: facing a high risk of extinction within the medium-term future. (5) Near Threatened: lower risk; rank below vulnerable; not the subject of a specific conservation program, without which the taxon would be at risk.

including all turtles, is protected. This protection is defined by a variety of laws, regulations, and international agreements that address the sustainable use of natural resources. Each state also has laws addressing matters related to flora and fauna.

It is illegal to collect, export, or remove from natural habitat any wildlife in Mexico without the proper and necessary permit. As time passes, rules and regulations change and tend to become more stringent. Permits to collect (or even to handle) protected animals become progressively more difficult to obtain, and the penalties for violating the law become greater. It is also illegal to eat, buy, transport, or trade all species of turtles, living or dead, without a permit. Heavy penalties, including incarceration, may be imposed for violations.

In order to collect or do research on turtles in Mexico one must obtain a scientific collecting permit. Permits are issued by SEMARNAT (Secretaría de Meioambiente y Recursos Naturales), the Mexican environmental authority. Application forms are available by writing to:

Dirección General de Vida Silvestre
Av. Revolución 1425, Nivel 1,
Col. Tlacopac San Ángel,
C.P.01040,
México, D.F.
or by Internet from:

http://www.semarnat.gob.mx/vs/tramites_vs_cofemer.shtml

Applications must be made in Spanish, the reason for collecting must have a scientific goal, and the applicant should be associated with a recognized scientific/academic institution and should include a resident Mexican scientist in the research plan. Applications for "general collecting" would almost certainly fail. Permits are issued for 1 year and may be processed in as little as 3 weeks.

Despite the well-meaning stringency of existing laws, current levels of enforcement are insufficient, and the law is frequently ignored. Some species covered in this book could be virtually extinct before they can be properly studied (say, by 2050). Effective protection will require a combination of grass roots cooperation by all levels of the Mexican populace with law enforcement and conservation agencies. Human cooperation and pride in this kind of conservation can work (see account of *Gopherus flavomarginatus*).

Import and Export Permits

Permits are necessary to legally import collected specimens into another country. In the United States, permits from USFWS (United States Fish and Wildlife Service) must be obtained. CITES permits (Convention on International Trade in Endangered Species of Wild Flora and Fauna) must be obtained for export of listed species from Mexico and to another country; if to a member country, a CITES permit for importation must also be obtained. CITES permits are not needed for unlisted species. CITES was established in 1974 and has 164 member nations. Mexico joined CITES in 1991.

The intent of CITES is to curtail and monitor international trade of threatened species (e.g., the illegal pet trade). It may take more than a year to obtain a CITES import permit to the United States for legitimate research (Vogt, pers. obs., 2005). CITES categories, listings, and criteria are summarized in Table 1.

Field and Laboratory Techniques

Much of the information presented in this book is based on large collections of turtles. Some of these (e.g., U.S. National Museum, American Museum of Natural History, Museum of Comparative Zoology, Harvard) were built up over time as parts of general herpetological collections. The largest collections of turtles (e.g., University of Utah, University of Florida, Tulane University, and Carnegie Museum of Natural History) were made chiefly in the second half of the 20th century, a time when turtles from anywhere in Mexico were a desideratum. The following treatment of collecting and preservation techniques is based on these specialized collections and our own personal experiences.

There is no need to duplicate existing collections—only to supplement them. Prospective collectors should question whether collecting is necessary at all and what might be learned by simple observation or capture, photography, and release at site of capture. The landmark studies of Rick Shine (e.g., 1994) and Henry S. Fitch (1982 and 1985) bear elegant testimony to the dissection of museum specimens in studies of diet and reproduction in squamate reptiles. We have harvested data in a like manner from museum specimens of turtles. There is probably good reason to obtain further material in poorly known regions of Mexico, provided existing collections have been surveyed and utilized to the extent possible. Series of specimens should contain adults of both sexes, juveniles, and at least some skeletal material.

In our experience with aquatic turtles, if one can scan a body of water and see turtles, one can catch at least some of those turtles. Baited hoopnet traps (Legler, 1960d) are probably the most common mode of collection, and canned sardines (despite their current high cost) are probably the most commonly used bait. Legler has used a combination of ripe bananas and sardines effectively in neotropical environments, but any aromatic animal or plant material can be used in the manner of "stinkbait."

A basic hoopnet trap is a collapsible tube of netting supported by lightweight rigid hoops and held open and rigid by adjustable rods. A funnel-shaped entrance (throat) at each end permits easy entry and impedes escape (Figure 5.2). The bait is folded into a square of galvanized hardware cloth and suspended in the center of the trap. Turtles easily find their way in but usually do not get out. It is uncertain how many turtles escape from traps. Traps are set parallel to shore or to objects that will naturally canalize their movements toward or into the trap. Most collectors make their own traps with netting purchased commercially. Nets made from chicken wire obviate the need for stiffening rods. Traps can be made in any size, but a diameter of 48 cm and length of 84 cm is practical and effective; one person can easily carry at least 20 traps of this size (at 0.9 kg each, including stiffeners) in a canvas backpack (Figure 5.3). Although the described trap has been referred to as a "Legler trap," the principle is thousands of years old.

Any hoopnet can be equipped with leads—long nets extending diagonally from the throat that direct the movement of turtles to the baited or unbaited trap. Setting such nets ("Fyke nets") is more complex than setting simple hoopnet traps and usually requires a boat. Turtles (and fish) can be startled into moving from one area to another by loud surface disturbances; Vogt has used an inverted funnel attached to a pole (a "carp horn" used by commercial fishermen in Wisconsin) to effectively move turtles into the leads of a Fyke net and into trammel nets. Turtles can also be frightened off basking logs into basketlike basking traps (Moll and Legler, 1971).

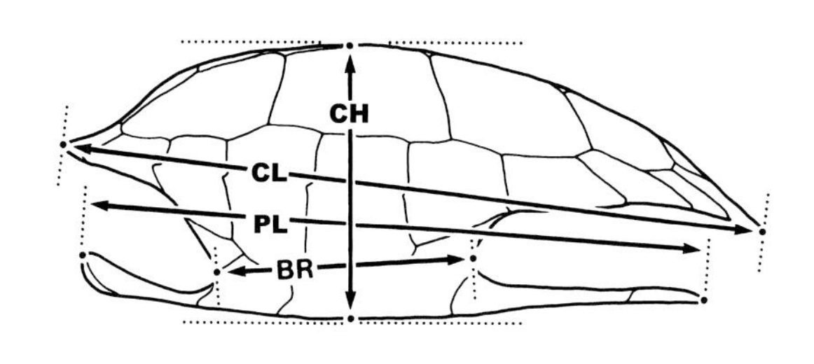

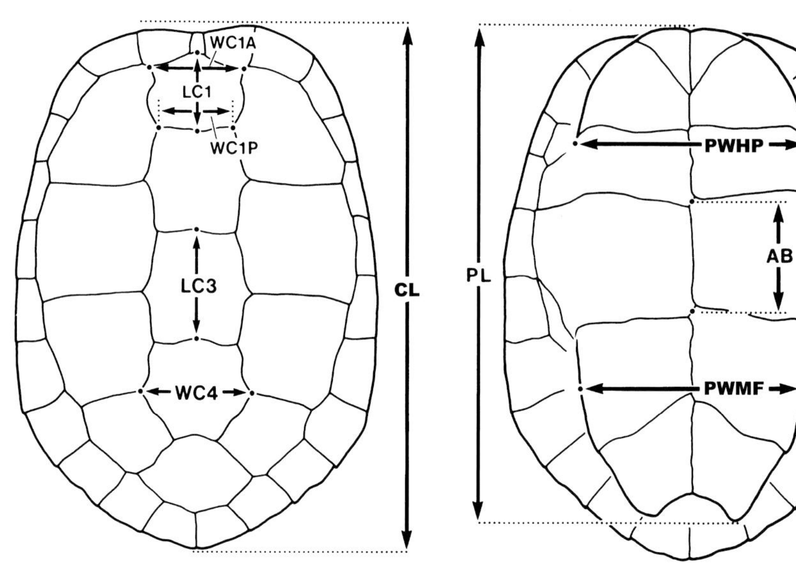

FIGURE 5.1 Dorsal, ventral, and lateral views of a *Trachemys scripta* shell showing method of measurement. Note that a single measurement expressed the mean length of right and left plastral scutes. Width of central scutes was measured from points common to three scutes. Central lengths were measured along midline. CL and PL were maximal and measured in a plane sometimes slightly tangential to true frontal plane. Abbreviations: AB, abdominal scute length; BR; bridge length; CH, shell height; CL, carapace length; LC1, length of central 1; LC3, length of C3; PL, plastron length; PWHP, plastral width at humeropectoral suture; PWMF, plastral width at midfemoral; WC1A, anterior width of central 1; WC1P, posterior width of central 1; WC4, width of C4.

FIGURE 5.2 Removing a *Kinosternon* (Chihuahua) from a standard baited trap (48 × 84 cm) with four aluminum tubing hoops. This trap was thrown from shore and tended via a long line.

FIGURE 5.3 Setting and checking traps by wading. Note backpack that accommodates 20 collapsed standard traps, designed and made by Avis J. Legler.

FIGURE 5.4 Diver about to catch a small *Trachemys*. Captured specimens are placed in a lightweight nylon backpack or in a tending boat. Just touching a swimming turtle on the edge of the shell perturbs its hydrodynamics and facilitates catching.

FIGURE 5.5 Students tending a trammel net. Note the two large-mesh, coarse outer nets and the small-mesh, fine net between them.

Trammel nets (Figure 5.5) and gill nets can be set in areas where turtles are likely to be moving; the turtles simply become tangled in the net (as does everything else). Turtles are pulmonates and can drown. Any submerged device must be checked frequently to prevent this.

Trapping tells one nothing about the animals that avoid a trap or escape from it. Diving with a face mask, swim fins, weight belt, and snorkel is one of the most effective (and enjoyable) ways to catch and observe turtles in clear water that is free of crocodilians and sharks (Figure 5.4). Also, one can identify and count all the turtles that escape. The technique was pioneered by Marchand (1945). It is ideal for exploring undercut banks, brush piles, or areas of thick aquatic vegetation. Two divers can work together and, by comparing observations, make fairly accurate estimates of turtle populations in streams. In places where the bottom can easily be seen, a diver simply moves slowly at the surface until a turtle is spotted and then uses the speed generated by the initial thrust of the dive to overtake the turtle or outmaneuver it.

Most of the foregoing techniques address the catching of aquatic turtles. Terrestrial chelonians (e.g., tortoises and box turtles) may require more luck than skill. In Veracruz, Vogt used a device (locally named a *chuzo*) to locate aestivating turtles (chiefly *Kinosternon acutum*) in piles of loose ground litter or dense brush. The device is essentially the sharp metal point of an 80 mm (tenpenny) nail projecting from the end of a wooden pole of broom handle proportions. When probing ground litter with a chuzo, contact with a turtle produces a distinctive sound (a "thunk"). Drift fences and natural rock walls produce more box turtles and tortoises than open areas. In a few fine-sand areas *Gopherus agassizi* can be tracked for long distances. Legler (1960c) quickly found that *Terrapene ornata* could be located with greater ease from horseback than on foot. Dogs have been trained to locate terrestrial chelonians. Carr (1952) mentions the detection of box turtles by bird dogs; Devaux and Buskirk (2001) record it for *Terrapene nelsoni* near Alamos, Sonora; Vogt has used the technique for *Kinosternon* in Veracruz; and Dodd (2001) records its extensive use for *Terrapene carolina* in the United States. The dogs may be perceiving the odor of musk gland secretions.

As long as turtles occur within the bounds of human civilization, they will continue to cross roads and be struck by vehicles. Every road kill has potential scientific value if harvested in a timely and legal manner. The simple presence of enlarged follicles or shelled eggs is important information.

In areas where people catch or buy turtles to eat, or where turtles are butchered for market, whole shells are often thrown away. Locating refuse heaps in villages can be a good source of shells. Looking through large markets (especially around Easter) one can often find live turtles of known provenance for sale.

Methods of Preparation

All preservation techniques require the sacrifice of a live animal. It behooves the preparator to do this quickly, efficiently, decorously, and out of the public eye. The term *euthanize* is incorrectly applied to this process but satisfies the need to avoid the word *kill*. We have used injections of sodium pentobarbital solution and ethanol successfully. Either substance, injected accurately into the spinal cord (subarachnoid space) between the last cervical and first dorsal vertebra, is virtually instantaneous.

The necessity for sacrificing live turtles to obtain dietary and reproductive information data has been virtually obviated by two techniques developed since the mid 20th century: stomach flushing (Legler, 1977, see "Diet"; Figure 5.8) and hormonal induction of oviposition (Ewert and Legler, 1978; see "Reproduction").

Whole injected specimens are the most common manner of preparation. Any part of the specimen can be dissected

FIGURE 5.6 The resulting parts of an S&P (shell and soft parts) dissection. The soft skin is removed, which contains the entire head and skeletal elements of the feet and tail. Eggs and stomach contents are removed; gonads and gut are tied in gauze; and remaining skeletal parts are fleshed and dried. All parts of specimen receive the same catalogue number. (Shell and skeletal elements not shown.)

at some later time (e.g., a skull prepared, gonads examined, etc.). Legler's technique of preference is injection of organs and body cavities with 10% formalin solution and immersion in 70% alcohol. Vogt and C. J. McCoy favored immersion in formalin for a time and then in alcohol. Colors can be preserved longer with alcohol immersion.

Specimens should be weighed prior to injection. Deep slits are made in the skin of the undersides of the limbs and neck with a very sharp blade. A small slit is made just behind the mandibular symphysis and the mouth is propped open with a small piece of swab stick—in a manner that makes everything inside the mouth visible without deforming the tongue. The specimen is then suspended by a double hook (fashioned from a paper clip and inserted in the symphyseal slit) until rigor sets in, and then immersed in liquid preservative.

Injection is done with a continuous-pump syringe and a long, large-diameter (#15) needle via the axillary and inguinal pockets. Efficient injection punctures the lungs and gut in several places, and injection fluid flows from the mouth when injection is complete (rather than inflating the body cavities).

Dissections

Of the approximately 12,000 specimens of turtles in the University of Utah collections, about 20% were prepared as dissections in the field. Dried shells are stored in specimen cases. The remainder of the specimen is prepared in two basic ways. A "shell and soft parts" ("S&P") preparation consists of the soft skin of the animal in two pieces, one containing the head and fore feet and the other the tail and hind feet, neatly tagged and bundled together (Figure 5.6). This is accomplished by severing the skin of the inguinal and axillary pockets where it joins the shell. The skin is then pulled back to the cervical-occipital articulation and the wrist and ankle joints; these joints are severed and the rest of the limb skeleton is left attached to the shell. The entire tail skeleton remains attached to the pelvis. The remaining limb skeletons and girdles are separated from the shell, fleshed out, dried in the field, and later cleaned with dermestid beetles. The viscera are bundled in cheese cloth and preserved in formalin. The shell is scraped clean and dried, and becomes a finished museum specimen in the field. The four component parts are labeled with the same field number. A complete skeleton differs from an S&P specimen in that only the soft skin is discarded. Drying of specimens in the field, even in the humid tropics, can be done overnight with a Coleman-type lantern inside a collapsible cylinder made with 1/4-inch hardware cloth and lined with aluminum foil.

Preparing specimens, especially large ones, in separate parts is well suited to fieldwork in a remote areas where all gear and specimens must be stored in a field vehicle and perhaps a small trailer. The preservation of visceral material is essential to further studies of diet and reproduction. In short, the technique makes the most of a sacrificed specimen.

FIGURE 5.7 Eggs, resulting from hormonal induction or dissection, incubating at controlled temperature in sealed, reused food containers. Incubation medium of damp lab tissues keep numbered eggs free of debris for easy inspection with transmitted light.

FIGURE 5.8 Stomach flushing of a small turtle. Copper screening covers top of used food container and is overlain by square of gauze to catch flushings (see Legler, 1977). Flexible plastic canula connected to continuous pump syringe is inserted into stomach via esophagus, and stomach contents are displaced by water. Particulate matter on gauze represents most recent feedings. Oldest meal is in a mucus-coated bolus.

Observation and Keeping Records

In the preparation of this book we have relied heavily on our own notes, catalogs, field journals, and photos made in the field over almost half a century. "Remembering" doesn't work. These notes contain localities stated either in terms of latitude and longitude or in distance from named places appearing on most maps. Most of these things are now simplified by modern electronic devices such as laptop computers, digital cameras, and global positioning devices. Even a primitive GPS will tell you almost precisely where you are on Earth. But nothing will replace notes kept in a known language.

We have never watched the undisturbed behavior of a turtle, in the wild or in captivity, without learning something new. Simple observation, with or without binoculars, from a vantage point (JML has frequently used a stepladder) is productive if recorded in a notebook or with a dictating machine. The huge aquaria in Snow Hall, University of Kansas, faced into a semidarkened hallway were the site of many observations of turtle behavior (Taylor, 1933; Legler, 1955).

Modern Taxonomic Studies and Techniques

Early alpha taxonomic descriptions of turtles, although often diagnostic, were brief and seldom very informative. Size was quantified subjectively ("moderate") or with a single measurement (assumed to be carapace length). Our own scientific careers span the time of purely morphological analysis (e.g., Legler, 1959) to the advent of molecular and morphological analysis combined (e.g., Vogt and McCoy,1980).

Morphological Studies

At mid 20th century some cheloniologists used a suite of measurements (e.g., length, width, and height of carapace) to quantify shape. The use of "ratios" (Milstead, 1969)—the expression of one dimension as a percentage of another—was popular. Klauber (1941) introduced descriptive statistics and tests of significance to herpetology (in his studies of rattlesnakes). Access to mainframe computers in the 1960s and 1970s facilitated sophisticated the statistical analyses of numeric data (Feuer, 1966; Berry, 1978; Vogt, 1993; Legler, 1990; e.g., canonical and multidiscriminate analyses of large series of carefully examined and measured specimens). Simple descriptive statistics are now expected in every study of morphometrics or ecology. Minimally these are mean, standard deviation, and extremes, all of which tell one something about the mean. Most meaningful analyses of turtle classification were based on morphology until the 1980s.

The best morphologically oriented studies have been done by workers who did their own fieldwork and who understood something about the natural history of the taxon, examined large series of specimens, and analyzed their results with sophisticated statistical protocols. These workers were able to acquire a ken for the animals they studied—a kind of knowledge that is difficult to quantify.

Molecular Studies

Molecular techniques in taxonomy harvest the hereditary information in biological macromolecules (proteins and nucleic acids) to address the same questions addressed by traditional morphological techniques. Examples of molecular targets used in cheloniology are mitochondrial DNA (Lamb, Avise, and Gibbons, 1989); allozyme electrophoresis (Georges and Adams, 1992; Morafka, Aguirre, and Murphy, 1994); and cytochrome b and base pairs of ribosomal DNA (Shaffer, Meylan, and McKnight, 1997). Molecular analysis of any kind requires special, often expensive techniques and lab facilities.

Molecular studies begin with tissue samples frozen in liquid nitrogen or preserved in 95% ethanol (Spinks, Shaffer, Iverson, and McCord, 2004; Simmons, 2002). Liver, cardiac muscle, and whole blood are favored tissues. Properly prepared tissues (Vogt, 1993) have been used for DNA comparisons after at least 31 years of proper storage (pers. comm., H. B. Shaffer). Voucher specimens must be placed in recognized museum collections.

Serological studies involving turtles were being conducted at mid 20th century (Dessauer and Fox, 1956; Frair, 1964). Iverson (1991b) successfully combined morphological data (27 characters) with molecular data (11 characters) in constructing a phylogeny of kinosternine turtles. Prior studies of the group had been early experiments with molecular analysis (Seidel and Smith, 1986; Seidel and Lucchino, 1981, Seidel, Reynolds, and Lucchino, 1981). Serious challenges to long-standing concepts of interfamilial relationships were made successfully by Shaffer, Meylan, and McKnight (1997) using a combination of molecular (cytochrome b gene sequence, ribosomal DNA) and morphological characters. More recently Spinks, Shaffer, Iverson, and McCord (2004) addressed the confused relationships within the subfamily Geoemydinae. The journal *Molecular Phylogenetics and Evolution* was founded in 1992.

No technique, new or old, is a panacea. Studies based only on morphology or only on molecular data are at risk of assessing only part of the genotype. Combining these data sets in a single phylogenetic analysis may circumvent this problem.

Analysis

A phenetic analysis is based on overall similarity of the taxa compared. A good phenetic analysis should be based on a suite of carefully selected characters. It is argued that overall similarity may not correspond to genetic relatedness. A phylogenetic analysis infers the phylogeny of the group and expresses it in a tree of hypothetical relatedness (cladogram) based on shared derived characters. Each approach has the potential to yield more or less the same results. Phylogenetic analysis originated in the mid 1960s (Hennig, 1966).

Further information on the foregoing subjects can be found in Wiley (1981); Wiley, Siegel-Causey, Brooks, and Funk (1991); and Avise (2004), and in the elegantly succinct account of "Systematics—Theory and Practice" in Zug, Vitt, and Caldwell (2001).

ACCOUNTS OF TAXONOMIC GROUPS

A Key to the Families and Genera of Mexican Chelonians (Excluding the Sea Turtles)

1A. Forelimbs flipperlike or paddlelike and bearing 2 or fewer claws (sea turtles of the families Dermochelyidae and Chelonidae—not covered in this book). . . . .

1B. Forelimbs with typical well-defined digits and three to five claws 2

2A. Shell covered with unbroken soft skin; peripheral bones lacking, imparting flexibility to posterior, leathery margin of carapace; snout in form of a distinctive proboscis . . . . **Family Trionychidae, genus *Apalone*, p. 185**

2B. Shell covered with large, typical epidermal laminae; snout not bearing a proboscis; no part of shell leathery or flexible 3

3A. Plastron cruciform but akinetic; abdominal scutes lie on narrow bridges and do not meet on midline; tail about as long as plastron at all ages and bearing a dorsal crest of large scales . . . . **Family Chelidridae, genus *Chelydra*, p. 355**

3B. All paired plastral scutes in midline contact; tail of adults much shorter than plastron and lacking a dorsal crest of large scales. **4**

4A. Twelve scutes in marginal series; 12 or more plastral scutes **5**

4B. Eleven scutes in marginal series; 11 or fewer scutes on plastron . . . . **Family Kinosternidae, 12**

5A. Eleven to twelve plastral scutes; three to five (or more) inframarginal scutes on bridge, separating marginals from plastral scutes; an elaborate vegetation chopping jaw armature . . . . **Family Dermatemydidae, genus *Dermatemys*, p. 69**

5B. Twelve plastral scutes; inframarginals usually consisting of an axillary and inguinal, not separating plastral scutes from marginals; jaw armature variable, never elaborate **6**

6A. Carapace high domed, plastron extensive, and bridge long, tending to reduce the anterior and posterior orifices of shell; a premaxillary beak; triturating surface of maxilla with or without ridging; palate vaulted to accommodate a large muscular tongue; hyoid skeleton moderate to small. Limbs modified for walking with plastron held clear of substrate; hind limbs elephantine and columnar; forelimbs flattened, antebrachia often covered with thick osteodermal scales. Skin on head divided into scales. Spurlike tubercles commonly present on rear of thigh. Extremely short tail in adults and young . . . . **Family Testudinidae, genus *Gopherus*, p. 204**

6B. Carapace shape ranges from relatively low and streamlined to high, domed, and (convergently) tortoiselike (e.g., *Terrapene*). Plastron extensive but bridge short. Limbs generalized, feet usually with interdigital webbing and adapted to swimming; hind feet never elephantine. Palate flat. Tongue simple, flat, and small. Skin on head smooth or with small scales posteriorly. Tail longer than testudinids, especially in young. Premaxillary bones form a beak, a notch, or a notched beak **7**

7A. Posteriormost marginal (M12 or "supracaudal") scutes relatively high, higher than M11 and overlapping suprapygal bone; a single articular socket between the fifth and sixth cervical centra; a strong lateral tuberosity on the basioccipital bone (the "batagurine process" of McDowell, 1964); angular bone excluded from contanct with Meckel's cartilage by a longitudinal process of the articular . . . . **Family Geoemydidae, genus *Rhinoclemmys*, p. 337**

7B. Posteriormost marginal (M12 or "supracaudal") scutes relatively low, lower than M11 and not overlapping suprapygal bone; a double articular socket between fifth and sixth cervical centra; no strong lateral tuberosity on basioccipital; angular bone forms floor of canal for Meckel's cartilage . . . . **Family Emydidae, 8**

8A. A single, transverse, ligamentous, kinetic hinge on plastron permitting anterior and posterior plastral lobes to tightly close the shell; digits minimally webbed. Small, terrestrial, tortoiselike emydids . . . . ***Terrapene*, p. 305**

8B. No transverse plastral hinge **9**

9A. Head and neck having a bright pattern of dark-bordered, parallel, yellowish or orangish stripes on a darker background; fully aquatic basking pond turtles **10**

9B. Head and neck not striped; markings at best a pale mottling on a darker ground; occurs in northern Baja California Norte. . . . . ***Actinemys*, p. 232**

10A. Upper jaw with median notch flanked by prominent cusps; carapace smooth on its surfaces and edges (no keels, no serrated edges); plastron orange with bold, black concentric markings. . . . . ***Chrysemys*, p. 236**

10B. Upper jaw without prominent notch or cusps **11**

11A. Maxillary triturating surface well developed with median alveolar ridge bearing conical denticles or tubercles; rear margin of carapace usually serrated; adult males with elongated foreclaws; ventral surface of lower jaw flat in cross-section . . . . ***Pseudemys*, p. 241**

11B. Maxillary triturating surface bearing median alveolar ridge smooth or with finer denticles; adult males (south of Rio Grande and west of Pecos) with elongated snout and unmodified foreclaws; ventral surface of lower jaw rounded in cross-section. . . . . ***Trachemys*, p.246**

12A. Plastron reduced or cruciform, covering half or less of ventral surface and bearing less than 10 scutes (usually seven or eight); entoplastron present **13**

12B. Plastron covering half or more of ventral surface and bearing 10 or more scutes; entoplastron absent **14**

13A. Axillary and inguinal scutes well developed; plastral bridge joins bridge by an akinetic joint; a single, kinetic, anterior plastral hinge . . . . ***Staurotypus*, p. 88**

13B. Axillary and inguinal scutes absent or vestigeal and reduced; an extremely narrow plastral bridge forms kinetic joint with carapace; no transverse plastral hinges; carapace weakly or not carinate . . . . ***Claudius angustatus*, p. 83**

14A. Plastron with two movable hinges; if six pairs of plastral scutes, scute number one never shortest; interlaminal seams of plastron do not contain soft tissue . . . . ***Kinosternon*, p. 103**

14B. Plastron with a single (anterior) movable hinge; plastral interlaminal seams containing soft tissue or not **15**

15A. Plastral interlaminal seams as in other *Kinosternon* or containing small amounts of soft tissue; a prominent hooked premaxillary beak; smallest adults >110 mm; no stripes on head and neck; occurs in Gulf drainages of northeastern Mexico . . . . ***Kinosternon herrerai*, p. 103**

15B. Plastral interlaminal seams often containing broad areas of soft skin; no premaxillary beak; most adults <100 mm CL; lateral stripes on head and neck of juveniles and young adults; once occurred in the internal El Sauz drainage of Chihuahua . . . . ***Sternotherus odoratus*, p. 180**

SUPERFAMILY TRIONYCHOIDEA

The trionychoids constitute a strange assemblage of widely varying taxa that seem, at first, to be so unusual as to be incorrect or fanciful, but the alliance is supported clearly by shared derived characters (molecular and morphologic). These close relationships have been commented on since the time of Baur (1891) and Boulenger (1889). Williams (1950), followed by Webb (1962) and others, broke this bubble of logic temporarily by including the family Chelydridae in the trionychoids, but most modern classifications place the snapping turtles closer to the Testudinoidea.

Four families are included, as follows: Dermatemydidae (p. 67), *Dermatemys mawi* being the sole extant representative; large, aquatic neotropical turtles with a "typical" emydidlike shell without plastral hinges. Kinosternidae (p. 77, 4 genera and 18 species in Mexico); medium to small aquatic or semiaquatic turtles with a highly modified shell—one or two plastral hinges or a kinetic articulation between bridge and plastron. Trionychidae (p. 183, one genus and two species in Mexico); medium to small completely aquatic turtles with a flattened, disklike shell having leathery, flexible edges. The Family Carettochelyidae with one extant species (*Carettochelys insculpta*) occurs only in New Guinea and northern Australia.

All trionychoids except *Dermatemys* have reduced the margin of the shell to 10 peripherals and 11 marginals, and Mexican trionychids (*Apalone* sp.) have no peripherals or epidermal scutes at all. The homologies of the plastral scutes are distinctive in *Dermatemys* and all Kinosternids (see taxonomic accounts).

Family Dermatemydidae

Genus *Dermatemys* Gray, 1847

Dermatemys mawi Gray, 1847

"Tortuga Blanca" is the most common name in Mexico and alludes to the pale plastron; Central American River Turtle (Iverson, 1992a); Giant River Turtle (Iverson and Mittermeier, 1980); Hachac (Mayan; pers. comm. Niko Chin).

ETYMOLOGY

Gr., derma, skin; dermatinos, leathery; Gr., emys, freshwater turtle. Probably alluding to the leathery appearance of the thin epidermal covering of the carapace, on which the scutes may be poorly defined in large adults; *mawi* is a patronym honoring the donor, a British naval officer named Solomon Maw (see the following).

HISTORICAL PROLOGUE

Two live specimens, presumably from a port on the Gulf Coast of Mexico, were brought to the London Zoological Society Gardens by a Lt. Maw (British Royal Navy) in 1833. The turtles lived until some time in the 1840s and eventually became specimens in the museum. One of these became the holotype of *Dermatemys mawi*. There has been confusion in the literature about the spelling of the specific name and the name of the collector, resulting in several emendations (e.g., Cope, 1865; Neill and Allen, 1959; Smith and Smith, 1979).

In the original description, Gray (1847) stated the name of the donor as "Mawe" and formulated the patronym "*Dermatemys Mawii*"; he used the name "Mawe" twice more (1855a and 1873) without mentioning an alternative spelling. However, all of the British Museum records and labels associated with the type specimen use the spelling "Maw." The following information was supplied by Dr. E. N. Arnold and Dr. Colin McCarthy of the British Museum (Natural History). The type specimen is BMNH 1947.3.4.12, a complete shell, CL 388 mm (420 mm measured on curve), and is the specimen illustrated by Gray (1855b: pl. XXI). Associated with or inscribed upon the type are "*Emys Mawi* Lt. Maw R.N. Corr. Memb. S. America 1833." Written three times on the inside of the shell, in the same hand, is "Lieuft Maw 1833." Records of the Zoological Society of London list a "Solomon Maw" as a corresponding member for 1833–1834.

Smith and Taylor (1950b) restricted the type locality to Alvarado, Veracruz.

The recent studies of Polisar (1996, 1992a, 1992b), Vogt (1992, 1998, 1988, 1987, and unpub. data), Vogt and Flores-Villela 1992a,b), and D. Moll (1986, 1989) have elevated our knowledge of the natural history of this species from a virtual mystery at mid 20th century to the status of reasonably well known at present.

DIAGNOSIS OF FAMILY DERMATEMYDIDAE AND *DERMATEMYS*

The following combination of characters distinguishes *Dermatemys* from all other Mexican chelonians: a hard-shelled carapace with a full compliment of carapace scutes, 11–12 plastral scutes and 3–5 (or more) inframarginal scutes on each bridge. Epidermal laminae discrete but thin; interlaminal seams may become obscure in old adults. Triturating surfaces of maxilla and mandible elaborately ridged (see the following).

FIGURE 6.1 Subadult male *Dermatemys* preparing to breathe at surface with only tip of snout exposed.

FIGURE 6.2 Dimorphism in head coloration in adult *Dermatemys*: male (left) and female (right).

Since *D. mawi* is the only extant member of its genus and family, no other diagnoses are given.

GENERAL DESCRIPTION

Dermatemys mawi is the largest freshwater turtle in Mexico (650 mm CL, 22 kg, rivaled in mass only by *Chelydra* sp.). It is seemingly the most generalized living trionychoid turtle.

Carapace is uniform medium brown, gray, or olive; plastron is distinctly paler, cream or yellowish (but often brown when stained by water deposits). The limbs and neck are neutral brownish or dark gray above and paler brown below.

Nuchal costiform processes are well developed, evident in young, and buried gomphotically in P1, P2, and C2 in adults. The plastron is connected to the carapace by a broad bridge. Carapace is low, broad, and evenly domed in profile and cross-section. Shell is smooth in adults; shell margins are rounded, not sharp, except in young; a median keel occurs in juveniles. Six or seven neurals in a continuous series posteriorly from the nuchal, the posteriormost being a much-reduced triangle with apex posterior and separated from pygal by the meeting of costals 7 and 8 or C6–8; some shells (ca. 20%) have a small plate (a neural or vestigial suprapygal) between the pygal and the dorsally joined C8 (Boulenger, 1889:28); an occasional specimen has eight neurals in a continuous series from nuchal to pygal. There are 11 peripheral bones and 12 marginal scutes on each side. The plastron has a small anterior, usually unpaired scute (extragular) followed by five pairs of larger scutes (gular, humeral, abdominal, femoral, and anal (*fide* Hutchison and Bramble, 1981). (See account of Family Kinosternidae and Figure 7.1.) Three to six inframarginals on each bridge (usually 4 or 5, but as many as 23) separate the marginals from the main plastral scutes.

BASIC PROPORTIONS

(RCV Database, Chiefly Veracruz and Tabasco)

CW/CL: males, 0.71 ± 0.028 (0.56–0.76) $n = 84$; females, 0.71 ± 0.037 (0.52–0.76) $n = 63$; im. unsexed, 0.77 ± 0.052 (0.59–0.88) $n = 203$.

CH/CL: males, 0.37 ± 0.026 (0.25–0.41) $n = 83$; females, 0.37 ± 0.031 (0.27–0.42) $n = 63$; im. unsexed, 0.37 ± 0.034 (0.24–0.53) $n = 351$.

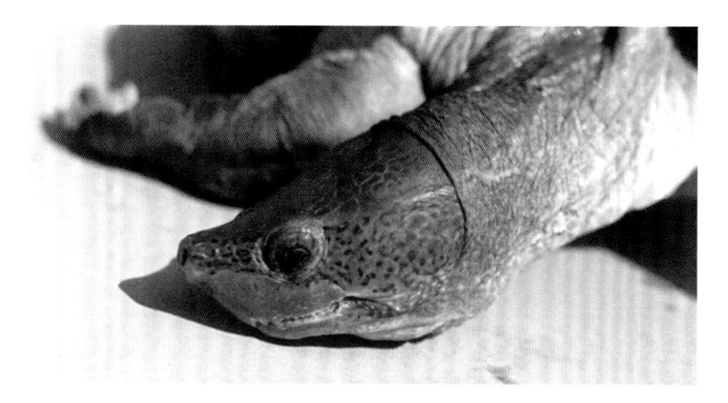

FIGURE 6.3 Adult *Dermatemys* showing typical drooping posture of head and neck when out of water.

CH/CW: males, 0.52 ± 0.036 (0.39–0.62) n = 83; females, 0.53 ± 0.049 (0.37–0.75) n = 63; im. unsexed, 0.48 ± 0.045 (0.35–0.67) n = 204.

As an herbivorous river turtle, *Dermatemys* shares many of its adaptations convergently with river turtles of distant relationship (e.g., *Batagur baska, Elseya dentata*). The elongate snout and the jaw ridging are examples of these characters.

Dermatemys has two musk glands, one at each extreme end of the sternal cavity; ducts passing through osseous canals in hyo- and hypoplastral buttresses; an axillary orifice in the soft skin of the axillary pocket at M3-4; and an inguinal orifice in seam between the posterior inframarginal and M8 (Waagen, 1972).

Winokur (1982a) reported a pair of very small tubercles in the gular region in 4 of 15 specimens and suggested they could be vestigial barbels.

Diploid chromosomes number 56, with 7 pairs of metacentric to submetacentric macrochromosomes, 5 pairs of telocentric or subtelocentric macrochromosomes, and 16 pairs of microchromosomes. The G-banded macrochromosomes are identical to those of *Chelonia mydas*, suggesting that this is the primitive karyotype for all nontrionychoid cryptodiran turtles (Carr, Bickham, and Dean, 1981).

THE JAW ARMATURE OF *DERMATEMYS*

In the largest adult skulls of *Dermatemys* (ca. 75 mm condylobasilar length; Figure 6.8) we see the epitome of palatal and mandibular ridging. In palatal aspect there are, on each side from lateral to medial, four structures of consequence to jaw armature: respectively, the tomial edge and, on the alveolar plate, three denticulate ridges—two on the alveolar shelf itself and one forming the raised medial edge of the secondary palate. The ridges parallel one another to form three distinct grooves. The mandible bears a tomial edge and two alveolar ridges, forming two grooves. These three ridges occlude with the three maxillary-palatal grooves (Figure 2.7), forming a rather elegant and efficient series of shearing-crushing-chopping devices.

There is also a pair of sharp-edged diagonal ridges on the anterior palatal plate of the maxillary, just posterior to the premaxillary-maxillary suture. They converge on the palate at a point common to the maxillaries, premaxillaries, and vomer. Corresponding diagonal ridges occur on the anterior mandible and form an additional shearing mechanism. They may be used to nip small bites from streamside vegetation (blades of grass and leaves) or to gain purchase on clumps of vegetation that are then pulled into the water. A *Dermatemys* can gently bite a piece from a floating leaf with scarcely any surface disturbance (JML, pers. obs.). However, we have made no detailed observations of *Dermatemys* feeding behavior.

Dermatemys has a further feature of jaw armature that is unique. The mandible bears an accessory ridge on the lower third of its lateral surface, here named the lateral dentary ridge. It forms one side of a groove that receives the tomial edge of the maxilla, further enhancing the shearing/cutting potential of the aforementioned ridging. Typically stomachs contain bite-sized chunks of leaves, grasses, and flowers. Seeds and fruit (e.g., figs) are consumed whole or in pieces.

The degree of alveolar ridging undergoes ontogenetic change that could account for differences of detail in published skull drawings (Bienz, 1896; MacDowell, 1964; present work; Figure 6.8). In juveniles (CL 58–66 mm) there is a single alveolar ridge on the maxilla, forming a groove with which the mandibular tomium occludes; the anterior diagonal ridging is evident but incipient. All other ridges and grooves develop much later. In its fullest development, the shearing/chopping capacity of the jaw ridging is approximately tripled in the course of ontogeny.

COLOR IN LIFE

The eye of females and immatures is dull—an unmarked bluish gray iris surrounding the black pupil. The eye is brighter in mature males—a yellowish iris variously marked with darker flecks that are sometimes concentrated in patches to suggest a horizontal eye stripe or a stellate pattern.

The top of the head is always darker than the sides or throat, is dull brown in females, and is yellowish in males (see Growth and Ontogeny). Jaw sheaths and the circumnarial area of the snout are uniformly pale and contrast with other head colors. Skin papillae on the neck, limbs, and tail are pale or dark but are always in sharp contrast to ground color.

FOSSILS

The Family Dermatemydidae is represented by 19 fossil genera from the Jurassic of Europe and the Cretaceous of North America, Europe, and East Asia (Hutchison and Bramble,1981; Iverson and Mittermeier,1980; Romer, 1956). *Dermatemys mawi* is not known as a fossil.

FIGURE 6.4 *Dermatemys* nest site in shallow cavity under overhanging river bank at water's edge; eggs are barely visible (arrow); inset on lower right shows detail of open nest cavity. Parts of at least five eggs are visible.

GEOGRAPHIC DISTRIBUTION

Map 5

Dermatemys mawi occurs in large rivers from the Rio Papaloapan drainage of central Veracruz, along the Gulf Coast to the Rio Champoton of Campeche, across the base of the Yucatán Peninsula in the lowlands of the Rio Usumacinta drainage of Tabasco and Chiapas, to southern Quintana Roo, Belize, and the Atlantic drainages of northern Guatemala and adjacent Honduras (Iverson, 1992a; Smith and Smith, 1979). It does not occur in the karst areas of the Yucatán Peninsula that lack external surface drainage. *Dermatemys* occurs sympatrically with *Trachemys scripta venusta* in lakelike habitats, the two species often being captured in the same traps and nets.

HABITAT

Dermatemys inhabits large, deep rivers and associated oxbow lakes throughout the year. In the rainy season they enter flooded forests and also travel up smaller tributaries to nest.

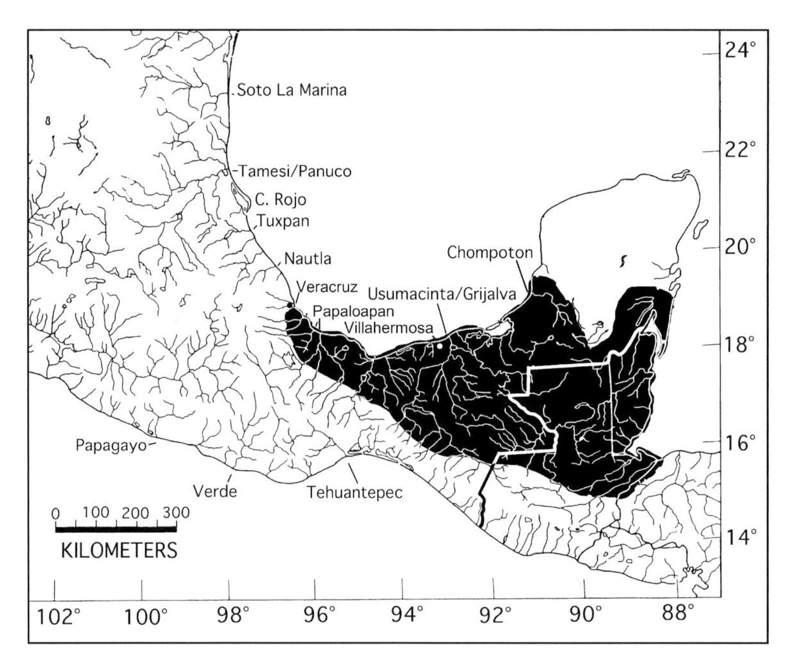

MAP 5 Geographic distribution of *Dermatemys mawi*.

FIGURE 6.5 Emerging *Dermatemys* hatchling. Note pale yellowish patch barely visible beneath caruncle and distinct pale supraorbital stripe tapering posteriorly and extending a variable distance onto the neck.

Moll (1986) reported a population in mangrove swamps in Belize. Vogt did not find the species in the mangrove areas he studied in Veracruz (Lago de Sontecomapan and Rio de Água Dulce) or in Tabasco (Lago de Cometa). As flood waters recede the turtles may become trapped in oxbow lakes until the next rainy season. No natural overland movements have been observed. Males seemingly never leave the water, and females do so only briefly to nest (see Reproduction).

They are fast, strong swimmers and capable of swimming up rapids. Usually they congregate on the bottom of deep (15 to 20 m) pools below the rapids, often in association with submerged tree trunks imbedded in the river bottom. Larger turtles may bury themselves partly in soft bottom substrate.

In the dry season (April and May), adult turtles congregate in deep holes; juveniles and subadults are found chiefly along the shorelines of smaller tributaries in deadwood snags or mounds of accumulated detritus. The younger turtles are easy to catch by diving off a boat, but subadults and adults usually are too fast for a diver unless they are grasped during the initial plunge. Scuba divers associated with Vogt explored deep holes in the Rio Lacantún and Rio Tzendales in May 1986. The divers saw many subadults and adults motionless on the bottom. The turtles were too wary to be approached closely. When three divers explored these same holes in May 1992, no concentrations of turtles were found.

DIET

Dermatemys is completely herbivorous from hatchling to adult (Vogt and Flores-Villela, 1992a). Information on diet was obtained from 220 stomach flushings of adults, subadults, and juveniles from the Rio Lacantún and Rio San Pedro, Reserva de Montes Azules, Chiapas, by Vogt and associates. Flushings were made in all calendar months from 1984 to 1997.

Diet varies seasonally and among populations. During high water periods they feed on shoreline vegetation (leaves, grasses, fruits, and flowers). When water levels rise (3–8 m in some habitats) the turtles have access to a greater variety of these foods in flooded forests and feed within the branches of some trees. Turtles in seasonally flooded marshlands often feed exclusively on grasses. Fruits are eaten opportunistically when available (e.g., figs). The bulk of the diet of riverine *Dermatemys* is the leaves of terrestrial trees, lianas, and grasses. Moll (1989) found estuarine populations in Belize to be feeding on three species of mangrove leaves.

Foraging usually occurs at night (Álvarez del Toro, Mittermeier, and Iverson, 1979; Moll, 1986, 1989). *Dermatemys* were captured in trammel nets and fyke nets mostly during the evening (1900–0100 h). Less than 5% were caught in traps during the day. In areas under tidal influence, Moll (1989) noted activity during a rising tide regardless of the time of the day. Foraging is also stimulated by rising water levels in freshwater rivers, perhaps because of increased access to fresh leaves and fruits. When water levels drop, *Dermatemys* feeds on submergent aquatic vegetation and on underwater accumulations of dead leaves that build up behind log jams and in calm backwaters. These natural stockpiles of leaves allow uninterrupted feeding throughout the year. Presence of *Dermatemys* in a river can be detected by the characteristic ragged bite marks made in the leaves of terrestrial plants that have been inundated. In lakes, *Dermatemys* feces packed with grasses can be found floating in the grass beds along shorelines (e.g., Laguna Oaxaca). In 2002 neither kind of telltale sign was present in the Laguna Oaxaca or the Rio Tzendales, in the Montes Azules Biosphere Reserve, Chiapas.

Laboratory experiments (Vogt, 1987) have shown that natural gut micro-organisms aid in the digestion of plant material much like that reported for *Pseudemys nelsoni* by Bjorndahl and Bolten (1990). Hatchlings (n = 12) were fed a diet of mosote (Compositae, *Melanopodium*) leaves, stems, and flowers. Half were allowed to inoculate themselves by feeding on adult feces during the first few days after hatching. Those inoculated were 30% heavier at the end of the first year of growth than noninoculated hatchlings on the same diet (means 159 g ± 20.3 [144–187] n = 6 versus 120 g ± 17.1 [95–142] n = 6, respectively). Although the organisms involved in the digestive process were not identified, they are perhaps similar to those described by Bjorndal and Bolten (1990) for *Pseudemys nelsoni*, an herbivorous species endemic to Florida. Under natural conditions the hatchlings might inoculate the gut by ingesting the nest substrate; females of most turtles discharge fluid from the cloaca in the process of nesting, and hatchlings of many species are found with nest substrate in the gut (Vogt, pers. obs.). Hatchlings could also simply inoculate

FIGURE 6.6 *Dermatemys* juveniles in first year of growth.

the gut by feeding on adult feces (as in hatchling green iguanas; Troyer,1982). See also the account of *Gopherus flavomarginatus*.

The small intestine of adults is usually packed with nematodes. Experiments need to be undertaken to determine if these nematodes act as parasites or commensals. Lopéz-León (2001) did not find any significant differences in the digestive efficiency of *Dermatemys*, *Rhinoclemmys areolata*, and *Trachemys scripta venusta* fed the same plant material and under the same laboratory conditions.

HABITS AND BEHAVIOR

Dermatemys does not bask in nature, either in or out of the water (Vogt, pers. obs.). References to aquatic basking in popular accounts (e.g., Álvarez del Toro, Mittermeier, and Iverson, 1979) may refer to captive turtles that are (pathologically) unable to submerge normally. Vogt observed this in captive adults at Los Tuxtlas, Veracruz, and Nacajuca, Tabasco, on at least three occasions.

Basking in emydids (e.g., *Trachemys*) raises body temperature (Moll and Legler, 1971) and seems to accelerate digestion. *Dermatemys* can digest plant material at the ambient water temperatures of the neotropics. Body temperatures of 37 *Dermatemys* in the Rio Tzendales and the Laguna Oaxaca ranged from 23.5°C to 32°C and approximated those of the water.

Vogt and students conducted repeated dry season floats of ca. 3 hours and 6 km in the Rio San Pedro (tributary to Rio Tzendales, Montes Azules Biosphere Reserve, Chiapas) from 1986 to 1992. Juveniles (15–30 per float) could easily be seen in the clear water among the submerged branches of fallen trees. Juveniles were not seen in the faster water of Rio Tzendales proper.

Dermatemys moves slowly and clumsily on land. Individuals out of water have difficulty holding the head off the substrate (as revealed in most illustrations [Holman, 1963]) and repeated observations by us (Figure 6.3). Adult *Dermatemys* cannot right themselves when inverted on flat ground.

Well-oxygenated water is preferred. The buccopharyngeal mucosa bears vascular papillae (Winokur, 1988) that can extract oxygen from water drawn in through the mouth and forced out the nostrils (R. M. Winokur, pers. comm.), permitting long periods of submergence without surfacing to breathe. Breathing is accomplished typically with only the tip of the snout exposed (Figure 6.1).

Legler and Elias Alfaro (a local Mayan resident) had occasion to observe *Dermatemys mawi* under natural conditions on the Rio Mopan, a major tributary of the Belize River, near Succotz, Belize (550 m elevation, 17°07′N, 89°08′W). The river is a medium-sized, clear-water stream with large pools up to 15 m deep, separated by narrower rapids that flow over a rocky bed. In early June we watched a large pool from a boat. Several large adults surfaced to breathe. We could see them rising rapidly to the surface when they were still at least 3 m deep. All were extremely wary of our presence and would either abort the rise or quickly dive again after surfacing. Local people caught the turtles in pools like this for household consumption using a hand line baited with mango. A trammel net set at the inlet to this pool in current fast enough to bow the net captured an immature male. Baited hoop net traps set in various places along the river caught *Trachemys s. venusta* and *Staurotypus triporcatus* but no *Dermatemys*. Vogt tested various baits (fresh and fermented bananas, plantains, yucca, mangos, figs) and caught no more turtles in the baited fyke nets than in the unbaited ones.

In April 1992, six adult females and two adult males were fitted with transmitters and followed for nearly a year. They moved up and down the Rio Tzendales from the first waterfall to the confluence with Rio San Pedro, often moving 1.5–2.0 km per day, and had not left the Rio Tzendales by December 1992. They stayed in wide, slow areas with abundant food resources. Two other adult females were fitted with transmitters in an area of the larger Rio Lacantún where food resources were less favorable. They swam out of range within 2 days and were never located again. These extensive movements were judged to reflect rapid travel through areas of relatively sparse food availability.

FIGURE 6.7 *Dermatemys* juvenile with orange patch on snout.

REPRODUCTION

MATING AND COURTSHIP

Observations of courtship and mating in nature have not been published. Copulation was documented in captivity at the Nacajuca (Tabasco) experimental turtle farm on 17 July 2005 at 0800 h (Claudia Zenteno-Ruiz, pers. comm.). The previous day 14 adult males and 32 adult females were transferred from one earthen pond to another with fresher, cooler water. Fourteen copulating pairs were seen floating on the surface, for 2–3 hours. The transfer may have simulated conditions at the beginning of the rainy season. Polisar (1996) mentioned verbal reports from hunters who reported seeing males and females "near each other" in April and early May.

Vogt housed captive adults in small (1.5 × 1.5 m) concrete tanks in 0.5 m of flowing water at Los Tuxtlas, Veracruz (Vogt, 1988). One male was placed with four females in each tank. Males persistently and aggressively pursued females, biting at the anterior and posterior margins of the carapace, inflicting bleeding wounds. Mounting was attempted but unsuccessful. The circumstances of captivity constrained the escape of females and the males had to be removed. Wounds were never noted in wild females.

Spermatogenesis is classified as prenuptial by Meylan, Schuler, and Moler (2002).

Vitellogenesis begins in June and July. Preovulatory follicles of 34.4 mm were noted in early July 1992. The earliest oviducal eggs were noted on 25 August (Rio Tzendales). Vogt found oviducal eggs, preovulatory follicles or corpora lutea, as late as early March in Chiapas. Fresh corpora lutea are 9–14 mm in diameter.

Polisar (1996) observed the following in Belize: Preovulatory follicles were 32–36 mm in diameter; ovarian follicles >25 mm were present from September to December; in 18 dry season ovaries (March–June) 67% of the ovaries had follicles 15–24 mm and ca. 13% of follicles >15 mm were atretic; regressing corpora lutea are still apparent in late April.

NESTING

Nest construction has yet to be observed. Nesting occurs chiefly during the rainy season (September–December) when water levels are at their annual peaks. A few females may be laying their last clutch of the season in late February or early March, but this is unusual. Reports of nesting in March and April (Lee, 1969; Moll, 1986) are based on gravid turtles from markets with eggs held in oviducts because of unfavorable conditions.

All nests observed (ca. 40) were within 3 m and most were within 1.5 m of a shoreline. None was more than one vertical meter above water, and all were low enough to be flooded. Nests are often constructed under an overhanging river bank immediately at the water's edge (Figure 6.4). Nests have also been found on clay banks and on clay mounds surrounded by patches of grasses on river flats. Captives in earthen ponds dig nest cavities into the bank at the waterline.

Nesting is apparently usually solitary. In nature the nests are scattered along shorelines that are continually changing with the rising and falling of water levels. Nests often go undetected by predators because they are covered first by water and subsequently by mud left at high water. The inundation of nests by flood waters has fueled the rural belief that eggs are laid underwater.

Vogt found a rare concentration of nests in September 1985, 20 m below a 6 m high waterfall on the Rio Tzendales; *Trachemys* occurred above the waterfall, but *Dermatemys* did not. The nests (26) were concentrated along ca. 4 m of shoreline on both sides of the downstream end of a small island. The vegetational association was a high, closed canopy evergreen rainforest on both sides of the river. At the time, this was clearly the upstream limit of opportunity for females seeking nesting sites. The island provided the only satisfactory sites near an otherwise rocky shoreline.

The earliest nests of the season may be inundated for weeks with no apparent negative effect on egg survivorship. Vogt transferred a clutch of 12 eggs that had been underwater for 6 weeks to the laboratory and incubated the eggs at 28°C with a 90% hatching success. Polisar (1996) documented a 80–100% hatching success in eggs subjected to more than 30 days of submersion in water.

EGGS

Data on 121 eggs in 12 clutches from the Rio Tzendales and Rio Lacantún were L, 61.2 ± 3.8 (56–67); W, 36.4 ± 2.9 (32–41); WT, 49.4 g ± 10.1 (37–72). Data on the 12 gravid females were CL, 466 mm ± 34.5 (420–502); WT with eggs, 12,050 g ± 2,545 (9,500–17,200); clutch WT, 514.1 g ± 107 (294–690); relative clutch mass, 0.043 ± 0.00981 (0.0294–0.0590).

Gonadal examination of eight mature females suggests a mean potential of 2.6 ± 1.06 (1–4) clutches per year in Chiapas. Number of eggs in 18 clutches from Chiapas and Veracruz was 14.8 ± 3.37 (10–24). Data on 44 eggs from Belize are L, 61.6 mm ± 3.9 (54.1–72.0); W, 35.8 mm ± 3.2

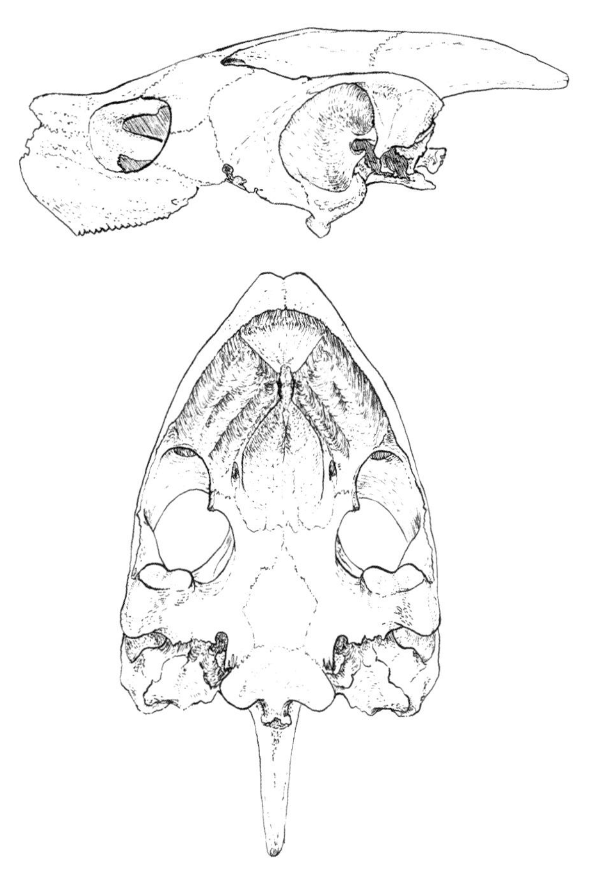

FIGURE 6.8 Lateral and ventral views of *Dermatemys mawi* skull; adult male; UU 13349; Playes De Catazajá, Chiapas; condylobasilar length, 76.4 mm. (Drawn by Elizabeth Lane.)

(32.4–49.8); WT, 50.2 g ± 9.8 (34.3–70.0); yolk diameter, 30.2–32.8 mm (Polisar,1996).

The eggs have brittle shells at oviposition. The outer calcareous layer cracks and pulverizes as the eggs expand and become turgid in a moist substrate, resulting in a shell that is flexible or "leathery." This accounts for the incongruence in the literature regarding the consistency of the eggshell (Smith and Smith, 1979; Holman, 1963; Álvarez de Toro, Mittermeier, and Iverson, 1979).

INCUBATION

Vogt conducted incubation experiments in the laboratory, shifting eggs between different moisture and temperature regimes. Incubation time to pipping in the laboratory varied from 115 to 223 days, at temperatures of 25–28°C. This variation can be attributed to embryonic diapause and embryonic aestivation (see Introduction). Sex determination is temperature dependent. Laboratory incubation temperatures above 28°C produced all females and temperatures of 25–26°C produced all males. Threshold temperatures have not been determined (Vogt and Flores-Villela, 1992b). There is an embryonic diapause in response to unfavorable environmental conditions (e.g., cool temperatures and flood conditions). Development after oviposition does not commence immediately if the substrate has a high moisture content. Diapause is broken by warmer temperatures and drying of the nests. Fully developed turtles aestivate in the eggshell until hatching is stimulated by moisture from the first summer rains. Diapause and embryonic aestivation synchronize the hatching of clutches laid over a 4-month period. The physiological mechanisms of this embryonic diapause require investigation.

Polisar (1992a) published the following information on reproduction in Belize. Incubation was 217–300 days from oviposition to emergence. A maximum of four clutches and 47 eggs per annum was observed. First clutches in recently matured females are small. There is a positive correlation of female body size with reproductive output. Of 21 adult females 7 produced one clutch; 7 produced two clutches; 6 produced three; and 1 produced four clutches per year. Eggs per clutch averaged 10.98 ± 3.72 (2–20) n = 49. Clutches over 15 were rare, and 8–15 eggs was common.

Polisar found that 3% of the adult females in Belize skipped seasons of reproduction. Skipped seasons were not observed in Mexico.

GROWTH AND ONTOGENY

In Mexico, females reach maturity at 395–420 mm and males at 365–385 mm.

Size data on 358 specimens from Veracruz and Tabasco (RCV database) were as follows: males, 424 ± 2.2 (366–479) n = 85; females, 434 ± 58.6 (188–512) n = 64; im. unsexed, 234 ± 8.6 (98–386) n = 209. Adult weight (kilos) was as follows: males, 9.6 ± 1.7 (6.5–17.3) n = 85; females, 10.6 ± 3.2 (2.3–17.2) n = 63.

Females grow larger than males. The head of mature males is sharply bicolored—pale yellow-ochre, yellowish orange, or bright orange above a line joining the nostril, top of orbit, and top of tympanum (Figure 6.2) and darker below. The pale dorsal colors become brighter in the breeding season (October and November) and often have darker speckling laterally. The head of females is paler than that of mature males and only slightly more yellow than the color of the legs. The anal notch of the plastron is deeper, and the tail is longer and thicker in males. Snout elongation occurs in juveniless and adults of both sexes but is exaggerated in old males.

Size and weight of 40 *Dermatemys* hatchlings from laboratory-incubated eggs obtained in Veracruz was CL, 52.8 mm ± 4.58 (41.8–62); WT, 31.2 g ± 5.7 (22.8–45.7).

Hatchlings and young juveniles have a distinct pale-yellow postorbital stripe, tapering posteriorly and extending a variable distance onto the neck; there is also an orangish field on the front of the snout extending from the tops of the external nares to the caruncle. Dark flecks are scattered across a pale-gray ground color on the carapace. These juvenile markings gradually fade and are lost completely by the age of 3 years.

Shedding of scutes occurs in the dry season (late April and May) in Chiapas. In some old turtles where growth has stopped, the scutes become fused to one another and are sloughed in a single sheet. Epizoic algae are shed with the scutes. The carapace is blue-black, soft, and easily damaged after shedding.

Growth rings are not valuable for age estimation in *Dermatemys*. Growth rings in juveniles represent different spurts of growth within the first year and are lost with the first ecdysis. Hatchlings raised in captivity on a diet of fresh plants reached carapace lengths of as much as 147 mm and weights as high as 475 g in the first year of growth. A 300 mm, 800 g juvenile captured in 1985 was recaptured in 1988 at 340 mm and 2,300 g. Another grew from 340 mm and 3,200 g in 1985 to 380 mm and 6,000 g in 1988, and a third from 340 mm and 5,200 g in 1985 to 420 mm and 8,400 g in 1989. Growth in adults does not slow abruptly at maturity.

A subadult male (CL 398 mm) collected in the Rio Lacantún in April 1992 weighed 8,500 g; it was dissected into weighed component parts as follows (wet weights): striated muscle, 2,350 g; viscera, 443 g; appendicular skeleton, 565 g; shell, 1,446 g; gut contents, blood, and other body fluids, 3,700 g.

Dissection of a subadult female (CL 358 mm) weighing 5,452 g yielded the following components and weights: striated muscle, 680 g; gut, 512 g; other organs, 266 g; appendicular skeleton, 263 g; skin, 165 g; fat, 50 g; shell, 1,940 g; gut contents, blood, and other body fluids, 1,576 g.

A juvenile from Belize (CL 120 mm) is near the end of its first full season of growth. A juvenile from Guatemala (CL 65.7 mm) with a closed umbilicus is in the early stages of areolar growth (one zone, ca. 1 mm). A cleared and stained specimen of the same size shows virtually no development of the dermal costal elements and only a bare outline of ossified plastral elements defining a single, large plastral fontanelle that includes most of the bridge.

PREDATORS, PARASITES, AND DISEASES

Several species of intestinal parasites have been described for *Dermatemys* (Caballero y Caballero, 1942; Caballero y Caballero and Rodriguéz, 1960; Thatcher, 1963; Caballero y Caballero, 1961).

Nest and egg predators include coati mundi (*Nasua nasua*) and large birds—gray-necked wood rails (*Aramides cajanea*), and limpkins (*Aramus guarauna*). Night herons (*Butorides virescens*, *Nyxticorax nycticorax*, and *Nycticorax vilacea*) are known to prey on hatchlings (Moll, 1986). Hatchlings and medium-sized individuals are eaten by crocodiles (*Crocodylus moreletii* and *C. acutus*) (Álvarez del Toro et al., 1982; Moll, 1986; Smith and Smith, 1979). Otters (*Lutra longicaudis*) are a major predator at all life stages, consuming eggs, hatchlings, and adults. The ravaged shells of turtles consumed by otters were a common sight along the Rio Lacantún in the 1960s and 1970s when turtle populations were high (Álvarez del Toro and Smith, 1982).

POPULATIONS

An intensive mark and recapture program was made in the Reserva de Montes Azules (Rios Tzendales, Lacantún, and San Pedro) from 1984 to 1992 to estimate populations in 26 river kilometers. Using the Jolly-Seber method of population estimation (Begon, 1979), the mean population estimate was 415 (90–2,266), as calculated from captures of 285 individuals in 23 5-day sampling periods, using one fyke net at each of 24 localities.

Vogt (1998) published the following data on numbers, size, and weight in two populations of *Dermatemys* in Chiapas. The data for relatively healthy populations (1984–1992) are followed by censusing results in 1996, after decimation by market hunters.

LAGUNA OAXACA (1985–1992)

10 adult males—CL, 421 mm ± 30 (358–455);
WT, 9,077 g ± 1,663 (6,500–12,000).

64 adult females—CL, 453 mm ± 32 (368–512);
WT, 11,912 g ± 2,591 (5,850–17,200).

78 subadults—CL, 189 mm ± 69 (113–338);
WT, 1,193 ± 1,243 (100–5,200).

In 1996 only one adult male (CL, 350 mm; WT, 5,100 g) and one adult female (CL, 410 mm; WT, 7,500 g) were captured.

RIO TZENDALES (1984–1992)

58 adult males—CL, 427 mm ± 28 (358–466);
WT, 9,422 g ± 1,707 (5,000–12,000).

119 adult females—CL, 437 mm ± 43 (320–500);
WT, 11,025 g ± 3,226 (4,000–16,500).

108 subadults—CL, 154 mm ± 48 (129–295);
WT, 558 g ± 514 (100–2,750).

Only 12 *Dermatemys* were captured in 1996.

Two adult males—CL, 372 mm ± 52 (335–408);
WT, 5,750 g ± 3,182 (3,500–80,007).

Seven adult females—CL, 377 mm ± 53 (323–453);
WT, 6,857 g ± 3,078 (2,500–11,000).

Three subadults—CL, 143 mm ± 32 (113–176);
WT, 474 g ± 434 (123–960).

Two of the adult females were recaptures.

In 1996 Vogt captured, measured, marked, and released 43 *Dermatemys* from a relatively undisturbed population in the Rio Blanco, a tributary of the upper Rio Papaloapan, 18°42´N, 95°52´W, as follows:

Three adult males—CL, 392 mm ± 53 (335–440);
WT, 6,900 g ± 2,551 (4,300–9,400).

Seven adult females—CL, 375 mm ± 60 (316–480);
WT, 6,471 g ± 3,176 (4,000–13,000).

33 subadults—CL, 221 mm ± 53 (145–312);
WT, 1,479 ± 101 (400–3,400).

CONSERVATION AND CURRENT STATUS

Humans are clearly the major predator of *Dermatemys*. Hunter-gatherers of the region have probably exploited this food source for centuries. The chief threat to an endangered *Dermatemys* is overcollecting by professional tortugueros (see Introduction). Habitat degradation has played only a minor role in the demise of *Dermatemys*.

There has been an official "closed season" in Veracruz, Tabasco, Chiapas, and Campeche since 1975, but it has not prevented overcollecting. Several captive reproduction programs exist in Tabasco and Veracruz. None of these operations is sufficiently productive to curb the decline of *Dermatemys* in Mexico within the next 50 years.

In April 1985 Vogt measured a series of 80 adult *Dermatemys* confiscated by the game wardens on the road leading out of the Selva Lacandona on 26 March 1985. The turtles had a mean weight of 10,580 g ± 2,356 (5,400–17,300) and mean CL of 431 mm ± 31.2 (338–512). The turtles had been captured in the Rio Salinas, a tributary of the Usumacinta on the frontier between Chiapas and Guatemala, and hidden under a load of corn en route to market in Villahermosa. This was probably the last large confiscation of illegal *Dermatemys* in Mexico. By 1990 one never saw more than five turtles in one place, whether being held from confiscation or for sale.

In 1984 the population in the Rio Tzendales, Selva Lacandona, Chiapas, was unperturbed by hunting. At that time one could catch and release 40–45 adults in 5 days of trapping with 20 fyke nets. By the mid 1990s this population had been decimated by turtle hunters, even though the area was completely within the Montes Azules Biosphere Reserve. In April 2003 no *Dermatemys* could be captured or detected by any means at this site. From 1990 to 2003 most of the *Dermatemys* entering the markets in Chiapas and Tabasco were coming from Guatemala and Campeche (pers. comm. Victor Gonzáles, Jefe de Pesca, Catazaja, Chiapas).

Currently the only viable populations of *Dermatemys* in Mexico exist in inaccessible areas far from civilization (areas we are loath to detail). Recuperating populations have a high percentage of subadults. At least 80% of the population should be adults in unperturbed populations. If there are few adults and subadults and a high percentage of hatchlings, the population can be regarded as fragile.

Family Kinosternidae

The Family Kinosternidae (Gray) 1968 comprise a unique and varied family of trionychoids with an extensive distribution in the Western Hemisphere. They range from the northeastern United States and southeastern Canada (45°N) through the eastern and southeastern United States, and through Mesoamerica to Bolivia and northern Argentina (27°S). There are 31 named taxa within the family representing 22 full species and 4 genera; 24 of these occur in Mexico, and 17 are endemic to Mexico and adjacent Mesoamerican countries. Three extant genera occur in Mexico: *Kinosternon*, *Staurotypus*, and *Claudius*. *Sternotherus* probably occurred in northern Mexico at one time but is now extinct there. The Mexican taxa display virtually the total range of variation present in the family. The authorship of family and subfamily names is somewhat clouded. We follow the extensive discussion of Smith and Smith (1979).

DIAGNOSIS

(Based on Hutchison, 1991; Waagen, 1972; and JML pers. obs.)

All kinosternids share the same derived pattern of peripheral and marginal reduction: 10 peripheral bones and 11 marginal scutes. The abdominal scutes (= scute 4 of the familiar testudinoid plan) have been lost; humeral scute greatly enlarged and divided into two separate scutes by straddling the anterior hinge (plastral scutes 3 and 4). Nuchal costiform processes present. Maximum of 7 neurals; inframarginal scutes 2 or fewer; plastron with one or two hinged lobes or with kinetic articulation to carapace. Two musk glands in sternal cavity; axillary duct lying in distinct axillary groove along inferior borders of marginal scutes and detectable in fossils and recent skeletons.

SYSTEMATIC RELATIONSHIPS

Two distinct subfamilial sister groups are recognized: the Staurotypinae and the Kinosterninae. Bickham (1981) ranked staurotypines and kinosternines as full families. Bickham and Carr (1983) considered staurotypines to be intermediate between the Chelydridae and the Platysternidae on the basis of chromosome morphology. Hutchison (1991) reviewed the strong evidence that the subfamilies are sister taxa and most closely related to the family Dermatemydidae (Hutchison, 1991; Iverson, 1992a). We follow this classification and its phylogenetic hypotheses. However, the subfamilies could easily be elevated to full families on the basis of morphological distinction and without damage to phylogenetic logic.

FOSSILS

Two fossil staurotypine genera are recognized: *Xenochelys* (Hay, 1908a, Oligocene and Eocene) and *Baltemys* (Hutchison, 1991, Eocene). All post-Oligocene records of kinosternines are referable to extant taxa. Ancestral kinosternids may have been present in late Cretaceous and earliest Tertiary times (Hutchison and Archibald, 1986). Staurotypines and kinosternines had differentiated by early Eocene times (Hutchison, 1991).

ORIGIN OF KINOSTERNIDS

In the most recent revision of the Kinosternidae, Iverson (1991b), following Hutchison and Bramble (1981) and Dunn (1931, 1940), posits that the kinosternines evolved from an unhinged *Xenochelys*-like ancestor in North America. Berry (1978) and Savage (1966) had predicted an origin in northern Central America, the region with the highest present diversity of kinosternid taxa (see account of *K. scorpioides* group).

GENERAL DESCRIPTION

The following are brief accounts of structure, function, and behavior that are uniquely kinosternid.

All kinosternids lay eggs with brittle shells that never become flexible. All have dorsal, cornified head shields and gular barbels of some sort. Kinosternids are referred to as "mud turtles" (Iverson 1981a, 1992a) in allusion to their preferred bottom-dwelling habits and also as "musk turtles" in allusion to their secretion of copious and pungent musk (witness the vernacular "stinkpot" for *Sternotherus odoratus*).

HOMOLOGIES OF PLASTRAL SCUTES IN KINOSTERNIDS

Books on turtles (e.g., Pope, 1939; Carr, 1952; Ernst et al., 1994) commonly identify the scutes of the plastron with a diagram of an emydid plastron that, like nearly all testudinoids, bears six pairs of scutes. Kinosternines also have six pairs of scutes (but only if one counts the unpaired "gular" as a pair). It was assumed that they were homologous to the testudinoid plan (from anterior to posterior—the gular, humeral, pectoral, abdominal, femoral, and anal scutes).

Yet more than half the turtles considered in this book (all kinosternids, *Dermatemys*, and *Chelydra*) have a derived plastral scute pattern that is only partly homologous with the testudinoid plan. The shell of trionychids is so specialized that it is considered separately in the account of that family. It had always been clear that staurotypines had only four pairs of scutes and that *Dermatemys* had a single anterior scute followed by five pairs of scutes. Bramble and Hutchison (1981) demonstrated that the scutes of the kinosternine plastron constituted a derived condition and that most of the homologies implied by comparison to the testudinoid plan were incorrect. They hypothesized that: (1) the pattern seen in *Claudius* and *Staurotypus* is ancestral; (2) the ancestral abdominal and pectoral scutes had been lost entirely in kinosternids but the abdominals retained in *Dermatemys*; (3) the two scutes bracketing the anterior hinge of kinosternines are subdivisions of an enlarged, "true" humeral scute; (4) the "true" gular scutes are the second pair of kinosternines and *Dermatemys* and the anteriormost scutes in staurotypines; and (5) a small unpaired scute (intergular, formerly "gular") is neomorphic and was added (independently) to the anterior end of the kinosternine and *Dermatemys* plastron as it was secondarily enlarged.

In summary, all trionychoid turtles with definable plastral scutes (*Dermatemys* and the kinosternids) have lost the pectoral scutes, and all kinosternids have also lost the abdominal scutes. The reader should refer to Bramble and Hutchison (1981) for complete explanations.

The forgoing hypotheses were substantiated with logical arguments and fossil evidence. They have subsequently been either ignored (Lemos-Espinal, Smith, and Chiszar, 2004), acknowledged but not used (Ernst et al., 1994), or accepted without discussion (Iverson, 1991). At this point, utility and scientific truth have collided in shattering confusion. We accept the work of Hutchison and Bramble. Figure 7.1 shows the plastra of *Dermatemys* and three genera of kinosternids with the new terminology on one side and the old on the other side. However, any hypothesis is subject to refutation. We seek to embrace logic and solve the problem of confusion by not using names and simply referring to the scutes, when necessary, by numbers, "1" being anteriormost in any given situation.

The following developmental events add ontogenetic credence to the phylogeny proposed by Hutchison and Bramble. Examination of *Kinosternon flavescens* embryos (UU, JML notes) showed: at 39% of development there was a single large scute (the humeral., sensu stricto) covering the future anterior hinge but all other seams were normally developed; at 45% of development the humeral was almost completely divided over the presumptive anterior hinge, being deficient only near the midline. Hatchlings had fully developed transverse interhumeral seams except in a few individuals with slight deficiency at the lateralmost parts of the seam; the two hinged lobes were slightly motile.

The humeral division occurs rather precisely across the areola of original scute, the division corresponding to the level of the anterior hinge. Growth patterns of plastral scutes (elsewhere on the plastron and in other groups of turtles) are usually anteromediad; the anterior component of the humeral (scute 3) retains this pattern whereas, in the posterior

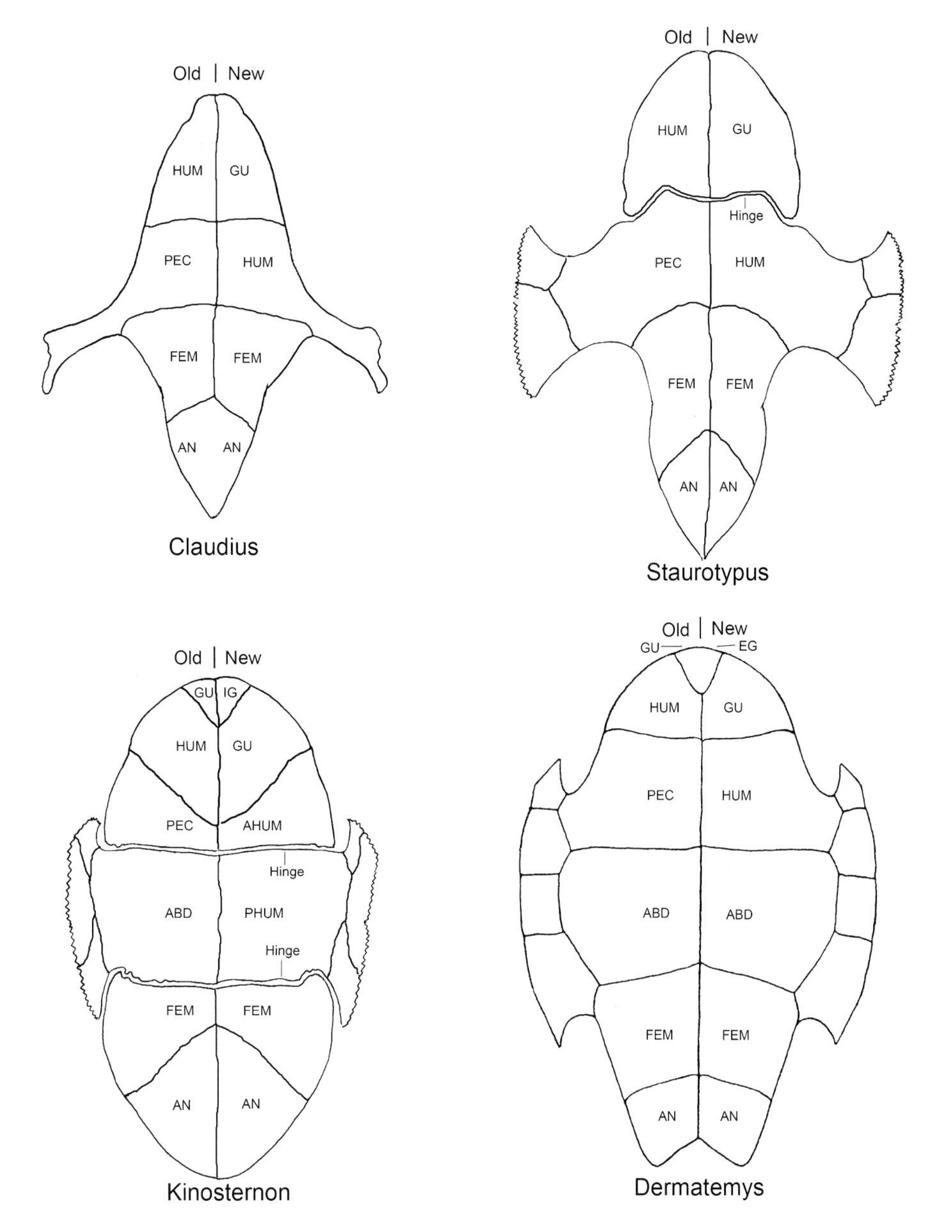

FIGURE 7.1 Plastral scute names in three genera of kinosternids and in *Dermatemys mawi*. Conventional ("Old") names are shown on the left of each plastron, and revised ("New") names, reflecting the homologies established by Hutchison and Bramble (1981), are on the right. Specimens (based on mature adults) are all shown at the same plastral length for comparison. All trionychoid turtles with definable scutes have lost the pectoral scutes; all kinosternids have also lost the abdominals (see account of *F. Kinosternidae*). Abbreviations: ABD, abdominal; AHUM, anterior humeral; AN, anal; EG, extragular; FEM, femoral: GU or GUL, gular; HUM, humeral; IG, intergular; PEC, pectoral; PHUM, posterior humeral.

component (scute 4), growth is posteromediad. Growth of either part of the humeral into the seam-hinge is minimal.

CLASPING ORGANS IN KINOSTERNID TURTLES

Clasping organs are opposed patches of cornified, spadelike epidermal scales on the hind limbs of male kinosternids (weakly developed in females of some species) (Figure 7.3) They appear at puberty and are unique to kinosternids. They are absent in both sexes in the *scorpioides* group of *Kinosternon* (7 species), a derived character state correlated with extensive plastral coverage fide Iverson (1981a, 1991b).

The patches lie immediately proximal and distal to the popliteal space (the depression at the back of the knee joint). The spades are angled slightly above the main plane of the skin and point anteriorly and (variably) into the popliteal space. Ergo, the posterior withdrawal of anything clasped by a flexed knee would be impeded by the apposition of the patches.

The patches are elliptical and aligned with the long axis of the hind limb. Each patch consists of 25–30 scales with heavily cornified epidermis. In preserved specimens the cornified layer of the pad can be prized away in one piece with a probe, leaving soft spades and normally cornified skin below.

The specific epithet "*Kinosternon hirtipes*" (Wagler, 1830) is derived from Latin "hirtus," rough, and "pes," foot, and alludes to the clasping organs of this species. Agassiz (1857a: 429) commented on their wide occurrence in "male Cinosternids." Since that time the structures have been mentioned by a confusing variety of other names, varying terminology sometimes used in the same work (stridulating organs, vincula, clasping patches, friction patches, curious patches, clasping organs, tilted scales, sexual scales, grasping organs, horny keeled tubercles, etc.).

Although the function of these structures has been hypothesized to be some sort of clasping during mating (e.g., Risely, 1930; Pope, 1939; Carr, 1952; Legler, 1965), clasping was not actually observed until Mahmoud's work in 1967. He observed mating in several kinosternines and reported that the male, flexing one or the other knee, grasped the tail of the female with his clasping organ and held it to one side during copulation. Figure 3 (labeled "kinosternids" but probably *Sternotherus sp.*) clearly shows this. Lardie (1975) observed clasping of the posterior edge of the female carapace on one side by male *Kinosternon flavescens*, the tail of the female being too short to be a significant obstruction to penetration.

These two short accounts and what has been reported herein for staurotypines suggest substantial variation in kinosternid mating behavior. Perhaps male kinosternids clasp whatever can enhance the copulatory union. Clasping organs (and the prehensile tail) might also be used in nonmating activities (e.g., stabilizing the turtle in an underwater brush pile).

The term "stridulating organs" addressed an old legend—that the pads (presumably those of same limb) made a strident sound when rubbed together. Pope (1939) attributed the idea to "reptile-men." The sounds generated were compared (E. G. Boulenger, 1914) to those made by grasshoppers. DeSola (1931) claimed actually to have heard cricketlike sounds emanating from a mating pair of *Kinosternon subrubrum* in New York. Evans (1961) provides the most recent entry in this speculation and claims to have heard ". . the scraping whistle of the male *S. minor* made by stridulating organs, in the presence of females" (and further suggests the possibility of the horny tail tip strumming against one of the clasping patches).

The anatomy of the tetrapod knee joint seems to have been ignored in more than a century of speculation. The knee is a modified hinge joint and moving the ventral surface of the leg over that of the thigh is only barely possible. One should envision, for a human knee, a patch of sandpaper attached just above and another just below the popliteal space. Rubbing the sheets together borders upon anatomical impossibility (JML, pers. obs.).

THE TAIL CLAW IN KINOSTERNIDS

The males of most turtles have a longer and more heavily developed tail than females. This is especially pronounced in kinosternids, in which the male tail assumes prehensile proportions and functions (Figure 7.2). In nearly all known kinosternids (exceptions are *Kinosternon angustipons* and *K. dunni*) the terminal scale of the tail is enlarged, cornified and clawlike in appearance. The shape of the claw varies among the genera. In kinosternines it bears close similarity to a phalangeal claw, with a curved sharp tip and a cross section that is convex dorsally and concave ventrally and is often as long as (but stouter than) the longest claw on the hind foot. In *Staurotypus* the cross section is more nearly triangular with the apex forming a dorsal ridge. In *Claudius* the tail claw is sharp-edged, cornified, vertically oriented and disc-shaped, the ventral and posterior edges of the disc bearing sharp edges. There is no reason to question the homology of tail claws in kinosternids or their analogous function in mating.

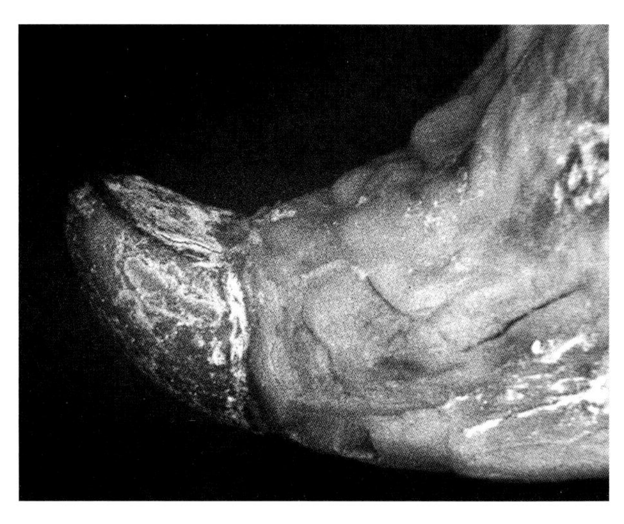

FIGURE 7.2 The kinosternid tail claw of a male *Kinosternon herrerai* (holotype, USNM 61249), ventral surface up.

The terminal scale is present in females, but never assumes the proportions or appearance of a claw. The tail claw has been reported to have a role in kinosternid mating, seemingly positioning the female tail and cloacal orifice prior to intromission (Lardie, 1975a and 1978; Mahmoud, 1967 and account of *Staurotypus salvini* in this book). Only Carr and Mast (1988) have actually described the use of a tail claw in mating (see account of *Kinosternon herrerai*). A tail claw is present in four of the five species of African *Testudo* (sensu stricto) (Loveridge and Williams, 1957) but has not been reported elsewhere among chelonians.

SEXUAL DIMORPHISM IN KINOSTERNINES

Clasping organs and tail claws have already been mentioned. Nearly all kinosternids show an increase in male head size, with age and size, which is easily expressed by maximum width of head as a percentage of CL. This is evident in profile, dorsal aspect and cross section. Included is a relatively massive development of the jaw sheaths and, in extremes, a bosslike development of the snout. These features can be discerned in the skulls alone (see figures for *Dermatemys, Staurotypus, Trachemys*). This dimorphism and its possible functional significance is largely unstudied.

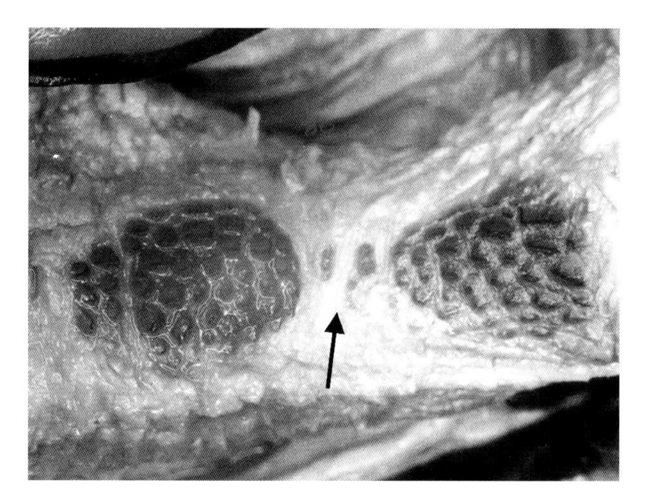

FIGURE 7.3 Kinosternid clasping organs (*Kinosternon herrerai* holotype, USNM 61249), posterior surface of left hind limb, proximal to left, distal to right. Specialized scales are separated by popliteal space (arrow). Note that free edges of spadelike scales will oppose each other when knee is flexed.

SOUNDS PRODUCED BY KINOSTERNIDS

The notes of Berendt (in Cope, 1865) described two distinct sounds made by *Staurotypus triporcatus* from Tabasco: an expiratory sound from which the name "Guau" or "Huau" may have been derived; and a squeaking noise. The first sound is made as (or shortly after) the head is withdrawn and usually by captives which are not used to handling. The sound varies from a hiss to a groan depending upon how widely the animal is gaping and how much air passes through the nostrils.

The "squeaking" sound may be unique to kinosternids; we have heard it in both species of *Staurotypus*, in *Claudius*, in *Kinosternon leucostomum*, and in *K. scorpioides*; Carr (1952) mentions it for Florida *Kinosternon* sp. and likens it to the sound of a *Terrapene* closing its shell. The sound is made by grating the anteriormost part of the mandibular sheath against that of the maxilla and ranges from squeaks to quacklike noises. The sound is usually made by individuals in a circumstance of duress. The sound and its palpable vibration are both startling. Since the jaws usually close without grating together, the sounds are probably made by simultaneously protracting and abducting the mandible.

This behavior could explain references to sounds of unknown origin emanating from tanks containing kinosternids (Risley, 1933) and sounds made during mating in *Staurotypus salvini* (Schmidt, 1970).

KEY TO SUBFAMILIES, GENERA AND SPECIES OF THE FAMILY KINOSTERNIDAE IN MEXICO

Keys to subspecies are imbedded in species accounts:

1A. Plastron reduced or cruciform, covering half or less of ventral surface and bearing fewer than 10 scutes (usually 7 or 8); entoplastron present . . . . **Staurotypinae, 2**

1B. Plastron covering half or more of ventral surface and bearing 10 or 11 scutes; entoplastron absent . . . . **Kinosterninae, 4**

2A. Axillary and inguinal scutes well developed; plastral bridge joins bridge by an akinetic joint; a single, kinetic, anterior plastral hinge . . . . ***Staurotypus*, 3**

2B. Axillary and inguinal scutes absent or vestigial and reduced; an extremely narrow plastral bridge forms kinetic joint with carapace; no transverse plastral hinges; carapace weakly carinate or noncarinate. . . . . ***C. angustatus***

3A. Head boldly marked with irregular dark and pale reticulations. Movable plastral lobe straight-sided and broadly rounded anteriorly. Second plastral scute 19–23% of PL; bridge width usually 16–20% of PL; CW/CL .57–.69. Three massively developed carapace keels in adults (Figures 9.2, 9.5, 9.11). Maximum adult size ca. 400 mm. Occurs in Gulf and Caribbean drainages. . . . . ***S. triporcatus***

3B. Top of head dark, usually unicolored, lacking bold pattern. Sides of movable plastral lobe convergent and terminating in a blunt apex. Second plastral scute 14–19% of PL. Bridge 12–16% of PL; CW/CL .61–.75. Carapace keels less massive in adults (Figures 10.1 and 10.2). Maximum adult size ca. 250 mm. Occurs in Pacific coastal drainages . . . . ***S. salvini***

4A. Keratinized plastral scutes usually separated by softer skin rather than midline seams; posteriormost scales of plastral forelobe four-sided (not triangular); once occurred at El Sauz, Chihuahua . . . . ***Sternotherus odoratus***

4B. Keratinized plastral scutes join along midline by typical interlaminal seams—not separated by softer skin; posteriormost scutes of anterior lobe usually triangular or nearly so . . . . ***Kinosternon*, 5**

5A. Ninth marginal scute much higher than eighth . . . . ***K. flavescens* Group, 6**

5B. Ninth marginal scute approximately same height as eighth **8**

6A. Plastral scute 2 relatively short, less than half length of forelobe (40–42%). Plastral scute 6 70–82% of poster lobe. Plastral scute 1 shorter than 6. Occurs (in Mexico) in tributaries of upper and lower Rio Grande (Conchos, Sabinas, Salado, and San Juan and northward flowing internal drainage systems) in the states of Chihuahua, Coahuila, Nuevo León and Tamaulipas and on the Gulf coast to Cabo Rojo, and the extreme upper Rio Yaqui in northeastern Sonora . . . . ***K. flavescens***

6B. Plastral scute 2 greater than half length of anterior plastral lobe *7*

7A. Plastral scute 2 52–60% of forelobe length; plastral scute 6 60–70% of posterior lobe; C1 and M2 not in contact; width of C1 < 88% of forelobe; plastral scute 5 < PS2; CW < 86% PL. Occurs in southwestern Arizona (Rio Sonoita) southward to the upper reaches of the Yaqui, Matape, Magdalena, Sonora, and Guaymas drainages in Sonora . . . . ***K. arizonense***

7B. Plastral scute 2 60–63% of anterior lobe; plastral scute 6 58–64% of posterior lobe; C1 usually in contact with M2; width of C1 > 85% of forelobe; plastral scute 5 > PS2; CW > 83% PL; width of C1 > 80% L forelobe. Occurs in the lower Rio Nazas and southwestern part of Bolsón de Mapimí, southwestern Coahuila, southeastern Chihuahua, and northeastern Durango . . . . ***K. durangoense***

8A. Posterior plastral hinge lacking; posterior lobe akinetic . . . . ***K. herrerai***

8B. Posterior plastral hinge present; posterior lobe kinetic 9

9A. Head shield furcate (notched, V-shaped) posteriorly . . . . ***K. hirtipes***

9B. Head shield not furcate posteriorly **10**

10A. Anterior pair of chin barbels very long, length subequal to orbital diameter . . . . ***K. sonoriense***

10B. Anterior pair of chin barbels not so long, never approaching orbital diameter 11

11A. Plastron with distinct posterior notch 12

11B. Plastron without distinct posterior notch. 14

12A. Width of plastral forelobe at hinge more than 67% of greatest CW; greatest width of hind lobe greater than 59.5% of greatest CW in males and greater than 62% in females; plastral scute 5 < 46% of bridge length and less than12% of maximum plastron length . . . . ***K. integrum***

12B. Width of plastral forelobe at hinge less than 67% of greatest CW; greatest width of hind lobe less than 59.5% of greatest CW in males and less than 62% in females; plastral scute 5 > 38% bridge length and more than 9% of maximum plastron length 13

13A. Carapace relatively narrow, depressed, and strongly tricarinate; both plastral lobes freely movable; bridge relatively wider—19.4–24.4% CL males, 22.7–26.4 females; PS1 longer, 15.2% CL males, 17.1 females; PS 3 shorter, 4.2% CL males, 4.1 females; axillary and inguinal scutes in narrow contact or not at all; C1 contacts M2. Occurs in Pacific coastal Guerrero and Oaxaca (ca. 99°40′–95°35′W) . . . . ***K. oaxacae***

13B. Carapace relatively broad and weakly tricarinate; anterior plastral lobe freely movable, posterior lobe slightly movable. Bridge narrower 15.3–20.6% CL males, 20.2–23.4 females; PS1 shorter, 13.9% CL males, 14.2% females; PS3 longer, 6.7% CL males, 6.1% females; axillary and inguinal scutes in broad contact; C1 narrow and seldom (12% of adults) contacts M2. Occurs in coastal Jalisco and Colima . . . . ***K. chimalhuaca***

14A. Plastral scute 1 broader on dorsal than on ventral surface; adult males with clasping organs; often a single, broad, pale (but variable) postorbital stripe . . . . ***K. leucostomum***

14B. Plastral scute 1 not broader dorsally than ventrally; adult males lack clasping organs; no single, pale, broad postorbital stripe 15

15A. Carapace lacking keels. . . . . ***K. alamosae***

15B. Carapace at least weakly keeled (one to three keels). 16

16A. Three obvious longitudinal keels on carapace . . . . ***K. scorpioides***

16B. One obvious longitudinal keel on carapace 17

17A. Anterior margin of posterior plastral lobe straight. . . . . ***K. acutum***

17B. Anterior margin of posterior plastral lobe not straight, bowed or angled posteriorly . . . . ***K. creaseri***

Subfamily Staurotypinae (Gray) 1869

Entoplastron present; seven neural bones; N1 in contact with nuchal; N7 reduced or not; plastron reduced to cruciform proportions; four pairs of plastral scutes and an occasional supernumerary (vestigial) anterior "gular"; fourth pair of scutes often fused; seams between scutes 1 and 2 lie on or close to hyoplastral-epiplastral sutures; scute 2 (humeral) not divided by a hinge; an anterior transverse plastral hinge only in *Staurotypus;* never a posterior hinge. Musk glands juxtaposed in center of restricted sternal cavity; anterior musk duct groove weakly impressed in peripherals; axillary groove extending to P1. One pair of fleshy gular barbels. Diploid chromosome number 54 (see account of *S. triporcatus* for remarks on sex chromosomes). Content: *Claudius* and *Staurotypus* (extant); *Baltemys* and *Xenochelys* (fossils).

Genus *Claudius* Cope, 1865

No transverse plastral hinges; bridge has kinetic articulation with carapace that permits tilting of entire plastron on long axis; N7 present but reduced and separated from suprapygal by midline contact of posterior costals; hyoplastron and hypoplastron co-ossified to form a single element on each side; inframarginals absent, vestigial or fragmented; carapace weakly tricarinate and bone of shell thin. Anterior musk gland with two ducts, one to an axillary orifice at M2, the other to an inframarginal orifice at M6; inguinal duct with orifice at M7. No special modification of jaw armature for crushing. Content: *Claudius angustatus* Cope 1865.

FIGURE 8.1 Head of adult *C. angustatus*: left, with mouth closed (rare in presence of humans); right, in typical gaping pose.

Claudius angustatus Cope, 1865

Chopontil (Veracruz, Vogt 1997a); Taiman (Chiapas; Vogt 1997a); Joloque (Tabasco; Vogt 1997a); Narrow-bridged mud turtle (Iverson,1992a).

ETYMOLOGY

Cope was not informative about the derivation of *Claudius* (*L. claudus*, crippled, limping, defective; Brown, 1954). There is no reason to think reference was to a Roman emperor. It has been assumed (Smith and Smith, 1979; Iverson and Berry, 1980) that reference was to the reduced or "defective" plastron (which would have been reiterated by *angustatus*). The root *claudus* (e.g., "claudication") is more often used to connote a crippled, limping, or lame condition . If either Cope or Berendt knew of the characteristic defense posturing of *Claudius*—a limping, thumping maneuver in response to a source of annoyance (see behavior)—the name could well refer to it.

L. angustatus, narrowest, in reference to the narrow bridge.

HISTORICAL PROLOGUE

The status of *Claudius* as a remarkably distinct monotypic genus has not been questioned nor has it acquired a nomenclatorial burden since its discovery. Bocourt (1868) described *Claudius megalocephalus*, which was quickly synonomized by Dumeril (1870). Iverson (1992a) ranks *Claudius* as the sister taxon of *Staurotypus*, but, in our opinion, *Claudius* may be distinctive enough to be placed in its own subfamily. Only anecdotal information was available for this species until Vogt and associates initiated natural history studies near Alvarado, Veracruz, in 1982.

DIAGNOSIS

The fused hyoplastron and hypoplastron, and extremely narrow bridge, in combination with a complete lack of transverse plastral hinges distinguish *Claudius* from all other North American chelonians. Pritchard (1971) is credited with the discovery of this fusion of plastral elements (Smith and Smith, 1979). Actually, Cope (1865) covered it in the first sentence of the type description.

GENERAL DESCRIPTION

Claudius is the smallest staurotypine: males, CL 116 mm ± 10.4 (98–140) *n* = 101; WT 269g ± 87.2 (140–500) *n* = 99; females, CL 106 ± 8.9 (89–122) *n* = 65; WT 194g ± 52.8 (105–400) *n* = 64. (Vogt database). It has the most reduced plastron and narrowest bridge of any kinosternid. The general character and appearance of this animal are dominated by its relatively huge head and its aggressive demeanor. Humans rarely are able to observe a *Claudius* out of water without facing a widely gaping maw.

Maximum CL 165 mm, 400 g (Vogt database).

BASIC PROPORTIONS FOR ADULTS

(HW = head width)

CW/CL: males, .63 ± 0.02 (.57–.68) *n* = 43; females, .64 ± 0.03 (.59–.76) *n* = 35.

CH/CL: males, .38 ± 0.02 (.35–.41) *n* = 43; females, .39 ± 0.02 (.34–.43) *n* = 35.

CH/CW: males, .61 ± 0.03 (.53–.71) *n* = 43; females, .60 ± 0.04 (.49–.68) *n* = 35.

HW/CW: males, .43 ± 0.02 ± (.38–.50) *n* = 43; females, .26 ± 0.02 (.22–.34) *n* = 35.

Carapace elongate in dorsal view, low in cross section and profile. First central scute significantly broader than long, usually making contact with the second marginal scute. Eleven marginal scutes on each side; M10 and sometimes the M11 distinctly higher than M9; 10 peripheral elements on each side. Carapace slightly tricarinate but usually worn smooth in old adults.

Color of carapace varies from pale gray to almost black. Plastron yellow or cream in adults; a midplastral pattern of darker gray lines following the seams in hatchlings. Color

FIGURE 8.2 Student trapping *Claudius* in typical habitat—marshy areas near Lerdo de Tejada, Veracruz.

FIGURE 8.3 Local men hunting by hand and sight in swamp habitat near Lerdo de Tejada, Veracruz.

of soft skin varies from yellow, cream, or pale gray ventrally to gray, brown, or black dorsally. Sides of neck, head, and nasal scute sometimes mottled.

Plastron cruciform and bridge narrow; number of plastral bones reduced to seven by fusion of hyo- and hypoplastral elements; plastron flexible overall but technically akinetic for want of intrinsic hinges; a ligamentous attachment of bridge to carapace permits tilting of plastron on long axis. Eight plastral scutes in pairs; posteriormost scutes sometimes fused.

Modal plastral formula 1 > 4 > 3 > 2 or 1 > 4 > 2 > 3.

There is a small oval head shield on the snout extending from the anterior edges of the prefrontal bones only to the level of the anterior third of the orbit (Figure 8.1). The usually retracted head and gaping visage mask the relatively very wide head (40–49% of CL); as opposed to the same ratio in both species and both sexes of *Staurotypus*—mean 25% (18–29). Maxillary and mandibular beaks are hooked, the former flanked by a pair of long, sharp cusps. In all but the largest adults the head shield is distinctively colored—a pattern of small pale dots or blotches on a darker ground (Figure 8.1). Papillae present on both dorsal and ventral surfaces of neck; one pair of chin barbels.

Color in life, based on three adult females and one adult male from "Rio El Limon," near Alvarado, Veracruz (JML field notes). Plastron pale straw yellow, darkened somewhat along seams; anterior tip brownish and callused. Ligamentous hinge on bridge cream. Ventral surfaces of limbs and tail yellowish cream (smallest) to slate cream (largest). Neck about same but with some dusky mottling and, in largest specimen, a slight suffusion of pink. Ground of chin and throat yellowish cream with black flecks on dirty dusky suffusion. Carapace brownish with faint suggestion of black radial pattern. Top of head and neck dark olive brown . A fairly distinct division between darker dorsal and paler ventral color extending along side of neck from just above tympanum to shoulder. Horny jaw sheaths dark straw to golden yellow with brown to black vertical barlike markings. Side of head from top of tympanum to jaw angle pale to dark yellow, with or without dusky flecking or suffusion. Dorsal surfaces of limbs and tail slate gray. Papillae and tubercles distinctly pale in smallest specimen.

Iris yellowish olive to amber, lacking ornamentation. Male differs from females chiefly in having greater flecking of black or gray on pale ventral surfaces of soft parts—also on chin and side of head. Neck not so distinctly bicolored.

Neonates have a dark slate colored plastral "pattern" on a pale yellow ground (Figure 8.4); the slate color extends to the edge of the plastron on the interlaminal seams, typically isolating 6 to 8 pale yellowish blotches (as in many other kinosternid neonates). Yellow patches of various sizes occur on the dark skin of the ventral head and neck, the thighs and the brachia. The carapace is a neutral dark color with narrowly darkened seams. Most of the plastral pattern is lost in the first year but traces remain in juveniles of 85–111 mm CL. Photos of gaping hatchlings show a black tongue (Figure 8.4).

KARYOTYPE

Sex chromosomes are lacking. The male karyotype is indistinguishable from the female karyotype of *Staurotypus* (Bull et al., 1974).

FOSSILS

Claudius is not known as a fossil. Archeological remains are known from Veracruz (Iverson and Berry, 1980).

GEOGRAPHIC DISTRIBUTION

Map 6

The Gulf coastal lowlands, from the Actopan and Antigua drainages just N of Cd. Veracruz through Tabasco, Chiapas, and Campeche, across the base of the Yucatán Peninsula

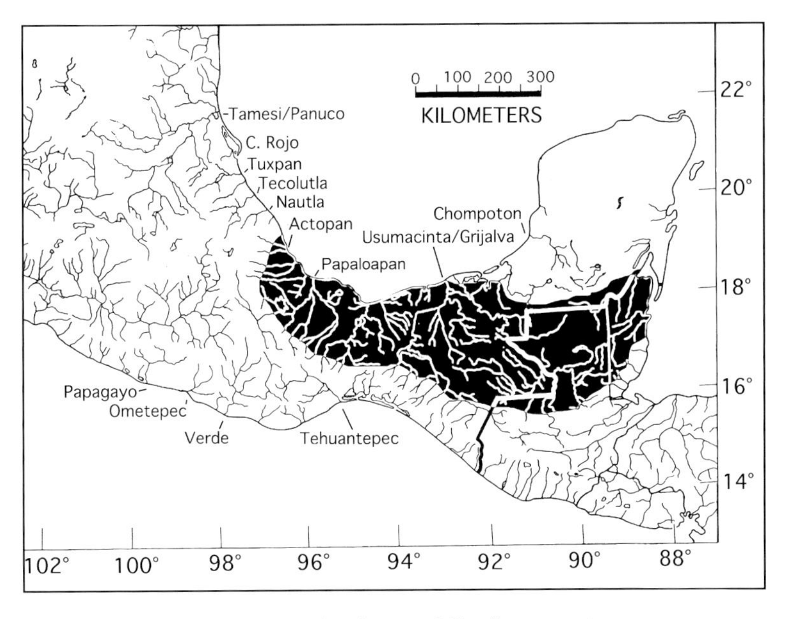

MAP 6 Geographic distribution of *Claudius angustatus*.

(including southern Yucatán and Quintana Roo), Belize, and northern Guatemala, at elevations below 300m.

HABITAT

Claudius angustatus is a bottom-crawling species that thrives in seasonal ponds, flooded fields, flooded forest, marshlands, and occasionally along the margins of shallow lakes. At the beginning of the dry season (March) they bury themselves in the loose soil or mud, often under vegetation as the water is receding, and remain in aestivation until summer rains flood their seasonal microhabitat again in mid-June to late July. Legler found *Claudius* in flooded forest pools, with *Kinosternon acutum*, in Belize (late June, 1966).

Most of the ecological data presented in this account are from a population at Lerdo de Tejada in wetlands inundated by the Rio Augustin (Papaloapan drainage, 19°39′N, 95°34′W), near Alvarado, Veracruz. The area is 30 m above sea level. Water quality is polluted by a sugarcane refinery. The area includes two small fishing villages, Sombrieta and Tecolopia, where numerous subsistence and commercial fishermen enter the wetlands in pursuit of fish, crustaceans, waterfowl, and turtles. Other turtles occurring in various parts of the wetland are *Kinosternon leucostomum, Staurotypus triporcatus, Chelydra rossignoni, Dermatemys mawi, Trachemys scripta venusta*, and *Rhinoclemmys areolata*. The wetland is ultimately confluent with the extensive Papaloapan drainage. This vast area has moderated the effect of long-term intensive market hunting. Although *Claudius* may be "threatened" here, it is not on the brink of extermination.

In September 1982, Vogt and associates negotiated with market hunters to salvage data and viscera from a series of 180 *Claudius*. This information was reported in the following papers and reports: Aguirre-León et al., 2002; Espejel Gonzáles, 2004; Espejel Gonzáles et al., 1998; Flores-Villela, 1986; Flores-Villela and Vogt, 1984; Flores-Villela and Zug, 1995; Vogt, 1997b).

DIET

Claudius angustatus is here regarded as an opportunistic carnivore, its diet varying with season and availability. Crustaceans, insects (chiefly Coleoptera, Hymenoptera, and Homoptera, but especially larvae) and arachnids comprised the bulk of identifiable gut contents from the mentioned visceral examinations. The only notable difference between males and females was the presence of ostracods in the stomachs of females. The high percentage of lepidopteran larvae and other terrestrial insects was probably a result of recent flooding.

Stomach flushing (Legler, 1977) was used to analyze two more samples of *Claudius* that were captured, marked, and released at Lerdo de Tejada. In July 1990, 101 individuals (63 males and 38 females) lacked lepidopteran larvae, but crustaceans (shrimp) and snail opercula and soft parts occurred in most stomachs. Trace amounts of vegetation were attributed to incidental ingestion when striking at moving prey. Principal contents of 78 stomachs flushed from July to December 1997 were crustaceans (chiefly freshwater shrimp) in 48 of 57 stomachs, and only a few guts (7) contained insects (Lopez-Luna et al., 1998).

Hatchlings feed readily and immediately after emergence from the nest on small (6 mm) crabs (*Avotrichodactylus constrictus*). They slowly stalk the crab, hold the neck retracted with the mouth closed, then strike with speed and accuracy, opening the mouth just before contact, grasping and partly crushing the crab.

The beak and flanking maxillary cusps function to puncture and hold crustaceans. Smaller prey items are

FIGURE 8.4 Neonate *Claudius* showing plastral pattern and black tongue.

FIGURE 8.5 Juvenile *Claudius* in year of hatching. Note angle formed by carapace and plastron. (Photo by B. Horne.)

swallowed whole, but larger animals are ripped apart with the fore claws while being held tightly in the jaws. Punctures were usually evident in exoskeletons of crabs and freshwater shrimp found in guts.

Diet probably varies with what is available locally and seasonally. Legler (field notes) found the opercula of large snails (*Pomacea*) in the stomachs of *Claudius angustatus* taken in the Rockstone Pond area of Belize; presumably, the turtles stalked the snails and then struck when the soft body was exposed.

HABITS AND BEHAVIOR

The seasonal activity pattern is strongly influenced by the dry season (see Habitat). Chopontils are most vulnerable to collection when emerging from aestivation. They are captured chiefly by hand in recently flooded areas in water 20–90 cm deep. Hunters stalk the turtles in these flooded areas at dawn and watch for movement of the grass stems to detect a turtle (Figure 8.3). They then thrust their hands blindly into the murky water to capture the turtle. Hundreds are captured daily in this manner for the first several weeks after emergence. This method of capture becomes less efficient as the water rises, the vegetation grows taller, and the turtles disperse.

Chopontils vie with snapping turtles (*Chelydra*) for pugnacity. This begins at emergence from the egg. Vogt has observed striking and biting movements, within the egg, in the last third of development. At hatching, in lieu of typical pipping with a caruncle, chopontils bite out chunks of the eggshell (Vogt pers. obs.).

The following stereotypic defense behavior of an adult captive was observed on a lab bench by Legler (see also Dodd, 1978). The head is usually pulled in and the mouth gapes maximally, depressing the anterior tip of the plastron. The carapace is tilted laterally, toward any source of annoyance, at an angle as great as the fully extended contralateral limbs can achieve. This juxtaposition is maintained by rapidly reorienting the long axis of the shell with limping, thumping movements. A strike can be elicited by any movement within the visual field. The strike is violent. Most strikes generated enough inertia to launch the entire animal forward and to produce an audible thump as the plastron struck the substrate, the entire turtle having been momentarily airborne. The neck is long enough to reach a handhold nearly anywhere on the shell.

Transmitters on two males and four females at Lerdo de Tejada in September showed that turtles remained in the same general area all year long, maintaining a home range that varied in diameter from 380–460 m. Aestivation sites were within the home range. As the water level dropped, the turtles buried themselves 8–12 cm deep in the bottom substrate and remained there until disturbed. Aestivation is solitary. Disturbed turtles showed no sign of torpor (Espejel Gonzáles, Vogt, and Lopez Luna 1998).

REPRODUCTION

Ledig (1988) noted that captives began mating in June and continued to do so for several weeks. One male mated randomly with the females present. Pauler (1981) reported the following for captives. Courtship included the touching of snouts and the male probing or sniffing the female from end to end, concentrating on the region of the bridge. Mounting and copulation proceeded after about a minute and lasted more than an hour. The female then concealed herself and attempted to avoid the male on subsequent mating attempts. If the animals were separated and subsequently placed together, mating could occur two to three times in a week. The female refused mating attempts near the time of laying.

FIGURE 8.6 Dorsal and ventral views of color and pattern, juvenile *Claudius* in year of hatching. (Photos by M. Ewert.)

FIGURE 8.7 Subadult *Claudius* showing radial carapace markings.

Most of the following data on reproduction and sexual maturity come from Vogt (unpublished data), Vogt and Flores-Villela (1986), and Flores-Villela and Zug (1994). All are based on the same population that Vogt studied (1982–1999) at Lerdo de Tejada, Veracruz.

Spermatogenesis is continuous and classified as acyclic by Meylan et al. (2002).

Our estimates of reproductive timing, reproductive potential, and size at sexual maturity are based on a grand total of 817 *Claudius* observed over a period of 15 years.

Vitellogenesis begins in June or July after the rains have stimulated emergence from aestivation. Females (10) from Tecolapia, Veracruz, examined during aestivation (20 May–20 June) contained no enlarged ovarian follicles. Females examined in July and August had enlarged follicles. The first clutches may be laid as early as September but usually not until November. Mean number of clutches per year is 2.5 ± 1.04 (1–5), based on palpation of oviducal eggs, enlarged follicles, and corpora lutea present in dissected females (n = 71) from 1982 to 1997. Mean number of eggs per clutch is 2.4 ± 0.81 (1–6) n = 92 clutches. The last clutches are produced as late as early March. All of the females examined produced at least one clutch. In November and December, all females examined (57) had oviducal eggs; 2 of 9 females bore oviducal eggs in January and February (one in each month) (Flores-Villella et al., 1995).

Females emerge from the water to dig nests. Nests are often dug in sandy areas elevated well above the high-water line, often along roadsides. Nests may also be excavated in loosely packed soil and leaf litter under shrubs within 4 m of the water. Three radio-tracked females were found aestivating in hard-packed soil at a depth of 8 cm with a final clutch of eggs directly beneath them (Espejel Gonzáles et al., 1998). Flores-Villella et al. (1995) reported unburied clutches of eggs hidden in vegetation and stated that no nest is constructed.

The eggs are oblong and have thin brittle shells that never become flexible. Data on 221 eggs are as follows: length, 30.7 mm ± 1.84 (26.4–39.1); width, 17.6 ± 0.7 (16.2–19.7:); weight 6.0 g ± 0.6 (4.8–7.9) (Vogt, 1997a).

The eggs are laid during the fall rainy season and remain in diapause until stimulated to begin developing by the drying of the nest substrate. Incubation time for laboratory-incubated eggs at 25°C was 95–229 days. This range of incubation times was influenced by diapause and embryonic aestivation, both of which were influenced by laboratory manipulations of humidity in the nest substrate.

Embryos remain at a gastrula stage until diapause is broken as the nest substrate dries and the environmental temperature lowers. The embryos then proceed to develop normally to term, at which time they undergo a period of quiescence or aestivation within the egg. Hatching is stimulated by the onset of the summer rains and rising humidity in the nest substrate. Pouring water on laboratory nests with fully developed embryos resulted in hatching within 24 hours. There is a 1:1 sex ratio at all laboratory-incubation temperatures in all natural nests and in all (52) wild-caught hatchlings (Vogt and Flores, 1992a). The sex ratio for 286 adult turtles captured by hand in natural habitat was 2.6 males to one female. This skewed sex ratio could be the result males moving more than females or higher predation on nesting females (Vogt and Flores, 1992).

Flores-Villella and Zug (1994) reported eggs per clutch as 2.4 ± 0.09(1–5) n = 58 collected in 1982–84; examination of corpora lutea showed 1 clutch of 6. On 29 November 1983, 80 female chopontils from Lerdo de Tejada were induced to oviposit in the laboratory by injection of 1 unit of oxytocin per 100g total weight of the turtle. The females laid 229 eggs, mean 2.8 ± 0.08 (1–5) eggs per clutch (Vogt, 1986). Hausmann (1968) reported 2–8 eggs in captives. Total clutch mass is 5.2–27.2 g . Eggs per clutch and clutch mass are positively correlated with female CL (Flores-Villella et al., 1995).

GROWTH AND ONTOGENY

Males reach carapace lengths of 165 and females 150 mm, with maximum weights of 400 and 434 g, respectively.

FIGURE 8.8 Large adult male *Claudius* gaping. Note width of head in relation to width of shell. This character persists throughout ontogeny.

These sizes would rank among the medium- to large-sized *Kinosternon* and are much smaller than either species of *Staurotypus*. Adults are sexually dimorphic, males are larger than females and have proportionately much larger heads, longer and thicker tails, and clasping organs on the hind limbs. Males mature at ca. 98 mm and females at 89–114 mm (Flores-Villella et al., 1995).

Size and weight of hatchlings are, respectively: 25.8 mm ± 1.2 (24–30) n = 69; and 3.8g ± 0.5 (2.3–5.2) n = 71(Vogt, 1997). Albinism (shell and skin completely white, eyes pink) occurred in one of 71 hatchlings hatched in the laboratory after incubation at 25°C (Vogt, 1986).

Espejel Gonzáles (2004) recaptured 11 turtles after a mean of 304 days ± 53.7 (261–449) and was able to show through photographic analysis that they add one growth ring per year, a new ring appearing 2–3 months after the turtles emerge from aestivation. Growth rings were countable on 90% of the turtles captured and on individuals up to 134 mm CL (maximum, 13 rings).

PREDATORS, PARASITES, AND DISEASES

Caracaras (Aves, *Caracara plancus*) are the major predator of nesting females. Raccoons (*Procyon*) prey on nesting females and nests. Many Chopontils are killed on highways as they disperse from aestivation or search for nest sites (Espejel, 2004).

POPULATIONS

Genetic sex determination (GSD) occurs in *Claudius* (Vogt and Flores-Villela, 1992b). Flores-Villella and Zug (1994) reported a sex ratio of 28 males to 11 females in a hand-captured sample from Lerdo de Tejada. Espejel Gonzáles (2004) captured and marked 254 individuals in an area of 17 ha at the same locality, in 2000–2001, in fyke nets with leads. In this year-long study, as many as 29% of the turtles captured in any month were recaptures. The sample was comprised of 120 males, 110 females, and 24 juveniles. Mean CL of these groups was 101 ± 20.0 (62–140), 92 ± 14.6 (61–132), and 58 ± 8.3 (39–75), respectively. The population was estimated at 286 individuals (32–960) and mean biomass at 2.8 kg/ha (maximum 9.4 kg/ha) using a Jolly-Seber model (MacDonald et al., 1980). Using a Schnabel model (MacDonald et al., 1980), the estimate was 505 turtles (± 35) for the entire period, a density of 29.7/ha and a biomass of 4.9 kg/ha.

Growth-ring studies showed males to be mature at 4 or 5 years and females at 5 years, 70% of the males and 50% of the females studied were sexually mature. Because there is GSD in this species, there appears to be a higher mortality of adult females. These data are similar to the hand-capture data, except that the sex ratio of hand-captured turtles was highly skewed. Sex ratios of hand captures were 370 males and 103 females in 12 capture periods from 1983–1997 at Lerdo de Tejada. Because there was a high number of immature turtles and a paucity of adults older than 10 years, this population was regarded as stable but perturbed by continued market hunting. With recruitment, the population seemingly can withstand collecting pressure.This population could serve as a model to test the resilience of other populations.

CONSERVATION AND CURRENT STATUS

Although *Claudius* is intensively hunted (hooked, trapped, and netted) in Mexico, there are still areas where it is abundant in Veracruz, Chiapas, and Tabasco—but most populations have been perturbed. Marshes of formerly high abundance near Villahermosa, Tabasco, Coatzacoalcos, and Minatitlan in Veracruz have been decimated by overhunting and petrochemical pollution since the 1960s. Local residents born after 1975 are unlikely to have ever seen the species.

The flesh of *Claudius* is savory and popular in southeastern Mexico. Growth is fast, and the ratio of edible parts to total body mass is the highest of any Mexican turtle studied by Vogt (unpublished data). This attribute would be favorable in a turtle farm environment. Three attempts have been made to raise *Claudius* in captivity. At La Florida, Veracruz (see Introduction), turtles are maintained in earthen tanks, allowed to aestivate naturally, and fed a natural diet of insects and fresh crustaceans (scraps from markets). Reproductive output of 39 captive females versus 17 wild caught females was not significantly different (Espejel-González, 2004). This facility is judged to be successful. Captive breeding at two other localities (Estación de Biología Los Tuxtlas, Veracruz, and the Nacajuca turtle farm in Tabasco) that involved concrete tanks and lower-quality food were judged as unsuccessful (Espejel-González, 2004).

Genus *Staurotypus* Wagler, 1830

A kinetic joint along epiplastron-hyoplastron suture; suprapygal usually in contact with N7 [Boulenger (1889): Figure 10] shows shell of *salvini* with seven neurals, the posteriormost separated from suprapygal]; hyoplastra and hypoplastra separate, not co-ossified; inframarginal scutes well developed; carapaco-plastral suture akinetic; carapace strongly tricarinate and bone of shell thick; posterior costals usually not in contact. Axillary musk duct orifice from mid M2 to M1–2; inguinal duct penetrating hypoplastral buttress with orifice at M8; orifice of short blind duct at mid M10 (Waagen, 1972). Massive maxillary crushing plates braced to roof of cranium by hypertrophied dorsal processes of palatine bones (JML, unpublished data).

Content: *Staurotypus triporcatus* (Wiegman) and *S. salvini* Gray, two heretofore poorly known sister species.

Staurotypus triporcatus (Wiegmann) 1828

Tres Lomos (Veracruz) (Liner,1994); Guao (Tabasco and Chiapas); Galapago (Veracruz); Mexican Giant Musk Turtle (Iverson,1992a).

ETYMOLOGY

Gr. *stauros*, cross; Gr. *typos*, shape; referring to shape of plastron. L. *tri*, three, and L. *porcatus*, ridged; alluding to the three distinctive keels on the carapace.

HISTORICAL PROLOGUE

In the early 1970s, Legler and E. O. Moll prepared an account of the genus *Staurotypus* ["Biosystematic studies of the genus *Staurotypus* (Chelonia: Kinosternidae)"]. The nearly completed manuscript lay fallow at the University of Utah for three decades but is freely cited in the current accounts for *Staurotypus* as "Legler and Moll (unpublished ms)." Intermittently from 1984 to 2001, Vogt studied a population of *Staurotypus triporcatus* in Laguna Oaxaca, an oxbow lake on the Rio Lacantún, Selva Lacandona, Chiapas. The population was extirpated before the final samples were made because of the building of a road that offered easy access. See account of *Dermatemys* for details of this study area. Vogt also conducted long-term mark-and-recapture studies of this species in Veracruz, near Lerdo de Tejada, Laguna Escondida, Rio de Agua Dulce, and Rio Las Margaritas from 1981–2000.

RELATIONSHIPS AND TAXONONOMIC STATUS

Staurotypus and *Claudius* comprise the subfamily Staurotypinae. This small assemblage itself displays more morphological variation than the entire subfamily Kinosterninae. All living staurotypines are nuclear Mesoamerican organisms that have previously been poorly studied.

DIAGNOSIS

Staurotypus triporcatus is the largest kinosternid turtle; maximum adult size and weight are at least 400 mm and 10 kg. Plastron not distinctly narrow, pointed only at posterior end, not distinctly cruciform. Second plastral scute 20–24% of plastron length. Sides and top of head boldly marked with irregular pale spots on a dark background and dark streaks on jaw sheaths. A movable epiplastral-hyoplastral hinge; hyoplastron not fused to hypoplastron.

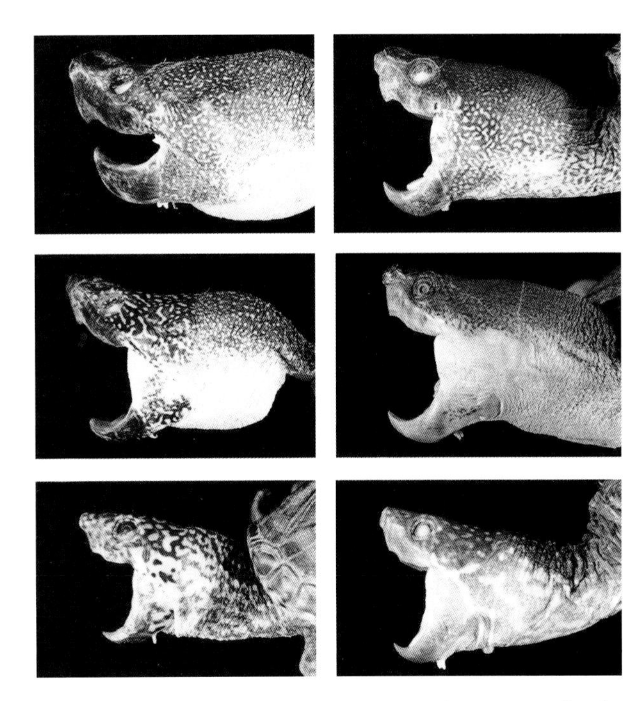

FIGURE 9.1 Lateral head profiles of *Staurotypus* showing sexual and ontogenetic differences. Left, *S. triporcatus* (region of Alvarado, Veracruz), from top to bottom: adult male, UU 6416, CL 303; adult female, UU 6413, CL 242; juvenile, KU 48968, CL 66. Right, *S. salvini* (El Salvador), from top to bottom: adult male, UU 6426, CL 186; adult female, UU 12132, CL 206; juvenile, UU 6424, CL 82.

GENERAL DESCRIPTION

(Comparative terms pertain to differences from *S. salvini.*)

BASIC PROPORTIONS FOR ADULTS

CW/CL: males, 62 ± 04 (.38–.66) n = 68, females, .63 ± .03 (54–.69) n = 60.

HT/CL: males, .34 ± .02 (.30–.40) n = 51, females, .36 ± .03 (.29–.42) n = 55.

HT/CW: males, .56 ± .05 (.46–.70) n = 65, females, .57 ± .05 (.46–.67) n = 54.

Modal plastral formulae: 1 > 4 > 3 > 2 (59%), 1 > 4 > 2 > 3 (22%).

Three longitudinal keels on carapace well defined at all ontogenetic stages but especially prominent in adults. Movable plastral lobe straight-sided and broadly rounded anteriorly. Bridge broad, usually 16–20% of plastron length. Shell relatively higher in adults (see cross sections (Figure 9.5). Carapace relatively narrow and tends to be straight-sided or sometimes slightly wider anteriorly.

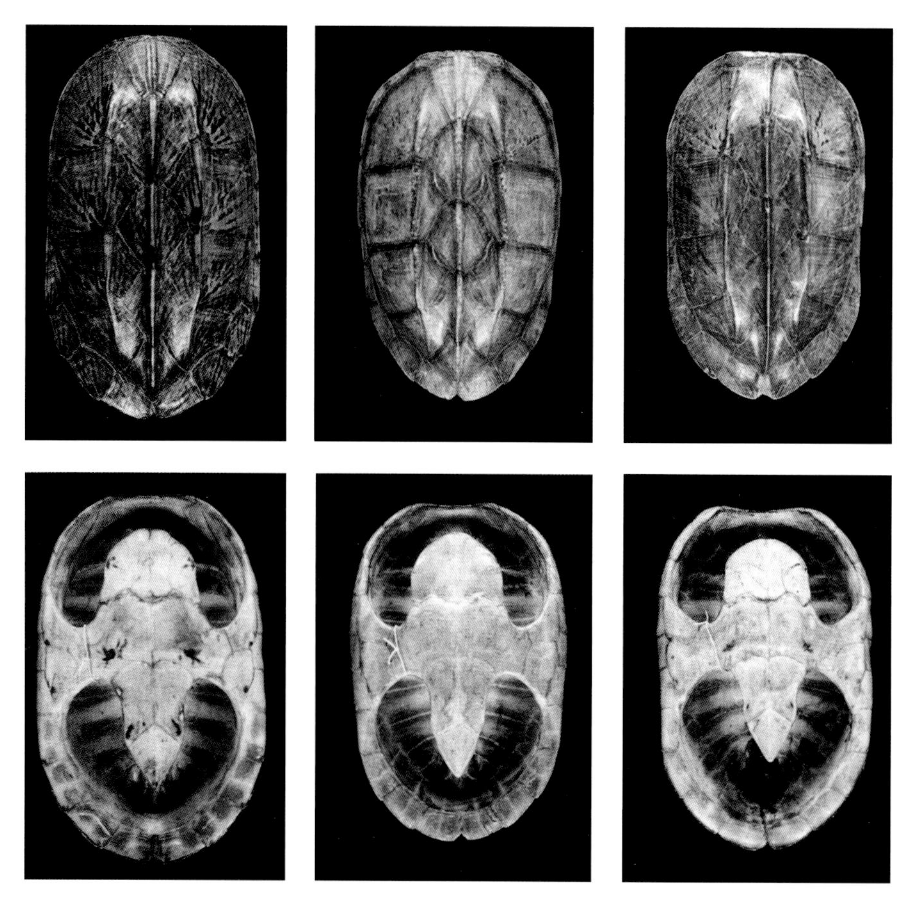

FIGURE 9.2 Dorsal and ventral views of *Staurotypus triporcatus* (region of Alvarado, Veracruz) showing sexual and ontogenetic differences. Left: UU 6413, subadult female, CL 242; middle: adult female, UU 6412, CL 314; right: adult male, UU 9781, CL 312.

Plastron longer and bridge wider; this in combination with narrower carapace make plastron seem less distinctly "cruciform" (Figures 9.1, 9.2, 9.3). Large axillary and inguinal scutes in contact and spanning the bridge.

Head shield variable, often ill-defined; confined to top of snout; trifurcate (as described for *salvini*) or a broad triangle with well-defined posterolateral lobes and extending posterior to orbits (Figure 9.8).

Snout of mature individuals of both sexes longer, upturned, and bosslike. Greatest width of skull on temporal bar, anterior to tympanic opening (Figure 9.6). Dorsal line of head profile, from highest point on snout to highest point on supraoccipital crest, a more or less continuous concave curve. Rostrum longer, at least equal to horizontal diameter of orbit. Supraoccipital crest comprised of two straight lines that form an obtuse angle near parieto-supraoccipital suture, more pronounced in males than in females. Two large gular barbels.

Carapace pale gray to dark brown, often with dark flecks or radial markings (Figure 9.11). Plastron pale and unmarked in adults. Jaw sheaths vertically streaked with dark color; head markings bold and contrasting in general. The broad head is dark brown to black with white reticulations.

Males of both *Staurotypus* species have sex chromosomes (Moon, 1974; Bull et al., 1974; Sites et al., 1979). Males are heterogametic (XY), and females are homogametic (XX). Diploid chromosome number 54 with 26 macrochromosomes and 28 microchromosomes (Moon, 1974).

FOSSILS

Staurotypus has no fossil record. Two fossil staurotypine genera are recognized: *Xenochelys* Hay 1906 (Oligocene and Eocene, South Dakota) and *Baltemys* Hutchison 1991 (Eocene, Wyoming) (Hutchison, 1991).

Legler has examined a large, nearly complete mandible of *S. triporcatus* from an archeological excavation in Belize (Peabody Museum 431), dated Late Classic Period, ca. A.D. 700–900 (pers. comm. with G. R. Willey). The mandible is from an adult of 400–430 mm CL with a head width of ca. 92 mm.

GEOGRAPHIC DISTRIBUTION

Map 7

Gulf lowlands (below 300 m) from the approximate latitude of Cd. Veracruz in central Veracruz (19°20′ N), southeastward through Tabasco, eastern Chiapas, and Campeche, across the base of the Yucatán Peninsula and thence, in

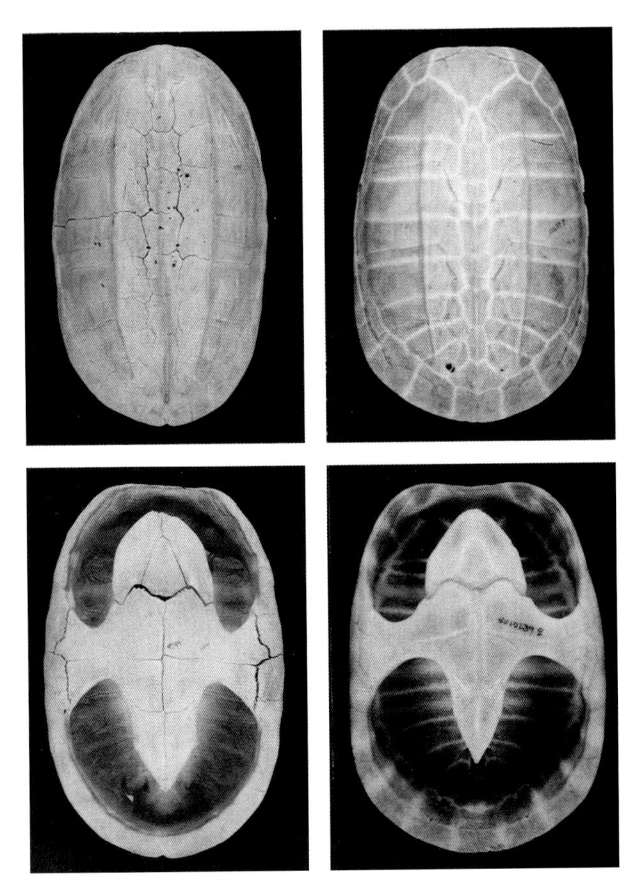

FIGURE 9.3 Scuteless shells of adult female *Staurotypus*, dorsal and ventral views. Left: *S. triporcatus* (Campeche), UU 9794, CL 292; right: *S. salvini* (El Salvador), UU 10239, CL 157. Larger punctures on dorsal surface of *S. triporcatus* shell may be spear wounds.

Caribbean drainages of Quintana Roo, Belize, northern Guatemala, and adjacent western Honduras (UU 6417, Masca, Honduras).

HABITAT

Staurotypus triporcatus occurs in a wide variety of permanent aquatic habitats—lakes, ponds, large rivers, oxbow lakes, mangrove swamps, and marshes. It is most abundant in rivers with slow current at depths of 1–2 m, usually along shorelines where they forage. They occur also in fast water but are not excellent swimmers.

In the Rockstone Pond region of Belize, other kinosternids (*Claudius* and *Kinosternon*) can be found on land and in temporary forest rain pools in the wet season; *S. triporcatus* was never found in these situations.

DIET

Holman (1963) mentioned mollusk fragments in the feces of a recent captive of *S. triporcatus*. Mollusks and large seeds were present in most of the guts examined by Legler and Moll (op. cit.). Some of the seeds were attached to partly digested fruit. Apple snails (*Pomacea flagellata gigantia*) and bivalves (*Psoronaias semigranosus* and *Lampsilus tampicoensis*) were identified. *Pomacea* and other gastropods were represented by opercula, attached soft bodies, and usually parts of the central shell spiral. Bivalve remains included most of the shell in large fragments. Fragments of snail shells were uncommon and never numerous enough to account for even one complete shell.

The massive crushing surfaces of the jaws (Figure 2.7) and their bracing (via a posterior palatine process) to the cranium in *Staurotypus* bespeak adaptation to a durophagous diet. Yet, in the absence of staple foods, this turtle can still be an opportunistic omnivore. Guts from the Rio Mopan in Belize contained shrimplike crustaceans, fish bones, a ventral scale from a snake, parts of an aquatic bird foot, peripherals and costals of a *Kinosternon*, plant stems, composite flowers, and fungus pellets (JML, field notes).

D. Moll (1990) documented *S. triporcatus* as regularly preying on *Kinosternon leucostomum* in Belize (frequency 40–100% from January through April). The consumption of most mollusks could be considered as carnivorous grazing. If, on the other hand, large *Pomacea* snails were consumed after pulling them out of their shells, a certain amount of stalking could be implied. Certainly the incidence of

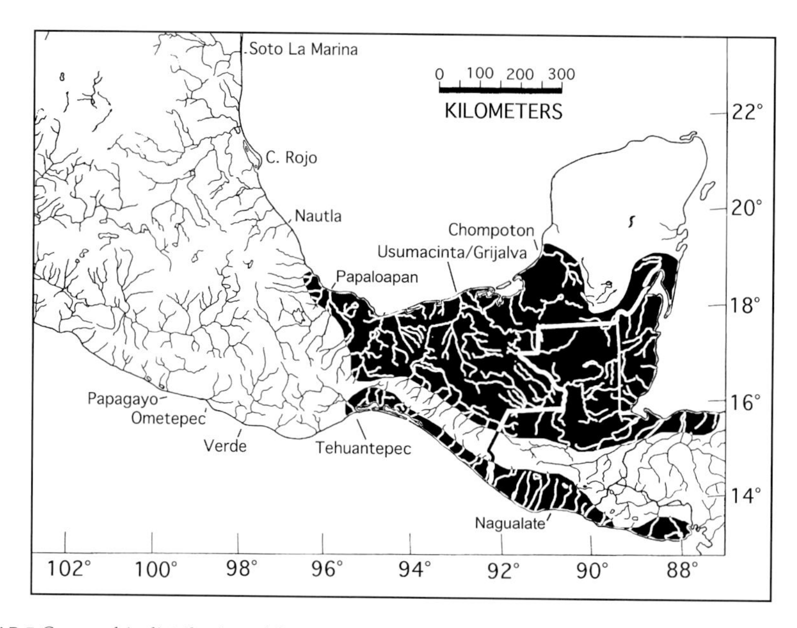

MAP 7 Geographic distribution of *Staurotypus triporcatus* (Atlantic) and *Staurotypus salvini* (Pacific).

Kinosternon (and other turtles) in the diet of a Belizian population qualifies *S. triporcatus* as an active predator.

We have never observed a *S. triporcatus* eating a mollusk, but certain logical assumptions on feeding technique can be made from the appearance of gut contents. It was common to find two to five snail opercula (*Pomacea*) in a single gut. These opercula ranged from 20 to 60 mm in greatest diameter and represented snails with greatest outside diameters of approximately 30–90 mm. Because muscular feet often remained attached to opercula, it is assumed that the turtles either employed some means of egesting most of the shell as the snail was crushed (as *Malayemys* and *Siebenrockiella* do), or they may have stalked the snail, bitten the exposed body, and then ripped it out of the shell with the forefeet. Remains of bivalves suggest that they are ingested whole and summarily crushed with most of the shell swallowed.

The feeding of captive juveniles (both *Staurotypus* species observed) is voracious, active, and aggressive. Small soft items are ingested in the suck-and-gape manner. Observed juveniles would bolt food of any size that would pass into the buccal cavity. Dissection of a small juvenile that died soon after feeding on small pieces of liver showed the stomach to be enormously distended, one piece of food in the esophagus, and another large piece in the pharynx (JML, pers. obs.). It is unknown if this kind of binge feeding occurs in nature.

Stomach flushing revealed that *Staurotypus* of all sizes were chiefly carnivorous in all but one of the sites studied by Vogt. Variations in diet in different habitats were noted as follows.

1. Rio Las Margaritas (tributary to Lago de Catemaco, Veracruz): 48 individuals, 100% apple snails.
2. Lerdo de Tejada (see discription under *Claudius*), $n = 64$: apple snails, crabs, shrimp (*Cambarus* and *Macrobracium*), and the hard seeds of the Jobo tree (*Spondias mombin*) with scraps of the fleshy fruit.
3. Laguna Escondida, a cool deep lake near Los Tuxtlas, Veracruz, $n = 85$: apple snails and *Macrobracium alcantura* made up 90% of the diet; *Kinosternon leucostomum* were abundant but never found in flushings.
4. Laguna Oaxaca, Chiapas: apple snails 50%, fish (most likely fish that had died in traps) 35%, and Jobo seeds 10%. Five adult males contained fragments of juvenile *Staurotypus*.
5. Rio de Agua Dulce, a mangrove estuary stream near Catemaco, Veracruz, $n = 47$: no apple snails present because of salinity. Diet was 85% plant material, mainly large hard seeds, the remainder being fishes and crustaceans. *Kinosternon leucostomum* and *Trachemys scripta venusta* were abundant, but there was no sign that *Staurotypus* were preying on them. Growth rates in this population were extremely slow; one subadult female turtle marked in June 1981 was recaptured in September 1995 with a growth increment of only 25mm CL.

Quality of diet is probably critical for growth. At the Nacajuca turtle farm (Tabasco; see Introduction), reproduction in *Staurotypus triporcatus* was poor from 1978 to 1990 when the turtles were fed chiefly vegetational market refuse (e.g., bananas, carrots, cabbage). Fifty females produced a total of about 20 small (4–6 eggs) clutches per year. When diet was changed to fish carcasses (chiefly *Tilapia* after filleting), most of these females began producing 1–2 clutches each year.

It is of interest that large seeds in the gut can result from either the digestion of the fleshy part of a fruit or

FIGURE 9.4 Dorsal and ventral views of juvenile *Staurotypus* showing ontogenetic change in form and pattern. Left: *S. salvini* (El Salvador), UU 6424, CL 82; middle: *S. triporcatus* (Belize), CL 82; right: *S. triporcatus*, UMMZ 12205, CL 43.

from seeds (chiefly legumes) dropping into the water from seedpods. Some seeds are crushed but most pass from the stomach to the feces with little or no alteration. The same phenomenon has been observed in short-necked chelids in tropical Australia (Legler, pers. obs.). The phenomenon requires study.

HABITS AND BEHAVIOR

Part of the population undergoes a period of aestivation during the dry season (April into June) when water temperatures reach 30°C. Individuals (all ages and sizes) leave the water and dig into a bank or crawl beneath forest leaf litter. Emergence from aestivation coincides with the advent of summer rains in late May or early June. These rains also stimulate the end of aestivation for hatchlings in nests of the previous season. Captives aestivate at the same times of the year in response to rising water temperature even when water levels in their tanks are constant.

Aerial basking has not been observed in *Staurotypus triporcatus*. Individuals have been seen floating just below the surface with only the nostrils protruding.

Both species of *Staurotypus* and *Claudius angustatus* gape for long periods when at rest on the bottom of an aquarium (Legler and Moll, op. cit.; Holman, 1963). Prolonged underwater gaping is uncommon in turtles (observed by us elsewhere only in *Macrochelys*).

Numerous observations of aquatic gaping were made (in Utah) during the day and at night. Some individuals simply gaped while sleeping; others were awake and presumably alert, terminating the behavior when human activity distracted them. In several instances, gaping was accompanied by tongue movements. Approximately every 2–3 minutes the entire tongue was depressed then slowly raised with a waving motion. Depression of the tongue was sometimes accompanied by an opening of the glottis.

Kinosternids lack the rich endowment of vascular papillae in the pharynx that occurs in other trionychoids (e.g., *Dermatemys* and *Apalone*). However, Winokur (1988) thought *Staurotypus* could be achieving respiratory exchanges with the buccopharyngeal mucosa.

Aquatic gaping with tongue movements was impressively similar to angling in *Macrochelys*, but live fishes placed in *Staurotypus* tanks were seldom eaten. Neither the presence of live fish in a tank nor the usual human activity associated with animal room feedings elicited gaping behavior in *Staurotypus* (as it did in *Macrochelys*). We conclude that gaping is probably not a feeding mechanism in kinosternids. However, respiratory exchange is known to occur through the general buccopharyngeal mucosa in aquatic turtles.

FIGURE 9.5 Optical cross-sections of *Staurotypus*. Left: *S. triporcatus* (from top to bottom) adult male, KU 40127, CL 378; adult female, USNM 51073, CL 353; juvenile, KU 48968, CL 66. Right: *S. salvini (from top to bottom)* adult male, USNM 109103, CL 176; adult female, USNM 109184, CL 192; juvenile, USNM 109100, CL 80. Cross-sections are presented at same size for comparison.

Gaping may be a respiratory mechanism used during sleep and other periods of minimal activity that circumvents the need to rise for air and the muscular effort of buccopharyngeal pumping. Gaping would substantially augment the known gaseous exchanges of the general cutaneous surface reported by Bagatto et al. (1997).

We do not suggest a close relationship of aquatic gaping in kinosternids and angling in *Macrochelys*, but the gaping behavior is a grade of morphological and behavioral organization from which angling could evolve.

When placed on a hard surface out of water, juveniles were clumsy and deliberate in their movements. Forward locomotion was accomplished with the plastron barely clearing the substrate and with the chin frequently scraping it. When approached by the observer, the turtle would withdraw limbs and neck, open the mouth widely, and release pungent musk from the inguinal glands. A touch on the carapace caused the turtle to lower the side toward the stimulus and raise the opposite side as high as possible.

Captive adults were chiefly inactive at the bottoms of their tanks. Juveniles were active almost constantly when the room was lighted. Individuals of all ages respond to intimidation by gaping with the head retracted but seem loathe to strike or snap defensively (in contrast to *Claudius*).

REPRODUCTION

Onset of vitellogenesis is coincident with the end of the dry season (late May to early June). Mature follicles were not found in the ovaries until late August. Females with shelled eggs ($N = 28$) were found from late August to mid-March in Veracruz, with a peak in October. In Laguna Oaxaca, Chiapas, Vogt collected 42 gravid females from 30 October 1984 to 11 November 1988: 13 in October, 11 in November, 7 in December, 10 in January, and 1 in February. There is no evidence that females skip seasons of reproduction.

Holman (1963) described an incomplete mating of *S. triporcatus* as follows: "The male was mounted on the posterior part of the female, with his forefeet grasping for purchase on her dorsolateral keels. His hind legs and tail were thrust under the posterior part of her carapace. The performance ended when the female moved away suddenly." No other observations have been made.

Nesting takes place in Veracruz from late August at the beginning of the monsoon rainy season and continues through the northerly storms of winter until the last clutches are laid by the middle of March. Nests are made near the water. Nests near Tlacotlalpan, Veracruz, were dug into riverbank soil just above the high water line less than 0.5 m from the river's edge. Females do not wander about in search of nesting places. In some 20 years of traveling roads through good *Staurotypus* habitat, Vogt has never seen females crossing roads or as road kills whereas this was commonplace with other freshwater turtles (*Chelydra, Claudius, Kinosternon,* and *Trachemys*).

The earliest published observations on reproduction may be those of Dr. K. H. Berendt in Cope (1865); 10 to 30 eggs were reported in Tabasco. Siebenrock (1907) reported 10 to 20 eggs.

For our data, the overall number of eggs per clutch is 4–17. For natural populations in Veracruz, number of eggs per clutch is 9.8 ± 1.82 (6–17) $n = 18$. Mean dimensions of 176 eggs were 42 mm ± 1.62 (38.6–44.2 mm) × 24 mm ± 1.24 (18.2–24.2) and weighed 14 g ± 0.91 (12–17 g).

Combined data for 59 gravid females and 468 eggs from Laguna Oaxaca, Chiapas (1983–1988), were: eggs per clutch 8.1 ± 1.96 (4–12); L 40.3 mm ± 2.1 (32.6–44.); W 23 mm ± 1.44, (14.1–25.1); egg weight 13g ± 1.14 (9–15.2). Mean clutch weight was 105 g ± 27.4 (49.2–156.6). Relative clutch mass was .043 ± .0154 (.014–.093). CL for the females was 279 mm ± 35.1(210–368) and weight was 2786 g ± 954.6 (1245–5150). There was no correlation (Pearson index) between clutch size and the weight or CL of the female. There was a significant positive correlation between the total clutch weight and the weight and CL of the female. There was also a significant correlation between the mean weight of the eggs in a clutch and the weight of the female—i.e., heavier females produced heavier eggs. There was not a correlation between the weight, length, or width of the eggs and the CL of the female. Both the length of the eggs and the width of the eggs were positively correlated to the weight of the female.

Data on two clutches of *S. triporcatus* eggs from Alvarado, Veracruz (15 September), with female size were: 317 mm CL, 15 eggs, L 36.4 ± 0.51 (35.4–37.3); W 21.9 ± 0.32 (21.4–22.4); volume 9.2 cc; 314 mm CL, 13 eggs, L 44.0 ± 0.55 (43.1–45.1);

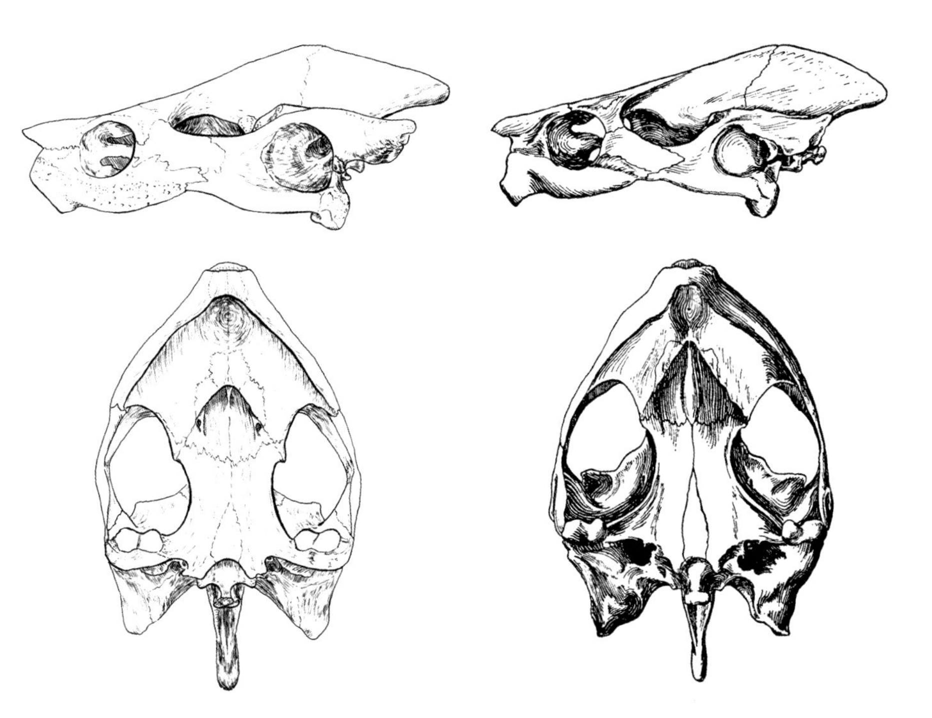

FIGURE 9.6 Lateral and ventral views of *Staurotypus* skulls. Left: *S. triporcatus*, adult female, UU 6411 (Rio Papaloapan near Alvarado, Veracruz), CL 315, maximum width 68, condylobasilar length 83. (Drawing by Elizabeth Lane.) Right: *S. salvini*, adult, MCZ 4989 (Tehuantepec, Oaxaca), maximum width 32 mm, condylobasilar length 42 mm. (Modified from Gaffney, 1979, Figure 167). Skulls are presented at same size for purposes of comparison.

W 25.9 ± 0.40 (25.4–26.9), mean volume 16.2 cc. Yolk diameter in one of the larger eggs (43.9 × 25.6 mm) was 23 mm but 17 mm in one of the smaller (36.4 × 21.9) eggs, being 90 and 78% of egg width, respectively. The discrepancy in egg size for two females of the same size, place, and date is of interest.

The eggs of both *Staurotypus* species have thick, brittle shells that never become flexible. The calcareous part of the shell is thickened into a raised band around the middle of the egg (Figure 9.7). Width of the band is approximately 20–30% the length of the egg. In some eggs, the band is incomplete for approximately one-third the perimeter of the egg. These bands are seemingly a kinosternid phenomenon that is especially well developed in *Staurotypus*. There is a slight tendency toward band formation in the eggs of *Kinosternon* (*angustipons*, *leucostomum*, *scorpioides*, and *subrubrum*), whether or not they have been held for long periods in the oviduct. The bands are most distinct in freshly laid eggs and are opaque, in sharp contrast to the translucent pinkish color of the rest of the shell. The bands remain visible and palpable throughout incubation (Legler and Moll, op. cit.). The bands bear many visible pores with jagged to craterlike orifices. Most of the pores penetrate the calcareous layer of the shell and may be necessary for gas and water exchange. It remains to be demonstrated why these pores lie on a thickened peripheral band and how the band is formed (Legler and Moll, op. cit.).

Incubation time varies in response to moisture and temperature (see Introduction). Embryonic diapause and embryonic aestivation synchronize developmental rates so that most eggs are ready to hatch at the beginning of the rainy season in June. Eggs incubated under controlled temperature (25–32°C) and humidity in the laboratory hatched in 180–260 days; variation in humidity caused embryonic diapause and embryonic aestivation to terminate differentially, thus making the range of incubation times extremely wide.

Sex is determined genetically in both species of *Staurotypus* by heteromorphic sex chromosomes (Bull et al., 1974; Vogt and Flores, 1992). Genetic sex determination can be overridden by hormones. Eggs treated with estradiol (at stage 17 of Yntema, 1968) produced 85% females rather than the 50% produced naturally. Induced gender was permanent, as indicated by later dissections (Vogt, 1991).

One to six clutches (determined by RCV dissection) are produced per reproductive season. Number of eggs per clutch is 8.4 ± 1.92 (4–17) n = 77 (natural populations in Veracruz and Chiapas). Mean number of clutches per year is 3.7 ± 1.09 (1–6) n = 26. Estimated annual reproductive potential is 31 eggs.

Goode (1994) recorded laying dates in captives over a 336-day period (18 June through 20 May, with 94% of layings from 18 August to 24 April. Number of eggs per clutch was 4–18, and 1–5 clutches were laid per year. Individual females laid 8–42 eggs per season.

The two September clutches mentioned above were first clutches of the season. The smaller female contained 62 enlarged follicles in 4 groups and the larger 68 enlarged follicles in 3 groups. These gonadal data would suggest a maximal ARP of 83 and 75 eggs in 4 or 5 clutches. Females examined in April (Veracruz) and June (Belize) contained

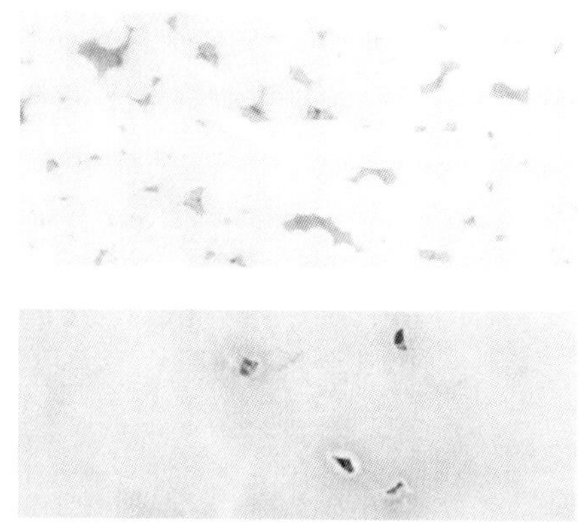

FIGURE 9.7 Left: Egg of *Staurotypus triporcatus*, length 36 mm (UU 6411). Note thickened porous band around narrow diameter of shell. Top right: Close up of pores in thickened band. Bottom right: Pores on thickened band on egg from another clutch (UU 6412).

FIGURE 9.8 Head of adult male *S. triporcatus* (Veracruz) in profile. It is unusual that the mouth is closed in the presence of humans.

old corpora albicantia and only a few (3–10) enlarged follicles (10–11 mm), suggesting that they had finished producing eggs for the year.

GROWTH AND ONTOGENY

The following sexually dimorphic characters occur in both *Staurotypus* species. Females are larger than males in all populations studied. The tails of males are longer and thicker than those of females. Males have clasping organs; females do not. Mature males have a short terminal claw at the tip of the tail; it consists of a terminal scale that is highly keratinized and is triangular in cross section, the apex being dorsal and the base ventral, with a sharp tip. The terminal claw becomes more pronounced with age. Females have a short stubby tail that has a terminal scale that never becomes keratinized or clawlike. Goode (1994) states that captive males develop secondary sexual characteristics at ages of 3–5 years and carapace lengths of 110–150 mm.

On the basis of all information available, we estimate that maximum size for *S. triporcatus* is in the range 400–500 mm. Size at sexual maturity varies, but most mature females are > 220 mm and most mature males > 180 mm.

Data on size and ontogenetic stage from JML database are as follows: adult males, 250 ± 52 (124–378) n = 36; adult females, 295 ± 53 (167–402) n = 22; im. males, 145 ± 17 (127–173) n = 5; im. females, 198 ± 25 (153–229) n = 7; all other im., 92 ± 32 (43–146) n = 15.

Size (CL) of all adults from our combined databases is males 258 ± 47 (183–360) n = 68; females 290 ± 43 (167–402) n = 59.

Goode's (1994) studies of a captive colony of *S. triporcatus* produced the following information. Captive reared females produced first clutches at known ages of 8–10 years and sizes of 248–319 mm, and a wild-caught captive produced eggs at 220 mm.

The two smallest preserved hatchlings of *S. triporcatus* (JML database, Veracruz) have carapace lengths of 43 and 44 mm. Goode (1994) gives hatchling size as 35–48mm. Size data on 24 lab-raised hatchlings from Veracruz (ambient lab incubation temperature) were CL 36 mm ± 1.44 (33–38) n = 24; CW 28 mm ± 1.93 (23–30), and weight, 8.4 g ± 0.77 (6.8–9.5) n = 24 (Vogt, unpublished data).

The three dorsal keels are prominent in hatchlings and become increasingly so with age The shell is boldly patterned in juveniles. The ground color of the carapace is dark brown to black, and each scute bears an irregular, ragged pale mark that covers as much as 20% of the scute. The marks do not extend onto other scutes, but they may extend along interlaminal seams. The overall effect is an optical disruption of outlines—a breaking up of the familiar outlines of scutes and often the margin of the

FIGURE 9.9 A clutch of newly hatched *S. triporcatus* (Veracruz) showing typical neonate carapace pattern.

FIGURE 9.10 Dorsal and ventral views of *S. triporcatus* juvenile that is near the end of the first year of growth.

FIGURE 9.11 Typical large, old male *S. triporcatus* (Veracruz). Pale patch on snout is seemingly scar tissue.

shell, much in the manner of boldly camouflaged military clothing. This neonatal pattern is lost by the end of the first full year of growth. The plastron is black with white reticulations. Traces of the plastral pattern persist in juveniles of 80 mm CL (Figures 9.4, 9.9, 9.10).

We have never observed scute shedding in *S. triporcatus* of any size class.

Growth rings were used in a few individuals (37 of 101) to approximate age up to about 13 years. Based on our combined data for largest immature and smallest mature individuals, the frequency of both categories was highest in year 6.

The data in Table 2, although subject to error, demonstrate considerable variation in size at any given age. In nearly all the captive age groups of both species of *Staurotypus* reported by Goode (1994), sizes were larger than those reported in our growth-ring analysis (see also account of *Staurotypus salvini*).

In only one instance was the growth of an individual actually followed by us: a juvenile male obtained in Veracruz, April 1962—80 mm CL and in its second full year of growth. It was fed regularly in captivity (at UU); aside from periodic obesity, it appears to have undergone "normal" growth. CL was 239 mm late in its 12th year of growth (February 1972), a growth record slower than any shown in Table 2.

PREDATORS, PARASITES, AND DISEASES

Crocodiles (*Crocodylus moreleti*) feed on juveniles and subadults but seemingly have difficult crushing the shells of large adults. Most of the large adults in Laguna Oaxaca, Chiapas, had tooth marks on the carapace that were surely inflicted by crocodiles. The heavy keels on the carapace may make it more difficult for crocodiles to actually crush the shell. However, large crocodilians do not have to crush a mass that can be swallowed whole. Berendt (in Cope, 1865) comments on the presence of whole *Staurotypus* in the stomachs of "alligators" in Tabasco (probably *Crocodylus moreleti*). *Crocodylus porosus* in northern Australia commonly swallow whole adult chelid turtles.

Berendt (op. cit.) also relates a rural belief that a swallowed *Staurotypus* will chew its way out of a crocodile, killing it in the process. Vogt was unable to confirm this in more than 20 years of work in southern Mexico. However, a *Staurotypus* could logically chew its way into a dead crocodile.

Many species of small mammals prey on the eggs (e.g., raccoons, opossums, skunks, and coatis).

POPULATIONS

Data on all individuals (with evident secondary sexual characteristics) from an undisturbed population (Laguna Oaxaca) in Chiapas are as follows: males, CL 226 mm ± 42 (122–346) n = 249; weight, 1574g ± 815 (200–4100) n = 249; females, CL 253 mm ± 43 (133–372) n = 352; weight 2208 g ± 1058 (210–6050) n = 352. The turtles were collected in unbaited fyke nets. It is not clear why significantly more females were collected than males. Perhaps females move more and are more susceptible to capture in the nesting season.

A population in the Rio de Las Margaritas, a protected tributary stream to Lago de Catemaco, produced 49 *Staurotypus* in two weeks in October 1996. The population consisted of

TABLE 2

Growth in *Staurotypus triporcatus* Expressed as Calculated Length of Carapace (mm) at Various Ages from Hatchling to Adulthood

Based on 55 Carapace Lengths of 37 Individuals

Ages[a]	Carapace Lengths (Estimated and Actual)[b]	Carapace Lengths (Captive)[c]
Year H	39 ± 5.27 (30.9–44.0) *n* = 5	[Goode data lacking]
Year 1	48 ± 6.40 (42.4–57.7) *n* = 5	[68 (54–90 *n* = 36)]
Year 2	82 ± 13.9 (61.9–98.0) *n* = 7	[92 (68–127) *n* = 24]
Year 3*	123 ± 36.4 (81.4–186) *n* = 7	[131 (75–180) *n* = 13]
Year 4*	121 ± 24.0 (104–138) *n* = 2	[152 (117–192) *n* = 12]
Year 5*	149 ± 15.9 (138–173) *n* = 4	[189 (153–222) *n* = 10]
Year 6*	220 ± 58.5 (153–246) *n* = 6	[217 (182–253) *n* = 9]
Year 7*	223 ± 16.4 (193–242) *n* = 6	[245 (214–278) *n* = 9]
Year 8*	228 ± 13.7 (211–242) *n* = 5	[263 (244–292) *n* = 7]
Year 9	252 ± 16.9 (240–264) *n* = 2	[301 (278–318) *n* = 7]
Year 10	287 ± 12.7 (280–302) *n* = 3	
Year 11	303 *n* = 1	
Year 12	No data	
Year 13	304 ± 13.4 (295–314) *n* = 2	

a. An asterisk (*) indicates the years in which sexual maturity probably was occurring. "YEAR H" signifies areolar growth in the season of hatching. "YEAR 1" is the first full year of growth.

b. Carapace lengths calculated from our measurement of plastral growth rings are combined with actual measurements of carapace length (on left).

c. Data in brackets on right are from a captive population studied by Goode (1994).

The forgoing data, although subject to error, demonstrate considerable variation is size at any given age. In nearly all the captive age groups of both species of *Staurotypus* reported by Goode (1994), size was larger than those reported in our growth ring analysis (see also account of *Staurotypus salvini*).

80% large, mature adults (Vogt, 1997b). In 2000, commercial turtle trappers decimated this population. The population was again sampled from September 2002 to March 2003. Of 83 turtles collected, only 10 adult males and 3 adult females were found, and 90% of the population was subadult. None of the turtles marked in 1996 were found. Mean CL had decreased to 116 mm ± 90.7 (49–320mm) and mean weight to 390 g ± 1351(49–4000g) (De la Torre-Loranca, 2004). The population still had a good stock of juveniles and, if not further exploited, might be able to recover in 20–30 years.

CONSERVATION AND CURRENT STATUS

Humans are the major predators of *Staurotypus triporcatus*. The species is still present in most of its natural range, but most populations have been severely affected by human predation. The population in Laguna Oaxaca, under study for 9 years by Vogt, had virtually ceased to exist by 1992. A new road allowed market hunters to decimate the population by at least 90% in a period of less than 6 months. Sampling in April 2003 yielded only 8 turtles in 5 days (vs. as many as 125 in the same time period with the same techniques before the hunting). None of the large, old (>39 years) adults from the 1980s were recaptured.

Despite extensive hunting pressures, there are still places in Mexico where *S. triporcatus* is abundant (places we are loathe to reveal).

The acceptability of *Staurotypus triporcatus* as a food varies in different parts of the range. The presence of *S. triporcatus* bones in pre-Columbian archeological sites bespeaks early use of the species for food by hunter-gatherers. Fragments and shells in rural kitchen middens in southern Mexico attest to moderate present use.

Mayan Indians in Belize prefer *Dermatemys* to all other turtles, and these preferences are reflected by the market in Belize City, where *Staurotypus* is seldom sold. Non-Mayan hunters seem to utilize *Staurotypus* for food whenever they can catch them. In the markets of Tabasco, Campeche, and Veracruz, *S. triporcatus* enjoys equal popularity with *Trachemys* and is sold for about the same price.

Staurotypus salvini Gray, 1864

Crucilla (Liner, 1994); Pacific giant musk turtle (Iverson, 1992a). The name Crucilla alludes to the cruciform plastron.

ETYMOLOGY

Gr. *stauros* (cross) and Gr. *typos* (shape), referring to the cruciform plastron.

The species name is a patronym honoring Osbert Salvin (1835–1898), who collected in Guatemala.

HISTORICAL PROLOGUE

See *Staurotypus triporcatus*.

DIAGNOSIS

Plastron is distinctly narrow, pointed at both ends, and cruciform. Second plastral scute is 16–20% of plastron length. Sides and top of head are unicolored or only dimly marked. Occurs in Pacific drainages.

GENERAL DESCRIPTION

(Comparative terms refer to *S. triporcatus*.)

Staurotypus salvini and *S. triporcatus* are allopatric sister species, differing chiefly in size and having relatively minor differences in proportion. Maximum adult size is 250 mm.

BASIC PROPORTIONS FOR ADULTS

CW/CL: males, 0.67 ± 0.04 (0.61–0.75) *n* = 10; females, 0.68 ± 0.03 (0.63–0.72) *n* = 13.

HT/CL: males 0.34 ± 0.01 (0.31–0.36) *n* = 10; females 0.37 ± 0.01 (0.35–0.39) *n* = 13.

HT/CW: males 0.52 ± 0.02 (0.47–0.55) *n* = 10; females 0.55 ± 0.03 (0.50–0.61) *n* = 13.

Modal plastral formula: 1 > 4 > 3 > 2 (93%).

Carapace is of uniform dark color or dimly mottled. Plastron is pale and unmarked in adults (Figures 10.1 and 10.2). Top of head is dark, usually unicolor, and lacking bold pattern; jaw sheaths are not streaked. Three longitudinal keels on the carapace are evident but weakly developed in comparison to *S. triporcatus*. Carapace is relatively wide and is wider posteriorly than anteriorly; CW (at widest point)/CL: males, 0.67 ± 0.036 (0.61–0.75) *n* = 10; females, 0.68 ± 0.032 (0.63–0.72) *n* = 13. Shell is relatively lower in adults (Figure 9.5).

Movable plastral lobe with convergent sides terminates in a blunt apex. Bridge is narrow, and its width is usually 12–16% of plastron length (PL). Entire plastron is relatively shorter and bridge is narrower, creating the illusion of a distinctly cruciform plastron. First plastral scute is usually longest, and scute 2 or 3 is shortest; second plastral scute is 14–19% of PL. Large axillary and inguinal scutes are in contact and span the bridge.

Head shield trifurcate behind, comprised of narrow band along each supraorbital rim and a narrow median band, all emnating from prefrontal region and extending to level of mid orbit (ill-defined in all age groups studied). Two large gular barbels.

Snout is only slightly upturned in adults and is not bosslike. Greatest width of skull is across the anterior part of the tympanum. Dorsal line of head profile is not or is only slightly concave; it is more often straight, slightly concave, or interrupted by a frontoparietal bulge. Rostrum is usually shorter than the horizontal diameter of the orbit. Supraoccipital crest is evenly curved from the frontoparietal suture to the tip in females and is slightly angular in males (Figure 9.6).

Karyotype and sex chromosomes are as described in *S. triporcatus*. Sex determination is genetic.

Neonates (Figure 9.10) have a distinctive plastral pattern of one or more dark blotches (each scute) on a pale ground, traces of which remain in juveniles of CL 82–89 mm. The ventral surfaces of the head and neck are marbled, and the small tubercles of the ventral thigh and brachium are evident as distinct pale points on a dark ground. The general aspect of the neonate carapace is medium to dark brown with vague markings.

GEOGRAPHIC DISTRIBUTION

Map 7

Found in the Pacific coastal lowlands of Mexico from the region just west of Salina Cruz, Oaxaca, through Chiapas and Guatemala into El Salvador, at low elevations (at least to the Rio Lempa and probably to Laguna Olomega). James R. Buskirk, who has explored the coastal lowlands west of the Rio Tehuantepec, has found no evidence of *S. salvini* in that region (pers. comm.).

The type specimen (British Museum of Natural History [BMNH] 1946.1.22.79) is a preserved female collected by "O. Salvin Esq., Huamuchal, Guatemala" (Boulenger, 1889:32). Many variations of "Huamuchal" exist (e.g., Huamuchil, Huamanchal, Guamuchal). "Huamuchal" is listed (*NIS Gazetteer*) at 14°04′N, 91°34′W, Suchitepequez Province, Guatemala, and is shown as a "village" on ONC chart K-25 (1:1,000,000), immediately west of the mouth of Rio Nagua. We consider this to be the type locality of *S. salvini*, in the immediate vicinity of the type locality of *Trachemys scripta grayi*.

HABITAT

In Chiapas and Oaxaca the species occurs in coastal lagoons (with *Trachemys scripta grayi* and *Kinosternon scorpioides cruentatum*). Other habitats are shallow lakes, backwaters, estuaries, and temporary pools. Habitat is more likely to be in turbid water than clear. *S. salvini* thrives in disturbed

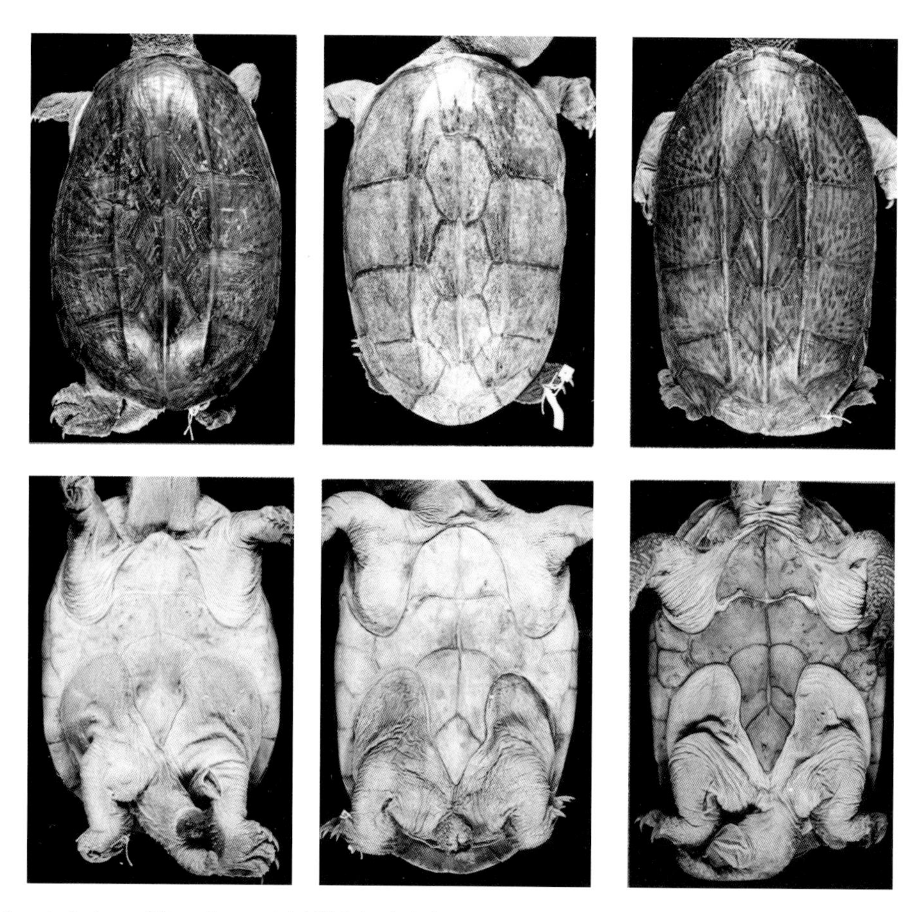

FIGURE 10.1 Dorsal and ventral views of *Staurotypus salvini* (El Salvador) showing ontogenetic change in form and pattern: left—UU 6430, young adult male, CL 135 mm; center—UU 12132, large adult female, CL 206 mm; right—UU 6426, older adult male, CL 186 mm.

habitats near highways and railroads (El Salvador), some of which contain sewage (Legler and Moll, unpublished manuscript; Dean, 1980).

DIET

Staurotypus salvini is an opportunistic omnivore but tends toward vegetation as a staple, exploiting various windfalls when they are available. Its jaw armature is virtually identical to that of *S. triporcatus*, and it has the potential for durophagy.

Dean (1980) is the only author to present an analysis of diet by frequency and volume. Unfortunately, the food item that accounted for 60% of the volume and was found in all stomachs could not be identified. Stomach contents of 29 *S. salvini* from Chiapas consisted of 23.4% plant material by volume. Diptera, Homoptera, Coleoptera, and Hymenoptera were found in 66% of the stomachs but made up only 6% of the volume. Cicadas were the most frequently ingested insect. Fish (*Dormitator latifrons*) were found in 62% of the stomachs but only accounted for 6% of the volume. Crabs and shrimp (*Macrobrachium*) were found in 34% and 31% of the stomachs but accounted for only 3.2% and 0.7 % of the volume, respectively. *Iguana* was found in 10% of the stomachs, for 1% of total volume.

Guts (n = 4) and fecal material (n = 12) from El Salvador consisted of ca. 50% plant material (monocot leaves and stems, fungus pellets) and 50% arthropods. The latter were represented chiefly by terrestrial insects of the families Scarabaeidae, Passalidae, Chrysomelidae, Alleculidae, Tettigoniidae, and Pentatomidae. A few spiders, a shrimplike crustacean, bones of *Smilisca baudini*, and parts of a small unidentified turtle were also present. A specimen from Acacoyagua, Chiapas, contained 100% vegetation (chunks of firm, fleshy fruit or fungus). Other specimens from El Salvador contained 50–90% vegetation—stems and leaves of streamside plants chopped into 30 to 40 mm lengths. Other items present were winglike seeds, large black beetles, large shrimplike crustaceans, much thin chitin, insect wings, and *Basiliscus* feet. There was no trace of mollusks in any of the *S. salvini* examined.

HABITS AND BEHAVIOR

In coastal Chiapas, Dean (1980) found *S. salvini* to be nocturnal and active from the beginning of the rainy season in May until October. Aestivation occurred in the dry season

FIGURE 10.2 Lateral and ventral views of adult *Staurotypus salvini* from Pacific coastal Chiapas (from a captive colony in Tuxtla Gutiérrez).

FIGURE 10.3 Dorsal and ventral views of young *Staurotypus salvini*, coastal Oaxaca. (Courtesy of M. Ewert.)

from November to April. The turtles were never seen basking, and cloacal temperatures were essentially the same as water temperatures (27–34°C) during the season of activity and 20°C in January (when the turtles were aestivating). Movements between recaptures were slight (maximum of 80 m). Mean home range was estimated at 1,200 m^2.

REPRODUCTION

Mating behavior has been observed in captivity by Legler and Moll (unpublished manuscript) and Schmidt (1970). Courtship is initiated by the male prodding the female's cloacal region with his snout. During this prodding the male vibrated his hyoid apparatus, as if he were sniffing for an olfactory cue. The male also bit the posterior margin of the female's carapace. This precoital phase lasted 2–5 minutes. The coital phase began when the male lunged on top of the female and attempted to clasp her with his forefeet under the margins of the carapace, just anterior to the bridge. Once mounted, the male curled his tail under that of the female. When the cloacal openings were in apposition, the male stabilized the union by pushing his tail claw into the base of the female's tail. Intromission lasted for up to 15 minutes.

Schmidt (1970) observed numerous matings by a captive pair of *S. salvini*. His observations differ from the foregoing in that the male used his hind limbs in clasping, the female displayed gaping behavior during coitus, and both sexes displayed "sniffing" behavior, sometimes in unison, during and after coitus. The author described these movements to be more nearly a rapid chattering of the jaws (*"klappert wiederholt mit dem Unterkiefer in schneller Folge"*). (This sound may be made by the jaw sheaths as described in the account of Family Kinosternidae.)

Nesting has not been reported for *S. salvini* but probably occurs in the later part of the rainy season (ca. October) with hatching and emergence at the beginning of the next rainy season (ca. March to April). By contrast, its congener, *S. triporcatus*, nests (like other neotropical endemic species in Gulf Coastal drainages) from September to March, and hatching is coincident with the thunderstorms of June. These differences result from differing rainfall patterns on the Pacific and Gulf sides of the Isthmian region (see Topography and Climate of Mexico).

The eggs are slightly smaller than those of *S. triporcatus*. Schmidt (1970) gave the size of 12 eggs as follows: length, 40.2 mm (38–43); width, 19.3 mm (18–21); and weight, 14 g (12–19). Mean data for 127 eggs from Goode (1994) were as follows: length, 37 mm (26–44); width, 21 mm (19–26); and weight, 10 g (5–14). Ewert (1979) reported that the egg of *S. salvini* weighs 11 g and the hatchling 6.1 g; the egg shell is 0.24 to 0.38 mm thick, with 75% of the thickness due to the mineral layer and 25% to the fibrous layer. He also noted that the eggshell was minutely pitted with pores near the lesser circumference.

Incubation period in captivity has been reported to vary between 207 (Schmidt, 1970) and 145 days (Sachsee and Schmidt, 1976). Nothing is known about incubation in nature. The long incubation period, with embryonic diapause and aestivation (Ewert, 1981) are tropical adaptations that synchronize hatching with the beginning of the rainy season. Like those of other tropical species, hatchlings are probably stimulated to leave the egg by the rise in humidity

TABLE 3

Growth in *Staurotypus salvini* Expressed as Calculated Length of Carapace (mm) at Various Ages from Hatchling to Adulthood

Based on 52 Carapace Lengths of 12 Individuals

Ages[a]	Carapace Lengths (Estimated and Actual)[b]	Carapace Lengths (Captive)[c]
Year H	41.9 ± 4.10 (35.0 – 47.8) n = 7	
Year 1	58.4 ± 7.8 (51.0 – 70.2) n = 7	
Year 2	73.3 ± 10.8 (54.9 – 90.9) n = 8	[89 (58 – 136) n = 26]
Year 3*	82.2 ± 7.7 (79 – 100) n = 8	[103 (75 – 138) n = 12]
Year 4*	103.4 ± 5.2 (98 – 112) n = 5	[109 (90 – 134) n = 8]
Year 5*	122.0 ± 11.7 (109 – 136) n = 6	[123 (114 – 145) n = 6]
Year 6	131.2 ± 10.9 (114 – 139) n = 5	[152 (131 – 161) n = 5]
Year 7	144 ± 8.8 (135 – 153) n = 4	[167 (143 – 184) n = 3]
Year 8	148.5 ± 10.6 (141 – 156) n = 2	[181 (164 – 198) n = 2]

a. An asterisk (*) indicates the years in which sexual maturity probably was occurring. "YEAR H" signifies areolar growth in the season of hatching. "YEAR 1" is the first full year of growth, etc.

b. Carapace lengths calculated from measurements of plastral growth rings are combined with actual measurements of carapace length.

c. Data in brackets on right are from a captive population studied by Goode (1994).

of the nest substrate caused by the onset of rain. Ewert and Wilson (1996) suggested that a chilling period of 25 days at 22.5°C is necessary to break diapause in *S. salvini* during laboratory incubation.

The smallest *S. salvini* we have seen are two juveniles (UMMZ 107,878 and USNM 109,100) with carapace lengths of 80 mm. Calculations from growth rings (Table 3) indicate a carapace length of 35.0–47.8 mm at hatching. Schmidt (1970) reported two hatchlings of *S. salvini* with carapace lengths of 41 and 42 mm at an age of 4 weeks. Hatchlings varied from 22 to 36 mm CL at emergence in a captive population studied by Goode (1994).

Goode (1994) maintained a captive colony of *S. salvini* for 13 years at the Columbus Zoo. Incubation times at uncontrolled temperatures ranged from 130 to 415 days (mean = 247, n = 127 eggs). Data were recorded for 105 clutches from five females purchased as adults and from 24 clutches by two females raised from hatchlings. First clutches were produced at ages of 6–7 years (CL 188 and 146 mm, respectively). The nesting season for the colony lasted 312 days (24 June–2 May), but 120 of 130 clutches were laid in a 209-day period (8 August–5 March), similar to the natural nesting seasons of *S. triporcatus*. Nesting seasons (n = 10) had a mean of 225 days, with a mean of 137 days for the intervening non-nesting season. Intervals between individual nestings ranged from 20 to 114 days, mean 58 days (n = 18).

Captive females use resources equivalent to 16.29% of their body mass to produce an average of 17.79 progeny per year (Goode, 1991).

Legler and Moll (unpublished manuscript) found 6–19 enlarged follicles in all females collected in El Salvador in June; these follicles represented two to three clutches of 6–10 eggs during the nesting season. Corpora lutea in one turtle retained from the previous reproductive season suggests that they nest annually.

Dean (1980) examined 10 reproductive tracts containing a mean of 11.3 ovulatory follicles. He found no corpora lutea. The range in the follicle size suggested two distinct size classes representing two clutches with 12 or more eggs. Sachsee and Schmidt (1976) reported a captive female laying 7–10 eggs per clutch for seven clutches. Schmidt (1970) reported that a female collected in Guatemala laid three clutches of eggs over a 3-month period (nine eggs on 18 September, seven in early November, and six on 27 December).

GROWTH AND ONTOGENY

Females are larger than males. Males have thicker and longer tails, clasping organs, and a tail claw. Females also have a higher domed carapace and tend to be wider (Figure 9.5). Dean (1980) noted that males had a greater interorbital breadth than females of the same size.

Adults of *S. salvini* are smaller than those for *S. triporcatus*. Data based on gonadal examination are as follows: males, 143 mm ± 30 (99–186) n = 10; females, 152 mm ± 32 (80–206) n = 13 (Legler and Moll, unpublished manuscript). Median size at puberty was calculated at 117 mm for males and 132 mm for females. Medians were based on one-half the difference between the largest immature (males 99 mm, females 136 mm) and the smallest mature specimens of each sex (males 135 mm, females 128 mm).

Dean (1980) found up to six growth rings on the plastra of this species, the rings being worn smooth in older individuals. The greatest growth increment occurred in the second growth period (probably the first full season of growth).

PREDATORS, PARASITES, AND DISEASES

Dean (1980) noted an 8% incidence of shell injuries in *S. salvini* and suggested these could be from attempted predation by *Caiman crocodylus*. As in *S. triporcatus*, the high-domed, tricarinate carapace may protect adults from crocodile predation. Similar injuries were noted in 40% of the *Trachemys* population. Epizoic algae (*Basicladia* sp.) grew on 78% of the population, 28% of these having a heavy algal growth.

POPULATIONS

The population studied by Dean (1980) was estimated to have 63.6 turtles per hectare. Sex determination is genetic, and the sex ratio was 1.36 males to 1 female, based on 78 turtles. Species in the study area with temperature-dependent sex determination had skewed sex ratios: *Trachemys scripta grayi* (0.52 males per female) and *Kinosternon scorpioides* (2.33 males per female).

ECONOMIC IMPORTANCE AND CONSERVATION

Legler was told that *S. salvini* is eaten in El Salvador but saw only *Rhinoclemmys pulcherrima* and *Kinosternon scorpioides* on sale in markets. Neither was it seen in the market at Tapachula, Chiapas, around Easter.

Subfamily Kinosterninae (Gray) 1869

Entoplastron is absent; there are six or fewer neurals; N1 may or may not be in contact with nuchal; posteriormost neural is not in contact with suprapygal; posterior costal bones are in contact; extent of plastron is variable but never reduced to extreme cruciform proportions; there are six pairs of plastral scutes; pair 6 is never fused. Seam between scutes 1 and 2 is well anterior to the epiplastron-hyoplastron suture; plastron always has one anterior hinged lobe usually also with a kinetic posterior lobe (exceptions are *Kinosternon herrerai* and *Sternotherus*); seam between scutes 2 and 3 lies on epiplastron only; scutes bracketing the anterior hinge are comprised of a single, divided humeral scute forming scutes 3 and 4; seam between plastral scutes 4 and 5 (humeral-femoral) corresponds to the posterior hinge or lies on the posterior half of the hypoplastron. Musk glands are at extreme ends of the sternal cavity; axillary groove extends to P2; anterior axillary orifice is at M3; duct of inguinal gland penetrates the hyoplastral buttress; inguinal orifice ranges from mid M7 to mid M8. There are one or more pairs of fleshy barbels. Diploid chromosome number is 56.

Content: *Kinosternon* and *Sternotherus*

Genus *Kinosternon* Spix, 1824

Plastral scute 1 is never shortest and is always present. Plastral scutes usually are separated by typical interlaminal seams. Carapace is moderately to weakly tricarinate. Posterior buttresses terminate in the posterior two-thirds of P7. Plastral lobes are moderately to very broad, often completely closing shell orifices; there are two kinetic plastral hinges (except in *K. herrerai*). Anal notch may or may not be present. Anterior musk duct orifice is at M3 to M3–4; posterior orifice is at M7 or M8.

Content: ca. 15 species, 12 of which occur in Mexico.

GENERAL REMARKS

There is an overall likeness of the recognized species of *Kinosternon* that masks the real distinctions between them. Consequently, to all but those who study them, they may all look very much alike. Iverson (1991b), in a major revisionary work, assumed that all of the extant specimens in museum collections would be misidentified, and found that about half of them were. This obfuscation of individuality is compounded by a preference for bottom dwelling and the accumulation of various encrustations, discolorations, dirt, and algae that may conceal the true character of the shell. This circumstance partially explains the ignoble vernacular name "mud turtle." The scrubbing of some large series of *Kinosternon* in the UU collection took almost as much time as their capture and careful preservation.

Thus, identifying *Kinosternon* has its problems. For a single specimen with no data it may be difficult. With a specimen of known provenance it is usually possible. Trenchant diagnostic characters may exist, but it was customary in the 1970s and 1980s to express them numerically in terms of "ratios" that were often identified with arcane and undefined abbreviations that are suitable for use perhaps in a multidiscriminate analysis but awkward for use by the nonspecialist. We attempt to use the results of these studies (e.g., Berry, 1978; Iverson, 1981a), but always to express them subjectively and quantitatively in terms defined in this book (e.g., "shell relatively wide and low, CW/CL 0.73, CH/CL 0.38"). Nonetheless, accurate knowledge of geographic provenance narrows the field and makes identification far simpler.

Kinosternon herrerai Stejneger, 1925

Herrera's musk turtle; casquito de Herrera (Liner, 1994); Herrera's mud turtle (Iverson, 1992a).

ETYMOLOGY

The name *herrerai* is a patronym honoring Dr. Alfonso L. Herrera (1869–1942), director of the National Museum of Mexico, founder of the Chapultepec Zoo, and founder of the Instituto de Biología, UNAM. Herrera donated the type specimens to the U.S. National Museum.

HISTORICAL PROLOGUE

Four specimens were donated to the U.S. National Museum in December 1918. The type locality was initially stated as "Xochimilco, Valle de Mexico"; it was later assumed that the specimens had been imported for sale in the local market. Smith and Taylor (1950b) first restricted the type locality to La Laja, Veracruz. Smith and Brandon (1968) subsequently restricted it to Tampico, Tamaulipas, a logical choice. Little has been written about this species, and no synonyms have been used. The information on natural history in the present account comes from a large series collected by James L. Christiansen in 1962 (75 specimens, UU, Veracruz), a report on a population at the northern limit of the range (Carr and Mast, 1988), and two graduate theses (Aquino-Cruz, 2003; Cazares-Hernandez, 2004).

DIAGNOSIS

Kinosternon herrerai is the only species of *Kinosternon* (sensu stricto) to lack a kinetic posterior plastral lobe. The following characters, in combination, further distinguish *K. herrerai* from all other species of *Kinosternon*: (1) adult carapace with single median keel, often obscure in old individuals; (2) first central scute narrow and not in contact with M2; (3) M10 and M11 elevated above M9; (4) at least three central scutes as wide or wider than long; (5) axillary scute always in contact with inguinal scute; (6) plastral scute 1 almost always less than half the length of anterior plastral lobe; (7) plastral scute 3 less than 10% of maximum PL; (8) plastral scute 4 (defining mid plastral lobe) 20–30% of maximum PL and shorter than other plastral lobes; (9) bridge narrow, 20–30% of PL (Berry and Iverson, 1980; and Smith and Smith, 1979).

GENERAL DESCRIPTION

Maximum recorded sizes and weights are as follows: males, 172 mm CL, 780 g; females, 157 mm CL, 580 g (Berry and Iverson, 1980; Carr and Mast, 1988; Cazares-Hernández, 2004). The largest adults rank well above the median for Mexican *Kinosternon* (JML database).

BASIC PROPORTIONS

(JML Database)

CW/CL: males, 0.67 ± 0.02 (0.63–0.73) n = 24; females, 0.69 ± 0.03 (0.64–0.77) n = 35; im. males, 0.72 ± 0.02 (0.69–0.76) n = 7; im. females, 0.72 ± 0.03 (0.68–0.78) n = 7; im. unsexed, 0.79 ± 0.05 (0.75–0.84) n = 3.

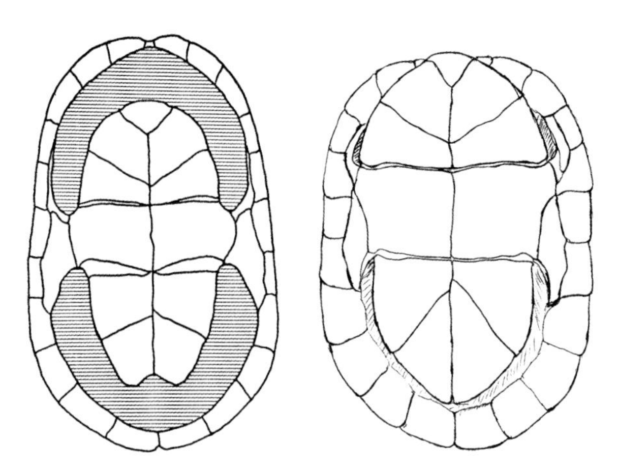

FIGURE 11.1 Plastral views of adult male *Kinosternon herrerai* (left) and K. *leucostomum* (right) demonstrating the range of plastral coverage in Mexican *Kinosternon*.

FIGURE 11.2 Adult male *K. herrerai* entering water from bank of small stream (Jalapa, Veracruz; courtesy of B. Horne). Note development of premaxillary beak and rostral boss.

CH/CL: males, 0.36 ± 0.01 (0.33–0.39) n = 29; females, 0.38 ± 0.02 (0.34–0.45) n = 35; im. males, 0.35 ± 0.01 (0.34–0.37) n = 7; im. females, 0.36 ± 0.02 (0.33–0.39) n = 7; im. unsexed, 0.38 ± 0.00 (0.38–0.39) n = 3.

CH/CW: males, 0.54 ± 0.03 (0.49–0.59) n = 29; females, 0.55 ± 0.04 (0.48–0.64) n = 35; im. males, 0.49 ± 0.02 (0.47–0.54) n = 7; im. females, 0.51 ± 0.05(0.45–0.58) n = 7; im. unsexed, 0.49 ± 0.03 (0.46–0.51) n = 3.

BASIC PROPORTIONS

(Iverson, 1991b)

PL/CL: 0.811 in males, 0.859 in females.

FL/CL: 0.292 in males, 0.294 in females.

HL/CL: 0.307 in males, 0.342 in females.

BL/CL: 0.164 in males, 0.190 in females.

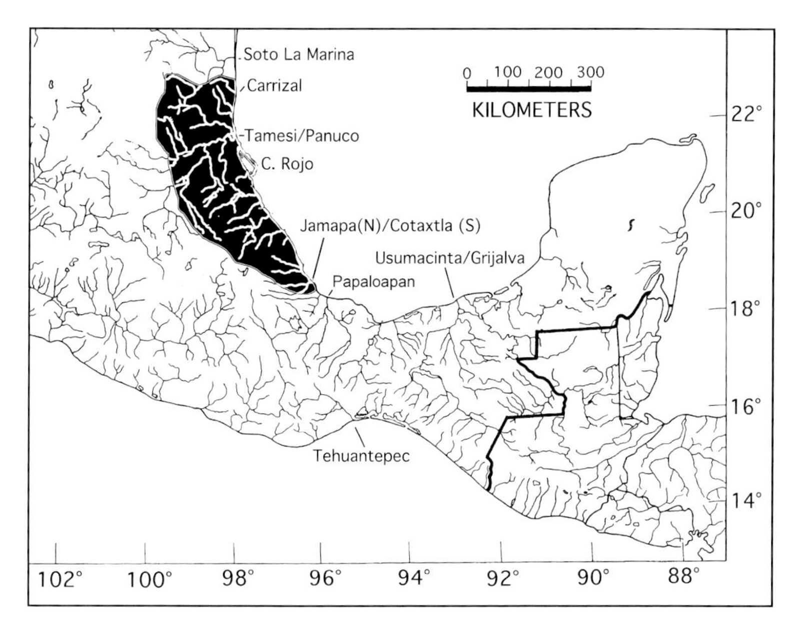

MAP 8 Geographic distribution of *Kinosternon herrerai*.

PS1/CL: 0.125 in males, 0.124 in females.

PS6/CL: 0.167 in males, 0.206 in females.

Modal plastral formula: 4 > 6 > (2 < = > 1) > 5 > 3.

The species is characterized by its distinctively reduced (but not cruciform; Figures 11.1 and 11.3) plastron, covering only about 50% of the shell opening. An anal notch is present. Carapace is relatively long, is only slightly domed, and is widest posteriorly. Immature individuals with CL as long as 57 mm have an indistinct dorsolateral keel on each side. Carapace is tan or brown to olive. Plastron is yellow to pale brown; the seams are often darker. Skin is gray or brown to yellowish; limbs are darkly spotted. Head is spotted or reticulated with brown; the ventral neck is nearly immaculate. Jaws are streaked with brown. There is a large, hooked premaxillary beak. Dorsal head shield bifurcates posteriorly. There is one pair of barbels on chin and one or two pairs on throat.

RELATIONSHIPS

Kinosternon herrerai is thought to be the most primitive kinosternine and is not closely related to any other living species (Iverson, 1991b).

GEOGRAPHIC DISTRIBUTION

Map 8

Kinosternon herrerai occurs in Gulf drainages from the Rio Tamesi and Rio Panuco in Tamaulipas to the Rio Actopan in Veracruz (just north of the Rio Papaloapan). Altitudinal occurence is from sea level to 1,540 m in headwater tributaries. The species has been recorded from the states of Tamaulipas, Veracruz, San Luis Potosi, Hidalgo, and Puebla. The only record for the state of Mexico is that of Mata-Silva, Ramirez-Bautista, Paredes-Flores, and Espino-Ocampo (2002).

HABITAT

Permanent, shallow, often rocky streams with fast water, as well as seasonal bodies of water are occupied by this species below 1,540 m. Vogt (pers. obs.) collected an adult male in a flooded field in a marshy coastal area near Nautla, Veracruz.

Carr and Mast (1988) studied the species in a small, clear, mud-bottomed stream flowing under closed forest canopy at Arroyo La Coma near Rancho Nuevo, Tamaulipas, near the northern limits of the range (23°11′N, 97°47′W). Most pools contained exposed rocks, with one pool having an entirely rock bottom and depth of 15–20 cm. The lowest and largest pool was at the base of a 1 m waterfall and was ca. 6 m wide, 1 m deep, and 15 m long.

Brackish water is tolerated long enough for barnacles to attach to the shell (see "Parasites" below). Brackish water habitats are probably subject to tidal influence.

DIET

Like most kinosternids, *K. herrerai* qualifies as an opportunistic omnivore, its diet varying with the seasonal availability of food.

Aquino-Cruz (2003) studied diet in a population near Jalapa, Veracruz (La Bomba, an intermittent stream in the Rio Actopan headwaters, ca. 19°31′N, 96°52′W, 1,137–1,151 m). Stomachs of 58 individuals were flushed, some from each month from October 1999 to September 2000. Forty-six of these (21 males and 25 females) contained food, as follows: animal matter—mollusks (*Opisthobranchia* and *Pulmonata*), decapods (*Procambarus*), isopods, diplopods,

FIGURE 11.3 Lateral and ventral views of adult male *K. herrerai* (Nautla, Veracruz).

ephemeropteran nymphs, Odonata (larvae), Orthoptera, Hemiptera, Coleoptera (terrestrials), dytiscids, Trichoptera (larvae), fish (Poecilidae), anurans (*Bufo*, *Eleutherodactylus*, *Rana* eggs, tadpoles, and adults), and unidentified animal material; plant matter—leaves, roots, stems, and figs; and miscellaneous items such as stones and soil.

Dragonfly larvae (Odonata) were found in 36% of the females. The seasonal availability of anuran eggs and tadpoles as well as tree fruits (*Ficus*) was reflected by differing relative percentages at different times of the year. Decapods, plant material, anurans, figs, dragonfly larvae, and tadpoles were the most often consumed food items. Frog eggs were consumed in large quantities only in autumn. Both sexes consumed more animal than plant material, with about 50% containing plants and more than 80% of each sex containing animal material.

Carr and Mast (1988) examined the stomach contents and feces in a small sample of *K. herrerai* collected in June 1981. The turtles were feeding on figs, other plant material, Coleoptera, Odonata, Orthoptera, and Diplopoda.

HABITS AND BEHAVIOR

Kinosternon herrerai is active throughout the year near Jalapa, Veracruz (Aquino-Cruz, 2003). Activity occurs at all times of the day and night (Carr and Mast, 1988; Reese, 1971; Smith and Brandon, 1968; J. B. Iverson, pers. comm.). Reese (1971) and Carr and Mast (1988) have commented on how fast *K. herrerai* can move on land and in the water, respectively.

REPRODUCTION

Courtship and copulation occur underwater. The male grasps the female with all four feet, utilizing the terminal tail claw to facilitate and maintain intromission (to the extent that the female pericloacal area was abraded and bleeding). During copulation the male arches his neck over the head and anterior shell rim of the female and moves it rapidly from side to side in several bouts of activity punctuated by rest. The lateral movements involve the rubbing of the male chin against the female shell margin. Copulation lasts about 5 minutes (Carr and Mast, 1988).

Carr and Mast (1988) found females with enlarged follicles and oviducal eggs in July and August. One of these females was maintained in captivity and laid eggs the following March. They reported egg size as 35 × 18 mm and 7.1 g, and mean number of eggs per clutch as three (two to four), and they found evidence for multiple clutches.

One of the paratypes (USNM 61250, 123 mm CL) was dissected by Legler. It contained three eggs: 30.3 ± 1.34 (28.8–31.3) × 17.6 ± 1.02 (16.4–18.3). The ovaries bore 10 nonatretic follicles of 1–3 mm and one of 7 mm; there were also seven atretic follicles in the range of 5–7 mm. The corpora lutea were regressing, suggesting that the specimen had been held in captivity. The egg shells were thick and chalky white. The shell surface was typically kinosternid—most of shell was finely pebbled, but there was a raised circumferential band on the middle third of the egg with distinct pores (as in *K. angustipons*, *Staurotypus*, *Claudius*, and probably all kinosternids). The pores have rounded orifices leading to deep pits with angular cross-sections. One can detect successive laminae of shell deposition on the walls of the rounded pores.

A series of 34 females obtained by J. L. Christiansen near Nautla and Tecolutla, Veracruz, 1–4 July 1965 was dissected by Vogt and Legler. The 32 mature females bore a total of 168 enlarged ovarian follicles with a mean diameter of 8.6 mm ± 2.6 (6–16). No corpora lutea were present. The largest yolked follicles would probably have ovulated in July and the others in August. Estimates of reproductive potential based on this series were as follows: number of eggs per clutch 3.7 ± 1.5 (2–8); two females had follicles sufficient for one clutch, 29 for two clutches, and one for three clutches. There was no evidence of skipped reproductive seasons.

GROWTH AND ONTOGENY

Plastral coverage of females, in general, is greater than that in males (Cazares-Hernandez, 2004; pers. obs.); the posterior lobe of the plastron and plastral scute 6 are much longer in females than those in males (Iverson, 1991b).

FIGURE 11.4 Adult and juvenile *K. herrerai* (Veracruz).

Females have short tails and lack a terminal claw, and males have long, thick tails with a well-developed terminal claw. Clasping organs are well developed in males but are small and uncornified in females. The head of males is proportionately longer and wider than that in females. The plastron is slightly concave in males. Male secondary sexual characteristics are well defined at CL 91 mm (Cazares-Hernandez, 2004).

Data on size (CL; gender based on gonadal examination; Nautla, Veracruz): males, 145 mm ± 13.72 (113–166) n = 29; females, 127 mm ± 8.99 (112–145) n = 35; im. males, 107 mm ± 8.23 (95–121) n = 7; im. females, 103 mm ± 7.98 (88–111) n = 7; im. unsexed, 76 mm ± 17.62 (57–92) n = 3.

One of the paratypes (USNM 61252, not dissected) is a seemingly mature female at 99 mm CL. The holotype (USNM 61249) is a mature male of 142 mm CL.

Hatchlings have not been described. The smallest specimen in the JML database is a juvenile of 52 mm CL. Ages were estimated for two juveniles in the UU collection as 57 mm CL, beginning of third year (dorsolateral ridges barely evident); 78 mm CL, beginning of fifth year (dorsolateral ridges not evident).

PREDATORS, PARASITES, AND DISEASES

Carr and Mast (1988) noted the empty shell of a balanomorph barnacle on the posterior carapace of an adult male, the first report of a barnacle on a kinosternid. They also found leeches (*Placobdella* of two species) and an epizoic alga (*Basicladia*) on adults.

POPULATIONS

In the population studied by Aquino-Cruz (2003) near Jalapa, Veracruz, 46 adults were measured and weighed. Data on carapace length and mass were as follows: 25 females —CL, 123 mm ± 14.7 (87–140), Wt, 265 g ± 77 (114–445); 21 males—CL, 130 mm ± 19 (91–157), Wt, 350 g ± 141 (100–620). Carr and Mast (1988) captured 7 males and 10 females. The series of 75 specimens obtained in baited traps by J. L Christiansen in late June and early July 1965 (coastal Veracruz, 21–22°N) was comprised of 26 adult males, 34 adult females, 6 immature males, 4 immature females, and 5 juveniles of undetermined sex.

CONSERVATION AND CURRENT STATUS

The actual status of this species is poorly known. Populations were formerly abundant in the region of Tampico. Many were dried, varnished, and shipped throughout Mexico in the 1960s to be sold as curios (Figure 4.2). Seemingly nothing has been done to conserve populations of this species, and long-term population studies are lacking. The benchmark studies of Aquino Cruz (2003) and Cazares-Hernandez (2004) have begun a trend that will hopefully be continued.

The *Kinosternon hirtipes* Species Group

As sister species, *Kinosternon hirtipes* and *K. sonoriense* comprise this group. Within their combined ranges of variation there are other *Kinosternon* that might be confused with them. In northern Mexico they could possibly be confused with *K. flavescens* group taxa or with *K. integrum* (sensu stricto). Trenchant characters distinguish both from the *hirtipes* group: the elevation of the ninth marginal in the *flavescens* group taxa, the lack of clasping organs in males of *integrum*, and all members of the *scorpioides* group.

Kinosternon sonoriense LeConte, 1854

Casquito de Sonora (Liner, 1994); Sonora mud turtle (Iverson, 1992a).

ETYMOLOGY

The specific name is a toponym referring to the Sonoran Desert.

HISTORICAL PROLOGUE

Iverson (1981a) produced the most recent revision of the *hirtipes* group, based on 1,298 museum specimens plus data taken from many other individuals captured and released in the field. We follow his taxonomic conclusions.

In the 150 years since *K. sonoriense* was described it has only accrued one synonym, *Kinosternon henrici*, described in 1859 by LeConte. *Kinosternon punctatum* (= *Kinosternon subrubrum*) was described by Gray in 1855a based on a specimen with collecting data "East Florida." Smith and Taylor (1950b) incorrectly synonomized *punctatum* with *sonoriense*.

Kinosternon sonoriense is easily and often confused with other species of *Kinosternon* (e.g., *hirtipes* and *flavescens*) and the literature is replete with erroneous identifications

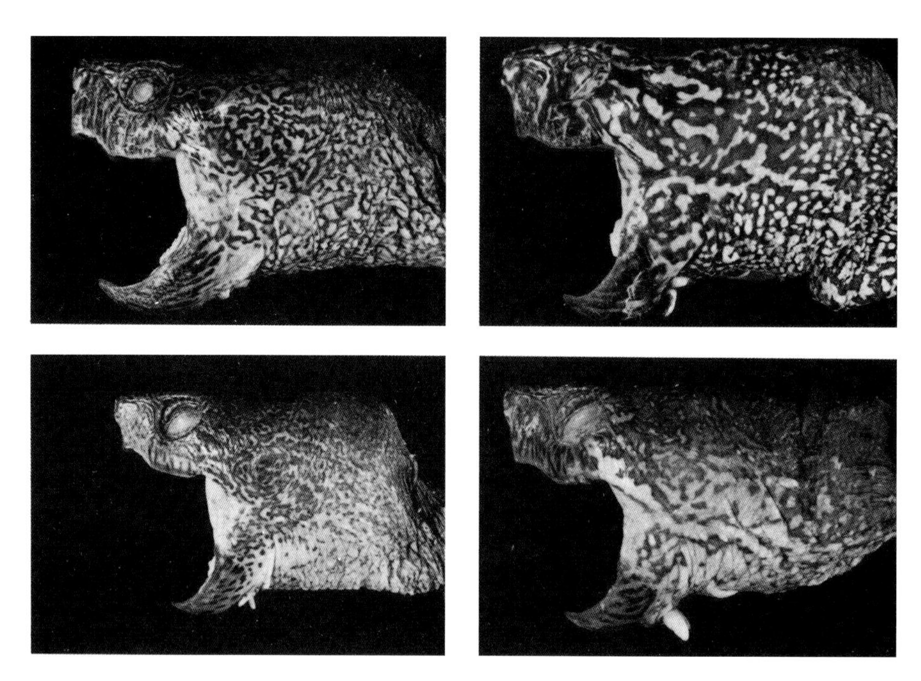

FIGURE 12.1 Comparison of adult heads of *Kinosternon hirtipes murrayi* (left) and *Kinosternon sonoriense* (right); males (top; KU 51320 and 51311, respectively), females (bottom; KU 51307 and 51313, respectively). Note the large barbels in *K. sonoriense*. (Modified, with permission, from Conant and Berry, 1978.)

and statements of geographic distribution (e.g., Carr, 1952; Pritchard, 1967; Pope, 1939; Stebbins, 1954; Brown, 1950).

Conant and Berry (1978) established an early plateau of scientific logic for the *hirtipes* group in northwestern Mexico and the adjacent United States. In a well-illustrated paper they redefined diagnostic characters, defined geographic distributions, and resolved many of the misunderstandings on identification and occurrence (Figures 12.1 and 12.3).

DIAGNOSIS

(See also diagnosis of *Kinosternon hirtipes*.)

A *Kinosternon* of the *hirtipes* species group has the following combination of characters: (1) Adult head shield is large (larger than any *hirtipes* in northern Mexico) and triangular, rhomboidal, or bell shaped (not furcate behind, little geographic variation); (2) there are three or four pairs of relatively long chin and posterior neck barbels, the longest of which are more than half the orbital diameter; (3) male plastron is relatively wide, with a maximum width of plastral forelobe of 47.2% (42–53) of CL; (4) adult females are usually larger than the males.

GENERAL DESCRIPTION

Figure 13.3

Size is generally smaller than *hirtipes* and larger than *flavescens*: Maximum CL is 169 mm in males and 176 in females.

BASIC PROPORTIONS FOR ADULTS

(JML Database: Arizona, Chihuahua, Sonora)

CW/CL: males, 0.65 ± 0.03 (0.59–0.71) n = 25;
females, 0.66 ± 0.03 (0.60–0.70) n = 17;
im. females, 0.72 ± 0.01 (0.71–0.73) n = 2;
im. unsexed, 0.81 ± 0.04 (0.73–0.85) n = 6.

CH/CL: males, 0.33 ± 0.02 (0.30–0.36) n = 24;
females, 0.37 ± 0.02 (0.33–0.41) n = 17;
im. females, 0.36 ± 0.01 (0.35–0.36) n = 2;
im. unsexed, 0.40 ± 0.02 (0.37–0.42) n = 6.

CH/CW: males, 0.51 ± 0.03 (0.46–0.58) n = 25;
females, 0.56 ± 0.05 (0.48–0.66) n = 17;
im. females, 0.50 ± 0.01 (0.49–0.50) n = 2;
im. unsexed, 0.49 ± 0.02 (0.46–0.52) n = 6.

PLASTRAL PROPORTIONS

(Iverson, 1991b)

PL/CL: 0.902 in males, 0.945 in females.

FL/CL: 0.318 in males, 0.321 in females.

HL/CL: 0.333 in males, 0.353 in females.

BL/CL: 0.215 in males, 0.248 in females.

PS1/CL: 0.167 in males, 0.173 in females.

PS6/CL: 0.188 in males, 0.223 in females (Iverson, 1991b).

Modal plastral formula: 4 > 6 > 1 > 2 > 5 > 3.

Carapace of adults is tricarinate with the mid-dorsal keel most pronounced; keels are variable with age, ranging

FIGURE 12.2 Dorsal and ventral views of hatchlings, *Kinosternon hirtipes murrayi* (left; BYU 14134, CL 27 mm) and *Kinosternon sonoriense* (right; BYU 14134, CL 27 mm). (Modified, with permission, from Conant and Berry, 1978.)

to completely smooth shells. Alga-covered shells have a rough, pocked appearance (Hulse, 1976b). Hatchlings and juveniles have three weak keels. C1 is broad (24.4% of CL in males and 25.5% in females) and usually in contact with M2. M9 is not elevated above preceding marginals; M10 is higher than M9; M11 may or may not be elevated to height of posterior M10. There are six neural bones; N1 is sometimes in contact (38%) with nuchal; N6 is not in contact with pygal element. Axillary and inguinal scutes are nearly always in broad contact; inguinal scute contacts M8.

Plastral coverage is extensive in comparison to *K. hirtipes*: maximum W of plastral forelobe is greater than 60% of CW at same level; maximum W of hindlobe (HL) is greater than 65% of CW at same level. Third interlaminal seam is 0.045 of CL in males and 0.038 in females. (See also plastral proportions in previous text.) There is a small anal notch.

Carapace is brown, olive, or gray, and seams are darker. Plastron is yellow to brown with darker brown seams. Bridge is black or dark brown. Adults have a contrasting reticulate-vermiculate pattern on the sides of the head that is much bolder than that of *K. h. murrayi* and is especially bold in males. In hatchlings the carapace is brown with a yellow margin, and the head striping is pale yellow. At older ages, the skin is dark gray, and the head and neck bear cream-colored mottlings that tend to form at least one pair of stripes on each side: one from the orbit above the tympanum and onto the neck and another from orbit to jaw angle. These stripes are very distinctive and yellowish in hatchlings. The shell of hatchlings is colored and patterned like that of *K. hirtipes murrayi*. Agassiz (1857b, Vol. ii, Pl. V, Figures 8–11) shows superb drawings of a hatchling (CL 28.5 mm).

GEOGRAPHIC VARIATION

Two subspecies have been described and characterized as follows.

Kinosternon sonoriense sonoriense LeConte, 1854

Occurs in New Mexico, Arizona, Sonora, and western Chihuahua. It is characterized by a long fifth plastral seam (19.5% of CL in both sexes), first central scute of medium width (24.4% of CL in males and 25.5% in females), and a relatively wide first plastral scute (20% of CL in males, 19.4% in females) (Figures 13.1 and 13.2).

Kinosternon sonoriense longifemorale Iverson, 1981a

"Longifemorale" alludes to the distinctively long fifth plastral scute (femoral).

Occurs in the Rio Sonoyta basin of Sonora, including Quitobaquito Springs in Organpipe Cactus National Monument, Pima County, Arizona. It is characterized by a short sixth plastral seam (14.4% of CL in males and 18.5% in females); a long fifth seam (12.8% of CL in males, females 13.5%), a wide C1 (28.9% of carapace length in males, 28.8% in females), and a narrow first plastral scute (17.7% of CL in males and 17.8% in females).

RELATIONSHIPS

There are similarities among *K. flavescens*, *K. arizonense*, *K. durangoense*, *K. hirtipes*, and *K. sonoriense* that have confused some workers. *Kinosternon flavescens* is unique in having M9 much higher than M8. *Kinosternon sonoriense* is distinguished by its full head shield (never notched or concave posteriorly) and its long barbels.

Kinosternon sonoriense seemingly evolved from a *hirtipes*-like ancestor that was isolated in the Sonoran Desert after migrating across the Sonora-Chihuahua Desert "filter barrier" in Arizona, New Mexico, and adjacent Mexico (Iverson, 1981a).

FOSSILS

Pleistocene (Rancholabrean) remains are known from Arizona and Sonora (Moodie and Van Devender, 1974; Van Devender, Rea, and Smith, 1985).

FIGURE 12.3 Comparison of three species of *Kinosternon* occurring in northern Mexico, lateral and ventral views: top row—*K. flavescens*, male (Comal Co., Texas); center—*K. hirtipes murrayi*, immature male (Rio Conchos, Chihauhua); bottom—*K. sonoriense*, male (Yavapai Co., Arizona). (Adapted, with permission, from Conant and Berry, 1978. Photos by E. H. Conant.)

GEOGRAPHIC DISTRIBUTION

Map 9

Kinosternon sonoriense (overall). From the lower Colorado and Bill Williams rivers of Arizona and California eastward in the Gila River drainage to New Mexico, southward in Sonora to the Rio Yaqui drainage, and eastward to the Rio Casas Grandes drainage of northwestern Chihuahua. The geographic ranges of *K. sonoriense* and *K. hirtipes* are separated by a scant 35 km in western Chihuahua (the distance between the Rio Piedras Verdes and the Rio Santa Maria).

HABITAT

Several studies of natural history have been conducted in central and extreme southeastern Arizona and adjacent New Mexico (Hulse, 1974, 1976a, 1976b, 1982; Stone, 2006; Van Loben Sels, Congdon, and Austin, 1997; Van Loben Sels, Congdon, Austin, and Austin, 2006; Van Loben Sels, Congdon, Hollett, Cameron, and Dickson, 2008). The last of these is a continuing study as of 2008, and we have cited abstracts of papers that describe some of its progress. Information on natural history for Mexico is sparse and anecdotal (Figure 13.4).

Enough information has been published to show significant intraspecific variation of life history parameters in different populations (reproductive potential, age and size at maturity, annual activity cycle). Much of this can be attributed to differing elevations and environmental temperatures for different populations. The principle studies cited hereinafter are as follows: Hulse (1974), 610 m and 1,200 m, Tule Cr., Yavapai Co., and Sycamore Cr., Maricopa Co., Arizona; Stone (2001, 2006), 1,700 m, Peloncillo Mts., Hidalgo Co., extreme SW New Mexico; van Loben Sels et al. (1997, 2006, 2008), 1,675 m, West Turkey Creek, Cochise Co., extreme SE Arizona.

The species inhabits permanent streams, springs, ponds, and stock tanks up to 2,042 m in elevation. In the northern part of the range they often inhabit streams in wooded areas. For initial capture of 580 individuals at West Turkey Creek, habitat preference was 90% in stock tanks; 5% in stream pools; and 5% on land.

In July 1959 Legler and companions were camped at the Cherry Ranch (shown as El Riito on some maps) on a small tributary of the Rio Sirupa, Sonora, elevation 2,100 m; this was a clear, cool, flowing stream with deep pools and intermittent stretches of flat rock over which the water flowed at a depth of ca. 50 mm. One could see two or three *K. sonoriense* at a time moving about in this shallow water

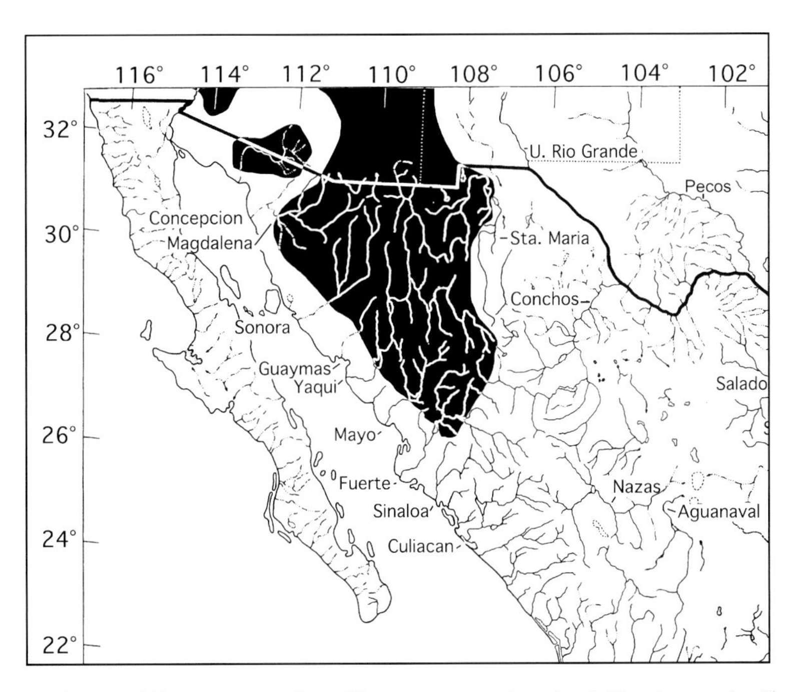

MAP 9 Geographic distribution of *Kinosternon sonoriense*. *Kinosternon s. sonoriense* (east); *Kinosternon s. longifemorale* (northwest).

with the tops of their shells exposed above the surface (possibly foraging). Seven were captured with baited traps in deep holes. This locality (ca. 28°48′N, 108°02′W) was in the upper southern headwaters of the Rio Yaqui drainage (Figures 13.2 and 13.4).

Until recently it was thought that *K. sonoriense* (like *K. hirtipes)* was dependent on permanent aquatic habitats (Iverson et al., 1991a; Ernst, Lovich, and Barbour, 1994; Hulse, 1976a,b). Stone (2001) and Ligon and Stone (2003b) have pointed out that terrestrial aestivation occurs in the Peloncillo Mountains of New Mexico. Populations have thrived and reproductive parameters have remained stable in the complete absence of surface water for substantial periods.

Aestivation occurred in shallow depressions (often with the posterior part of the shell exposed) under clumps of vegetation or beneath large rocks. Turtles left isolated canyon pools (not necessarily ones that were drying) and traveled 1–79 m to sites where they aestivated for 11–39 days, returning to water only when the pools were connected by flowing water.

Stone (2006) has also reported a previously unobserved behavior involving terrestrial activity. Turtles hand captured in rapidly drying stock tanks left the stock tanks within 30 minutes after release and pursued a direct overland course up the nearest and steepest slope. This was termed a "terrestrial flight reaction" and remains under study by Stone and at West Turkey Creek.

DIET

Hulse (1974) studied the feeding behavior and diet of *K. sonoriense* in three steams in central Arizona. Plant material (aquatic angiosperms, chlorophyta, and *Chara*) made up 18.3% of the total volume of material consumed. Animal material comprised 81.7% of the total volume of the stomach contents overall. Plant versus animal material varied with altitude—31% plant at Tule Stream (612 m) and 6% at Sycamore Creek (1,200 m). Choice of food reflected benthic foraging (Anisoptera, *Physa* (gastropoda), Tricoptera, Diptera, Coleoptera, and Ephemeroptera) comprising 63% of volume. Other animals were consumed in lower quantities: fish 4.5%, Hemiptera 4.3%, *Rana pipiens* 2.8%, Zygoptera 2.2%, Megaloptera 1.4%, *Procamberus* 0.6%, and ostracods 0.6%.

Gut contents, in general, reflected opportunistic feeding on what was available in quantity. Opportunistic feeding was evident in the Tule Creek population. When caddis flies were at their peak in March, 66% of the stomachs contained them. When the ostracod population was at its peak from May through July, 66–80% of the intestines contained them. In many instances the intestinal boluses contained only ostracods, suggesting careful selection and separation from algae while feeding. No age or sexual differences in diet were noted.

Two events of predation on vertebrates deserve mention. Stone, Babb, Stanila, Kersey, and Stone (2005) observed a female of 102 mm CL to attack, kill, and begin eating a black-necked garter snake (*Thamnophis cyrtopsis*) in a small, shallow pool (Hidalgo Co., New Mexico). Ligon and Stone (2003a) recorded two instances of predation on *Bufo punctatus* in small pools in the same general region; in one instance the attack was actually witnessed, and in the other the toad carcass was found. In both cases the ventral mass of the toad had been consumed, and all that remained were the head, vertebral column, parts of the limb girdles, and the dorsal integument, including all parts bearing

FIGURE 13.1 *Kinosternon s. sonoriense*. Adult male entering small spring in southern Chiricahua Mountains. (Courtesy of Justin Congdon.)

FIGURE 13.2 *Kinosternon s. sonoriense*. Adult wading in shallow (ca. 20 mm) water flowing over solid rock in small stream near Cocomorachic, Chihuahua—a common occurrence. Individuals could also be found under rocks.

toxic tubercles. These incidents suggest that any animal of appropriate size attracted to the pools could be potential prey for the turtles.

HABITS AND BEHAVIOR

Hulse (1974) observed *K. sonoriense* crawling along the bottoms of streams with necks fully extended (swimming was used only to move to another foraging area). As the turtle moves forward it swings the head from side to side, presumably searching for food. When a food item is located, the head is retracted and the turtle lunges forward, often moving the entire body. Small food items are swallowed whole. Large ones are masticated and torn apart with the foreclaws before ingestion. Algae are often ingested incidentally. Foraging turtles surfaced to breathe every 5–10 minutes. Foraging sites ranged from densely vegetated to open areas but are always on the stream bottom. Feeding is diurnal in spring and autumn but becomes increasingly nocturnal with warming summer temperatures, occurring diurnally only on cooler, cloudy days. Occasional aerial basking occurs. Smaller individuals in particular were often seen basking on branches and small floating logs in Quitobaquito Springs from June to August 1968 (Vogt, pers. obs.). The species is said to be shy and will withdraw when handled (Ernst, Lovich, and Barbour, 1994).

REPRODUCTION

At Tule Stream, Arizona, males are mature at sizes of 76–82 mm CL and 5–6 years of age, and females are mature at 93 mm CL. Mature females of 96 to 99 mm CL were 8–9 years old (Hulse, 1982). Rosen (1987) reported that the same population was maturing in 5 years. At higher altitudes (1,200 m) females mature at 130 mm CL and 12 years (Hulse, 1982). At other sites in Arizona, Rosen (1987) found females to mature at 112 mm CL in 6 years. At West Turkey Creek the minimum female size and age at sexual maturity are 105 mm CL and 5 years.

Courtships were observed in an artificial pond in March, and four copulations were observed in nature, in April, at a water temperature of 21°C (Hulse, 1982) and in May at another Arizona site (Iverson, 1981a). Courtship is similar to that of other kinosternids and involves males chasing females until they become submissive.

The testes reach maximum size in June, maintain this size through August, and then regress in September. Females from Tule Stream maintained enlarged ovarian follicles (5.0–9.9 mm) throughout the year and bore oviducal eggs from June to September. According to Hulse (1982), nesting in Arizona begins in late May and continues through September at lower elevations, but terminates in July at higher elevations (1,200 m).

The following is from van Loben Sels et al. (1997, 2006, and 2008) and pertains to the West Turkey Creek population (1,675 m). Laying begins at the start of the summer monsoon season (July) and continues through August and early September. Fifty-eight nests were observed. Ten females were tracked with transmitters and thread bobbins. Some females left the water several days before nesting, traveled up to 440 m, and remained on land for up to 5 days, whereas others nested and returned to water within 12 hours. Once a site was selected, nesting took at least 3 hours. Nestings were observed from 1,100 to 2,000 hours. Digging was done with the rear legs, but some females initiated the process with the forelimbs. Completed nests were well concealed and difficult to see. Nest temperatures from laying to hatching ranged from 0.7 to 36.5°C. Embryonic development remained in diapause for ca. 9 months and resumed in May when nest temperatures exceeded 20°C. Hatchling emergence coincided with early monsoon storms, in the year following laying. The emergence of

FIGURE 13.3 *K. s. sonoriense:* Dorsal and ventral views of series El Riito, Chihuahua. Left group, M, M, F, Juv on left, F, F on right. Right group, ventral views of same specimens on left. All specimens July 1959, KU. (See also Figs. 12.1 and 12.2.)

FIGURE 13.4 *Kinosternon s. sonoriense.* Typical habitat in high-altitude headwaters, Papigochic drainage of Chihuahua: left, El Riito; right, near Cocomorachic.

young may therefore overlap the subsequent nesting season.

Females at Tule Creek laid up to two clutches of two to four eggs (Hulse, 1982). Rosen (1987) reported up to four clutches for this same population.

Data for the West Turkey Creek population (1,675 m) follow: Mean eggs per clutch for 231 clutches was 6.7 (2–12), as determined by radiography. The eggs are laid in two to four clutches per year. Mean egg width is 16.9 (13.6–20.0) (van Loben Sels et al., 2006).

The eggs are brittle shelled. Egg size and weight given by Iverson (1992c) are 28–35 × 13.8–19.0 mm and 4.2–5.6 g. Size and weight of 21 eggs from the West Turkey Creek study were L, 28.9 (27.0–31.8); W, 16.9 (15.8–18.2); and Wt, 5.03 (4.7–5.5). There is a positive correlation between egg size (width) and clutch size with female size (Hulse, 1982; van Loben Sels et al., 1997).

Legler (pers. obs.) noted clutches of four and five oviducal eggs in early July at El Riito and Cocamorachic, Chihuahua.

The period from laying to hatching is extremely long because of embryonic diapause. Eggs under laboratory conditions remained in diapause for 265 days and hatched after 80 days of incubation for a total of 345 days from time of laying. Sex is determined by incubation temperature (Ewert, 1991).

GROWTH AND ONTOGENY

Data on size (JML database, Arizona, Chihuahua, Sonora): males, 127 ± 13.5 (101–152) n = 25; females, 131 ± 21.4 (87–173) n = 17; im. females, 83 ± 3.5 (80–85) n = 2; im. unsexed, 54 ± 16.2 (39–79) n = 6.

Data from the West Turkey Creek population are as follows: Adults ranged from 90 to 168 mm CL and 191 to 784 g weight. Maximum adult size was 169 mm for males and 160 mm for females. Mean size and weight of hatchlings (overall): 24 mm (17–27) and 2.9 g (1.0–4.5). Mean size and weight of four hatchlings (with caruncles) captured in a nesting area were 22.2 mm CL and 3.0 g (van Loben Sels et al., 1997, 2006.). Growth rates of juveniles from hatching to age 6 averaged 17.5 mm CL per year. Survivorship was 0.45/year in juveniles and 0.89/year in adults.

FIGURE 13.5 *Kinosternon sonoriense longifemoral.* Dorsal and ventral views of adult male (Quitobaquito Springs, Organpipe Cactus National Monument, Pima Co., Arizona). Note the diagnostic elongation of fifth plastral scute (femoral); compare to Figure 13.3.

Females are larger than males and reach 175 mm CL (Iverson, 1981a); males reach 169 mm CL (van Loben Sels et al., 1997). Females have short, stubby tails with a small claw; males have long, thickened tails with a hypertrophied tail claw. Males have clasping organs; females do not. The plastron is slightly concave in males.

Growth rate varies by population and gender. At Tule Stream the mean annual increment of carapace length was 1.57 mm for mature males and 1.50 mm for mature females. Growth rate during the first 4 years does not differ between the sexes; subsequently, males grow at a slower rate and reach a smaller maximum size than females (Hulse, 1976b). A Sonoran mud turtle lived for 36.5 years in the Baltimore Zoo (Snider and Bowler, 1992).

In the West Turkey Creek population sexual dimorphism in size was not evident at 5 or 10 years of age. Overall, however, adult females are judged to be larger than males (see "General Description" and Iverson [1976]).

PREDATORS, PARASITES, AND DISEASES

Bullfrogs (*Rana catesbeiana*) eat hatchlings of *K. sonoriense*. The frogs were introduced in the West Turkey Creek drainage in about 1983 (Rosen, Schwalbe, Rarazek, Holm, and Lowe, 1994) and are now numerous enough to potentially reduce hatchling/juvenile survivorship. Hulse (1976a) found evidence of minor external shell damage on 10 individuals and suggested that it occurred when turtles were stepped on by cattle in stock tanks. Ernst, Lovich, and Barbour (1994) mention archaeological evidence that *K. sonoriense* was eaten by Native Americans.

POPULATIONS

Hulse (1982) estimated a density of 740–825/ha for *K. sonoriense* at Tule Stream, Arizona (1.2 km of small stream, 0.4 ha). Iverson (1982) estimated the biomass of this population to be 100.3 kg/ha. Hulse marked 190 Sonoran mud turtles in this population: 79 (41.5%) were mature males, 40 (21%) were mature females, and 71 (37.5%) were immature. Estimates based on mark and recapture data for actual numbers of mature males, females, and juveniles were 136, 69, and 123, respectively.

In the West Turkey Creek study from 1990 to 1994 (van Loben Sels et al., 1997), 573 live captures consisted of 18.5% juveniles, 35.85% adult males, and 45.7% adult females. Adult sex ratio was 1 male per 1.29 females. If data from all sexable immature animals were considered, sex ratio was 1 male to 1.1 females. Egg/nest survivorship is seemingly unknown.

CONSERVATION AND CURRENT STATUS

Kinosternon sonoriense is listed as vulnerable by the International Union for Conservation of Nature (IUCN) Redlist 2006, but it is not listed in Norma Oficial Mexicana Special, CITES 1996, or USFWS 2006. Its IUCN listing is predicated on probable increased water use and habitat desiccation.

Since this species is not consumed by humans, it is not of interest to the pet trade, and populations are abundant in many streams today. There appears to be little need for management or conservation planning for this species.

Kinosternon hirtipes (Wagler) 1830

Mexican rough-legged mud turtle; casquito de pata rugosa (Liner, 1994). We take the liberty of supplanting the name "roughfoot" (Iverson, 1992a) because the clasping organs are on the thigh and leg—not the foot.

ETYMOLOGY

L. *hirtus*, rough; L. *pes*, foot; alluding to the apposed patches of spadelike scales on the thigh and leg (herein termed clasping organs—see Family Kinosternidae).

FIGURE 14.1 *Kinosternon hirtipes murrayi*. Rio Papigochic drainage, Cd. Guerrero, Chihuahua, ca. 2,411 m; top—adult head patterns (female on left); bottom—shells of specimens from top figures (KU) (male on right in each pair).

HISTORICAL PROLOGUE

The type locality was listed as "America" by Wagler (1830) and later restricted by Wagler (1830) to "Mexico." Schmidt (1953) restricted the type locality to "Lakes near Mexico City." The exact provenance of the type is still unclear but was almost certainly in the Valley of Mexico (Smith and Smith, 1979).

The species remained monotypic until the description of *Kinosternon murrayi* by Glass and Hartweg (1951). The extensive work of Iverson on *Kinosternon* (1978–present) has defined a total of six subspecies of *hirtipes* (including *murrayi*). The distribution and relationships of these taxa require further study. The following diagnoses, descriptions, and information on natural history pertain chiefly to *K. h. murrayi* and are based largely on Iverson (1981a, 1985, 1991b, 1992a); Iverson, Barthelmess, Smith, and de Rivera (1991); and Smith and Smith (1979).

DIAGNOSIS

Because of the wide variation in *Kinosternon hirtipes* (sensu lato), a trenchant diagnosis is difficult and the characters distinguishing the geographic populations are found in the more subjective general description.

A *Kinosternon* of the *hirtipes* species group is distinguished by the following combination of characters:

1. Adult head shield variable but notched or concave posteriorly in all populations except *K. h. hirtipes* (in which it is large and triangular or bell shaped).
2. Usually two pairs of relatively short gular barbels and none posteriorly on ventral neck; all barbels less than half orbital diameter in length.
3. Plastron relatively narrow, especially in males; maximum width of plastral forelobe 43% (36–50) of CL.
4. Adult males usually larger than females.

The type specimen is anomalous in lacking a precentral scute. It is typically present.

GENERAL DESCRIPTION

Kinosternon hirtipes ranks with *leucostomum*, *integrum*, and *oaxacae* in achieving sizes greater than 170 mm CL (maximum sizes in all cases are for males). Maximum CL: 186 mm males, 160 mm females (Iverson, 1981a, 1991b; JML database). However, judging size in terms of maximum CL may be misleading (see Iverson, 1991b:26, Table A). In a database of 2,547 Mexican *Kinosternon* (JML) the largest 300 individuals were nearly all male *K. leucostomum* (exceptions were four *hirtipes*, four *integrum*, and two *oaxacae*). When all *leucostomum* were removed from the database, 739 cases remained. Of these, the largest 50 individuals were 25 *hirtipes*, 15 *integrum*, 8 *oaxacae*, and 2 *sonoriense*—nearly all of which were males.

BASIC PROPORTIONS

(*K. h. murrayi*, Conchos and Papagochic Drainages of Chihuahua, JML Database)

CW/CL: males, 0.64 ± 0.02 (0.57–0.69) n = 43;
females, 0.70 ± 0.03 (0.66–0.76) n = 48;

FIGURE 14.2 *Kinosternon hirtipes murrayi.* Rio Papigochic drainage, Minaca, Chihuahua, ca. 2,100 m. Top—adult heads; two males above, two females below. Bottom—dorsal and ventral views; two males left, two females right.

im. males, 0.70 ± 0.04 (0.65–0.72) $n = 3$;
im. females, 0.73 ± 0.02 (0.70–0.77) $n = 10$;
im. unsexed, 0.76 ± 0.03 (0.72–0.78) $n = 4$.

CH/CL: males, 0.37 ± 0.01 (0.34–0.40) $n = 43$;
females, 0.40 ± 0.03 (0.35–0.46) $n = 48$;
im. males, 0.37 ± 0.01 (0.37–0.38) $n = 3$;
im. females, 0.39 ± 0.02 (0.36–0.41) $n = 10$;
im. unsexed, 0.40 ± 0.01 (0.39–0.42) $n = 4$.

CH/CW: males, 0.58 ± 0.03 (0.54–0.66) $n = 43$;
females, 0.57 ± 0.04 (0.49–0.64) $n = 48$;
im. males, 0.53 ± 0.03 (0.51–0.57) $n = 3$;
im. females, 0.53 ± 0.02 (0.49–0.57) $n = 10$;
im. unsexed, 0.53 ± 0.02 (0.50–0.56) $n = 4$.

The shell of *hirtipes* is wider and higher relative to CL than that in *K. sonoriense*.

PLASTRAL PROPORTIONS FOR ADULTS (OVERALL)

(Iverson, 1991b)

PL/CL: 0.860 in males, 0.925 in females.

FL/CL: 0.311 in males, 0.327 in females.

HL/CL: 0.308 in males, 0.344 in females.

BL/CL: 0.200 in males, 0.236 in females.

PS1/CL: 0.142 in males, 0.155 in females.

PS6/CL: 0.185 in males, 0.236 in females.

Modal plastral formula: 4 > 6 > 1 > 2 > 5 > 3.

Carapace is elongated and tricarinate to varying degrees; carapace keels are almost never absent; mid-dorsal keel is always evident at least posteriorly. C1 is broad (18–32% of CL) and usually is in contact with M2; M9 is not elevated above preceding marginals; M10 is higher than M9; M11 may or may not be equal in height to the posterior edge of M10. Marginals posterior to the bridge are flared. There are six neural bones; N1 is rarely (10.2%) in contact with nuchal; N6 is not in contact with pygal element. Axillary scute is almost always in contact with the inguinal scute.

Plastral coverage is reduced, and the plastron much smaller than the shell orifices (more pronounced in males than in females); maximum W of forelobe is less than 48% of CL in males, less than 52% in females, and less than 65% CW at same level; maximum W of hindlobe is less than 50% of CL and less than 60% CW at same level. Third plastral seam is 0.071 CL in males and 0.061 in females. Bridge length is 20.1% (16–24) of CL in males and 23.6% (18–29) in females (see also plastral proportions in previous text). There is a distinct anal notch.

Head shield is typically large and posteriorly furcate but variable among subspecies, ranging from a small crescent anterior to orbits to a large rhomboid extending posterior to orbits (see subspecies accounts). There are two pairs of gular barbels, the longest (anterior pair) being less than half the orbital diameter.

Carapace is pale to dark brown and may be almost black. Interlaminal seams are darkened in paler specimens. Plastron usually is yellow with darkened seams and is occasionally reddish, brown or black, or (rarely) orange; bridge is brown. Skin of soft parts varies from cream to gray or black; head may be spotted, mottled, or reticulated with dark brown or black.

Adults of *K. h. murrayi* have a reticulate-vermiculate pattern on the side of the head that is accentuated in males (Figure 14.1); this pattern is much less bold than that of *K. sonoriense*. Hatchlings have a dark central plastral figure that occupies about two-thirds of the plastron and follows the interlaminal seams to the bridge and free plastral edges (scarcely different from *K. sonoriense*). The carapace of hatchlings bears a dark spot on the areolar epicenter of each scute but little else. The plastral and carapacal markings are lost at an early stage—both carapace and plastron are essentially unmarked in posthatchling stages except for darkened interlaminal seams.

Substantial geographic variation within the subspecies *Kinosternon h. murrayi* is shown in Figures 14.2 and 14.4–14.7.

GEOGRAPHIC DISTRIBUTION OVERALL

Map 10

Kinosternon hirtipes (sensu lato) occurs from southwestern Texas (Presidio Co.) and northern Chihuahua (Santa Maria, Carmen, and Conchos drainages) southward and southeastward on the Mexican Plateau to the Chapala, Zapotlan, San Juanico, Pátzcuaro, and Valle de Mexico basins of the Sierra Volcanica Transversal in southern Mexico at elevations of about 800–2,600 m. Within this region there are at least 21 disjunct drainage basins, usually isolated by desert or volcanic mountain ranges (Iverson, 1981a). Iverson (1981a) identifies errors in the literature regarding distribution records from misidentified specimens. Maps are shown in Iverson (1981a, 1985, and 1992a).

CORRIGENDA The distribution map in Iverson (1985:361.2) shows the distributions of the various races properly but is incorrectly labeled. Correct labeling is as follows:

1. *K. h. hirtipes*
2. *K. h. chapalaense*
3. *K. h. magdalense*
4. *K. h. megacephalum*
5. *K. h. murrayi*
6. *K. h. tarascense* (Iverson, pers. comm.)

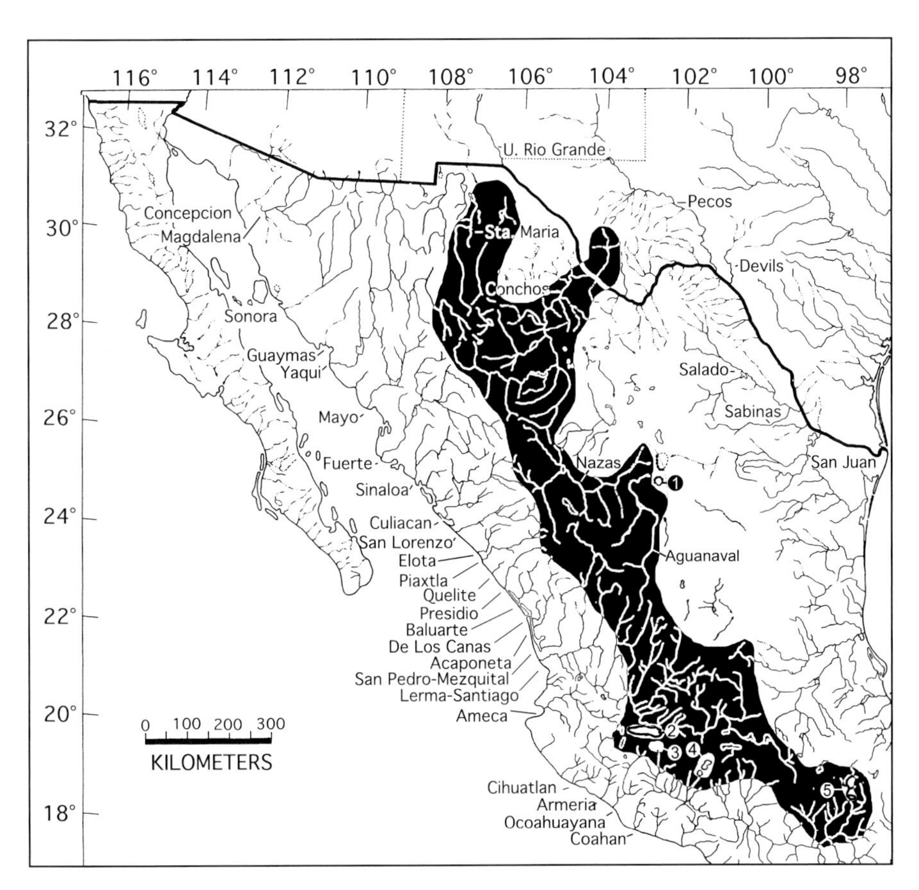

MAP 10 Geographic distribution of *Kinosternon hirtipes*. Most of the blackened area is the range of *K. h. murrayi*. Ranges of subspecies with small geographic ranges are as follows: (1) *K. h. megacephalum*; (2) *K. h. chapalaense*; (3) *K. h. magdalense*; (4) *K. h. tarascense*; and (5) *K. h. hirtipes*.

FIGURE 14.3 *Kinosternon hirtipes murrayi*. Top: high altitude habitat in Chihuahua; above, Rio Papigochic near headwaters. Bottom: an artificial impoundment on private property near Minaca. Both had large populations of *K. hirtipes*. Turtles were observed basking in both places.

GEOGRAPHIC VARIATION

Iverson (1981a) recognized six subspecies of *Kinosternon hirtipes*, as summarized in the following. All occur on the Mexican Plateau from the northern parts of the Basin and Range Province to a series of restricted ranges in the Neo Volcanic Cordillera (see Introduction). Two occur in the northern Basin and Range Province: *K. h. murrayi* has the most extensive range and barely enters the United States, and *K. h. megacephalum* occurred formerly in a single isolated drainage system. The remaining four subspecies have relatively small geographic distributions in isolated basins and are known from relatively few specimens. Known differences between them are small, and Iverson, in some contexts, refers to them as "races."

Kinosternon hirtipes hirtipes (Wagler) 1830

Known only from three localities in the Valley of Mexico (and may be extinct). Maximum adult size: 140 mm in males, 140 mm in females. Adult head shield is triangular, rhomboidal, or bell shaped and is not indented posteriorly; there is a mottled head pattern (typically organized into a pale streak extending posteriorly from the angle of the jaw with a similar pale postorbital streak variably evident); bridge (BR) is short—BR/CL is 17.6% in males and 21.7% in females; PS5 is short—6.9% of CL in males and 7.1% of CL in females; PS6 is long—20.0% of CL males and 25.8% of CL in females.

Kinosternon hirtipes murrayi Glass and Hartweg, 1951

murrayi: a patronym honoring Dr. Leo T. Murray (1902–1958), a well-known naturalist and former director of the Strecker Museum.

Ranges from the Big Bend region of Texas and Chihuahua southward to northern Jalisco, northern Michoacan, and the eastern portion of Estado de Mexico. This is the largest subspecies: Maximum adult size is 182 mm in males and 157 mm in females. Adult head shield is furcate posteriorly; there is an extremely variable mottled to reticulated head pattern; bridge is long—20.0% of CL in males and 23.7% of CL in females; PS1 is long—14.7% of CL in males and 15.8% of CL in females.

Kinosternon hirtipes chapalaense Iverson, 1981a

chapalaense: a toponym referring to Lago de Chapala, Jalisco, where the type series was collected by Norman Hartweg in 1947.

Occurs only in the Chapala and Zapotlan basins of Jalisco and Michoacan. A medium-sized subspecies: Maximum adult size is 152 mm in males and 149 mm in females. Head shield is reduced, crescent-shaped, and lies anterior to the orbits; there is a boldly mottled to reticulate head pattern with two dark postorbital lines but with very little dark pigment present dorsally. Bridge is long—20.2% of CL in males and 25.3% of CL in females; PS6 is long—19.1% of CL in males and 25.2% of CL in females.

Kinosternon hirtipes megacephalum Iverson, 1981

megacephalum: Gr. *mega*, large; Gr. *kephale*, head; in reference to the large head of this extinct taxon.

The type series of seven specimens was taken in or near the Laguna Viesca, the terminal lake of the Rio Água Naval, an internal drainage system that has been disrupted by anthropogenic works (but is always subject to the vicissitudes of drying). The internal Rio Nazas terminates in the Laguna Mayran slightly W and N of the Laguna Viesca. At times of higher water these two basins were probably confluent. The types were collected in 1961 when drying was occurring; specimens of *Trachemys scripta* were taken at the same time. Legler has seen good photos of these specimens and they are *Trachemys s. hartwegi*. Both basins are currently dry. *Kinosternon h. megacephalum* has been assumed to be extinct.

Maximum adult size is 99 mm in males and 117 mm in females. The head is distinct among *Kinosternon* in being wide and in having the triturating surfaces of both jaws expanded to crushing plates and an hypertrophied mandibular adductor musculature—a condition ("megacephaly" or "macrocephaly") found in several other unrelated

FIGURE 14.4 *Kinosternon hirtipes murrayi*. Dorsal and ventral views, adult male (left) and adult female (right); Rio Santa Maria near Galeana, Chihuahua (UU).

turtle groups (both suborders) and nearly always associated with durophagy, usually mollusk crushing (Legler, 1989; Legler and Georges, 1993).

Head shield is V shaped, and head pattern is similar to that of *murrayi*. Plastron seemingly is the most reduced of any *K. hirtipes*—W of hindlobe/CL is 28.2% in males and 31.8% in females; the bridge and the first and sixth plastral scutes are relatively short—BR/CL, 17.3% in males and 23.9% in females; PS1/CL, 11.0% in males and 12.8% in females; PS6/CL, 15.9% in males and 20.9% in females.

Kinosternon hirtipes tarascense Iverson, 1981a

tarascense: honors the Tarascas, an indigenous human population inhabiting the Lake Pátzcuaro region in Michoacan.

Occurs only in Lago de Pátzcuaro, Michoacan.

Adult size is small to medium: Maximum CL is 136 mm in males and 132 mm in females. Adult head shield is furcate posteriorly; there is a finely mottled or spotted head; bridge is short—18.0% of CL in males and 21.4% of CL in females; PS1, 10.6% of CL in males and 12.6% of CL in females; PS3 is long—10.1% of CL in males and 8.5% of CL in females.

Kinosternon hirtipes magdalense Iverson, 1981a

magdalense: a toponym referring to the Valle de Magdalena, Michoacan.

Endemic to the Magdalena Valley of Michoacan (San Juanico Basin).

Among the smallest adult sizes attained by *K. hirtipes*: 136 mm in males and 132 mm in females. Adult head shield is furcate posteriorly; there is a finely mottled to spotted head pattern with very little or no jaw streaking; there is a narrow plastron—W of plastral forelobe/CL is 41.9% in males and 43.5% in females; bridge is short—BR/CL, 18.0% in males and 21.4% in females; PS1 is short—9.9% of CL in males and 11.0% of CL in females; PS3, 8.7% of CL in males and 11.0% of CL in females.

RELATIONSHIPS

The *Kinosternon hirtipes* group evolved on the Mexican Plateau. Its various geographic races have remained there even though numerous streams drain the plateau, either naturally or by stream piracy. By contrast, *Kinosternon integrum* has migrated both upstream and downstream in several major drainages (e.g., Rios Mezquital and Balsas).

Kinosternon hirtipes is very similar morphologically and ecologically to *K. sonoriense*. Iverson (1981a) speculated that *K. herrerai* might be the closest relative of *K. hirtipes*, based on shared characters.

HABITAT

Iverson, Barthelmess, Smith, and de Rivera (1991) studied a dense population of *K. h. murrayi* in the Santa Maria drainage of Chihuahua over a period of 13 years (ca. 650 individuals) that accounts for most of the knowledge of natural history that follows.

The rough-legged mud turtle inhabits small permanent bodies of water in arid grasslands in the northern part of its range. It is also found in lakes, ponds, streams, and rivers flowing into lakes, artificial stock ponds, and marshy areas (Figure 14.3). Iverson, Barthelmess, Smith, and de Rivera (1991) found dense concentrations beneath undercut banks in a small, artesian-fed spring in Chihuahua. Legler found dense concentrations in isolated, cold, spring-fed pools of the channel of the Rio Santa Maria in the same general region (see account of *Chrysemys picta*).

DIET

Diet is reported to be carnivorous, being comprised of insects, other invertebrates, amphibians, and fish (Ernst, Lovich, and Barbour, 1994). Iverson, Barthelmess, Smith,

FIGURE 14.5 *Kinosternon hirtipes murrayi*. Camargo, Chihuahua, ca. 1,485 m. Above: beads of adult male (left) and adult female. Bottom: dorsal and ventral views, group of five—three adult males below, two adult females above.

and de Rivera (1991) found them to forage on mollusks, tricopteran larvae, and "aufwuchs" (presumably algae and other organisms on underwater objects).

HABITS AND BEHAVIOR

Activity is diurnal in areas without human disturbance and nocturnal in areas frequented by humans (Iverson, Barthelmess, Smith, and de Rivera, 1991). Stream temperatures are warm enough to permit activity throughout the year in some areas. Dependence on permanent bodies of water may be related to it having a higher evaporative water loss than more terrestrial *Kinosternon* (Seidel and Reynolds, 1980). Iverson (1981a) recorded aerial basking.

Legler (field notes, 1959) was able easily to catch large series of *K. h. murrayi* in an artificial impoundment on a small tributary of the Rio Papagochic (3.2 km W of Minaca, Chihuahua, 2,070 m); aerial basking was common. Some of the adult males were of maximum size (170 to 186 mm CL). From a nearby high bank overlooking the Rio Papagochic itself, a few much larger *Kinosternon* sp. (possibly 250 mm) were seen basking on logs and rocks; these were probably *K. h. murrayi*.

REPRODUCTION

In general, most females larger than 100 mm are mature at the ages of 6–8 years, based on ovarian examination. The smallest mature females were 99 and 97 mm CL, and the largest immature females were 102, 97, and 94 mm CL. One copulation event was observed on 2 August at 0800 hours (Iverson, Barthelmess, Smith, and de Rivera, 1991).

Ovulation and nesting occur from at least early May to late September and perhaps as early as late April. Females captured in early May had oviducal eggs or enlarged follicles near ovulatory size. A female (CL 119 mm) had two sets of corpora lutea in May, suggesting that she had laid her first clutch in late April. Eight females collected in

FIGURE 14.6 *Kinosternon hirtipes murrayi.* Rio Nazas below Lázaro Cardenas Dam, Durango, 1,524 m. Top: adult heads, clockwise from lower left, M, F, F. Bottom: dorsal and ventral views—one adult female (left) and two adult males (right).

June had at least one set of corpora lutea and at least one set of preovulatory follicles, suggesting that at least two clutches are laid by all females by early July. Four of six females collected by Iverson in July 1982 did not have corpora lutea, although each had ovulatory-sized follicles indicating that they would soon be ovulating. Nesting in Chihuahua may cease in the hottest part of the year (July) and resume in August (Iverson, Barthelmess, Smith, and de Rivera, 1991).

Up to four clutches of eggs are produced per year and the mean number of eggs per clutch is 3 ± 0.82 (1–6) n = 34 (Iverson, Barthelmess, Smith, and de Rivera, 1991), based on egg counts and gonadal examination.

Ergo, annual reproductive potential could be as high as 24 eggs, but actual annual production is estimated at 12 eggs. Moll and Legler (1971) reported four to seven eggs per clutch in Durango and Chihuahua, and Iverson (1981a) reported two to six eggs on the Mexican Plateau.

Size data for 74 eggs (means and extremes) were 28.9 mm (24.2–33.2) × 16.3 mm (14.6–18.6) (Iverson, Barthelmess, Smith, and de Rivera, 1991). Similar data were reported for Durango and Chihuahua (28 × 17 mm; Moll and Legler, 1971) and across the Mexican Plateau (29–32 mm × 16–18.9 mm; Iverson, 1981a). Egg size and number of eggs per clutch are directly correlated to female size, with larger females laying larger eggs and more eggs per clutch (Iverson, Barthelmess, Smith, and de Rivera, 1991). Relative clutch mass was 7.1% and is independent of body mass.

Two hatchlings (20 and 16 mm CL) with fresh umbilical scars but lacking caruncles were found in early August. Incubation time at a constant temperature of 29°C ranged from 196 to 201 days, suggesting embryonic diapause (Rudloff, 1986). Manner of sex determination is unknown.

GROWTH AND ONTOGENY

Data on size (JML database, Conchos and Papagochic drainages of Chihuahua): males, 157 ± 11.95 (114–186) n = 43; females, 124 ± 10.46 (100–140) n = 48; im. males, 101 ± 13.32 (90–116) n = 3; im. females, 94.5 ± 10.1 (73–108) n = 10; im. unsexed, 69 ± 7.23 (60–75) n = 4. The largest adults are from the highest elevations.

Mean adult mass in Chihuahua (Iverson, Barthelmess, Smith, and de Rivera, 1991) was as follows: 295 males, 212 g (range 57–380); 255 females, 203 g (range 92–325).

Among populations sexual size dimorphism varies from those with females slightly larger than males to those where males are larger than females. Males of *K. h. murrayi* grow faster and are larger than females by or before 5 years of age. Sexual dimorphism increases with age.

Adult females have paler chins and short, stubby tails. Males have darker chins and long, thick tails. Only males have clasping organs. A tail claw is present in both sexes but is much more robustly developed in males.

Growth occurs from March through October in northern Chihuahua; one growth ring is formed per annum

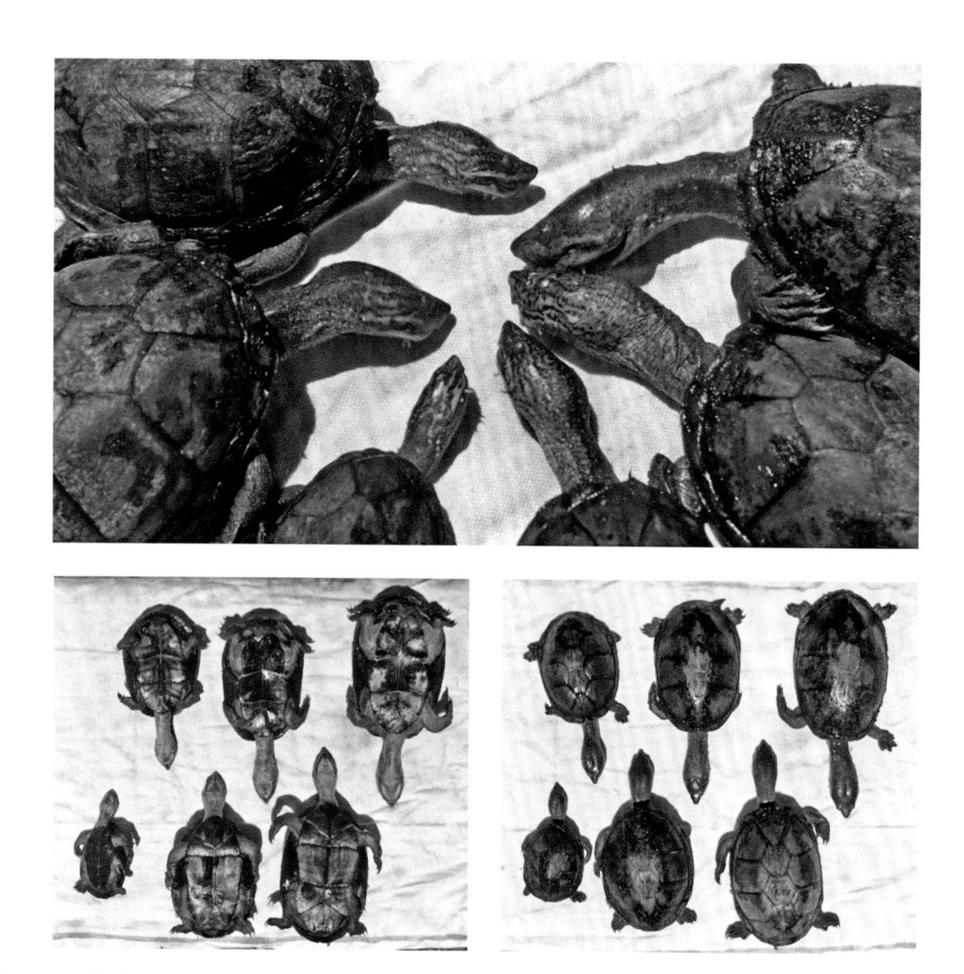

FIGURE 14.7 *Kinosternon hirtipes murrayi*. Rio Tunal, ca. 1,890 m; a small tributary of the Rio Mesquital near Cd. Durango. Top left (top to bottom)—two adult females, one juvenile; top right—two adult males, one immature male. Bottom—dorsal and ventral views, immature male and two adult males above, juvenile and two adult females below. This is a distinctive population of small adult size (male adults are as small as 80 mm CL).

(in winter) (Iverson, Barthelmess, Smith, and de Rivera, 1991).

PREDATORS, PARASITES, AND DISEASES

Freshly eviscerated shells were common along the stream studied by Iverson, Barthelmess, Smith, and de Rivera (1991), but predators were not identified.

Parasites were reported by Caballero y Caballero (1939, 1940); Caballero y Caballero and Cerecero (1943); Hughes, Higgenbotham, and Clary (1941, 1942); Yamaguti (1958); Thatcher (1963); Lamothe-Argumedo (1972); and Ernst and Ernst (1977).

POPULATIONS

The Chihuahuan population studied by Iverson, Barthelmess, Smith, and de Rivera (1991) was dense; turtles were hand collected from beneath deeply undercut banks. No population density estimates were given, but an experienced person could obtain 75 individuals per hour. Similarly high population densities have been observed elsewhere—in the Conchos, Nieves, and Mesquital drainages of Chihuahua, Durango, and Zacatecas (Legler, pers. obs., and UU specimens).

Iverson, Barthelmess, Smith, and de Rivera (1991) marked and released 604 *K. hirtipes* in the Chihuahua study; of these, 585 were sexable adults—310 males and 275 females. *Kinosternon hirtipes* was the only turtle present.

In the Rio Grande at Lajitas, Brewster Co., Texas, and in the Rio Conchos near Ojinaga, Chihuahua, *K. hirtipes* was found to be rare, *Apalone spinifera* abundant, and *Trachemys scripta gaigeae* uncommon (Legler, 1960b), whereas in the Rio Conchos at the mouth of the Rio San Pedro (Ojinaga, Chihuahua) *K. hirtipes* was the most abundant turtle.

Kinosternon hirtipes occurs with *Chrysemys picta* in the Rio Santa Maria proper (near the Iverson study site), and both species may occur in abundance in dry season pools (see remarks on relative abundance under *Chrysemys picta*).

CONSERVATION AND CURRENT STATUS

Kinosternon hirtipes is wide ranging and its habitat requirements, in most cases, demand little more than permanent water, ranging from clear, cold springs to stock tanks.

Most reported populations appear to be dense. In general, there has been scant anthropogenic impact on *Kinosternon hirtipes*. The Huichole people are known to have eaten this species (Malkin, 1958), and it is eaten locally in Pátzcuaro, Michoacan, where it is captured by local fishermen (Flores-Villela, 1980). They have been imported and sold as pets in Mexico City markets. In the region SW of Camargo, Chihuahua, Legler was told the turtles were not eaten, but that their blood was rubbed on the skin to treat arthritis. Overall, the species may need no special conservation effort.

The *Kinosternon flavescens* "Group" in Mexico

Three species of Mexican *Kinosternon* (*flavescens*, *arizonense*, and *durangoense*) comprise a natural group that includes *K. bauri* and *K. subrubrum* in the United States. We take the liberty of referring to this as a "group" in Mexico. In the following we diagnose and describe *K. flavescens* in detail and then focus on the characters by which the other two (derived) taxa differ from *flavescens*. Our treatment follows the most recent revisions (Iverson, 1979b, 1998), but these three taxa have been referred to also as subspecies of *flavescens* (Houseal, Bickham, and Springer, 1982) and the matter is likely to remain contentious for some time.

Iverson (1998) submitted mitochondrial DNA sequence data to phylogenetic analysis alone and together with morphological and protein data to determine relationships within the *Kinosternon* group. He found that *flavescens*, *arizonense*, *durangoense*, *baurii*, and *subrubrum* are closely related to each other and separate from other members of the genus. They are also similar ecologically and morphologically, nearly parapatric, and could be considered as a distinct genus. If future analysis supports the monophyly of this group, the generic name *Thyrosternum* (Agassiz, 1857b) is available for use (type species *Thyrosternum pensylvanicum* = *K. subrubrum*).

Kinosternon flavescens achieves the smallest adult size within the group, but all members of the group in Mexico rank above the median CL for Mexican *Kinosternon*. *Kinosternon arizonense* reaches maximum sizes of 181 in males and 167 in females, ranking near the larger species *integrum* and *hirtipes*.

The *flavescens* group in Mexico can be distinguished from all other Mexican *Kinosternon* by the following combination of characters (in adults): M10 is higher than either M9 or M11, with M9 forming a distinct high peak where it meets the seam between L3 and L4; bridge bears a distinctive longitudinal crease or groove. Seam between third plastral scutes is extremely short or nil; third plastral scutes are often separated by posterior apices of second pair. Plastral coverage is reduced, especially posteriorly (as reflected in lengths of hindlobe and bridge), with a greater reduction in males than in females.

Head shield is reduced, always distinctly indented posteriorly, and often consists of three separate parts: a nasal shield and a narrow crescent over each orbital brow. Clasping

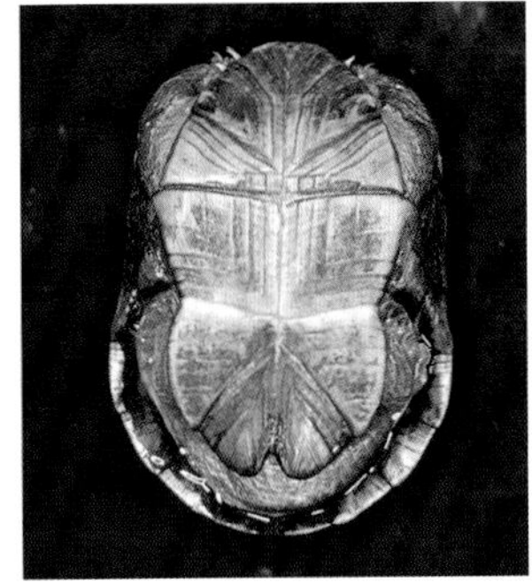

FIGURE 15.1 *Kinosternon flavescens*. Adult male; La Laca, Tamaulipas, UU 9857.

organs are present in adult males and absent in females; tail claw is present in males and absent in females. There is one pair of mental barbels and one pair on the throat.

There are statistical differences in shell proportions, with *durangoense* having the lowest, widest cross-section and *arizonense* having the narrowest (see Basic Proportions for each species).

The characteristic visage of an adult member of the *flavescens* group is a smooth-shelled turtle that is low and wide, with a tannish- to olive-colored carapace marked only with fine, dark interlaminal seams. A narrow interorbital breadth imparts a wide-eyed appearance (Figures 15.1 and 15.3).

Iverson (1979b) recognized the diagnostic value of the groove on the bridge, but it has not been adequately defined nor is it always evident without close scrutiny. The groove is a narrow, elongate trough pointed anteriorly and lying on the posterior two-thirds of the bridge on plastral scute 4 over the hyo- and hypoplastral contributions to the bridge. Its posterior end forms the medial one-fourth of the posterior edge of the bridge.

Diagnoses rely heavily on the extent of plastral coverage and the relative lengths of plastral scutes and lobes. We base the following diagnoses on Iverson (1979b, 1998), Berry and Berry (1984), the original data sheets of James F. Berry, and our own observations of specimens.

Kinosternon flavescens (Agassiz) 1857

Yellow mud turtle (Iverson,1992a); Casquito amarillo (Liner, 1994).

ETYMOLOGY

L. *flavus*, yellowish, alluding to the coloration on the ventral head and neck.

HISTORICAL PROLOGUE

Agassiz (1857b) described *Platythyra flavescens* from five syntypes, three of which bore localities that were

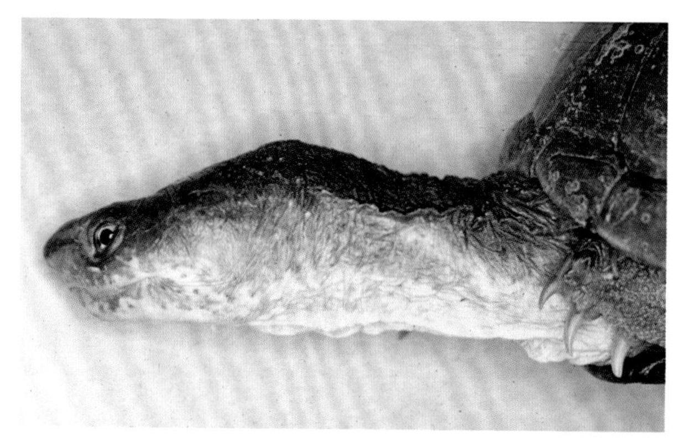

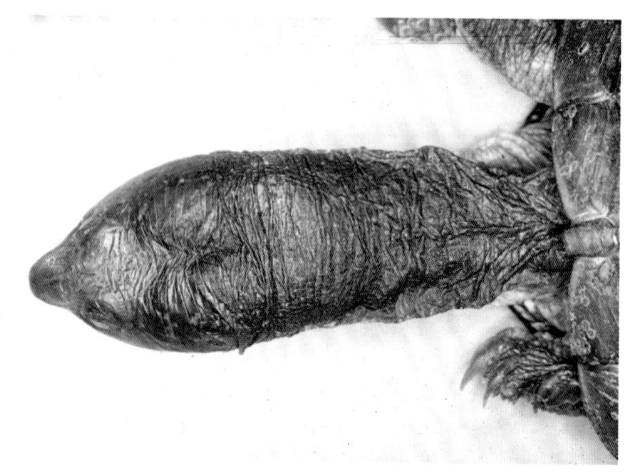

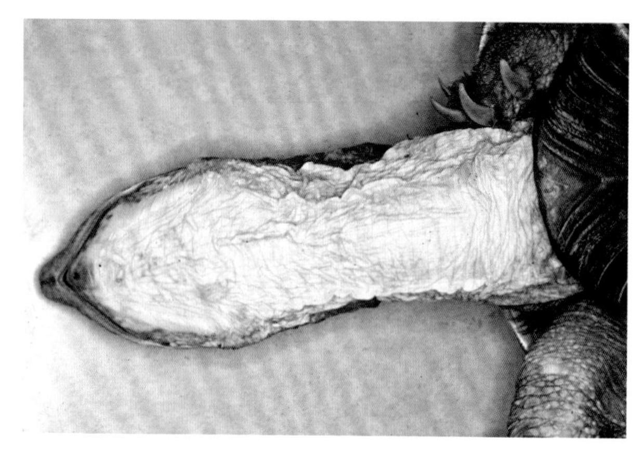

FIGURE 15.2 *Kinosternon flavescens,* adult male. Lateral, dorsal, and ventral views of head and neck of specimen in Figure 15.1.

indefinite or dubious. Two (U.S. National Museum) had the reasonably precise locality data "Texas, near San Antonio," where the species is known to occur. Iverson (1978) nominated USNM 50, a well-preserved adult male, as lectotype and restricted the type locality to "Rio Blanco, near San Antonio, Texas" (near San Marcos, Comal Co.). See Iverson (1978 and 1979b) and Smith and Smith (1979) for details.

Most of the natural history information available for this species group has been gathered for the nominate species at the northern part of its range in the United States (Iverson, 1991c; Mahmoud, 1969). We include some anecdotal information gathered by JML in northern Mexico.

DIAGNOSIS

(Diagnosis and description based mainly on Iverson, 1991c; Berry and Berry, 1984; and JML databases.)

The smallest adult size in *flavescens* group, maximum CL 144 in males, 128 in females. M10 is variable and may or may not be higher than M11. Both plastral lobes are relatively wide, and the anterior hinge is 61–80% of CW; maximum width of hindlobe is 51–70% of CW. Plastral scute 1 is less than half the length of the forelobe (40–42%). Plastral scute 6 is relatively long, 70–82% of posterior lobe. Axillary and inguinal scutes are nearly always (>90%) in contact. C1 is in contact with M2 in about half of specimens.

GENERAL DESCRIPTION

BASIC PROPORTIONS

CW/CL: males, 0.73 ± 0.05 (0.66–0.98) n = 41; females, 0.74 ± 0.07 (0.45–0.85) n = 29; im. males, 0.77; im. females, 0.78 ± 0.01 (0.77–0.79) n = 2; im. unsexed, 0.8.

CH/CL: males, 0.38 ± 0.04 (0.33–0.58) n = 41; females, 0.41 ± 0.03 (0.35–0.47) n = 29; im. males, 0.41; im. females, 0.43 ± 0.02 (0.42–0.44) n = 2; im. unsexed, 0.41.

CH/CW: males, 0.52 ± 0.03 (0.41–0.59) n = 41; females, 0.56 ± 0.08 (0.47–0.91) n = 29; im. males, 0.54; im. females, 0.55 ± 0.02 (0.54–0.56) n = 2; im. unsexed, 0.52.

PLASTRAL PROPORTIONS

(Iverson, 1991b)

PL/CL: 0.897 males, 0.953 females.

FL/CL: 0.335 males, 0.342 females.

HL/CL: 0.316 males, 0.345 females.

BR/CL: 0.191 males, 0.216 females.

PS1/CL: 0.138 males, 0.141 females.

PS6/CL: 0.218 males, 0.274 females.

Modal plastral formula: 6 < = > 4 > 2 > 1 > 5 > 3.

Cross-section is flattened and shell often is almost as wide as long in dorsal aspect. M9 is much higher than M8 and is peaked where it meets the seam between L4 and L5; it is at least somewhat higher than M10 and M11. Carapace is devoid of evident keels except in smallest juveniles (Figures 15.4 and 15.5). Plastral coverage is reduced (Figure 15.1), especially posteriorly, as reflected in lengths of hindlobe (HL) and bridge (BR)—a greater reduction than in other members of the group.

Carapace is yellowish olive to dark brown; plastron is yellow with darkly pigmented seams; head is yellow to gray with paler jaws, often with dark spots; limbs and other soft parts are yellow to gray. Marking of iris varies from nil to a complete horizontal stripe transecting the eye, to a vague stellate pattern (see Conant and Berry, 1978, and figures in Vetter, 2004, 2005). The iris in a live male adult (La Laca, Tamaulipas) was very pale yellow with a well-defined, dark brown, horizontal stripe (Figure 15.2). There was a small patch of brown pigment at the bottom of the iris but not enough to suggest a stellate pattern.

FOSSILS

Fossils are known from the early Pliocene (Hemphillian and Blancan) and Pleistocene of Nebraska, Kansas, and Texas (Fichter, 1969; Holman and Schloeder, 1991; Holman, 1986; Rogers, 1976; Holman and Winkler, 1987; Parmley, 1992). Late Holocene remains (150 years before present [YBP]) are known from Texas (Parmley, 1990).

GEOGRAPHIC DISTRIBUTION

Map 11

Kinosternon flavescens has a broad distribution in the central United States and northern Mexico.

Overall: the Mississippi and other Gulf drainages southward from NW Nebraska through Kansas, Oklahoma, Texas, and extreme southeastern Arizona to northern and Gulf Coastal Mexico. Isolated populations occur in the northern part of the range in the United States.

In Mexico: the tributaries of the upper and lower Rio Grande (Conchos, Sabinas, Salado, and San Juan) in the states of Chihuahua, Coahuila, Nuevo León, and Tamaulipas, thence southward in the Gulf Coastal drainages of Tamaulipas and Veracruz to the latitude of Cabo Rojo (ca. 21°33′N). Distribution in northern Chihuahua includes the northward flowing internal drainage systems of the Rios Guzman, Santa Maria, and Carmen and the extreme upper reaches of the Rio Yaqui in northeastern Sonora (Iverson 1979b, 1992a; Smith and Smith, 1979; Houseal, Bickham, and Springer, 1982; Berry and Berry, 1984; Conant and Berry, 1978).

HABITAT

Stagnant or slow-moving waters are the preferred habitat throughout the range—ponds, lakes, reservoirs, cattle tanks, rivers, swamps, and backwater sloughs. Bottom substrate is typically mud or sand. Iverson (1991c) has found *K. flavescens* to be abundant in ponds in the Sand Hills of Nebraska. In these habitats the turtles can easily burrow below the frost line.

In early September 1960 Legler found *K. flavescens* to be present and usually common in several spring-fed impoundments within a few kilometers south of the U.S. border near Palomas, Chihuahua. An especially dense population was observed in a 0.2 ha pond fed by a protected spring in the town of Guzman. As many as 10 individuals were seen basking on a log in the pond (from which 18 individuals were trapped in 45 minutes). The water flowing into all these impoundments was clear, and the ponds had a grayish mud bottom. Except for tules and other semiaquatic vegetation growing on the periphery of the ponds, the area was one of mesquite desert. It was clear (pers. obs. and unanimous agreement of residents) that *K. flavescens* was the only turtle present. Within an 80 km radius of

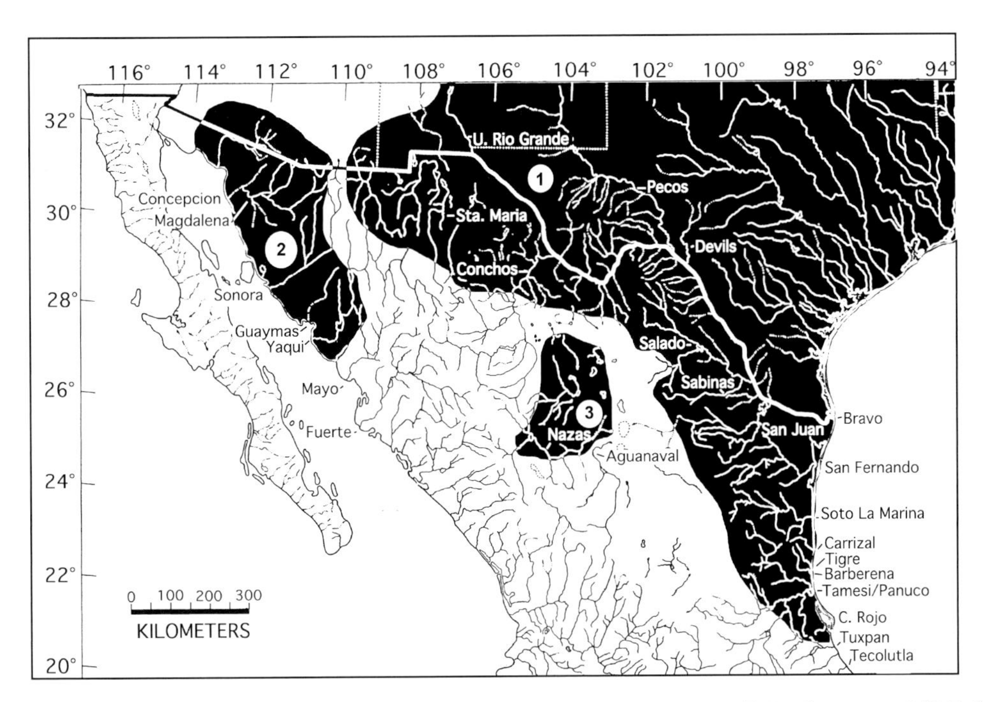

MAP 11 Geographic distribution of the *Kinosternon flavescens* "group": (1) *Kinosternon flavescens*; (2) *K. arizonense*; and (3) *K. durangoense*.

FIGURE 15.3 *Kinosternon flavescens*. Shell of adult male from Chihuahua.

FIGURE 15.4 *Kinosternon flavescens*. First-year juvenile in process of losing bright coloration of plastron and texture on carapace.

Guzman (in the same period) we found only *Kinosternon sonoriense* in flowing tributaries of the Rio Casas Grandes and *Kinosternon hirtipes* (with *Chrysemys picta*) in Rio Santa Maria.

DIET

In the season of activity feeding occurs mostly along pond bottoms. Diet varies among populations but, in general, bespeaks opportunistic omnivory—feeding on whatever is present and in abundant supply. Studies in the central United States have reported planarians, nematodes, oligochaetes, isopods, crayfish, insects, snails, amphibians, carrion, fish, tadpoles, earthworms, duckweed, algae, and other aquatic vegetation (Christiansen, Cooper, Bickham, Gallaway, and Springer, 1985; Mahmoud, 1968; Iverson, 1975; Punzo, 1974). It is generally assumed that feeding does not occur during underground aestivation. Christiansen, Cooper, Bickham, Gallaway, and Springer (1985) found little evidence of new growth during aestivation in July in Illinois. D. Moll (1979) suggested that earthworms might be eaten during underground aestivation.

HABITS AND BEHAVIOR

The season of activity varies with latitude: 15 April to 15 October, New Mexico (Christiansen and Dunham, 1972); 100–128 days, late April to mid July, and late August to September or October, Missouri and Iowa (Christiansen, Cooper, Bickham, Gallaway, and Springer, 1985).

Activity is chiefly crepuscular, occurring from 0800 to 1000 and 1700 to 2000 hours. In the season of activity (Oklahoma) they are active at water temperatures of 18–32°C. Preferred body temperature is 25°C (Mahmoud, 1969).

As temperatures cool in autumn in Oklahoma, the turtles leave the water and bury themselves on land, usually in association with stumps, shrubs, logs, brush piles, or leaf litter. In sandy soil they dig burrows. In some areas they will hibernate in muskrat houses or in the mud at the bottom of ponds. They sometimes over-winter with other turtles (e.g., *Terrapene ornata* and *T. carolina*) (Carpenter, 1957).

At the end of an activity period in Texas the turtles leave the water to aestivate on land as deep as 25 cm, usually in sandy soil. As an adaptation for aestivation this species has the highest carcass-lipid index of any turtle studied (Long, 1985), the stored fat providing energy and insulation during aestivation and hibernation.

In Oklahoma, Mahmoud (1969) observed extensive overland movements in the season of activity. Mean distance between first and last capture for more than 100 days of activity was 258 m for females and 135 m for males (extremes 3.3 and 435 m). The areas utilized were 0.11 ha for males and 0.13 ha for females. In Iowa, Christiansen. Cooper, Bickham, Gallaway, and Springer (1985) used radio telemetry to record movements of 100–450 m from water to elevated sandy dunes, where the turtles burrowed to depths of 5–15 cm. Nocturnal movements were stimulated by rainfall. Overland migrations were crepuscular in the Nebraska Sand Hills (Iverson, 1990b).

Aggressive male-to-male combat was observed by Lardie (1983); males attacked other males whether a female was present or not, seemingly defending a small aquatic territory.

REPRODUCTION

Males reach reproductive maturity at 80 to 90 mm CL, in the fifth and sixth years of growth; females mature at 80- to 125 mm CL at ages ranging from 4 to 16 years (Mahmoud, 1967; Christiansen and Dunham, 1972; Iverson, 1991c). Females mature at older ages (11–16 years) in the northern part of the range (Iverson, 1990b).

Courtship behavior is nearly identical to that of *Sternotherus odoratus* and usually occurs in water, but copulating pairs have been found on land. Coitus lasts from 10 minutes to 3 hours (Mahmoud, 1967). Lardie (1978) described the courtship behavior as consisting of three distinct parts: tactile, initial mounting, and copulating. Detailed descriptions can be found in Mahmoud (1967)

FIGURE 15.5 *Kinosternon flavescens*. Newly hatched siblings from a clutch of four eggs, Comal County, Texas. Note virtual closure of umbilici and lack of caruncles. (See also Figure 3.8.)

and Lardie (1978) or summarized in Ernst, Lovich, and Barbour (1994).

Nesting occurs chiefly from May through July, but some females continue to nest into August. In Nebraska, females move quickly, purposefully, and directly over distances of 21–191 m from the water to sparsely vegetated south-facing slopes of sand hills and bury themselves to depths of about 13 cm. When buried the female excavates a nest cavity with her hind limbs, deposits her eggs, and voids fluid into the cavity. The nests are 17–23 cm below the surface. The female remains with the eggs for at least 1–38 days. Some nesting females may not return to the water until the following spring. Iverson (1990b) hypothesized this behavior bespeaks parental care, the presence of the females near a nest increasing nest humidity and discouraging nest predators.

Christiansen and Dunham (1972) made the following observations on the male reproductive cycle in New Mexico. The testes reach maximum size in August and September; thereafter, regression takes place until near minimum size is attained in November. Hibernation occurs from November to April. Thereafter further regression produces a minimum size in May. Spermatogonia and primary spermatocytes appear with the initiation of testicular growth in May and June. Spermatogenesis begins in July and sperm pass into the epididymides from September to November. Some sperm were present in the epididymides throughout the year, but the volume decreased in June and July.

Christiansen and Dunham (1972) reported a mean of four (one to six) eggs per clutch in New Mexico. The ovarian cycle in New Mexico begins with vitellogenesis in September or October; nearly ovulatory-sized follicles are present in early May; and ovulation occurs in mid May until mid June. The ovaries are quiescent in August, but more than 50% of the turtles they examined had enlarged follicles in August, suggesting that these follicles are retained for ovulation the following year. Data from corpora lutea and oviducal eggs confirm that only one clutch is laid per year in New Mexico, suggesting an annual reproductive potential of four. Long (1972) also noted that the three different follicle sizes found in *K. flavescens* did not suggest three clutches per season, but rather follicles that would be ovulated in future reproductive seasons. Neither study showed any evidence of follicular atresia.

In Nebraska Iverson (1991c) reported a mean of 6.5 eggs (four to nine) in 49 clutches studied. Nesting occurs once per year in Nebraska, but may occur twice in the southern part of the range. In Nebraska 75–95% of the females produce eggs in a good year, and as few as 41% do so in a harsh year (as determined by capture and palpation and radiography rather than gonadal examination).

The eggs are hard shelled from laying to hatching. The egg shell of *K. flavescens* is thinner (0.18 mm) than other *Kinosternon* studied (*minor, bauri*, and *hirtipes*) (Packard, Hirsch, and Iverson, 1984) (see also "Eggs" in Introduction).

Data on 265 eggs from 49 clutches (Iverson, 1991c) are as follows: L, 26.6 ± 1.84 (22.7–31.4); W, 15.4 ± 0.59 (14.1–18.3); Wt, 4.1 g ± 0.6 (3.0–5.8). Clutch mass in Nebraska averages 10.9% of the weight of the female and is correlated with both female size and clutch size.

Incubation temperature determines sex, females being produced at both high and low temperatures and males at intermediate temperatures (Vogt, Bull, McCoy, and Houseal, 1982). Iverson (1991c) found the overall adult sex ratio in Nebraska to be 1:1. Elsewhere sex ratios vary among populations. Mahmoud (1969) reported sex ratios of 0.7–1.5 males to 1 female in three Oklahoma populations.

Fully developed young remain within the egg until rain stimulates hatching. In most areas studied (Iowa, Nebraska, Texas) hatching takes place in the fall, and hatchlings burrow farther into the sand and emerge the following spring (Christiansen and Gallaway, 1984; Long, 1986; Iverson, 1991c).

GROWTH AND ONTOGENY

Adult size data: males, 115 ± 16.7 (81–140) *n* = 41; females, 104 ± 13.11 (78–124) *n* = 29; im. males, 82; im. females, 98 ± 4.24 (95–101) *n* = 2; im. unsexed, 83.

Males are larger than females and have a slightly concave plastron. Females have a flat plastron. Other sexually dimorphic characters are covered in the diagnosis of the *flavescens* group.

Mahmoud (1969) studied the growth rate at various ontogenetic stages in a natural Oklahoma population of *K. flavescens*. Individuals with CL of 21–40 mm increased 7.7 mm per growing season; those of 41–60 mm increased 7.9 mm; those of 61–80 mm increased 3.1 mm; and those greater than 80 mm increased 2.7 mm.

Hatchlings and juveniles are an exception to most proportional characters stated in the diagnosis and general description. They are virtually round in the dorsal aspect and have a barely evident mid-dorsal keel (Figures 15.4 and 15.5), and M9 and M10 are equal to or slightly lower than M8. M9 and M10 become higher than M8 at a CL of ca. 67 mm (Cahn, 1937).

Examination of eight freshly hatched individuals (JML, notes) from two clutches of eggs (Taylor Co., Texas; Figure 15.5 hatchling) are as follows: Carapace is medium to dark brown with vague, irregular, small black flecking; interlaminal seams are narrowly darkened; and most marginal scutes have pale-orangish edges. Ground color of plastron is pale orangish with a black seam following plastral figure that is widest on mid lobe and narrowest on forelobe. Head and neck have irregular pale markings that usually define at least a vague postorbital stripe and a stripe beginning at jaw angle, both running onto neck.

Females are larger than males at hatching. After 3 years males are larger in all age classes (Iverson, 1991c). The annual growth period is 90 days in Oklahoma (Mahmoud, 1969).

Iverson has studied his Nebraska population since 1981. His complete life tables show a mean cohort generation time of 28 years. This is an important contribution that has rarely been calculated for any species of turtle.

HATCHLINGS AND EMBRYONIC DEVELOPMENT

Two clutches of five eggs each from Texas (see "Growth and Ontogeny") were incubated at ambient lab temperature in Salt Lake City. Hatching occurred after 98 days. In the late stages of development the eggs cracked and vitelline blood and albuminous fluid oozed through the cracks (Figure 3.8). Each umbilicus was closed and yolk sacs were completely internalized at hatching. All of the hatchlings bore atypical, minuscule, barely detectable caruncles. All had virtually transparent lower eyelids through which details of the eye could be seen (see remarks on accessory ocular structures in Introduction and in Legler, 1993b). It was evident that these hatchlings had spent an indeterminate time of quiescence within the egg before actually emerging. Little was known of diapause and embryonic aestivation at the time (in 1975).

PREDATORS, PARASITES, AND DISEASES

Many predators have been documented. Hatchlings and juveniles are taken by black hawks (*Buteo gallus*), fish, water snakes (*Nerodia*), and other turtles. Raccoons (*Procyon lotor*), skunks (*Mephitis mephitis*), rodents, and hognose snakes (*Heterodon nasicus*) are known to be nest predators (Christiansen and Galloway, 1984). Perhaps more adults are killed by traffic than predators as the turtles attempt to cross roads. Populations are adversely affected by pond and swamp drainage, and pollution from agricultural chemicals (Flickinger and Mulhen, 1980).

POPULATIONS

Densities vary among populations, whether these are due to actual differences attributable to habitat quality, food availability, or differential predation is unclear. It is probable that differences in collecting methods and analyses are also causes of variation.

With the exception of a long-term study in the Nebraska Sand Hills (Iverson, 1991c), other estimates of population density or biomass are interesting but probably do not reflect the actual state of these populations. They may be indicative only of ease of capture at a particular time of the year and the period of time over which these methods were used.

Three sites in Missouri (Kangas, 1986) had populations (Schnabel estimates) of 932 (655–1,610), 61 (43–101), and 3. In a 1,050 ha site in Iowa Christiansen, Gallaway, and Bickham (1990) estimated the population to contain 2,925–3,207 individuals. As with many studies this was estimated by extrapolation; only about 33% of the area was sampled. Population density for the entire area was estimated to be three turtles per hectare but could have been as high as nine turtles per hectare if the whole area had been sampled and the turtles were uniformly distributed.

Survivorship in Nebraska from egg to hatchling emergence in spring is 19% (Iverson, 1991c). This includes loss due to egg predation, infertility, over-wintering mortality, and death during migration from nest to water. Survivorship of Nebraska juveniles is 30–70%, 90% at 6 years and 95% for turtles 8 years or older.

CONSERVATION AND CURRENT STATUS

Stable and abundant populations of *Kinosternon flavescens* exist throughout its wide range; there is no known commercial exploitation; and preferred habitat is not limited. Iverson (1991c) has been studying a population estimated to be in the thousands on a wildlife reserve in Nebraska since 1981 (1,702 marked, with 3,640 recaptures as of 1988).

The chief threats to local populations are those from industrial, agricultural, and chemical pollution, as in the populations formerly called *K. f. spooneri* (Smith, 1951) in Iowa. A chemical plant was wary of environmental and legal sanctions if they polluted the habitat of a unique species with a limited range. Iverson (1979b) had concluded that *spooneri* was a unique taxonomic entity. Two additional studies were commissioned by the chemical company (Houseal, Bickham, and Springer, 1982; Bickham, Springer, and Galloway, 1984). Both concluded that *spooneri* was not a unique taxon.

The National Academy of Sciences was asked to make a ruling on the issue. They requested that yet another taxonomic study be undertaken, and this was the well-cited work of Berry and Berry (1984), who also concluded that *spooneri* was not a valid entity. Ergo, the chemical plant proceeded apace and taxonomic science was favored with a rare firm majority (see Dodd 1982, 1983, for details). Much later a molecular analysis (Serb, Phillips, and Iverson, 2001) again showed that *spooneri* was not unique but that the three remaining subspecies (*flavescens*, *arizonense*, and *durangoense*) were distinct enough to be recognized as full species.

Kinosternon arizonense Gilmore, 1922

Arizona mud turtle (Iverson, 1992a).

ETYMOLOGY

The specific name *arizonense* refers to the state of Arizona where the type specimen was obtained.

HISTORICAL PROLOGUE

Gilmore (1922) first described *K. arizonense* from late Pliocene (Blancan) fossils—two nearly complete shells from a quarry ca. 3 km S of Benson, Cochise Co., Arizona. The holotype (USNM 10463) is an adult male. Iverson (1978, 1979a, and 1979b) established that *arizonense* was conspecific with extant populations of *flavescens* in the southwesternmost part of the range and treated *Kinosternon flavescens stejnegeri* (Hartweg, 1938) as a synonym of *arizonense*. The taxa *arizonense* and *durangoense* were considered to be subspecies of *Kinosternon flavescens* until the molecular analysis of Serb, Phillips, and Iverson (2001) justified their elevation to species rank. Nearly all the life history information for *arizonense* results from data collected in a 3-year period by Iverson (1989b). See also the account of *K. flavescens*.

DIAGNOSIS

(Diagnosis and description based mainly on Iverson, 1991c; Berry and Berry, 1984; and JML databases.)

FIGURE 16.1 *Kinosternon arizonense*. Habitat, southern Arizona. (Photo by P. P. van Dijk, used with permission.)

FIGURE 16.2 *Kinosternon arizonense*. Adult surfacing in pond (Figure 18.01). (Photo by P. P. van Dijk, used with permission.)

FIGURE 16.3 *Kinosternon arizonense*. Juvenile in first year of growth; Quijotoa, Pima Co., Arizona (Photo by Cecil Schwalbe).

Adult size is intermediate to large: maximum CL is 181 mm in males and 167 mm in females (see "Growth and Ontogeny," following). M10 is variable and may or may not be higher than M11. Anterior plastral lobe is wide at hinge and is 68–79% of CW. Posterior plastral lobe is narrow with a maximum width of 59–67% of CW. Plastral scute 1 is of intermediate length, 52–60% the length of the forelobe. Plastral scute 6 is relatively long, 60–70% of the posterior lobe. Axillary and inguinal scutes are nearly always (>90%) in contact. C1 and M2 are not in contact.

GENERAL DESCRIPTION

BASIC PROPORTIONS FOR ADULTS

CW/CL: males, 0.69 ± 0.04 (0.65–0.75) $n = 6$;
females, 0.70 ± 0.02 (0.66–0.76) $n = 23$.

CH/CL: males, 0.37 ± 0.01 (0.36–0.39) $n = 6$;
females, 0.40 ± 0.02 (0.37–0.45) $n = 23$.

CH/CW: males, 0.54 ± 0.04 (0.47–0.57) $n = 6$;
females, 0.57 ± 0.03 (0.51–0.64) $n = 23$.

PLASTRAL PROPORTIONS

(Iverson, 1991b)

PL/CL: 0.911 males, 0.950 females.

FL/CL: 0.337 males, 0.341. females.

HL/CL: 0.284 males, 0.345 females.

BR/CL: 0.227 males, 0.255 females.

PS1/CL: 0.284 males, 0.345 females.

PS6/CL: 0.227 males, 0.228 females.

Modal plastral formula: 4 > 6 < = > 1 > 2 > 5 > 3.

Adults differ from *flavescens* and *durangoense* by having a narrower carapace, narrower C1 (21–27% of CL), and high-domed cross-section. Plastral coverage is more extensive than either *flavescens* or *durangoense* (as reflected in the lengths of the hindlobe and bridge).

Shell is medium dark, usually olive, brown, or yellow-brown. The marginals are often marked with yellow. The plastron is yellow, often with black along the seams. The dorsal surface of the head is gray or brown. The sides of the head are plain yellow or cream, distinguishing it from the sympatric *K. sonoriense*, which has reticulations on the head and neck.

FOSSILS

Fossils (other than the types) are known from the Pleistocene of Cochise Co., Arizona (Gilmore, 1922). Van Devender, Rea, and Smith (1985) reported Pleistocene fossils from Sonora.

GEOGRAPHIC DISTRIBUTION

Map 11

The species ranges from southwestern Arizona (Rio Sonoyta), southward to the upper reaches of the Yaqui, Matape, Magdalena, Sonora, and Guaymas drainages in Sonora, at elevations below 1,000 m (Iverson, 1989b, 1992a; Berry and Berry, 1984). The species no longer occurs in the immediate vicinity of the type locality (San Pedro river and valley) or in the Gila drainage. Sandy, porous soils near the coast are too well drained to support temporary ponds. *Kinosternon arizonense* has been found in microsympatry with *K. alamosae* in Sonora, and Vogt found them with *K. sonoriense* at Quitobaquito Springs, Pima Co., Arizona (Vogt, pers. obs., 1968).

HABITAT

Kinosternon arizonense occurs and thrives in temporary ponds, cattle tanks, and roadside borrow pits at elevations of from 200 to 800 m in the Sonoran Desert. They are sometimes found in permanent lentic situations, but according to Iverson (1989b) they are rarely found in permanent streams and rivers. Historically they probably only occurred in pools, arroyos, riverbed oxbows, and playas that held water only during the wettest parts of the year. They now thrive in various manmade ponds, stock tanks, and roadside ditches.

DIET

Arizona mud turtles are carnivorous, often occupying the top of the food chain in small, temporary pools. They feed on adult and larval anurans; dytiscid, hydrophilid, and other aquatic beetles; dragonfly nymphs; and fairy shrimp. They probably feed diurnally (Iverson, 1989b).

HABITS AND BEHAVIOR

Activity is chiefly diurnal. Iverson (1989b) collected few at night. The activity period in the Sonoran Desert is from early July to mid August, but they become active whenever torrential summer rains drench their terrestrial aestivation sites. They often migrate overland between water bodies. Vogt (pers. obs.) often found them crossing roads during summer rains in Pima Co., Arizona. They frequently bask, particularly juveniles, out of the water on logs or branches at air temperatures is as high as 45°C (Iverson, 1989b).

When temporary aquatic habitats dry up, they aestivate until the next heavy rains occur, a period that could exceed 1 year (Iverson, 1989b). Details of aestivation have not been recorded.

REPRODUCTION

Females reach reproductive maturity at sizes of 120 to 130 mm CL (115 to 125 mm PL) and ages 6–10 years (Iverson, 1989b).

The activity and reproductive cycles of *K. arizonense* differ distinctly from those of *flavescens*. They are most active from July to the middle of October. A copulating pair was collected in July 1955 in Quitobaquito Springs, Organpipe Cactus National Monument, Pima Co., Arizona, by P. W.

Smith and M. Hensley (1957). Iverson (1989b) also reports courtship and copulation of this species in July. Nesting occurs between 1 July and 15 August, during which two to three clutches are produced (gonadal data, Iverson, 1989b).

Mean number of eggs per clutch is 4.7 eggs (three to seven). *Kinosternon arizonense* eggs are larger than those of *flavescens*, as follows (means and extremes): L, 31.6 ± (29.2–35.2) $n = 24$; W, 17.9 ± (16.2–19.5) $n = 24$; Wt, 6.3 g (5.8–6.8) $n = 8$ (Iverson, 1989b).

One egg from a five-egg clutch was incubated at 25°C and hatched after 320 days (Ewert, 1991). Other eggs incubated at 25°C completed embryonic development in 122–134 days. Arizona mud turtles may complete development, aestivate within the egg, and then hatch only when stimulated by summer rains in the following year, thus emerging a year or slightly more after laying (Ewert, 1985, 1991). This hatching pattern would be synchronous with natural weather patterns, where eggs laid in July and August of 1 year would hatch at the beginning of the rainy season (July to August) the following year.

Females invest about 4.9% of their gravid body weight in a clutch of eggs (Iverson, 1989b).

GROWTH AND ONTOGENY

Data on adult size (Berry and Berry, 1984) are as follows: males, 140 ± 25.1 (98–161) $n = 6$; females, 136 ± 11.8 (113–158) $n = 23$.

Adult sizes vary considerably among populations. Iverson (1989b) found the mean CL of four populations to vary from 140 to 165 mm in males (extremes 117 and 181 mm, $n = 52$). There is less variation in females (means, 131–136).

Males average larger size than females. Length of plastral hindlobe is shorter than that in other members of the group (males 28% of CL, females 35% of CL). The two smallest individuals collected by Iverson (1989b) were 24.8 and 26.9 mm CL. Hatchlings from lab incubated eggs were 27.0 mm CL (24.9–28.4), with a maximum weight of 5.1 g (Iverson, 1989b).

Iverson (1989b) provides the only data on growth in this species. The following data are extracted from his Table 2 and are based on plastral length (mean PL in millimeters followed by percentage increment from previous season of growth):

Age 2, males 41, 136%	females 46, 165%
Age 4, males 71, 19%	females 74, 21.1%
Age 6, males 87, 9.3%	females 88, 10.0%
Age 8, males 107, 12.1%	females 102, 2.9%
Age 10, males 122, 3.0%	females 112, 1.1%

In his 3-year study Iverson (1989b) found that adults greater than 124 mm PL grew 0.53–1.25 mm/year. One female of 104 mm PL grew exceptionally fast to 117 mm at a rate of 4.4 mm/year.

POPULATIONS

Iverson (1989b) trapped 25 turtles in a 0.15 ha pond in 45 minutes, calculating a biomass of 58.3 kg/ha. However, the actual total weight of the turtles captured was only 8.8 kg.

The high densities observed and calculated for desert-adapted species reflect the crowding of small bodies of water as they begin filling; densities would, of course, be lower as the pools enlarge and would return to nil when they dry up. Although of interest, these densities cannot be compared with those for turtles inhabiting permanent bodies of water.

CONSERVATION AND CURRENT STATUS

Iverson (1989b) found Arizona mud turtles to be much more common than previously thought, and that populations are not endangered either in Arizona or Sonora, Mexico. In many areas turtles are probably more common now than they were in the past. Optimum habitat has increased with the abundance of earthen stock tanks that retain water during the rainy season.

Kinosternon durangoense Iverson, 1979b

Durango mud turtle (Iverson, 1979b); casquito de Durango (Liner, 1994).

ETYMOLOGY

The specific name is a toponym referring to the State of Durango, from which the species was first described.

HISTORICAL PROLOGUE

See account of *K. flavescens*.

DIAGNOSIS

Diagnosis and description based mainly on Iverson (1979b, 1991b), Berry and Berry (1984), and JML databases.

The largest adult size in the *flavescens* group: maximum CL is 211 in males and 148 in females. M10 is usually (86–88%) higher than M11. Anterior plastral lobe is wide at hinge and is 64–69% of CW. Posterior plastral lobe is 55–67% of maximum CW. Plastral scute 1 is 60–63% of anterior lobe. Plastral scute 6 is 58–64% of posterior lobe. Axillary and inguinal scutes are usually (75–86%) in contact. C1 is usually (>86%) in contact with M2.

GENERAL DESCRIPTION

BASIC PROPORTIONS FOR ADULTS

CW/CL: males, 0.74 ± 0.02 (0.71–0.77) $n = 8$;
females, 0.78 ± 0.04 (0.73–0.81) $n = 6$.

CH/CL: males, 0.35 ± 0.01 (0.33–0.37) $n = 8$;
females, 0.39 ± 0.02 (0.36–0.42) $n = 6$.

FIGURE 17.1 *Kinosternon durangoense*. Head of subadult male. (Courtesy of J. B. Iverson.)

FIGURE 17.2 *Kinosternon durangoense*. Adult male. (Courtesy of J. B. Iverson.)

CH/CW: males, 0.47 ± 0.02 (0.43–0.51) n = 8;
females, 0.50 ± 0.04 (0.47–0.56) n = 6.

PLASTRAL PROPORTIONS

(Iverson, 1991b)

PL/CL: 0.898 males, 0.920 females.

FL/CL: 0.314 males, 0.320 females.

HL/CL: 0.328 males, 0.357 female.

BR/CL: 0.208 males, 0.237 females.

PS1/CL: 0.328, 0.353.

PS6/CL: 0.162, males, 0.205 females.

Modal plastral formula: 4 > 6 > 1 > 5 > 2 > 3.

The carapace is relatively wider and C1 is wider (27–31% of CL) than that in *K. arizonense* adults. *Kinosternon durangoense* is morphologically closer to *arizonense* than to *flavescens*. Iverson (1979b) suggested this could be a convergent adaptation to similar xeric environments. Plastral coverage is slightly greater than that in *K. flavescens*, as reflected in lengths of hindlobe (HL) and bridge (BR). Carapace is horn-colored; head is same color as that in *flavescens* (Figures 17.1–17.3).

FOSSILS

Van Devender, Rea, and Smith (1985) reported Pleistocene fossils from Sonora.

GEOGRAPHIC DISTRIBUTION

Map 11

The species is an isolated population occurring in the lower Rio Nazas and the southwestern part of the Bolsón de Mapimí (east of Ceballos), in the tri-state area where SW Coahuila, SE Chihuahua, and NE Durango share a common point, at elevations of 1,000–1,600 m (Berry and Berry, 1984; Iverson, 1979b; Iverson,1992a). Distribution maps exist in Iverson (1979b) and Houseal, Bickham, and Springer (1982).

HABITAT

Kinosternon durangoense occurs in the grasslands of the Chihuahuan Desert (Aguirre-León, Morafka, and Adest, 1997). This is a moderately warm region with harsh alkaline soils, low relative humidity and low precipitation, high solar radiation, and wide temperature fluctuation. A strong seasonality includes dry winters with evening frost; little precipitation from November through March; a dry, warm spring (April through May); and hot, moist summers (June–October) (Morafka, 1982). Average annual rainfall is 271 mm, 74% of which occurs in September.

Vegetation is a patchy distribution of thorn scrub. Dominant plant species are creosote bush (*Larrea tridentata*), prickly pear cactus (*Opuntia rostrata*), mesquite (*Prosopis glandulosa*), tar bush (*Flourensia cernua*), tobosa grass (*Hilaria mutica*), mallow (*Sphaeralcea angustifolia*), and grama grass (*Bouteloua* sp.) (Morafka, Berry, and Spangenberg, 1997; Figure 17.4).

Kinosternon durangoense has been collected in or near temporary playa lakes, but only during the height of the rainy season—mid July to early November—when there is standing water in the playas (Iverson, 1979a). This season of activity differs markedly from *flavescens* in the north, which is active in spring and early summer but aestivates in mid summer and early fall (Mahmoud, 1969; Christiansen and Dunham, 1972).

Little else is known of life history, but similarities to that of *K. arizonense* might be expected. Iverson (1979b) predicts that *durangoense* aestivates for much of the year.

FIGURE 17.3 *Kinosternon durangoense*. Plastron of large adult male. (Courtesy of J. B. Iverson.)

FIGURE 17.4 Wet season habitat of *Kinosternon durangoense* near Pedriceña, Durango. (Courtesy of J. B. Iverson.)

GROWTH AND ONTOGENY

Adult size data: males, 139 ± 13.4 (118–158) *n* = 8; females, 120 ± 17.0 (96–148) *n* = 7 (Berry and Berry, 1984).

CONSERVATION AND CURRENT STATUS

Stable and abundant populations of *Kinosternon durangoense* exist within its small range, and it is not known to be exploited by humans or affected by pollution. The generalized nature of its habitat is not limited, and no special concerns for conservation or management need be addressed.

The *Kinosternon scorpioides* Group

This assemblage includes 7 of the 16 species occurring in Mexico and is distinguished by a shared, derived character—the lack of clasping organs in both sexes (see Family Kinosternidae). The loss of clasping organs may be correlated with increased plastral coverage (Iverson, 1981a, 1988b, 1991b; Berry,1978). The *scorpioides* group includes the species *scorpioides, integrum, alamosae, chimalhuaca, oaxacae, acutum,* and *creaseri*. All occur in Mexico and all, except *scorpioides*, occur only in Mexico and adjacent Guatemala and Belize.

Berry (1978) presented a series of logical hypotheses on the origin and dispersal of *Kinosternon scorpioides* and *K. integrum* (and their derivatives) from a common ancestral stock in the mesic lowlands of southern Mexico/ northern Central America in middle Cenozoic times. Physiographic and climatic events (uplift and aridity in Miocene) divided the ancestral stock at the Isthmus of Tehuantepec, "forming a southern Mexican group and a nuclear Central American group." The SE Mexican population became ancestral *K. integrum* in the southern Pacific coastal region. This stock moved northwestward along the coastal plain and colonized most Pacific drainages as far as the Yaqui basin. Concomitantly, the ancestral stock colonized the Mesa Central via the extensive Balsas, Lerma-Santiago, and San Pedro-Mezquital systems (Berry, 1978, Figure 35). Ultimately this stock reached the headwaters of the Tamesi-Panuco and Papaloapan drainages, by overland migration and/or by headwater stream piracy. It remained in the upper reaches (above 800 m) of these Gulf drainages. Thus far there seem to have been no exchanges with Gulf Coastal streams entering the sea between the Tamesi-Panuco systems and the R. Papaloapan.

Following the generally northward migration of *K. integrum*, small populations were isolated along the Pacific versant and were differentiated into three species: *K. alamosae, K. chimalhuaca,* and *K. oaxacae*—all of which were discovered and described after 1980.

Kinosternon alamosae is the oldest and most distinct of these derivatives and has adapted to xeric conditions in northwestern Mexico. It is also the only one of the derivatives that occurs in sympatry with the parent species, *integrum*. Its zoogeographic history remains unclear. *Kinosternon chimalhuaca* and *K. oaxacae* were isolated from the main *integrum* stock on relatively small stretches of the Pacific coastal plain as interglacial sea levels rose.

The nuclear Central American group of the *scorpioides* stock was later able to disperse into South America and cross the Isthmus of Tehuantepec, gaining access to the southern Pacific coastal plain, the Mexican Gulf Coast, and the Yucatán Peninsula, giving rise to *K. s. cruentatum, K. creaseri,* and *K. acutum*. Subsequent dispersal northward to the Soto la Marina drainage in Tamaulipas involved circumvention of Punto del Morro at low sea levels.

Kinosternon scorpioides (Linnaeus) 1766

Red-cheeked mud turtle (Iverson, 1992a); scorpion mud turtle (Berry and Iverson, 2001b); tortuga de pecho quebrado (Liner, 1994).

ETYMOLOGY

L. *scorpio*; Gr. *skorpion*; seemingly alluding to the tail claw and its fancied similarity to that of a scorpion.

DIAGNOSIS AND DEFINITION

Kinosternon scorpioides (sensu lato)

Overall maximal CL is 205 mm in males and 174 mm in females; M10 is higher than M11; carapace is variably tricarinate; there are five neural bones, and N1 and N5 are isolated from nuchal and pygal contact by mid-dorsal articulation of costal elements; central scutes 1–4 are notched on posterior edges; plastral lobes are extensive enough to close or nearly close the shell; plastral scute 6 is long, 29–31% of CL in adults and almost as long as posterior plastral lobe; plastral scutes 3 and 5 are short, often barely touching at midline or narrowly separated; clasping organs are lacking; both sexes have a tail claw; premaxillary beak is strongly hooked in adult males; there is an anterior pair of large barbels and 2–3 pairs of smaller barbels posteriorly (Berry, 1978; Smith and Smith, 1979).

PLASTRAL PROPORTIONS

(Iverson, 1991b)

PL/CL: males 0.947, females 0.972.

FL/CL: males 0.318, females 0.323.

HL/CL: males 0.362, females 0.371.

BL/CL: males 0.293, females 0.310.

PS1/CL: males 0.159, females 0.166.

PS6/CL: males 0.303, females 0.325.

Modal plastral formula: $6 > 4 > 1 > 2 > 5 > 3$.

HISTORICAL PROLOGUE

Kinosternon scorpioides is a polytypic species that reaches the northeasternmost extent of its range in Mexico. Four subspecies are recognized in the most recent review (Berry and Iverson, 2001b). The nominate subspecies, *K. s. scorpioides*, occurs from western Panama into South America discontinuously, as far as 10°S in easternmost Brazil and to about 25°S in the upper Parana drainage of northern Argentina, with a disjunct population at 15°S in northern Bolivia. *K. s. albogulare* occurs from eastern Panama to northern Nicaragua. *Kinosternon s. cruentatum* and *Kinosternon s. abaxillare* are the subspecies occurring in Mexico and the ones considered in the following accounts.

Siebenrock (1906) considered *K. integrum* and *K. scorpioides* to be conspecific. Berry (1978) unraveled some of the complexities of this group in an extensive study of variation in the *Kinosternon scorpioides* and *K. leucostomum* groups; he demonstrated that *scorpioides* and *integrum* were distinct species and that *cruentatum* and *abaxillare* were subspecies of *scorpioides*. He also observed that there were at least three other identifiable species in the *scorpioides* group that were later described as *K. alamosae*, *K. chimalhuaca*, and *K. oaxacae* (see separate accounts).

GEOGRAPHIC VARIATION

Kinosternon scorpioides (sensu lato) is quite variable over its entire range and within its relatively small range in Mexico. It is a taxon with which we have little firsthand experience. For Mexico, most of our observations are based on a series of specimens obtained in Yucatán and Quintana Roo by Legler in 1966, and on Iverson (2010).

Kinosternon scorpioides cruentatum (Duméril and Bibron) 1851

Red-cheeked mud turtle (Liner, 1994); *chachagua* local name.

ETYMOLOGY

L. *cruentus*, *cruentatas*, made bloody or spotted with blood; in reference to the bright red or orange coloration on the head and limbs of some individuals, particularly males (Figure 18.3).

GENERAL DESCRIPTION

Adults are smaller in Mexico than farther to the SE. The largest *K. scorpioides* recorded for Mexico is a female of 148 mm CL (extreme northern end of range, Tampico Embayment) and a male of 166 mm CL (Veracruz lowlands; Iverson, 2010). Most of the adult *K. s. cruentatum* ($n = 268$) in the JML database rank from slightly above the median (Pacific coastal drainages) to well below the median in a database of 2,765 Mexican *Kinosternon*. The smallest Mexican *K. scorpioides* occur on the lower Yucatán Peninsula. There is a general increase in adult size through Central America and thence into South America.

BASIC PROPORTIONS

(Yucatán Peninsula, JML Database)

CW/CL: males, 0.64 ± 0.02 (0.60–0.69) $n = 55$;
females, 0.67 ± 0.03 (0.62–0.77) $n = 28$;
im., 0.68 ± 0.03 (0.64–0.72) $n = 5$.

CH/CL: males, 0.40 ± 0.02 (0.35–0.43) $n = 55$;
females, 0.43 ± 0.02 (0.37–0.47) $n = 28$;
im., 0.40 ± 0.03 (0.37–0.44) $n = 5$.

CH/CW: males, 0.62 ± 0.03 (0.56–0.69) $n = 55$;
females, 0.64 ± 0.03 (0.57–0.71) $n = 28$;
im., 0.59 ± 0.05 (0.54–0.65) $n = 5$.

FIGURE 18.1 Dorsal, ventral, and lateral head views of *Kinosternon scorpioides cruentatum* from vicinity of Tonalá, Chiapas: left—UU 7631, male, CL 116 mm; right—UU 7633, female, CL 103 mm. (From Berry and Legler, 1980, with permission.)

BASIC PROPORTIONS: PACIFIC COASTAL CHIAPAS

(Courtesy N. Lopéz-León)

CW/CL: males, 0.67 ± 0.023 (0.61–0.73) n = 117; females, 0.71 ± 0.023 (0.66–0.80) n = 138; im. unsexed, 0.72 ± 0.027 (0.69–0.76) n = 10.

CH/CL: males, 0.43 ± 0.023 (0.34–0.50) n = 117; females, 0.48 ± 0.024 (0.42–0.56) n = 138; im. unsexed, 0.43 ± 0.036 (0.37–0.48) n = 10.

CH/CW: males, 0.65 ± 0.033 (0.53–0.74) n = 117; females, 0.68 ± 0.036 (0.59–0.80) n = 138; im. unsexed, 0.60 ± 0.069 (0.50–0.68) n = 10.

Pacific populations (Figures 18.1, 18.6, and 18.7) are larger, relatively wider, and relatively higher in cross-section and have brighter red markings on the head compared to those on the Yucatán Peninsula (Figures 18.3–18.5).

Modal plastral formula: 6 > 4 > 1 > 2 > 5 > 3.

Carapace is moderately to strongly tricarinate in all but largest individuals. Shell is relatively high; CH/CL is 43% in males and 48% in females. Carapace is oval in the dorsal aspect. C1 is wider than long; central scutes 1–4 have distinct notches at their tapered posterior edges; M10 is always higher than M9 and usually (73%) higher than M11. Margins of carapace are distinctly flared outward in some populations (e.g., Isla Cozumel) but not in others (Veracruz, Chiapas).

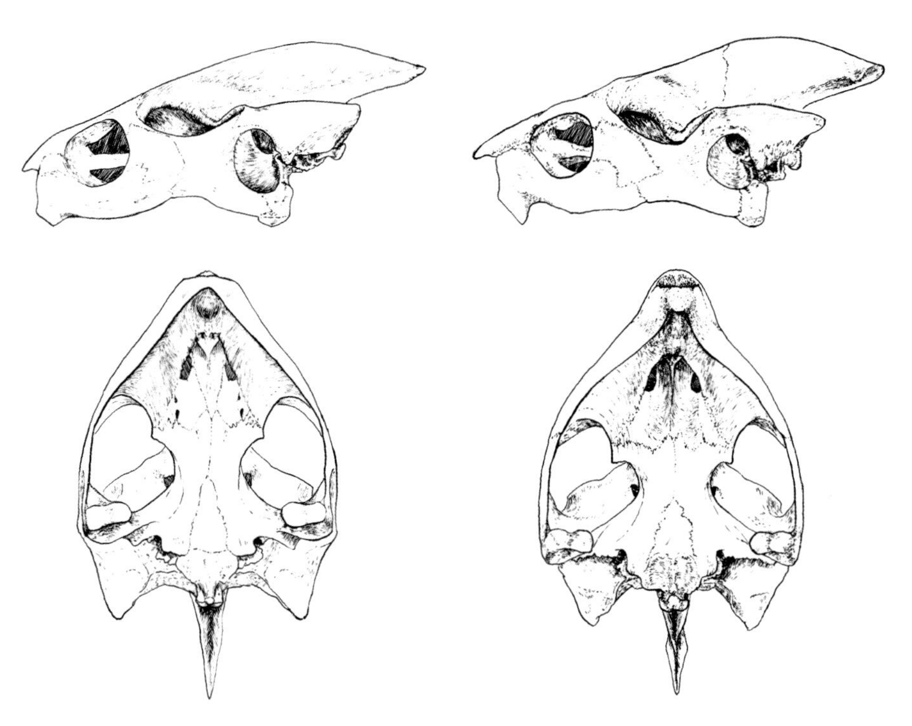

FIGURE 18.2 Lateral and ventral views of adult *Kinosternon scorpioides (albogulare)* skulls (Dominical, Puntarenas Prov., Costa Rica): left—female, UU 7552, CBL 43 mm; right—male, UU 7567, CBL 33 mm. (Drawings by Elizabeth Lane.)

Width of posterior plastral lobe is 47% of CL in males and 48% of CL in females; plastron is extensive and usually completely closes the shell. Bridge length is 29% of CL in males and 31% of CL in females; axillary scute is present and rarely in contact with inguinal. Plastral scute 6 is long, ranging from 19% to 39% of CL (mean 30%); seam between third plastral scutes is very short or absent, with the scutes not or barely meeting on the midline. Length of plastral scute 4 is 26% of CL in males and 28% of CL in females. Anal notch is minuscule or absent (Berry and Iverson, 2001b).

Color of carapace varies widely from population to population—usually horn colored or pale brown with darker seams in Veracruz and Chiapas, and olive to black on Cozumel. Plastral color varies from population to population and may be yellow, orange, brown, gray, or black, with seams usually darker. Degree of plastral closure varies from complete (Veracruz) to partial (Cozumel). Head shield is bell shaped, rhomboidal, or triangular, but rarely V-shaped. Maxillary sheaths are hooked, more strongly so in adult males than in females and juveniles.

Color pattern of the head, neck, and limbs is extremely variable. Head markings are red, orange, yellow, or with paler dots or reticulations on a darker ground (brown, gray, or olive). Limbs usually are gray, brown, or black and may be spotted with small darker spots. Barbels are present in three or four pairs, with the anterior pair being largest.

FOSSILS

Kinosternon scorpioides is known from the Pleistocene of Yucatán (Kuhn, 1964; Langebartel, 1953).

GEOGRAPHIC DISTRIBUTION

Map 12

Kinosternon scorpioides cruentatum occurs in Gulf drainages from the Rio Soto la Marina in Tamaulipas southward to just south of the Rio Papaloapan, along the Pacific coastal plain from the Rio Tehuantepec (Oaxaca) into Guatemala and El Salvador and, on the northern two-thirds of the Yucatán Peninsula and adjacent to Guatemala and Belize, thence to Honduras. There is a paucity of records for the Usumacinta-Grijalva lowlands of Tabasco and Chiapas (Smith and Smith, 1979; Mertens and Zilch, 1952; Iverson,1992a). In Mexico its range overlaps that of *Kinosternon creaseri* and *Kinosternon acutum*, but it is unlikely that microsympatry with these taxa occurs. *K. scorpioides* occurs in some of the same Gulf Coastal drainages as *K. integrum* but occupies the lower reaches, whereas *integrum* occurs in tributaries on the Mesa Central. Overall the ranges of *scorpioides* and *integrum* are almost exactly mutually exclusive in Mexico.

HABITAT

Kinosternon scorpioides lives in permanent and ephemeral water bodies at low elevations, including small streams,

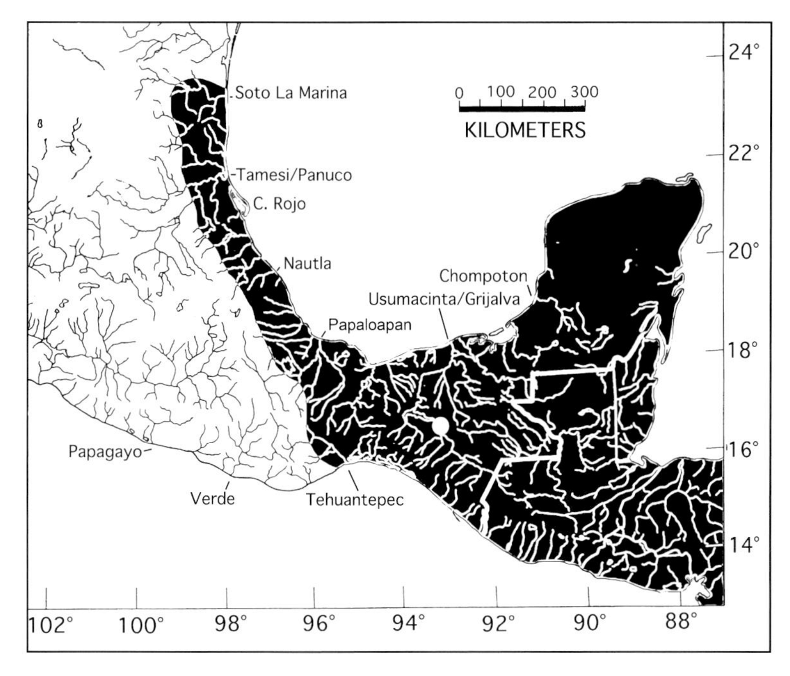

MAP 12 Geographic distribution of *Kinosternon scorpioides* in Mexico. Most of the blackened area is for *Kinosternon s. cruentatum*; the large white dot represents the immediate environs of Tuxtla Gutierrez, Chiapas, and the entire range of *Kinosternon s. abaxillare*.

oxbow lakes, marshes, and ponds. In Yucatán, Quintana Roo, and Belize they also occur in open cenotes (see account of *Trachemys scripta venusta*). On Isla Cozumel *Kinosternon scorpioides cruentatum* was the most abundant turtle in marshy ponds (Vogt, pers. obs., 1979–2010).

DIET

The diet is omnivorous and includes fruits, seeds, algae, frogs, shrimp, crabs, and fish (Álvarez del Toro, 1960). Moll (1990) took samples of stomach contents from 80 adult *K. s. cruentatum* in a slow-moving river (Belize) from January to April 1984. By volume there was slightly more animal (65%) than plant material. Identifiable food items were insects (50.2%), snails (4%), fish (1.2%), *Paspalum peniculatum* (12.2%), and *Elodea densa* (2.6%). Insects were found in all of the stomachs, snails in 76.3%, *Elodea* in 50%, *Paspalum* in 23.8%, and fish in 10%.

Legler noted that adults from the lower Yucatán Peninsula, which were not eating mollusks, seemed to have narrower heads than those from Belize that were eating mollusks. James F. Berry (pers. comm., June 1978) demonstrated statistically that this was true.

HABITS AND BEHAVIOR

Aestivation occurs in the bottom mud of ponds when water levels drop. The turtles emerge when rain fills the ponds. One can often see trails in the mud when the water is only several centimeters deep and follow these trails across mud flats to more distant refugia (Vogt, pers. obs.). When threatened by predators *K. s. cruentatum* closes the plastron tightly and remains closed until danger has passed. They are not prone to bite when handled.

REPRODUCTION

In an analysis of *K. s. cruentatum* on the Yucatán Peninsula (Iverson, 2010), the three largest immature females were 98, 98, and 88 mm CL. The four smallest gravid females were 104–109 mm CL. Age of two immature subadults (86 and 98 mm CL) was estimated (from growth rings) to be 8 years. The foregoing data suggest that females in this region mature at ca. 100 to 105 mm CL and 9–10 years of age (Iverson, 2010).

In a series of 88 *K. scorpioides* from the Yucatán Peninsula (JML, UU) females were mature at 104 mm and males at 101 mm CL.

Courtship and copulation have been observed in October and November in Venezuela (see Sexton [1960] for detailed account). Nesting has been reported in Northern Guatemala from March to May with 6–10 eggs per clutch (Campbell, 1998). Nesting occurs in March and April in Chiapas (Álvarez del Toro, 1982).

Kinosternon scorpioides has been observed active on the Yucatán Peninsula in every month of the year except September. It is seemingly capable of activity whenever standing water is available. The nesting season extends over at least 10 months of the season of activity. Females (n = 26) collected by JML (12–31 July 1966) bore no corpora lutea, shelled eggs, or ovulatory follicles; only slightly enlarged (2–3 mm) follicles were present.

FIGURE 18.3 *Kinosternon scorpioides cruentatum*. Libre Union, Yucatán, Mexico, UU: dorsal and ventral views of adults—top, males; bottom, females.

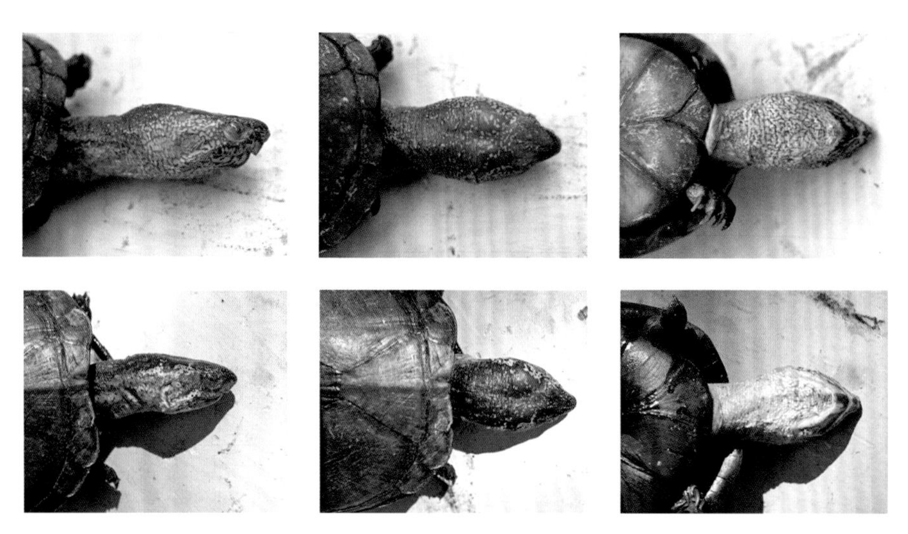

FIGURE 18.4 *Kinosternon scorpioides cruentatum*. Lateral, dorsal, and ventral views of adult heads (Libre Union, Yucatán): top, male; bottom, female.

These data fill a gap in data reported by Iverson (2010), who predicted that *K. s. cruentatum* would begin vitellogenesis in July. He found oviducal eggs, fresh corpora lutea, or ovarian follicles of preovulatory size in nearly all months from August to May. The preceding combined data therefore suggest that laying begins as early as August and extends at least until May, and that ovarian activity may be suspended in June and July. Iverson's data further indicate a typical production of two to four (possibly five) clutches per annum.

Modal number of eggs per clutch was two eggs (21 of 37 clutches); means were 2.22 ± 0.44 (2–3) n = 9, based on oviducal eggs and corpora lutea and 2.04 ± 0.88 (1–4) n = 23 based on follicular size classes (Iverson, 2010).

Size and weight of all eggs from southeastern Mexico to Costa Rica (subspecies *cruentatum* and *albogulare*; compilation of data from Castillo-Centeno, 1986, and Iverson, 2010) were as follows: L, 33.6 ± 2.57 (24.2–39.7) n = 348; W, 17.9 ± 1.05 (14.8–20.1) n = 345; Wt, 6.72 g ± 1.13 (3.82–9.45) n = 345.

Size and weight of 19 eggs from the Yucatán Peninsula (Iverson, 2010) were as follows: L, 31.4 mm ± 1.8 (29.6–34.8); W, 16.9 mm ± 0.6 (15.8–17.5); WT, 5.41 g ± 0.21 (5.19–5.86) n = 8. In general, the eggs of *K. s. cruentatum* from Mexico are smaller than those of *K. s. scorpioides* in South America and are small for *Kinosternon* in general (Iverson, 2010: Table 3).

Ewert (1979) gave the components of a 7 g *K. scorpioides* egg by dry mass as follows: fibrous layer, 16%; mineral

FIGURE 18.5 *Kinosternon scorpioides cruentatum.* Dorsal and ventral views of immature specimens (Libre Union, Yucatán, Mexico, UU): top, juvenile in year of hatching; bottom, juvenile in second or third year of growth.

layer, 84%. The brittle egg shell had a thickness of 0.35–0.51 mm.

Ewert (1979) reported incubation times of 176 days at 25°C and 91 days at 30°C. He also reported that the recently hatched neonate of 3.7 g represented 60.9% of the egg weight.

Sex is determined by incubation temperature (Ewert and Nelson, 1991). Incubation is complex and involves embryonic diapause and aestivation (Ewert, 1991).

CAPTIVE STUDIES

Goode (1994) studied reproduction of Mexican *K. scorpioides* in captivity (a population founded by *K. s. albogulare* from Honduras fide Iverson [2010]). Size of reproductive females ranged from 122 to 146 mm CL. Mean size of hatchlings was 33 mm (27–36) n = 67. Minimum size and age at first reproduction of these hatchlings was 132 mm and 5.32 years of age, which is larger than in the smallest-sized adults Goode received initially. Mean eggs per clutch was three (one to eight) and one to three clutches were laid per season with an inter-nesting interval of 32–35 days. Data on size and weight for 476 eggs (160 clutches) from captive turtles were as follows: L, 32 mm (27–37); W, 18 mm (11–22); Wt, 6.4 g (3.8–9.1). RCM (4.9%) was low compared to that of wild *Kinosternon scorpioides*.

GROWTH AND ONTOGENY

Data on size, Yucatán and Quintana Roo (JML database): males, 114 ± 5.2 (101–128) n = 55; females, 116 ± 8.69 (94–128) n = 28; im., 92 ± 11.37 (79–106) n = 5.

Iverson (2010) recorded a female of 141 mm CL from the Yucatán Peninsula.

Data on size, Pacific drainages (Lopéz-León, 2008): males, 120 mm ± 6.4 (98–134) n = 118; females, 121 mm ± 6.7 (103–138) n = 139; im. unsexed, 77 mm ± 20.2 (38–102) n = 11.

The plastron of adult males is smaller than that of females, particularly the posterior lobe. Anal notch, if present, is deeper in males than females. Color patterns are more brilliant in males. The tip of the tail in both sexes has a terminal claw, which is small in females, and there is a heavily developed claw at the end of a muscular, prehensile tail in males. Both sexes lack clasping organs. Adult size is seemingly not significantly dimorphic (Iverson, 2010).

HATCHLINGS

Mean CL for eight neonates (Yucatán Peninsula) was 27.3 mm ± 3.2 (23–34). Two of these showed no evidence of growth and were collected on roads during rains with larger numbers of *Kinosternon creaseri* (Iverson, 2010).

GEOGRAPHIC VARIATION IN SIZE WITHIN MEXICO

We have been aware that adult size and head coloration are geographically variable in *K. scorpioides* in Mexico, and small adult size and relatively dull coloration is characteristic of populations on the Yucatán Peninsula. Iverson (2010) has quantified this in a very useful comparative table to which we have added recent data from our own databases and that of Lopéz-León (2008).

TAMPICO EMBAYMENT (IVERSON, 2010)

Males, 127.9 mm ± 6.8 (120–139) n = 9; females, 129.8 mm ± 18.5 (93–148) n = 13.

VERACRUZ LOWLANDS (IVERSON, 2010)

Males, 130.3 mm ± 24.3 (108–166) n = 5; females, 116.3 mm ± 8.9 (107–131) n = 8.

MEXICAN PACIFIC CHIAPAS (IVERSON, 2010)

Males, 115.6 mm ± 8.4 (98–130) n = 28; females, 119.9 mm ± 9.4 (91–135) n = 47.

MEXICAN PACIFIC (BIOSPHERE RESERVE, ENCRUCIJADA, CHIAPAS ON PACIFIC COASTAL PLAIN BETWEEN PIJIJIAPAN AND MASTEPEQUE; LOPÉZ-LEÓN, 2008)

Males, 120 mm ± 6 (98–134), Wt, 282 g ± 43 (140–380) n = 118; females, 117 mm ± 14 (103–138), Wt, 308 g ± 82 (195–442) n = 153.

FIGURE 18.6 *Kinosternon scorpioides cruentatum*. Tonalá, Chiapas: adult male showing bright head coloration.

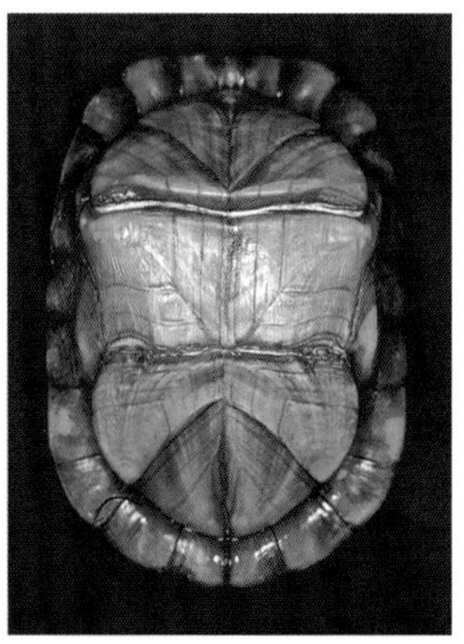

FIGURE 18.7 *Kinosternon scorpioides cruentatum*. Tonalá, Chiapas, three views of adult male, UU.

YUCATÁN PENINSULA (INCL. BELIZE) (IVERSON, 2010)
Males, 108.0 mm ± 9.7 (96–132) n = 45; females, 110.1 mm ± 10.9 (97–141) n = 53.

YUCATÁN PENINSULA (JML DATABASE)
Males, 114 mm ± 5.2 (101–128) n = 55; females, 116 mm ± 8.69 (94–128) n = 28; im., 92 mm ± 11.37 (79–106) n = 5.

K. S. ABAXILLARE, CHIAPAS (IVERSON, 2010)
Males, 122.4 mm ± 15.7 (88–149) n = 21; females, 118.6 mm ± 14.6 (84–153) n = 36.

CAPTIVE STUDIES

Goode (1994) reported on growth and maturity of a breeding captive colony from hatching to maturity. Most hatchlings reached maturity in 5.32 years at a minimum CL of 132 mm. Some of the hatchlings reached this size in 3–4 years but did not reproduce until year 5. Growth slowed after the 5th year, and total growth increments were no more than 4 mm between the ages of 5 and 9 years. Sexually dimorphic characters became apparent at 118 mm CL and 4.2 years in captive males and at 122 mm CL in a male from the wild. Although females did not increase in length after the 5th or 6th year, growth continued in the form of changes in shell shape (e.g., height and width) that increase total volume, bulk, and the capacity to accommodate greater numbers of eggs. The carapace keels diminished with age.

PREDATORS, PARASITES, AND DISEASES

D. Moll (1990) found *Staurotypus triporcatus* to be one of the principal predators of *Kinosternon scorpioides* and *K. leucostomum* in Belize, comprising 10–57% of the volume of total stomach contents in 27 of 40 adult *Staurotypus*.

POPULATIONS

Iverson (1982) reported on a population of *abaxillare* in a Chiapas pond where he found a density of 272.4/ha and a biomass of 59.2 kg/ha (number of turtles and size of pond not given). The turtles had a mean carapace length of 111 mm and a mean mass of 217 g. Dean (1980) also reported on a population of this species in a shallow coastal swamp ("estero") of 1.5 ha in Chiapas. Based on 21 captures, population was estimated at 48.1/ha using the Schnabel method and a biomass of 10.2 kg/ha. The turtles had a mean CL of 110 mm and a mean mass of 210 g.

D. Moll (1990) found *K. s. cruentatum* to be the second most abundant turtle in a 4 ha slow-moving stream community of four species of turtles in Belize. Population size estimates ranged from 35 to 92 individuals between January and April 1984 (similarly, the population size of *K. leucostomum* in the same area and time span varied from 28 to 135 turtles). The populations of these two species were higher in April than in January, and Moll attributed this to turtles migrating into the stream from nearby temporary ponds that had dried due to lack of rain.

Populations of both *K. s. cruentatum* and *K. s. abaxillare* seem to exhibit female-biased sex ratios, whereas those of *K. s. albogulare* and *K. s. scorpioides* appear to exhibit no such bias (Iverson, 2010: Table 1).

Legler's trapping data from the Yucatán Peninsula are as follows: Aguada Sayusil, Yucatán, 34 males, 21 females, 2 immature males, 6 juveniles = 63 total. Rock Stone Pond

FIGURE 18.8 *Kinosternon scorpioides abaxillare.* Tuxtla Gutiérrez, Chiapas: left—subadult male, UU 12423, CL 112 mm, dorsal,ventral and lateral head views; right—dorsal and ventral views of shell, USNM 7519, subadult male paratype, CL 122 mm; and detail (arrow) of axillary-abdominal seam, immature male, USNM 7527, CL 113 mm. (From Berry, 1978, with permission.)

#1, Belize: 8 males, 21 females, 1 immature male, 1 immature female = 31 total.

CONSERVATION AND CURRENT STATUS

Kinosternon scorpioides abaxillare and *Kinosternon scorpioides cruentatum* are protected in Mexico (see "Conservation" in Introduction).

Kinosternon scorpioides cruentatum is eaten locally throughout its range in Mexico, but its small size and usually low population densities preclude commercial hunting. On the Atlantic coastal plain of Brasil (e.g., Belem) *Kinosternon scorpioides* is hunted specifically for consumption as a delicacy in restaurants.

From 1980 to 1986, the market in Alvarado, Veracruz, openly sold live turtles and the restaurants sold cooked *Kinosternon* sp. by the dozen. Among several thousand adult *Kinosternon* (chiefly *leucostomum*) Vogt found only one *Kinosternon scorpioides cruentatum.* Legler (pers. obs.) saw *K. s. cruentatum* on sale at Tapachula, Chiapas, in April 1962.

Acuña Mesén (1990) reported the effects of fire and drought on a population of *Kinosternon scorpioides albogulare* in Costa Rica. A 3-day fire killed 140 individuals from a marked population of 206. Another 24 died during the rest of the year from

FIGURE 18.9 *Kinosternon scorpioides abaxillare.* Bottom, male; top, female (Tuxtla Gutiérrez, Chiapas).

the effects of the fire or drought-related causes. Mortality was 45% in juveniles, 37% in females, and 18% in males.

Fires are also a problem in Mexico, particularly in Tabasco, but also in Veracruz. Market hunters set fire to marshes and grasslands in April and May, hoping to drive aestivating *Kinosternon* sp., *Claudius*, and nesting *Trachemys* into the open for easy capture. However, more turtles are burned in place than driven to the periphery, and habitat destruction is substantial. The practice will cease only when there is no market for the turtles or when the turtles no longer exist.

Kinosternon scorpioides abaxillare (Baur)

Central Chiapas mud turtle (Iverson, 1992a)

Dr. George Baur based a description of the taxon on a series of 12 shells at the U.S. National Museum in the period 1890–1892. The specimens were from Tuxtla Gutiérrez, Chiapas, obtained by Dr. C. H. Berendt in 1863 or 1864. Baur died (at age 39) after leaving the National Museum and before the manuscript was published. Stejneger (1925) published the description and credited Baur with the discovery.

The type locality and seemingly nearly the entire geographic range of *K. s. abaxillare* is "near" Tuxtla Gutiérrez, Chiapas, and is completely surrounded by the range of *K. s. cruentatum.* The taxon was considered a full species until Berry (1978) showed it to be, at best, a subspecies of *K. scorpioides.*

ETYMOLOGY

L. *ab-*, away, and connoting the lack (or loss) of an axillary scute.

DIAGNOSIS AND DESCRIPTION

Adult size (Iverson, 2010): males, 122 mm ± 15.7 (88–149) $n = 21$; females, 119 mm ± 14.6 (84–153) $n = 36$.

Kinosternon scorpioides abaxillare is a member of the *scorpioides* group, which is known to differ from *cruentatum* and other *scorpioides* in only one character—the lack of a discrete axillary scute, a variation that occurs also in other subspecies of *K. scorpioides.* Iverson (pers. comm.) mentions that they are very different in shape, and the carapace is longer and narrower and not as high domed as in *cruentatum.*

Mean proportions may differ slightly from *cruentatum* as follows: CH/CL is slightly lower, 38% in males and 39% in females; posterior plastral lobe is about the same, 48% of CL in both sexes; bridge is longer, 28% of CL in males and 30% of CL in females; fourth plastral scute is longer, 31% of CL in males and 33% of CL in females; head markings are less colorful, consisting of yellow, cream, or pale gray dots or reticulations on a gray or olive background (Berry and Iverson, 2001b).

A population study of *K. s. abaxillare* (Sanchez-Montero, Romero, Vogt, Lopez-León, and Dadda, 2000) was conducted in 1998–1999 in permanent ponds on an intermittent stream, Piedra Parada, in the municipality of Ocozocoautla, Chiapas. They collected 21 males and 49 females. The population was estimated to have 94 individuals (± 6.42, Schnabel; Macdonald, Ball, and Hough, 1980) based on captures of 25 marked turtles over 312 trap hours. Males were significantly larger in head width, carapace length, and plastron length, but there were no differences in carapace height or width. Most females had five growth rings, and fewer were found with one, eight, and nine rings. Most males were in the four- or five-ring age class, and fewer were found with 3, 7, 8, and 10 rings. This suggests that the population is recuperating from heavy predation on mature adults (Sanchez-Montero, Romero, Vogt, Lopez-León, and Dadda, 2000).

Ecological studies of all *Kinosternon scorpioides* are needed in Mexico. Intensive sampling is needed in the region of Tuxtla Gutiérrez, Chiapas, to determine the true range of *abaxillare.* If *cruentatum* and *abaxillare* appear to represent separate gene pools, tissue samples should be taken for molecular analysis.

Kinosternon integrum (LeConte) 1854

Mexican mud turtle (Iverson, 1992a); casquito de burro (Liner, 1994).

FIGURE 19.1 Dorsal, ventral, and head profile views of adult *Kinosternon integrum* (Rio San Lorenzo, 2 km N of El Dorado, Sinaloa, Mexico): left—male, UU 7778, CL 176 mm; right—female, UU 7783, CL 163 mm. (After Berry and Legler, 1980, with permission.)

ETYMOLOGY

L. *integer,* whole, complete; in probable reference to the large plastron and its complete closure of the shell.

HISTORICAL PROLOGUE

LeConte (1854) described this species from a specimen in the Academy of Natural Sciences, Philadelphia, collected by a Mr. Pease. No catalog number was assigned, and the holotype has been lost (Iverson, Young, and Berry, 1998). The type locality was given as "Mexico." Smith and Taylor (1950b) restricted the type locality to Acapulco, Guerrero. No lectotype has been designated. Because of the similarity of this species to *Kinosternon hirtipes* and *K. scorpioides* (it was at one time considered a subspecies of *K. scorpioides*) and the recent descriptions of *K. alamosae*, *K. oaxacae*, and *K. chimalhuaca*, much of the literature prior to 1978 contains abundant identification errors. Most of these have now been clarified (see "Relationships," following).

Siebenrock (1907) considered *K. integrum* to be a subspecies of *K. scorpioides.* This classification was followed by Wermuth and Mertens (1961), Casas Andreu (1967), Pritchard (1967, 1979), and Morafka (1977a). We subscribe, without discussion, to the classification of Berry, Seidel, and Iverson (1997); Berry and Legler (1980); Iverson (1992a); and Iverson (1991b) in treating *K. integrum* and three closely related taxa (*alamosae*, *oaxacae*, and *chimalhuaca*) as full species within the *Kinosternon integrum* "group."

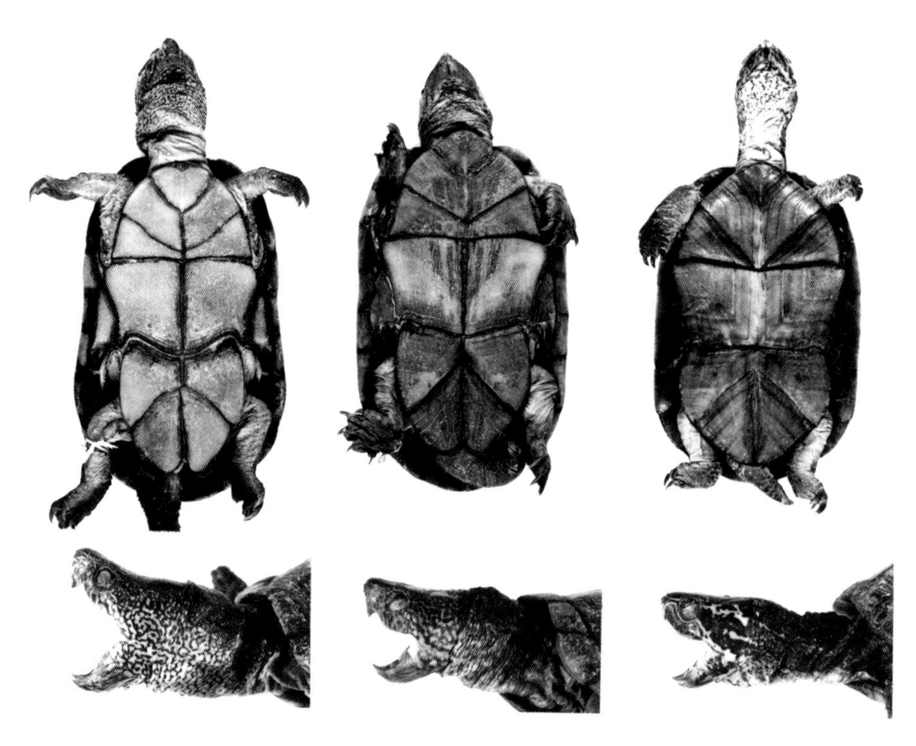

FIGURE 19.2 Comparison of adult males of three species of *Kinosternon* with ranges in Oaxaca—plastral and lateral head views. Left to right: *K. oaxacae*, UCM 48857 (holotype); *K. integrum*, UU 7849; *K. scorpioides cruentatum*, UU 7631. (From Berry and Iverson 1980, with permission; see also accounts of all three species.)

Aside from anecdotal accounts by us and others, most of what we present on natural history is from Iverson (1999) and Macip-Ríos (2005).

DIAGNOSIS

Carapace is weakly tricarinate, and keels are frequently lost in old individuals; C1 contacts M2; 11th marginal is flared outward, not vertical; there are six neural bones, and the first and last are isolated from contact with nuchal and suprapygal by dorsal articulation of costal bones. Plastron is intermediate to large, often completely closing shell; both plastral hinges are freely movable; length of plastral scute 6 is greater than 66% of posterior lobe; anal notch is usually present but variable, being deeper in males than females; plastral scute 4 (mid lobe of plastron) is 0.265 of CL in males and 0.279 in females. Axillary and inguinal scutes are narrowly in contact or separated; inguinal is barely in contact with M5. Clasping organs are lacking in both sexes; tail claw is present in both sexes but is more developed in males; there are two or more pairs of gular barbels plus two or more pairs on throat.

GENERAL DESCRIPTION

Maximum adult size: males, 210 mm; females, 196 mm. *Kinosternon integrum* and *K. hirtipes* rank as the largest Mexican *Kinosternon*. In a large database (JML), 92% of the Mexican *Kinosternon* of 160 mm CL or larger were of these species.

BASIC PROPORTIONS

CW/CL: males, 0.63 ± 0.02 (0.59–0.69) $n = 34$; females, 0.66 ± 0.04 (0.61–0.89) $n = 45$; im. males, 0.69 ± 0.01 (0.69–0.70) $n = 2$; im. females, 0.72 ± 0.07 (0.66–0.94) $n = 16$.

CH/CL: males, 0.36 ± 0.01 (0.33–0.40) $n = 34$; females, 0.40 ± 0.04 (0.35–0.62) $n = 45$; im. males, 0.38 ± 0.00 (0.38–0.38) $n = 2$; im. females, 0.39 ± 0.04 (0.36–0.52) $n = 16$.

CH/CW: males, 0.58 ± 0.02 (0.53–0.63) $n = 34$; females, 0.60 ± 0.03 (0.53–0.70) $n = 45$; im. males, 0.55 ± 0.01 (0.55–0.56) $n = 2$; im. females, 0.55 ± 0.03 (0.51–0.62) $n = 16$.

PLASTRAL PROPORTIONS

(Iverson, 1991b; Iverson, Young, and Berry, 1998)

PL/CL: 0.921 males, 0.954 females.

FL/CL: 0.323 males, 0.331 females.

HL/CL: 0.336 males, 0.350 females.

BL/CL: 0.261 males, 0.268 females.

PS1/CL: 0.161 males, 0.166 females.

PS6/CL: 0.219 males, 0.253 females.

Modal plastral formula: 4 > 6 > 1 > 2 > 5 > 3.

FIGURE 19.3 Comparison of adult females of three species of *Kinosternon* with ranges in Oaxaca—plastral views. Left to right: *K. oaxacae* (AMNH 88884, allotype), *K. integrum* (UU 7877), and *K. scorpioides cruentatum* (UU 5150). (From Berry and Iverson, 1980, with permission; see also accounts of all three species.)

Carapace is elongated; usually three carapacal keels are present and are most prominent posteriorly. Keels usually wane with age. C1 is usually in contact with M2; M9 is not elevated above anterior marginals. M10 is higher than M9, and M11 is usually not elevated to level of posterior edge of M10. Posterior width of forelobe is 47% (42–54) of CL in males and 53% (45–57) in females. Head shield is large, triangular, or bell shaped, and posterior margin is convex. Skin is smooth.

COLOR

Carapace is highly variable, ranging from pale tan to dark brown with darker interlaminal seams. Plastron is yellow to yellow-orange, usually with darkened seams. Head is spotted, mottled, or reticulated with cream or yellow markings on a dark brown to black ground. Iris usually has a complete or partial stellate pattern.

RELATIONSHIPS

This species is widespread and variable, but no subspecies are recognized. Berry's (1978) extensive analysis supports this premise.

There are numerous errors in the literature concerning the identification of *K. integrum*, most often confusing it with *K. hirtipes*. These mistakes are usually in areas of sympatry, and most of them have been corrected by Berry (1978), Berry and Legler (1980), Iverson (1981a), and Smith and Smith (1979).

FOSSILS

Mooser (1980) reported fossil *K. integrum* from the Pleistocene of Aguascalientes.

GEOGRAPHIC DISTRIBUTION

Map 13

The Mexican mud turtle is one of the most widely distributed and most often encountered turtles in Mexico, occurring from sea level to 2,545 m in western, central, and southern Mexico. Occurrence along the Pacific coastal plain is from the Rio Matape (Yaqui drainage, 28°44'N, 110°21'W) in central Sonora to coastal Guerrero and the Rio Verde in Oaxaca. *Kinosternon integrum* occurs throughout the San Pedro-Mezquital, Lerma-Santiago, and Balsas systems, the headwaters of which come into close proximity to those of the Tamesi- Panuco and the Rio Papaloapan on the Mesa Central where populations exist. The range extends to ca. 24°N on the western side of the plateau (Durango) and to 17°N on the eastern plateau in Tamesi-Panuco headwater tributaries. Distribution on the eastern side of the plateau is all above 800 m.

The northern limit of distribution in Sonora corresponds to that of the tropical thorn scrub community as do many other species (Stuart, 1964; Berry, 1978; see also account of *Terrapene nelsoni klauberi*). Specimens reported to the north of this are other species of *Kinosternon*.

Kinosternon integrum is replaced at lower elevations on the Pacific coast by *K. chimalhuaca* in southwestern Jalisco and in western Colima and by *K. oaxacae* in southern Guerrero and southwestern Oaxaca, and is replaced by *K. scorpioides* in the Rio Tehuantepec and on the Gulf Coastal plain.

Kinosternon integrum occurs in sympatry with *K. hirtipes* on the southern Mexican Plateau and with *K. alamosae* and *K. sonoriense* in northwestern Mexico, but not with *K. chimalhuaca*, *K. oaxacae*, or *K. scorpioides*. The populations in the Valley of Mexico are supposedly introduced (Iverson, Young, and Berry, 1998).

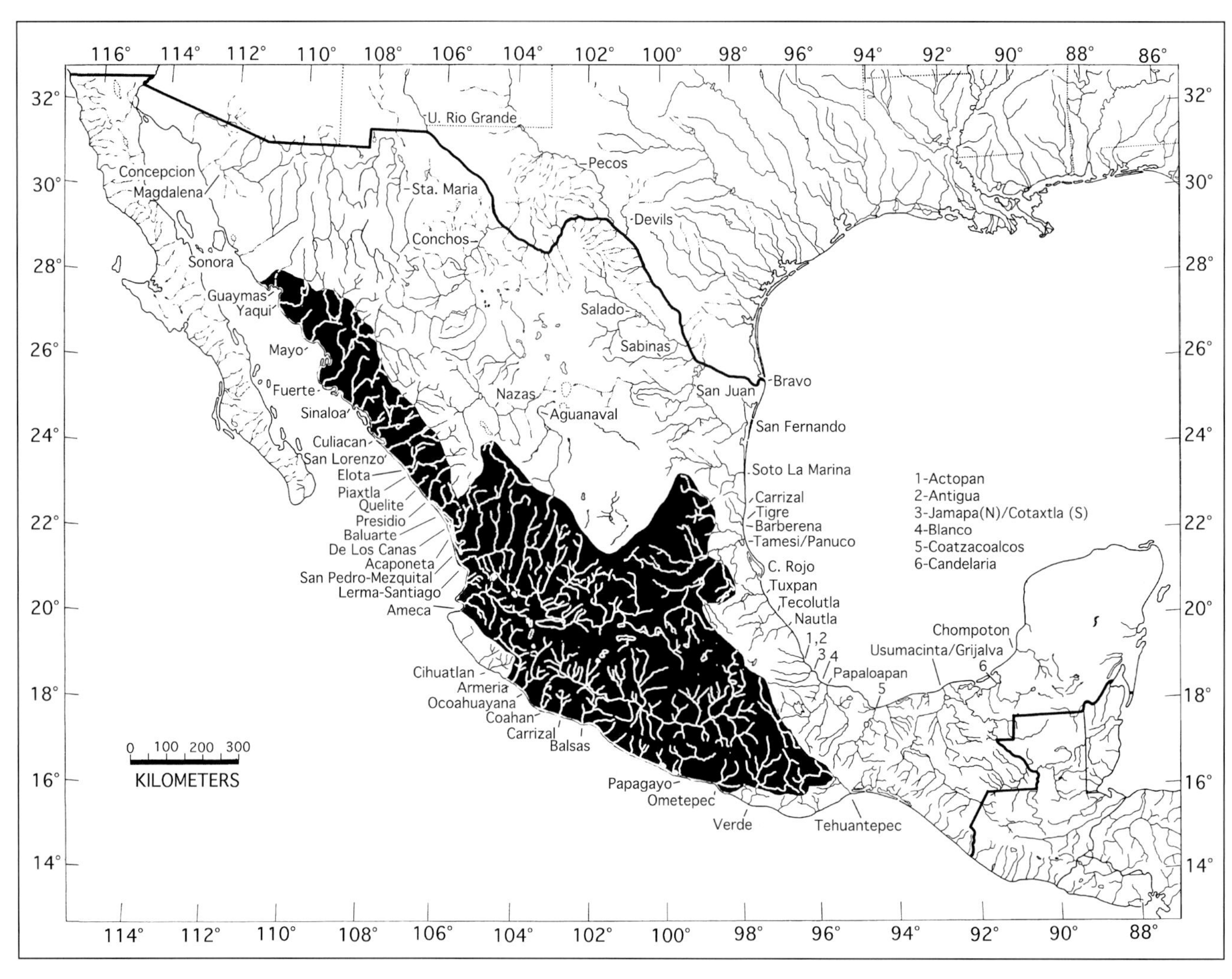

MAP 13 Geographic distribution of *Kinosternon integrum*.

HABITAT

Kinosternon integrum occurs in a wide variety of usually lentic temporary and permanent aquatic habitats, including slow-moving streams and temporary ponds from sea level to 2,500 m and in association with diverse plant communities except the most xeric or montane. Its distribution includes the states of Sinaloa (Hardy and McDiarmid, 1969), Michoacan (Duellman, 1965), Colima (Duellman, 1958b), Guerrero (Davis and Dixon, 1961), and Sonora (Heringhi, 1969). A collection of this species was made in intermittent pools near the headwaters of the Rio Grande de Santiago in Durango in a mountainous area of pine-oak forest at 2,545 m (Webb and Hensley, 1959). Aestivation occurs in areas where aquatic habitats dry in winter.

Flannery (1967) found an old adult male (July 1962) buried in the bottom mud of a small spring-fed pond in the Tehuacan Valley of Puebla; it bore a thick growth of epizoic algae on the carapace.

The Laguna Rio Viejo (2 km N of El Dorado, Sinaloa), an old oxbow of the Rio San Lorenzo (400 × 100 m, 1 to 2 m deep; Figure 48.2), seemed to be a stable, mature habitat when Legler and companions visited it in late March–early April 1964. The entire lake was bordered by tules, and most of them appeared to be on floating mats. The water was darkly stained and the bottom muddy. Hyacinth covered 5–10% of the lake and drifted with the wind. A few rooted tree stumps occurred in the water. *Kinosternon integrum* was the most common turtle; *Trachemys scripta ornata* occurred in lesser numbers (see "Populations" below). Otters were observed in the lake.

DIET

Like most other musk turtles, *K. integrum* is an opportunistic omnivore. Accounts of diet are nearly all of mixed plant and animal matter: grass, seeds, filamentous algae, coleopterans, dragonflies, dipterans, snails, leeches, and amphibians.

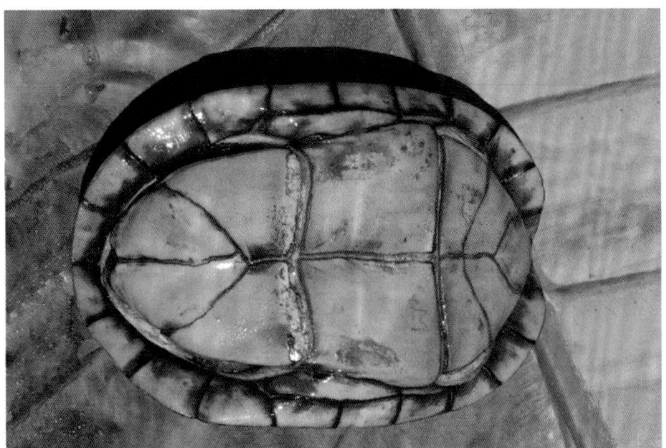

FIGURE 19.4 Adult female *Kinosternon integrum* near Mazunte, Oaxaca.

Macip-Ríos (2005) found that juveniles ate a higher percentage of plant materials than adults. During the dry season juveniles consumed a lower diversity of prey items and in lower numbers than adults. When resources were not limited in the rainy season, juveniles and adults were eating the same things.

In a dissected female, Flannery (1967) found mud, a large water beetle (probably *Belostoma*), parts of ichneumon flies, remains of medium-sized and small aquatic snails, termite wings, and parts of the leafy vegetation that surrounded the pool. He was told by local residents that the turtles left the water to feed on carrion.

Stomach flushing of freshly caught specimens from the Rio Mayo and Fuerte drainages (various habitats) of Sonora contained small insect larvae, algae, bite-sized succulent stems, hymenopteran larvae, ants, pieces of leaves, and coleopteran parts (JML field notes, May 1978).

Legler (field notes, March 1964) observed the following with binoculars from the Rio Sinaloa bridge, Guasave, Sinaloa: A large adult in a stagnant pool was chasing small aquatic animals (probably tadpoles); it demonstrated speed and agility in maintaining close pursuit, abruptly and deftly changing course when necessary.

HABITS AND BEHAVIOR

Based on trapping with bait, hand collecting, stomach flushing, and diving, most activity, especially feeding, was judged to be nocturnal or crepuscular near Alamos, Sonora, and El Dorado, Sinaloa.

Kinosternon integrum is more active in the summer and autumn rainy season than in the spring or winter months. Activity, as assessed by the abundance of turtles captured, was correlated to mean monthly temperatures and precipitation (Macip-Ríos, 2005). Iverson (1999) found an individual aestivating under a rock in oak woodland (Jalisco, May) where no surface water was present.

REPRODUCTION

A copulating pair was observed under natural conditions on 26 August in Michoacan (Duellman, 1961).

Iverson (1999) produced the only definitive study of reproduction by dissecting 52 females from various localities over the wide geographic range (Guerrero, Jalisco, Michoacan, Morelos, Puebla, Queretaro, and Sonora) from 1978 to 1985, in the months of April through August.

The mean CL of 38 mature females was 147 ± 8.7 (131–168). The smallest mature females ranged from 132 to 136 mm CL and 270 to 352 g weight; the largest immature females had 133 to 136 mm CL and 339 to 353 g weight. Ergo, maturity was logically predicted at sizes of 130 to 140 mm CL. There was no evident geographic variation in size at maturity (Iverson, 1999).

Number of clutches per year was 2.26 (1–4) $n = 38$. Eggs per clutch based on corpora lutea was 5.8 ± 2.3 (3–12) $n = 19$; based on ovarian follicles it was 5.0 ± 1.69 (2–10) $n = 39$.

Data on size and weight of 66 eggs: L, 29.5 ± 3.1 (23.1–35.4) $n = 66$; W, 16.7 ± 1.2 (13.9–19.1) $n = 66$; Wt, 5.8 g ± 0.97 (4.4–7.8) $n = 23$.

Relative clutch mass was 0.060 ± 0.017 (0.036–0.097) $n = 14$, and is the smallest reported for the genus *Kinosternon*, supporting the hypothesis that low RCM is correlated with larger body size.

The predicted nesting season for *K. integrum* is May–September (similar to that of *K. hirtipes*), possibly beginning prior to the rainy season. However, most turtles can retain shelled oviducal eggs for variable periods, and nesting has not actually been observed in the wild. Webb (1984) reported a captive female laying in early October. Females bore preovulatory follicles (15–17 mm) in all months sampled. Females collected in February in Sonora had smaller enlarged follicles (4.5–11.5 mm, $n = 3$) and lacked corpora lutea. The earliest date for the presence of oviducal eggs was early May. Some females from early April bore preovulatory follicles. The earliest occurrence of two sets of corpora lutea was 19 July. Females in early August had one or two sets of enlarged follicles.

Despite the seeming lack of variation in known life history parameters, Iverson (1999) reported a significantly

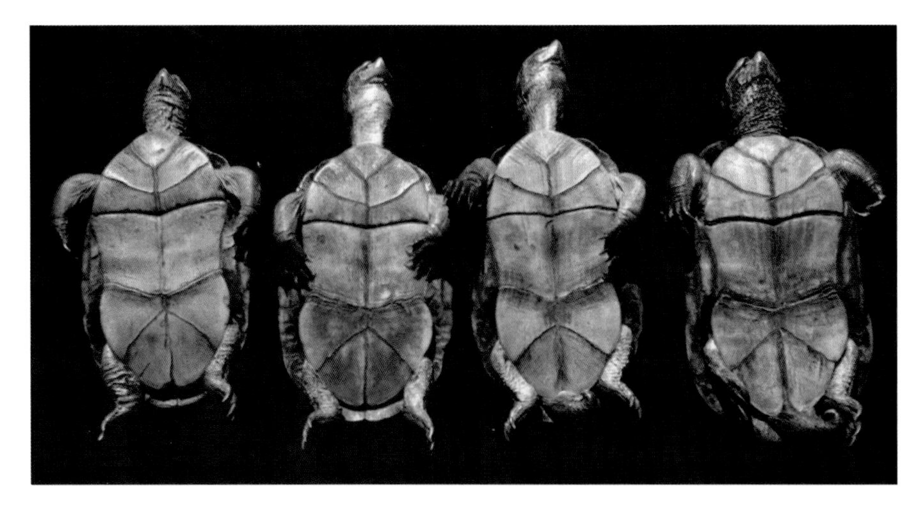

FIGURE 19.5 Series of adult *Kinosternon integrum*, Alamos, Sonora; two females on left, two males on right (KU).

divergent population producing larger clutches of smaller eggs in a heavily polluted (garbage and feces) stream draining a poor suburb near Cuernavaca, Morelos.

Macip-Ríos (2005) reported the nesting season to be late June to late October in the State of Mexico. Number of eggs per clutch was 4 ± 1.8 (1–8) n = 20 clutches and directly correlated to body size. Egg size was as follows: L, 30.4 ± 2.2 (23.9–35.0) n = 78; W, 16.4 ± 1.0 (13.0–18.3) n = 75; Wt, 5.1 g ± 0.6 (3.4–6.5) n = 57. Mean egg length was inversely correlated to clutch size.

Newly emerged hatchlings have been found in Sinaloa from late July to early September (Hardy and McDiarmid, 1969) and on 17 June in Oaxaca (Iverson, 1999). Since *K. integrum* has embryonic diapause and embryonic aestivation these hatchlings could represent eggs from the previous year.

GROWTH AND ONTOGENY

Adult males are larger than females. Macip-Ríos (2005) reported maximum weights as follows: males, mean 336.8 g, maximum 774.1 g, n = 50; females, mean 281.4 g, maximum 654.2 g, n = 89.

Data on size (a large series from 2 km N of El Dorada, Sinaloa, early April 1964; JML notes): males, 152 ± 11.4 (130–170) n = 34; females, 146 ± 8.1 (128–165) n = 45; im. males, 113 ± 7.8 (107–118) n = 2; im. females, 106 ± 20.6 (62–128) n = 16.

PREDATORS, PARASITES, AND DISEASES

Parasites have been recorded in wild-caught *K. integrum* from many localities (Bravo-Hollis, 1944; Bravo-Hollis and Caballero Deloya, 1973; Caballero y Caballero, 1938, 1940; Caballero y Caballero and Herrera-Rosales, 1947; Herrera Rosales, 1951; Pérez Reyes, 1964; Yamaguti, 1958).

Macip-Ríos (2005) attributed the paucity of hatchlings to predation by herons, snakes (*Drymarchon corais* and *Masticophis mentovarius*), skunks, procionids (ringtails and raccoons), and opossums. There was no indication of human use in his study area, but domestic animals (dogs, cats, and pigs) were also predators.

POPULATIONS

Macip-Ríos (2005) gave the following data for specimens obtained between October 2003 and November 2004 using baited funnel traps and seines in small ponds 20–300 m in diameter. In 14 capture sessions he collected 204 turtles and recaptured 118 of them. He found immature turtles (65) to make up the largest percentage of the population, while hatchlings were rarely found. The adult sex ratio was 50 males to 89 females. The mean population size was calculated to be 197 turtles (128.28–415.94). The mean monthly survivorship was calculated to be 0.9574.

Legler and students (field notes, May 1978) made two samples in northwestern Mexico: (1)The Laguna Rio Viejo (see description in habitat)—155 captured in 3 days (19 baited traps, 764 trap hours, in all parts of the lake); 105 saved and cataloged (UU) were 35 males, 58 females, 1 immature male, 6 immature females, and 5 juveniles. (2) A small (10 × 10 m, maximum depth 1.5 m), cool, spring-fed pond near Alamos, Sonora—82 specimens by hand and baited traps: 5 males, 17 females, 11 immature, 49 juveniles.

CONSERVATION AND CURRENT STATUS

There are no special projects involving the conservation or management of *K. integrum*. Populations appear to remain abundant even among dense human populations (Macip-Ríos, 2005).

The species has been used as a source of food (chiefly by hunter-gatherers) in the Tehuacan Valley (Rio Salado and its tributaries, the headwaters of the Rio Papaloapan) at least since 5000 BC. It is the only chelonian now living in the valley (Flannery, 1967).

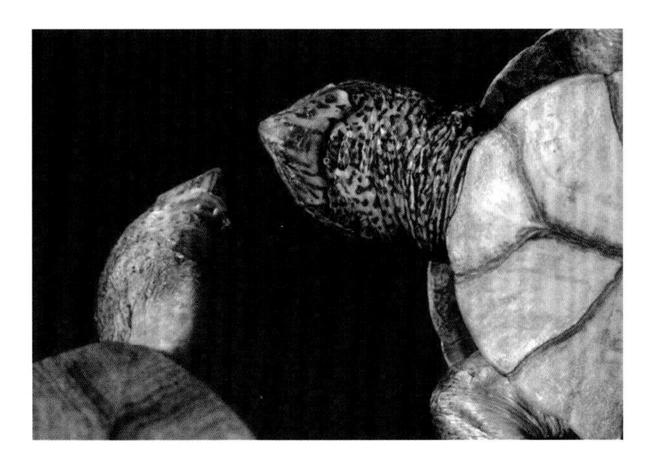

FIGURE 19.6 Throat patterns of two adult *Kinosternon integrum* (female left, male right). Alamos, Sonora (KU).

Kinosternon alamosae Berry and Legler, 1980

Alamos mud turtle (Berry and Legler, 1980); casquito de Alamos (Liner, 1994).

ETYMOLOGY

The specific name is a toponym referring to the lovely colonial town of Alamos, Sonora, near the type locality.

HISTORICAL PROLOGUE

Berry (1978), in a detailed analysis of the taxonomy of the *Kinosternon scorpioides* and *Kinosternon leucostomum* complexes, noted that most of the specimens identified as *Kinosternon hirtipes* from the region of Alamos, Sonora (Heringhi, 1969), represented an unrecognized species. Berry and Legler (1980) published the description of *Kinosternon alamosae* based on a total of 33 specimens from 15 localities, which included a small body of reproductive information. Subsequently, Iverson (1989a) obtained 56 additional specimens from 25 localities (1981–1986) and contributed a substantial amount of natural history information on the species. Virtually all the information in this account comes from these two papers and from Iverson (1990a).

Smith and Smith (1979) (with permission) published the name "*Kinosternon alamosae* Berry and Legler nomen nudem" in a manner that would insure its validity when published a few months later. Pritchard (1979) obtained the name *alamosae* from an unpublished source, misspelled it as "alamose," and included it in a semipopular book (see Melville, 1983; International Commission on Zoological Nomenclature, 1985).

RELATIONSHIPS AND TAXONONOMIC STATUS

The species is a member of the *Kinosternon scorpioides* complex (including *scorpioides, integrum, creaseri, acutum, alamosae, oaxacae,* and *chimalhuaca*). The last three of these species were named and described since 1980. The group is characterized by an extensive, virtually fully closable plastron, a bell-shaped head shield, a basically tricarinate carapace, and lack of clasping organs in both sexes.

Kinosternon alamosae is phenetically closest to *K. integrum* (Berry and Legler, 1980; Iverson 1989a, 1991b).

DIAGNOSIS

(Description and diagnosis chiefly from Berry and Legler [1980] and Iverson [1991b]).

Carapace is noncarinate, broadly rounded, or flat-topped in cross-section. C1 usually is not in contact with M2; there are six neural bones, the first is in contact with nuchal, and the last is isolated from the suprapygal. Plastral lobes are extensive, closing or nearly closing the shell orifices; anal notch is lacking or minuscule; both plastral hinges are freely movable; plastral scute 4 (on mid lobe of plastron) is 0.301 and 0.318 of CL in males and females, respectively. Posterior margin of carapace is nearly vertical in profile, straight or evenly curved, and never recurved or flared outward; bridge is longest of *integrum* group (26–33 % of carapace length). Axillary and inguinal scutes are widely separated, inguinal scute is narrowly in contact with M6 but never in contact with M5. Clasping organs are lacking in both sexes; tail claw is present in both sexes but is more heavily developed in males. There is one pair of gular barbels, which are almost never longer than their basal width; there are no barbels on the throat.

GENERAL DESCRIPTION

Kinosternon alamosae is a medium-sized species; maximum CL is 135 mm in males and 113 in females.

BASIC PROPORTIONS

CW/CL: males, 0.64 ± 0.03 (0.61–0.68) $n = 5$;
females, 0.67 ± 0.03 (0.62–0.71) $n = 8$.

CH/CL: males, 0.36 ± 0.01 (0.34–0.37) $n = 5$;
females, 0.40 ± 0.02 (0.37–0.42) $n = 8$.

CH/CW: males, 0.56 ± 0.02 (0.52–0.58) $n = 5$;
females, 0.59 ± 0.02 (0.55–0.61) $n = 8$.

PLASTRAL PROPORTIONS

(Iverson, 1991b)

PL/CL: 0.932 males, 0.963 females.

FL/CL: 0.295 males, 0.290 females.

HL/CL: 0.336 males, 0.347 females.

BL/CL: 0.286 males, 0.307 females.

PS1/CL: 0.165 males, 0.170 females.

PS6/CL: 0.204 males, 0.236 females.

Modal plastral formula: 4 > 6 > 1 > 5 > 2 > 3.

FIGURE 20.1 Dorsal, ventral, and lateral head views of *Kinosternon alamosae*. Left—UU 14299, male paratype, 0.8 km W of Alamos, Sonora, CL 126 mm; right—female holotype, LACM 127639, Rancho Carrizal, 7.2 km N and 11.5 km W of Alamos, Sonora, CL 122 mm. (From Berry and Legler, 1980, with permission.)

Carapace is relatively narrow, oval in dorsal aspect, flat-topped in cross-section, and lacking mid-dorsal keel; large females may have slight concavity in mid-dorsal region. M1–9 are about the same height; M10 is abruptly higher than M9, but only slightly higher than M11. Carapace is slightly flared laterally in the region of marginals 8–10, but the extreme posterior edge of the carapace is straight. Plastral lobes are large and almost completely close shell orifices; plastral lobes are evenly rounded.

Rostrum is short and broad dorsally; rostral pores are well developed. Premaxillary hook is lacking or weakly developed. Head shield is well developed and covers the entire surface of the underlying frontal and prefrontal bones with blunt extensions over postorbital arches and parietal region; posterior border is straight, never concave or V-shaped. Anterior part of tongue has papillae.

COLOR IN LIFE

Carapace is olive to brown with darker interlaminal seams and is paler at margins. Plastron is yellow with dark-brown seams. Skin of head is pale gray above with numerous dark spots; sides of the head are mottled pale gray and yellowish cream with less dense dark spotting. An indistinct pale stripe extends from the posterior ventral edge of the orbit, above the maxillary sheath to the jaw articulation. Jaw sheaths are yellowish cream to pale gray, and those of males have faint brown vertical streaks. Dorsal surfaces of neck are pale gray

FIGURE 20.2 *Kinosternon alamosae*: live adult male, J. B. Iverson 858, 27.3 km W and 2.9 km S of Alamos, Sonora. Width of head is 21 mm. (From Berry and Legler, 1980, with permission.)

grading to immaculate yellowish-cream on the ventral surfaces of the head and neck. Skin of legs and tail is pale gray and is slightly darker above. Iris typically is pale to dark yellow and marked with a dark horizontal stripe.

GEOGRAPHIC DISTRIBUTION

Map 14

The type locality is Rancho Carrizal, 7.2 km N and 11.5 km W of Alamos, Sonora. Known geographic range is the Pacific coastal lowlands (to 1,000 m) of Sonora and Sinaloa from the Rio Sonora drainage southward at least to the region of Guasave (Rio Sinaloa drainage). The northern limit of the range coincides with that of thorn scrub forest and the northern limit for many tropical vertebrates (Stuart, 1964; Berry, 1978). Iverson (1989a) reported one specimen (Hwy. 15, 62 km S of Nogales) some 200 km N of the main range and regarded it as "extralimital." However, this seasonal species evaded discovery until 1980 because even good turtle collectors were in the right places at the wrong time. There are two reasonable explanations for the out-of-range specimen, in order of likelihood: (1) a collector without proper permits released it there before crossing the border; (2) the species occurs there.

Kinosternon alamosae occurs in broad sympatry with *K. arizonense*, *K. sonoriense* (to the north), and *K. integrum* (to the south). It has been found in microsympatry with *arizonense* in two impoundments east of Hermosillo, Sonora (Iverson, 1989a).

Most of the following natural history information is based on Iverson (1989a).

HABITAT

Alamos mud turtles inhabit temporary ponds, including arroyos, cattle tanks, other artificial impoundments, and roadside ditches, and are found chiefly during the rainy season (July to September). None have been collected in permanent water habitats. On sunny days it is possible to document the presence of the species by scanning a pond for heads. They readily enter baited traps set for as little as one daytime hour.

DIET

Kinosternon alamosae is an opportunistic omnivore and eats much of what is available in seasonal ponds. Iverson (1989a) examined the gastrointestinal tracts of 25 *K. alamosae*—6

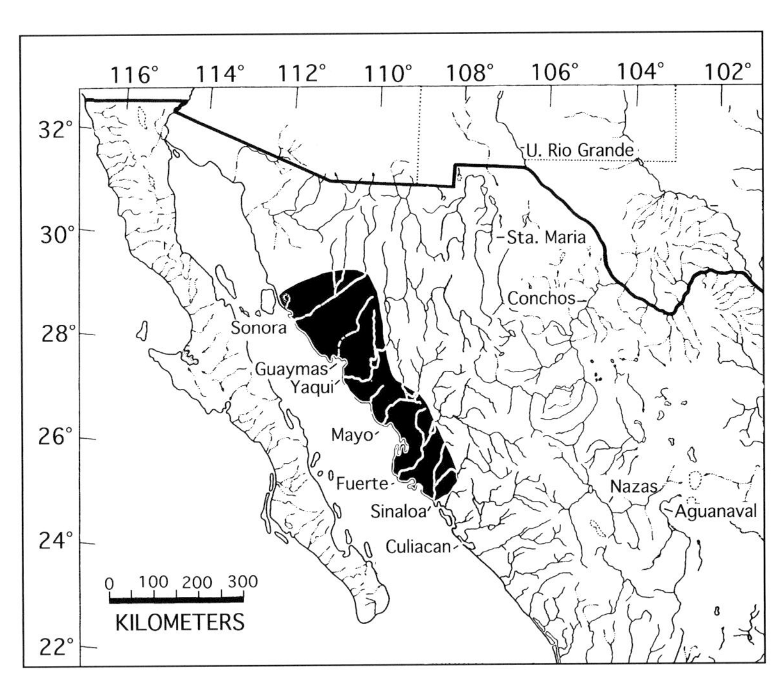

MAP 14 Geographic distribution of *Kinosternon alamosae*.

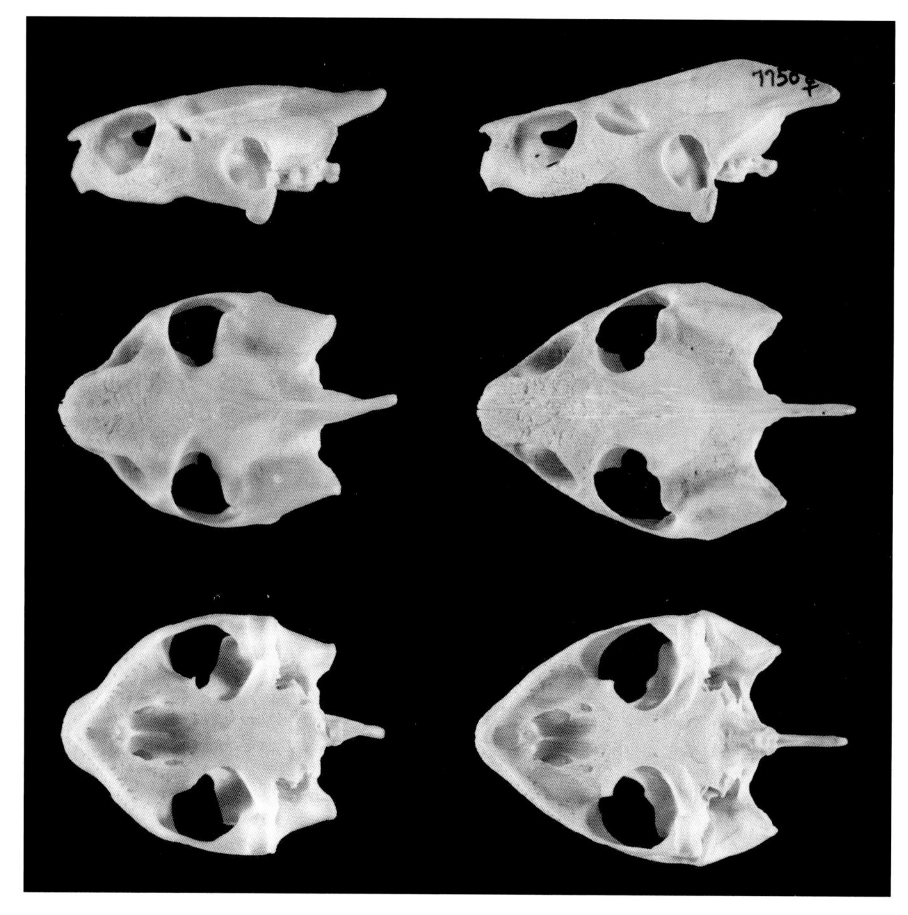

FIGURE 20.3 Comparison of *Kinosternon* skulls, Left—*K. alamosae*, UU 14281, female paratopotype, condilobasilar length 26.2 mm; right—*K. integrum*, UU 7750, female, Laguna Rio Viejo, 2 km N of Eldorado, Sinaloa, condylobasilar length 35.1 mm. (From Berry and Legler, 1980, with permission.)

were empty; 17 contained the following in decreasing order of occurrence: Coleoptera, Hymenoptera, bivalves, shrimp, centipedes, Orthoptera, and scorpions. The gastrointestinal tracts of two individuals also contained anuran eggs and another a frog bone; 10 of the 19 tracts contained plant material (mainly seeds) and two contained only plant material.

HABITS AND BEHAVIOR

Alamos mud turtles are shy; when handled they withdraw and do not attempt to bite.

The annual cycle of activity seems to correspond to seasonal rains. Of Iverson's series, 42 were collected between 5 July and 12 September, and more than half of these in the month of July. These data suggest an annual season of activity that could be as short as 2 months in an average year, interspersed with long periods of aestivation (presumably in the mud of drying pools). In times of drought, aestivation might be extended to 18 months. Some *Kinosternon flavescens* have been shown to have a similarly abbreviated annual activity cycle (Christiansen, Cooper, Bickham, Gallaway, and Springer, 1985). *Kinosternon alamosae* has been shown to have the lowest rate of evaporative water loss among four desert-adapted species of *Kinosternon* (Seidel and Reynolds, 1980).

Aerial basking has been observed (Iverson, 1989a). Thermal tolerance is extremely high. Turtles were hand collected and trapped in shallow water as warm as 42°C and uncomfortable to human skin.

REPRODUCTION

Mounting and copulation behavior have been observed in the field on several dates in July (Iverson, 1989a; Wiewandt, Lowe, and Larson, 1972). No details were given.

Four females with oviducal eggs were found from 20 July to 2 September. All had at least one set of 8 to 11 mm follicles and one set of corpora lutea. These data suggest that *alamosae* may lay up to three clutches per year. Timing of vitellogenesis is unknown; enlarged follicles may be retained into the next breeding season.

Mean number of eggs in four clutches was 4.0 ± 0.8 (3–5). Mean egg dimensions were as follows: L, 27.2 (24.6–29.5); W, 15.2 (13.5–16.5). Mean egg weight was 3.66 g (Iverson, 1989a). Nesting has not been observed.

FIGURE 20.4 *Kinosternon alamosae*. Old male, 1 km W of Alamos, Sonora. (Courtesy of J. F. Berry.)

GROWTH AND ONTOGENY

Data on size: males, 113 ± 7.5 (105–125) n = 5; females, 115.38 ± 9.6 (101–131) n = 8; im. male, 89.

Males are larger than females. Males have a less extensive plastron than females (PL is 93.2% of CL in males and 96.3% of CL in females) and a relatively narrower shell. The plastron is slightly concave in males and flat or slightly convex in females. Tail of male is elongate and prehensile and is longer than one-half the length of the posterior plastral lobe; tail of female is much shorter than one-half the length of the posterior lobe.

Sexual maturity (females) is attained between 95 and 100 mm CL (90 to 95 mm PL) at 5–7 years of age. The largest immature females were 92 and 93 mm at 5 years, and the smallest mature female was 100 mm at 5 years. Males begin to exceed females in size in the fifth year of growth.

Iverson (1989a) based the following information on gonadal examination and growth-ring analysis in a large series of specimens. Size more than doubles (256%) by the end of the second year of growth, is ca. 30% in the following season, and decreases to 6–11% by year 6.

PREDATORS, PARASITES, AND DISEASES

Nematodes were found in 8 of 25 stomachs by Iverson (1989a). The usual range of predators is assumed.

POPULATIONS

Sex is probably determined by incubation temperature, as in all other *Kinosternon* studied. Iverson's collecting in 1989 produced more males (28) than females (12).

Kinosternon chimalhuaca Berry, Seidel, and Iverson, 1997

Jalisco mud turtle (Berry, Seidel, and Iverson, 1997)

ETYMOLOGY

The specific name, *chimalhuaca*, alludes to a tribe of indigenous people, the Chimalhuaca Indians, who are said to have lived along the southern Pacific coast of Mexico at the time of European colonization (Jennings, 1980; Willey et al., 1964; West and Augelli, 1976).

HISTORICAL PROLOGUE

Berry (1978) concluded, in a multiple discriminate analysis of 36 external morphological characters, that this population of *K. integrum* was phenetically distinct from *integrum*, *oaxacae*, and *scorpioides*. However, Seidel, Iverson, and Adkins (1986), using electrophoresis of 13 protein systems, found no differences between *K. chimalhuaca* ("new sp.") and *K. integrum*. Iverson (1991b) concurred with Berry after producing a cladistic analysis of all known *Kinosternon* based on a series of 27 morphological characters and 11 protein loci. He concluded that the taxa now known as *chimalhuaca*, *integrum*, and *oaxacae* were closely related, confirming that *K. chimalhuaca* is a member of the *K. scorpioides* complex but distinct from all other members of the complex.

Rogner (1996), seemingly using unpublished information, used the species name "*Kinosternon chimalhuaca*" (with a photograph) before the original description was published. Thus the name became available as of Rogner (1996; prior to the actual scientific description). Technically, therefore, Rogner is the author of the name, according to the rules of Zoological Nomenclature. We choose to ignore this technicality.

DIAGNOSIS

Diagnosis and description based largely on Berry, Seidel, and Iverson (1997) and Iverson (1991b, 1998).

Carapace is weakly tricarinate and depressed; C1 seldom (12% of adults) contacts M2; there are five neural bones, and the first and last are isolated from contact with the nuchal and suprapygal bones by dorsal articulation of the costal bones.

Plastron is relatively small (smaller in adult males than in adult females and juveniles) and does not completely close shell orifices; width of plastral hindlobe is 50.5–57.6% of CW in males and 56.7–73.1% of CW in females; there is an anterior plastral lobe that has a straight hinge and is freely movable and a posterior lobe that has a curved hinge (bowed posterioly) and is only slightly movable; there is a distinct anal notch that is deeper in males than in females; plastral scute 4 (on mid lobe of plastron) is of moderate length, 0.241 of CL in males and 0.258 of CL in females; bridge is shortest of *integrum* group (15–21% of CL in males and 20–23% of CL in females). Axillary and inguinal scutes are in contact; inguinal is not or is barely in contact with M5.

Clasping organs are lacking in both sexes; tail claw is present in both sexes but is more heavily developed in males; there are two or more pairs of gular barbels plus two or more pairs on the throat.

FIGURE 21.1 Dorsal, ventral, and head profile views of *Kinosternon chimalhuaca* type specimens. Left—adult male holotype, CM 140,201, CL 147 mm; right—adult female allotype, CM 140,202, CL 138 mm. Note unintentional anterior and posterior truncations on male images. (From Berry et al., 1997, with permission.)

GENERAL DESCRIPTION

Maximum size is 160 mm CL in males and 130 mm CL in females.

BASIC PROPORTIONS

CW/CL: males, 0.66 ± 0.02 (0.60–0.70) n = 37; females, 0.70 ± 0.02 (0.65–0.75) n = 42; im. males, 0.71 ± 0.02 (0.68–0.73) n = 6; im. females, 0.72 ± 0.02 (0.70–0.76) n = 9; unsexed juveniles, 0.74.

CH/CL: males, 0.34 ± 0.02 (0.29–0.38) n = 37; females, 0.38 ± 0.02 (0.34–0.42) n = 42; im. males, 0.36 ± 0.02 (0.34–0.39) n = 6; im. females, 0.38 ± 0.02 (0.36–0.41) n = 9; unsexed juveniles, 0.46 ± 0.13 (0.37–0.55) n = 2.

CH/CW: males, 0.52 ± 0.03 (0.45–0.56) n = 37; females, 0.54 ± 0.03 (0.48–0.60) n = 42; im. males, 0.51 ± 0.03 (0.47–0.56) n = 6; im. females, 0.53 ± 0.02 (0.51–0.57) n = 9; unsexed juveniles, 0.51 ± 0.02 (0.50–0.52) n = 2.

FIGURE 21.2 Dorsal and lateral views of head of adult male *Kinosternon chimalhuaca*, Rio Purificatión, Jalisco, CM 15235. (After Berry et al., 1997, courtesy of A. Rhodin and John Iverson.)

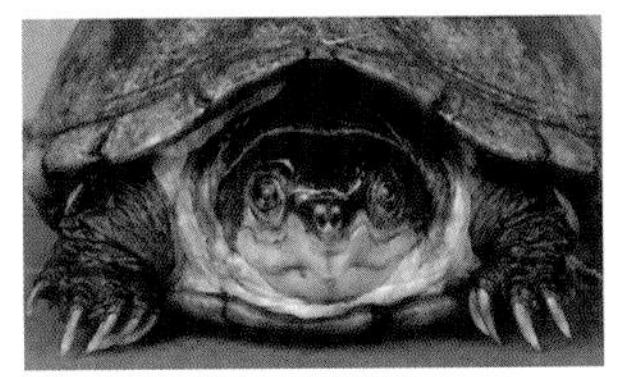

FIGURE 21.3 Three views of adult female *Kinosternon chimalhuaca* (CL 123 mm) from near Chamela (19°32′N), coastal Jalisco. (Photos courtesy of P. P. vanDijk.)

PLASTRAL PROPORTIONS

(Iverson, 1991b)

PL/CL: 0.858 males, 0.908 females.

FL/CL: 0.287 males, 0.292 females.

HL/CL: 0.330 males, 0.355 females.

BL/CL: 0.188 males, 0.221 females.

PS1/CL: 0.139 males, 0.142 females.

PS6/CL: 0.189 males, 0.216 females.

Modal plastral formula: 4 > 6 > 1 > 5 > 2 > 3.

Carapace is relatively wide and low; there are three low, longitudinal keels; carapacal scutes are imbricate; growth rings are evident even in some large adults. M1–9 are aligned dorsally, and M10 is abruptly higher than M9; M11 is lower than M10 but higher than M9. Carapace is flared laterally at M8–10. Plastral coverage is the least extensive of the *integrum* group, and relative lengths of PS6 and PS1 are shortest. Head shield is V shaped or bell shaped. Upper jaw is strongly hooked, especially in large adult males. There are four to eight rows of papillae on the lateral and dorsal surfaces of the neck.

COLOR

Carapace is dark brown, olive, or tan, often with dark mottling; interlaminal seams are darkened. Head is dark green to brown and mottled with bright yellow to orange-yellow or pale brown. Limbs and tail are brown above and yellow ventrally. Plastron is yellow to brown with darker seams.

RELATIONSHIPS

Kinosternon chimalhuaca, *integrum*, and *oaxacae* are closely related and are members of the *K. scorpioides* complex. The range of *chimalhuaca* lies about 700 km SE of *alamosae* and about 500 km NW of *oaxacae*; it is virtually surrounded by the range of *K. integrum*.

GEOGRAPHIC DISTRIBUTION

Map 15

Kinosternon chimalhuaca occurs along ca. 150 km of Pacific coast, immediately southeast of Cabo Corrientes in Jalisco and Colima, from approximately the Rio Tuito (19°58′N) southeastward to the Rio Cihuatlan (19°09′N), ca. 26 km NW of Manzanillo, Colima. It is common in the R. Purificación and R. Cihuatlan.

Kinosternon integrum replaces *chimalhuaca* in the Rio Armería and Rio Ameca to the north and east, respectively. The restricted coastal distribution suggests that it, like *K. alamosae* and *K. oaxacae*, is an isolated descendant of a *K. integrum*–like ancestor (Berry, 1978; Iverson, 1986a, 1989a). The poeciliid fishes, *Poecilia chica* (restricted to the Rio Cuitzmala, Purificación, and Cihuatlan basins) and *Poeciliopsis turneri* (known only from the Rio Purificación and Cihuatlan basins) are also endemic to the region (Miller, 1975, 1983).

FIGURE 21.4 Adult *Kinosternon chimalhuaca* surfacing in a natural pond near Chamela, Jalisco. (Photo courtesy of P. P. vanDijk.)

HABITAT

Specimens were collected in ponds with emergent and submergent vegetation. The type locality is a clear 1-acre pond, possibly spring fed, in a hardwood swamp. Specimens were also obtained from muddy ponds. They were not found in nearby clear water streams.

DIET

Stomach contents contained mollusks, insects, crustaceans, and plant material. Berry, Seidel, and Iverson (1997) suggested that, because of its generalized jaw structure, the diet is primarily carnivorous but opportunistic.

REPRODUCTION

The smallest mature male had 105 mm CL and the largest immature male had 97 mm CL, suggesting that males reach sexual maturity at about 100 mm CL at the ages of 5–7 years.

Analyses of ovaries and growth rings in 30 females from 8–9 May and 11 females from 29–30 June showed that the smallest females with ovarian follicles greater than 5 mm had 99 to 110 mm CL and the largest immature females had 100 to 106 mm CL, suggesting that females reach maturity at 99 to 107 mm CL at the ages of 7–8 years (Berry, Seidel, and Iverson, 1997).

Seven of the foregoing females were considered to be sexually mature. None had fresh corpora lutea or oviducal eggs, but all had maturing follicles (2–8) in the largest size class of 6–15 mm. Although all the mature females had the follicular potential to produce a second clutch of eggs, Berry, Seidel, and Iverson (1997) predicted that this species might have the same reproductive pattern as *K. oaxacae*—one clutch per year in July or August (Iverson, 1986b).

Captive turtles laid three clutches of two, four, and five eggs. Size of eggs was as follows: L, 33.4 (29.0–36.6); W, 17.5 (16.8–18.2).

GROWTH AND ONTOGENY

Data on size: males, 119 ± 17.38 (88–160) $n = 37$; females, 104 ± 9.47 (84–127) $n = 42$; im. males, 87 ± 6.57 (78–95) $n = 6$; im. females, 92 ± 6.53 (80–99) $n = 9$; im. unsexed, 65, $n = 2$.

The plastron is smaller in adult males than females and never completely conceals the fleshy parts of the turtle

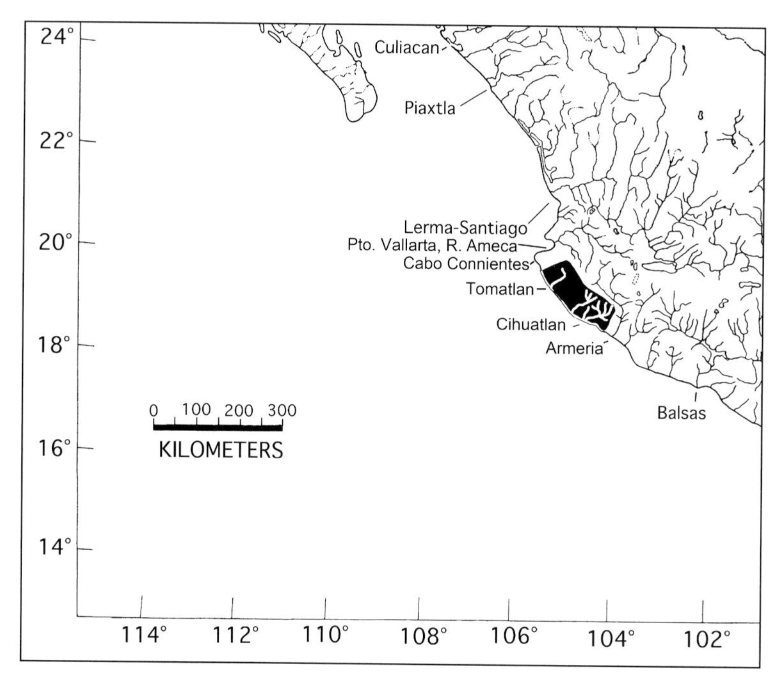

MAP 15 Geographic distribution of *Kinosternon chimalhuaca*.

FIGURE 22.1 Dorsal, ventral, and lateral head views of *Kinosternon oaxacae*. Top row—adult male holotype, UC 48857, CL 159 mm; middle row—adult female allotype, AMNH 88884, CL 130 mm; bottom row—male on left, female on right. Both specimens from ca. 12 km N of Pochutla, Oaxaca. (Modified from Berry, 1978; see also comparisons in account of *K. integrum* [Figures 19.2 and 19.3]; note unintentional truncations of male snout and tail.)

in either sex. The plastron is concave in males and flat to slightly convex in females. Anal notch is larger in males than in females. The head is larger in males than in females.

Kinosternon oaxacae Berry and Iverson, 1980

Oaxaca mud turtle (Iverson, 1983b); casquito de Oaxaca (Liner, 1994).

ETYMOLOGY

The specific name is a toponym referring to the state of Oaxaca.

HISTORICAL PROLOGUE

Berry and Iverson (1980) described *Kinosternon oaxacae* from a series of eight specimens collected from five localities. While the paper was in press Pritchard (1979), seemingly by access to a manuscript, published a description under the name "*Kinosternon oaxacae*" and legally occupied the name. The International Commission of Zoological Nomenclature (Mellville, 1983) suppressed Pritchard's work after being petitioned by Pritchard and Pronek (1982).

Kinosternon oaxacae is one of the least known species in Mexico. Since the original description, one natural history note has been published and 36 specimens from 10 new localities have been reported (Iverson, 1986b).

FIGURE 22.2 *Kinosternon oaxacae*. Adult female, region of Mazunte, Oaxaca.

DIAGNOSIS

Carapace is strongly tricarinate and depressed; C1 contacts M2; there are six neural bones, and the first and last are isolated from contact with the nuchal and suprapygal by dorsal articulation of the costal bones.

Plastron is relatively small (more so in males than in females) and does not completely close the shell; width of hindlobe is 57–63% of CW; both plastral hinges are straight or only slightly curved, and both are freely movable; there is a distinct anal notch; plastral scute 4 (on mid lobe of plastron) is relatively long, 2427% of CL and less than 26% of PL; bridge is short, 22–25% of CL. Axillary and inguinal scutes are in contact; inguinal scute contacts M5.

Clasping organs are lacking in both sexes; tail claw is present in both sexes but is more heavily developed in males. There are two or more pairs of gular barbels plus two or more pairs on the throat.

GENERAL DESCRIPTION

Kinosternon oaxacae is one of the larger species in the *scorpioides* complex: maximum size of males is 175 mm and females is 157 mm. Average body size is exceeded only slightly by *K. integrum* (Iverson, 1986b).

BASIC PROPORTIONS

CW/CL: males, 0.63 ± 0.01 (0.62–0.65) $n = 3$;
females, 0.67 ± 0.06 (0.62–0.73) $n = 3$.

CH/CL: males, 0.34 ± 0.02 (0.32–0.36) $n = 3$;
females, 0.36 ± 0.01 (0.36–0.37) $n = 3$.

CH/CW: males, 0.54 ± 0.02 (0.52–0.55) $n = 3$;
females, 0.54 ± 0.05 (0.49–0.59) $n = 3$.

PLASTRAL PROPORTIONS

(Iverson, 1991b)

PL/CL: 0.875 males, 0.926 females.

FL/CL: 0.295 males, 0.312 females.

HL/CL: 0.332 males, 0.351 females.

BL/CL: 0.225 males, 0.249 females.

PS1/CL: 0.152 males, 0.171 females.

PS6/CL: 0.223 males, 0.225 females.

Modal plastral formula: 4 > 6 > 1 > 5 > 2 > 3.

L4 is in contact with M10; M10 is elevated above the level of other marginals. Width of plastral forlobe is less than 70% of CW; width of plastral hindlobe is less than 60% of CW in males and less than 62% of CW in females. Head shield is V shaped or bell shaped.

COLOR

Carapace is brown, black, or mottled brown and black. Seams are darkened in pale individuals. Plastron and bridge are usually yellowish with darker brown seams but sometimes have darker environmental stains. Skin varies from gray to black. Head is mottled with dark on paler ground and is dark brown to black dorsally and ventrally cream to yellow with a few dark spots. Young turtles often have an orange suffusion on the sides of the head.

GEOGRAPHIC DISTRIBUTION

Map 16

Kinosternon oaxacae has a small range in Pacific drainages of Oaxaca and eastern Guererro, at elevations of 100 to

FIGURE 22.3 *Kinosternon oaxacae*. Adult male (left) and female (right), region of Mazunte, Oaxaca.

FIGURE 22.4 *Kinosternon oaxacae*. Adult head, region of Mazunte, Oaxaca.

800 m. These drainages enter the sea between the approximate longitudes of 99°40′W (Rio Papagayo) and 95°35′W (small unnamed streams). The distribution map includes all records based on specimens or reliable sight records (Iverson, 1992a,b; Carr, 1993). This species is known from 44 specimens and 15 localities (Iverson, 1986b). The geographic range of *K. oaxacae* lies ca. 500 km SE of the range of *K. chimalhuaca*.

HABITAT

Oaxacan mud turtles commonly inhabit small seasonal forest ponds with turbid water and muddy bottoms but are also found in permanent ponds, open water at the edge of marshes, and in deeper (up to 1 m) pools in fast-moving clear water streams (Iverson, 1986b). Martha Harfush showed RCV a dry, rocky river bed near Mazunte, Oaxaca, in which *K. oaxacae* was common in rainy season pools during July.

DIET

Like many species of *Kinosternon* this species appears to be an opportunistic omnivore. Iverson (1986b) reported that plants comprised the bulk of the material in 19 gastrointestinal tracts. Seeds and/or fruits were found in 15 of the stomachs. Animals were found in 10 stomachs and included Coleoptera, freshwater shrimp, tadpoles of *Bufo marinus*, and fish bones.

HABITS AND BEHAVIOR

Activity is seasonal. During the rainy season (June–October) terrestrial activity increases. Individuals appear and migrate to ephemeral ponds, presumably to find food and mates. Presumably aestivation occurs, but it has not been documented. Aerial basking was observed in April and June but considered to be uncommon. When handled *K. oaxacae* is shy and not aggressive. Individuals rarely opened their mouths and always kept their heads maximally retracted when handled (Iverson, 1986b).

REPRODUCTION

Vitellogenesis begins at the end of the dry season (April and May) and continues into the rainy season (June and July). Mating and ovulation probably occur in June and July (Iverson, 1986b). Iverson found two females in April with enlarged ovarian follicles (>8 mm) and three others with numerous follicles, none larger than 5 mm. Maximum

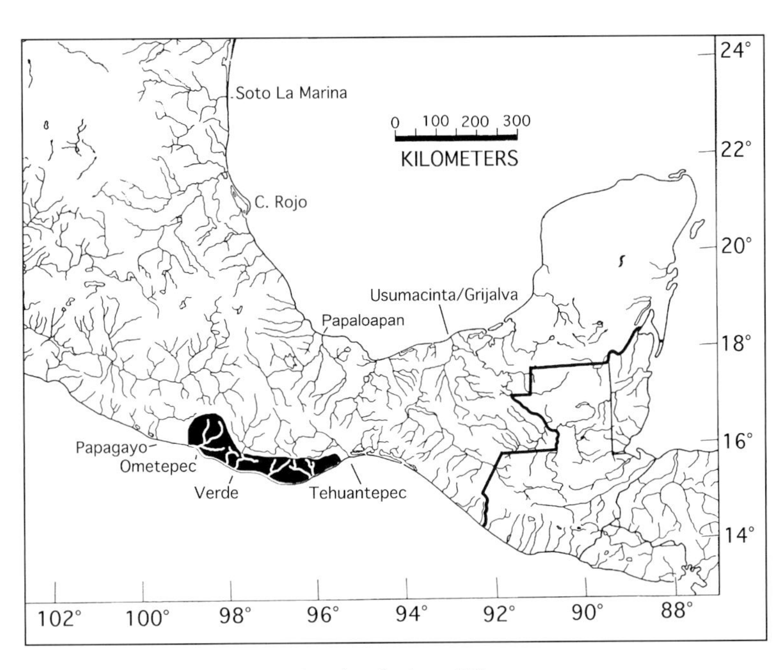

MAP 16 Geographic distribution of *Kinosternon oaxacae*.

FIGURE 22.5 *Kinosternon oaxacae*. Dorsal and ventral views of juvenile, region of Mazunte, Oaxaca.

follicle size from mid June was 11 mm, but most females had none greater than 8 mm.

Females with oviducal eggs or corpora lutea were not found. Nesting probably begins in mid July. Estimated clutch size for five females (115 to 141 mm CL) was two to five eggs. Iverson's data suggest a reproductive potential of two clutches of two to six eggs per clutch (Iverson, 1986b).

GROWTH AND ONTOGENY

Males are larger than females. Coloration of the chin is paler in females. Data on size (Iverson, 1986b:121): males, 141 (93–175) $n = 18$; females, 127 (95–157) $n = 21$. Based on gonadal examination, males of 113 to 125 mm CL were mature at 7–10 years. Females mature at 115 mm CL and 8–9 years. Calculated growth of *K. oaxacae* was judged to be more rapid than that of any other *Kinosternon*.

POPULATIONS

Iverson (1986b) found the density and biomass of *K. oaxacae* to be greater than that of any other kinosternid. He calculated a biomass of 523 kg/ha and a density of 1,750 turtles/ha (!) on the basis of 20 turtles from an ephemeral pool 6 × 20 m and less than 50 cm in depth. This may reflect the propensity of newly emerged turtles to congregate in small pools prior to dispersal, with the first pools formed at the beginning of the rainy season having unusually high densities.

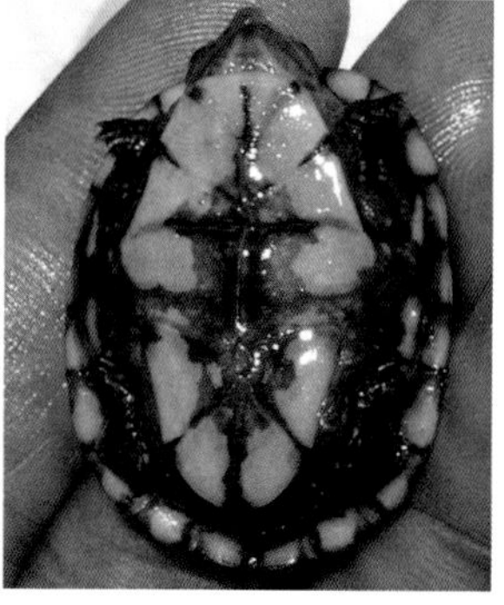

FIGURE 22.6 *Kinosternon oaxacae*. Neonate, region of Mazunte, Oaxaca.

CONSERVATION AND CURRENT STATUS

Vogt visited the range of *K. oaxacae* at Mazunte, Oaxaca, in April 2006. Martha Harfush (Centro Mexicana de Tortugas) has found the species to be abundant in the intermittent streams near Mazunte and has observed them to be active during the rainy season. There is little human interest in this species.

Kinosternon acutum Gray, 1831

Montera, chechagua de Monte (Liner, 1994); Tabasco mud turtle (Iverson, 1992a).

ETYMOLOGY

L. *acutus*, sharp, pointed; referring to the posterior tip of the plastron, which is pointed rather than notched.

HISTORICAL PROLOGUE

Vogt discovered a large population of *Kinosternon acutum* near Lerdo de Tejada, Veracruz, in 1995. Until then the species was considered to be "rare" (ca. 45 specimens in collections worldwide; Smith and Smith, 1979; Emys System Electronic Database, 2008). The basis for most of this account stems from studies of this population by Vogt and students from 1995 to 1997. All aspects of life history were studied, and data were taken from some 419 marked and released individuals.

The following diagnosis and description are adapted chiefly from Iverson (1980, 1988a, 1991b). Since *K. acutum* and *K. creaseri* are often confused, diagnostic and descriptive characters are often stated in terms of a comparison of these two species.

DIAGNOSIS

Kinosternon acutum and its sister species, *K. creaseri*, are members of the *scorpioides* group of *Kinosternon*, an assemblage of seven species occurring in Mexico and distinguished by a shared derived character—the lack of clasping organs in both sexes.

There is one usually evident mid-dorsal keel that becomes less distinct with age (dorsolateral keels are evident only in young); carapace is flat topped in cross-section and is never tricarinate. Plastral coverage of shell orifices is nearly complete, and PL/CL is 0.930 in males and 0.960 in females. Posterior plastral hinge is transverse, nearly straight, and is not angled or bowed posteriorly; posterior plastral lobe is bluntly pointed or spade shaped and is not notched. Premaxillary beak is not heavily developed and slopes posteroventrally rather than almost vertical. Adult head shield is large and covers the rostrum, and extends posteriorly over supraorbital ridges and a short distance down the postorbital bar, extending posteriorly as a triangle with a blunt apex at the base of the supraoccipital spine (same in a juvenile of 59 mm). There is one pair of gular barbels and one pair on the throat.

GENERAL DESCRIPTION

Kinosternon acutum is the smallest member of the genus occurring in Mexico; maximum size for males is 106 mm and for females is 116 mm (Iverson, 1991b). The largest adults observed in the Lerdo de Tejada population were as follows: males, CL 103 mm; females, CL 111 mm.

BASIC PROPORTIONS

(Vogt database)

CW/CL: males, 0.63 ± 0.02 (0.56–0.69) *n* = 125;
females, 0.63 ± 0.03 (0.56–0.76) *n* = 107;
im. males, 0.70 ± 0.02 (0.67–0.75) *n* = 12;
im. females, 0.68 ± 0.07 (0.50–0.78) *n* = 13.

CH/CL: males, 0.37 ± 0.02 (0.31–0.45) *n* = 125;
females, 0.39 ± 0.02 (0.34–0.44) *n* = 107; im.
males, 0.37 ± 0.03 (0.30–0.41) *n* = 12; im.
females, 0.40 ± 0.06 (0.35–0.58) *n* = 13.

CH/CW: males, 0.59 ± 0.04 (0.50–0.70) *n* = 12;
females, 0.62 ± 0.04 (0.50–0.72) *n* = 107;
im. males, 0.53 ± 0.04 (0.43–0.59) *n* = 12;
im. females, 0.53 ± 0.04 (0.43–0.59) *n* = 12.

PLASTRAL PROPORTIONS

(Iverson, 1991b)

PL/CL: 0.930 males, 0.960 females.

FL/CL: 0.293 males, 0.296 females.

HL/CL: 0.351 males, 0.368 females.

BL/CL: 0.302 males, 0.317 females.

PS1/CL: 0.181 males, 0.174 females.

PS6/CL: 0.305 males, 0.324 females.

FIGURE 23.1 *Kinosternon acutum*. Young adult female, Lerdo de Tejada, Veracruz.

C1 is in contact with M2; L4 is usually in contact with M11; M10 and M11 are elevated above M9. Head shield covers most of the dorsal surface of the head. The following lengths of plastral scutes are given as a percentage of CL: PS1, 9.1% males, 1.1% females; PS6, 28.7% males, 29.7% females. Axillary and inguinal scutes are in contact or narrowly separated.

Modal plastral formula: 6 > 4 > 1 > 2 > 6 > 3.

Carapace is dark brownish or slate to black with paler ground colors and darkened interlaminal seams. Plastron and bridge are yellow to pale brown with darker brown seams. Ground color of skin is grayish. Head (and forelimbs) are patterned with red, yellow, and black mottling from temporal region onto neck. Chin color varies from yellow to white and is mottled with brown or black.

GEOGRAPHIC VARIATION

In specimens from Veracruz M9, M10, and M11 are elevated above M8; in specimens from Tabasco, Chiapas, Belize, and Guatemala only M10 and M11 are so elevated (Vogt and Iverson, 2011).

RELATIONSHIPS

Kinosternon acutum and *K. creaseri* are considered sister species (Iverson, 1988a, 1991b); they are similar in size, shape, and ecology, and may replace each other geographically. Fide Iverson (1988a) the ranges narrowly approach each other and may actually overlap in southern Campeche and Quintana Roo. Difficulty in identification of these species is common (see Iverson, 1988a). The two species can, however, be distinguished by the usually straight posterior hinge in *acutum* (it is curved and bowed posteriorly in *creaseri*) (Iverson, 1988a). However, even this diagnostic character must be used in combination with others. *Kinosternon creaseri* also has a proportionately larger head, larger beak, and longer anterior plastral lobe.

FIGURE 23.2 *Kinosternon acutum*. Older adult female, Lerdo de Tejada, Veracruz.

Coloration of head and limbs in *creaseri* is less vivid than in *acutum*.

Corrigendum—We feel it necessary to note that a published plastral view of *"acutum"* in Iverson (1980:261.1, Figure 2) is actually *Kinosternon creaseri* (SM 11448). The error was acknowledged by Iverson (1988a).

GEOGRAPHIC DISTRIBUTION

Map 17

Kinosternon acutum occurs in the Gulf Coastal lowlands of central Veracruz and adjacent northern Oaxaca in the Rio Papaloapan drainage, below 300 m, from ca. 18°45′N eastward to southwestern Campeche, northern Guatemala, and northern Belize, excluding most of the Yucatán Peninsula where it is replaced by *Kinosternon creaseri*. Records for Laguna Escondida, Veracruz (Smith and Smith, 1979; Pérez-Higareda, 1978), are in error.

HABITAT

In the Rio Papaloapan drainage *K. acutum* is not conspicuos. It occurs in small pools in open areas of thorn scrub with sandy soil. They are semi-terrestrial, concealing themselves in leaf litter at the bases of shrubs and fallen logs by day and foraging in shallow pools at night. Understanding of these habitat characteristics was a chief factor in the discovery of large, healthy populations by Vogt in 1995. The population studied lived in a thorn scrub-riparian community that has been used for grazing since 1958 (Augustin Lara, pers. comm.). This pattern of land use is probably conducive to maintaining the open habitat where *K. acutum* thrives.

Elsewhere (e.g., Chiapas, Selva Lacandona [Vogt, pers. obs.], Guatemala [Duellman, 1963], and Belize [Legler, pers. obs.)] *Kinosternon acutum* is found in temporary forest pools after rains and sometimes in permanent pools (e.g., Rockstone Pond #1, Belize). Monteras do not favor fast moving water. Vogt twice observed adults struggling on the surface of flowing water (Chiapas, Rio Lacantun, and Rio San Pedro). The turtles were thrashing about clumsily in an attempt to reach the shore. Both incidents occurred after torential rains, suggesting that the turtles had been washed in by floods.

DIET

In Veracruz *K. acutum* feeds on insects and other invertebrates that are captured nocturnally while foraging in shallow pools. Stomach flushing of adults revealed larvae

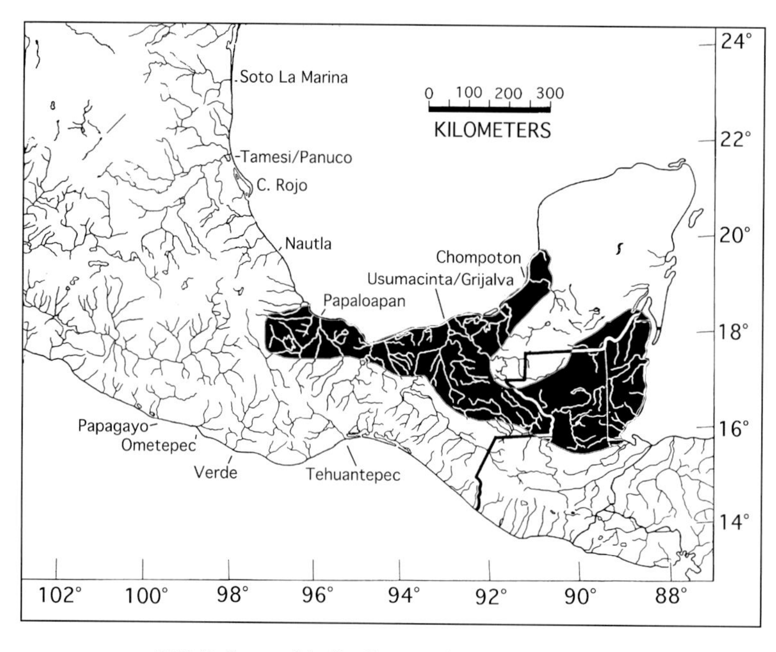

MAP 17 Geographic distribution of *Kinosternon acutum*.

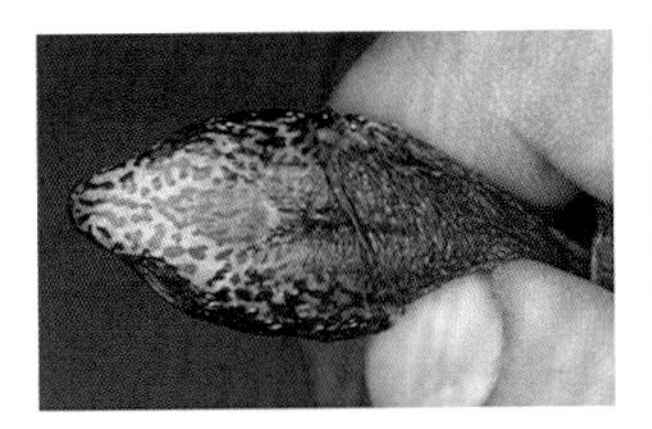

FIGURE 23.3 *Kinosternon acutum*. Top—dorsal and lateral views of adult male head; bottom—hatchling. Lerdo de Tejada, Veracruz.

FIGURE 23.4 *Kinosternon acutum*. Adult male, northern Belize.

of Coleoptera, Lepidoptera, Odonata, dyticids, and small snails, and no plant material. Soil occurred in some stomach flushings and could have resulted from terrestrial feeding.

HABITS AND BEHAVIOR

During the rainy season (July-March) they move into rain-filled pools and forage nocturnally. After the pools dry there is an inactive period from April until the rains of July-August fill the pools again. Legler observed three males and one juvenile *K. acutum* and three male *Claudius angustatus* in flooded forest—a small area of interconnected rain pools—at Rockstone Pond #1, Belize in June 1966.

Male-female pairs are often found together, in pools or leaf litter. Radio-telemetry showed that individual males followed the same female over a 6-month period. Usually, the turtles were within 6 cm of each other or actually in contact. Only rarely were the turtles more than 2 m apart. Mean home range for 10 turtles with radio transmitters was 120 $m^2 \pm 99.7$ (29–314). Females had larger home ranges than males, 169.50 $m^2 \pm 138.3$ (29-314) $n = 4$, versus 99 $m^2 \pm 67.2$ (52–198) $n = 4$, respectively. Two juvenile turtles with transmitters had the smallest home ranges, 61 $m^2 \pm 9.2$ (55–68). There was apparent selection for specific sites within the home range; individuals moved directly from a resting site to a feeding pool and back with little deviation (Vogt et al., 2000).

Male–female pairs are often found together in pools or leaf litter. Radiotelemetry showed that individual males followed the same female over a 6-month period. Usually the turtles were within 6 cm of each other or actually in contact. Only rarely were the turtles more than 2 m apart. Mean home range for 10 turtles with radio transmitters was 120 $m^2 \pm 99.7$ (29–314). Females had larger home ranges than males: 169.50 $m^2 \pm 138.3$ (29–314) $n = 4$ versus 99 $m^2 \pm 67.2$ (52–198) $n = 4$, respectively. Two juvenile turtles with transmitters had the smallest home ranges: 61 $m^2 \pm 9.2$ (55–68). There was apparent selection for specific sites within the home range; individuals moved directly from a resting site to a feeding pool and back with little deviation (Vogt, Dath, Espejel, and López-Luna, 2000).

REPRODUCTION

Females reach reproductive maturity at sizes of 77 to 97 mm CL in 5–8 years and mature males are as small as 74 mm.

At Lerdo de Tejada (see *Claudius* account for a detailed description of the area) egg laying commences at the beginning of the fall rainy season (September) and continues until the beginning of the dry season (March).

Mean number of eggs per clutch was 1.6 ± 0.76 (1–3) $n = 13$. Two to four clutches are laid. We do not know if individual females lay eggs every year.

Data on size and weight for 20 eggs are as follows: L, 33 ± 3.05 (30.0–38.7); W, 17 ± 1.22 (15–19); Wt 6.2 g $\pm$ 1.45 (4.4–9.1).

Incubation time was 98–180 days at 25°C. It varied with moisture and temperature regimes that influence embryonic diapause and embryonic aestivation. Sex is determined by incubation temperature, with the threshold temperature being about 27°C. Mostly males are produced at 25°C and mostly females above 29°C.

A nest was found on 12 February 1997 with three eggs. The eggs were covered with a mixture of loamy soil and dry leaves. The nest had a diameter of 60 mm, was 80 mm deep, and was in shade 1 m from the trunk of an *Acacia* tree. Another nest was found 2 June 1997 under similar conditions. Nest temperatures were taken with a data logger during the remainder of the incubation period; mean daily temperatures ranged from 31°C in July to 19°C in February.

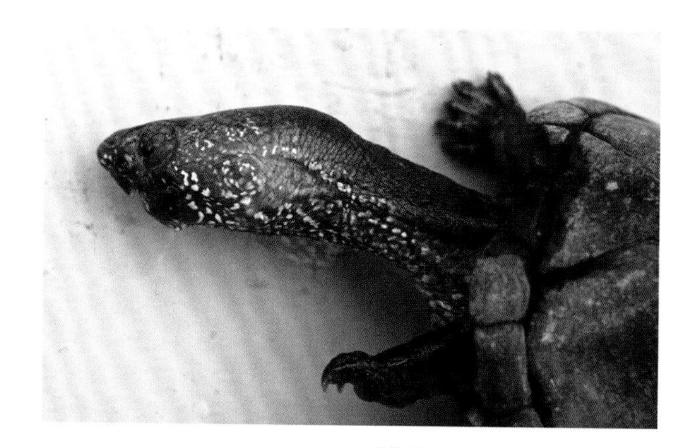

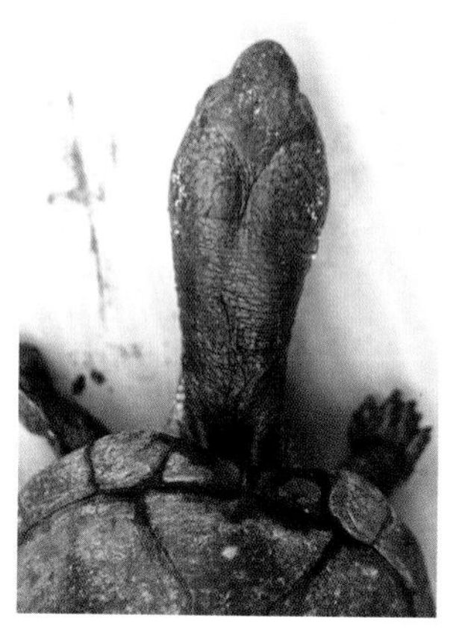

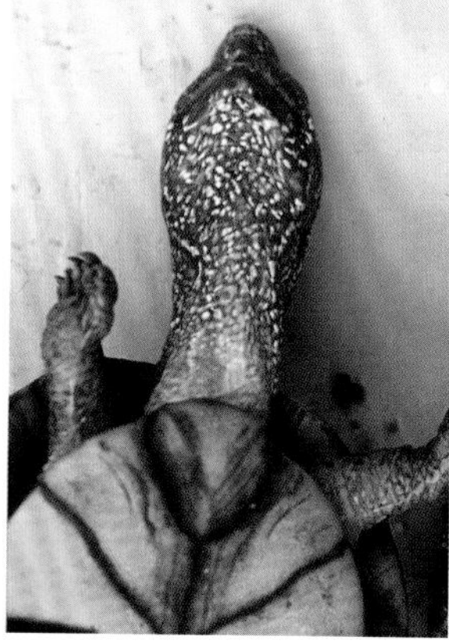

FIGURE 23.5 *Kinosternon acutum*. Three views of head, adult male, northern Belize.

FIGURE 23.6 *Kinosternon acutum*. Dorsal and plastral views of immature individual, northern Belize.

GROWTH AND ONTOGENY

Males have bolder coloration and longer, stouter tails than females, and concave plastra. Both sexes have a terminal tail claw and lack clasping organs. Females are slightly larger than males and have a more extensive plastron, a longer bridge, and a longer sixth plastral scute.

Data on size: males, 93 ± 5.9 (74–103) n = 126; females, 97 ± 7.35 (77–111) n = 107; im. males, 68 ± 6.34 (59–76) n = 21; im. females, 68 ± 7.62 (52–76) n = 13; im. unsexed, 62 ± 17.48 (26–72) n = 6.

Weight data: males, 100 ± 18.05 (52–185) n = 125; females, 117 ± 26.5 (54–171) n = 107; im. males, 41 ± 11.9 (23–55) n = 12; im. females, 46 ± 22.1 (17–95) n = 13; im. unsexed, 37 ± 18.7 (17.50–95.00) n = 13.

Data for 12 hatchlings were as follows: CL 26 mm ± 1.4 (25–29); Wt, 3.8 g ± 0.76 (2.8–4.7) n = 12.

Recaptures of marked turtles showed that one major growth ring is formed per year and that old turtles stop growing in length. Complete sets of growth rings were counted on 257 individuals (134 males, 117 females, 5 juveniles). In each age class there was an equal number of females and males. The relationship among growth rings, size, and weight seemed not to differ greatly between the sexes in any of the groups below.

Mean size and weight of each age group, as estimated by growth rings, are given as follows:

3 rings: CL, 69 mm ± 9.3 (54–82); Wt, 40.7 g ± 15.6 (22–64) n = 7.

4 rings: CL, 78 mm ± 14.6 (52–103); Wt, 62.9 g ± 36.4 (18–140) n = 14.

5 rings: CL, 87 mm ± 16.4 (62–102); Wt, 91 g ± 28.9 (28–147) n = 23.

6 rings: CL, 89 mm ± 9.6 (60–108); Wt, 93 g ± 26.3 (26–142) n = 49.

7 rings: CL, 90.5 mm ± 10.8 (64–104); Wt, 94 g ± 29.4 (30–157) n = 47.

8 rings: CL, 95.5 mm ± 5.5 (78–108); Wt, 110 g ± 23.4 (62–185) n = 36.

9 rings: CL, 95.5 mm ± 6.6 (76–107); Wt, 107 g ± 24.6 (46–150) n = 31.

10 rings: CL, 99 mm ± 4.7 (91–110); Wt, 116 g ± 18.3 (91–154) n = 21.

11 rings: CL, 99 mm ± 7.99 (75–111); Wt, 124 g ± 28.7 (51–171) n = 17.

12 rings: CL, 100 mm ± 4.2 (93–106); Wt, 130 g ± 21.5 (102–165) n = 10.

14 rings: CL, 95 mm ± 6.29 (90–99); Wt, 101 g ± 15.2 (90–113) n = 2.

PREDATORS, PARASITES, AND DISEASES

Raccoons (*Procyon lotor*) prey on eggs and adults.

POPULATIONS

Vogt and students collected and marked 419 *acutum* during the study; 287 were captured between January and June, and of these, 96 were recaptured in leaf litter by probing with *chuzos* (see Field and Laboratory Techniques).

Fyke nets were used from October to December 1997, collecting 132 turtles in small, rain-filled pools; 61 of these were recaptured. The population size for this 5 ha study area over a 12-month period was estimated to be 756, based on marking and recapture (Vogt, Dath, Espejel, and López-Luna, 2000). The overall sex ratio at Lerdo de Tejada was not significantly different from 1:1. The size and age classes collected in this study formed a normal curve, unlike most other species studied elsewhere.

CONSERVATION AND CURRENT STATUS

The market in Villahermosa was still selling locally caught and mummified *K. acutum* in the 1980s, but increased vigilance and enforcement seems to have stopped this. This species is small and not a desideratum for human food, nor does it seem important to the pet trade. Its populations in Mexico may be among the least disturbed by anthropogenic works. Populations are high and the habitat vast along the riparian community of the Rio Papaloapan. Matters of conservation and management for *K. acutum* would seem to have a low priority in comparison to other turtles in the same general area (e.g., *Claudius angustatus*).

Kinosternon creaseri Hartweg, 1934

Casquito de Creaser (Liner, 1994), Creaser's mud turtle (Iverson, 1988a).

ETYMOLOGY

The name is a patronym honoring the collector, Dr. Edwin P. Creaser (1907–1981).

HISTORICAL PROLOGUE

Kinosternon creaseri is one of the few turtle species in Mexico that has been referred to by the same name since the original description. Until recently biological and distributional information about this turtle was an enigma. Only 70 specimens were known from 18 localities. They had been reported to inhabit cenotes. J. C. Lee's seminal work (1980) on the Yucatán Peninsula reported only two specimens found during 9 months of fieldwork. Himmelstein (1980) found only one specimen of this species during six trips (totaling 53 days across 5 different months). Iverson (1988a) finally discovered that the preferred habitat was in temporary forest pools. Most of the natural history information in this account is based on Iverson (1988a).

DIAGNOSIS

Diagnosis and description are adapted chiefly from Iverson (1983a, 1988a, 1991b).

FIGURE 24.1 *Kinosternon creaseri*. Top—old male, Puerto Morelos, Quintana Roo; bottom—young adult male, Yucatán Peninsula. (Courtesy of J. B. Iverson.)

Kinosternon creaseri is a member of the *scorpioides* group of *Kinosternon* in which clasping organs are absent in both sexes, and tail claws are present in both sexes but are more heavily developed in males. It is distinguished from other Mexican *Kinosternon* by the following combination of characters: one obvious mid-dorsal keel that becomes less distinct with age (dorsolateral keels are evident only in young); carapace is never tricarinate or flat topped; cross-section of shell is roof shaped in young. Plastral coverage of shell orifices are nearly complete, PL/CL is 0.945 in males and 0.961 in females; posterior plastral hinge is curved, bowed, or angled posteriorly; rear edge of posterior lobe is evenly rounded and not pointed or notched. Premaxillary beak is heavily developed and almost vertical in profile in adult males. Adult head shield is extensive and covers the entire dorsal exposure of snout, extending posteriorly over the orbits with short, blunt extensions over postorbital bars then tapering posteriorly to a point near the middle of the supraoccipital crest (bluntly furcated in juveniles—only the portion over the snout and orbits is cornified). There is one pair of gular barbels and one pair on the throat.

GENERAL DESCRIPTION

Maximum CL is 125 mm in males and 121 mm in females, larger on average than *K. acutum*.

FIGURE 24.2 *Kinosternon creaseri.* Adult plastra: female left, male right; Puerto Morelos, Quintana Roo.

BASIC PROPORTIONS

(JML Database)

CW/CL: males, 0.66 ± 0.01 (0.64–0.68) n = 6; females, 0.69 ± 0.03 (0.66–0.74) n = 6; im. males, 0.67 n = 1; unsexed juveniles, 0.82 ± 0.02 (0.81–0.84) n = 4; hatchlings, 0.83 ± 0.02 (0.79–0.85) n = 11.

CH/CL: males, 0.35 ± 0.04 (0.28–0.38) n = 6; females, 0.39 ± 0.01 (0.37–0.41) n = 6; im. males, 0.37, n = 1; unsexed juveniles, 0.41 ± 0.01 (0.40–0.42) n = 4; hatchlings, 0.44 ± 0.02 (0.42–.47) n = 11.

CH/CW: males, 0.53 ± 0.06 (0.42–0.58) n = 6; females, 0.57 ± 0.03 (0.52–0.60) n = 6; im. males, 0.56, n = 1; unsexed juveniles, 0.50 ± 0.00 (0.49–0.50) n = 4; hatchlings, 0.53 ± 0.02 (0.49–0.57) n = 11.

PLASTRAL PROPORTIONS

(Iverson, 1991b)

PL/CL: 0.945 males, 0.961 females.

FL/CL: 0.307 males, 0.312 females.

HL/CL: 0.358 males, 0.372 females.

BL/CL: 0.278 males, 0.280 females.

PS1/CL: 0.195 males, 0.186 females.

PS6/CL: 0.320 males, 0.331 females.

C1 is in contact with M2; L4 usually is in contact with M11; M10 and M11 are elevated above the level of the preceding marginals. Length of plastral scutes, given in terms of mean percentage of CL, are as follows: PS2—8.6% in males, 9.8% in females; PS4—27.9% in both sexes. Axillary and inguinal scutes are in contact or narrowly separated.

Modal plastral formula: 6 > 4 > 1 > 2 > 6 > 3.

Carapace is usually dark (brown to black); if paler then interlaminal seams are darkened. The plastron and bridge are pale yellow to pale amber and unmarked except for darker interlaminal seams. Head is dark brown to black dorsally and is mottled with bright yellow laterally and light gray with darker speckles ventrally. The head and limb coloration of *creaseri* is less vivid than that of *acutum.* Iris is solidly black to tannish-orangish with at least two dark blotches contributing to a horizontal stripe; more often there are four blotches forming a stellate pattern.

Coloration varies geographically. Turtles from Quintana Roo are usually black to dark brown with paler yellow pigmented areas on the head and neck, whereas turtles from near Piste in central Yucatán are much paler with bright yellow markings on the neck, sides of head, and anterior antebrachium.

RELATIONSHIPS AND PHYLOGENY

The Yucatán Peninsula was submerged beneath warm, shallow seas from the beginning of the Cenozoic; emergence began in the Oligocene and ". . . continued until the Pleistocene" (Ward, 1985). The common ancestral stock of *K. acutum* and *K. creaseri* probably invaded the peninsula N of 18°N at some time during the Pleistocene and differentiated into *Kinosternon creaseri.*

Kinosternon creaseri and *K. acutum* are parapatric sister taxa of similar size and bear an overall similarity that has confused most of the scientists who have studied them (Duellman, 1965; Iverson, 1980; Legler, field notes). The foregoing diagnoses and general descriptions combined with geographic provenance should distinguish them (see also remarks under Behavior).

Iverson (1988b), using discriminate analysis, found significant differences between populations of *K. creaseri* in Yucatán and those in Quintana Roo and Campeche but made no suggestions for taxonomic change.

GEOGRAPHIC DISTRIBUTION

Map 18

Restricted to the mainland of the northern and central Yucatán Peninsula, in the states of Quintana Roo, Yucatán, and Campeche (N of 18°). Vogt (pers. obs.) did not find the species on Isla Cozumel despite intensive searching in March 1979 and July 1982.

An extralimital record from Tabasco (between Teapa and Tenosique; Flores-Villela, García, and Montes Oca, 1991) has been examined by Vogt and is not *creaseri.*

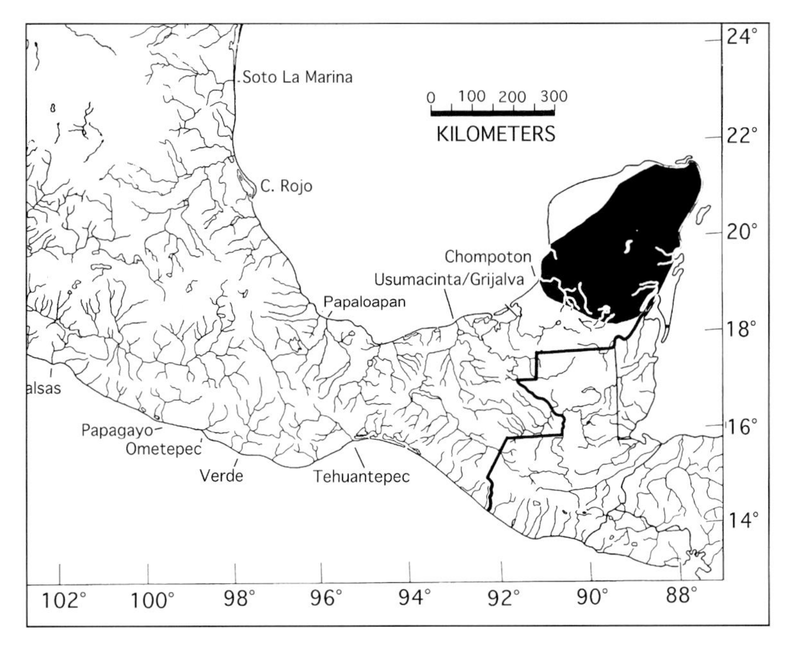

MAP 18 Geographic distribution of *Kinosternon creaseri.*

The general ranges of *K. acutum* and *K. creaseri* may overlap at the base of the Yucatán Peninsula (southern Campeche and Quintana Roo; Lee, 1996) but the species have not been found in microsympatry despite similarity of habitat preference. Both species may occur in microsympatry with *Kinosternon scorpioides* (which is tricarinate and easily distinguishable).

HABITAT

Modern cheloniologists came to expect that all "aquatic" turtles can be found, caught, or observed in water, and that terrestrial turtles should be sought on the surface of Earth (reference is chiefly to the work of Iverson, Legler, and Vogt, all of whom worked intensively in Mexico in the period 1958–2002). Turtle species not detected by these means were judged to be absent or "rare." The work of Iverson (1988a) and Morales-Verdeja and Vogt (1997b) has shown clearly that several species are cryptic in that they secret themselves in terrestrial habitats for part of the year—chiefly during the neotropical dry season. This fact accounted for the seeming "rarity" of several species. *Kinosternon creaseri* and *K. acutum* are in this category (but the same principles apply to terrestrial species that spend a large portion of their lives underground (see accounts of *Terrapene ornata luteola, Gopherus flavomarginatus,* and the *Kinosternon flavescens* group).

Kinosternon creaseri was initially thought to occur in cenotes but could not be found there. Actually they inhabit shallow temporary pools less than 1.0 m deep in undisturbed forest, temporarily filled roadside ditches, and small pools in shallow limestone depressions. However, they are truly "rare" in permanent water bodies such as cenotes and aguadas (Iverson, 1988a; JML, pers. obs.).

Legler and family were camped in the central park/playing field of Pueblo Nuevo, X-Can, Quintana Roo, on 31 July 1966 after the roads had reached impassability during a day-long heavy rain. A local resident, knowing of our interest in turtles, presented us with a series of 20 juveniles of *K. creaseri,* all of which were the young of the year or in their first full year of growth. These had been found in shallow rain pools formed in limestone depressions (referred to locally as *sardinales*). At that time the forest in that region had not been significantly disturbed—one could walk less than 1 km from a side road and observe spider monkeys.

DIET

Iverson (1988a) identified snail shells, insect parts, and palm seeds in feces. Captives ate only animal matter. Under natural conditions they probably feed on anything that comes into, lives in, drops into, or is forced from a subterranean refugium by standing water. This is a perfect dietary scenario for binge feeding on anything that is in an ephemeral pool (and then retires again to a terrestrial refugium). There is no evidence *K. creaseri* leaves the water to feed.

HABITS AND BEHAVIOR

The turtles probably move about through a mosaic of rain pools during the wet season (June to December). The only terrestrial movements actually observed were by juveniles on roads during rains.

Iverson (1988a) hypothesized that *creaseri* spends most of the year buried underground and emerges only in the

FIGURE 24.3 *Kinosternon creaseri.* Young adult male showing head and neck coloration and characteristic beak, Puerto Morelos, Quintana Roo.

wet season, when there is sufficient rain to form pools. Aestivation sites are unknown, but the turtles probably bury into leaf litter at the bases of shrubs and logs much like *K. acutum* in Veracruz (Vogt, pers. obs.).

Kinosternon creaseri differs from *acutum* (and perhaps other *Kinosternon*) by its extremely aggressive behavior—in the presence of other *creaseri* and when being handled. Rather than withdrawing into its adequately protective shell it remains exposed and bites viciously at the hands of a captor or at any other *creaseri* in the same pool. Iverson observed this behavior under natural and captive conditions and reported damage to the shell margins resulting from it. They also secrete a more pungent musk than other neotropical *Kinosternon*. Iverson (1988a) thought the powerfully developed beak could be an adaptation for aggression rather than feeding.

With a few exceptions, only one individual of *K. creaseri* (of any gender, age, or size) was found in any single pool or pond, and only twice was *creaseri* found in a pool with any other species of chelonian. Adult males were found alone in all but two instances—one a mating pair (17 August) and the other with a female close by.

Since only one male was ever found in a given pool, Iverson interpreted this as territoriality. The only male in a pool would have exclusive access to any female entering the pool.

REPRODUCTION

Females reach reproductive maturity at sizes of 110–115 mm and ages of 10–15 years (Iverson, 1988a). Two females of 106 and 108 mm CL were immature, with no follicles larger than 2 mm in diameter.

A female of 116 mm CL and 207 g (25 August) with 17 plastral annuli contained one large egg (38.2 × 19.1 mm; 8.9 g; Iverson, 1988a). The ovaries bore one corpus luteum and two enlarged follicles of 19 and 13 mm. A dissected female from Pueblo Nuevo X-Can, QR (112 mm, 31 July 1966) contained the following: one corpus luteum, 8 mm; three enlarged follicles, 18.0, 13.4, and 10.3 mm; three smaller-yolked follicles (8, 6, and 5 mm); and numerous follicles smaller than 3 mm (JML, pers. obs.).

These scant data suggest a potential for three to four one-egg, wet season clutches and is consistent with the neotropical reproductive "pattern" (Moll and Legler, 1971) that is shared with other turtles in this region of Mexico—multiple clutches, each with a few large eggs.

Relative clutch mass is low for *Kinosternon*—4.3% versus more nearly 10% in other species. But this figure in a single egg clutch represents high energy expenditure per egg in comparison to other *Kinosternon* (ca. 2%).

Nests and nesting have not been reported.

GROWTH AND ONTOGENY

Kinosternon creaseri is larger than *K. acutum* but is still one of the smaller *Kinosternon* in Mexico. Males are larger than females and have a heavily developed, seemingly prehensile tail terminating in a large claw.

Data on size (JML and RCV databases): males, 117 ± 3.67 (113–122) n = 6; females, 108 ± 4.38 (100–112) n = 6; im. males, 0.90, n = 1; unsexed juveniles, 43.2 ± 3.87 (39.7–46.7) n = 4; hatchlings, 33.0 ± 1.95 (30.3–37.1) n = 11.

Iverson (1991b) reported the following adult size data: males, 113.4 (96.0–124.9) n = 15; females, 106.8 (89.1–121.4) n = 11.

Juvenile growth is rapid. Males grow faster than females by their second year. Six juveniles captured in the first year of growth attained sizes of 88.3 mm CL (84 -93) and 80.9 mm PL (78–86) during 2 years of captivity. This rate was faster than that for wild turtles (Iverson, 1988a).

The series of young specimens obtained at Pueblo Nuevo X-Can, Quintana Roo (31 July 1966 UU), provide most of the following data on juvenile stages.

Individuals of 40 to 47 mm CL were classified as juveniles; each had a wide major growth zone preceded by a distinct growth ring at the edge of uninterrupted areolar growth; the wide growth zone contained several minor growth rings (perhaps each resulting from a separate emergence).

Individuals of 30–34 mm, classified as "hatchlings," were clearly older and larger than would be expected in freshly hatched young; the umbilicus was closed in all, and only one bore a worn caruncle. All showed slight areolar growth (detectable only by scute surface texture), and there were no growth rings within this areolar growth zone.

At all ages the plastron is pale yellowish to straw colored and unmarked except for darkened interlaminal seams. Zones of new growth in juveniles are darker. In all hatchlings and juveniles the carapace is solidly reddish brown with black interlaminal seams; on each carapace scute there is a darker brown spot (small but twice the width of the dark interlaminal seams) at the epicenter of each

FIGURE 24.4 *Kinosternon creaseri*. Series of 18 neonates and first-year juveniles taken after a heavy day-long rain, Pueblo Nuevo X-Can, Quintana Roo.

carapacal areola. The pale field of the plastron is vaguely specked with small, gritty-looking black marks in about half the young specimens.

Hatchlings and juveniles have a much less developed head shield than adults, extending from the snout over the orbits in two prongs separated by a wide notch. Top of head is mostly unicolored pale brown; there are irregular pale marks on the side of the head and adjacent neck that are usually organized into an irregular orbitocervical stripe and another such stripe running from the inferior border of the orbit to the jaw angle.

PREDATORS, PARASITES, AND DISEASES

Iverson (1988a) found ticks on the neck of a large adult near Dzibalchen, Campeche.

POPULATIONS

Considering new information on the cryptic nature of habitat, Iverson (1988a) suggested that *Kinosternon creaseri* could actually be the most abundant turtle on the Yucatán Peninsula. This species is ripe for a thorough study of life history using modern techniques of mark and recapture, including radio telemetry, stomach flushing, and trained dogs.

CONSERVATION AND CURRENT STATUS

The paved highways to tourist attractions on the northern Yucatán Peninsula were under construction in 1966. Quarries for road-base material were numerous and level, and we (JML and family) used them to park our large travel trailer. The borrow pits contained turtles even then, and these pits remained in mature form as ditches when Iverson (1988a) worked the area some 20 years later. Ditches may constitute some of the most enduring turtle habitats in Mexico—they are virtually eternal. Although Iverson (1988a) found *creaseri* in ditches, prime natural habitat consists of wet season pools in mature forest. It is the waning of the forest (cutting, desiccation, and utilization of low cleared areas for crops) that may threaten *Kinosternon creaseri* populations.

The two-lane highway with roadside pools has been replaced by a four-lane, high-speed highway from Chetumal and Cancun to Merida. Iverson (pers. comm.) wondered what effect this traffic has had on the population of *K. creaseri*. RCV and sons spent ca. 8 hours on this road between Cancun and Playa del Carmen during and after rainstorms in July 2010 and again ca. 14 hours in the section from Cancun to Chetumal in July 2011. No roadside pools and no turtles of any kind were seen on the highway, day or night. Further study of *K. creaseri* must now be done in the realtively undisturbed habitats closer to the center of the Yucatán Peninsula.

ADDENDUM

The events of life history between emergences into temporary pools remain unknown. Until they are, the following hypothesis and order of events seem worthy of testing.

Emergence is stimulated by heavy rains (the ephemeral pools may actually inundate refugia); feeding and mating occur in the pools; nesting occurs possibly in the refugia; when pools dry the turtles return to the refugia; the little known about the reproductive cycle would provide for the laying of one large, thick-shelled egg during each of several emergences; a major growth ring is formed after each emergence; if it does not rain, there is no emergence, no feeding, and possibly no growth. (None of this is fanciful; all events hypothesized are known to occur in other turtles.)

Kinosternon leucostomum (Duméril and Bibron) 1851

Pochitoque (Veracruz, Liner, 1994); chechahua (Alvarado, Veracruz); white-lipped mud turtle (Iverson, 1992a).

HISTORICAL PROLOGUE

Duméril and Bibron (1851) gave the type locality as "N. Orleans: Mexique Rio Sumasinta (Amer. Central)." Schmidt (1941) restricted it to the Rio Usumacinta, El Peten, Guatemala. This was a locality where the collector of the type specimens, Morelet, had received other specimens of reptiles. The Rio Usumacinta forms the border between Chiapas and Guatemala in this region.

ETYMOLOGY

Gr. *leukos*, white; Gr. *stoma*, mouth; referring to the pale maxillary jaw sheaths in many populations.

RELATIONSHIPS AND TAXONONOMIC STATUS

Two subspecies are now recognized by most authors (Berry, 1978; Iverson, 1992a; Smith and Smith, 1979): *K. l. leucostomum*, which ranges from Veracruz southward to the Mosquito Lowlands of Nicaragua; and *K. l. postinguinale*, ranging from Nicaragua to Ecuador. *Kinosternon spurelli* (Boulenger, 1913) is a synonym of *K. l. postinguinale* (Berry and Iverson, 2001a). Other authors have ignored subspecific designations because of the great amount of variation occurring in the species. *Kinosternon l. leucostomum* is the subspecies that occurs in Mexico.

There is much (thus far unevaluated) geographic variation, chiefly in color, in Mexico and elsewhere. There is also substantial geographic variation in some aspects of life history. We wish to make it clear that the taxon we report on here is *Kinosternon leucostomum leucostomum* (Duméril and Bibron, 1851) and that we hereinafter refer to it simply as "*Kinosternon leucostomum*." We do not place *K. leucostomum* in any "group."

DIAGNOSIS

The following combination of characters distinguishes *Kinosternon leucostomum* from all other Mexican *Kinosternon*: carapace is unicarinate in juveniles and becomes smooth in adults; carapace is highly domed in adults. Movable plastral lobes are large and almost entirely close the shell orifices. C1 usually (88.5%) contacts M2. Axillary and inguinal scutes are usually (81.5%) separated. Clasping organs are present in males and absent in females. Plastral scute 4 is short, ca. 23% of CL. Posterior plastral lobe is rounded. Anal notch is small or lacking. Length of second plastral scute is less than 60% of anterior plastral lobe.

GENERAL DESCRIPTION

Kinosternon leucostomum is a medium- to large-sized turtle over most its range and ranks well above median size within the genus. It attains the largest adult size for any *Kinosternon* in Mexico (maximum CL and Wt are 214 mm and 1,214 g in males and 208 mm and 1,087 g in females). In a database containing 2,811 total specimens of *Kinosternon* sp. (amalgamation of JML, RCV, and J. F. Berry data), median adult CL was 96 mm; 79% of those above the median were *K. leucostomum*. Size varies greatly between populations.

BASIC PROPORTIONS

(Overall for 1,352 Adults from Mexico)

CW/CL: males, 0.63 ± 0.027 (0.55–0.79) $n = 763$;
females, 0.66 ± 0.029 (0.56–0.81) $n = 589$.

CH/CL: males, 0.41 ± 0.026 (0.31–0.52) $n = 763$;
females, 0.44 ± 0.037 (0.34–0.53) $n = 587$.

CH/CW: males, 0.65 ± 0.043 (0.50–0.85) $n = 763$;
females, 0.67 ± 0.052 (0.53–0.81) $n = 587$.

FIGURE 25.1 *Kinosternon leucostomum*. Dorsal, ventral, and lateral head photographs: top row—adult male, UU 9521, CL 124 mm; middle row—adult female, UU 9526, CL 116 mm; bottom row—male on left. Both specimens from region of Frontera, Tabasco. (From Berry, 1978, with permission. Snouts of top row inadvertently cropped in original publication.)

PLASTRAL PROPORTIONS

(Iverson, 1991b)

PL/CL: 0.907 males, 0.925 females.

FL/CL: 0.323 males, 0.325 females.

HL/CL: 0.356 males, 0.371 females.

BL/CL: 0.245 males, 0.255 females.

PS1/CL: 0.137 males, 0.139 females.

PS6/CL: 0.280 males, 0.298 females.

Modal plastral formula: 6 > 4 > 2 > 1 > 5 > 3.

C4 is usually in contact with M11. M10 and M11 are subequal in height. Fourth plastral scute is 27% of the plastral length and less than 80% of the forelobe length; bridge length is less than 28.5% of CL. Plastral lobes completely close shell orifices, much as in box turtles (*Terrapene*). Clasping organs are present but poorly developed in males and are absent in females. Tail claw is present in both sexes but is more heavily developed in males. There is one pair of gular barbels.

In adults the carapace is gray, straw colored, brown, or black, and the plastron and cream, yellow, brown, or black. The dark coloration can be due to natural melanin production or staining from chemicals in aquatic habitats (Figure 3.7).

Head pattern varies geographically, locally, and ontogenetically within and between populations (Figure 25.9)

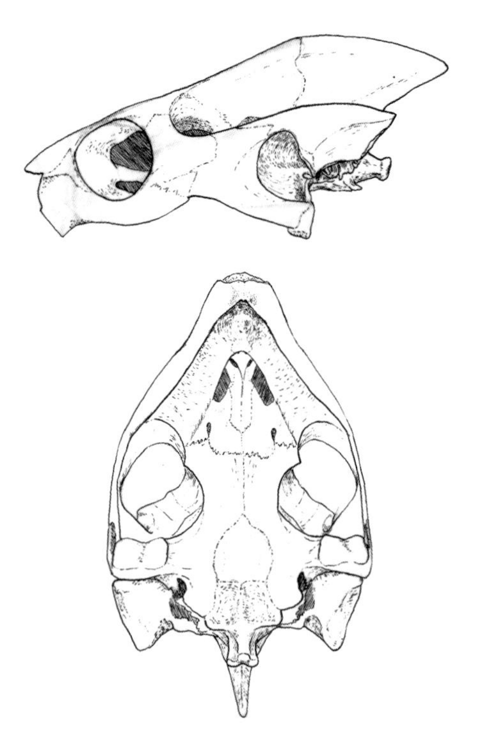

FIGURE 25.2 *Kinosternon leucostomum*. Adult female skull (*K. l. postinguinale*), UU 4233, CBL ca. 25 mm, Gamboa, Canal Zone. (Drawing by Elizabeth Lane, from Legler, 1965, with permission.)

but has never been properly analyzed or quantified. A common pattern consists of two pale stripes on the side of the head, one on the upper jaw and the other extending posteriorly from the orbit. Some populations have only the jaw stripe, and the remainder of the head is mottled brown. Other populations are composed of individuals with a mottled head that lacks stripes. In other populations (e.g., Laguna Zacatal and Laguna Escondida, Los Tuxtlas, Veracruz), for example, patterning of soft parts is lost with age and replaced with unicolored dark brown or black. Most hatchlings have a dark plastral figure, but the plastron may be uniformly pale (orangish, yellowish, or cream) in the same population (Figure 25.9).

The basic pattern in Veracruz is a pair of pale stripes on the side of the head, both beginning on the snout—one passing from the snout over the orbit and widening on the temporal region, the other passing from the snout along the maxilla at least to jaw angle (Figure 25.3). The stripes vary in width and define a narrow, dark, intermediate stripe on each side of the snout and a much wider dark stripe or patch from the eye onto the neck. The extent of pale (yellowish) areas and dark (brownish) areas can produce patterns ranging from almost solidly pale to solidly dark with various degrees of mottling, vermiculation, and speckling (Figures 25.3, 25.4, 25.7, and 25.8).

GEOGRAPHIC DISTRIBUTION

Map 19

Kinosternon l. leucostomum ranges from the lowlands of central Veracruz just south of Punta del Morro (18°55′N), in the Gulf drainages from the Rio Papaloapan to the Usamacinta, and thence across the Yucatán Peninsula to the Caribbean drainages of Quintana Roo, Belize, Guatemala, and Honduras, and southward to the latitude of Lake Nicaragua (at altitudes of <300 m in Mexico). The northern limit of the range corresponds to that of many other neotropical species.

The species also occurs in both Pacific and Caribbean drainages of the Mesoamerican isthmus from southwestern Nicaragua southeastward to the Atrato and Magdalena drainages of Colombia and to west-central Ecuador at

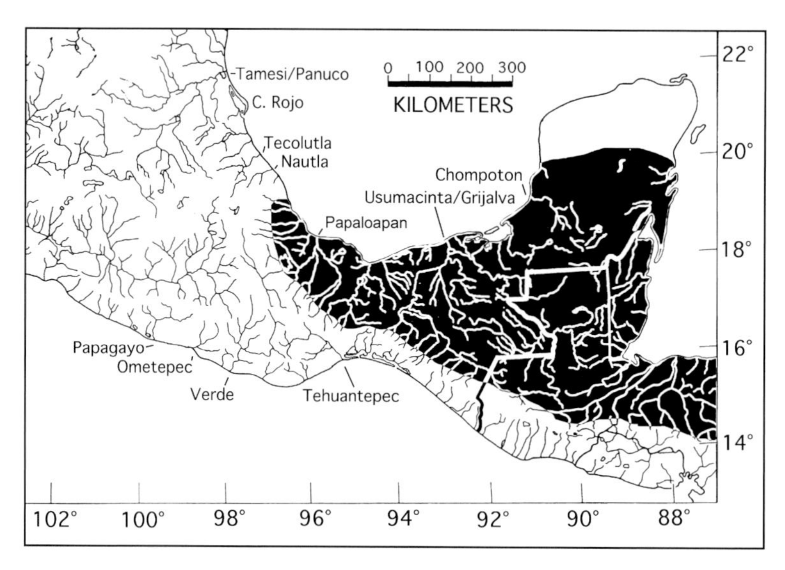

MAP 19 Geographic distribution of *Kinosternon leucostomum*.

FIGURE 25.3 *Kinosternon leucostomum*. Adult female, Villahermosa, Tabasco. This demonstrates the basic shape, shell colors, and head patterns that show great variation in other populations. Note dull carapace, contrasting yellowish plastron, temporal speckling, and the pale "lips" alluded to in the specific name. (Photo courtesy of Claudia Zenteno-Ruiz.)

FIGURE 25.4 *Kinosternon leucostomum*. Adult, Jesus Carranza, Veracruz, Coatzacoalcos drainage near Oaxacan border. (Photo courtesy of B. Horne.)

elevations below 750 m (Berry and Iverson 2001a; Meyer and Wilson, 1973). Records for northwestern Peru (Pritchard, 1984; Pritchard and Trebbau, 1984) are unsubstantiated (Carr and Alemendriz, 1990). The photos published in Bartlett and Bartlett (2003) are, in fact, a color morph of *Kinosternon scorpioides* common in the state of Acre, Brazil, on the frontier with Peru.

GEOGRAPHIC VARIATION

Within *Kinosternon leucostomum*, and even within the subspecies *K. l. leucostomum*, there is a tremendous amount of variation in color, body form, and characteristics of life history. Attempts have been made to analyze this variation (Legler, 1958, research notes; Berry, 1978), but much critical study, including molecular analysis, is needed to explain it fully.

Vogt has gathered life history data on some 1,330 marked individuals of *K. leucostomum* in central and coastal Veracruz. Most of the life history information in what follows is based on this unpublished work unless otherwise cited.

In this account we describe some differences in life history characteristics that occur in two narrowly separated populations in Veracruz. It is possible that head patterns and shell proportions are influenced by diet or habitat preference. For example, individuals that live in small streams often do not grow larger than 120 mm or 500 g. Individuals in these populations usually have the distinctive white jaws. Other populations live in temporary lakes; are active for only part of the year; have a highdomed, unmarked, smooth carapace; and grow to a 210 mm CL and 1,000 g. In what follows, we offer some insight into some of this variation and describe it where possible, but are unable satisfactorily to explain it.

HABITAT

Kinosternon leucostomum occurs in a wide variety of aquatic habitats that include temporary ponds, roadside ditches, marshes, edges of large lakes, small streams, rivers, mangrove swamps, and spring pools. They populate and utilize nearly every aquatic niche present in the neotropics. Populations are highest in small ponds and marshes where *Staurotypus triporcatus* and crocodiles are not abundant. This kind of habitat, in the Rio Papaloapan drainage near Lerdo de Tejada, has the richest biodiversity of freshwater turtles in Mexico.

In rivers and large lakes pochitoques occur chiefly within 3 m of the shore at depths of 0.5–2.0 m and avoid water deeper than 5 m. Therefore, extrapolations of population density can not and should not be made from shoreline trapping or observation. Densities in small ponds are predictably high and should not be used to calculate populations in larger bodies of water.

Pochitoques feed along the shoreline, crawling along the bottom and surfacing frequently for air. They can often be found by muddling in shallow water among the aquatic vegetation, in bottom mud, or under fallen trees. Some populations aestivate by crawling out of a drying pond and burying themselves on the forest floor, under boulders or tree trunks, at the bases of buttressed trees, and in the leaf litter. They also utilize the burrows of small mammals, such as armadillos (*Dasypus*) and pacas (*Cuniculus*), as refugia.

DIET

Kinosternon leucostomum is omnivorous, feeding on fruits (e.g., *Ficus*), shoots, flowers, insects, fish, crayfish, tadpoles, and mollusks. The proportion of the diet composed of plant or animal material varies among populations, depending on the resources available. Turtles in Laguna de Zacatal, where 75% of the diet was animal material, had faster growth rates than turtles in a nearby deep lake, Laguna Escondida, where the diet was 75% plant material

FIGURE 25.5 *Kinosternon leucostomum*. Shell shape, pattern, and coverage in specimens from Alvarado, Veracruz (top), and Jesus Carranza (bottom); males on right.

(Vogt and Guzman, 1988). See comparative data following the account of reproduction.

At Lerdo de Tejada, where *K. leucostomum* lives in sympatry with *Claudius*, *Kinosternon acutum*, *Chelydra*, *Dermatemys*, *Staurotypus triporcatus*, and *Trachemys scripta venusta*, they feed on terrestrial and aquatic insects, snails, crustaceans, and to a lesser extent fruits.

HABITS AND BEHAVIOR

Even in small lakes, where they are the only species of turtle, pochitoques utilize only the shoreline habitat (Morales-Verdeja and Vogt, 1997a). Populations that inhabit temporary ponds leave the ponds as water levels drop during the dry season in January and February, and aestivate for up to 5 months buried beneath the leaf litter in the forest. The ability to close the shell tightly provides protection from predators and impedes desiccation. Studies are lacking concerning the use of energy stores while aestivating.

Populations within 2 km of each other in Veracruz differ in that one in a temporary lake aestivates during the dry season while the population in the permanent lake does not. Those in the permanent lake continue producing clutches of eggs until late May, whereas those in the temporary lake (Laguna de Zacatal) stop nesting by mid March. Summer rains (June–July) stimulate emergence from aestivation and the return to the temporary lake.

Hatchlings and juveniles can frequently be seen basking on branches of trees and shrubs that overhang the water. Basking is communal, on the same log or on branches of the same shrub. This seems to be an aggregation behavior that could serve to reduce predation. Basking occurs mainly between 0700 and 0900 hours, and usually in shrub tangles. Adults bask under the same conditions, but represent less than 10% of the individuals seen basking by Vogt in 20 years.

Transmitters were attached to 11 females and 3 males to document activity cycles and movements in and around Laguna de Zacatal (Morales-Verdeja and Vogt, 1997b). A trained dog helped find aestivating turtles once the general area was located by the radio receiver. The dog produced distinguishable barks for turtles and snakes (e.g., fer-de-lance, *Bothrops asper*).

Turtles had linear home ranges, 50 200 m long, around the shoreline. Activity was concentrated in 10–50% of total home range. Turtles moved up to 600 m from the home range to nest or aestivate. In 2 years of study turtles were found to return to the same activity areas in the lake after aestivation in the surrounding rain forest.

Turtle density was estimated as 29 turtles/ha, considering the entire lake as suitable habitat. However, using a 3 m band of shoreline as potential habitat, an estimate of 449 turtles/ha is probably more realistic.

REPRODUCTION

(RCV unpublished data unless otherwise cited.)

MATING

Vogt observed males trailing and mounting females in August at Laguna de Zacatal within 2 weeks after the summer rains began to fill the lake. All mating events (18) were in clear water less than 1 m deep in the grassy end of the lake. Ten pairs were seen between 0800 and 1100 hours and eight between 2300 and 0200 hours. Actual copulation was not observed. There does not appear to be any stereotyped courtship ritual in this species; a male chases and bites at the head of a female and mounts when she becomes quiescent.

MALE REPRODUCTIVE CYCLE

Spermatogenesis occurs in Panama from April to October. Specimens obtained in January were at the end of the cycle—mature sperm were found in the epidydimides tubes but not in the seminiferous tubules (Moll and Legler, 1971).

Morales-Verdeja (2000) studied the testicular cycle in a population of *K. leucostomum* at El Jicacal, southern Veracruz (ca. 18°35'N, 94°34'W). Habitat was a stream

TABLE 4
Data from Mature Female *K. leucostomum* from Lerdo de Tejada

Date	Number of Females	Clutches per Year[a]	Eggs per Clutch	ARP[b]
06 July 1990	25	1–3	1–3	4–10
09 July 1983	3	3–4	2–3	6–10
30 July 1983	8	2–4	1–2	4–7
24 Sept. 1983	3	3–5	1–4	7–10
09 Nov. 1983	18	2–4	1–4	5–12
19 Nov. 1983	7	3–4	2–4	7–12
07 Dec. 1983	4	3	2–4	6–9
17 Dec. 1983	9	3–6	1–5	5–13
04 Feb. 1984	6	2–4	1–5	4–13
19 Mar.1984	9	4–7	1–3	6–12
23 Mar. 1983	2	3–4	2–3	6–8

Data were calculated by measuring and counting follicles >4 mm, corpora lutea and shelled oviducal eggs in 94 mature females.

a. Clutches are the number of potential clutches per annum as estimated from size grouping of follicles, corpora lutea and egg counts.

b. Reproductive output is stated in terms of a potential annual reproductive potential (ARP) that may or may not be realized.

flowing through a marsh and pasture land ca. 200 m from the ocean shore. The study was based on dissections and histological sections from individuals obtained monthly from February 1992 through January 1993.

Stage of the spermatogenic cycle was determined from histological sections (as in Moll and Legler, 1971); testis size alone was not a reliable indicator. Testicular regression began in the fall and winter (October–January) in synchrony with rainfall and lower temperatures. Spermatogenesis was maximal from late summer to early autumn (August and September). Testicular regression began in late summer and early autumn (August and September), and partial gonadal quiescence began in early winter (November–December).

Aestivation may or may not occur, depending on environmental conditions and the particular population studied. During aestivation (in the hottest part of the year, May–June) the testes are in a state of quiescence. Spermatogenesis is probably initiated In April, 1 month before the dry season begins, permitting males to enter aestivation at an advanced stage and be ready to inseminate females upon emergence in July or August.

This pattern of spermatogenesis is termed "postnuptial" by Meylan, Schuler, and Moler (2002) (see Introduction). The pattern described by Morales-Verdeja (2000) is essentially continuous spermatogenesis for a given population, since the beginning and end of the process overlap among individuals. The actual duration of the cycle is about 11 months.

Males of 99 167 mm CL were studied. Males of 99, 121, and 126 mm showed neither spermatogenesis nor spermatozoa in the epididymis and were judged to be immature.

FEMALE REPRODUCTIVE CYCLE

Vitellogenesis probably occurs throughout the year, but ovulation (in Veracruz) occurs from late August to May in all populations and habitats. Of 136 females dissected, some females contained enlarged ovarian follicles in all calendar months.

Moll and Legler (1971) commented that *K. leucostomum* reproduced throughout the year in Panama and Costa Rica, based on the presence of shelled eggs in 9 months of the year (no samples in December, February, and March).

In Veracruz, *K. leucostomum* is unusual in retaining follicles of preovulatory size into the next nesting season without reabsorption. It is possible (but undocumented) that some reabsorption occurs during aestivation. In any case, Vogt (unpublished data) documented that females enter and leave aestivation with enlarged follicles and, in populations that do not aestivate, bear enlarged follicles in all months of the year.

NESTING

Nesting occurs from late August through late May in Veracruz and September–December in Tabasco (Vogt, field notes). Nesting is crepuscular, nocturnal, or occurs on

FIGURE 25.6 *Kinosternon leucostomum*. Twenty-four-hour-old neonate, Los Tuxtlas, Veracruz. Note caruncle below nares.

overcast days. Nests are often dug adjacent to fallen trees or within the form where the female is aestivating.

Most nests (ca. 90%) were dug in soil or leaf litter. However, eggs are often found lying among leaves on the forest floor with no evidence of nest excavation or attempts to conceal the egg. Random surface oviposition of single eggs has not been noted in other *Kinosternon*, but Moll and Legler (1971) described a similar scenario for *Rhinoclemmys funerea* in Panama.

Vogt (1981b) found the following proportion of gravid to nongravid adult females in his study areas near Los Tuxtlas, Veracruz: January 6/2, February 54/35, March 25/27, April 5/47, May 5/118, June 0/122, July 0/145, August 5/7, September 14/39, October 14/14, November 28/2, December 4/0. Enlarged follicles were present in the months in which no shelled eggs were observed.

Mean number of eggs per clutch for 289 females (from all study areas in central Veracruz) was 2.26 ± 0.94 (1–5). Data on size and weight of the gravid females were as follows: CL, 147 mm ± 18.47 (106–178); Wt (before laying), 491 g ± 220 (204–998).

EGGS

From one to five brittle-shelled eggs are laid in up to six clutches per reproductive season. There is no evidence for skipped seasons of reproduction. Data on overall size and weight for 482 eggs from three localities in central Veracruz are as follows: L, 34.5 ± 2.75 (27–43); W, 18.6 ± (12–27); Wt, 7.65 ± 1.65 (3.9–12.9). Number of eggs per clutch is correlated with female size and particular population (see the following). Data on eggs and hatchlings from 12 clutches (central Veracruz) are as follows: egg L, 36 ± 3.4 (32–42); egg Wt, 9.5 g ± 1.76 (5.5–12.8) $n = 19$; hatchling CL, 30.2 mm ± 3.14 (22–35); hatchling Wt, 5.8 g ± 1.24 (3.1–8.3).

Data for *K. leucostomum* in Panama and Costa Rica were as follows: one to two eggs per clutch (5 clutches of two eggs and 16 of one); mean egg size 37 × 19 mm (35 × 18 to 40 × 21 mm); incubation time at temperatures of 20–33°C was 126–148 days; hatchling CL was 32.7 mm; nests were made in leaf litter (Moll and Legler, 1971).

INCUBATION

Incubation time at 25°C was 125 days ± 26.6 (88–168). Lab incubation times varied from 90 to 265 days at 28°C, depending on humidity and its influence on embryonic diapause and embryonic aestivation (see Introduction).

If eggs were maintained on a moist substrate for 1–4 months they remained in diapause and ceased development. Eggs left on a dry substrate for 2–3 weeks and then put in a moist substrate began to develop. Lowering incubation temperature to 23°C for 2–3 weeks and then raising it to 28°C also stimulated development. See other details and comments presented under "Embryonic Diapause And Synchronous Emergence" in Introduction.

Sex is controlled by incubation temperature: Males are produced at 25–27°C and females at 28°C and above (Vogt and Flores-Villela, 1992a). *K. leucostomum* may be in a stage of evolving genetic control of sex determination; 100% of either sex was never produced at any temperature. Adult sex ratios in some populations are 1:1; others have been found to be 2:1 in favor of males.

Larger eggs produce heavier hatchlings but not necessarily greater CL. Heavier hatchlings were also produced from eggs maintained under moister incubation conditions in the laboratory at Los Tuxtlas.

VARIATION IN NATURAL HISTORY PARAMETERS

Most of the foregoing data on diet, reproduction, and other parameters of life history (e.g., egg size) are stated in overall terms for the three populations studied intensively in central Veracruz. Actually, these populations varied significantly and we deem it worthwhile to summarize these differences below.

LAGUNA DE ZACATAL, VERACRUZ

Summary data: Turtles have larger size, feed on more animal material, have faster growth rates, and lay larger clutches of larger eggs, and all turtles aestivate.

Laguna de Zacatal (18°35′N, 94°04′W) is a seasonal volcanic crater lake of 7.5 ha, 7 m average depth, and maximum depth 13 m. It begins to fill with the summer rains of July–August. Water temperature is 25-31°C. Its overflow enters the L. Esdondida ca. 2 km to the southeast. Water levels fall rapidly in January, and the lake dries by late March or April. *Kinosternon leucostomum* leave the lake to aestivate in surrounding rain forest until the lake fills again. The dry lake bed is covered with low seasonal vegetation, including grasses. When the lake fills, these provide forage for tadpoles, turtles, and apple snails. Freshwater shrimp (*Macrobrachium*) migrate into the lake at high water. *Kinosternon leucostomum* is the only species of turtle, and there are no large fish (a few cyprinids may swim

FIGURE 25.7 *Kinosternon leucostomum.* Relatively dull adult shell and head patterns, Alavarado, Veracruz (top, females), near northern edge of range.

upstream in the overflow channel). No turtles occur in the connecting stream. Diet is 75% animal material. Adult size is larger, and growth rates are faster (Vogt and Guzman, 1988).

Kinosternon in Zacatal have a higher-domed carapace than other populations and hold the head high and the plastron off the substrate while traversing land.

SIZE AND WEIGHT

CL: males, 174 mm ± 17.5 (100–214) n = 265;
females, 164 mm ± 12.6 (108–192) n = 205;
im. unsexed, 82.9 ± 32 (36–138) n = 62.

WT: males, 797 ± 213 (136–1,214) n = 265;
females, 764 ± 155 (190–1,087) n = 205; im.
unsexed, 119 ± 108 (7.50–365) n = 62.

BASIC PROPORTIONS

CW/CL: males, 0.63 ± 0.028 (0.55–0.79) n = 265;
females, 0.66 ± 0.026 (0.58–0.81) n = 205;
im. unsexed, 0.74 ± 0.053 (0.62–0.83) n = 62.

CH/CL: males, 0.42 ± 0.022 (0.32–0.52) n = 265;
females, 0.47 ± 0.027 (0.40–0.53) n = 203;
im. unsexed, 0.43 ± 0.025 (0.38–0.48)
n = 62.

CH/CW: males, 0.67 ± 0.039 (0.51–0.85) n = 265;
females, 0.71 ± 0.045 (0.58–0.81) n = 203;
im. unsexed, 0.58 ± 0.041 (0.50–0.71)
n = 62.

Females gravid 26 August–20 April.
Clutches per year: one to five.
Eggs per clutch: 3.0 ± 1.08 (1–5) n = 66 clutches.

EGGS PER CLUTCH VERSUS MEAN WT OF FEMALE

1 egg: 680 g ± 65.5 (372–820) n = 7.

2 eggs: 757 g ± 54.6 (683–909) n = 15.

3 eggs: 759 g ± 55.8 (678–858) n = 27.

4 eggs: 843 g ± 56.7 (715–935) n = 11.

5 eggs: 830 g ± 54.6 (731–935) n = 6.

Clutches of four and five eggs occur near the end of the nesting season; nesting stops when aestivation begins; the last clutch may be laid in aestivation refugium.

Egg size and weight of 103 eggs: L, 37.5 ± 2.6 (27–43); W, 2.7 ± 1.36 (17–27); Wt, 9.8 ± 1.34 (4.7–12.9)

Hatchling size and weight: CL, 31.4 mm ± 2.7 (22–35) n = 20; Wt, 6.4 ± 0.91 (3.9–8.3) n = 20.

Mean annual growth increments in the first 5 years are extrapolated from plastral scute measurements and expressed as a percentage of size at the end of the previous growth season: 68% (n = 20), 34% (n = 17), 21% (n = 14), 19% (n = 10), 10% (n = 7). Total increase 356% in 5 years.

LAGUNA ESCONDIDA, VERACRUZ

Summary data: Turtles are smaller in size; consume more plant material; have slower growth rates, fewer eggs per clutch, and smaller eggs; and are active in water all year.

Laguna Escondida is a permanent lake of 18.3 ha, 2 km SE of L. Zacatal, with a mean depth 13 m and temperature of 22–26°C. Water flows from Zacatal to Esondida via a small stream in which no turtles of any kind occur. No evidence of migration of marked turtles from one lake to the other was detected in a 20-year period. *Kinosternon leucostomum* remain in the lake all year. *Trachemys scripta venusta* and *Staurotypus triporcatus* also occur in the lake and there is a resident fish population. Diet is 75% plant material (Vogt and Guzman, 1988).

Kinosternon in Escondida have a lower-domed shell and extend the head and neck just above the substrate and drag the plastron while traversing land.

Most of the perimeter of both lakes is surrounded by high evergreen rain forest; many fruits and leaves fall into the water, particularly figs and palm fruits.

SIZE AND WEIGHT

CL: males, 148 mm ± 14.7 (81–189) n = 367;
females, 139 mm ± 13.0 (100–172) n = 231;
im. unsexed, 69 ± 18.5(36.4–98.0) n = 55.

Wt: males, 465 ± 141 (62–978) n = 366; females,
415 ± 122 (117–810) n = 231; im. unsexed,
50 ± 33 (8.2–140) n = 55.

FIGURE 25.8 *Kinosternon leucostomum*. Variation in adult head pattern: top—Estación de Biologia Tropical Los Tuxtlas, Veracruz; bottom left—male, Balzapote, Veracruz; bottom right—Jesús Carranza, Veracruz. (Bottom photos courtesy of B. Horne.)

BASIC PROPORTIONS

CW/CL: males, 0.64 ± 0.026 (0.55–0.77) n = 367;
females, 0.66 ± 0.026 (0.59–0.75) n = 230;
im. unsexed, 0.72 ± 0.037 (0.65–0.82)
n = 55.

CH/CL: males, 0.41 ± 0.020 (0.35–0.48) n = 367;
females, 0.44 ± 0.023 (0.39–0.51) n = 230;
im. unsexed, 0.40 ± 0.024 (0.36–0.46)
n = 56.

CH/CW: males, 0.65 ± 0.036 (0.55–0.77) n = 367;
females, 0.67 ± 0.037 (0.57–0.76) n = 230;
im. unsexed, 0.57 ± 0.030 (0.51–0.63)
n = 55.

Females are gravid 4 September–26 May.
Clutches per year: one to four.
Eggs per clutch: 1.3 ± 0.3 (1–2) n = 45 clutches.

EGGS PER CLUTCH VERSUS WT OF FEMALE

1 egg: 427 g ± 65.5 (254–682) n = 21.

2 eggs: 509 g ± 45.6 (451–646) n = 13.

3 eggs: 434 g, n = 1.

Egg size and weight (20 eggs): L, 36.8 ± 2.31 (32–41); W, 19.9 ± 0.91 (18–22); Wt, 9.18 ± 1.25 (7.4–12.3).
Hatchling size and weight: CL, 30.2 mm ± 2.3 (24–34) n = 20; Wt, 5.7 ± 0.92 (3.1–7.2) n = 27.

Mean growth annual increments in first 5 years: 41% (n = 33), 33% (n = 33), 27% (n = 28), 12% (n = 25), 11% (n = 23). Total increase 293% in 5 years.

LERDO DE TEJADA, VERACRUZ

Summary data: Turtles are smallest in size of three populations compared and have a lower silhouette. They have an intermediate clutch size, the smallest eggs, and the smallest hatchlings.

The Lerdo de Tejada study area (ca. 126 km^2) is part of a vast wetland in the lower Rio Papaloapan drainage near Alvarado, Veracruz (18°39′N, 95°34′W). It supports the highest diversity of nonmarine turtles in Mexico. Diet is variable and tends to be opportunistically omnivorous. Other turtles occurring in various parts of the wetland are *Claudius angustatus, Staurotypus triporcatus, Chelydra rossignoni, Dermatemys mawi, Trachemys scripta venusta*, and *Rhinoclemmys areolata*.

SIZE AND WEIGHT

CL: males, 136 mm ± 10.8 (107–162) n = 78;
females, 126 mm ± 10.6 (104–208) n = 162; im. males, 121 ± 6.9 (110–130) n = 6; im. females, 114 mm ± 8.5 (102–120) n = 4.

Wt: males, 332 ± 70.6 (210–509) n = 78; females, 314 ± 62 (156–558) n = 237; im. males, 226 ± 26.7 (178–257) n = 6; im. females, 217 ± 28.3 (187–254) n = 4.

 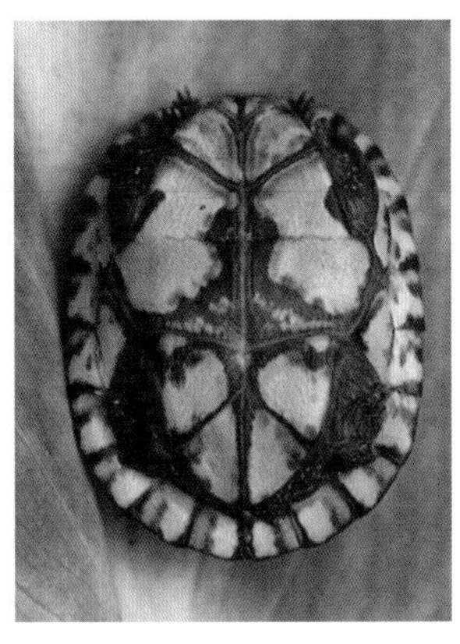

FIGURE 25.9 *Kinosternon leucostomum*. Extreme variation in neonatal plastral color and pattern in the immediate vicinity of the Estación de Biologia Tropical Los Tuxtlas, Veracruz.

BASIC PROPORTIONS

CW/CL: males, 0.63 ± 0.02 (0.56–0.70) n = 78;
females, 0.66 ± 0.024 (0.61–0.76) n = 160;
im. males, 0.66 ± 0.02 (0.62–0.68) n = 6;
im. females, 0.65 ± 0.02 (0.63–0.68) n = 4.

CH/CL: males, 0.39 ± 0.02 (0.35–0.43) n = 78;
females, 0.44 ± 0.029 (0.37–0.55) n = 161;
im. males, 0.39 ± 0.02 (0.36–0.41) n = 6;
im. females, 0.40 ± 0.02 (0.38–0.42) n = 4.

CH/CW: males, 0.62 ± 0.02 (0.56–0.69) n = 78;
females, 0.66 ± 0.045 (0.58–0.89) n = 160;
im. males, 0.59 ± 0.01 (0.58–0.60) n = 6;
im. females, 0.61 ± 0.04 (0.58–0.67) n = 4.

Gravid females: observed in all months except June, July, and August, but enlarged follicles present in all months of the year. May produce clutches all year.

Clutches per year: two to six.

Eggs per clutch: 2.3 ± 0.8 (1–5) n = 170.

EGGS PER CLUTCH VERSUS WT OF FEMALE

1 egg: 297 g ± 54.6 (204–417) n = 21.

2 eggs: 324 g ± 55.7 (204–482) n = 79.

3 eggs: 343 g ± 55.4 (225–465) n = 32.

4 eggs: 377 g ± 58 (301–470) n = 13.

5 eggs: 350 g, n = 1.

Egg size and weight (359 eggs): L, 33.5 ± 2.04 (28–38); W, 17.9 ± 1.10 (12–22); Wt 6.95 ± 1.08 (3.9–10.1).

Hatchling size and Wt: CL, 28.1 mm ± 1.9 (24–31) n = 20; Wt, 4.4 ± 0.9 (2.6–6.5) n = 23.

There are no data on growth rate.

GROWTH AND ONTOGENY

Adult males reach sizes of 214 mm and weights of 1,200 g, and females 163 mm and 1,050 g (Vogt, field notes). Clasping organs are weakly developed in males and absent in females; the tail claw is more heavily developed in males. Large females are relatively broader and higher domed than males.

Age and size at sexual maturity have not been determined precisely. Overall CL of sexable individuals is as follows: males, 152.5 mm ± 25.1 (81–214) n = 763; females, 139.3 ± 25.8 (86–208) n = 591. These data are based chiefly on secondary sexual characters (tail size, tail claw, clasping organs, etc.) and there is little doubt as to their gender. Relatively few of the 1,354 individuals were dissected.

Immature males studied by Morales-Verdeja (2000) ranged from 99 to 126 mm. Moll and Legler (1971) reported the smallest mature male and female to have 118 mm and 102 mm CL, respectively (Panama). Berry (1978, and original data sheets for Mexico) examined adult males of 108 mm and adult females in the range 102–109 mm. We think it possible that maturity could occur at less than 100 mm CL but that the minimal carapace measurements shown previously represent immature individuals ("subadults").

POPULATIONS

The population of pochitoques in Laguna Zacatal from 1981 to 1988 was comprised of 313 males and 234 females. During that same period, in Laguna Escondida, 365 males and 231 females were captured. No concentrated mark and recapture study was conducted at Lerdo de Tejada.

PREDATORS, PARASITES, AND DISEASES

Staurotypus triporcatus and Morelet's crocodiles (*Crocodilus moreleti*) are major predators of *K. leucostomum*. The presence of large populations of these two predators has a negative effect on *Kinosternon* populations. This was noted particularly in Laguna Oaxaca, Chiapas, where both predators were common and only six pochitoques were collected over the entire study period from 1984 to 1993.

Vogt observed an American bittern capture an adult male *K. leucostomum* (CL ca. 120 mm) in a rain pool at the Chajul Field Station in Chiapas (September 1984). The bird struck at the turtle once and swallowed it whole, the entire process taking about 1.5 minutes. Vogt collected a 2-m-long adult *Drymarchon corias* in the dry lake bed of Laguna de Zacatal in May 1982; it had swallowed three *K. leucostomum* eggs. The eggs were removed from the snake and incubated in the laboratory, and two of them hatched. Mammals such as opossums (*Didelphis*), raccoons (*Procyon*), and coatis (*Nasua*) also prey on the eggs. Otters (*Lutra*) were observed preying on eggs and juveniles, but were not able to open large, thick-shelled mature adults. Deep grooves made by canine teeth were noted on the carapace of 14 adults in Laguna de Zacatal. Otters were frequently captured in turtle traps but usually chewed their way out.

CONSERVATION AND CURRENT STATUS

Pochitoques are still abundant throughout most of their range even though they are heavily preyed upon for human consumption. Many populations exist within reserves. This virtually ubiquitous species in southern Mexico seems not to require more than moderate protection (see Economic Uses and Conservation). Until the 1990s *Kinosternon leucostomum* was openly sold in urban markets and restaurants in Alvarado, Veracruz, and Villahermosa, Tabasco. It was common to see tables with piles of whole boiled turtles in the cantinas of Lerdo and Alvarado. However, the open selling of turtles virtually ceased by 2000 because of increased law enforcement. Clandestine capture, sale, and consumption still occur in rural areas.

Sternotherus odoratus Latreille, 1801

Common musk turtle (Iverson,1992a); tortuga amizclera, (Liner, 1994).

Although there are two authors to the work (Sonnini de Manoncourt and Latreille, 1801) in which the description appears, Harper (1940) credits Latreille with the name.

ETYMOLOGY

Gr. *sternon*, chest, Gr. *thairos*, hinge; alluding to a hinged plastron; L. *odoratus*, odoriferous.

FIGURE 26.1 *Sternotherus odoratus* (FMNH 1404), young adult male, taken by Seth Meek at El Sauz, Chihuahua, 29 May 1903. Specimen examined and photographed by Legler in 1963. Basic measurements: CL, 83 mm; CW, 62 mm; Ht, 39 mm.

FIGURE 26.2 Arroyo El Sauz, 20 June 1965. Sierra del Nido barely visible in background. View is to SW. No *Sternotherus* have been captured in Mexico since 1903.

GEOGRAPHIC DISTRIBUTION

The species is widely distributed in the eastern half of the United States without significant geographic variation (Zug, 1986; Reynolds and Seidel, 1983).

Sternotherus odoratus once occurred in Mexico.

Seth Meek collected two specimens of *Sternotherus* (*"Aromochelys"*) *odoratus* at El Sauz, Chihuahua (FMNH 1403 and 1404) on 29 May 1903. The former has seemingly never been examined and has been missing at least since 1948. Norman Hartweg examined and identified FMNH 1404 in 1950. The provenance of this specimen is unequivocal. It bears three virtually identical data labels, sewn through the skin of the ventral base of neck, the left forelimb, and the right hind limb. This record of *Sternotherus odoratus* lies some 300 km SW of the known range. *Kinosternon hirtipes* is the only other chelonian known from this drainage (UU 8549–53).

Specimen FMNH 1404 was examined and photographed by Legler in 1963 (Figure 26.1) and independently by Moll and Williams (1963); it is a young adult male with basic measurements of CL 83 mm, CW 62, CH 39.

Meek also collected four species of fishes at El Sauz: *Pantosteus plebeius*, *Notropis lutrensis*, *Cyprinodon eximius*, *and Gambusia affinis*. Meek traveled by railroad in Mexico. El Sauz was the last collecting locality on an epic 3-month expedition that formed the basis of Meek's (1904) monograph on Mexican fishes.

El Sauz lies ca. 40 km NNW of Cd. Chihuahua on the main railroad line and may have been little more than a railroad pumping station in 1903. The Rio Sauz is a small internal drainage system (also referred to as "Encinillas") that springs from the eastern slope of the Sierra del Nido and flows intermittently to the desert floor near Sauz and thence intermittently northward, paralleling the S. del Nido and ultimately disappearing ca. 100 km NNW of Cd. Chihuahua (Iverson, 1981a).

Like other basins in the Chihuahuan desert, the Sauz drainage suffers the vicissitudes of periodic drying. Hubbs (in Hubbs and Springer, 1957) and Miller (1961) reported that the entire Sauz valley became dry in 1947.

Meek described the habitat as a small stream about 15 miles in length situated a short distance north of the city of Chihuahua. At the middle of its course in Sauz it contains a small amount of running water during the dry season with holes up to 6-feet deep, a sandy bottom, and much aquatic vegetation (mostly algae).

The status of *Sternotherus* in Mexico has been discussed as follows: Zug (1986; "unconfirmed"); Reynolds and Seidel (1982; "presumably valid"); Ernst, Lovich, and Barbour (1994, "doubtful"); Iverson (1981a, "problematical"), Conant and Berry (1978, "valid"); Moll and Williams (1963, "valid"). Subsequent visits to the Sauz drainage by competent scientists have failed to produce other specimens.

Minckley and Koehn (1965) collected three of the four fish species reported by Meek (no *Pantosteus*) from a permanent impoundment in the Sauz basin in 1964. No turtles were taken. E. O. Moll (journal, 1964) set baited traps on the east side of the town of El Sauz on 20 June 1964 (28 trap hours) and caught *Kinosternon hirtipes murrayi* and one suckerlike fish (presumably *Pantosteus*). At that time there was no flow in the stream, but isolated pools were bordered by good stands of willows, cottonwoods, and grasses (lusher than the brief scenario presented by Meek). Depth was 0.3–1.0 m and the bottom was firm mud to gravel (Figure 26.2). *Thamnophis eques* and *Rana* sp. were common. Iverson (1981a) and associates trapped permanent waters near Sauz in May 1977 and reported neither turtles nor fishes.

Fishes from El Sauz are said to be more closely related to those in the Rio Conchos than to those in other drainage systems to the west (Hubbs and Springer, 1957; Smith and Miller, 1986). Moll and Williams (1963) proposed that *Sternotherus* may have colonized the El Sauz drainage by dispersal up small tributaries of the Rio Conchos.

Minckley and Koehn (1965) review the climatic vicissitudes of the Sauz basin. Hubbs (1954, in Hubbs and Springer, 1957) reported that no permanent water had existed in the basin since 1947. R. R. Miller visited Meek's locality in 1964 and found only a moist area in the river bed (pers. comm. cited by Minckley and Koehn, 1965).

We conclude that *Sternotherus odoratus* is probably extinct in Mexico. However, if fishes and other kinosternids can survive in the basin, so might *Sternotherus*.

We extend special thanks to Mary Anne Rogers and Kathleen M. Kelly of the Field Museum of Natural History for assistance with this investigation.

Family Trionychidae

Softshell Turtles

L. *tres*, G. *treis*, three; G. *onyx*, -ychos, fingernail, talon, claw; alluding to the three claws on each manus and pes.

The Family Trionychidae Fitzinger, 1826 are here regarded as the most morphologically specialized of nonmarine turtles. They have achieved these specializations without altering the basic chelonian Bauplan.

The term "softshell turtle" is at once descriptive but misleading. Only the periphery of the shell (lacking peripheral bones) is "soft" and flexible. The rest of the carapace, which is underlain by dermal bone, is as stout and almost as protective as that of typical chelonians.

CLASSIFICATION OF THE TRIONYCHIDAE

The work of Meylan (1987) is here accepted as the most recent revision of the Family Trionychidae. He offers an expanded taxonomy of trionychids that contains 2 subfamilies, 6 "tribes," 14 genera, 30 species, and 39 terminal taxa with a wide distribution, chiefly in the Northern Hemisphere. Older classifications recognize approximately seven genera.

DIAGNOSIS OF FAMILY

(Based chiefly on Meylan, 1987.)

Recent trionychids can be diagnosed by the following combination of characters (many of which alone are diagnostic): peripheral bones are absent (present only in reduced form in *Lissemys*); there are no discrete epidermal laminae anywhere on the shell at any ontogenetic stage—rather, a continuous layer of skin of varying texture; there are three claws on manus and pes; there is hyperphalangy in digit 4 (4–6 phalanges) and digit 5 (2–4 phalanges); pygal and suprapygal bones are lacking; there are seven to eight neural bones; the posteriormost pair of costals often meets on the dorsal midline; there are no pygals or suprapygals. Dermal bones of shell have a distinctive pitted or pebbled appearance (shared only with *Carettochelys*).

Bony plastron is fenestrate; posterior plastral elements are separated on the midline; plastral fenestrae may be covered by epithecal callosities in adults; entoplastron is present, a boomerang-shaped element with its concavity posterior and its apex anterior, and its posterior arms articulating with the hyoplastra and flanking a fenestra in which there are no bony structures (Figure 27.1). Cervical centra narrow and all are opisthocoelous except the 8th, which lacks a ventral process and a central articulation to the first thoracic vertebra. Body of hyoid is comprised of six or eight ossifications; ilia are curved posteriorly; pectineal processes lie in a single plane and are in broad contact with plastron, equal to or wider than the interpubic contact. Quadratojugal is not in contact with postorbital; jugal is in contact with parietal; premaxillae are fused and excluded from external narial aperture by maxillae. A narrow postorbital bar separates extensive dorsal and ventral emarginations of temporal roofing; premaxillary bones are fused; squamosal are isolated from parietal and postorbital bones; stapes are enclosed by quadrate (Figure 27.2).

Two musk glands are on each side; anterior axillary gland is beneath the anterior margin of the carapace immediately anterior to rib 2, with a short duct to the orifice on the adjacent margin of the carapace; inguinal gland is immediately anterior to rib 6 with a short duct passing ventrally between the lateral bifurcation of hypoplastron; osseous

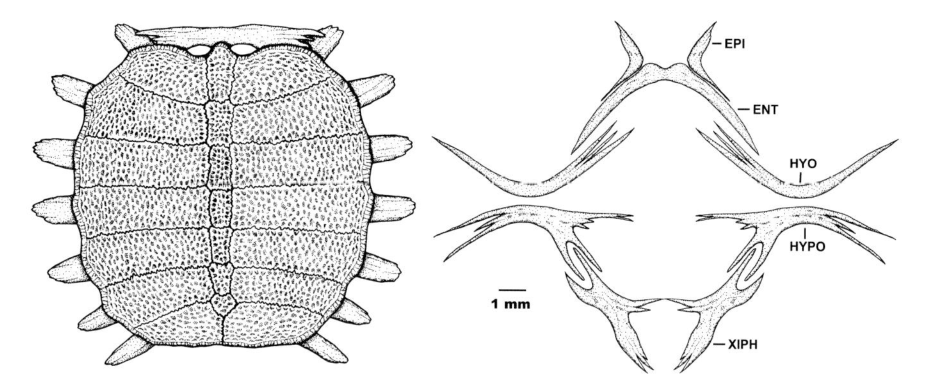

FIGURE 27.1 Skeletal carapace of *Apalone spinifera*. Left—carapace, KU 90876, small adult, length ca. 90 mm; anterior transverse bar is homologous to fused costiform processes; neural and costal elements are clearly shown; peripherals are lacking; note pitted texture of dermal elements. Right—plastral bones of embryonic *Apalone spinifera* (ventral view, KU 290151, stage 21, total length ca. 12.2 mm). Abbreviations: EPI, epiplastron; ENT, entoplastron; HYO, hyoplastron; HYPO, hypoplastron; XIPH, xiphiplastron. (Images modified from Sheil, 2003, with permission.)

canal is present in dermal bone of adults (Waagen, 1972). Cloacal bursae are lacking; buccopharyngeal mucosa are vascular and papillose. Fleshy upper and lower lips conceal all but the anteriormost parts of the jaw sheaths. There are external nares at the end of a long proboscis-like snout.

There are two extant subfamilies (Cyclanorbinae and Trionychinae), and about 220 fossil species are recognized. A third subfamily (Plastomeninae) is sometimes invoked for some of the fossil taxa. Extant members of the family are completely aquatic turtles occurring in Africa, Asia, the Indo-Australian archipelago, and North America. The oldest fossils are from the late Jurassic (see account of *Apalone*).

Subfamily Trionychinae Lydekker, 1889

Common Softshell Turtles (Iverson, 1992a)

Nuchal bone is a horizontal bar, its horizontal axis at least three times its medial longitudinal length; anterior and posterior costiform processes of nuchal are fused; peripheral elements are completely lacking; depressions for articulation of ilia are absent on 8th costals; bridge is narrow; there are usually two or more ossifications in the second branchial horn of the hyoid; emargination of edge of external narial aperture is variable; epipterygoid is fused to pterygoid in adults.

The subfamily contains four monophyletic species groups ("tribes") with relationships that are not entirely clear. North American taxa are all placed in the "tribe" Trionychini and the genus *Apalone* (Rafinesque, 1832) with the genera *Rafetus* and *Trionyx* (sensu stricto, *T. triunguis* only) of S. Asia and Africa, respectively.

THE PLASTRAL BONES OF TRIONYCHIDS

The anterior plastral bones of trionychids are unusual, and their homologies have been debated. A pair of elements articulate with the sides of a boomerang-shaped element and project, pronglike, anterolaterally. The boomerang-shaped element has historically (and logically) been termed the entoplastron and the anterior pair the epiplastra—assumptions made chiefly on proximity (Figure 27.1).

An obscure paper by Lane (1909) demonstrated the boomerang-shaped element to be paired in a 14 mm CL embryo of *Apalone spinifera*. Williams and McDowell (1952) subsequently homologized the boomerang-shaped element to paired, fused epiplastra and the anterior pronglike elements to neomorphic bones they termed "preplastra." Meylan (1984) referred to the boomerang-shaped bone, without comment, as the entoplastron; in 1987 Meylan rejected the Williams and McDowell hypothesis and cited a manuscript (Bramble and Carr, unpublished) in support of this premise.

Bramble and Carr (unpublished) concluded that the homologies of trionychid plastral elements are the same as those in other turtles, based on the attachments of anterior trunk musculature (see also Bramble, Hutchison, Carr, and Legler, 2009). Lane did not mention his techniques. His demonstration of paired entoplastra (as shown in his Figures 1 and 2) has not been repeated. However, cleared and stained specimens of juvenile *Apalone* and *Pelodiscus* (Legler, pers. obs., and Sheil, 2003) show median constrictions of the entoplastron that suggest a possible dual origin. The two most recent and authoritative accounts of plastral bone development (Sheil, 2003; Sanchez-Villagra et al., 2009) don't mention the controversy of plastral homology and the latter work (Figures 3b and 4a) clearly show a single entoplastral anlage at stage 21 (a stage comparable to 14 mm CL).

GENERAL DESCRIPTION

Practical identification of North American softshells is quite simple. They are completely aquatic turtles with a flattened, disk-shaped shell that has flexible, boneless edges; a long, proboscis-like snout; and only three claws on each foot.

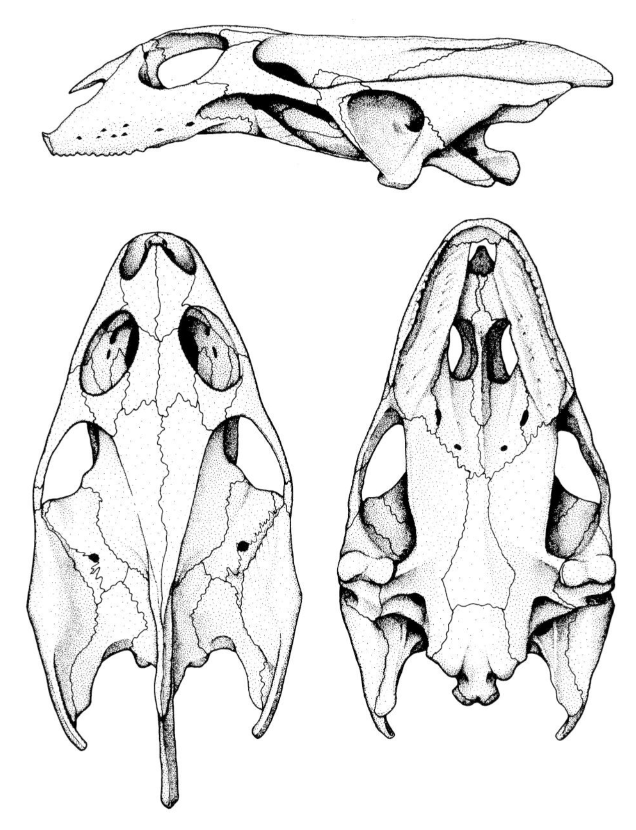

FIGURE 27.2 Skull of adult *Apalone spinifera* (KU 2757, CBL 58.4, greatest breadth 33.6) demonstrating general skull elongation among trionychids. (Modified from Sheil, 2003, with permission.)

CONTENT

Apalone is the only New World genus, is widely distributed in the eastern United States, and barely enters northern Mexico. The genus has shown a remarkable ability to survive and prosper after being purposely introduced for human consumption in the Hawaiian Islands, on Mauritius (Ernst, Lovich, and Barbour, 1994; Webb, 1980; McKeown and Webb, 1982), and possibly in the Sacramento River of California. Its presence in the Colorado River drainage of the United States and Mexico is thought to result from accidental introduction (Webb, 1962; Miller, 1946).

Genus *Apalone* Rafinesque, 1832

North American softshell turtles.

ETYMOLOGY

Gr. *apalos*, soft to the touch, tender.

TAXONOMIC STATUS

According to the classification of Meylan (1987), the native North American species of trionychids (*mutica*, *spinifera*, *atra*, and *ferox*) should be placed in the resurrected genus *Apalone*. The species of *Apalone* are monophyletic and distinct from other trionychids. "*Trionyx*" is an old generic name that has been used as a taxonomic dumping ground for anything vaguely trionychid. Meylan's classification of the trionychids has come into general use (Iverson, 1992a; Shaffer, Meylan, and McKnight, 1997). Webb (1990) and Ernst, Lovich, and Barbour (1994) continued to use *Trionyx* (despite the use of *Apalone* by Ernst and Barbour in 1989).

FOSSIL RECORD

The North American softshells diverged from their closest relative during the Cretaceous (Gardner, Russell, and Brinkman, 1995). The oldest known trionychid fossil is *Sinaspideretes wimani* from the late Jurassic of China (Ernst and Barbour, 1989). Rapid diversification of trionychids occurred throughout ". . . fluviolacustrine environments of Laurasia beginning in the early Late Cretaceous" and they are "mostly absent from southern continents" (Head, Aguilera, and Sanchez-Villagra, 2006).

Fossil remains of *Apalone*-like softshells have been found in northeastern South America (Venezuela) in Miocene and early Pliocene deposits (saline and marginally marine; Head, Aguilera, and Sanchez-Villagra, 2006) and in the lower Pliocene of southeastern Costa Rica (Laurito et al., 2005). There is controversy about whether softshells reached these localities by way of dispersal or by southeastward coastwise migration of northern populations. Head, Aguilera, and Sanchez-Villagra (2006) propose that softshells may have colonized fluvial and coastal habitats in northeastern South America for most of the Neocene period.

DIAGNOSIS AND DESCRIPTION

All New World trionychids (*Apalone*) lack peripheral bones and individual epidermal laminae, replacing them with a continuous covering of undivided skin. The form of the combined shell is low and disclike, round, and with sharp flexible edges, reminiscent of the form usually assumed for "flying saucers." The head is equally distinctive, bearing a long proboscis-like snout with nostrils on its tip. Fleshy lips ("flaps" in some accounts) almost completely conceal the maxillary and mandibular jaw sheaths. There are four or more plastral callosities in adults. There are seven to nine neurals (usually 8); old males have seven plastral callosities; adults are 200 mm or smaller; and eighth pair of costals is reduced or absent.

Karyotype: $2n = 66$; 12 macrochromosomes and 54 microchromosomes (Ernst, Lovich, and Barbour, 1994).

UNIQUE STRUCTURES, FUNCTIONS, AND BEHAVIOR

The unique morphology of trionychids facilitates unique underwater behavior patterns. Softshells are completely

FIGURE 27.3 Captive *Apalone ferox* in final phases of concealment (sand and fine gravel substrate): left—head withdrawn to point where skin of neck partially covers eyes; right—completely withdrawn.

aquatic, emerging only to bask and to nest. They are fast, maneuverable swimmers. Divers have difficulty catching them in a straight chase. Swimming skills are probably aided by the hydrodynamic form of the shell and the extensive webbing of the feet.

When resting underwater they conceal themselves in bottom vegetation or, more often, in bottom substrate (mud, sand, gravel, or simply debris). When so concealed they can sustain life with alternative, nonpulmonary respiration. The concealment seems to be chiefly for protection, but it could also function in ambushing prey.

When under duress, a softshell may dive, at speed, into a silty bottom, causing a cloud of silt that temporarily conceals subsequent movements. The process of settling into the bottom is stereotypic. The turtle may first shift from side to side to make a small depression. Sand or other substrate is then thrown onto the shell with flipping movements of the feet until the shell is almost covered. There is then a stereotypic final movement of the entire trunk. This is a brief "shuddering"—a rapid, almost vibratory movement of the entire shell from side to side. At this point the shell or the entire turtle is buried or at least covered with a thin layer of bottom substrate. The turtle, thus concealed, may then remain at rest for long periods on the bottom. One can sometimes detect these refugia by slight bulges of the bottom substrate. Turtles may explode from these places and swim away at speed if disturbed. However, Marchand (1942), as quoted by Carr (1952), observed *A. ferox* to be remarkably unconcerned in the presence of a diver, tolerating some human contact while buried. The low, flattened, sharp-edged carapace facilitates bottom concealment, and the elongated snout is an adaptation for aquatic respiration during such concealment.

While concealed at rest, the proboscis-like snout ultimately projects above the bottom substrate (Figures 27.3 and 27.5). In extremely shallow water the tip of the snout may project above the surface of the water for normal pulmonary respiration. However, repeated changes in buoyancy would be likely to disturb the substrate in which the turtle is concealed. More often, the snout projects just far enough above the substrate to avoid being fouled by it. Under these circumstances the turtle switches to an alternative mode of respiration by passing water over vascular buccopharyngeal papillae and fimbriae. All of this has been observed (chiefly in captives) by both of us and by Winokur (1988).

Nearly all aquatic turtles can achieve some kind of alternative respiration when resting underwater. In all cases this involves the passage of oxygenated water over a vascular surface (e.g., cloacal bursae, plastral skin, or modified skin of the tongue and buccopharyngeal lining). Vascular surface area is increased with papillae and fimbriae.

Gage (1884) and Gage and Gage (1886) were the first to record the alternative respiratory functions of *Apalone*. They observed and counted hyoid movements of a captive in an aquarium and then measured oxygen and carbonic acid in the water at the end of the experiment. Vascular buccopharyngeal villi were mentioned.

Members of the genus *Apalone* have well-developed lungs (pers. obs. and Gräper, 1931), despite the brief comments of Agassiz (1857) to the contrary. Additionally, they have evolved highly vascular buccopharyngeal fimbriae (analogous to gills) capable of extracting oxygen from the water (Carr, 1952; Winokur, 1988; Dunson, 1960).

The mechanics of buccopharyngeal respiration involve the same musculoskeletal structures as gape and suck feeding (see Structure and Function). The hyoid apparatus is lowered, volume in the buccopharyngeal cavity is increased, pressure decreases, and water flows into the cavity through the nares and or mouth and is expelled by the same route when the process is reversed.

Buccopharyngeal and/or other nonpulmonary respiration may suffice to support a turtle *at rest* for hours, days, or months, depending on environmental conditions (e.g., water temperature). Under conditions of strenuous activity (e.g., being chased by a human diver) they must rely on pulmonary respiration. Legler and James F. Berry crudely tested this idea in northern Florida in May 1975. We swam together on the surface until we observed a large adult female of *Apalone ferox* fleeing from a place of concealment. Thereafter we worked as a team, one man underwater

FIGURE 27.4 *Apalone atra*. Adult female showing extent and occlusion of fleshy lips characteristic of trionychids. Note particularly the occlusion of lips on left side of head and central exposure of jaw sheaths.

FIGURE 27.5 Adult *Apalone atra*, Tio Candido, Cuatro Ciénegas, shortly after burial in bottom silt with only head protruding. Vague depressions show position of shell. The species is rarely seen active during day.

in chase and the other swimming at the top. The turtle became progressively slower and weaker and ultimately simply rose to the surface in complete exhaustion and bobbed up and down in a period of recovery using pulmonary respiration. It could be approached and handled with impunity. In the same period we tried the same experiment with a large adult *Chelydra serpentina*. *Chelydra* has relatively small, simple cloacal bursae. The results were essentially the same. These two incidents are in sharp contrast to certain Australian chelids (e.g., *Rheodytes*; Legler and Cann, 1980) with huge cloacal bursae who, under chase, swim with the cloaca wide open (exposing a red interior lining of long vascular fimbriae) and can sustain a chase longer than a team of divers.

The "fleshy" lips of trionychids (Figures 27.4 and 28.5) have received remarkably little comment in the literature on turtles except as a diagnostic character. They constitute a problem in functional morphology that remains to be scrutinized empirically. The materials and speculative remarks in the following are based on our own observations of captives and personal communications with D. M. Bramble, R. M. Winokur, and P. A. Meylan.

The occlusion of the jaw sheaths is imperfect in all non-trionychid turtles we have observed. One can observe currents and debris passing out the sides of a closed mouth during feeding. Trionychid lips are in the form of elongate, bulbous, or flattened masses of skin and subcutaneum. They are attached to the upper and lower jaws where the jaw sheaths meet the general soft skin of the head. This attachment is a long, narrow isthmus through which any communicating vessels or nerves must pass. The lips are usually reflected away from the jaw sheaths in preserved specimens. When apposed to the jaw sheaths they appear to seal the sides of the buccal cavity. Examination of sections of whole heads shows flattened or stepped occlusal surfaces on the lips (Figure 27.4). Macrosections of the lower lip in a large *Apalone ferox* (preserved) have a spongy appearance in which there are some large cavities. Gentle pressure on the lip causes copious flow of liquid from the cut surface.

These simple observations suggest a hydrostatic system of controlled turgidity that could enhance the sealing of the mouth when turgid and keep the lips away from the tomia when flaccid.

If trionychid lips, when adpressed, could seal most of the buccal cavity, the anteriormost part might be sealed by the tip of the tongue and an apposed pad of tissue in the premaxillary region. The occluded lips could therefore canalize the flow of water in and out of the buccal cavity.

Softshells are known to feed on aquatic invertebrates, many of which are small. The negative pressure created in the buccal cavity by hyoid movements would create an increased velocity of flow at a relatively narrow anterior orifice and act as an enhancement to suction feeding. R. M. Winokur (pers. comm.) has seen captive *Apalone* rapidly "sucking in" an earthworm in this manner.

More importantly, the sealing and canalization imparted by the lips enhances buccopharyngeal respiration by directing a nearly continuous flow of clean, well-oxygenated water over the vascular pharyngeal papillae. Dunson (1960) counted 16–25 hyoid pulsations per minute in experimental *A. spinifera*. When a softshell is buried and the snout projects above the substrate, water passes, alternately, in both directions through the snout. Bidirectional flow through the nostrils has been verified (in *A. ferox*) with particulate matter by Marchand (1942, in Carr, 1952). Furthermore, if water were taken in through the mouth by a turtle buried in silt, a plume of silty water would come out the nostrils.

In small specimens of *A. spinifera* and *A. ferox*, preserved with the mouth closed, thumb pressure on the throat region at the level of the mandibular joint causes major expulsion of liquid from the nostrils and virtually nothing from the sides of the mouth (JML, pers. obs.).

Winokur described erectile tissue and smooth muscle in the snouts of trionychids and *Carettochelys* (1981 and 1982) and proposed that these structures controlled the turgidity and angle of the snout and the internal diameter of the long narial passageways.

CLOACAL PUMPING

There are repeated references to flow of water in and out of the cloaca (e.g., Carr, 1952; Cahn, 1937; Dunson, 1960), and most aquatic turtles probably practice it. Dunson (1960) demonstrated that softshells could harvest some oxygen in this manner. Yet softshells and all other trionychoids lack cloacal bursae (see Structure and Function). It must be assumed that an unmodified cloacal wall accomplishes this. Ward (1970) has demonstrated that virtually any part of the gut of experimental turtles has the ability to extract oxygen from water.

NATURAL HISTORY

There seems to be general agreement that *Apalone* favors a diet of aquatic invertebrates; nearly every account of diet emphasizes the consumption of crayfish and aquatic insects, but the most complete accounts allow that *Apalone* retains the ability to be an opportunistic omnivore.

All *Apalone* lay spherical eggs.

KEY TO THE SPECIES OF *APALONE* IN MEXICO

1A. Carapace dark, gray to smudgy black in adults; rear dorsal surface of carapace with distinctive longitudinal corrugations, sharp edge often ragged; posterior one-third of carapace lacks small pale tubercles; a posterior vestige of pale, circumferential band on carapace present or not; occurs only in Basin of Cuatro Ciénegas, Coahuila . . . . ***Apalone atra***

1B. Carapace generally pale tan to brown with a complete yellowish circumferential band; posterior one-third of carapace with small, pale tubercles; occurs in the Rio Grande, northern Gulf, and Colorado drainages . . . . ***Apalone spinifera emoryi***

Apalone atra (Webb and Legler) 1960

Tortuga de concha blanda negra (Liner, 1994); black spiny softshell (Liner, 1994). Local residents near Cuatro Ciénegas refer to softshell turtles as *tortuga blanca* and make no distinction between *A. spinifera emoryi* and *A. atra* (Legler, field notes).

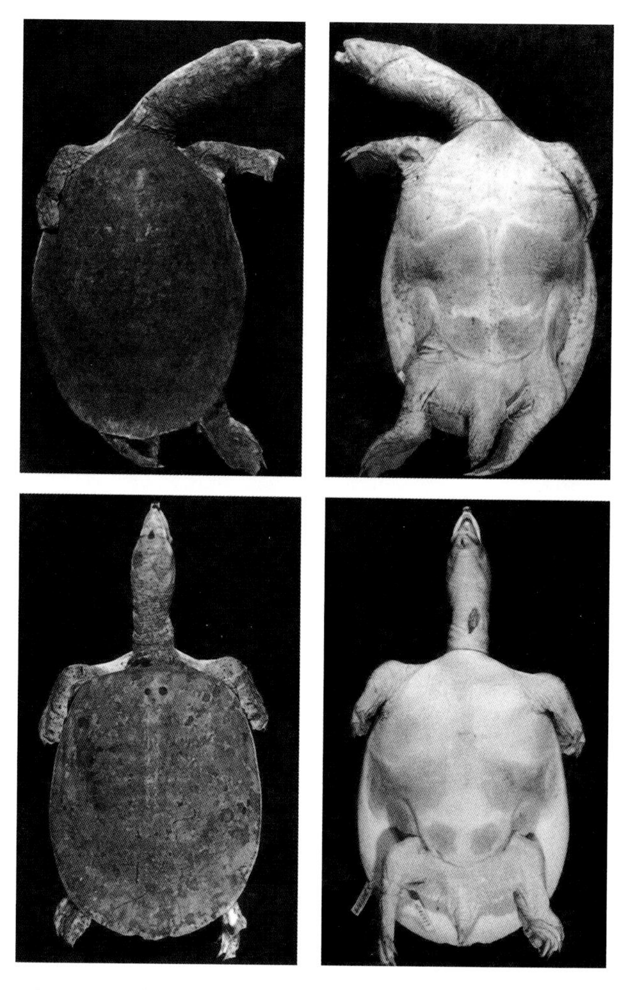

FIGURE 28.1 The two species of Mexican *Apalone*. Top—*Apalone atra*, holotype (adult female, KU 46903, CL 243 mm); bottom—*Apalone spinifera emoryi* (adult female, KU 46913, CL ca. 184 mm) from Rio Chiquito. (From Webb and Legler, 1960, with permission.)

ETYMOLOGY

L. *atra*, black; referring to the overall dark coloration of the carapace and skin.

HISTORICAL PROLOGUE

Apalone atra was described as a full species that occurred in Tio Candido (the type locality), a pond 16 km SW of Cuatro Ciénegas de Carranza, central Coahuila, Mexico, in sympatry with *A. spinifera emoryi* (Webb et al., 1960). Webb (1962, 1973) continued to rank it as a full species. Various other works have ranked the taxon as little more than a "race" of *A. spinifera* (Morafka, 1977a), as a subspecies of *A. spinifera* (*emoryi*), or as a full species, but few of these rankings were based on actual studies or knowledge of the Basin of Cuatro Ciénegas. Winokur (1968; MS thesis) conducted a thorough study of variation in most of the aquatic habitats in the Cuatro Ciénegas basin and concluded that, although morphological intermediacy occurred, *atra* deserved full

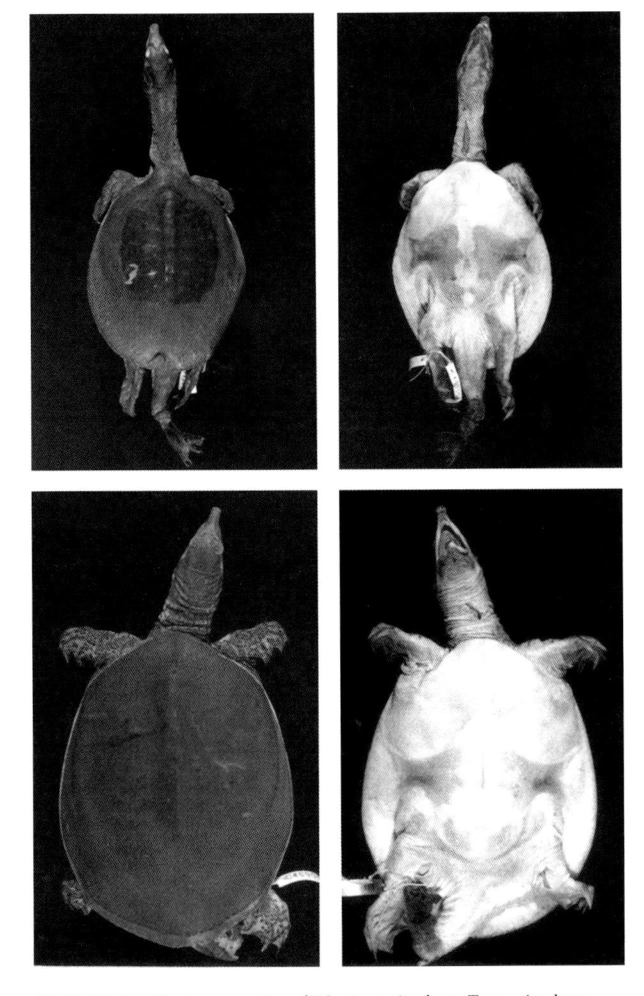

FIGURE 28.2 The two species of Mexican *Apalone*. Top—*Apalone atra*, allotype (adult male, KU 46911, CL 134 mm); bottom—*Apalone spinifera emoryi* (adult male, KU 46907, CL ca. 144 mm) from type locality of *A. atra*. (From Webb and Legler, 1960, with permission.)

species rank. Smith and Smith (1979) thought *atra* showed "clear evidence" of being absorbed by *A. spinifera emoryi* and suggested that *atra* was an extinct taxon of recent age, and that all softshells in the basin should be considered hybrids of *Apalone atra* × *A. spinifera emoryi* (no citations given).

The most recent study (McGaugh and Janzen, 2008) is based on molecular and morphological evidence and substantial fieldwork; the authors refer to "*Apalone atra*" and "*Apalone spinifera emoryi*" throughout the text but conclude that *atra* "is not a separate species."

We see no evidence that *atra* is extinct or that it has been swamped by *emoryi*. Cerda-Ardura, Soberón-Mobarak, McGaugh, and Vogt (2008) report records of several "pure" *atra* being taken from Tio Candido in 1983 and a "pure" adult *atra* female and an *A. spinifera emoryi* being taken in the same trammel net in 1986.

We regard *atra* and *spinifera* as full species on the premise that (1) they occur in sympatry in some places, and (2) their hybridization results in part from anthropogenic phenomena. Their distinction nests in a blurred and confusing combination of phenotypic characters.

DIAGNOSIS

(Based on Webb and Legler [1960], Webb [1962, 1973], Winokur [1968].)

Apalone atra is a medium-sized softshell turtle that can be distinguished from its nearest relative, *A. spinifera emoryi*, by the following suite of characters: (1) Dorsal posterior edge of carapace bears fine, parallel, longitudinal, cutaneous corrugations; (2) pale border on adult carapace is lacking or indistinct; (3) there is uniform to lichenlike blackish pigmentation on carapace; (4) ventral surfaces are speckled with dark pigmentation; (5) narial septal ridges are present but are often reduced in males; (6) females are lacking small white tubercles on posterior half of carapace.

Winokur (1968) found no single unique character that would diagnose *A. atra*. Webb (1962) considered the cutaneous corrugations on the rear of the carapace alone to be diagnostic. McGaugh and Janzen (2008) found the most useful characters for field identification to be carapace pigmentation, dark plastral speckling, and the corrugations on the posterior carapace, in all of which they perceived a minimum of ontogenetic change.

In fact, all characters considered in this account, "diagnostic" or not, are subject to substantial variation and general ontogenetic change (see Reproduction).

GENERAL DESCRIPTION

(Webb and Legler, 1960; Webb, 1962, 1973; Winokur, 1968; pers. obs., JML.)

Apalone atra is a softshell approximating the size of *A. spinifera emoryi*. Maximum recorded CL is 247 mm for males and 294 mm for females. It most closely resembles its nearest geographic neighbor, *Apalone spinifera emoryi*, with which it occurs sympatrically in parts of the Basin of Cuatro Ciénegas.

BASIC PROPORTIONS

CW/CL: males, 0.78 ± 0.042 (0.72–0.83) *n* = 9;
females, 0.74 ± 0.028 (0.69–0.81) *n* = 21;
im. females, 0.76 ± 0.056 (0.69–0.82) *n* = 31;
juveniles, 0.85 ± 0.012 (0.84–0.87) *n* = 3.

HT/CL: males, 0.28 ± 0.038 (0.25–0.31) *n* = 2;
females, 0.27 ± 0.031 (0.25–0.31) *n* = 3;
juveniles, 0.28 ± 0.024 (0.26–0.31) *n* = 3.

HT/CW: males, 0.35 ± 0.039 (0.31–0.38) *n* = 2; females, 0.36 ± 0.057 (0.31–0.43) *n* = 3; im. females, 0.34 ± 0.022 (0.30–0.36) *n* = 10; juveniles, 0.33 ± 0.024 (0.31–0.36) *n* = 3.

WH/CL: males, 0.16 ± 0.0008 (0.16–0.17) *n* = 4;
females, 0.14 ± 0.0008 (0.12–0.15) *n* = 18;
im. females, 0.15 ± 0.015 (0.10–0.19) *n* = 27;
juveniles, 0.22 ± 0.0002 (0.22) *n* = 3.

FIGURE 28.3 Tio Candido, type locality of *Apalone atra* and *Trachemys scripta taylori*. Top view (May 1976) is northeastward and includes only southern end of poso. Diver in foreground stands on bottom. *Apalone spinifera* and *A. atra*/*A. spinifera* hybrids have been reported here, but none were taken in 1976. Compare to Minckley (1992: Figure 14, pt. 2, p. 111). Bottom view—Tio Candido from western shore, 2005. (Photo by S. McGaugh, used with permission.)

Snout W/WH: males, 0.24, $n = 2$; females, 0.22 ± 0.018 (0.19–0.24) $n = 17$; im. females, 0.22 ± 0.032 (0.17–0.29) $n = 19$; juveniles, 0.23 ± 0.014 (0.22–0.24) $n = 3$.

Data presented as "basic proportions" suggest that *A. atra* differs from *A. spinifera emoryi* in smaller adult size, a carapace more nearly ovoid than round, higher shell profile and cross-section, slightly wider head, and a relatively broader, shorter snout.

The differences in snout shape between *emoryi* and *A. atra* (Webb, 1962; Winokur, 1968) have never been satisfactorily investigated. The snout of *emoryi* has a long, thin taper, is widened perceptibly at the tip and flared where it meets the main mass of the head. In mature *atra* the snout flares somewhat at its proximal end but otherwise has the thick, straight-sided proportions of a modern beverage can.

Apalone atra has a uniform or patchy, lichenlike carapace that is black, dark gray, or slate colored. The dorsal surfaces of the limbs, neck, and head are uniform in coloration, sometimes with obscure patterns. Tubercles on the anterior edge of the carapace vary in size and number but are blunt and never spinous.

No function has been ascribed to the diagnostic cutaneous corrugations on the posterior carapace. A "ragged" posterior edge in females might be ascribed to the nipping behavior observed by Legler (1955) during courtship of *A. spinifera*.

In preserved specimens, Winokur (1968) found the choanal flaps of *atra* to be almost always horizontally oriented and those of *emoryi* (within the Basin) to be vertical. The choanal flaps of *atra* have papillose edges and cover about half the choanal orifice, whereas those of *emoryi* have serrated edges and cover a smaller area in paletal views of heads).

McGaugh (2008) carefully explored the phenomenon of dorsal color matching bottom substrate color in softshells in various habitats in the Basin. Her results were impressive in most cases but did not explain how the dark carapace of *A. atra* could act as camouflage on the silver-gray bottom sediment at Tio Candido (Figure 27.5). Softshells there rely chiefly on concealment in bottom vegetation or burial in bottom substrate rather than on camouflage.

MOLECULAR STUDIES

(McGaugh and Janzen, 2008)

One mitochondrial marker (cytochrome b) is known to show strong resolving power in *Apalone* (Weisrock and Janzen, 2000; McGaugh et al., 2008). Three nuclear loci are known to show species resolution power in turtles generally (Engstrom, Shaffer, and McCord, 2004; Krenz, Naylor, Shaffer, and Janzen, 2005) but showed little to no divergence between *A. atra* from the type locality and *A. s. emoryi* from the Rio Grande (McGaugh and Janzen, 2008).

As such, McGaugh and Janzen (2008) agreed with the current checklist designations (e.g., Bickham et al., 2007) of *atra* as a subspecies of *A. spinifera*. However, the lack of differences in one genetic system does not necessarily imply that these are not distinct species. Failure to detect genetic differentiation does not necessarily negate the distinction of *atra* as reckoned by unique phenotypic characters and ecological preferences (Cerda-Ardura, Soberón-Mobarak, McGaugh, and Vogt, 2008).

As shown with starch gel electrophoretic studies (Vogt and McCoy, 1980; Vogt, 1993), closely related species of turtles often appear to be very similar genetically at many of the loci that are good indicators of species in other groups of animals. This does not mean that the populations studied are conspecific. A more thorough genetic evaluation using a genomewide scan must be done to resolve this issue (Cerda-Ardura, Soberón-Mobarak, McGaugh, and Vogt, 2008).

EVOLUTIONARY SCENARIO

(See also accounts for *Trachemys scripta taylori* and *Terrapene coahuila*.)

We propose that a population of *Apalone* (*A. spinifera* or whatever occurred in the Rio Salado drainage at the time)

FIGURE 28.4 *Apalone atra*. Gravid adult female captured on land, Tio Candido (CL, 233 mm; Wt, 2,067 g; UU 17569; May 1976).

FIGURE 28.5 *Apalone atra*. Head of specimen shown in Figure 28.4.

was isolated in the Cuatro Ciénegas Basin by closure of one or both of the narrow portals exiting the valley. This isolated stock differentiated in warm, still-water habitats (limnocrines) to become *Apalone atra*. This probably occurred at some time during the current interglacial period (e.g., in the last 11,000 years). When the basin was subsequently reconnected to the Rio Salado, *A. s. emoryi* entered the basin, colonized fluviatile habitats, and eventually entered still-water habitats occupied by *A. atra*. This may have happened in relatively recent, perhaps historic, times.

The population of *A. s. emoryi* in the basin has an extensive, continuous distribution outside the basin. *Atra* occurs only in parts of the basin. Differentiation of *atra* within the basin was insufficient to produce complete genetic isolation, and the two species hybridized in some of the places where they occurred together. This sympatry did not include all aquatic habitats in the basin.

Webb (1962) suggested that *spinifera* could have reached isolated *atra* habitats by overland dispersal during rains or via known subterranean connections. For turtles as thoroughly aquatic as *Apalone* we consider overland dispersal to be unlikely. Wendell L. Minckley and Legler experienced several heavy rains in the Basin in September 1958; *Terrapene coahuila* were moving about almost everywhere, but no other turtles were observed on land. The possibility of turtle dispersal via underground waterways is also considered unlikely (until it can actually be observed).

Whatever the means of dispersal for *A. s. emoryi*, distributional overlaps were accelerated within historic times (late 19th century) by human works that connected formerly isolated aquatic habitats. Thus, the present drainage patterns in the basin are not a completely natural phenomenon.

Isolation and reconnection of the basin occurred more than once (Minckley, 1969). A subsequent invasion event by *A. spinifera* might have complicated the scenario if and when *A. spinifera emoryi* hybridized not only with *A. atra* and but also with existing hybrid individuals of *atra* × *spinifera*. As of yet, we know of no experiments to identify or test the fertility of hybrid *Apalone*. The evolutionary events described for *Apalone atra* parallel those for *Trachemys scripta taylori*, *Terrapene coahuila*, and many other aquatic organisms in the basin. Thus, the taxonomic status of *atra*, its relationship to *spinifera* in the basin, and the interesting problems of how the present situation came to be are equivocal. The aforementioned events have produced a rather blurred phenotypic picture.

We choose to rank *atra* as a species because (1) in places it does occur in microsympatry with *A. spinifera*, and (2) seemingly "pure" individuals of both forms are still found (especially *atra* at the type locality). Furthermore, a hybridization event does not necessarily invalidate a taxon, especially when associated with an anthropogenic factor. Viable offspring have been produced by hybridization between *Glyptemys insculpta* and *Emydoidea blandingi* in a seminatural captive situation (Harding and Davis, 1999). Also, *Apalone spinifera* has been observed to mate with *A. mutica in* captivity (Legler, 1955).

GEOGRAPHIC DISTRIBUTION

Map 20

Apalone atra is endemic to permanent warmwater ponds in the Cuatro Ciénegas Basin of Coahuila, Mexico. These ponds are confined to an area near the northernmost extension of the Sierra de San Marcos. This includes the following localities: Tio Candido (type locality), El Mojarral, Posos de la Becerra, Rio Mesquites, Laguna Churince, Rio Chiquito, Saca del Fuente Canal, Rio Puente Chiquito, La Angostura Canal, Rio Canon, Anteojo, Tierra Blanca, and Laguna Grande (Winokur, 1968).

HABITAT

See description of the Basin of Cuatro Ciénegas in account of *Trachemys scripta taylori*.

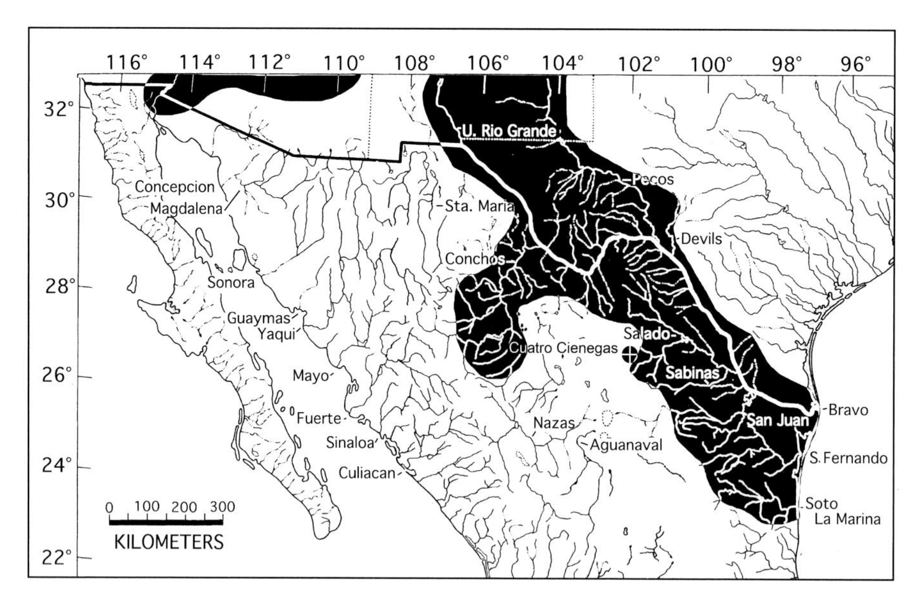

MAP 20 Geographic distribution of *Apalone spinifera* (entire darkened area) and *Apalone atra* (endemic to Basin of Cuatro Ciénegas, where both species occur).

Morphologically typical *Apalone atra* occur chiefly in—and probably evolved in—permanent, spring-fed, clear-water ponds in the Cuatro Ciénegas Basin (figure 30.03), an area of about 600 km^2, elevation 720 m. The basin is marshy with dry, sandy slopes. Individuals have been observed, netted, and trapped at depths of 2–3 m and, to our present knowledge, are absent or less common in fluvatile habitats and playa lakes (Winokur, 1968; McGaugh and Janzen, 2008).

The bottoms of the preferred ponds consist of gray silt and are partially covered with dense, submerged aquatic vegetation, mainly *Chara*. Water lilies grow in the shallow parts, and dense stands of cattails (*Typha*) and *Eleocharis* border the ponds. The ponds may be separated from potential nesting sites on the sandy slopes by up to 100 m of flat, marshy grassland. Water temperature is ca. 27–29°C (Webb and Legler, 1960). *Apalone spinifera emoryi is* found in rivers and playa lakes (Winokur, 1968; McGaugh and Janzen, 2008). Little is known about the ecology, population dynamics, reproduction, or behavior of these two taxa in the Cuatro Ciénegas Basin.

Although *A. atra* and *A. s. emoryi* may occur together, may hybridize, and probably compete at several places in the Basin, competition and hybridization seem to be minimal or nonexistent in the lagoons where *atra* evolved (Cerda-Ardura and Soberon-Mobarak, unpublished data; Cerda-Ardura, Soberón-Mobarak, McGaugh, and Vogt, 2008).

DIET

Apalone atra feeds selectively on aquatic insect larvae. Webb and Legler (1960) identified 23 larvae of long-eared leaf beetles (Chrysomelidae; *Donacia* sp.) and many small pieces of the roots of rushes, possibly ingested while capturing the insect larvae. Winokur (1968) examined the stomach contents of four adult females from Posos de la Becerra and four from Tierra Blanca and found large numbers of *Donacia* sp. larvae and pupae, parts of fish (*Ictalurus lupus* and *Cichlosoma* sp.), unidentified seed pods, and the lower parts of sedges. One stomach contained the shed skin of a large snake (*Nerodia* sp.).

HABITS AND BEHAVIOR

Legler (Webb and Legler, 1960) observed the type locality systematically with binoculars and was able to see heads at the surface only at dusk. No turtles were seen on the surface or in the crystal clear water during daylight hours, nor were any free-swimming turtles seen by divers in 1976 (JML, field notes). Careful searching revealed *atra* and *Trachemys scripta taylori* under dense vegetation or in sediment at the bottom of the pond. *A. atra* were captured in baited traps between dusk and dawn, or on overcast days. These observations were repeated by Legler in 1976 (field notes). Little aerial basking has been reported. Perhaps warm water temperatures obviate the need for basking as a thermoregulatory measure.

REPRODUCTION

Sexual maturity has been confirmed by dissection for lengths of 134 to 150 mm CL for males and 255 to 280 mm CL for females. Sex is genetically determined in *Apalone spinifera*, in other trionychids (Vogt and Bull, 1982; Greenbaum and Carr, 1996), and probably in *atra* (see "Sex Determination" in Introduction).

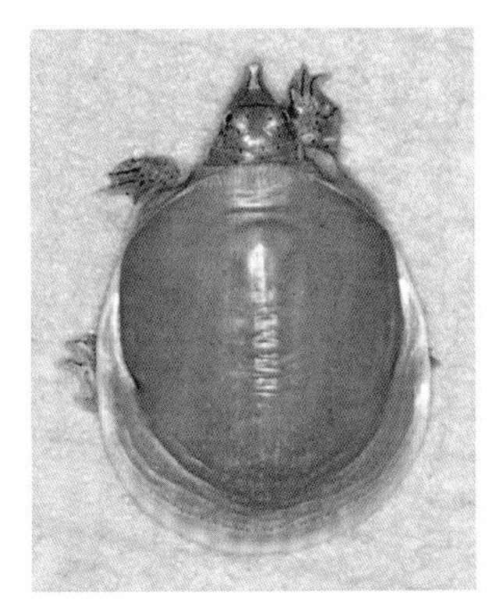
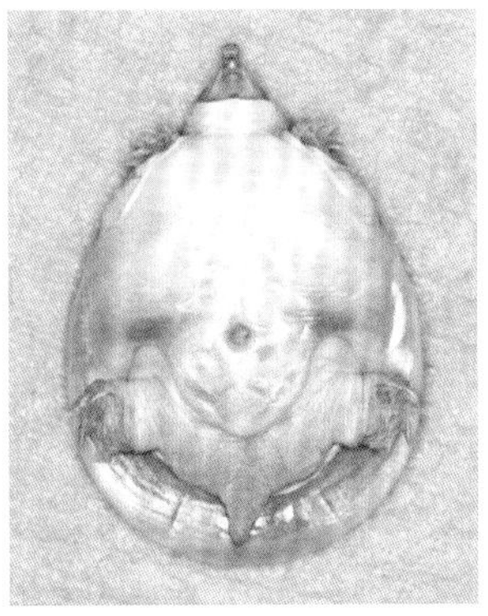
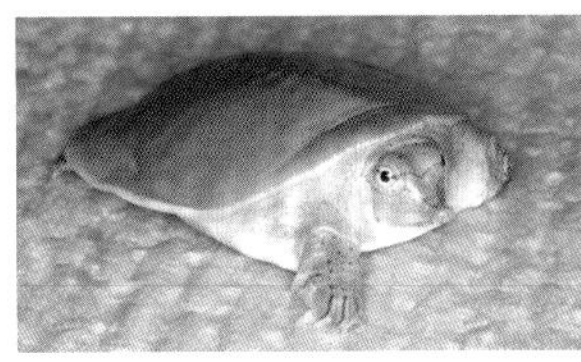

FIGURE 28.6 *Apalone atra*. Dorsal, ventral, and oblique views of hatchlings from eggs produced by female shown in Figures 28.4 and 28.5.

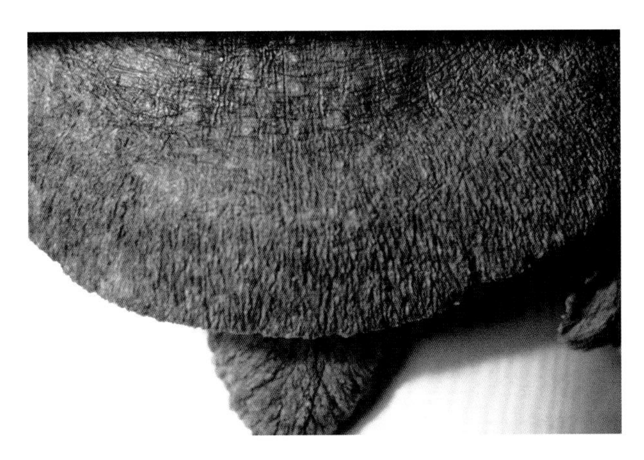

FIGURE 28.7 *Apalone atra*. Posterior edge of carapace of old male showing longitudinal rugosities. (Courtesy of R. M. Winokur.)

The second and third largest female paratypes (KU 46906 and 46908, 231 and 213 mm CL) were dissected and found to have immature ovaries. The slightly larger holotype (243) may be immature as well. Larger adults may have been unable to enter hoop nets, the largest of which had openings slightly less than 250 mm wide.

Courtship behavior is unknown for *atra*. A note by Legler (1955) describing a complete courtship/copulation event between a male *A. spinifera* and a female *A. mutica* could provide some clues. The "nipping" of the carapace edges by the male might account for the sometimes ragged or "notched" posterior edge of the carapace in female *atra*.

No turtle nests of any kind have been found in the Basin. There are no sand banks or mud banks associated with the habitat favored by *A. atra*. Matted vegetation is abundant and is a possibility for nesting sites.

A preovulatory female (CL, 233 mm; Wt, 2,067 g; UU 17569; Figures 28.4 and 28.5) was captured by hand at Tio Candido (25 May 1976) and transported to Utah. It was palpated repeatedly and deemed gravid on 11 June. Seven eggs were obtained on that date by hormonal induction (Pitocin) and incubated at 30°C. Four of the eggs developed normal chalk marks and three of these hatched. Incubation period was 80.5 days ± 12 hours. The eggs were almost perfectly spherical and hard shelled; diameter and weight were, respectively, 29.7 mm ± 0.92 (27.8–30.4) and 14.2 g ± 0.83 (13.0–15.6) (Legler and J. F. Berry, pers. obs.).

The hatchlings (Figure 28.6) were kept in captivity for an unrecorded period; all had traces of an umbilical scar but no trace of caruncles when preserved. Carapace length was 43.8 mm ± 2.3 (41.7–46.2). Salient characters of these juveniles were as follows: anterior edge of carapace has only a trace of tuberculation; there is a pale-yellowish peripheral band from anterior one-third of dorsal carapace that increases in width to 3.3 mm on posterior edge; choanal flaps are horizontal with weakly papillose edges; carapace is tan and otherwise unmarked; plastron is pale and unmarked; head has transverse line across interorbital region connecting the anterior one-third of each eyelid; there is another fine, dark line from the angle of the eye along the side of the snout almost to the tip; there are fine, dark lines suggesting a triangle with an anterior apex; septal processes are distinct; iris has an incipient, dark transverse stripe that does not fully reach the pupil. The differences between these juveniles and their maternal parent represent an enormous amount of ontogenetic change. Webb (1962) reported a hatchling of *atra* that was indistinguishable from *emoryi* (without further comment).

A larger juvenile of undoubted *Apalone spinifera emoryi* (UU 4140; CL, 79 mm; Val Verde Co., TX) shows the following characters: there are fine white dots on the posterior one-third of carapace; carapace is tan and otherwise unmarked; plastron is pale and unmarked; there are no discrete septal processes (rather, an amorphous lump on each side of nasal septum); there is a discrete, pale, complete circumferential band from anterior to posterior on the carapace, less than 0.5 mm wide anteriorly and 5.7 mm wide posteriorly; choanal flaps are horizontal with serrated edges.

GROWTH AND ONTOGENY

Data on CL and ontogenetic stage (S. McGaugh and Legler databases): males, 190 mm ± 56 (115–247) n = 9; females, 252 mm ± 15.6 (232–294) n = 21; im. females, 198 mm ± 26.3 (136–231) n = 32; unsexed juveniles, 44.4 mm ± 1.7 (43.0–46.3) n = 3.

Weights associated with CL (S. McGaugh, pers. comm.) are as follows: five males (218 to 247 mm CL), 1,416 g ± 263 (1,160–1,739); females, 1,554 g (245 mm CL), 2,534 g (282 mm CL), and much greater than 2,000 g (294 mm CL); five immature females (189 to 218 CL), 633–1,202 g. The

FIGURE 28.8 Hybrids, *A. spinifera* × *A. atra*, Rio Churince, Basin of Cuatro Ciénegas. Top, male; bottom, female. (Courtesy of R. M. Winokur.)

two heaviest females were taken in June and were probably gravid.

Males are smaller than females and have an elongated preanal region with the cloaca extending beyond the posterior edge of the carapace.

Most of the characters distinguishing *atra* from *emoryi* develop in the course of ontogeny and reach full phenotypic expression in adults. There is a striking resemblance among all hatchling and juvenile *Apalone*.

There have been too few neonates described from the basin to make accurate statements on ontogenetic change in diagnostic characters. McGaugh and Janzen (2008) thought that none of their diagnostic characters underwent ontogenetic change. The events we describe under "Reproduction" would indicate that nearly all diagnostic characters undergo substantial ontogenetic change.

POPULATIONS AND RELATIONSHIPS WITHIN THE BASIN

Webb and Legler (1960) pointed out that *atra* was relatively abundant at Tio Candido. *Trachemys scripta taylori* trapped there at the same time outnumbered *A. atra* by 4.2:1.

Winokur (1968) confirmed the existence of intermediates but concluded that hybridization was restricted and that *atra* was maintaining its genetic integrity as a full species. He examined specimens from 11 localities in the Basin; both species occurred at eight of these—*atra* alone at one locality and *emoryi* alone at two localities. Identity was determined on the balance of all phenotypic characters. Results are shown in Table 5.

McGaugh and Janzen (2008) surveyed populations of *Apalone* at nine sites in the Cuatro Ciénegas Reserve using baited hoop-net traps; their results are based on 62 days of trapping in 2003 and 2004 (mid May to early July). Three criteria for identification were selected: dark pigmentation on the carapace, speckled pigmentation on the plastron, and cutaneous corrugations on the posterior carapace. They captured six *A. atra* but no *spinifera* at Tio Candido. Four *A. atra* were trapped in Antiguos Mineros Grande and three from Poso de las Tortugas. The authors characterized these sites and the softshells in them as follows:

Lagoons
- Tio Candido: *atra* characters only (source of "pure *atra*" specimens); no *emoryi* caught.
- El Mojarral Este: *atra* with some *emoryi* characters.
- Antiguas Mineros Grande: *atra* with some *emoryi* characters.
- Anteojo: *atra* and *emoryi* characters in same individuals.

Rivers
- Rio Mesquites: *atra* and *emoryi* characters in same individuals.
- Rio Canon: A preponderance of *emoryi* characters (margins not rugose, no dark pigmentation on carapace).

Wastewater pond
- Ejido San Marcos: predominantly *emoryi* characters.

Playa lakes
- Laguna Grande: predominantly *emoryi* characters.
- Los Gatos: predominantly *emoryi* characters.

In summation, *Apalone atra* favors lagoon habitats and exists in Tio Candido in "pure" form and in sympatry with a few *A. s. emoryi* (Winokur, 1968). Other lagoons contain *atra* with some *emoryi* characters. Riverine habitats show an intermediacy of *atra* and *emoryi* or a preponderance of *emoryi* characters. Playa lakes show a preponderance of *emoryi* characters and virtually no *atra* characters. After half a century, the status at Tio Candido shows little or no change.

Trapping may not accurately reflect population structure or natural sex ratios. This could be attributed to method of capture, differing activity cycles, or the idiosyncrasies of softshell behavior. Therefore, to say that populations of *A. atra* are "low" is simply a guess. It may take nocturnal diving or circumferential trammel netting to solve these problems.

PREDATORS, PARASITES, AND DISEASES

Minckley (1966) found the remains of two *A. atra* (with the remains of nine *Trachemys*) near a site where a coyote (*Canis latrans*) was actually seen to stalk and kill a *Trachemys*

TABLE 5
Relative abundance of *Apalone* sp. at 11 localities in the basin of Cuatro Ciénegas

Locality	*A. atra*	*A. s. emoryi*
Tio Candido, 14 km S Cuatro Ciénegas type locality	13 female, 1 male	1 male
El Mojarral, 9 km S Cuatro Ciénegas	4 female	1 female, 2 male
Anteojo, 8.5 km WSW Cuatro Ciénegas	1 female	1 hatchling
Posos de la Becerra, 14 km SW Cuatro Ciénegas	4 female, 1 male	2 female
Rio Churince, 0.75 km NW Laguna Churince	2 female	1 male
Rio Churince, small unnamed laguna, 1 km NW Laguna Churince	1 female	2 male
Rio Mesquites, ca._7 km WSW Cuatro Ciénegas	1 female	None
Rio Mesquites, 10 km S Cuatro Ciénegas	1 female	6 female
Rio Canon, 5.3 km NNW Cuatro Ciénegas	None	1 female
Saca del Fuente Canal, 9.6 km E Cuatro Ciénegas	1 female	1 male, 1 hatchling
La Angostura Canal, 6.6 km WNW Cuatro Ciénegas	None	1 female

scripta taylori (9 km SW of Cuatro Ciénegas). All the turtle remains were crushed near the anterior end of the carapace. Nesting females would be extremely vulnerable to mammalian predators.

Dissection of an immature male from Rancho Orozco by James F. Berry (UU 17568) revealed the outside of the stomach and part of the small intestine to be covered with small pustules.

CONSERVATION AND CURRENT STATUS

Populations of *atra* are said to be low, perhaps critically so, but the taxon is neither extinct nor invalid (Cerda-Ardura, Soberón-Mobarak, McGaugh, and Vogt, 2008). A long-term capture-release study should be done at Tio Candido and other selected sites in the Basin. These sites must receive absolute protection from human and domestic animal use, and from industrial or agricultural use of groundwater. The study should include the protection and restoration of potential nesting sites, X-ray examination of females for gravidity, and a facility to study captive behavior under seminatural conditions.

At present the following factors place the survivorship of *atra* at risk: (1) competition from or introgression with *A. s. emoryi* (which we contend is an invasive species); (2) industrial and agricultural use of water affecting water levels and temperature and establishing connections between formerly isolated habitats with canals; (3) degradation of ponds and potential nesting sites by feral and domestic livestock; and (4) at least occasional human use for food (Webb, 1962).

Present conservation measures must be reviewed. The Cuatro Ciénegas Basin was proposed as a National Park in 1987 and as a Biosphere Reserve in 1989, but was not officially protected until 1994 when the 84,000 ha Cuatro Ciénegas Protected Area for Flora and Fauna was designated a Natural Protected Area. This area was expanded to 840,000 ha in 2007 and was designated an international RAMSAR wetland and included in the Nature Conservancy Parks in Peril program. In 2006 it became a UNESCO Man and the Biosphere Reserve.

Although the foregoing proposals and credentials are impressive, they only hint at success. Conservation and protection cannot exist without enforcement (see remarks under "Economic Uses and Conservation" in Introduction).

Apalone spinifera emoryi (Agassiz) 1857

Tortuga de concha blanda Texana (Liner, 1994); Texas spiny softshell (Iverson, 1992a).

ETYMOLOGY

Emoryi is a patronym honoring William Hemsley Emory (1811–1887), Major, U.S. Army Corps of Topographical Engineers, who was associated with the Mexican Boundary Survey.

HISTORICAL PROLOGUE

Three species of *Apalone* occur in the United States (and eastern Canada)—*mutica*, *spinifera*, and *ferox*. A fourth species, *atra*, occurs only in Mexico. The species *A. spinifera* (Le Sueur, 1827) has the widest geographic distribution, chiefly in the United States. It is of intermediate size—larger than *mutica* and smaller than *ferox*. Six subspecies are recognized: *spinifera*, *hartwegi*, *aspera*, *pallida*, *guadalupensis*, and *emoryi* (Web, 1973). Only *emoryi* occurs in Mexico. See the account of *Apalone atra* for comments on possible hybridization with *emoryi* in the Basin of Cuatro Ciénegas.

FIGURE 29.1 *Apalone* with preponderance of *A. spinifera emoryi* characters; male, Rio Churince, Basin of Cuatro Ciénegas. (Photo courtesy of R. M. Winokur.)

FIGURE 29.2 *Apalone spinifera emoryi*. Male; UU 11862; Las Norias, Tamaulipas; bluish marks correspond to dermal shell elements but are inexplicable.

DIAGNOSIS

The following suite of characters distinguishes *A. s. emoryi* from *Apalone atra* and all other subspecies of *Apalone spinifera*:

1. There is a pale circumferential band on rim of dorsal carapace, four to five times wider posteriorly than anteriorly.
2. Tubercles on anterior edge of adult carapace are low, rounded, and not spinous.
3. Posterior one-third of carapace often bears small, pale tubercles imparting a speckled appearance.
4. Postorbital stripe is pale and usually interrupted, leaving a conspicuous, pale, usually dark-bordered postorbital mark.
5. Carapace ground color is brownish to tan and unmarked except for the pale-yellowish circumferential band.
6. Plastron and ventral limb surfaces have few or no dark markings.
7. Fine, dark lines on head form a facial triangle with anterior apex, a base connecting the anterior angles of the eyes and sides converging near tip of snout.

GENERAL DESCRIPTION

Maximum CL for all specimens of *emoryi* from Mexico and adjacent United States is as follows: males, 273 mm; females, 329 mm. Degenhardt, Painter, and Price (1996) report CL maxima in southern New Mexico as 361 mm for males and 367 mm for females. Halk (1986) reported a seemingly record size for the species *spinifera* from northern Louisiana (a female—CL, 540 mm; CW, 406 mm; PL, 362 mm, Wt, 11.7 kg).

BASIC PROPORTIONS

(Excluding Basin of Cuatro Ciénegas)

CW/CL: males, 0.84 ± 0.038 (0.75–0.91) n = 31;
females, 0.77 ± 0.027 (0.69–0.81) n = 24;
im. females, 0.82 ± 0.076 (0.51–0.98) n = 34;
juveniles, 0.78 ± 0.155 (0.51–0.91) n = 6.

HT/CL: males, 0.24 ± 0.065 (0.22–0.30) n = 16;
females, 0.23 ± 0.021 (0.20–0.26) n = 7;
im. females, 0.25 ± 0.015 (0.22–0.27) n = 10.

HT/CW: males, 0.30 ± 0.03 (0.27–0.37) n = 16;
females, 0.29 ± 0.026 (0.26–0.34) n = 7; im. females, 0.30 ± 0.017 (0.27–0.34) n = 10.

WH/PL: males, 0.20 ± 0.016 (0.17–0.25) n = 29;
females, 0.16 ± 0.019 (0.11–0.20) n = 32;

im. females, 0.20 ± 0.014 (0.17–0.22) n = 34;
females, 0.25 ± 0.021 (0.23–0.30) n = 10;
juveniles, 0.17 ± 0.031 (0.13–0.21) n = 6.

Snout W/WH: males, 0.19 ± 0.023 (0.11–0.23) n = 41;
females, 0.21 ± 0.04 (0.15–0.38) n = 30;
im. females, 0.19 ± 0.023 (0.15–0.27) n = 33; juveniles, 0.19 ± 0.029 (0.15–0.22) n = 6.

The internarial septum bears a distinct horizontal ridge projecting into a large nostril on each side. Lips are yellowish with darker spotting. Diploid chromosome number is 66 (Stock, 1972).

Relationships are discussed in the account of *Apalone atra*.

FOSSILS

Apalone spinifera is known from the Pleistocene in Brazos and Knox Counties, Texas (Holman, 1969); Tillman County, Oklahoma (Gehlbach, 1965); and Sangamon County, Illinois (Holman, 1966).

GEOGRAPHIC DISTRIBUTION

Map 20

Apalone spinifera has an extensive distribution in the eastern United States. The subspecies *Apalone s. emoryi* occurs throughout the modern Rio Grande drainage system from northern New Mexico to Brownsville/Matamoros.

FIGURE 29.3 *Apalone spinifera emoryi*. Head and neck of UU 11862, Las Norias, Tamaulipas.

FIGURE 29.4 *Apalone spinifera emoryi*. Juvenile, northern Mexico, outside Basin of Cuatro Ciénegas.

This includes the drainages of the Pecos in Texas and the R. Conchos in Chihuahua. It also occurs in numerous smaller tributaries on both sides of the lower Rio Grande, notably the R. Salado, R. Sabinas, and R. San Juan of Coahuila, Nuevo León, and Tamaulipas. Northeast of the lower Rio Grande (in Texas) the range abuts that of *T. s. guadalupensis* and an intergrade population of *guadalupensis/pallidus* (Conant, 1975; Iverson, 1992a).

The range extends southward in Gulf drainages to the R. San Fernando and R. Soto la Marina in Mexico (ca. 23°48′N), which constitute the southernmost distributional limits of trionychids in the Western Hemisphere. See account of *A. atra* regarding distribution in the Basin of Cuatro Ciénegas. We discount the mention of *Apalone* occurring in the Rio Alamos near the town of Alamos, Sonora (Webb, 1962). Legler has visited the precise locality three times and was able to scan large clear-water pools with binoculars from a high vantage point. The only turtles observed were *Kinosternon* sp. and *Trachemys script hiltoni*.

A. s. emoryi was introduced in the Colorado–Gila drainage around 1900 after escaping from a privately stocked pond on the Gila drainage near Duncan, Greenlee Co., Arizona (Miller, 1946). Henceforth, it aggressively colonized most of that drainage system. It now occurs in Arizona, California, Nevada, Utah, and New Mexico. Rorabaugh, Montoya, and Gomez-Sapiens (2008) documented occurrences near the mouth of the Colorado River, establishing its presence in the states of Sonora and Baja California Norte.

Since there are few natural history data for this subspecies in Mexico, the following data are taken from the literature and our combined experience studying *Apalone s. spinifera* in Wisconsin, *A. s. aspera* in the southern United States, and *A. s. emoryi* in the American southwest. Unless otherwise stated, many remarks and observations are attributed only to "*Apalone spinifera*."

HABITAT

Spiny softshell turtles inhabit rivers and are commonly seen basking on sandbars (but may also bask on logs and rocks). They are usually more abundant in areas with soft sandy or muddy bottoms or with submergent aquatic vegetation. Smaller streams, lakes, and oxbow lakes are less favored. Females are thought to prefer more open, deep water than males (Williams and Christiansen, 1981).

Legler has collected *emoryi* in the Rio Grande and Rio Conchos in Chihuahua (streams bordered by harsh desert habitat) at times when these streams were flowing. The R. Conchos at Mesquita, Chihuahua (13 km S and 26 km W of Ojinaga), was narrow, clear, and flowing in June 1959. Careful scrutiny from the banks revealed no sign of turtles. Traps set during the day produced one adult female *A. spinifera emoryi*. Large cottonwood trees grew near the water but were replaced by mesquite/creosote bush habitat a few meters away. The sandy banks shortly gave way to rockier substrate. This probably represents an extreme of habitat tolerance by *Apalone*. By contrast, *emoryi* was observed and trapped (JML, field notes, 1966) in estuarine parts of the R. San Fernando and nearby drainages in Tamaulipas.

DIET

Apalone spinifera is an active predator that swims along the bottom and thrusts the snout into vegetation or under rocks and debris searching for prey. They sometimes actively chase small fish but more often use gape and suck behavior to swallow small prey items (thus the addition of plant material, sand, and gravel in the stomach contents). Large prey is held in the mouth and torn apart with the front feet before swallowing. By contrast, Vogt (pers. obs.) found acorns to be common in stomach flushings of *Apalone spinifera* from the Pearl River in Mississippi. These were found with such frequency that they must have been selectively eaten.

No particular dietary studies have been conducted for *emoryi*. Other subspecies in the Mississippi drainage are chiefly carnivorous, feeding on insect larvae, nymphs, and terrestrial adults (Ephemeroptera, Trichoptera, Odonata, Diptera, and Coleoptera). Other invertebrates include decapods, isopods, oligochaetes, gastropods, and pelecypods. Small fishes were found more often in female stomachs than in males, while males ate more dragonfly nymphs than females (Cochran and McConville, 1983).

Williams and Christiansen (1981) reported on the gut contents of 52 spiny softshells in Iowa as follows (proportion of total volume): large fish, 17.2%; small fish, 2.2%; crayfish, 24.2%; larval and nymphal insects, 27.9%; plant material, 26%. Lagler (1943) found the stomachs of 11 spiny softshells in Michigan to include crayfish (47%) and insects (52%).

HABITS AND BEHAVIOR

Annual activity cycles depend on temperature. Activity may occur all year in the southern United States and adjacent Mexico. Activity occurs from April through October in Kentucky (Ernst, Lovich, and Barbour, 1994) and from May to late September in Wisconsin (Vogt, 1981). However, they are active all year in a Wisconsin lake where an electrical plant discharges warm water and the lake does not freeze (Vogt, 1981).

Apalone spinifera is diurnal and spends the night buried in soft substrate or among branches. They are among the most aquatic of all turtles, foraging, floating on the surface, or buried in the sand with only the head and neck exposed. Well-oxygenated water is necessary, since several forms of alternative respiration are practiced (see account of genus). Captive hatchlings often die at night if the water they are kept in becomes anoxic (RCV, pers. obs.).

Spiny softshells are heliothermic and spend much time basking—on sandbars, if available, and less frequently on logs or rocks. When basking on sandbars they are usually poised on the edge of a slope with neck extended and facing the water. At the slightest provocation they sprint to the water at speed. Groups of up to eight individuals were often seen basking together on islands of sand resulting from dredging in the Mississippi River near Stoddard, Wisconsin.

REPRODUCTION

Little reproductive information pertaining to *emoryi* is available, and we attempt to summarize what is known about *A. spinifera* ssp. elsewhere as being of possible value.

Of the Mexican specimens in our database, CL of the smallest mature male was 96 mm and the largest immature male 103 mm; corresponding figures for females were 102 and 167 mm.

In Oklahoma *A. spinifera "emoryi"* males are mature at carapace lengths of 104–117 mm (converted from PL) and females are mature at 234 to 260 mm CL (Webb, 1962). Males of *A. s. spinifera* in Tennessee are mature at 130 g; females are mature at 1.5 kg (Robinson and Murphy, 1978).

Robinson and Murphy (1978) made a detailed study on the male and female reproductive cycles of *A. s. spinifera* in Tennessee, and the following data are from their study unless otherwise noted. In April spermatogonia and Sertoli cells are abundant in the walls of the seminiferous tubules, and the lumen has mature sperm. Spermatocyte production peaks in June–July. By August spermatocytes are fewer in numbers. Mature sperm are still present in the tubules in July and August when the weight of the testes is greatest. In September few spermatocytes are present. Sperm enter the epididymides in September and October where they are stored over winter. Mating occurs in April and May. Viable sperm can be stored in small tubules in the walls of the oviduct (Gist and Jones, 1989).

Vitellogenesis begins in September, and by mid September to mid October ovarian follicles are 12–22 mm in diameter at hibernation. The follicles are the same size at spring emergence and grow to ovulatory size in May. Ovulation occurs in May and June with nesting in June and July. Two clutches of eggs may be produced each year. Nesting usually occurs in bright sunlight at the highest edge of sand banks that slope into the river. Nests are usually within a few meters of the water's edge, but females may travel as far as 100 m or more on land to nest (Vogt, 1981). Spiny softshells have been found nesting in the rocky substrate of elevated railroad grades in Wisconsin. Unlike many species of turtles *Apalone spinifera* is wary when nesting. If disturbed while digging the nest cavity, they will abandon the nest and race to the water (Robinson and Murphy, 1978).

The brittle-shelled, spherical eggs are 28 mm (24–32 mm) in diameter and weigh 10–11 g. Four to thirty-nine eggs per clutch were reported by Vogt (1981, Wisconsin). The egg shell is 0.15 to 0.19 mm thick (nearly twice the thickness of a *Chrysemys picta* egg shell) and composed of 27% fibrous layers and 73% mineral layers (Ewert, 1979). *Apalone* eggs are much more resistant to desiccation than those of other aquatic turtles, but they exchange water at five times the rate of bird eggs (Packard, Taigen, Packard, and Shuman, 1979).

An egg of *emoryi*, illustrated by Agassiz (1857: Pl VIII, Figure 20) is 29 mm in diameter. Other *Apalone* eggs and their diameters in the same plate are as follows: *A. mutica*, 22 mm (Figure 21); *A. ferox*, 25 mm (Figure 22); and *A. spinifera*, 26 mm (Figure 23). The eggs of *emoryi* and *atra* probably rank as the largest among *Apalone*.

All species of *Apalone* studied demonstrate direct embryonic development and incubation times directly correlated with temperature: 95 days at 25°C and 58 days at 30°C in *A. spinifera* (Ewert, 1979). Eggs hatch in 82–84 days (late August) in natural nests in Wisconsin (Vogt, 1981). Sex determination is genetic, and a 1:1 sex ratio is produced at all incubation temperatures (Bull and Vogt, 1979; Bull, Vogt, and McCoy, 1982).

GROWTH AND ONTOGENY

Data on size for Mexican specimens, including Cuatro Ciénegas (S. McGaugh and Legler databases): males, 143 ± 37 (96–273) $n = 45$; females, 228 ± 48 (102–329) $n = 49$; im. females, 137 ± 23 (90–167) $n = 32$; unsexed juveniles, 41.7 ± 1.2 (41–43) $n = 3$. Size data for southern New Mexico (Degenhardt, Painter, and Price, 1996; CL mean and extremes): adult males, 157 mm (108–361); adult females, 215 mm (110–397).

The largest immature females we dissected (217 and 188 mm CL, Basin of Cuatro Ciénegas) contained a few follicles of 2–3 mm and one of 6.7 mm. Both had typically immature, narrow, ribbonlike oviducts. It is probable that they would have reproduced within a year.

Females develop a mottled or blotched pattern before they reach sexual maturity (Webb, 1956). Adult males retain the juvenile pattern and have long, thick tails with the cloacal opening near the tip. Females have a shorter tail that terminates anterior to the margin of the carapace. Adult females are much larger than males.

Males tend to retain the shell coloration of juveniles; females may develop a darker lichenlike carapace pattern. At larger immature sizes (>90 mm CL) individuals not demonstrating male secondary sexual characteristics can accurately be sexed as immature females.

The smallest *emoryi* reported by Webb (1962) were juveniles of 45 to 52 mm CL and a hatchling of 25 mm. Hatchling color pattern is similar to that of males—an unmarked carapace field of olive to tan that is speckled posteriorly with pale tubercles. The pattern is more distinct in hatchling males than females. Accurate sexing of hatchlings by external characters is a rare circumstance in turtles.

Breckenridge (1955) studied the growth of *A. spinifera* in Minnesota. Adult CL related to age was estimated as follows: females—250 mm, 10 years; 297 mm, 15 years; 333 mm, 20 years; 381 mm, 30 years; and 430 mm, 53 years; males—160 mm, 10 years; 170 mm, 15 years. Growth of hatchlings is rapid; a hatchling of 38 mm CL grew 48 mm in its first year but slowed to 42 mm in its second year.

A longevity record of 25 years, 2 months, and 17 days was reported for a female *A. spinifera* in the Racine, Wisconsin, Zoo (Snider and Bowler, 1992).

PREDATORS, PARASITES, AND DISEASES

Raccoons (*Procyon lotor*), red fox (*Vulpes fulva*), and skunks (*Mephitis mephitis*) raid nests on the Mississippi River in Wisconsin (Vogt, 1981). Young are eaten by fish, other turtles, wading birds, and mammals (Webb, 1962). Spiny softshells can move rapidy on land and in the water and may escape many predators by deft and speed. The long neck functions in circumspection as well as in feeding and defense. They strike and bite viciously when handled, and their sharp jaws can inflict injury.

POPULATIONS

Few studies have addressed population sizes or structure in this species. Cagle and Chaney (1950) reported spiny softshells making up less than 1% of more than 1,000 turtles taken from a drainage ditch near Jacob, Illinois; however, he caught 16% softshells in a collection of 214 turtles collected from a stream near Elkville, Illinois. They found *spinifera* to be more abundant in streams with current in Louisiana. Spiny softshells made up 53% of the turtles caught in the spillway of Caddo Dam, 25% in Lacassine Refuge, and 67% in the Sabine River. Areas without current had lower populations: Caddo Lake, 31%; False River, 9%; and Lake Providence, 13%. Adult sex ratios in the Cahaba River of Alabama, Chickasawhey, and Pearl Rivers of Mississippi were 1:1 (Vogt and Bull, 1982).

CONSERVATION AND CURRENT STATUS

No specific conservation measures have been undertaken for this species in Mexico, but since it is pursued occasionally for food, it deserves full protection.

SUPERFAMILY TESTUDINOIDEA

As covered here, the testudinoids are comprised of three large families (Testudinidae, Emydidae, and Geoemydidae) containing about 58% of the recognized 466 terminal taxa in the world (Bickham et al., 2007).

Testudinoids demonstrate most of the variation seen in cryptodires. Body form varies from high-domed tortoises with elephantine feet, to a depressed hydrodynamic shell and webbed feet, to box turtles with hinged plastra. Adult size varies from the tiny *Glyptemys muhlenbergi* to giant Galapagos tortoises.

Inframarginals are absent (except as axillary and inguinal scutes), permitting marginals and plastral scales to lie in contact on bridge; axillary and inguinal buttresses are well developed and in contact with costal elements except in hinged plastra; eighth cervical vertebra is biconvex; blade of ilium is laterally curved with double origin for iliotibialis muscle.

Family Testudinidae

Family Testudinidae Batsch, 1788—The testudinids are the "tortoises" in the strict sense of that vernacular term—chiefly medium to large, predominantly herbivorous, completely terrestrial chelonians. Most tortoises are the quintessence of slow, deliberate movement. They are docile and innocuous and are usually depicted as happy creatures in children's books. "The Tortoise" is a principal character in an Aesop fable (sixth century BC) and the international symbol for "slow" on lawn mowers.

The family is considered a natural group with 19 genera, 54 species, and 85 terminal taxa (Bickham et al., 2007). They share characters with geoemydids and emydids but are more closely related to the former (Hirayama, 1984; Gaffney and Meylan, 1988). *Manouria* is regarded as the most primitive testudinid genus and the sister taxon of all other tortoises (Crumly, 1994).

Present geographic distribution of the family is in temperate and tropical regions of the Eastern and Western Hemispheres and a few oceanic islands. The oldest known testudinids are from the late Paleocene of Mongolia; the earliest North American fossils are early Eocene (Holroyd and Parham, 2003). *Gopherus* is the only genus occurring in North America.

DIAGNOSIS OF FAMILY

Carapace is usually high domed and forms a long, akinetic, sutural bridge with an extensive plastron that tends to reduce the anterior and posterior orifices of the shell (Figure 30.1). General reduction of skeletal mass results usually in a thinner and lighter shell than other chelonians. Shell kinesis occurs in *Kinixys* (carapace), *Pyxis* (plastron), some *Testudo* (plastron) and in females of *Gopherus berlandieri* (plastron). Bones and scutes of shell are generalized—6 pairs of plastral laminae, 11 peripherals, and 12 marginals on each side; axillary and inguinal scutes are present. Nuchal bone lacks costiform processes. Rib heads are reduced and often lacking, essentially eliminating the epaxial space. Temporal roofing of skull is widely emarginated posteriorly and ventrally, forming narrow postorbital and zygomatic bars; columella is completely surrounded by quadrate; descending processes of prefrontal bones define a wide ethmoidal fissure; there is a premaxillary beak; there is a triturating surface of maxilla with or without ridging; palate is vaulted, accommodating a large, muscular tongue; hyoid skeleton is moderate to small (see "Feeding Mechanics" in Introduction; Figure 30.2).

Limbs are modified for walking with the plastron held clear of substrate (Figure 31.3); hind limbs are elephantine and columnar; forelimbs are flattened (becoming elephantine only in largest taxa), and antebrachia are often covered with thick osteodermal scales; all digits are shortened (almost never more than two phalanges in any digit); feet are webless; foreclaws in burrowing species are heavily spatulate. Trochanteric fossa of femur are narrowed by close approximation of the trochanters. Skin on head is divided into scales. Spurlike tubercles are commonly present on rear of thigh. Tail is extremely short in adults and young. Musk glands and cloacal bursae are lacking (based on Williams, 1950; Loveridge and Williams, 1957; Auffenberg, 1974; Gaffney, 1975a; Smith and Smith, 1979; Ernst, Lovich, and Barbour, 1994).

Genus *Gopherus* Rafinesque, 1832

Gopher tortoises.

ETYMOLOGY

The masculine *Gopheru*" is derived from "gopher," a phonetic spelling of the French *gaufre* (= waffle). The name was used by early French settlers for any small, burrowing animal. "Gopher" was first used by Bartram (1791). Holbrook (1842) mentions "mungofa" as a common name; "mungofa" is derived from a West African word *gofa*, meaning to dig (Auffenberg and Franz, 1978a).

DIAGNOSIS AND DESCRIPTION

Gopher tortoises are diurnal herbivores that are usually an evident part of a fauna. The following characters (in combination or individually) unequivocally distinguish *Gopherus* from all other testudinids (Crumly, 1994): premaxillary bones form a median ridge along their common ventral suture (shared only with the Oligocene fossil *Stylemys*); there is a variably enlarged otolith in the sacculus of the internal ear; prefrontal pits (a recess on the ventral surface of each prefrontal bone) are present at some ontogenetic stage; there is a small sessamoid bone (os transiliens) in the tendon of the adductor mandibulae muscle where it passes over a trochlear process formed by the quadrate and prootic bones; marginal scutes 4 and 6 are enlarged, and M4, 5, and 6 are in contact with second lateral scute; there is a pair of large (class 1) mental glands on the skin of the throat (see Introduction and account of *Gopherus agassizi*). The karyotype is the same for three of the recognized species (*berlandieri*, *agassizi*, and *polyphemus*): diploid number is 52: 26 macrochromosomes and 26 microchromosomes (Stock, 1972; Dowler and Bickham, 1982). No karyotype has been reported for *G. flavomarginatus*.

The fossil record of *Gopherus* extends from early Oligocene through the Pleistocene (Auffenberg and Franz, 1978a).

Reynoso and Montellano (2004) have recently described *Gopherus donlalio*, a giant tortoise (PL 520 mm) from the Rancholabrean Pleistocene of Tamaulipas (80 km N of Cd. Victoria). The skull is unique but resembles that of the *polyphemus–flavomarginatus* group.

RELATIONSHIPS

The genus is monophyletic but devisable into two groups based on morphology and molecular data (Crumly, 1994; Bramble, 1982; Auffenberg, 1976; Gaffney and Meylan, 1988; Lamb, Avise, and Gibbons, 1989; Lamb and Lydeard, 1994; Morafka, Aguirre, and Murphy, 1994).

The two groups (sometimes referred to as subgenera) have been allopatric except in the Pleistocene of Aguascalientes. Gene flow does not occur between the groups nor between *Gopherus polyphemus* and *G. flavomarginatus*, but might occur between *agassizi* and *berlandieri* (Morafka, Aguirre, and Murphy, 1994).

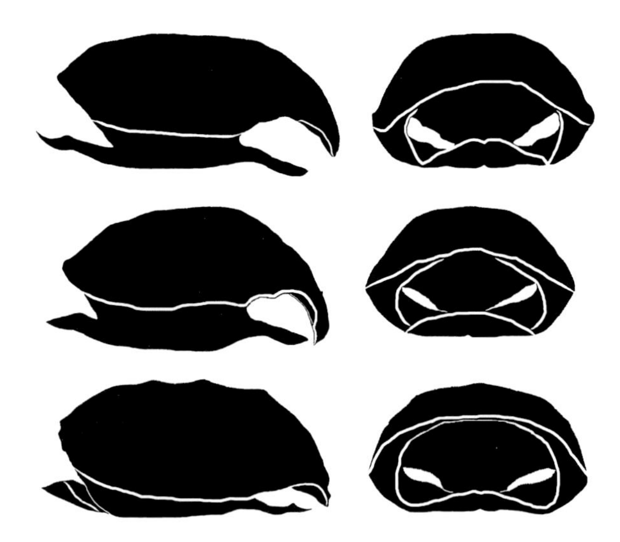

FIGURE 30.1 Typical profiles and optical cross-sections of the three species of *Gopherus* occurring in Mexico. From top to bottom: *Gopherus agassizi*, adult; *Gopherus berlandieri*, adult; *Gopherus flavomarginatus*, immature. (All images modified from Auffenberg, 1976, Figures 29–31.)

Bramble (1982) defined these groups morphologically and placed them into two separate genera, characterizing them as follows (see also Germano, 1993): *Gopherus* (containing *flavomarginatus* and *polyphemus*), characterized by specializations for digging and living in burrows and distinguished by a wider head, wider manus, and a distinctive inner ear and first dorsal vertebra; *Scaptochelys* (*agassizi* and *berlandieri*), a group more nearly adapted to simple walking than burrowing, with a narrower head and manus and lacking the inner ear and vertebral specializations.

Most remarkable among the features distinguishing *Gopherus* (sensu Bramble) is a large saccular otolith that impinges upon sensory hair cells in the inner ear and is classically regarded as a receptor of static equilibrium—a sense of position of the head. But it could and probably does act as a high-grade receptor of ground-borne and perhaps airborne vibrations (Adest, Aguirre-León, Morafka, and Jarchow, 1989a; Patterson, 1976; Bramble, pers. comm.). This feature is unique among tetrapod vertebrates (the saccular otolith of *agassizi* and *berlandieri* is small; Bramble,1982; Crumly, 1994).

Bour and Dubois (1984) demonstrated that *Scaptochelys* was a junior synonym of *Xerobates* (Agassiz, 1857). Crumly (1984, 1994) then proposed the suppression of *Xerobates* because it is not a natural group. We follow Crumly and include all four living species in the genus *Gopherus*.

A KEY TO THE SPECIES OF *GOPHERUS* (BASED ON AUFFENBERG AND FRANZ, 1978A)

1A. A wider manus; base of first claw on forefoot to base of third claw equal to base of first claw to base of fourth claw on hind foot. 2

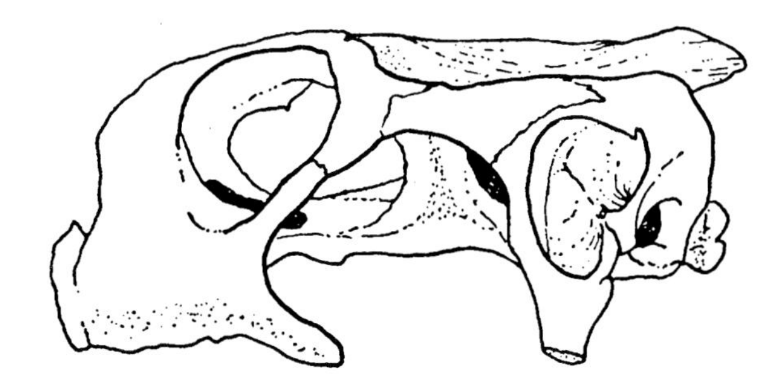

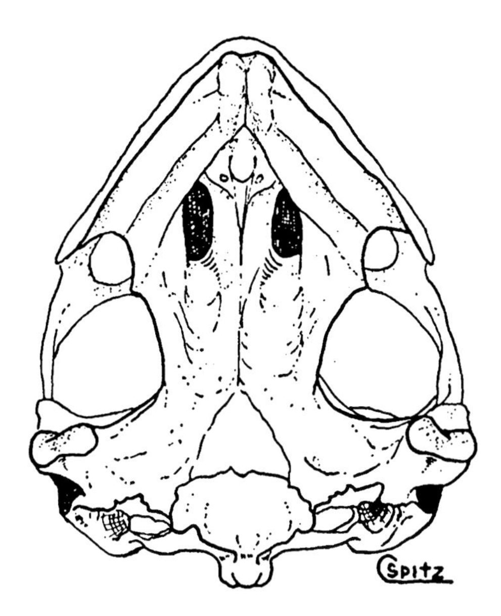

FIGURE 30.2 Skull of *Gopherus flavomarginatus*, lateral and ventral views (holotype, adult, USNM 60976, CL 246 mm, CBL 45 mm, greatest width of skull 39 mm). (Modified from Legler, 1959, with permission.)

1B. A narrower manus; base of first claw on forefoot to base of fourth claw on forefoot equal to same measurement on hind foot 3

2A. Carapace narrower posteriorly than anteriorly (not flared posteriorly); usually one enlarged femoral spur; axillary scale single, triangular (at least pointed posteriorly); precentral scute always single; ground color of carapace brown, laminae often with paler centers; plastron usually clear yellow (does not occur in Mexico) . . . . ***G. polyphemus***

2B. Carapace narrower anteriorly than posteriorly (flared posteriorly); at least two enlarged femoral spurs; a single enlarged axillary scute, wider posteriorly than anteriorly, not triangular, not pointed posteriorly; precentral scute often divided; carapace ground color yellow, laminae with dark brown centers; plastron yellow with darker (areolar) markings in some specimens (occurs only the Bolsón de Mapimí of Mexico) . . . . ***G. flavomarginatus***

3A. Snout wedged-shaped in dorsal aspect; axillary scales often paired; gular projections often divergent in large males; carapace ground color brown to black, laminae often with yellowish centers; head usually black with evident yellow temporal patches (occurs in southern Texas and northeastern Mexico southward to northern Veracruz) . . . . ***G. berlandieri***

3B. Snout rounded in dorsal aspect; axillary scales single; gular projections not usually divergent in adult males; carapace ground color brown, laminae often with yellowish centers; head usually brown, lacking paler temporal patches (occurs in Mojave and Sonoran deserts from SW Utah, SW Arizona, and SE California to N. Sinaloa . . . . ***G. agassizi***

Gopherus agassizi (Cooper) 1863

Desert tortoise (Iverson, 1992a); galapago del desierto (Liner, 1994).

ETYMOLOGY

A patronym honoring Jean Louis Rodolphe Agassiz (1807–1883), a Swiss zoologist at Harvard College, in particular his two-volume work (1857), *Contributions to the Natural History of the United States of America*, which contained a classic coverage of North American chelonians and turtle embryology.

HISTORICAL PROLOGUE

Gopherus agassizi has been well studied in the northern part of its geographic range. Ecological observations in Mexico have been made by Bury, Luckenbach, and Munoz (1978), Reyes Osorio and Bury (1982), and Fritts and Jennings (1994).

DIAGNOSIS

The following combination of characters is diagnostic among *Gopherus*: highest point is at middle of shell; upper anterior rim of shell in anterior view is nearly parallel to main dorsal arch of carapace; shell is relatively narrow; gular prongs of plastron are relatively short, wide, and blunt; there is a single trapezoidal axillary scale; there is one large femoral spur; snout is rounded in dorsal aspect; pes wide relative to head; hind foot diameter/WH is 0.85–1.15%; angle formed by maxillae is less than 60°.

Despite variations in structure and habits, *Gopherus agassizi* digs burrows throughout its range.

GENERAL DESCRIPTION

Gopherus agassizi is intermediate in size among its congeners in Mexico—larger than *berlandieri* and smaller than *polyphemus* or *flavomarginatus* (Germano, 1994a). All adults on Isla Tiburon had smaller than 300 mm CL with the exception of one abnormally oblong male of 371 mm. Median CL was 250 mm for females (180–290, n = 57)

FIGURE 31.1 Adult male *Gopherus agassizi* with fully developed, secreting mental glands (Washington Co., Utah).

and 260 mm for males (180–288, n = 68; Reyes Osorio and Bury, 1982). Maximum sizes for natural populations elsewhere are as follows: 286 mm in Nevada (Burge and Bradley 1976) and 316 mm in Utah (Woodbury and Hardy, 1948). The largest individual recorded (381 mm) is a captive named "Maximus" (St. Amant, 1976).

The following data on basic proportions and on size, growth and ontology are given separately for three allopatric populations (following Germano, 1993) that, along with others, may ultimately receive taxonomic rank.

BASIC PROPORTIONS

(D. Germano and Legler databases):
Mojave:

CW/CL: males, 0.78 ± 0.035 (0.69–0.84) n = 52; females, 0.79 ± 0.029 (0.72–0.85) n = 43; adults unsexed, 0.78 ± 0.027 (0.75–0.82) n = 6; im. unsexed, 0.77 ± 0.036 (0.71–0.82) n = 12; juveniles, 0.85 ± 0.035 (0.78–0.91) n = 28.

CH/CL: males, 0.46 ± 0.028 (0.39–0.52) n = 52; females, 0.45 ± 0.027 (0.40–0.51) n = 43; unsexed adults, 0.43 ± 0.037 (0.38–0.47) n = 4; im. unsexed, 0.48 ± 0.031 (0.44–0.53) n = 12; juveniles, 0.48 ± 0.030 (0.44–0.56) n = 28.

CH/CW: males, 0.59 ± 0.028 (0.50–0.64) n = 52; females, 0.58 ± 0.035 (0.51–0.64) n = 43; unsexed adults, 0.55 ± 0.051 (0.49–0.61) n = 4 ; im. unsexed, 0.62 ± 0.032 (0.56–0.67) n = 12; juveniles, 0.57 ± 0.031 (0.52–0.64) n = 28.

Sonoran:

CW/CL: males, 0.76 ± 0.037 (0.70–0.83) n = 14; females, 0.75 ± 0.024 (0.70–0.80) n = 16; unsexed adults, 0.78 ± 0.031 (0.74–0.82) n = 5; im. unsexed, 0.77 ± 0.044 (0.72–0.82) n = 5; juveniles, 0.81 ± 0.029 (0.77–0.86) n = 6.

CH/CL: males, 0.45 ± 0.023 (0.40–0.48) n = 14; females, 0.43 ± 0.030 (0.39–0.49) n = 16; unsexed adults, 0.41 ± 0.025 (0.38–0.43) n = 3; im. unsexed, 0.48 ± 0.036 (0.43–0.51) n = 5; juveniles, 0.48 ± 0.033 (0.41–0.50) n = 6.

CH/CW: males, 0.59 ± 0.037 (0.54–0.67) n = 14; females, 0.58 ± 0.038 (0.51–0.64) n = 16; unsexed adults, 0.51 ± 0.028 (0.48–0.54) n = 3; im. unsexed, 0.62 ± 0.019 (0.59–0.64) n = 5; juveniles, 0.59 ± 0.042 (0.51–0.62) n = 6.

Sinaloan:

CW/CL: males, 0.72 ± 0.029 (0.65–0.76) n = 10; females, 0.73 ± 0.033 (0.68–0.78) n = 9; im. unsexed, 0.75 ± 0.041 (0.70–0.80) n = 5; juveniles, 0.81 ± 0.065 (0.74–0.90) n = 5.

CH/CL: males, 0.42 ± 0.024 (0.38–0.46) n = 14; females, 0.42 ± 0.036 (0.38–0.48) n = 6; im. unsexed, 0.43 ± 0.029 (0.39–0.46) n = 4; juveniles, 0.45 ± 0.031 (0.41–0.48) n = 4.

CH/CW: males, 0.59 ± 0.029 (0.54–0.63) n = 14; females, 0.57 ± 0.036 (0.54–0.63) n = 6; im. unsexed, 0.56 ± 0.031 (0.52–0.59) n = 4; juveniles, 0.55 ± 0.067 (0.45–0.61) n = 4.

Color in desert tortoises (and Texas tortoises) is highly variable and probably related to age, size, and wear. Pale markings are more distinct and contrasting in young specimens. In general, the shell of *agassizi* is paler and smoother than that of *berlandieri*, the ground colors of adults being best described as various shades of neutral browns, tans, duns, and horn colors with varying amounts of contrast by pale or dark laminal centers. The head is brownish to tan, and the soft nonosteodermal skin is a neutral yellowish-gray color. Mental glands and two to five rostral pores are well developed in males.

The only other terrestrial chelonians that occur within the range of *agassizi* are box turtles (*Terrapene ornata* in southern Arizona and northern Sonora and *T. nelsoni* in Sonora and Sinaloa), both of which are easily distinguished by the plastral hinge and smaller size.

RELATIONSHIPS

There is general agreement that *Gopherus agassizi* is most closely related to *G. berlandieri*. The Sonoran desert population of *G. agassizi* is probably the "inceptive lineage" of

FIGURE 31.2 *Gopherus agassizi.* Typical adult female feeding on grass shoots soon after spring emergence (Lytle Ranch Preserve, SW Utah). (Courtesy of Dr. D. Kessler.)

G. berlandieri, the 500 km gap between the two populations having been closed as recently as 20,000 years ago (Morafka, 1977b; Morafka, Aguirre, and Murphy, 1994; Lamb, Avise, and Gibbons, 1989). Hybridization has been recorded between captives of *G. agassizi* and *G. berlandieri* (Woodbury, 1952; Mertens, 1956).

Gopherus agassizi was once thought to be a wide-ranging species with wide habitat tolerance (Woodbury and Hardy, 1948; Smith and Smith, 1979). Classical morphological and fossil studies (Auffenberg, 1976; Bramble, 1982; Weinstein and Berry, 1987; Germano, 1993), molecular studies of allozymes, plasma protein markers and mitochondrial DNA (Rainboth, Buth, and Turner, 1989; Lamb, Avise, and Gibbons, 1989; Lamb and Lydeard, 1994; Glen et al., 1990), and ecological studies (Germano, 1994a,b) are congruent in suggesting that *Gopherus agassizi* is polytypic. It is probably best to think of these findings as valuable information on the evolution of desert tortoises rather than as a need for immediate taxonomic change.

Lamb. Avise, and Gibbons (1989) recognized five different mtDNA clones within *Gopherus agassizi* (sensu lato). These clones comprise major assemblages east and west of the Colorado River. The western assemblage contains three distinct clones, the most common and widespread genotype being found throughout the Colorado and Mojave Deserts of California and extending into southern Nevada. Two smaller clones occur in the northeastern Mojave (southern Nevada and southwestern Utah) and define the northern edge of the geographic range of *G. agassizi*. Of the two clones east of the Colorado R., one has an extensive distribution in the Sonoran desert from west central and southern Arizona to central Sonora. The fifth clone occurs in southern Sonora and northwestern Sinaloa.

Edwards et al. (2009a,b) are currently studying relationships of *Gopherus agassizi* in Mexico. They recognize the following groups (in addition to the Mohave group) and their association with habitat types: (1) Sonoran, occurring in Sonoran desert scrub from approximately Hermosillo, Sonora, to Kingman, Arizona; (2) Sonoran, from approximately Cd. Obregon to southern Sonora in a transitional zone of foothill thornscrub; (3) Sinaloan, in the general region of Alamos, Sonora, and adjacent Sinaloa in tropical deciduous forest. They base their classification on fixed differences in STR alleles, autosomal STR motifs, and mtDNA haplotypes that clearly distinguish a unique "Sinaloan *Gopherus*" that diverged some 5–6 MYA from an ancestral stock that gave rise also to the Sonoran and Mojave lineages. Our use of "Sonoran" follows that of Germano (1993) in reference to these groups. The juxtapostion of sister groups of tortoises on opposite sides of the Colorado R. can logically be explained by interruption of an ancestral stock by the Bouse embayment, a Pliocene extension of the Gulf of California northward to southern Nevada (ca. 5.3–5.7 MYA by K-Ar decay dating; see discussions in Lamb, Avise, and Gibbons [1989] and Lamb and Lydeard [1994]).

Thus, a combination of molecular and morphologic data gathered from about 1976 to present, once collated, will probably define three to five taxa now included in the species *Gopherus agassizi*.

FOSSILS

Pleistocene fossils of *G. agassizi* are known from Arizona, California, New Mexico, and Nevada. Remains are known from archeological sites in Arizona, southern California, and southern Nevada (Ernst, Lovich, and Barbour, 1994). Crumly and Grismer (1994) record fossils of "testudinids" at four places on the Baja California Peninsula.

GEOGRAPHIC DISTRIBUTION

Map 21

G. agassizi are found in extreme southwestern Utah, the southern tip of Nevada, and extreme southeastern California (Mojave Desert) southward through southwestern Arizona and most of Sonora (Sonoran Desert) to regions of subtropical thorn scrub and deciduous forest in northern Sinaloa (west of the Sierra Madre Oriental). This contiguous latitudinal range (ca. 26°30′N–37°00′N) is greater than that of any other *Gopherus* species. Known localities are listed by Smith and Smith (1979), Patterson (1982), and Fritts and Jennings (1994). The southern third of the range lies in Mexico and is known to extend to El Fuerte, Sinaloa, but could logically extend farther southward in the Sinaloan foothills. Tortoises are absent from the flat agricultural land between Guaymas and Alamos, Sonora.

Bogert and Oliver (1945) extended the known range to its present southern limits (near Alamos, Sonora) and referred to it as a partly differentiated population not worthy of taxonomic recognition.

Ottley and Velazques Solis (1989) described a new Gopher tortoise (*Xerobates lepidocephalus*) from the region of La Paz, BCS. Crumly and Grismer (1994) placed the name in the synonymy of *Gopherus agassizi* and suggested that the types were released captives. Grismer (2002) later suggested the

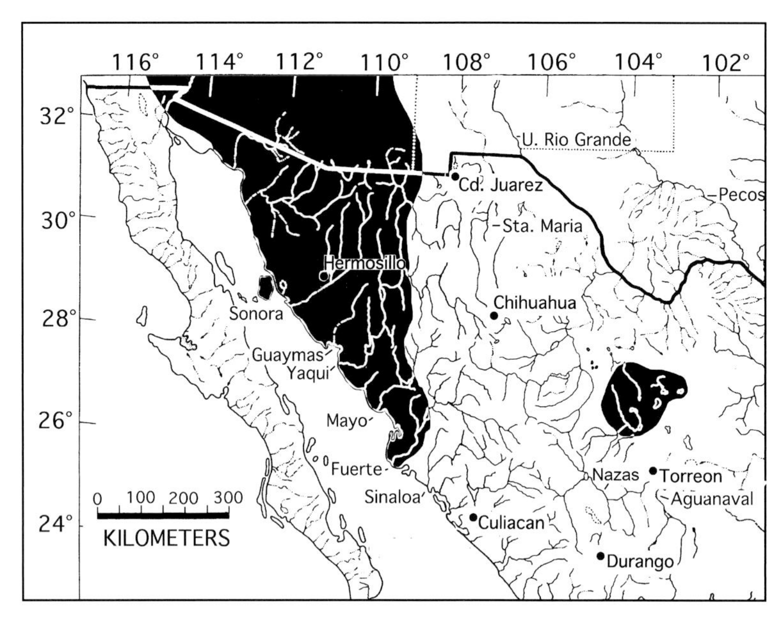

MAP 21 Geographic distribution of *Gopherus agassizi* (west) and *Gopherus flavomarginatus* (east).

possibility of a minuscule but natural population of *G. agassizi* near La Paz. Grismer (2002) further stated the high probability that *G. agassizi* once occurred in extreme northern BCN (based on Clavigero, 1789), although no specimens or modern records exist. Bury, Germano, Van Devender, and Martin (2002) regard the northern population as an introduction.

HABITAT

Gopherus agassizi occupies diverse habitats within its extensive range (Germano, Fritts, and Medica, 1994). In the north, tortoises occur in arid lands that have sandy or gravelly soil to altitudes of at least 1,070 m—desert alluvial flats, washes, valley floors, and hillsides. Characteristic vegetation is creosote bush, salt brush, grasses, and cacti (Berry, 1989; Schamberger and Turner, 1986; Barrett, 1990). Tortoises frequent rocky slopes and higher elevations in the eastern Mojave and in the Sonoran desert of Arizona (Bury, Esque, DeFalco, and Medica, 1994; Germano, Fritts, and P. A. Medica, 1994; Fritts and Jennings 1994). There are isolated colonies of tortoises in small mountain ranges in SW Arizona and extreme southeastern California (Patterson, 1982).

Altitudes of 1,300–1,500 m are reached in the Mojave Desert. The highest locality for Mexico is 1,050 m at Rancho La Palma, NE of Vavicora, Sonora (Fritts and Jennings, 1994). Aridity and sandy soils in the Colorado River delta seem to limit the distribution to the west and exclude the species from Baja California. The westernmost locality in Sonora is the Pinacate Lava Flow (ca. 31°45′N, 113°31′W).

Annual precipitation within the range (long-term averages over 60–100 years) ranges from 101 mm in the Mojave Desert to 664 mm in Sinaloan deciduous forest (Germano, Fritts, and Medica, 1994). Tortoises are not known in areas receiving less than 100 mm of rain per annum (Fritts and Jennings, 1994).

Habitats on the Mexican mainland range from xeric conditions near sea level to the lower edges of evergreen oak and juniper woodlands at 800 m elevation; fewer tortoises occur in coastal mountains and low elevations than at intermediate elevations of 300–500 m; at elevations above 500 m densities varied with local features of vegetation, soil, and exposure (Fritts and Jennings, 1994). Mexican habitats include slopes, bajadas, and the sides of arroyos.

Altitudinal distribution in northeastern Sonora is roughly defined by the 800 m contour. Eastern range limits in the south are sharply defined by the abrupt change from lowland Sinaloan thorn scrub to Madrean evergreen woodland at about 800–1,000 m; a few tortoises are found in the lower margins of this woodland vegetation zone (Fritts and Jennings, 1994).

In northwestern Sonora the boundary of tortoise distribution coincides closely with the 100 mm annual precipitation isopleth. In northern and Central Sonora tortoises occur in disjunct patches and narrow ribbons of Sinaloan thorn scrub associated with hills, isolated mountains, and mountain ranges. In eastern and southern Sonora there are larger, more continuous expanses of habitat; these probably support the highest densities of desert tortoises in Mexico.

In the foothills of the Sierra Madre Occidental in Sinaloa tortoises occur chiefly in arroyos, slopes, and bajadas, and are conspicuously absent from valley floors and other flat areas (Fritts and Jennings, 1994). Many of the hills and mountain slopes occupied by tortoises in Sinaloan thorn scrub have rainfall at the upper limits for desert scrub. In extreme southern Sonora and northern Sinaloa tortoises

FIGURE 31.3 *Gopherus agassizi*. Adult female walking in typical wash habitat (Ivanpah Valley, California). (Courtesy of P. Medica.)

occupy mesic and luxuriant habitats in Sinaloan deciduous forest.

HABITS AND BEHAVIOR

Gopherus agassizi is well studied in the United States, but little has been recorded on the natural history of desert tortoises in Mexico. The two principal studies are by Reyes Osorio and Bury (1982) for Tiburon Island and Fritts and Jennings (1994) for a general survey of distribution in Sonora and Sinaloa. General information on the species in the United States is included in the following where it has possible application to populations in Mexico (see summary by Ernst, Lovich, and Barbour, 1994).

The diet is herbivorous and opportunistic. Tortoises browse on various kinds of succulent vegetation (e.g., grasses, cacti, and blossoms) as they move through the environment, but may remain in a place where food is abundant. Desert annual plants are the most important dietary staple. Flowers are preferred, but the whole plant may be eaten. The availability of annual plants with flowers (April–June) coincides with a peak in tortoise activity. Tortoises acquire reserves for periods of quiescence during this relatively brief period (Ernst, Lovich, and Barbour, 1994). Little food is available to hatchlings in late summer, and they may enter hibernation before eating or drinking anything (Ernst, Lovich, and Barbour, 1994).

Esque and Peters (1994) record the ingestion of stones, soil, and weathered vertebrate bones (NE Mojave), and suggest that these substances could be dietary supplements. Females are known to eat soil with high calcium content during the nesting season (Marlow and Tollestrup, 1982).

Water is consumed whenever and wherever it is available. A small home range may not contain standing water, and rainwater is often the only water available. Tortoises occasionally construct small catchment basins during showers that are capable of holding water for up to 6 hours (Woodbury and Hardy, 1948; Nagy and Medica, 1977, 1986; Medica, Bury, and Luckenbach, 1980; Nagy, 1988). Drinking of water may produce a weight gain of as much as 43%, but the basic water supply comes from food. As fat is accumulated for hibernation, it is a source for metabolic water. Tortoises can gain weight in spring by eating 3–4% of their body mass in succulent annual plants each day, but water intake drops to 5 mL/day and metabolic rates decline in summer (Nagy and Medica, 1977, 1986; Ernst, Lovich, and Barbour, 1994).

Water retention and salt balance are problems in the desert. Water loss is by evaporation and urination. The urine is more nearly liquid when surface water is available but becomes semisolid during dry periods. Tortoises go for long periods without discharging any cloacal fluid. They binge on standing water during summer storms, void the concentrated urine, and accumulate dilute urine. This restores osmotic balance and allows a resumption of feeding on dry grasses. Tortoises begin feeding on succulents in late September and are in good physiological condition by mid November when hibernation begins (Nagy and Medica, 1986; Ernst, Lovich, and Barbour, 1994).

Throughout the known range, summer rains stimulate tortoise activity. There is a long annual period of activity. In the north, emergence from hibernation occurs from February to late April, and hibernation begins from late September to November, the timing being related to latitude, altitude, precipitation, and the particular population. In Mexico the period of greatest surface activity is in late summer and early autumn—coincident with the summer monsoons (Reyes Osorio and Bury, 1982; Fritts and Jennings, 1994). On Tiburon, activity is concentrated in the cooler season from early September to late November with occasional activity in December. Few tortoises are seen at other times of the year (Reyes Osorio and Bury, 1982). This contrasts to a spring peak of activity in the Mojave.

Desert tortoises spend most of their time in their burrows where temperature and humidity are relatively stable. Temperature and humidity are likewise important during the relatively little time tortoises spend outside the burrow. They leave their burrows during the cooler hours of morning and remain active until high temperatures force them to take cover in burrows or the shade of low vegetation, then they resume activity late in the day. Seemingly there is no fidelity to a particular burrow. They may remain active all day during the more moderate temperatures early and late in the season, but daily activity is bimodal in the hottest parts of the year (Luckenbach, 1982; Ernst, Lovich, and Barbour, 1994). Hibernation in burrows or dens begins in October or November throughout the range. It is common for several tortoises to congregate in a single shelter (two to four on Tiburon, and up to 23 tortoises in one huge den in Utah).

Forward motion is slow and deliberate and more easily expressed in hours per mile (3.3–7.3) than vice versa. There is broad variation in home ranges. Berry (1986) recorded home ranges of 1–268 ha in five areas of the United States.

FIGURE 31.4 *Gopherus agassizi*. Adult female in front of typical burrow at bottom of wash (Ivanpah Valley, California). (Courtesy of P. Medica.)

Home ranges overlap, and there is no evidence of territoriality. Males have larger home ranges than females. California desert tortoises know the locations of the burrows, mates, water sources, and mineral licks within their home ranges. Despite this familiarity, individuals may make long-distance excursions outside a home range of 16 days to 5 years for unknown reasons (Berry 1986). In Utah tortoises migrate from winter hibernacula to summer feeding grounds (Woodbury and Hardy, 1948).

Desert tortoises dig burrows throughout the known geographic range (Germano, Fritts, and Medica, 1994). Burrows are dug in gravelly or sandy soil with alternate movements of the forefeet, and loose soil is pushed to the entrance with the sides of the shell. The finished burrow has a semicircular cross-section.

Burrows are longer (up to 10 ms) and more varied in the Mojave part of the range, where there are smaller summer burrows and much larger winter "dens" (Woodbury and Hardy, 1948). Burrows in Mexico are shorter (0.2–1.5 m), sometimes barely concealing the tortoise, and are used as refugia all year long. Most of the burrows on Tiburon (78%) were dug into the sides of arroyos. Others were dug under the roots of trees or shrubs in upland desert areas, and none were in the open. By contrast, Mojave summer burrows are dug downward at angles of 20–40° on flats and benches. Woodrats (*Neotoma* sp.) are usual commensals in burrows. Desert tortoises also excavate shallower cavities called pallets that may provide some protection but are rarely large enough to conceal the whole tortoise.

An individual tortoise may habitually use up to 25 cover sites (dens, burrows, pallets, and other resting places) in a year of activity. These sites are commonly used by other tortoises or shared with them (especially the winter dens; Ernst, Lovich, and Barbour, 1994). No dens (burrows >2 m long) were found on Tiburon, but about half of the burrows contained multiple occupants (two to four) in various age and gender combinations (Reyes Osorio and Bury, 1982).

The winter dens in Utah are spectacular in that they are dug horizontally and deeply into the sides of high, steep arroyo walls, just below the edge of the gravel flat at the top of the slope. Distinct, well-worn paths lead to such burrows (Woodbury and Hardy, 1948; Legler, pers. obs.).

Medica, Lyons, and Turner (1986) described an ingenious method ("tapping") to entice desert tortoises from their burrows, avoiding the possible trauma of using hooks to remove them. A stout bamboo pole of 2 m length is used to lightly tap the shell of a tortoise (if seen) or to tap on the entrance to a burrow. The method is about 80% effective, and the authors don't know why it works.

Behavior of adults is usually docile. That of very young juveniles may be pugnacious by comparison (Ernst, Lovich, and Barbour, 1994). Barrett and Humphrey (1986) recorded two instances of adult females driving off *Heloderma* that were scratching about at the mouths of tortoise burrows (perhaps digging for eggs). Wild populations have dominance hierarchies related to the defense of burrows, water, and nests (Berry, 1986a). When two tortoises meet they bob their heads rapidly and sometimes touch snouts before passing. Males often fight when they meet. After preliminary head bobbing they separate and then rush toward one another with heads pulled in. They meet head on, butting each other with their gular prongs. Fights are usually innocuous, but one tortoise may be turned over. Righting is usually possible; if not, exposure to sun could result in mortality (Ernst, Lovich, and Barbour, 1994).

Campbell and Evans (1967) record two types of sounds made by tortoises: a short grunt and a drawn-out moan for which the frequency range is at least one octave; these sounds are not involved in courtship or combat, and their significance is unknown.

Cloacal temperatures of active animals range from 15.0°C to 38.3°C; critical thermal maxima range from 39.5°C to 43.1°C. The ability to recover from an inverted position is lost at about 39°C; small tortoises are active at lower temperatures than larger ones (Woodbury and Hardy, 1948; McGinnis and Voigt, 1971; Brattstrom, 1965; Hutchison, Vinegar, and Kosh, 1966; Berry and Turner, 1984).

Gopherus is unique among testudinids in having Type 1 mental glands (Winokur and Legler, 1975). These are large, paired, multilobed, holocrine integumentary glands on the throat skin with slitlike secretory orifices. The glands are inactive prior to sexual maturity, are larger in males than females, and are under androgenic control. Mental glands develop seasonally into huge protuberances that alter the outline of the head (Figures 2.10, 31.1; Winokur and Legler, 1975). Secretions consist of saturated and unsaturated free fatty acids (which are volatile), phospholipids, triglycerides, and steroids, plus a protein fraction that varies between and within species (Rose, 1970).

The precise function of mental gland secretions is not clear, but available data suggest they serve as chemical signals in close-range social interactions (Alberts, Rostal,

FIGURE 31.5 *Gopherus agassizi*. Juvenile (Washington Co., Utah).

and Lance, 1994). These secretions would qualify as pheromones according to the definition of Mason (1995).

There is evidence for *G. agassizi* that one tortoise can detect the secretions of another. This may serve to identify individuals in a population. When male mental gland secretions were smeared on the carapace of a female, a male who had previously courted that female treated her as though she were a male and initiated combat (Alberts, Rostal, and Lance, 1994).

In *G. polyphemus* both sexes can distinguish between "unfamiliar" and "familiar" members of the same species. Alberts, Rostal, and Lance (1994) suggested that being able to recognize individuals with whom dominance has been previously established would avoid potentially traumatic encounters (e.g., being tilted and overturned in hot weather).

There are also references to the odor of feces and urine serving as a pheromone. As yet there is little known about what imparts this odor, but it could be an as-yet-undiscovered gland opening into the cloaca. There is some, seemingly powerful, evidence of the pheromonal effect of cloacal secretions in other tortoises. A male *Geochelone carbonaria* mounted a dried shell on which "cloacal secretions" from a mature female had been rubbed; a male *Geochelone denticulata* attempted to copulate with a head of lettuce similarly treated. Urine from *G. berlandieri* spread in a burrow shared by 17 *G. agassizi* caused some of the tortoises to leave and sleep in the open. Fecal pellets from unfamiliar *G. agassizi* caused all the tortoises to leave. Fecal pellets from dominant males caused subordinate males to leave (Mason, 1995).

REPRODUCTION

Data on size (CL) presented in Growth and Ontogeny are from free-living animals preserved as museum specimens (n = 227). They suggest that both sexes mature in the range of 180 189 mm CL (after excluding one possibly anomalous male of 158 mm CL from Tiburon; Reyes Osorio and Bury, 1982).

Germano (1994a, 1994b) associated size and age in mature females in the eastern Mojave and Sonoran regions as mean CL of 189 mm, 15.4–15.7 years, and those in the western Mojave as mean CL of 176 mm, 14.4–15.7 years. Few data are available on males.

Although genetic factors are certainly involved in growth rates and maximum size, it is logical that well-nourished tortoises of the same genetic stock will grow faster, achieve larger sizes, and probably reach maturity sooner than their wild counterparts (Turner, Hayden, Burge, and Roberson, 1986). Captives on a regular, high quality diet grow rapidly and reach maturity at much younger ages than free-living animals. One captive male was mature (motile sperm detected) at 4 years and 236 mm (Jackson, Trotter, Trotter, and Trotter, 1978).

In the northern part of the range, mating and courtship begin after spring emergence (as early as late March and continuing until October; Tomko, 1972; Luckenbach, 1982; Medica, Lyons, and Turner, 1982). Courtship is usually initiated by a male approaching a female. There follows a display of mutual head bobbing, during which the male presses his snout to the female's body. The male then pursues the female, after which there is mutual circling of one by the other. Ultimately the male bites at the head and forequarters of the female and mounts after she retracts into her shell. Copulation is accompanied by vertical pumping movements and puffing—grunting noises (which seem to have no significance as vocal communication). Courtship and copulation may take as long as 80 minutes (Weaver, 1970; Ernst, Lovich, and Barbour, 1994). Courtship and mating have not been observed in Mexico.

Nests with eggs were found on Tiburon from mid September to early November, a cycle differing from northern populations that lay mainly in May and June (Reyes Osorio and Bury, 1982). Three nests were found: one in mid September with one egg; two in early November with one and five eggs, respectively (Reyes Osorio and Bury, 1982).

In the United States nesting occurs from mid May through July and usually takes place in early morning or late afternoon. Some nesting has been recorded as late as September and October (Luckenbach, 1982).

Females are said to become restless in the days prior to nesting. After nesting, they spend less time on the surface than do the males (Luckenbach, 1982). The eggs of a single clutch may be deposited in several different places. Nests are dug in sandy or friable soil, usually at the mouths of burrows or in the shade of a bush but seldom in the open (Ernst, Lovich, and Barbour, 1994). Nest excavation may require one to several hours; actual oviposition takes 15–30 minutes. The nest cavity is distinctive in being wider at the top than the bottom (Nichols, 1953; Ernst, Lovich, and Barbour, 1994).

The eggs are hard shelled, have a coarse texture, and range from elliptical to nearly spherical. Four eggs measured by

FIGURE 31.6 *Gopherus agassizi* from southern extent of range: top left—adult female, NE of El Fuerte, Sinaloa; top right—two immature near Alamos, Sonora, the smaller ca. 140 mm CL; bottom left—juveniles, northern Sinaloa; bottom right—thornscrub habitat NE of El Fuerte, Sinaloa (where mature female was photographed). (All images courtesy of J. R. Buskirk.)

Miller (1932) were 41.6–48.7 mm × 36.6–39.6 mm; weight was 33–34 g. Average length of seven eggs from Isla Tiburon was 40 mm.

Egg shells found on Tiburon in September were interpreted as a hatching event. A recently hatched tortoise (45 mm CL) with a soft shell was found in November (Reyes Osorio and Bury, 1982).

Ewert (1979) gives incubation times for eggs in natural nests (to pipping) of 84–120 days; no data are available for incubation at controlled temperatures. Emergence of hatchlings from nests occurs from mid August to October in the north (Ernst, Lovich, and Barbour, 1994).

Temperature-dependent sex determination (TSD) is of the "Pattern 1 a" type in the population near Las Vegas; temperatures above 32.5°C produce all females and below 30.5°C produce all males. A relatively high estimated threshold temperature of ca. 31.3°C produces a 1:1 ratio of sexes (Ewert, Jackson, and Nelson, 1994; Spotila et al., 1994; Rostal, Wibbels, Grumbles, Lance, and Spotila, 2002).

Known range of eggs per clutch is 2–15 (Ernst, Lovich, and Barbour, 1994), but clutches of four to six are the norm. Although females may produce up to three clutches of eggs in a season, they can also skip seasons of reproduction (Berry, 1978; Germano, 1994a; Turner, Hayden, Burge, and Roberson, 1986). Multiple clutches are usually produced by larger females (Turner, Hayden, Burge, and Roberson, 1986).

Mean number of eggs per clutch in the eastern Mojave and Sonoran deserts is 4.5 (1–8) in one to three clutches (mean 1.73) per year (Germano, 1994a). In California, mean number of clutches per year was 1.57–1.89 over a 3-year period, the first one or two clutches being laid in May and June (Turner, Hayden, Burge, and Roberson, 1986).

Turner, Hayden, Burge, and Roberson (1986) also provide some valuable information on the timing of eggshell formation and the interval between clutches for California populations. Shells of the second clutch were visible in X-rays 9–10 days after the first clutch was laid, and eggs of the second clutch were actually laid about 3 weeks after the shells of the first clutch were visible in X-rays.

GROWTH AND ONTOGENY

Data on size from three populations of *Gopherus agassizi* (D. Germano and Legler databases):

Mojave: males, 232 ± 29.9 (181–310) n = 52; females, 221 ± 20.3 (184–274) n = 43; unsexed adults, 239 ± 35.6 (184–288) n = 6; im. unsexed, 136 ± 24.3 (102–179) n = 12; juveniles, 69.5 ± 14.7 (43–97) n = 28.

Sonoran: males, 233 ± 24.9 (189–275) n = 14; females, 229 ± 17.0 (188–256) n = 17; unsexed adults, 238 ± 25.6 (204–271) n = 5; im. unsexed, 135 ± 31.1 (102–181) n = 5; juveniles, 71 ± 19.0 (50–99) n = 6.

Sinaloan: males, 208 ± 23.4 (158–244) n = 20; females, 228 ± 19.3 (186–252) n = 9; im. unsexed, 131 ± 24.5 (105–158) n = 5; juveniles, 69.4 ± 12.6 (57–84) n = 5.

There is less ontogenetic change in shell shape than in *berlandieri*. In lateral and dorsal aspects the shell form is more nearly hemi-elliptical than hemispherical. The top of the shell may be more nearly flat than arched in profile.

The posterior edge of the shell typically lies well dorsal to the posteriormost points of the plastron.

Males are slightly larger than females in the northern populations and are of approximately equal size in Sonoran populations, and females are larger than males in Sinaloan populations (Germano, 1994b). Males have thicker tails, longer gular prongs, concave plastra, more massive claws, and larger mental glands than females (Ernst, Lovich, and Barbour, 1994; Winokur and Legler, 1975).

Hatchlings have 36 to 48 mm CL and are slightly longer than broad. The osseous shell contains many fontanelles and remains soft for 5–10 years. At hatching there is a large yolk sac that impedes normal locomotion, but is absorbed in the first 2 days. The umbilicus remains visible for several months. The caruncle disappears gradually (Ernst, Lovich, and Barbour, 1994).

Growth rate is rapid for about 20 years but then slows dramatically. One-year-old turtles from the western part of the range are 51–59 mm, whereas those from eastern populations are 46–49 mm at the same age. Carapace lengths of 10-year-old individuals from the same regions were 139–140 and 122–125 mm, respectively (Ernst, Lovich, and Barbour, 1994). Utah juveniles are known to grow to 100 mm in 5 years (Woodbury and Hardy, 1948).

Annuli can be used to determine age in the first 20–25 years of life, but are better used to study growth rate and the number of years in which growth occurred. Germano (1994b) presents a table in which he compares calculated growth rates for four populations (western and eastern Mojave, Sonora, and Sinaloa). Tortoises from the western Mojave Desert and Sinaloa have relatively rapid growth rates and attain larger mean sizes, whereas those from the eastern Mojave and Sonoran Deserts have the slowest growth rates and attain smaller sizes. Berry (2002; Mojave and Colorado Deserts, California) found photographs of scutes to be unreliable in distinguishing major and minor growth rings and therefore inaccurate in determining age.

Only a few tortoises live to 50 years, and many as 29% live past 25 years; these percentages are greater in eastern populations than in western ones. The oldest age approximation for a free-living tortoise is 48–53 years (Germano, 1992). Ages of 67–80 years have been reported for captives (Jennings, 1981; Glenn, 1983).

PREDATORS, PARASITES, AND DISEASES

The following are known natural predators of life stages from eggs to adults: coyotes (*Canis latrans*), bobcats (*Lynx rufus*), ravens (*Corvus corax*), golden eagles (*Aquila chrysaetos*), and gila monsters (*Heloderma suspectum*). Many other predators (snakes, birds, and carnivores) are suspect (Ernst, Lovich, and Barbour, 1994). Predation by mountain lions in Nevada has been confirmed (P. Medica, pers. comm.).

High-density populations of ravens adversely affect tortoise populations by preying on juveniles smaller than 110 mm, judged by tortoise shells beneath raven perches (Boarman, 1993; Berry, 1991).

Although many natural predators (e.g., coyotes) are present on Isla Tiburon, predation on tortoises is seemingly minimal (the coyotes are probably preying on small mammals). Woodbury and Hardy (1948) suggested that predation on tortoises increases if rodent populations decrease.

Berry (1991) recorded a 50% decline in a protected California population over 19 years and attributed it partly to upper respiratory tract disease (URDT). The disease is caused by *Mycoplasma*, a small bacterium (Jacobson et al., 1991). Spread of the disease may have been caused by the introduction of nonnative tortoises (Ernst, Lovich, and Barbour, 1994).

Anthropogenic factors are probably the chief cause of mortality in many populations (see Conservation and Current Status).

POPULATIONS

Reyes Osorio and Bury (1982) identified 146 individual tortoises at six study sites in an area of 2.25 km^2 on eastern Isla Tiburon. Average estimated density was 65 tortoises/km^2 (0.65/ha; extremes 28.9–87.3 km^2). This high population density is exceeded only by certain populations in California. Tortoises with 180 mm CL and greater (92%) could be sexed externally. Among 126 adults, 55 (45%) were female and 69 (55%) were male. Composition of the population was as follows: 0.7% hatchlings and juveniles 100 mm or smaller; 3.8% juveniles 101–179 mm; 4.5% subadults 180–214 mm; 91% adults larger than 215 mm. These data resemble those for the Utah population studied by Woodbury and Hardy (1948) from 1935 and 1945 (90% adults)—adults with a sex ratio of 40% females to 60% males. Approximately 77% of the adults on Tiburon were larger than 215 mm, and most of these were in the range of 230–270 mm. Berry (1976) reported the population composition in four populations (southern California, Nevada, and Utah) to be 42–58% adults, 14–17% subadults, 18–33% juveniles, 5–10% "very small," and 1–2% hatchlings. Luckenbach (1982) reported sex ratios for California of 20% females to 80% males ($n = 124$).

Population densities vary widely in other populations. Woodbury and Hardy (1948) estimated a density of 1.62/ha in Utah. Berry (1986) found population densities to range from eight or fewer tortoises/km^2 (0.8/ha) to as many as 184/km^2 (1.84/ha) at 27 sites in southern California; numbers of adults and nonadults were about equal, and only 4 of the 27 populations had sex ratios differing from equality. Iverson (1982) calculated the biomass in southern California as 0.59–0.64 kg/ha (after Barrow, 1979) and 2.05 kg/ha in southern Utah (after Woodbury and Hardy, 1948).

Annual survivorship of eggs and hatchlings is low in all populations but can be as high as 100% in adults. In the Mojave Desert survivorship of eggs to hatching ranged from 46% to 93%; from hatching to 1 year, 51%; from 1 year to adulthood, 71–89%; and for adults, 75–100% (Germano, 1994a). Annual survivorship of adults in California is given as 88% by Turner, Medica, and Lyons (1984). Annual adult mortality in California rises from about 5% to 18% in dry periods (Luckenbach, 1982).

Germano and Joyner (1988) gave mortality rates of a population in southern Nevada during two periods from 1979 to 1987, respectively, as follows: from hatching to 14 years, 6.1%; 15–25 years, 9.3%; and more than 25 years, 10.3%. Significantly higher rates of mortality from 1979 to 1983 decreased mean age and size in the population but had no effect on density, probably because the lower juvenile mortality rate maintained the population.

CONSERVATION AND CURRENT STATUS

Desert tortoises in the United States are suffering population declines and local extirpation from respiratory disease and substantial habitat damage (Berry, 1989; Diemer, 1989; Morafka, Aguirre, and Murphy, 1994). The species is listed under the Endangered Species Act as threatened in all U.S. states where it occurs except Arizona. Mojavien populations north and west of the Colorado R. were listed as threatened (by USFWS) in April 1990 (Ernst, Lovich, and Barbour, 1994). The species is protected by state agencies in Arizona and Utah (Morafka, Aguirre, and Murphy, 1994).

Loss of habitat in the Mojave Desert has resulted from urbanization, road construction, utility lines, off-road vehicles, grazing and other agriculture, introduction of exotic plants, and fires caused by humans. Off-road vehicles may actually crush tortoises or collapse burrows, as well as destroy vegetation. There is a 40% overlap in the native vegetation used by tortoises and grazing animals (Ernst, Lovich, and Barbour, 1994). The species is used for food and in rituals by some American Indians (Schneider and Everson, 1989).

The only known insular population of *G. agassizi* occurs on Isla Tiburon, Sonora. The island has been little altered by human activity since 1955. It lacks feral and domestic animals and is protected by the Mexican government. Vegetation (creosote-mixed desert shrub and subtropical thorn scrub) is in good condition. This is one of the densest, healthiest, and best-protected populations of desert tortoises (Reyes Osorio and Bury, 1982; Bury, Luckenbach, and Munoz, 1978). The population merits continuing protection and further study.

Fritts and Jennings (1994) provide the following information on use of tortoises in mainland Mexico. Most residents are aware of a Mexican government ban on taking tortoises. Human consumption of tortoise meat is widespread but infrequent, the usual reason being the expense of other meat. Tortoise populations near large human settlements (e.g., Hermosillo, Guaymas, and Caborcas) have probably experienced long-term damage from human exploitation, habitat degradation, road kills, predation by domestic dogs, and collection as pets. However, the steep, rocky, and heavily vegetated hillsides favored by tortoises in this part of the range make it possible for populations to exist virtually undetected in such well-used places as rest stops on major highways. Tortoise gathering is opportunistic, and there is no evidence that commercial or other organized gathering of tortoises has occurred. Some tortoises have been collected near Alamos for sale to tourists or shipment out of Mexico.

Gopherus berlandieri (Agassiz) 1857

Berlandier's tortoise (Auffenberg and Franz, 1978c; Iverson, 1992a); Texas tortoise (Ernst, Lovich, and Barbour, 1994); galapago tamaulipas (Liner, 1994).

ETYMOLOGY

A patronym honoring Jean Louis Berlandier (1803–1851), a French naturalist who collected the type specimens.

HISTORICAL PROLOGUE

Gopherus berlandieri is the least known of the three Mexican species of *Gopherus*. Most existing information on natural history results from studies done in southern Texas (Auffenberg and Weaver, 1969; Rose and Judd, 1982; Bury and Smith 1986; Judd and Rose, 2000; Kazmaier et al., 2002) where the densest populations occur. Ernst, Lovich, and Barbour (1994) summarize existing knowledge on the species. To date there have been no studies made in Mexico. The life history characteristics of *Gopherus berlandieri* demonstrate extreme plasticity and variation from one local population to another (Auffenberg and Weaver, 1969; Rose and Judd, 1982). Auffenberg and Franz (1978c) and Smith and Smith (1979) compiled extensive literature summaries of the species.

DIAGNOSIS

Gopherus berlandieri can be distinguished from its congeners by the following combination of characters: highest point of shell posterior to middle (Figure. 30.1); a distinctive cross section in anterior view, upper anterior rim of shell peaked, not gently arched, not parallel with dorsal arch of carapace; shell wider relative to Cl (WC/Cl 0.84 ± 0.04(.75–.94); gular prongs long, pointed and often divergent in males; axillary scales usually paired, triangular or trapezoidal; femoral spurs poorly developed or absent; head narrow and snout pointed in dorsal aspect; pes narrow relative to head; hind foot diameter/WH 0.57–.0.89; angle formed by maxillae 65–73°.

GENERAL DESCRIPTION

Gopherus berlandieri is the smallest species of the genus and the most sexually dimorphic. Maximum adult size is ca. 220 mm. Auffenberg (1966) examined a specimen of 314 mm and commented on its record size.

FIGURE 32.1 *Gopherus berlandieri*. Adult male in prime condition (Brownsville, Texas). (Courtesy of P. Scanlan, Gladys Porter Zoo.)

BASIC PROPORTIONS

(Texas and northern Mexico, D. Germano and Legler databases).

CW/CL: males, 0.84 ± 0.033 (0.76–0.91) n = 48; females, 0.86 ± 0.041 (0.78–0.94) n = 36; unsexed adults, 0.85 ± 0.049 (0.78–0.92) n = 10; im. unsexed, 0.85 ± 0.046 (0.77–0.92) n = 9; juveniles, 0.88 ± 0.034 (0.82–0.93) n = 14.

CH/CL: males, 0.51 ± 0.032 (0.43–0.58) n = 48; females, 0.51 ± 0.033 (0.45–0.57) n = 36; unsexed adults, 0.48 ± 0.041 (0.39–0.53) n = 10; im. unsexed, 0.53 ± 0.024 (0.48–0.57) n = 9; juveniles, 0.54 ± 0.025 (0.48–0.59) n = 14.

CH/CW: males, 0.62 ± 0.032 (0.52–0.68) n = 48; females, 0.60 ± 0.034 (0.53–0.65) n = 36; unsexed adults, 0.56 ± 0.035 (0.48–0.61) n = 10; im. unsexed, 0.62 ± 0.023 (0.60–0.66) n = 9; juveniles, 0.61 ± 0.028 (0.55–0.65) n = 14.

Scute 4 is always longest; 3 or 6 is always shortest; modal plastral formula: 4 > 1 > 2 > 5 > 6 > 3.

There is much variation in color. In general aspect the shell is brown and roughly ridged with well-developed growth rings. The dark scutes have paler centers in younger individuals. The plastron has a variable combination of brown and paler colors, the latter tending to be on the posterior portions. The dorsal parts of the head are dark brownish and may be marked with paler patches. The soft skin of the limbs and neck are varying shades of yellowish cream or gray. Mental glands are well developed in males; there are two to four rostral pores (Winokur and Legler, 1974, 1975).

Within the range of *Gopherus berlandieri* the only other terrestrial chelonians are box turtles (*Terrapene ornata* and *T. carolina triunguis* in the north and *T. mexicana* in the south), all of which are clearly distinguishable by smaller size and a distinctively hinged plastron.

RELATIONSHIPS

Gopherus berlandieri is most closely related to the population of *G. agassizi* in the eastern Sonoran desert (see account of *G. agassizi*). The two stocks became separated sometime between late Pliocene and late Pleistocene (Lamb, Avise, and Gibbons, 1989). Hybrids between captives of *G. agassizi* and *G. berlandieri* have been recorded (Woodbury, 1952; Mertens, 1956). Populations of *G. berlandieri* are known to differ in size (Rose and Judd, 1982). No taxonomic division of the species has been suggested.

GEOGRAPHIC DISTRIBUTION

Map 22

Southern Texas, from a line connecting Del Rio, San Antonio (29°30′N), and Rockport (28°N) southward through eastern Coahuila, Nuevo León, and Tamaulipas to extreme eastern San Luis Potosi, Mexico, and extreme northern Veracruz (ca. 22°N). About half the range is in Mexico and lies north and east of the Sierra Madre Oriental. The records for San Luis Potosi and Veracruz suggest possible occurrence in northern Hidalgo and Queretaro (Auffenberg and Franz, 1978a; Smith and Smith, 1979; Rose and Judd, 1982, 1989; Judd and Rose, 2000; Iverson, 1992a,b; Pérez-Higareda, 1980). In Texas the range corresponds closely to the boundaries of the Tamaulipan Biotic Province (Kazmaier, Hellgren, and Ruthven, 2002).

HABITAT

Texas tortoises inhabit scrub forest in the humid, subtropical parts of Texas and semidesert scrub in Mexico. Substrates range from well-drained sandy soils to harder clay and caliche. Tortoises occur from sea level to 200 m in Texas and up to 884 m in northeastern Mexico (Carr, 1952; Judd and Rose, 2000; Ernst, Lovich, and Barbour, 1994).

Habitat in the coastal areas of Texas is chiefly lomas, clay dunes, or ridges that are islands of habitat surrounded by salt flats and marshes (Judd and Rose, 2000; Auffenberg and Weaver 1969; Rose and Judd, 1975; Judd and Rose, 1983; Bury and Smith, 1986). Populations on these islands are isolated, and there is seemingly little gene flow between them.

DIET

The staple diet is herbivorous. There is a preference for stems, fruits, and flowers of *Opuntia* cactus; grasses, violets, asters, and other plants are also eaten. Parts of crayfishes, terrestrial snails, and beetles have been recovered from feces. Feeding on feces of peccaries, rabbits, and other tortoises, and on dried bones (cows and rabbits) has been reported (Auffenberg and Weaver, 1969; Mares, 1971; Rose and Judd, 1982).

Texas tortoises are opportunistic in their consumption of water, and movement is stimulated by rains. There is

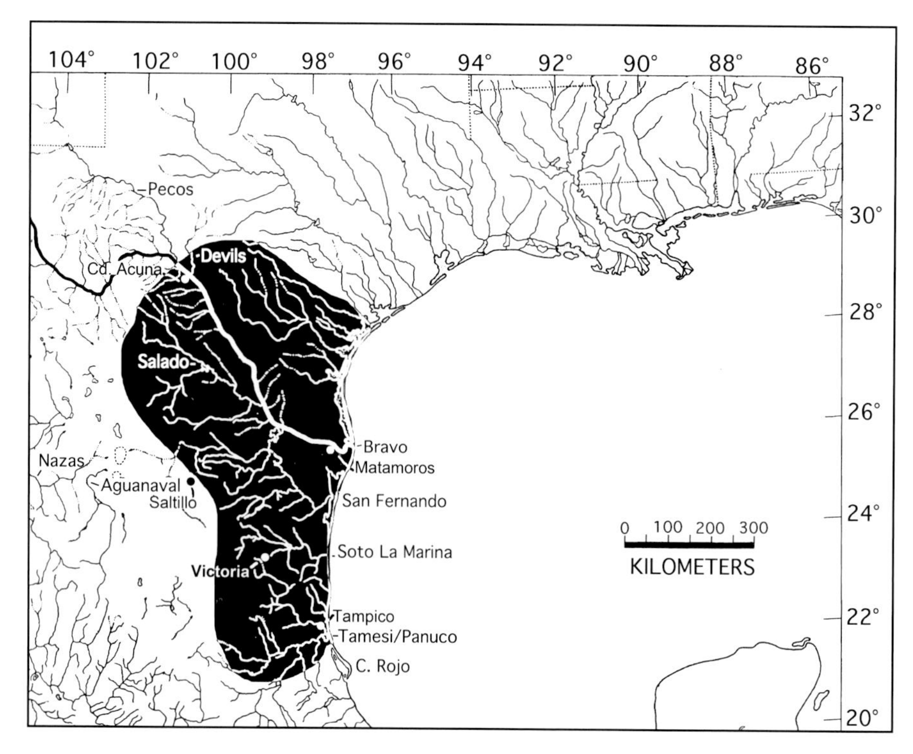

MAP 22 Geographic distribution of *Gopherus berlandieri*.

usually no access to standing water for long periods. The water content of *Opuntia* is high (stems, 91–95%; fruits, 84.5%; Rose and Judd, 1982). Water is reabsorbed from the bladder, and nitrogenous wastes are excreted as semi-solids. Ernst, Lovich, and Barbour (1994) review the physiological aspects of excretion, lipid metabolism, and water conservation.

HABITS AND BEHAVIOR

Auffenberg and Weaver (1969) found tortoises active in all months of the year (south Texas), with reduced activity in the hottest and coldest parts of year. Nearby populations on the Gulf Coast definitely hibernate. Rose and Judd (1975) found no tortoises above ground from December to February. In these periods the tortoises find whatever cover they can and wait passively for warmer weather, with at least half the shell exposed.

Periods of daytime activity are geared to temperature. Early and late in the season (April and November) there is one period of activity at midday. During warmer periods activity occurs in the morning and again in late afternoon. Tortoises retreat to shade in the hottest part of the day and resume activity when body temperature reaches about 28°C (Bury and Smith, 1986).

Populations studied by Auffenberg and Weaver (1969) were referred to as nomadic but maintain restricted home ranges for short periods. Distance moved per period of time was inversely related to food availability. Tortoises larger than 150 mm moved farther per unit of time than smaller ones. Judd and Rose (1983) and Rose and Judd (1975) recognized defined home ranges in their studies; mean home range was 2.38 ha for males and 1.4 ha for females. Mean distances between captures were 56.7 m for males, 42.0 m females, and 41.3 m for juveniles. Home ranges were closely associated with *Opuntia* cacti. Experiments demonstrate that homing ability exists. All tortoise activity was associated with lomas, and movement between lomas is uncommon.

In contrast to the loma habitats of southeastern Texas, Kazmaier, Hellgren, and Ruthven (2002) studied populations some 300 km to the northwest (Chaparral Wildlife Management Area) on the western Rio Grande plains of Dimmit and LaSalle Counties. Contiguous usable habitat for tortoises was much greater than that on lomas. Mean home ranges (in various microhabitats and by different means of estimation) were 7–46 ha for males and 3–9 ha for females, and there was evidence that home ranges were larger in ungrazed than in grazed pastures.

Texas tortoises are highly opportunistic in their use of shelter. They dig burrows only occasionally and where soil texture permits. Auffenberg and Weaver (1969) recorded a few tortoise-made burrows up to 1.3 m long and 0.3 m deep in areas west of the Gulf Coast, chiefly in sandy soils. Tortoises also enlarge and use burrows made by mammals. Most tortoises living on lomas used pallets for shelter

FIGURE 32.2 *Gopherus berlandieri*. Old male (Port Mansfield, Texas). (Courtesy of P. Scanlan, Gladys Porter Zoo.)

(shallow depressions in the ground, often in the shade of a bush or cactus). Pallets are at first large enough only to accommodate part of the plastron. Repeated use may enlarge them to a kind of incipient burrow, up to 330 mm long and 100 mm deep. Different tortoises occupy a pallet on a first-come basis. Pallets in thick brush are preferred in summer and those in more open situations in winter. Shelter is also sought beneath rocks, stumps, or debris (Auffenberg and Weaver, 1969).

Male combat prior to the breeding season consists of biting, ramming, and attempts to overturn another male. There seems to be a hierarchy among males, but the function of male combat is poorly understood. Two adult males may occur in close proximity without showing interest in each other.

Most active tortoises have temperatures of 30–35°C. Mean body temperature is 31.1°C in spring and 33.1°C in summer. Exposure to direct solar radiation is fatal in as little as 20 minutes. Heat stress begins at 37.5°C; the righting response is lost at about 40°C (mean 40.4°C, extremes 39.8–40.7°C) and the critical thermal maximum is 42.9°C (42.5–43.2°C). Tortoises hibernating in winter have mean body temperatures of 9.9°C (6.5–18.5°C; Hutchison,Vinegar, and Kosh, 1966; Judd and Rose, 1977; Grant, 1960a; Voigt and Johnson, 1976). The head is involved in thermoregulation; heating and cooling occur less rapidly when the head is withdrawn (Voigt and Johnson, 1977).

REPRODUCTION

Secondary sexual characteristics appear at 105–125 mm and sexual maturity is estimated to occur at 120–125 mm and the ages of 3–5 years in males and at larger than 150 mm in females, as judged by the presence of shelled oviducal eggs (Auffenberg and Weaver, 1969; Judd and Rose, 1989; Hellgren, Kazmaier, Ruthven, and Synatzske, 2000).

Mating occurs from June to September in Texas (Weaver, 1970). The male pursues a female for up to 1 hour, accompanied by head bobbing. Ultimately he overtakes, stops, and confronts her by biting and ramming. The pair then moves in a pattern of diminishing concentric circles until she presents herself by raising the rear of her shell. The male then mounts and copulation occurs. Copulatory thrusts are strong enough to move the female forward. Mating is completed in 10 minutes or less, after which the female simply walks away (see detailed description in Ernst, Lovich, and Barbour, 1994). There are tubules in the oviduct capable of storing sperm (Gist and Jones, 1989).

Nests are dug in areas lacking low grassy vegetation but under or near the drip zone of bushes. Nests of different females are usually found in small clusters, and these favored nesting sites are used repeatedly (Auffenberg and Weaver, 1969; Vermersch, 1992). Auffenberg and Weaver (1969) described nest construction in captives. A ramp is constructed by deepening the end of a pallet with the gular prongs; the female then backs into this depression and digs a flask-shaped cavity (50 to 57 mm deep with an orifice 33–45 mm in diameter and its greatest diameter near the bottom). Cloacal fluid is voided during nest excavation.

Auffenberg and Weaver (1969) found oviducal eggs or evidence of nesting in all months from April to November, and ovarian follicles greater than 10 mm in March and November (suggesting that vitellogenesis begins in autumn and is resumed in spring). They concluded that there were two periods of egg laying in extreme southeastern Texas (presumably two clutches, the first in late June or July and the second in August or September).

Tortoises X-rayed by Judd and Rose (1989) contained shelled eggs from April to July. Shelled eggs are retained for as long as 39 days. Females laid the eggs from a single clutch in several places at different times. This may explain why Auffenberg and Weaver (1969) reported only one egg in 38 of 60 nests.

The eggs are elongate and hard shelled, and have a granular surface. Auffenberg and Weaver (1969) give the measurements of nine eggs from Brownsville as follows: L, 43.6 ± 2.99 (43.6–57.7); W, 34.0 ± 1.35 (32.2–35.6); these dimensions are comparable to those given for 12 eggs from Port Isabel by Judd and Rose (1989): L, 41.5 ± 0.73 (37.0–46.9); W, 34.1 ± 0.60 (30.0–36.2); mean Wt, 26.9 g ± 1.06 (18.8–30.4).

The xiphiplastral prongs of the plastron and the posterior projection of the carapace define an osseous aperture that is often too small for the eggs to pass through. This gave rise to the notion that the eggs were soft shelled until laid (Grant, 1960a). Rose and Judd (1991) demonstrated that eggshells were hard before laying, and that a hinge between the xiphiplastron and hypoplastron permits lowering of the xiphiplastron by about 30° during oviposition and copulation (see remarks on plastral kinesis in Introduction).

FIGURE 32.3 *Gopherus berlandieri.* Immature (Brownsville Texas). (Courtesy of P. Scanlan, Gladys Porter Zoo.)

Eggs per clutch in various populations ranges from one to seven, with means of 1.42, 2.65, and 4.30 at different localities and an overall mean of 1.99. One or two clutches are produced annually (Germano, 1994a). About 33% of females skip seasons of reproduction (Judd and Rose, 1989).

Judd and Rose (1989) reported one to five eggs per clutch in southern Texas (mean, 2.65/clutch) and found no evidence for second clutches. However, the evidence they previously presented (Judd and Rose, 1983) contradicts this. They reported road-killed females bearing oviducal eggs with up to seven enlarged ovarian follicles (21.0–30.5 mm). Follicles of this size (62–90% of mean egg width) could reach ovulatory size in the current season (Legler, pers. obs.).

Eggs taken from nests made by captive females and transferred to flower pots hatched in 88–118 days (mean 105; Judd and McQueen, 1980). No investigation of temperature-dependent sex determination has been done in *G. berlandieri* (but might be expected in view of its occurrence in its closest relative, *G. agassizi*).

GROWTH AND ONTOGENY

Data on size (CL; Texas and northen Mexico; D. Germano and Legler databases): males, 169 mm ± 24.5 (105–220) n = 48; females, 142 mm ± 16.6 (107–183) n = 36; unsexed adults, 160 mm ± 27.1 (123–202) n = 10; im. unsexed, 112 mm ± 5.8 (101–120) n = 9; juveniles, 87.4 mm ± 11.9 (56–103) n = 14.

The species undergoes the greatest ontogenetic shell change seen in *Gopherus*. In adults the lateral and dorsal aspects of the shell form are nearly hemispherical (rather than hemi-elliptical). The rear of the carapace drops off sharply to a posterior edge that usually "curls" ventroanteriorad to where it almost contacts the posterior plastral edge and may even overlap it. The shell margin flares to accommodate the hind limbs. Adult males differ from adult females in the following characters: they are relatively longer and narrower; they have longer and more deeply forked gular projections; they have a concave plastron; a longer tail; and larger, more complex mental glands (Ernst, Lovich, and Barbour, 1994).

Adult size and the degree of sexual dimorphism decrease west of the Gulf Coast (Judd and Rose, 2000). Minimum adult size and age for females are given by Germano (1994a) as 140 mm and 13.3 years (extremes 11–17).

Captive hatchlings from Hidalgo Co., Texas, were nearly circular in dorsal aspect. They grew from hatching until winter and then resumed growth in spring. Average weight and CL for eight juveniles were 24.3 g and 42.2 mm at hatching and 80.1 g and 65.9 mm at the end of the first full year of growth (age 371–380 days; Judd and McQueen, 1980). The carapacal scutes of hatchlings are brownish with pale areolar centers. Dark pigment may cover the entire plastron or be restricted to the interlaminal seams; the head is brownish with various pale marks on the side, and the throat is chiefly pale (Auffenberg and Weaver, 1969).

Growth rings can be used to estimate age to sizes of about 180 mm CL. A maximal growth ring count of 18 was recorded at 201 mm. Lengths of 105–128 mm are attained at 3–5 years, and about 130 mm in year 6; thereafter, growth slows to a constant annual rate of about 5%. Average absolute annual growth of females (11.1 mm) and males (7.9 mm) suggests that females may grow faster than males (Auffenberg and Weaver, 1969). At the Chaparrral Wildlife Management Area all of the 1,700 tortoises marked could be aged accurately by growth rings (R. Kazmaier, pers. comm.).

The following weights and ages are reported by Olsen (1976): 6 years, 200 g; 9 years, 500 g; 10 years, 860 g; and 11 years, 875 g.

Known ages of captives exceed 60 years (Judd and McQueen, 1982; Judd and Rose, 1989; Germano, 1994a). It is reasonable to assume a life span of 30–50 years for wild individuals (Judd and Rose, 2000).

PREDATORS, PARASITES, AND DISEASE

There are few firsthand data on predators. Woodrats (*Neotoma micropus*) are probably egg predators. Woodrats also gnaw on the shells and feet of hibernating tortoises. Opossums are known to eat juveniles. Raccoons and foxes prey on captives (Judd and Rose, 2000; Auffenberg and Weaver, 1969). Judd and Rose (2000) suggest many potential predators of all life stages. There is little evidence that Native American peoples used the species for food (Judd and Rose, 2000).

Adults may be afflicted with nonfatal "necrotizing scute disease," a fungal (*Fusarium semitectum*) infection that causes the scutes to take on a whitish appearance leading to a complete degradation of the scute (Judd and Rose, 2000; Rose, Koke, Koehn, and Smith, 2001). The fungus degrades old keratin in the scute but does not degrade new keratin

FIGURE 32.4 *Gopherus berlandieri*. Young juvenile from captive colony. (Courtesy of F. Rose.)

growth in the most recently formed layers. The disease is relatively rare in the western part of the range.

Captive *berlandieri* have tested positive for URDT. Symptoms include a watery, bubbly exudate from the nares, lethargy, and swollen eyelids (Judd and Rose, 2000; see account of *G. agassizi* for etiology).

POPULATIONS

All of the information on population density comes from studies in extreme southern Texas (Cameron Co.). Auffenberg and Weaver (1969) reported densities for three vegetational communities on one loma as follows: 122/ha in a brush community, 33/ha in a *Baccharis* community, and 8/ha in a grass and cactus community. The brush community was so dense that ca. 5% of it was too thick to enter. Judd and Rose (1983) estimated density on a loma with grass, prickly pear, and scattered shrubs; mean estimates ranged from 10.0 to 22.9/ha over 5 years. They suggested that a maximum density of 16/ha was likely. Bury and Smith (1986) located 107 tortoises on about 3 ha of dirt roads and adjacent areas at Laguna Atascosa National Wildlife Refuge but made no density estimates. Iverson (1982) estimated biomass of a typical population at 54.5 kg/ha.

Auffenberg and Weaver (1969) reported the mean population structure on three lomas in southeastern Texas as follows: 31.4% males (20.0–46.6%); 50.2% females (40.0–57.6%); and 27.4% immature (14.7–40.0%). Judd and Rose (1983), a few miles to the northeast, reported 42% males, 29% females, and 26% immature. Bury and Smith (1986) reported 63% males and 37% females at Laguna Atascosa.

Survivorship from egg to hatching is ca. 60% in artificially incubated eggs (Judd and McQueen, 1980).

CONSERVATION AND CURRENT STATUS

Nearly all anthropogenic activities are deleterious to *G. berlandieri*. In the lower Rio Grande Valley of Texas (from Laredo downstream) land has been modified for agriculture or grazing since about 1900. In worst-case scenarios (e.g., "chaining" followed by plowing) virtually all the cover and food plants of the tortoises were destroyed. Introduced buffel grass (*Cenchurus ciliaris*) is taller than native buffalo grass and impedes the movements and vision of tortoises. These changes have reduced and fragmented the range and decreased the abundance of tortoises. The initial clearing devastates a tortoise population, but in resulting areas of managed pasturage overgrazing actually improves tortoise habitat by favoring more nearly open short grass–cactus associations (Auffenberg and Weaver, 1969; Judd and Rose, 2000).

About half of the range is in Mexico where virtually nothing is known about the species, but where the same kinds of habitat damage are occurring. In Tamaulipas 243,800 ha were modified for agriculture in 1953 versus an astounding 1,310,000 ha by 1988 (Judd and Rose, 2000).

Vehicles kill a significant number of tortoises on public and private roads throughout the range. The predilection of tortoises for roads is partly to blame for this.

Collecting for the pet trade has probably reduced some populations. Many specimens are said to reach California from northeastern Mexico via New Mexico in order to avoid Texas laws. Shipments of 4,000–8,000 specimens have been mentioned in the literature, and estimates of annual totals have been as high as 40,000. Judd and Rose (2000) concede that the practice exists but think its magnitude may be exaggerated.

In Mexico tortoises are killed, dried, and varnished for sale to tourists in border cities (from Cd. Acuñya to Matamoros). Of 346 mummified specimens reported by Judd and Rose (2000), 39 were *Gopherus berlandieri* (most were *Kinosternon* sp.).

The species is legally protected and considered threatened by agencies in Texas, but there is no state conservation plan. The species is not federally protected, and there is no federal recovery plan. The Convention on International Trade in Endangered Species (CITES), endorsed by the United States in 1973, endorses the requirement of export permits from the country of origin and places all *Gopherus* in Appendix II of the List of Endangered Species. Collecting of native fauna in Mexico requires a permit.

It is enforcement rather than the law that flags. Enforcement requires the ability to identify animals and a knowledge of the rules. Even if an illegal tortoise is confiscated, a law officer may not have the facilities to dispose of it properly. Tortoises confiscated by U.S. Customs at the United States–Mexican border are said to be released on the U.S. side of the border (Judd and Rose, 2000).

Thus far there seem to be no management plans or specific reserves for the species. *Gopherus berlandieri* may have suffered the least of all the *Gopherus*; much of its range occurs on blocks of private ranch land that are off limits to the public.

Populations studied by Auffenberg and Weaver (1969) are not protected on most of the lomas, some of which have been disturbed by military maneuvers, hunting, and residential development. Populations studied by Judd and Rose (1983, 1989) and by Bury and Smith (1986) are under private or federal protection (Judd and Rose, 2000).

Studies of basic life history should be made on several sites with differing habitats. This would be most feasible in Gulf Coastal areas where high population densities occur and where a resident investigator would not be necessary. Controlled burning of dense brush areas would provide an advantageous mosaic of dense and open vegetation. It would be helpful if road signs could warn against possession of tortoises and urge caution for animals on the road (Judd and Rose, 2000; Bury and Smith, 1986).

FIGURE 33.1 *Gopherus flavomarginatus*. Typical adult female (KU 53806). (Photo by J. K. Greer.)

Gopherus flavomarginatus Legler, 1959

Bolson tortoise (Iverson, 1992a); tortuga Llanera (Legler and Webb, 1961); other local names are tortuga del monte, tortuga del tierra, and tortuga grande (Morafka, 1982).

ETYMOLOGY

L. *flavus*, yellow; L. *marginatus*, a border; alluding to the distinctive yellowish color of the marginal scutes as seen in a lateral view of the shell (Figures 33.1 and 33.2).

HISTORICAL PROLOGUE

The late scientific discovery of this large, distinctive tortoise (Legler, 1959) can be attributed to its occurrence in a remote area. Taxonomic, morphological, and natural history notes (Legler and Webb, 1961) quickly followed the type description. Other early natural history notes were made in three popular articles by Pawley (1968, 1969, 1975) and Morafka (1982). Morafka, Adest, Aguirre, and Recht (1981) conducted baseline studies of natural history from 1970 to 1979; this work was followed by Morafka and McCoy (1988) and Lieberman and Morafka (1988). These studies are seminal; they began the scientific study of a threatened species that is now one of the best known chelonian species in North America. Studies have been based on two robust populations: Rancho La Ventura in the Cerro Emilio region of Chihuahua and the Mapimi Biosphere Reserve in Durango. Knowledge of other *Gopherus* has also grown rapidly in the same period (Bury and Germano, 1994; Van Devender, 2002; Rhodin, 2002). Useful reviews of literature were made by Bury and Germano (1994) and by Aguirre-León, Morafka, and Adest (1997).

DIAGNOSIS

Gopherus flavomarginatus can be distinguished from all other Mexican chelonians by the following combination of characters: carapace is narrower anteriorly versus posteriorly; there are at least two femoral spurs on the back of each thigh; there is a single enlarged axillary scute that is wider posteriorly versus anteriorly, not triangular, and not pointed posteriorly; intergular seam is longer than interhumeral; precentral scute is often divided; carapace ground color is yellow and laminae have dark centers; plastron is yellow with darker central markings in some specimens (based substantially on Auffenberg and Franz, 1978a).

GENERAL DESCRIPTION

Bolson tortoises attain the largest size of any *Gopherus* and are the largest terrestrial chelonians in the Northern Hemisphere. Adult size is commonly greater than 340 mm, and maximum recorded length is 371 mm (Legler, 1959). Anecdotal accounts allude to a much larger size in the past (Legler and Webb, 1961; Aguirre-León, Morafka, and Adest, 1997). All or most of the shell laminae have a pale ground coloration with darker, often sharply contrasting areolar centers (whereas the converse is usually true of other *Gopherus*). The most distinctive feature of a bolson tortoise is the pale yellowish coloration of the lateral marginal scutes (M3–M6) with sharply contrasting dark brown or black centers.

BASIC PROPORTIONS

(D. Germano and Legler Databases)

CW/CL: males, 0.83 ± 0.026 (0.79–0.88) n = 22; females, 0.78 ± 0.025 (0.74–0.83) n = 22; unsexed adults, 0.79 ± 0.024 (0.77–0.83) n = 8; im. females, 0.81; im. unsexed, 0.81 ± 0.024 (0.77–0.86) n = 11; juveniles, 0.89 ± 0.036 (0.85–1.00) n = 18.

CH/CL: males, 0.47 ± 0.019 (0.44–0.51) n = 22; females, 0.45 ± 0.021 (0.41–0.48) n = 22; unsexed adults, 0.46 ± 0.004 (0.46–0.47) n = 5; im. females, 0.44; im. unsexed, 0.48 ± 0.033 (0.43–0.55) n = 11; juveniles, 0.53 ± 0.089 (0.47–0.86) n = 17.

CH/CW: males, 0.57 ± 0.024 (0.51–0.59) n = 22; females, 0.57 ± 0.029 (0.51–0.63) n = 22;

FIGURE 33.2 *Gopherus flavomarginatus*. Subadult (with clean shell) showing natural colors; CL ca. 270 mm (Reserva de Mapimí, Durango). (Photo by M. A. Recht, in Bury et al., 1988, with permission of Carnegie Museum.)

unsexed adults, 0.58 ± 0.017 (0.55–0.60) n = 5; im. females, 0.54; im. unsexed, 0.58 ± 0.039 (0.54–0.68) n = 11; juveniles, 0.60 ± 0.096 (0.53–0.95) n = 17.

Scutes 4 or 1 are always longest; 3 or 6 are always shortest; modal plastral formula: 4 > 1 > 5 > 2 > 3 > 6.

GEOGRAPHIC VARIATION

The northernmost population of *G. flavomarginatus* (Diablo region of Chihuahua) is separated from the rest of the geographic range by about 100 km of dry playa basins. It differs from the main southern population in color and the degree of sexual dimorphism (southern males lack a plastral concavity; Morafka, 1982; Adest, Aguirre-León, Morafka, and Jarchow, 1989a; Trevino, Morafka, and Aguirre-León, 1995; Trevino, Morafka, and Aguirre, 1997). Morafka (1988) suggested that intermittent contact between the two populations could have occurred over a period of 10,000 years.

RELATIONSHIPS (SEE ALSO ACCOUNT OF *GOPHERUS*)

Gopherus polyphemus (southeastern United States) is the closest relative of *G. flavomarginatus*. This was first suggested by Legler (1959) and supported by subsequent studies (Bramble, 1982; Auffenberg, 1976; Lamb, Avise, and Gibbons, 1989; Lamb and Lydeard,1994; Morafka, Aguirre, and Murphy, 1994; Crumly, 1984, 1994).

FOSSILS

The fossil history of *Gopherus* dates from the Oligocene epoch and may be more complete than that for any other North American chelonian (Morafka, Aguirre, and Murphy, 1994). In Pleistocene times, *G. flavomarginatus* was more widely distributed (from Oklahoma, southern New Mexico, and Arizona to the edges of the Transverse Volcanic Ranges in Aguascalientes; Auffenberg and Franz, 1978a; Morafka, 1982). The species now occurs in only 10% of its former range (Morafka, 1982). *Gopherus huecoensis* (Strain, 1966; mid-Pleistocene of West Texas) is probably conspecific with *G. flavomarginatus*; the fossil site lies about 500 km N of the current range of *flavomarginatus* (Auffenberg, 1974; Auffenberg and Franz, 1978d).

GEOGRAPHIC DISTRIBUTION

Map 21

The entire geographic range is now in the Bolsón de Mapimí of northern central Mexico where the states of Chihuahua, Coahuila, and Durango share a common point (26°42′N, 103°09′W), ca. 200 km S of the Big Bend of the Rio Grande. The Bolsón de Mapimí is actually a series of interconnected closed basins lying in a small part of the southern Chihuahaun Desert Biotic Province (Morafka, 1982) and is a part of the Mesa Central del Norte of the Altiplano of Mexico (González-Trapaga, 1995).

The current range is fragmented into six discrete areas (see map of Bury, Morafka, and McCoy, 1988) partly due to human causes (Aguirre-León, Morafka, and Adest, 1997). Tortoises occur in colonies, and this contributes to a patchy distribution, even in good habitat (Aguirre-León, Morafka, and Adest, 1997).

The total area occupied by the species has been estimated variously at 40,000–50,000 km^2 (Morafka, 1982; Morafka, Adest, Aguirre, and Recht, 1981); 6,000 km^2 (Bury, Morafka, and McCoy, 1988); and as small as 1,000 km^2 (Aguirre-León, Morafka, and Adest, 1997). Including the gaps produced by extirpation, the diameter of the geographic range is less than 150 km. Populations are discontiguous and restricted to narrow belts below rocky outcrops and above bolson floors. These gaps are not shown in Map 21.

Morafka (1982) gave coordinates for extreme points of the range as follows: northernmost, Rancho La Ventura, Chihuahua (27°36′N, 104°00′W); westernmost, edge of Sierra de los Remedios, Chihuahua (26°55′N, 104°25′W); northeasternmost, Rancho (Ejido) El Socorro, Coahuila (27°20′N, 103°15′W); southernmost, the type locality, 48 to 64 km N of Lerdo and adjacent Sierra Banderas, Durango (26°25′N, 103°30′W). Lists of all known localities and reports by local inhabitants appear in Morafka (1982) and Morafka, Adest, Aguirre, and Recht (1981). The range of *Kinosternon durangoense* overlaps the range of *G. flavomarginatus* in the southwestern part of the Bolsón de Mapimí (east of Ceballos, Durango). No other chelonians occur in the basin.

HABITAT

Bolson tortoises occur in the Mapimian tobosa (*Hilaria mutica*) grasslands of the Chihuahuan Desert (Aguirre-León,

FIGURE 33.3 *Gopherus flavomarginatus*. Head of adult male, east of Ceballos, Durango. (Courtesy of R. G. Webb, A. D. MacEwen collection.)

Morafka, and Adest, 1997). This is a moderately warm region with harsh alkaline soils, low relative humidity and precipitation, high solar radiation, and wide temperature fluctuation. A strong seasonality includes dry winters with evening frost; little precipitation (November–March); a dry, warm spring (April–May); and hot summer (June–October; Morafka, 1982). Average annual rainfall is 271 mm, 74% of which occurs in September.

Habitat requirements include suitable substrates for burrowing and nesting, and sufficient plant cover for forage and shelter. Optimum habitat seems to be at altitudes of 1,000–1,300 m and chiefly on sandy slopes with a grade of 1.0–2.5%. Distribution may extend to 1,400 m (Morafka, 1982). Fewer tortoises occur on flats with a grade less than 0.5%. Burrows disappear abruptly where rocky soil is overlain by desert pavement.

Vegetation is a patchy distribution of thorn scrub. Dominant plant species are creosote bush (*Larrea tridentata*), prickly pear cactus (*Opuntia rostrata*), mesquite (*Prosopis glandulosa*), tar bush (*Flourensia cernua*), tobosa grass (*Hilaria mutica*), mallow (*Sphaeralcea angustifolia*), and grama grass (*Bouteloua* sp.; Morafka, Berry, and Spangenberg, 1997). An excellent photo of habitat in Chihuahua appears in Bury, Morafka, and McCoy (1988: Figure 5).

Although *G. flavomarginatus* now occurs in arid habitat, it is not regarded as a desert specialist either phylogenetically or ecologically. Most of its quaternary evolution occurred in nonarid grasslands (with other sympatric tortoises). Its habits, thermal requirements, and social behavior are more consistent with a burrow microhabitat rather than surface life in the Chihuahuan desert (Aguirre-León, Morafka, and Adest, 1997).

DIET

The dietary staple is tobosa grass, a wiry bunch grass. Tortoises feed opportunistically on other grasses, succulent herbaceous annuals, and *Opuntia* fruits. Despite these dietary preferences, tortoises are not limited to areas of tobosa grass and may be absent where solid carpets of it exist (Morafka, 1982).

Crop-and-move grazing occurs along well-defined grazing trails (Morafka, 1982; Appleton, 1986). Foraging ceases in early fall and resumes in late spring (Appleton, 1980). Captives of both sexes have been observed eating earth. Stones and woody stems have been found in scats (Lindquist and Appleton, 1982; Appleton, 1986).

Hatchlings eat fresh adult feces (Appleton, 1980), thereby insuring inoculation of the gut with the microorganisms necessary for herbivorous digestion. Feces of 1- to 2-year-old wild tortoises contain tobosa grass. Hatchlings require about twice the protein of larger animals and prefer broad-leafed herbaceous plants (with high phosphorus and high protein content) to grass (Tom 1994; Appleton, 1986). This knowledge was important in increasing the survivorship of nursery-reared animals (Morafka, Berry, and Spangenberg, 1997; Adest, Aguirre-León, Morafka, and Jarchow, 1989b). Tortoises drink when it rains and may sit in the water while drinking (Appleton, 1980; Adest, Aguirre-León, Morafka, and Jarchow, 1989a).

HABITS AND BEHAVIOR

Bolson tortoises spend most of their time in burrows. Even under favorable conditions in the season of activity, monitored individuals spent 80–90% of their time in burrows. If the period of winter quiescence is considered, as little as 1% of a tortoise's life is spent out of a burrow (Adest and Aguirre-León, 1995). Laboratory studies (Ayala Guerrero, Calderón, and Peréz, 1988) suggest that burrow time is spent sleeping.

The cold, dry period from November through March or April is spent in a period of semiquiescence. Some tortoises bask at burrow mouths or just outside burrows beginning in March, and occasional foraging may occur in this period. Sporadic emergence occurs in the hot, dry period of May and June, especially if there is rain. Surface activity is concentrated in the hot, wet period from June to October. Activity accelerates in July, August, and September. Intense grazing begins in July, after nesting has occurred. Monitored adults averaged 45 minutes of surface activity per day from June through September. Surface activity in summer occurs typically in the morning (0900–1100 hours) and a short period (1700–1900 hours) in afternoon. There is some activity prior to 0800 hours and some resting at burrow entrance after sunset.

Within the known geographic range tortoises occur as scattered individuals and in colonies of differing densities. Tortoise distribution is associated with sandy soils with high salt content, often at the bases of bajadas but above playas. Burrow density varies from 5/km^2 (0.05/ha) at Mapimi Biosphere Reserve (MBR) to 7/ha at Cerro Emilio,

FIGURE 33.4 *Gopherus flavomarginatus*. Head of adult female (UU 3572), east of Ceballos, Durango.

Chihuahua (Aguirre-León, Adest, and Morafka, 1984; Adest and Aguirre-León, 1995).

Burrow counting may not accurately estimate number of individuals. Each burrow may be used by more than one tortoise, each tortoise may occupy more than one burrow, and occasionally a burrow is simultaneously shared. Males use an average of 3.5 different burrows, females 2.7, and juveniles 1.8. Some males used up to eight different burrows. Tortoises of all ages spend 75% of their burrow time in a primary burrow and 25% in all other burrows (Aguirre-León, Adest, and Morafka, 1984). In the case of a shared burrow, one wonders how an underground head-to-head meeting would be resolved.

A nuclear area at MBR had a density of 3 tortoises/ha, which decreased to 1 tortoise/8–10 ha in a peripheral area. Within colonies, burrow distribution is clumped. Tortoises exhibit linear movements associated with foraging, exploration, and reproduction. Mean daily distance moved was 265 m for males, 165 m for females, and 165 m for juveniles. Mean maxima for adults were greater than 500 m, and 256 m for juveniles. All age and sex classes moved significantly more during wet versus dry periods (Adest Aguirre-León, 1995).

Home range area was estimated by two techniques (minimum modified polygon and 68% bivariant ellipse models, respectively) as follows: males, 3.1 and 4.1 ha; females, 2.5 and 3.1 ha; and juveniles 0.4 and 1.2 ha (Adest and Aguirre-León, 1995).

There is home range overlap with other individuals in 32% of males, 26% of females, and 9% of juveniles. No overlap occurs in the ca. 10% of home range used intensively for foraging (Adest and Aguirre-León, 1995).

Centers of activity are correlated with burrows, and distribution of burrows is correlated with high densities of tobosa grass. Long-distance movements (1,500–6,000 m) away from a home range occurred in 14% of adults and 18% of juveniles. Individuals moving this far away from a colony did not return during an 11-month study (Adest and Aguirre-León, 1995).

Gopherus flavomarginatus habitually digs and occupies sizable burrows that serve as refugia, a means of thermoregulation, and to retard evaporative water loss. Most aspects of tortoise life center on the burrow and its entrance. Habitual movement patterns suggest a knowledge of burrow locations (Morafka, 1982).

Burrows are 1.5 to 2.5 m deep, up to 10 m long, and have a single opening. The observation of a branched burrow system with multiple entrances (Pawley, 1975) has never been repeated. Captives often start burrow excavation using the purchase of a wall or foundation, and natural burrows may be started next to a mesquite bush or other substantial vegetation. Burrow mouths tend to face northeastward. An entrance mound 10 to 30 cm high is characteristic of active burrows and may serve as an impediment to flooding (some burrows face uphill). Abandoned burrows have rounded, eroded entrances. Active burrows have the same cross-section as a tortoise. The burrow mouth and all parts of the burrow are high enough to permit locomotion with the plastron raised off the substrate. Entrance height is 20% higher than the tortoise, and width is 25–40 % greater than carapace length, permitting a tortoise to rotate on its long axis in the entrance. Burrow diameter tapers to ca. 10% greater than carapace length.

A typical burrow has a sharp, short curve of 10–20° to one side then a steep slide downward at 30° or more for 60% of burrow length, and then becomes more nearly level. The burrow ends in a humid chamber large enough for a tortoise to rotate easily on its long axis. Fecal pellets are found on the flat floor of the burrow and on the outdoor entrance mound (Morafka, 1982). Captives bring up dried feces from burrows and spread them outside the entrance at the beginning of the wet season (Appleton, 1986).

Spiders and common tenebrionid beetles are the only commensals recorded for burrows (Morafka, 1982). Hatchlings dig their own burrows; a captive dug a burrow large enough to cover itself in 2 days (Lindquist and Appleton, 1982).

There is anecdotal evidence that tortoises in burrows react to light. Captive tortoises in a state of quiescence become active if they can see sunlight (Morafka, Adest, Aguirre, and Recht, 1981). Tortugueros may use a mirror to shine light down a burrow to get a tortoise to come out (Legler and Webb, 1961). A captive in Salt Lake City would come out of its burrow at night when a flashlight was shined into it (A. D. MacEwen, pers. comm.).

Most information on hatchlings in the wild comes from Tom (1994), who followed 10 hatchlings (average age 17 days) with radio transmitters for a period that included the warm, dry season (May and June) and the hot, rainy season (August–October). Hatchlings construct small burrows (length 6.0–19.0 cm) that provide full cover, and pallets, which are depressions providing partial cover. Hatchlings moved between pallet and burrow sites frequently in the rainy season but occupied only one

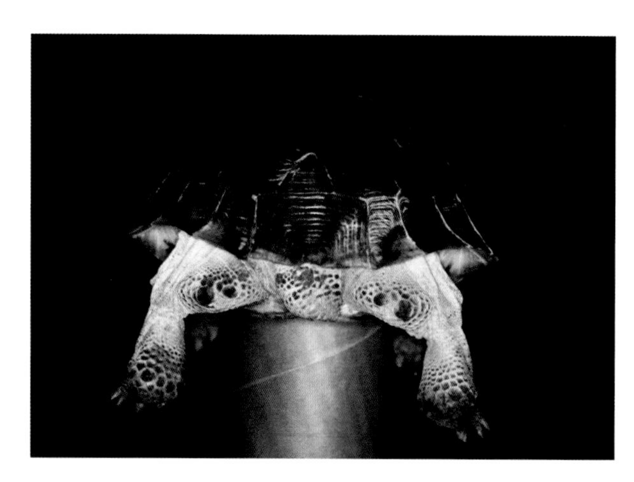

FIGURE 33.5 *Gopherus flavomarginatus*. Adult male showing two femoral spurs on back of each thigh.

burrow or pallet in the dry season. Ten hatchlings occupied a total of 21 burrows and 14 pallets. At all seasons the hatchlings preferred to dig in the shade of *Opuntia* cacti. Four of the ten hatchlings were alive after 11 months (four known dead and two presumed dead by predation). Hatchlings cannot right themselves when overturned, according to Appleton (1980).

On sunny mornings tortoises bask on entrance mounds, often with head and neck extended. Approaching humans are detected as far away as 100 m. When startled, the tortoise rotates its body to face the burrow and quickly dives down the relatively steep entrance slide (Morafka, Adest, Aguirre, and Recht, 1981). This behavior has evident selective advantage, but it could also account for human searches that revealed more burrows than live tortoises.

The idea of a large, cumbersome tortoise doing all of this quickly seems, at first, to be ludicrous. Legler had occasion to observe the fenced colony of bolson tortoises at the Research Ranch in Elgin, Arizona, guided by the caretaker, Ms. Chaun Copus, in late March 2003. This was on a sunny morning, early in the season of activity, and there were signs of grazing near the burrows. One of the largest tortoises ("Gertie," CL >300 mm) was seen from a distance with the head barely visible above an entrance mound (Figure 33.7). In this protected, captive colony the tortoises are conditioned to the presence of humans and some hand feeding. Cautious approach brought us within ca. 1.8 m, close enough to view the entire tortoise and see into the burrow. At this point the tortoise pirouetted 180°, and we witnessed the headlong dive into the burrow and an almost immediate 90° rotation within the burrow, tightly wedging the shell at right angles to the main axis and blocking it, at a depth of no more than 1 m. The speed and agility of the entire maneuver were judged fast enough to foil any mammalian predator.

Social behavior is loosely hierarchical. Tortoises sometimes defend burrows, but there is no strong evidence of territorial defense, of males defending females, or of harem formation (Alberts, Rostal, and Lance, 1994; Adest and Aguirre-León, 1995). Intraspecific signaling consists of head bobbing, neck extension, and posterior raising of the shell (Morafka, 1982). Head bobbing is used in challenge, recognition, and courtship, but is not a prelude to actual combat.

In captives, there is interspecific agonistic behavior among *Gopherus* but not with other genera of tortoises (Auffenberg 1969, 1976; Morafka, 1982). Aggression between males is common. In interactions of two males of differing size, the larger male was dominant and chased the smaller into its burrow. The larger male bobbed at the burrow entrance of the smaller male even when the burrow was empty. Adults did not defend burrows against other adults. When humans disturbed burrow mounds, the resident tortoise blocked the entrance and made hissing noises. Burrow exchanges usually involved tortoises of the same size or a smaller individual usurping a larger burrow (Lindquist and Appleton, 1982; Appleton, 1986).

Bolson tortoises regulate body temperature with elegant simplicity by moving in and out of burrows. The burrow constitutes a stable thermal gradient. When surface temperatures reach 40–42°C in July, temperatures on the floor of the burrow are in the range of 25–28°C. A gradient in the opposite direction could be predicted in winter (Morafka, 1982). Burrow temperatures are probably too low for efficient herbivorous digestion. Tortoises frequently bask on the entrance mound. Basking may include the hyperextension of limbs and neck (as in aquatic emydids) and is terminated when body temperature reaches 29–30°C. Typical behavior in summer is foraging along well-defined trails 5–20 m from the burrow. In this time, deep core body temperature may rise as high as 41°C, after which the tortoise quickly retreats a variable distance into a burrow. Most actively foraging tortoises have temperatures of 33–34°C (extremes, 28–40°C; Morafka, Adest, Aguirre, and Recht, 1981; Rose, 1983; Adest, Aguirre-León, Morafka, and Jarchow, 1989a; Adest and Aguirre-León, 1995).

Tortoises tolerate higher body temperatures in the wet season versus in the dry (Adest and Aguirre-León, 1995). The following estimates are based on 2,042 deep core body temperatures: critical thermal maximum, 43–45°C; average maximum, 32.5°C in hot, dry season and 36.4°C in hot, wet season (Adest, Aguirre-León, Morafka, and Jarchow, 1989a).

REPRODUCTION

Courting and copulation begin at emergence from hibernation (as early as 20 March) and continue until early October, with peaks in July and August (Adest and Aguirre-León, 1995). Sperm used in these inseminations are produced in the previous year. Inseminated females can store sperm. Testosterone levels in males are maximal in July. July testosterone levels of 1,329 ng/mL are the highest

FIGURE 33.6 *Gopherus flavomarginatus*. Typical active burrow of small adult, on level ground (Research Ranch, Elgin, Arizona).

reported for any vertebrate, fide Gonzáles-Trapaga, Aguirre, and Adest (2000).

Soft-shelled oviducal eggs can be palpated in late February and early March, soon after emergence. Hard-shelled eggs can be palpated from mid March to July with a peak in May. It is probable that follicles begin enlarging before hibernation, continue enlarging during hibernation, and reach ovulatory size at about the time of emergence (Adest and Aguirre-León, 1995; González-Trapaga, Aguirre, and Adest, 2000).

There is evidence (from captives and wild populations) that a relatively few males mate with a majority of the females (Bickett, 1980; Adest and Aguirre-León, 1995). The actively mating males do not exclude subordinate males. Males observed in copulation tended to occupy a greater number of burrows and travel greater distances than other males, but were not the largest males in the colony (Adest and Aguirre-León, 1995).

Consistent elements of courtship are head bobbing, trailing, circling, and ramming. There may be a male-female pursuit (Lindquist and Appleton, 1982). Mating behavior is usually initiated by a male bobbing near the burrow of a female or on the burrow apron; she may come out of the burrow or already be out. When ready, the female turns away to face downward toward the burrow. The female is sometimes the initiator (Bickett, 1980). Mating may occur frequently. A pair of captives mated 14 times in 40 days (Appleton, 1986).

In a copulation the male mounts and rocks back and forth, shifting weight from one hind limb to the other. During coital thrusting, the mouth is open and the head is repeatedly extended and retracted at intervals of 15 seconds or less; this is accompanied by rasping grunts (suggesting the exertion of intra-abdominal pressure).

Males often fall off (or are scraped off when the female dives into a burrow); they are able to right themselves and mount again (Appleton, 1986; Lindquist and Appleton, 1982). Males from Durango may lack a concave plastron and have difficulty remaining in mounted position (Bickett, 1980; Morafka, 1982).

Nesting begins in early April but is most common in May and June (Adest and Aguirre-León, 1995). Hatching and emergence are coincident with autumn rains, a period of about 4 months, peaking in September (Adest, Aguirre-León, Morafka, and Jarchow, 1989b; Adest and Aguirre, 1995). Hatchlings appear in late summer and early fall (Adest and Aguirre, 1995).

Of 13 natural nests observed at MBR, six were on an entrance mound or at a burrow entrance, three were located within 10 m of a burrow, and four were within 150 m of a burrow. Of those away from a burrow, three were in the open and four in shade (Adest and Aguirre, 1995). Trevino, Morafka, and Aguirre (1997) allude to ample evidence of nests at Rancho Sombreretillo, Chihuahua. Pawley's (1975) allegation that eggs are laid within the burrow has not been confirmed.

There are no reports of the nesting process under completely natural conditions. Nesting usually occurs in the afternoon. Most nests require 1–2 hours for completion. Trial nests are commonplace. The nest is dug with alternate movements of the hind limbs. As each egg is laid, the hind feet are placed alternately in the hole, seemingly to arrange the eggs. Soil is pulled back in with alternate semiencircling movements of the hind limbs and followed by tamping movements of the hind limbs. When the nest is filled, the loose earth is smoothed by rotating the plastron. The nest is completed and concealed by a final backward flipping of soil, grass, and other debris by both forelimbs and hind limbs. Some false nests include all these steps and lack only the eggs. The nest cavity is flask-shaped and angled into the ground, 15 cm in greatest depth and 20 cm long from lip of orifice to floor of cavity. The top egg may be covered by no more than 1–2 cm of soil. Tamping with the hind feet may damage, dent, or break the uppermost eggs (Appleton, 1986; Lindquist and Appleton, 1982).

There is some parental concern for a completed nest; captive females have repeatedly and aggressively approached humans disturbing the nests (Morafka, Adest, Aguirre, and Recht, 1981).

The eggs are roughly spherical with mean greatest and least diameters of 47.6 and 44.1 mm and mean weight of 54.98 g (Adest and Aguirre-León, 1995). Morafka, Adest, Aguirre, and Recht (1981) provide the following data on four shelled oviducal eggs from the same captive female: L, 45.7 mm ± 0.95 (44.9–46.6); W, 43.3 mm ± 0.14 (43.2–43.5); Wt, 50.8 g ± 1.26 (49–52). The mean ratio of width to length was 0.95. Bolson tortoise eggs are the largest of any *Gopherus*.

In a "seminatural" nest in Durango, two hatchlings emerged after 120 days; 19 days later two more hatchlings were found dead under compacted earth and egg shells. Since

FIGURE 33.7 *Gopherus flavomarginatus* (Research Ranch, Elgin, Arizona). Left—large adult female positioned 90° to main axis of burrow, alert to approaching humans; right—same individual after pirouetting then "diving" into borrow immediately after left photo was taken.

yolk sac retraction was complete, these hatchlings may have been trapped by the emergence of their sibs. Hatchlings from other seminatural clutches showed diminished strength and vigor in comparison to those hatched in incubators (Adest, Aguirre-León, Morafka, and Jarchow, 1989b).

Statements of incubation period in the literature range from 75 to 149 days (Morafka, 1982; Morafka, Berry, and Spangenberg, 1997). Appleton (1980) records an incubation period of 62–65 days for eggs from a natural nest soon after laying and placed in an incubator at 32°C. Seemingly there are no data on incubation time or time of emergence from a natural nest. Adest and Aguirre-León (1995) mention a clutch that overwintered and took 332 days to hatch.

Most of the useful information on incubation comes from the hatchery-nursery at the MBR in extreme northeastern Durango (Adest, Aguirre-León, Morafka, and Jarchow, 1989b). The results for three hatchery incubation protocols were as follows (incubation times and rate of success):

1. Temperature-controlled incubation (constant 30°C water bath incubator with eggs either buried in sand or exposed to air), 111.1 days ± 14.1 (92–150) $n = 17$; success rate, 72.1% ± 21.7.
2. Passive incubation (eggs on sand floor in an air chamber exposed to fluctuating ambient temperatures [22–37°C] approximating those of natural nests), 94.7 days ± 14.9 (74–111) $n = 21$; success rate, 64.7% ± 25.3.
3. Seminatural (eggs incubated outdoors under predator-proof screening), 115.9 days ± 16.0 (93–139) $n = 17$; success rate, 77.7% ± 27.3.

Temperature-dependent sex determination is known in *G. polyphemus* (Rostal, Wibbels, Grumbles, Lance, and Spotila, 2002) but has not been reported for *G. flavomarginatus*.

Most wild adult females (56%) do not lay eggs in any given year (Adest, Aguirre-León, Morafka, and Jarchow, 1989a). Captives in good condition skip fewer seasons and produce up to three clutches per season (California—Morafka, 1982; MBR—González-Trapaga, Aguirre, and Adest, 2000). Environmental conditions and the availability of food probably influence the reproductive output of wild animals (Morafka, 1982). Number of eggs per clutch ranges from 2 to 12 ($n = 60$; Adest and Aguirre-León, 1995). Wild females produce an average of 1.4 clutches of 5.4 eggs each per year (Aquirre et al., 1997). Of 60 wild females, 57 produced one clutch and three produced two clutches in the same year; no wild females were known to produce three clutches (Adest and Aguirre-León, 1995). Mean total clutch mass is 298 g (extremes, 85–708). Adest and Aguirre-León (1995) consider total hatchling mass to represent the best estimate of net reproductive output.

None of the foregoing information is based on ovarian examination. The following information from Legler and Webb (1961) has not been mentioned in recent literature.

Dissection of a mature female (UU 3571, 326 mm CL) from Ceballos, Durango, yielded the following information. The ovaries bore eight follicles 14–16 mm in diameter and numerous white, yolkless follicles less than 5 mm. There were 13 large spheroid bodies 6–7 mm in diameter with hollow centers lying beneath circular openings in ovarian epithelium; these were probably corpora lutea but could have been atretic follicles or follicles in the process of reabsorption. There were also 27 puckerings on the ovarian epithelium that were interpreted as older corpora lutea or corpora albicantia. Ergo, there were 48 sites on these ovaries that represented ovulations or potential ovulation events over a period of about 2 years (21 for year of capture and the rest for the previous year, probably two clutches in each year).

GROWTH AND ONTOGENY

The largest recorded specimen (CL 371 mm) is one of the paratypes (Legler, 1959) and was probably obtained in the late 1880s.

Date on size (CL; D. Germano and Legler databases): males, 308 ± 32.3 (220–356) $n = 22$; females, 346 ± 31.8 (222–382) $n = 22$; unsexed adults, 294 ± 13.5 (278–316) $n = 8$; im. female, 164; im. unsexed, 157 ± 48.3 (112–245) $n = 11$; juveniles, 70.1 ± 12.3 (53–94) $n = 18$.

There is a decrease in body size with time in the fossil record (Aguirre-León, Morafka, and Adest, 1997; Adest, Aguirre-León, Morafka, and Jarchow, 1989a). Size estimates (anecdotal and perhaps exaggerated) of 1 m CL and weights of 35–50 kg by rural people are common and suggest that the largest known individuals have never been seen by contemporary scientists or no longer exist (Legler, pers. obs.; Morafka, 1982). A large tortoise is said to have been displayed in 1967 at Jaco and Carillo, Chihuahua; it was heavy enough to require a mechanical hoist to get the tortoise into a truck bed (Morafka, 1982).

Gopherus flavomarginatus is the least sexually dimorphic gopher tortoise. Males are smaller than females (Germano, 1994a). Sexing live animals is difficult using external characters (Legler and Webb, 1961; Auffenberg and Franz, 1978b; Morafka, 1982). At the MBR sex determination is done by manual palpation for a penis or clitoris on the ventral floor of the cloaca (Adest, Aguirre-León, Morafka, and Jarchow, 1989b).

Mean adult size in a natural colony is reported as follows: females, 272.5 mm ± 51.9, $n = 24$; males, 262.2 mm ± 31.2, $n = 19$ (Morafka, 1981). Sexual maturity seems to be size (not age) dependent (Germano, 1994b). Estimates of CL at sexual maturity vary. Adest, Aguirre-León, Morafka, and Jarchow (1989a) give 230 mm, and Germano (1994a, 1994b) 280–285 mm. Legler and Webb (1961) observed two live captive males; the larger was 220 mm and behaved as an adult; the smaller was 164 mm and behaved as a subadult.

Mean age at sexual maturity is given by Germano (1994a) as 13.9 years (extremes, 12–17) for female captives. Wild females known to bear eggs had growth-ring counts of 16–21 and CL of 326 to 380 mm (Adest and Aguirre-León, 1995). Precise data on longevity for this species is not available.

Means and extremes for hatchling size are as follows: CL, 45.7 mm (33.7–54.4); CW, 41.4 mm (34.5–47.0); and Wt, 33.1 g (15.4–49.0; Adest and Aguirre-León, 1995). The hatchlings are dark brown to black (as predicted by Legler, 1959); new growth peripheral to the areola is yellowish and remains paler than the areola. Month-old hatchlings from the Sierra del Diablo region had black areolae surrounded by dark brown with only a narrow margin of yellow on the anterior edges of the lateral marginal scutes (Morafka, 1982).

Gopherus flavomarginatus maintains a higher rate of growth for a longer period (12 years) than its congeners. Growth rates are highest (18 mm/y) for individuals in the 8- to 12-year range (170 to 240 mm CL; Germano, 1994a). Growth slows dramatically at about 20 years as adult size is approached. A high quality diet produces dramatic growth rates in captives (Morafka, 1982). Growth rings can be used accurately to age regularly growing individuals and to assess growth rates (Germano, 1994b; Legler, 1960c; Moll and Legler, 1971). The four species of *Gopherus* reach adult size in approximately the same number of years regardless of final CL (Germano, 1994b).

Nearly all information on early juvenile growth comes from a hatchery-nursery at the MBR. Mean growth rates in the first year ($n = 23$) were 48.8% ± 5.7 for CL and 184.7% ± 37.9 for weight. These rates fell to virtually zero during the third year (Morafka, Berry, and Spangenberg, 1997; Adest, Aguirre-León, Morafka, and Jarchow, 1989b). In this period juveniles suffered from various illnesses, especially "knobby shell" (a pyramiding of scute and underlying shell bone) referred to as "severe nutritional osteodystrophy due to calcium deficient diets" (Jackson and Cooper, 1981; Adest, Aguirre-León, Morafka, and Jarchow, 1989b). This problem was solved by feeding immature tortoises fresh green shoots and fresh-cut alfalfa, increasing protein content from 8% or less to 16%. Thereafter, growth was sustained past the third year, survivorship increased markedly, and shell deformities were reduced (Morafka, Berry, and Spangenberg, 1997).

A dissected female (Ceballos, Durango) of 326 mm CL had 19 clear zones of growth and had been reproductive for at least 2 years. The growth record was clear and congruent on all scutes. Calculated hatchling size for that female was 55 mm. Growth increments were 15% in year of hatching, 36% at end of first full year of growth, and 21%, 15%, 10%, and 9% in the following 4 years; this was followed by increments of 3–9% for the 6th through 19th full years. In one year of this period of slower growth (1953), the increment was 16% (Legler and Webb, 1961).

The incidence of scute anomalies and supernumerary scutes in captive-reared juveniles (60%) is about the same as that in free-living individuals (66.5%; Adest, Aguirre-León, Morafka, and Jarchow, 1989b). Legler (1959) reported a high incidence of supernumerary scutes in the small type series.

PREDATORS, PARASITES, AND DISEASES

Humans are the chief predators of bolson tortoises and their eggs (Morafka, Adest, Aguirre, and Recht, 1981). Firsthand accounts of nonhuman predators are actually rare (Morafka, Adest, Aguirre, and Recht, 1981). Local predators at the MBR were listed as coyotes, foxes (*Urocyon*), badgers, skunks (*Mephitis*), road runners (*Geococcyx*), hawks (*Buteo*), and ravens (*Corvus*). Eggs, hatchlings, and other juvenile stages are the most vulnerable life stages. About 90% of the eggs laid are destroyed (Adest, Aguirre-León, Morafka, and Jarchow, 1989b). All the aforementioned animals plus procyonids, large rodents (e.g., *Rattus*), corvids, and snakes could be regarded as potential predators of eggs and young.

Healthy tortoises harbor substantial numbers of nematodes in the gut. This could constitute a symbiotic relationship, with movements of the worms breaking up "the food bolus" (Morafka et al., 1986). Adobe ticks (*Ornithodoros turicata*) have been found attached to interlaminal seams of the carapace and to the skin of the soft parts, including the eyelids (Morafka, 1982; Morafka, Adest, Aguirre, and Recht, 1981).

Common ailments of captives include respiratory infections, nasal and ocular discharges, abscesses, labored breathing, chronic weight loss, diarrhea, and lethargy. These may be bacterial infections. Most infections are produced by opportunistic gram-negative bacilli, some of which are present in the soil (Adest, Aguirre-León, Morafka, and Jarchow, 1989b). A detailed analysis of fecal flora of healthy adult tortoises is given by Adest, Aguirre-León, Morafka, and Jarchow (1989a).

POPULATIONS

Seeing and counting tortoises in natural habitat is difficult. Even large adults are active for only short periods, are wary, and, if near a burrow, can disappear rapidly. Small juveniles are cryptic and seldom seen at all. Trained scientists observed less than one tortoise per person per month outside of areas of known highest tortoise density (Bury, Morafka, and McCoy, 1988). Population composition is categorized by size as follows: juveniles, 117–198 mm; subadults, 199–249 mm; adults, greater than 250 mm. Populations consist chiefly of adults and juveniles, with fewer intermediate size classes. These percentages varied from year to year. Lumped data for several years showed 81% adults, 6.3% subadults, and 12.7% juveniles (Adest, Aguirre-León, Morafka, and Jarchow, 1989b). The preponderance of adults could result from enhanced survivorship in past small cohorts (Aguirre-León, Morafka, and Adest, 1997). Mortality is highest in eggs and small juveniles. An estimated 87% of natural nests are destroyed by predators; 67% of released hatchlings and 69% of 1- to 4-year-old juveniles die in the first year and up to 93% over 5 years. Many of the released tortoises are simply never seen again, making it difficult to account for their fate. Hatching success in a nursery (eliminating completely infertile clutches) is 65–77% (Morafka, Berry, and Spangenberg, 1997; Aguirre, Morafka, and Adest, 1997). At any given point, potential survivors weighed more than those that ultimately perished (Adest, Aguirre-León, Morafka, and Jarchow, 1989b). At least half of the mortality could be attributed to the unnatural aspects of the hatchery-nursery program. Adult survivorship is high—near 100% if human predation is removed from the equation.

No method has been developed for determining the sex of neonates. However, Morafka, Adest, Aguirre, and Recht (1981) reported an adult sex ratio of 44% males to 56% females and considered it not significantly different from 1:1.

CONSERVATION AND CURRENT STATUS

There is no question that bolson tortoises were once much more abundant than they are now. Interesting, compelling, and seemingly accurate anecdotes of this former abundance are reviewed by Morafka (1988). The species was probably endangered long before its scientific discovery in 1958.

Decline is coincident with increasing human populations in tortoise habitat (Bury, Morafka, and McCoy, 1988). Estimates from the two study areas suggest total living animals at no more than 10,000. Most populations are in decline. Human influence on Bolson tortoises is almost entirely negative via direct predation and habitat alteration. Population declines accelerated around 1900, coinciding with the increase in agriculture, highways, and railroads (Aguirre-León, 1995). Human predation has probably occurred since pre-Columbian times and was accelerated by the advent of metal tools used in burrow excavation (Morafka, 1982). Land reform plans have moved people to cooperative agricultural communities (*ejidos*) where they will eventually irrigate, raise crops, and practice grazing. But they begin as hunter-gatherers and take tortoises for food. Even where human populations are relatively sparse (four families in a plain 40 km wide), there is abundant evidence of tortoise excavation and butchering.

Tortugueros come in from substantial distances with horses and pickup trucks. Some Tortugueros use hooks on long poles to pull tortoises out of burrows (Morafka, 1982). Perhaps they will eventually use earth-moving equipment to excavate burrows.

Tortoises have been extirpated nearly everywhere west of highway 49 (to Cd. Chihuahua) and from 10 km strips on either side of the east–west railroad (they were in fact loaded onto railroad cars and shipped to the West Coast at one time). The railroad divides the known range into northern and southern halves (Morafka, 1982).

Some local residents are concerned with conservation. In 1958 Legler spoke to Bolson residents who were surprised that such a large animal had just been "discovered," concerned that numbers were clearly declining, but were hopeful that scientific information would be used to arrest the decline. Pawley (1975) stated that Bolson tortoises had been domesticated and harvested like a crop at Hacienda de los Remedios, Chihuahua, for many years. Morafka (1982) contended that this is not an ancient custom.

Although present reasons for decline are evident, this decline may be a final episode in a process that began at least 11,000 years ago over most of the former range—a process coincident with increasing populations of pre-Columbian humans (Mallouf, 1986; Adest, Aguirre-León, Morafka, and Jarchow, 1989b).

Little legal protection existed before 1978. *Gopherus flavomarginatus* was officially declared an endangered species on 27 November 1978 (Federal Register, Vol. 43 (223): 55314–19; Morafka, 1982). The species has been listed as endangered by the USFWS since 1987 (Morafka, Aguirre, and Murphy, 1994) and is legally protected in Mexico and at the U.S. border (Morafka, 1982). Bolson tortoises were given high priority by the IUCN-Species Survival Commission (SSC) Tortoise and Freshwater Turtle Specialist Group's Action Plan in 1991 (Aguirre-León, 1995).

Protection in Durango was initiated by a coalition of state officials, landowners, and residents, supported by the federal government and Mexican Fauna Silvestre, and managed by the Ecological Institute of Mexico (Morafka, 1982).

Despite evident progress, there is little enforcement of protective legislation even in the study areas. Protection in remote rural areas is difficult where local residents may regard tortoises as a source of food. The MBR provides a variable amount of protection to 10–20% of surviving tortoises (Aguirre-León, 1995).

There are two areas in which field studies are concentrated, where habitat is good to excellent, and in which tortoises are nominally protected. These areas include the two best areas of geographic range defined by Aguirre-León (1995), and both constitute binational (Mexico and United States) cooperative ventures.

An active Bolson tortoise management program was founded at the MBR in 1983 with the objective of achieving baseline studies and establishing a husbandry program whereby eggs and young are protected in predator-proof enclosures and eventually (at age 3–5 years) released in natural habitat. Eggs are harvested from as many females as possible to preserve genetic variability. This approach has the potential to accelerate recovery of an endangered population (Aguirre-León, 1995; Morafka, Berry, and Spangenberg, 1997). Active management through a hatchery-nursery program probably produces the equivalent of the one good year per decade that might be necessary for the recruitment of a cohort (Morafka, Berry, and Spangenberg, 1997).

The Laboratorio del Desierto, more often referred to as the Mapimi Biosphere Reserve (MBR), was established in the early 1980s in northern Durango. Its center is located 33 km E of Ceballos, Durango (26°20′N–26°52′N, 103°58′W–103°32′W; González-Trapaga, Aguirre, and Adest, 2000). The MBR has a well-equipped laboratory, a paid support staff, and facilities for resident and visiting investigators. The husbandry program and the hatchery-nursery facilities are located at MBR. Tortoise populations are estimated at 1,500–2,000 in densities from 0.1 to 3.0 tortoises/ha (Aguirre-León, 1995). There is voluntary human abstinence from hunting on the MBR, and this has been an important element of a slow recovery program (Aguirre-León, Morafka, and Adest, 1997). Management procedures are reviewed by Adest, Aguirre-León, Morafka, and Jarchow (1989b) and Adest, Jarchow, and Brydolf (1988).

Rancho Sombreretillo, established as a study area in 1985, is a privately owned ranch at the northern edge of the Bolsón de Mapimí and the northern edge of the geographic range of the species. The location is ca. 99 km N and 17 km E of Ceballos, Durango (coordinates, 27°25′N, 103°58′W; 1,254 m elevation), bounded by the Sierra de Diablo to the west, the Cerros Emilio and the Sierra de Almagre to the northeast, and the Sierra Mojada to the southeast (Trevino, Morafka, and Aguirre-León, 1995). The area contains 10,000 ha of prime tortoise habitat and a robust population of *G. flavomarginatus*, with medium- to high-density and ample evidence of nests, hatchlings, and other juvenile age classes (Aguirre-León, 1995; Trevino, Morafka, and Aguirre, 1997). There is a full-time caretaker. A consortium of conservationists provide technical assistance, a scientific advisor, a patrol vehicle, and improved communications equipment (Trevino, Morafka, and Aguirre, 1997).

Other research sites include various private tortoise keepers and the Research Ranch in Elgin, Arizona (Aguirre-León, Morafka, and Adest, 1997). The captive colony of 37 bolson tortoises at the Research Ranch (Appleton Ranch) was transferred, in September 2006, to the Armendaris Ranch in central (Sierra Co.) New Mexico with the objective of restoring wild populations from captive breeding. Edwards et al. (2009a) examined mtDNA and autosomal microstellite loci in this colony, in wild populations in Durango, and in captives at the El Paso Zoo with the objective of measuring genetic diversity as it pertained to management of the Research Ranch colony. They found that the colony possessed 97.5% of the "total genetic diversity" thus far known for the species—an important matter for maintaining diversity in a species that has been in decline for about a century. The Armendaris Ranch is ca. 01°35′N and 02°41′W of the old Reserch Ranch.

Aguirre-León, Morafka, and Adest (1997) found no evidence of recolonization in areas where extirpation had occurred, nor of recruitment in the very small populations west of highway 49, and stated that the post-Columbian extirpation is a trend "unlikely to be reversed."

Morafka (1982) predicted that the species could be totally extinct by the year 2000. As this book is published, the status of this species remains endangered, but its extinction has probably been arrested on the two reserves. We can probably be optimistic that *Gopherus flavomarginatus* will not become extinct if current efforts continue at present levels.

Family Emydidae

Pond Turtles

EMYDID CHELONIANS IN MEXICO

There is no satisfactory vernacular name that applies accurately to all members of the Family Emydidae Rafinesque, 1815. Emydids are a natural group but are far more diverse than the testudinids. Diet ranges over the entire scale of omnivory, and habits range from completely terrestrial to completely aquatic.

As treated here with full family status, the emydids are a chiefly western hemisphere group, ranging from Canada (ca. 52°N) to temperate South America (ca. 36°S; Legler, 1990). The single exception is the genus *Emys* (northern Africa and Eurasia). The family is comprised of 12 genera, 39 species, and a total of 93 terminal taxa (Bickham et al., 2007). The oldest fossils are from the Upper Cretaceous of Montana.

DIAGNOSIS AND DESCRIPTION OF FAMILY

(Based on Williams, 1950; Loveridge and Williams, 1957; Auffenberg, 1974; Gaffney, 1975a; Smith and Smith, 1979; Ernst, Lovich, and Barbour, 1994.)

The families Emydidae and Geoemydidae share the following characters, most of which distinguish them from the Testudinidae: Rib heads are well developed, with the proximal part of rib forming floor of a substantial epaxial space. No carapacal kinesis is known; there is well-developed plastral kinesis in some genera. Skeletal mass and shell thickness are usually greater than those in testudinids. Limbs are generalized, and the feet usually have interdigital webbing; hind feet are never elephantine; there are three phalanges in digits 2–4 of pes; osteoderms in soft skin are rare; there are no spurs on thigh.

Trochanteric fossa of femur are large. Palate is flat. Tongue is simple, flat, and small. Hyoid skeleton is well developed. Descending processes of prefrontal bones define a narrow ethmoidal fissure. Quadrate is not completely closed posteriorly and does not completely surround columella. Premaxillary bones form a beak, a notch, or a notched beak. Skin on head is smooth or has small scales posteriorly. Tail is longer than that of testudinids, especially in young.

Musk glands are present or absent; cloacal bursae are usually present. Nuchal costiform processes are variable, usually evident in young, and are seldom more than vestigial in adults; there are 11 peripheral bones, 12 marginal scutes, and 6 pairs of plastral laminae.

DIAGNOSTIC EMYDID CHARACTERS

Emydids can be distinguished from geoemydids as follows: There is a double articular socket between fifth and sixth cervical centra. Angular bone forms floor of canal for Meckel's cartilage. There is no strong lateral tuberosity (the "batagurine process" of McDowell, 1964) on basioccipital bone.

Five genera of emydids occur in Mexico: *Actinemys*, *Chrysemys*, *Pseudemys*, *Trachemys*, and *Terrapene*.

Genus *Actinemys* Agassiz, 1857

Pond turtles; tortugas de Charcos (Liner, 1994).

FIGURE 34.1 *Actinemys marmorata*. Head and forequarters of adult female, second from right in Figure 34.2 (Arroyo Santo Tomás, Baja California Norte). (Courtesy of T. Akre.)

ETYMOLOGY

Gr. *aktis, actinos*, ray or beam. Perhaps in reference to basking in the sun; more likely alluding to the pattern of radiating dark lines on carapacal scutes at younger ontogenetic stages ("radiating striae," Agassiz 1857:444; see also Buskirk, 2002; Vetter, 2004).

RELATIONSHIPS, TAXONOMY, AND NOMENCLATURE

Until 1964 the genus *Clemmys* (Ritgen, 1828) was perceived to include a variety of Old World forms that are now considered to be batagurines and have been placed in other genera (Boulenger, 1889; Carr, 1952; Loveridge and Williams, 1957; Wermuth and Mertens, 1961). McDowell (1964) achieved a more realistic partitioning of the genus using morphological characters (chiefly the skull). More recent works (Iverson, 1992a; Ernst, Lovich, and Barbour, 1994; Smith and Smith, 1979) have considered the genus *Clemmys* to include four North American species—*guttata, marmorata, insculpta*, and *muhlenbergi*. Even this more logical arrangement included a tremendous amount of morphological diversity.

It is now reasonably clear that these four species are not as closely related as once thought, and that *Clemmys* is paraphyletic (Bickham, Lamb, Minx, and Patton, 1996; Lenk, Fritz, Joger, and Wink, 1999; reviewed by Feldman and Parham, 2002). *Actinemys* is more closely related to *Emys* and *Emydoidea* than to the other three species of "*Clemmys*" (sensu lato; Holman and Fritz, 2001). The revised taxonomy is as follows: *Clemmys guttata*; *Actinemys marmorata*; *Glyptemys muhlenbergi*; *Glyptemys insculpta*. We follow Bickham, Lamb, Minx, and Patton (1996) as do most other authors since that time (e.g., Jones, 2010; Tuttle and Carrol, 2005).

Actinemys marmorata (Baird and Girard) 1852

Pacific pond turtle (Iverson, 1992a); tortuga de Charcos (Liner, 1994); tortugita (Grismer, 2002).

ETYMOLOGY

L. *marmor*, marble, alluding to the marbled or reticulate pattern often present on the shell and soft parts; L. *pallidus*, pale, alluding to the paler color of the southern subspecies.

HISTORICAL PROLOGUE

Actinemys marmorata is yet another example of a North American species that barely enters Mexico, has been virtually unstudied there (Grismer, 2002), but is fairly well known in the United States (Holland, 1994; Bury, 1972; Buskirk, 1990; Lovich and Meyer, 2002; Buskirk, 2002). Noble and Noble (1940) based a useful laboratory manual of chelonian anatomy on this species and included notes on the reproductive cycle and embryonic development. Grismer (2002) summarizes most of what is known of the species in Mexico.

DIAGNOSIS

The taxon is distinguished from all other chelonians in its Mexican range by having five foreclaws, nonelephantine hind limbs, epidermal laminae on the shell, and lack of bright coloration in the form of postorbital patches and/or stripes on head and neck (Grismer, 2002).

GENERAL DESCRIPTION

Actinemys marmorata is of medium size (slightly larger than *Chrysemys* and smaller than most *Trachemys*) and has the low, smooth shell profile and cross-section typical of other aquatic emydids in Mexico. Overall maximum size is 210 mm CL and ca. 1,200 g, but rarely exceeds 160 mm (Holland, 1994; Buskirk, 2002).

We have neither examined nor measured specimens from Mexico.

BASIC PROPORTIONS

Mojave R. California population, raw data courtesy of J. Lovich (see Lovich and Meyer, 2002):

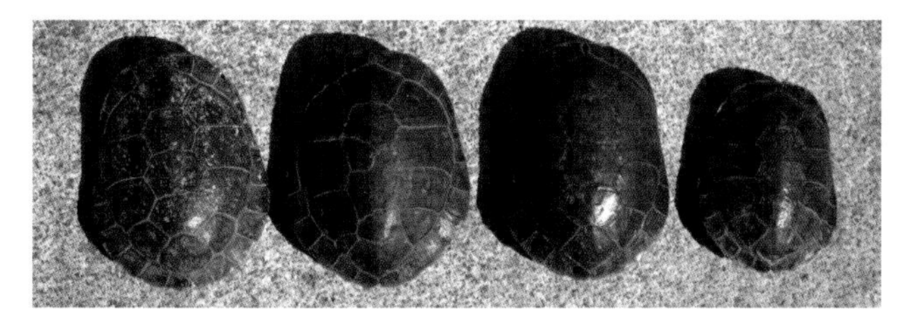

FIGURE 34.2 *Actinemys marmorata*. Series of four individuals from Arroyo Santo Tomás, Baja California Norte. Dorsal and ventral views from left to right: adult female, adult male, adult female, immature male. (Courtesy of T. Akre.)

CW/CL: males, 0.78 ± 0.047 (0.68–0.89) $n = 16$;
females, 0.79 ± 0.023 (0.75–0.84) $n = 18$.

CH/CL: males, 0.34 ± 0.012 (0.32–0.36) $n = 16$;
females, 0.40 ± 0.020 (0.36–0.44) $n = 18$.

CH/CW: males, 0.44 ± 0.026 (0.39–0.48) $n = 16$;
females, 0.50 ± 0.021 (0.46–0.55) $n = 8$.

Modal plastral formula (from photos): $6 > 3 > (4 = 5) > 1 > 2$.

Carapace is relatively broad and smooth and is widest behind bridge and highest at C2 or C2–3; mid-dorsal keel is present only in young. Plastron is extensive, nonkinetic, and slightly shorter than carapace; there is a shallow, obtuse pygal notch at posterior edge of carapace and an extremely wide, obtuse anal notch on plastron. Edges of shell are not serrated. There is an extensive bridge with plastral buttresses suturally united with carapace. Interanal seam is clearly longest on plastron; other plastral scutes are of more or less subequal length (Holland, 1994), axillary scute is small, and inguinal scute is absent in Mexican populations; pectorals and abdominals contact marginals 4–7 on a wide bridge.

Triturating surfaces of jaws are narrow and unridged; apex of maxillary sheath is notched but not hooked; jugal is not in contact with pterygoid; maxilla reaches border of temporal notch; frontal reaches orbital rim; posterior palatine foramina are enlarged and fenestra-like; caroticopharyngeal foramen is large, on or near pterygoid-basisphenoid suture.

There are five foreclaws and four hind claws; digits are fully webbed (to bases of claws). Head is rather large and flat topped; tail is moderately long and slender with irregular whorls of scales; tympanum is elliptical; skin of neck and gular region is granular with several loose folds.

Cloacal bursae are well developed; there are one to three rostral pores (Winokur and Legler, 1974); buccopharyngeal mucosa have short, glandular lingual papillae (Winokur, 1988); there are no mental glands (Winokur and Legler, 1975); there is one musk gland in anterior sternal cavity on each side with duct passing to one or two orifices beneath M3 and/or M4 (Waagen, 1972).

Karyotype: $2n = 50$; there are 16 macrochromosomes with median or submedian centromeres, 10 macrochromosomes with terminal or subterminal centromeres, and 24 microchromosomes (Bickham, 1975).

Top of head is dark and covered with reticulum of pale spots; ground of chin and throat is cream with varying degrees of dark longitudinal striping or spotting; upper surface of limbs is dark with weak indications of striping; undersides of limbs are black to cream with various intermediate patterns; tail is dark with black vertebral stripe; carapace is brown to olive green, with or without darker markings; many individuals have dark lines radiating out from areolae; plastron is olive green to cream; there are dark markings on plastron usually confined to interlaminal seams; entire central region of plastron is dark in some individuals. Specimens from low elevations in coastal regions (e.g., Arroyo San Telmo, Arroyo Santo Domingo, and Arroyo Grande) are dark in overall ground color, whereas individuals at higher elevations have a paler ground color. At higher elevations (e.g., Las Cruces Spring in Sierra San Pedro Martir) the paler ground color accentuates darker patterns on the limbs and tail (Grismer, 2002). Except where cited, diagnosis and description are adapted from McDowell (1964), Bury and Ernst (1977), Carr (1952), Mlynarski (1976), and Smith and Smith (1979).

Seeliger (1945) described two subspecies, the nominate *marmorata* in the north (western Washington to San Francisco Bay and western Nevada) and a new subspecies, *pallida*, in the south (south of San Francisco Bay), with integrading populations in central and southern California. Storer (1930) is seemingly the first to record *A. marmorata* from Mexico. Thus far, the Baja California populations have been referred to as *A. m. pallida*. However, Seeliger (1945) was unable to assign Mexican specimens to either of the two subspecies, and Janzen, Hoover, and Shaffer (1997), studying

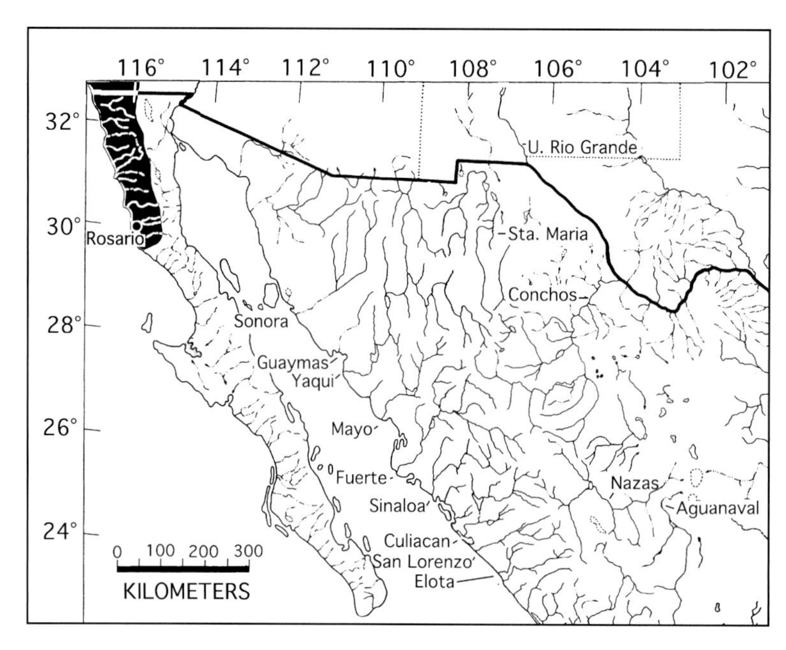

MAP 23 Geographic distribution of *Actinemys marmorata*.

cytochrome b, suggested that the Mexican populations could represent an undescribed taxon (Grismer, 2002).

FOSSILS

Fossil remains are known from Pliocene (Blancan) deposits in California and Oregon and the Pleistocene (Irvingtonean, Rancholabrean) of California and Washington. Remains in archeological sites in California suggest that American Indians ate these turtles (Ernst, Lovich, and Barbour, 1994).

GEOGRAPHIC DISTRIBUTION

Map 23

The species has been extirpated in extreme southwestern British Columbia and now occurs west of the Cascade, Sierra Nevada, and Mexican Peninsular ranges, from western Washington southward to about 31°N in northern Baja California (Bury, 1970; Iverson, 1992a; Ernst, Lovich, and Barbour, 1994). Isolated populations existed in the Truckee, Carson, and Humbolt Rivers of Nevada, but may now be extinct (Buskirk, 1990). An isolated population exists in the internal Mojave R. drainage of California as far north as Afton Canyon (Lovich and Meyer, 2002).

In Baja California Norte *A. marmorata* occurs naturally in Pacific streams draining the Sierra San Pedro Martir, at least from the Rio San Carlos southward to Arroyo Grande (31°04′N), east of El Rosario—the southernmost natural distribution of a wide-ranging species in western North America (Grismer, 2002). Recent studies have further detailed the natural distribution of this species in Baja California Norte (Lovich et al., 2005, 2007).

There are reports of former occurrence as far south as ponds below Mission San Fernando Velicata (ca. 129 km south of present known distribution); the ponds were silted in following a hurricane in 1978, and no turtles have been seen since. The same thing occurred in the lower ponds of Arroyo San Telmo in 1992–1993. The species has been introduced in ponds of the Arroyo Grande at Rancho Metate, 50 km E of Rosario near the present southern edge of range (Grismer, 2002). Populations still exist in remote riparian habitats at the base of the Sierra San Pedro Martir (J. R. Buskirk, pers. comm.; Figure 34.3).

Distribution was much more extensive in the Pliocene and Pleistocene. As the climate became drier in the last 10,000 years, the range contracted and shifted westward along the Pacific coast, leaving relict populations in the deserts of western Nevada, the Mojave River, and Baja California (Grismer and McGuire, 1993). Gene flow between isolated desert populations is probably minimal.

HABITAT

Various aquatic habitats are mentioned in the literature, but the species seems to prefer permanent pools in smaller streams and occurs to altitudes of about 1,830 m. It is known to occur occasionally in brackish estuarine habitats (Smith and Smith, 1979; Ernst, Lovich, and Barbour, 1994). Welsh (1988) found an *A. marmorata* in deep pools of a perennial stream in Baja California (Arroyo Santo Domingo, Rancho San Antonio, at base of the western scarp) near the southern edge of the range. This was an isolated aquatic habitat in distinctly Sonoran desert with large cacti occurring near the water. Desert populations are thoroughly aquatic and are as dependent upon water as are desert fishes.

Lovich and Meyer (2002) studied the species in the internal Mojave River drainage of San Bernardino Co., southern California (ca. 35°N, 117°W, 250 km NW of known Mexican

FIGURE 34.3 *Actinemys marmorata*. Habitat, Arroyo Santo Tomás, Baja California Norte. (Courtesy of T. Akre.)

populations), where mean annual precipitation is 112 mm. Riverside vegetation, once dominated by cottonwoods and willows, has been replaced by tamarisk. The turtles occur in permanent pools with maximum depth of 1.5 m and total surface area of 0.25 ha.

DIET

The species is generally recognized as an opportunistic omnivore, feeding on algae, various plants (including pads of water lilies), snails, crustaceans, isopods, aquatic insects, and carrion. Feeding on small organisms in the surface film (neustophagia) has been observed (Holland, 1994; Ernst, Lovich, and Barbour, 1994). Grismer (2002) observed feeding on aquatic vegetation and carrion in Baja California.

HABITS AND BEHAVIOR

Many reports indicate that pond turtles are wary and retreat from basking sites with little provocation, concealing themselves in bottom mud or beneath undercut banks. Grismer (2002) observed activity in Baja California from February to early December.

In the arid southern part of the range (e.g., Mojave R. and probably Baja California) turtles seldom leave the water except to nest and bask. In more northern parts of the range they commonly spend 1–2 months per year in terrestrial habitats, and some individuals in Oregon overwinter in terrestrial refugia for as long as 8 months (Holland, 1994; Lovich and Meyer, 2002).

Aerial basking is known throughout the range. Bury (1972) found that basking occurred after a period of foraging, but mostly in a period of 1 hour at mid morning. Temperatures higher than 34°C are avoided and basking is terminated near this temperature. Critical thermal maximum is around 40°C (Ernst, Lovich, and Barbour, 1994).

REPRODUCTION

Some females reproduce at a CL of 111 mm and an age of 6–7 years, but most do not reproduce until they are 120 mm and at least 8–10 years old. In the Mojave R. population females contained shelled eggs from 26 May to 14 July (Lovich and Meyer, 2002). Elsewhere nesting occurs from April through August (Ernst, Lovich, and Barbour, 1994). Holland (1988) provides a detailed description of a partial mating sequence in San Luis Obispo Co., California.

Shelled eggs may be retained in the oviducts for up to 3 weeks before being laid. Egg-laying movements in the Mojave R. drainage were oriented toward the dry river channel. Nesting females left the water for periods of 0.83–86 hours. One confirmed nest was made under a bush on 11 June less than 5 m from water. It was flask shaped and ca. 70 mm deep (Lovich and Meyer, 2002).

Data on eggs over the entire geographic range are as follows: L, 30–43 mm; W, 19–23 mm; Wt, 8.3–11.1 g. The yolk is 27% lipid. The egg shells are hard (Ernst, Lovich, and Barbour, 1994).

Natural incubation periods range from 80 to 122 days in California and 94 to 106 days in Washington (Holland, 1994). Recorded laboratory incubation times range from 60 to 130 days. Eggs incubated at 25–33°C (Lardie, 1975b; Feldman, 1982) hatched in 73–81 days. Temperatures in a natural nest cavity in July may range from 19°C to 30°C within 24 hours. Feldman (1982) observed that hatchlings did not leave the egg if the temperature exceeded 27°C; when moved to a cooler temperature they emerged within 2–3 hours. Pond turtles have TSD, with a threshold temperature at about 30°C under constant incubation temperatures in the laboratory; males are produced below this temperature and females above it (Ewert, Jackson, and Nelson, 1994). Eggs in moist incubation medium are said to "rupture," presumably from the inability of the hard shell to expand (Feldman, 1982; Holland, 1994; Holte, 1998; Buskirk, 2002).

Hatchlings in northern populations overwinter in the nest, shed the caruncle during that time, and emerge in spring. Of 22 hatchlings found in California and northern Mexico, 77% were found in April and May (Buskirk, 1990). Some growth could be expected to occur in the nest before emergence.

Number of eggs per clutch is 1–13 over the entire range of species and is positively correlated with carapace length. Mean eggs per clutch for 12 Mojave females was 4.58 ± 0.90 (4–6). Mean size of gravid females was 144 mm ± 0.98 (133–160) $n = 10$ (Lovich and Meyer, 2002). Many females skip seasons of reproduction. Some females in all populations have the potential to produce up to two clutches in the same year with known intervals between clutches of 38–41 days (Stebbins, 1985; Holland, 1994; Goodman, 1997; Ernst, Lovich, and Barbour, 1994).

GROWTH AND ONTOGENY

Data on size in the Mojave R. population: males, 140 ± 11.0 (118–155) $n = 16$; females, 144 ± 8.2 (132–160) $n = 18$;

hatchlings, 26.4 ± 1.30 (25.6–27.9) $n = 3$. Weights recorded for adults in the same population were as follows: males, 406 ± 79 (265–533) $n = 18$; females, 523 ± 80 (400–750) $n = 18$ (Lovich and Meyer, 2002).

Growth rings can be read with fair accuracy up to about 120 mm, at which point the shell becomes smooth (presumably because of scute shedding and wear). Most individuals in the 100 to 110 mm group are 4–5 years old but can range from 3 to 12 years at this size. Ages of 40–50 years for individuals in the wild have been recorded (Holland, 1994).

There is no sexual dimorphism in size (Buskirk, 2002). In males the plastron is concave, the tail is thick at base, the vent usually is posterior to the rear margin of the carapace, and the forelimbs and throat tend to be predominantly yellowish. Females have a flat or convex plastron, an unthickened tail base, a vent that is even with or anterior to the posterior edge of the carapace and forelimbs, and throat tending to be more nearly brownish or grayish.

Overall, hatchlings are 25 to 31 mm CL and weigh 3–7 g (Holland, 1994). Buskirk (1990) records a 27 mm hatchling (with caruncle) from Baja California. Fully developed but unhatched young in the Mojave R. population had mean CL of 26.4 mm ± 1.3 (25.6–27.9; (Lovich and Meyer, 2002). Small juveniles are nearly circular in dorsal aspect.

Neonates are essentially patternless (Agassiz, 1857, pl. III; Buskirk, 2002), having a dark unmarked carapace and a plastron with pale (yellowish) ground color bearing a dusky central nonconcentric figure that radiates variably along the plastral interlaminal seams; the undersides of the marginals are pale, and this color may be visible in dorsal and lateral views. There are vague blotches and stripes on the limbs and on the head and neck. As growth begins there develops a carapacal pattern of alternating pale and dark lines that radiate out from the areola of each scute, in the direction of epidermal growth. Buskirk (2002, Figure 15) illustrates the variation of this radial pattern from predominantly pale with narrower dark radiations to predominantly dark with fine yellow radiations on an essentially black ground. Age, wear, and probable secondary melanism may produce various combinations in which the markings are lost and the shell becomes unicolored pale or dark. The original plastral figure seems at first to become restricted to interlaminal seams, but adult plastra vary from nearly uniform pale to uniform black. Buskirk (2002) shows colored photos that depict most of this ontogenetic variation. In general, the species is the least colorful emydid in Mexico.

PREDATORS, PARASITES, AND DISEASES

There are few reports on natural predation in the southern part of the range. Lovich and Meyer (2002) recorded ants as nest predators. Holland (1994) gives an impressive list of known predators that includes fishes, bullfrogs, large birds, large insects, raccoons, skunks, coyotes, and red fox as nest and general predators. The species has been popular with the pet trade. Bury (1989) reported the removal (and shipment to Europe) of some 500 specimens from a southern California lake, and Buskirk (2002) chronicles its extermination, chiefly through habitat loss, in large areas of California (e.g., San Joaquin Valley). Pacific pond turtles were considered a delicacy for human consumption in U.S. West Coast cities at least until the 1930s (Storer, 1930; Holland, 1994).

POPULATIONS

In the Mojave River population density was estimated to be 50–62 turtles/ha at two different localities, and standing crop biomass to be 26.23 kg/ha. Sex ratio was essentially equal. All turtles observed were adults greater than 117 mm. Absence of immature stages in the population could signal either lack of recruitment or that juveniles are difficult to find. In other populations typical densities are in the range of 23–214 turtles/ha, and males are reported to outnumber females 1.7:1.0 (Ernst, Lovich, and Barbour, 1994; Holland, 1994). Survivorship in the first 3 years is 10–15%, but this becomes higher at sizes of greater than 120 mm. Annual mortality of adults is estimated at 3–5%. Up to 99% of the nests are predated (Holland, 1994).

CONSERVATION AND CURRENT STATUS

Populations are declining in southern California and over most of the northern part of the range, chiefly because of degradation or outright destruction of wetlands habitat. Reduction in numbers of turtles in areas of former abundance is estimated at greater than 90% (Buskirk, 2002). Welsh (1988) commented on the degradation of riparian habitats for agricultural purposes in Baja California. Some populations in Baja California survive by being remote.

At present, only northern California and southern Oregon support extensive populations. The species is protected by the California Department of Fish and Game as a "species of special concern" (Jennings and Hayes, 1994). The USFWS was petitioned to list *Actinemys marmorata* as an endangered species (Federal Register, 1992; Ernst, Lovich, and Barbour, 1994; Buskirk, 2002). In Mexico the species is protected by the same laws as all other reptiles and amphibians. The tenuous status of *A. marmorata* demands conservation action (Lovich and Meyer, 2002).

Chrysemys picta belli (Gray) 1831

Western painted turtle (Iverson, 1992a); tortuga pintada occidental (Liner, 1994).

ETYMOLOGY

L. *pictus*, painted. The subspecies name is a patronym honoring Thomas Bell (1782–1880) who was a contemporary of Gray.

FIGURE 35.1 *Chrysemys picta belli.* Typical head and neck pattern; adult from upper Mississippi drainage. (Courtesy of R. M. Winokur.)

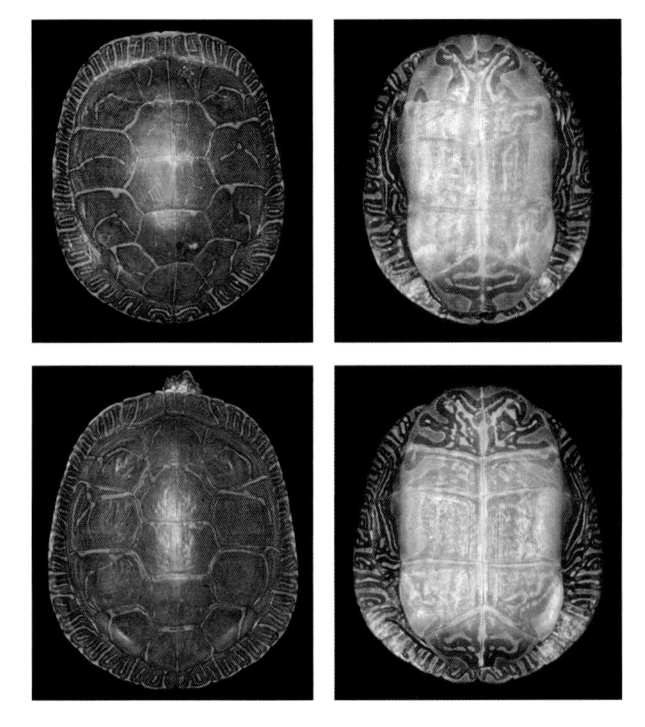

FIGURE 35.2 *Chrysemys picta belli.* Dorsal and ventral views of adult shells, Rio Santa Maria near Galeana, Chihuahua (UU). Top, male; bottom, female.

HISTORICAL PROLOGUE

This account concerns the painted turtles of the monotypic genus *Chrysemys* (sensu stricto). The single species, *Chrysemys picta* (Schneider, 1783), has several subspecies with wide distribution in Canada and the United States. *Chrysemys picta belli* is the taxon occurring in northern Mexico.

Gray (1831) first described this taxon as *Emys bellii* without stating a type locality. The type, purchased from Ashton Lever in 1806 and placed in the collection of the Royal College of Surgeons (London), was destroyed by wartime bombing in 1941. The type locality was restricted to Manhattan (Riley Co.) Kansas by Smith and Taylor (1950b) and later "designated" by Schmidt (1953) as "Puget Sound, Washington." The combination "*Chrysemys picta bellii*" was first used by Bishop and Schmidt (1931).

Strauch (1890) first recorded *Chrysemys* ("*Clemmys bellii*") from "Mexico." Stejneger and Barbour (1917) included "northern Mexico" in the range of "*Chrysemys marginata bellii*" and may have based this on Strauch's account. Smith (1939) established the first definite locality in Mexico by reporting a juvenile specimen (31 mm CL) obtained in 1902 by C. M. Barber for the Field Museum in the Rio Santa Maria (FMNH 2440; Smith and Smith, 1979).

DIAGNOSIS

Chrysemys is one of three closely related genera of aquatic emydid turtles in Mexico (*Chrysemys*, *Pseudemys*, and *Trachemys*), all of which have a bright pattern of pale parallel stripes on the head and neck. *Chrysemys* is the smallest of these taxa in Mexico. This bright pattern in combination with a hingeless plastron distinguishes the three genera from *Actinemys* and *Terrapene*, respectively. *Chrysemys* is unique in having a deep median notch in the maxillary beak that is flanked by a pair of sharp cusps.

GENERAL DESCRIPTION

(Diagnosis and description based chiefly on Carr, 1952; and Ernst, Lovich, and Barbour, 1994.)

Chrysemys picta belli attains a larger adult size (female CL up to 250 mm) than any of the other subspecies. There is a cline in adult size, with the largest adults occurring in the north, the smallest in the south.

BASIC PROPORTIONS

(Rio Santa Maria, Chihuahua; Legler Database)

CW/CL: males, 0.78 ± 0.02 (0.74–0.80) $n = 5$;
females, 0.79 ± 0.017 (0.75–0.82) $n = 12$;
im., 0.84 ± 0.026 (0.82–0.86) $n = 2$.

HT/CL: males, 0.35 ± 0.02 (0.33–0.38) $n = 5$;
females, 0.37 ± 0.014 (0.34–0.39) $n = 12$;
im., 0.35 ± 0.006 (0.34–0.35) $n = 2$.

HT/CW: males, 0.45 ± 0.028 (0.43–0.50) $n = 5$;
females, 0.47 ± 0.019 (0.43–0.49) $n = 12$;
im., 0.41 ± 0.006 (0.41–0.42) $n = 2$.

Scute 4 is nearly always longest; scute 2 is nearly always shortest.

Modal plastral formula: 4 > 6 > 1 > 3 > 5 > 2.

A large, dusky, concentric, central plastral figure covers more that one-half the plastral surface. Central and lateral laminae are staggered, and their interlaminal seams are not aligned; a mid-dorsal pale line is inconspicuous or lacking; carapacal scutes are lacking wide pale anterior margins; there is often a melanistic reticulation on carapace; there is less red or duller red on marginals than in other subspecies.

FIGURE 35.3 *Chrysemys picta belli*. Head and neck patterns of four adult males (left) and three adult females (right); Rio Santa Maria near Galeana, Chihuahua (UU).

The tail is longer in *Chrysemys* than in other Mexican emydids; this is especially evident in juveniles, but has never been quantified.

Karyotype: $2n = 50$; there are 26 macrochromosomes (16 metacentric, 6 submetacentric, 4 telocentric) and 24 microchromosomes (Stock, 1992; Killebrew, 1977a).

Chrysemys is closely enough related to *Trachemys* and *Pseudemys* to have generated substantial controversy over its generic standing (see account of *Trachemys*). The most recent molecular and morphological analyses (Seidel, 2002; Shaffer, Meylan, and McKnight, 1997; Feldman and Parham, 2002) agree on its allocation as a monotypic genus and the distinction of all three genera.

FOSSILS

No fossils are known for Mexico. Fossils from within the present range of *C. p. belli* are from the late Miocene of Nebraska, the Pliocene of Kansas and Texas, and the Pleistocene of Kansas (Ernst, Lovich, and Barbour, 1994).

GEOGRAPHIC DISTRIBUTION

Map 24

Chrysemys picta has a great total geographic range in Canada, the United States, and adjacent Mexico (ca. 30°N–52°N) and is the only species to occur naturally from coast to coast across the northern United States and southern Canada. Painted turtles and snapping turtles (*Chelydra serpentina*) occur near the northern limits for oviparous reptiles (Legler and Georges, 1993). Distribution to the south is essentially west of the Mississippi and east of the Rocky Mountains on the Great Plains. In the extreme southwestern parts of the range there are seemingly disjunct populations in Utah, Arizona, Colorado, and New Mexico. Bleakney (1958) proposed that the ancestral stock of *C. p. belli* had been isolated from the

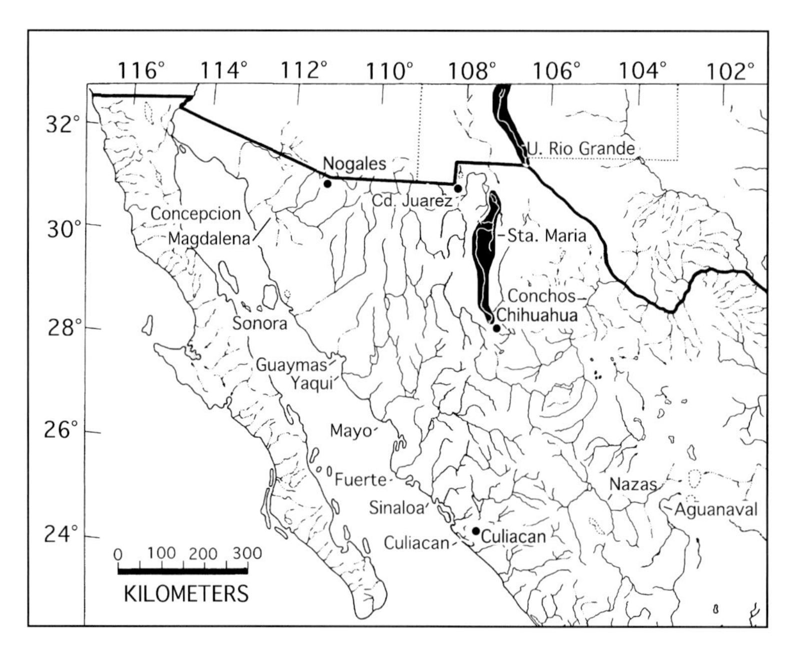

MAP 24 Geographic distribution of *Chrysemys picta belli*.

other subspecies during Wisconsin glaciation, with gene flow being reestablished by glacial recession.

Distribution in Mexico constitutes the southernmost part of the range and is restricted to the small drainage area of the Rio Santa Maria, terminating about 75 km S of Columbus, New Mexico. The Rio Santa Maria is an isolated internal drainage system lying between and closely paralleled by two other such drainages in the Laguna de Bustillos Basin—the Rio Carmen to the east and the Rio Casas Grandes to the west. Each flows northward to a terminal playa lake within 100 km of the Mexican-U.S. border. These internal drainage systems were part of the Pleistocene Upper Rio Grande drainage system (see account of *Trachemys scripta gaigeae* and Legler, 1990).

Known records for *Chrysemys picta* occur in the mid to lower reaches of the Rio Santa Maria (between the latitudinal intersects of 29°50′N and 30°50′N) and are concentrated near Galeana, Chihuahua. There are seemingly no records south (upstream) from Buenaventura nor any from the Laguna de Santa Maria. Trapping, observation, and substantial input from local residents in early September 1960 produced no evidence that painted turtles occurred in the Carmen or Casas Grandes drainages (Legler, pers. obs.). The record of Van Devender and Van Devender (1975) for the Rio Casas Grandes is in error according to Smith and Smith (1979).

This species has been taken or observed in the Rio Grande near El Paso (Smith and Smith, 1979; Dixon, 2000), raising the possibility that it could occur in the few small Chihuahuan tributaries of that river downstream from El Paso. Degenhardt, Painter, and Price (1996) show abundant records for *C. picta* in the Pecos River of New Mexico, but Dixon (2000) shows no records for the Pecos in Texas.

Chrysemys picta is the only aquatic emydid in the Santa Maria drainage of Chihuahua. The subspecies occurs in sympatry with a *Trachemys* and/or a *Pseudemys* in the upper Rio Grande and Pecos drainages (Ernst, Lovich, and Barbour, 1994).

HABITAT

Painted turtles prefer still water with mud bottom, basking sites, and abundant aquatic vegetation (Carr, 1952; Ernst, Lovich, and Barbour, 1994). They occur in ponds, lakes, and the open water of marshes, and can be seen basking on logs and other objects protruding from the water. In the Rio Grande drainage of New Mexico, they occur in rather dense populations in borrow pits along roads and railways paralleling the river, and seem to eschew the moving water of the river proper (Degenhardt, Painter, and Price, 1996; Legler, pers. obs.).

In early September 1960 the Rio Santa Maria (4.8 km N and 3.2 km W of Galeana) consisted of isolated pools of cool water (ca. 15.5°C), 30 to 100 m long and 9 to 15 m wide. The bottom consisted of about 30–45 cm of gray mud. No submergent or emergent aquatic vegetation was evident. One bank was high and stabilized by grass and tree roots. The pools were probably spring fed. Many large cottonwood trees had toppled into the water, forming ideal basking sites. In one such pool there was an evident abundance of painted turtles—as many as 50 individuals could be seen at one time, either basking on the logs or with their heads above the surface. Baited traps captured 54 *Chrysemys picta* and 24 *Kinosternon hirtipes* during a 2-hour period.

The Laguna de Santa Maria proper lies 150 km downstream from Galeana and was dry (could be driven on with vehicles) in early September 1960. There were two large, spring-fed marshes at the base of a rocky area near the northwesternmost point of the basin. These contained large areas (ca. 1 ha) of open water 1 m or less deep, abundant emergent and submergent vegetation, and a mud bottom. No turtles were seen or trapped in these marshes. At Buenaventura, 40 km upstream from Galeana, the river is clear and fast with rocky riffles and a few deep holes beneath undercut banks. No turtles were observed or caught. The source of the Santa Maria is ca. 160 km upstream from Buenaventura on the eastern face of the Sierra Madre Occidental (ca. 50 km NW of Cuauhtemoc).

DIET

All accounts agree that diet is opportunistically omnivorous, the emphasis changing with seasons and whatever is available in abundance (Carr, 1952; Ernst, Lovich, and Barbour, 1994).

HABITS AND BEHAVIOR

Chrysemys picta has been well studied over a wide latitudinal range in Canada and the United States and may be the best-known emydid species ecologically. This knowledge has produced a prodigous body of literature. The most complete modern review of the natural history is by Ernst, Lovich, and Barbour (1994). Although there is little first-hand information for Mexico, extrapolations can be made from studies in southern New Mexico.

Painted turtles are diurnal and rest or sleep at night on the bottom or on submerged objects. Basking is common, especially before and after foraging. In northern populations painted turtles may remain dormant for 5–7 months of the year, but farther south they may become active during warm periods in winter. In New Mexico lakes they are probably active all year long but do not eat between mid October and April (Christiansen and Moll, 1973). They are known to move long distances either from one water body to another or within streams and rivers. Homing ability has been demonstrated (Cagle, 1944b; Ernst, Lovich, and Barbour, 1994).

REPRODUCTION

The following account is based, where possible, on populations in the Rio Grande of New Mexico (ca. latitude 34°N) and chiefly on the surveys of Christiansen and Moll (1973),

Iverson and Smith (1993), and Iverson, Balgooyen, Byrd, and Lyddan (1993). Data not specific to New Mexico are from Ernst, Lovich, and Barbour (1994).

Vitellogenesis begins in September and ceases when females stop feeding in winter. Ovulation and oviposition occur from May through mid July, after an additional period of follicular growth in spring. Ovarian quiescence begins in July and lasts until September (Christiansen and Moll, 1973; New Mexico). Female reproductive cycles are prolonged at southern latitudes (e.g., 6 weeks in Wisconsin and 10 weeks in Louisiana; Christiansen and Moll, 1973; Ernst, Lovich, and Barbour, 1994).

Courtship and mating begin soon after spring emergence but may continue later in the season. Maximum testis size is not attained until August. Sperm used in spring matings are therefore produced in the previous season (Christiansen and Moll, 1973; New Mexico). Sperm can be stored for up to 165 days in the oviducts (Ernst, Lovich, and Barbour, 1994).

Chrysemys practice a complex Liebespiel that occurs, with modifications, also in *Trachemys*, *Pseudemys*, and some *Graptemys* (see *Trachemys* account, Figures 37.4 and 37.5, and Legler, 1990). Taylor (1933) first described this curious behavior, and Ernst, Lovich, and Barbour (1994) have updated it in some detail. The male pursues the female, overtakes her, turns to face her, and continues swimming backwards in front of her as he vibrates the backs of his elongated foreclaws against her face and chin with the forefeet turned (pronated) palm outward. Taylor observed essentially the same behavior in *Trachemys scripta elegans* housed in the same large aquarium but did not remark on any interspecific matings. Females respond by titillating the male's outstretched forelimbs with their foreclaws. The male may swim away from the female several times between such bouts of mating behavior. Ultimately the female sinks to the bottom where copulation occurs (Ernst, Lovich, and Barbour, 1994; Ernst, 1971).

Nesting begins in May over most of the range (Carr, 1952). The nesting process is typical of most nonmarine turtles—it occurs usually on a slope well exposed to the sun and fairly close to the water, and in late afternoon or evening. Nest excavation is by alternate movements of the hind limbs and facilitated by moistening of the soil with substantial quantities of fluid from the cloaca. Eggs are laid and arranged in the nest at intervals of about 30 seconds, and the soil is scraped back into the nest in the same order that it was removed. The female completes the nest by packing it with the plastron. The entire nesting process may take as long as 4 hours (Carr, 1952; Ernst, Lovich, and Barbour, 1994).

The egg shell has a well-defined calcareous layer and numerous pores (Packard, Packard, and Boardman, 1982). It is thin and rigid at laying and becomes flexible as the egg absorbs water and swells during incubation.

Mean dimensions of oviducal eggs for New Mexico are as follows: 31.4 × 18.4 mm (Christiansen and Moll, 1973); mean egg weight and clutch weight are 6.37 g and 56.7g, respectively (Iverson and Smith, 1993). Carrie Morjan (pers. comm.) examined eggs from 35 undisturbed nests along the Rio Grande near Socorro, New Mexico, in May and June 1999 and 2000. Mean number of eggs per nest was 9.2 ± 1.97 (6–14); mean egg weight for 159 eggs from 19 nests was 7.14 g ± 0.893 (5.05–8.95).

The only reproductive data from the Rio Santa Maria population are from John Iverson (pers. comm.). He obtained a gravid female (163 mm CL; Wt, 610 g) on 3 August 1989 and induced oviposition of five eggs with oxytocin. The dimensions of four unbroken eggs were as follows: L, 32.3 ± 1.054 (31.4–33.8); W, 18.9 ± 0.521 (18.35–19.50). The largest male from this series was 157 mm CL and 485 g.

Ewert (1985) recorded the following incubation times for *C. picta* at controlled temperature: 61–68 days at 25°C (mean 64.4, $n = 22$) and 47–56 days at 30°C (mean 50.3, $n = 16$). Sex determination is temperature dependent; incubation temperatures of 22–27°C usually produce only males and those of 30–32°C produce 100% females. Both sexes are produced at the pivotal temperatures of 20°C and 28°C (which is important in the far north where ground temperatures may never exceed 28°C; Ewert and Nelson, 1991; Ernst, Lovich, and Barbour, 1994). Application of estradiol benzoate on eggs causes the embryo to develop as a female (Gutzke and Bull, 1986).

The yolk is a rich source of energy. Of total egg lipids, 38% are used during embryonic development and 62% remain in the hatchling as sustenance (Congdon and Tinkle, 1982).

Multiple clutches (up to five) are produced over most of the range. A single large clutch is produced in the Pacific Northwest (Nussbaum, Brodie, and Storm, 1983). Total known range of eggs per clutch is 1–23. A variable percentage of females (20–60%) may skip seasons of reproduction in Michigan and Quebec, but the incidence of this decreases with age (Ernst, Lovich, and Barbour, 1994) and is seemingly unstudied in New Mexico.

Mean eggs per clutch in New Mexico is 9 or 10, depending on method of estimation. Half of the females produce two clutches and 14% produce three clutches per year. Average annual reproductive potential is 14.8 eggs (Christiansen and Moll, 1973). Iverson and Smith (1993) interpret the same data as follows: mean 1.6 clutches per year, averaging 8.9 eggs per clutch and an annual reproductive potential of 14.2 eggs.

The following correlates were obtained over the entire range of the species: Eggs per clutch, clutch weight, and egg width are positively correlated with female shell length (Ernst, Lovich, and Barbour, 1994); adults of both sexes mature later and at a larger size and attain larger adult size at more northern latitudes; the reproductive season is longer at more southern latitudes; more eggs are produced per clutch and there are fewer clutches per year in the north; the annual reproductive potential is about the same regardless of latitude (Iverson and Smith, 1993; Ernst, Lovich, and Barbour, 1994; Christiansen and Moll, 1973).

GROWTH AND ONTOGENY

Size in a small series from the Rio Santa Maria drainage of Chihuahua (UU, Legler database) is as follows: males,

139 ± 20.1 (113–164) n = 5; females, 149 ± 11.9 (124–171) n = 12; im. unsexed: 87.5 ± 3.5 (85–90) n = 2.

Adult size data from first-time captures of *Chrysemys picta* at Rio Grande Nature Center State Park, Albuquerque, Bernalillo Co., New Mexico, were as follows (Stuart, 2000): 29 males—CL, 150 mm ±1 4.0 (122–174), Wt, 406 g ± 102 (232–573); 23 females—CL, 166 mm ± 22 (101–202), Wt, 639 g ± 212 (140–1,064).

Sexual maturity in both sexes is correlated positively with size (Ernst, Lovich, and Barbour, 1994). Females in New Mexico mature on average in the fifth year at a CL of about 139 mm; average CL for adult females is 160 mm. Males mature in the third year at as small as 90 mm CL (Christiansen and Moll, 1973; Iverson and Smith, 1993). Longevity may be 30–40 years (Gibbons, 1968).

Mature males have extremely elongated foreclaws and a heavier, longer tail with the vent located beyond the posterior edge of the carapace. Females are larger than males and have unmodified foreclaws and a shorter tail with the vent anterior to the posterior edge of the carapace.

Hatchlings are about 25 mm CL, are essentially round in dorsal aspect, and have a mid-dorsal keel. The plastron is more intensely pigmented, and the tail is relatively longer than in the adult (Carr, 1952). Bone mass is 2% of body mass in hatchlings; this ratio increases to 20–27% at about 100 mm CL then stabilizes (Iverson, 1982; Ernst, Lovich, and Barbour, 1994).

PREDATORS, PARASITES, AND DISEASES

Various rodents, carnivores, birds, snakes, fish, and humans are known as predators. Raccoons (*Procyon*) are major nest predators. Nest predation is as high as 95–100%. Aquatic mammals, snapping turtles, snakes, raccoons, bullfrogs, large fishes, herons, and water bugs prey on young turtles (Ernst, Lovich, and Barbour, 1994).

CONSERVATION AND CURRENT STATUS

Chrysemys of all subspecies are probably too small to be taken seriously as a source of food for humans, but anthropogenic degradation of habitat is rife and is a chief factor in the decline of populations. John Iverson (pers. comm.) visited an area of springs (Ojos de los Reyes, ca. 9 km SE of Galeana, Chihuahua) six times from May 1977 to June 1990 and observed painted turtles in a marsh 200 m downstream from the spring heads on each visit. In 1990 the springs, rivulets, and marshes were being drained directly to the river by ditches to create pasturage for cattle. Iverson speculates that this habitat disturbance may have seriously reduced populations of *Chrysemys* and *Kinosternon hirtipes* at this particular locality. Propst and Stefferud (1994) surveyed 12 sites on the Santa Maria drainage for the declining Chihuahua Chub (*Gila nigrescens*), which occurred in all reaches of the river in the mid 1800s. Prime habitat is pools with large trees and root masses. They found few large riparian trees near developed areas. The fish were common only in a few remote and unmodified parts of the river. Habitat degradation has proceeded upstream from the terminal lake (dewatering) and downstream from the headwaters (logging in the 1950s), and the midsections of the river are likely to be the least modified by human works (David Propst, pers. comm.). Smith and Smith (1979) have suggested that a sanctuary of some kind be established in a remaining area of good habitat.

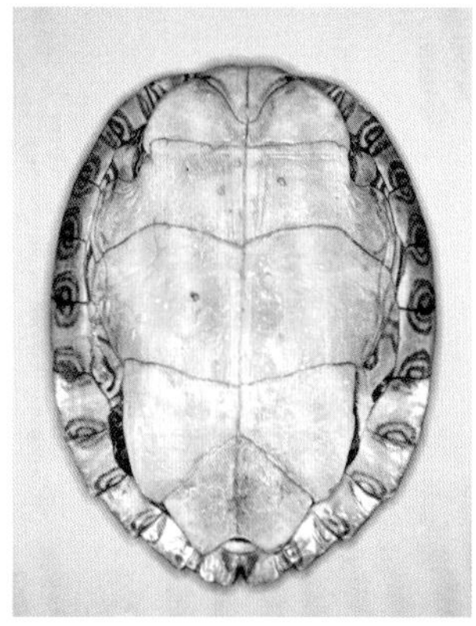

FIGURE 36.1 *Pseudemys gorzugi*. Dorsal and ventral views of adult female (Del Rio, Texas) showing typical markings and coloration. Note that only anterior "V" of plastral pattern remains. (Courtesy of J. J. Bull.)

FIGURE 36.2 *Pseudemys gorzugi*. Adult female in water (Black River, New Mexico). (Courtesy of C. W. Painter.)

Pseudemys gorzugi Ward, 1984

Rio Grande cooter (Ernst, 1990); Jicotea del Rio Bravo (Liner, 1994).

ETYMOLOGY

A patronym honoring George R. Zug (b. 1938), Curator Emeritus of Amphibians and Reptiles, National Museum of Natural History.

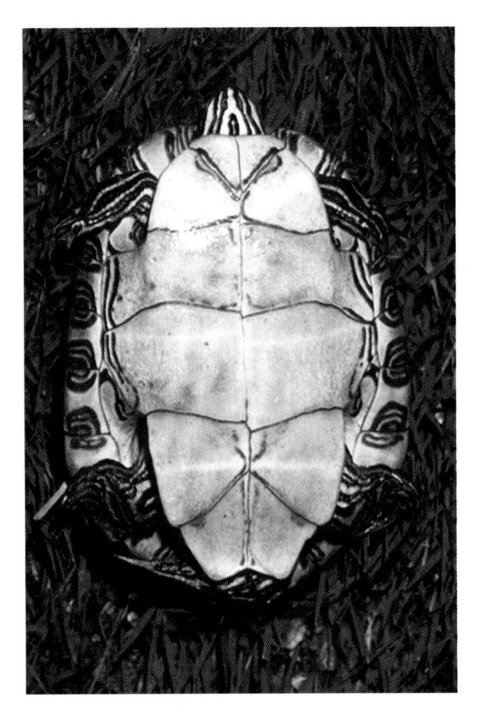

FIGURE 36.3 *Pseudemys gorzugi*. Subadult immature plastron (Black River, New Mexico). (Courtesy of C. W. Painter.)

FIGURE 36.4 *Pseudemys gorzugi*. Carapace pattern of juvenile in first year of growth (Black River, New Mexico). (Courtesy of C. W. Painter.)

HISTORICAL PROLOGUE

Ward (1984) described *Pseudemys gorzugi* as a subspecies of *Pseudemys concinna* (LeConte, 1830) from the Rio Grande and Pecos drainages and elevated *P. c. texana* to species rank. The type locality and holotype are as follows: Rio San Diego, 5.6 km W of Jimenez (29°15′N, 100°51′W), Coahuila, Mexico, 260 m elevation (KU 39,986, adult female skeleton). Ernst (1990) elevated *gorzugi* to full species rank, based on probable lack of gene interchange between *P. gorzugi* and the closest population of *P. concinna* (*P. c. metteri*). Much of the literature regarding *P. gorzugi* was published under the species names *P. concinna* or *P. floridana*. Little has been published concerning its natural history, and little is known about Mexican populations.

DIAGNOSIS

The following suite of characters distinguishes *P. gorzugi* from all other Mexican emydine turtles: carapace is elongate with serrated posterior margin and marked with contrasting black and yellow lines and blotches; first and/or second lateral scutes have large, reversed, C-shaped mark; most of adult plastron is unmarked or dimly marked on pale straw to yellowish ground; prefrontal part of maxillary tomium bears a shallow curved depression flanked (if at all) by shallow scallops (not an angular notch flanked by sharp tomial cusps); maxillary triturating surfaces bear large, sharp conical denticles.

GENERAL DESCRIPTION

Pseudemys gorzugi is of medium to small size by comparison to other members of the *P. floridana-concinna* group, but where it occurs in sympatry with other emydids (*Chrysemys picta* and *Trachemys scripta*), it is the largest emydid. Maximum recorded size is 284 mm in males and 372 mm in females. Specimens from near the type locality are smaller, but M. Forstner (pers. comm.) has observed much larger individuals in the Sabinas, Salado, and San Juan tributaries farther downstream in Coahuila.

Pseudemys texana differs from *gorzugi* in having a median notch, often flanked by toothlike cusps, on maxillary tomium and a plastron with a red-tinged rim.

BASIC PROPORTIONS

(164 specimens from Pecos and lower Rio Grande tributaries in Coahuila, Texas, and New Mexico; M. Forstner and Legler databases):

CW/CL: males, 0.74 ± 0.039 (0.58–0.85) $n = 96$;
females, 0.76 ± 0.040 (0.67–0.88) $n = 48$;
im. males, 0.92 ± 0.041 (0.87–0.97) $n = 4$;
im. females, 0.83 ± 0.046 (0.77–0.90)
$n = 12$; 1 juvenile, 0.87.

CH/CL: males, 0.37 ± 0.026 (0.25–0.48) $n = 84$;
females, 0.40 ± 0.026 (0.35–0.45) $n = 42$;
im. males, 0.49 ± 0.039 (0.46–0.53) $n = 3$;
im. females, 0.41 ± 0.026 (0.37–0.46) $n = 11$;
1 juvenile, 0.46.

CH/CW: males, 0.49 ± 0.034 (0.36–0.64) $n = 83$;
females, 0.53 ± 0.033 (0.46–0.60) $n = 42$;
im. males, 0.53 ± 0.053 (0.47–0.58) $n = 3$;
im. females, 0.49 ± 0.024 (0.43–0.52) $n = 11$;
1 juvenile, 0.52.

Modal plastral formula: $4 > 6 > 3 > 1 > 5 > 2$.

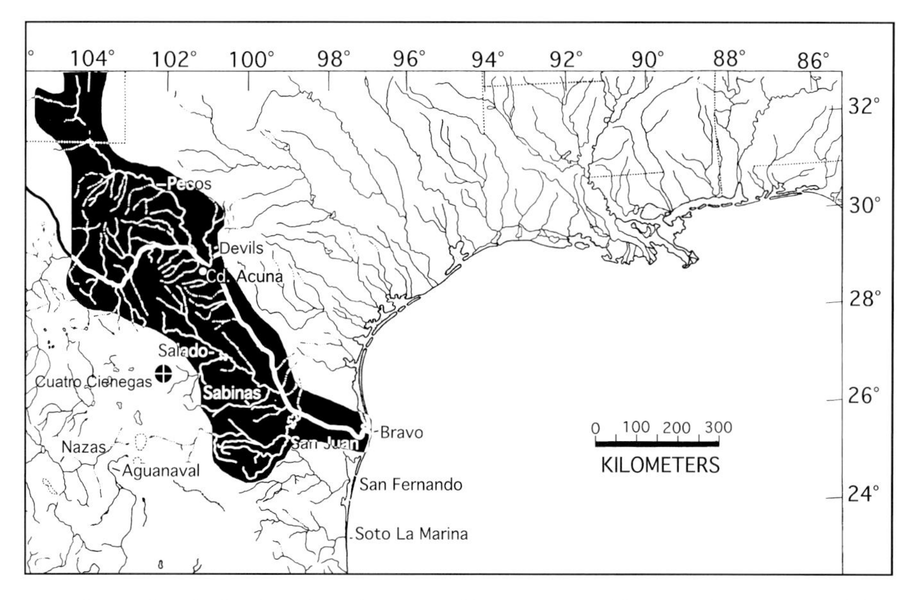

MAP 25 Geographic distribution of *Pseudemys gorzugi* in northeastern Mexico and adjacent United States.

Karyotype unknown but assumed to be same as *P. concinna*—$2n = 50$, 26 macro- and 24 microchromosomes (Ernst, Lovich, and Barbour, 1994).

Carapace is oval in dorsal aspect, low, and slightly keeled; it is highest at middle and widest posteriorly; posterior margin is serrated; surface of lateral scutes is shallowly rugose. Plastron is notched posteriorly.

COLORATION AND PATTERN

In hatchlings there is an indistinct, dusky, seam-following plastral pattern of narrow whorls on all interlaminal seams that quickly fades to a residual but conspicuous V-shaped vestige on the gular-humeral seams. The adult plastron is otherwise more or less unicolored straw to pale yellow with distinct and narrowly darkened seams. Neonates have a pair of dark horizontal bars crossing the bridge that fade to vestiges on the axillary and inguinal ends of the bridge.

The carapace pattern is complex and difficult to accurately define. It contains many of the components seen in other members of the *Pseudemys/Chrysemys* group. The pattern at first glance suggests a reticulum but is basically ocellar. The ocelli are crowded together on the lateral and marginal scutes. An ocellus that borders an interlaminal seam terminates abruptly at that seam and truncates the ocellus, accounting for the backward-facing "C-shaped" marks mentioned here and in other accounts (i.e., the open part of the C faces posteriorad and is precisely aligned with an interlaminal seam). Such marks are usually most evident on L1 and L2. A typical carapace pattern, unmodified by ontogenetic change, is seen in hatchlings (Figure 36.5).

The carapacal ocelli have a broad orangish-yellow border with well-defined dark edges. The ocelli and all marks within them are variously compressed and seldom circular. In some cases an entire lateral scute may have a yellow border. The pale ocellar centers are progressively obscured by melanistic blotches. In the oldest, largest adult in our small series (UU), the melanistic blotches are all that remain of the pattern (see Degenhardt, Painter, and Price, 1996; and Figures 36.2 and 36.7, present work).

The soft skin has a greenish to slate ground color with yellow stripes or other pale markings on the head, neck, legs, and tail. The limbs are marked with red, yellow, and black, especially in the webbing between the toes where there are black half-moon markings on a red background color.

The basic pattern of the head and neck seems to consist four to five parallel pale stripes, variously broken and colored, that extend posteriorad from the orbit onto the neck on each side (Figures 36.2 and 36.6).

Common elements in this variation are an unbroken stripe from the inferior orbital rim joined by a stripe from the mandibular ramus and continuing to the base of the neck, forming a lateral Y; a yellow stripe from the mid-posterior orbit broken into blotchlike segments in two or more places; an upper stripe, usually isolated from the orbit, that is widened and forms an orange postorbital mark like that seen in many *Trachemys* and *Pseudemys*, and a median stripe aligned with the mandibular symphysis and joined to a ventrolateral stripe on each side to form a "Y" (the symphyseal Y as discussed under *Trachemys*).

The pattern of the head and neck could logically have been derived from an ancestral pattern like that shown for the genus *Trachemys* (Figure 37.1). Ward (1984) discussed the clinal variation in head and shell markings.

GEOGRAPHIC DISTRIBUTION

Map 25

The Pecos River and tributaries (south of 35°N) in New Mexico and Texas, and the Rio Grande drainage from the

FIGURE 36.5 *Pseudemys gorzugi*. Hatchling with caruncle, showing the full juvenile plastral pattern (Black River, New Mexico). (Courtesy of C. W. Painter.)

FIGURE 36.6 *Pseudemys gorzugi*. Old adult showing melanism on shell and jaggedness of linear head pattern (Black River, New Mexico). (Courtesy of C. W. Painter.)

Big Bend region downstream to Brownsville in Coahuila, Nuevo León, and Tamaulipas (essentially the ancient Rio Grande; see account of *Trachemys scripta gaigeae*). The tributary streams of both systems tend to be small. Those of the Rio Grande begin downstream from Del Rio and Cd. Acuña and are not numerous. Legler (field notes) caught only one *Trachemys scripta elegans* in a small impoundment near Langtry, near the mouth of the Pecos. Lengthy scanning with binoculars revealed no other turtles. Few, if any, records are from the Rio Grande proper. The Rio Grande does not constitute a continuous homogeneous ribbon of habitat for any kind of aquatic chelonian. There was no sign of *P. gorzugi* in the Rio Grande or its tributaries at Ojinaga (mouth of Rio Conchos) or downstream at Lajitas (*Trachemys s. gaigeae*, *Apalone spinifera*, and *Kinosternon hirtipes* were common at both localities in 1959, [Legler, field notes]).

Baited traps in the flowing and muddy Pecos R. proper (Chaves Co., New Mexico, 1962) caught only *Apalone spinifera*. In 1962 Legler and party found the greatest abundance of *gorzugi* in smaller Pecos tributaries (Evans Creek, 25.1 km NW of Del Rio, Val Verde Co., Texas, and Black River, 11.3 km W of Malaga, Eddy Co., New Mexico, and in the Rio San Rodrigo near Moral, Coahuila (29°04'N, 101°41'W; 20.5 km downstream from Jimenez); these were shallow, clear streams with chiefly rocky or sandy bottoms and long, slow reaches between riffles. *P. gorzugi* could be easily observed from the bank in clear water or basking on emergent objects, as could the *Trachemys scripta elegans*, which were observed in roughly equal numbers at the same localities.

Of the 10 specimens of "*Pseudemys gaigeae*" reported from Coahuila by Schmidt and Owens (1944), eight were identified by Legler as *P. gorzugi* (from Las Rusias, San Juan Sabinas, Hermanas, Lampacitas, and Allende; R. Sabinas, R. Salado, and R. San Juan; see account of *Trachemys scripta taylori*).

The lower Rio Grande tributaries, downstream from the region of the type locality (ca. 29°15'N, 100°51'W), are the only waters in Mexico where two species of emydids with head striping occur in natural sympatry (see general account of *Trachemys scripta*).

Rio Grande Cooters are active at 12°C in December in the Black River of New Mexico (Degenhardt, Painter, and Price, 1996). Aquatic and aerial basking occur, the latter on logs and rocks along the river edge. Bailey, Forstner, Dixon, and Hudson (in press) monitored the surface temperature of *gorzugi* using temperature data loggers attached to the carapace for 3 months in the summer of 2004. The monitored turtles maintained temperatures at least 2°F above water temperature at a depth of 1 m. Between the hours of 0800 and 1600 temperature was gradually increased by aerial basking to a peak of 5°F above the water temperature. Body temperatures decreased after 1600 hours.

Movements are limited. In the Black River of New Mexico, turtles marked in deep pools moved a maximum of only 300 m between captures.

REPRODUCTION

Based on all sources of information (secondary sexual characteristics, gonadal examination, and clear growth-ring data), most males probably mature near 100 mm CL in the fourth or fifth full season of growth, and most females do so at about 160 mm CL in the fifth or sixth full season of growth (JML, pers. obs.).

As suggested by the long foreclaws in males, there is probably a stereotyped courtship behavior involving titillation with the foreclaws, similar to that described for *P. concinna* by Marchand (1944) and significantly different from allopatric emydids (*Trachemys* and *Chrysemys*). Courtship of female *P. gorzugi* by male *Trachemys scripta elegans* (but not

FIGURE 36.7 *Pseudemys gorzugi*. Ventral and dorsal views of three females and one juvenile (UU; Black River, New Mexico).

copulation) has been observed where they occur in sympatry (M. Forstner, pers. comm.).

In New Mexico nesting occurs in late May (based on dissection and robbed nests (Degenhardt, Painter, and Price, 1996). Recently emerged hatchlings (with caruncle) have been observed in mid August and late October. Hatchlings may overwinter in the nest along the Pecos River; hatchlings were found in a nest on 14 April (Degenhardt, Painter, and Price, 1996). Multiple clutches have not been documented, but the meager data on nesting dates suggest that they are produced.

A large female (240 mm CL, Black River, New Mexico) obtained on 23 May contained nine oviducal eggs with mean dimensions of 42 × 31 mm. Hatching occurred after 70 days of incubation under laboratory conditions (no temperatures were given). Mean measurements and weights of four surviving hatchlings were 33.9 mm CL and 10.1 g, respectively (Degenhardt, Painter, and Price, 1996).

Bailey, Forstner, Dixon, and Hudson (in press) reported on measurements and weights of 14 neonates hatched in captivity as follows: CL, 39.8 mm ± 2.1 (34.6–42.4); CW, 37.2 mm ± 1.8 (33.2–39.7); Wt., 14.4 g ± 2.7 (8–18). Four recently emerged hatchlings with caruncles (17 August, Black River, New Mexico) had a mean CL and weight of 37.8 mm and 11.3 g, respectively (Degenhardt, Painter, and Price, 1996).

No data on temperature-dependent sex determination are available, but the phenomenon is known in *Pseudemys concinna*.

GROWTH AND ONTOGENY

Data on size for 163 specimens from the Pecos and lower Rio Grande tributaries in Coahuila, Texas, and New Mexico (CL, M. Forstner and Legler databases): males, 197 mm ± 37.1 (109–284) $n = 98$; females, 257 mm ± 43.5 (130–372) $n = 48$; im. males, 84 mm ± 7.5 (74–92) $n = 4$; im. females, 114 mm ± 32.6 (67–166) $n = 13$.

Adult males have straight, elongated foreclaws and longer, thicker tails with the cloacal opening posterior to the edge of the carapace (anterior to edge in females). Old males often become melanistic (see General Description for ontogeny of color and pattern).

POPULATIONS

Bailey, Forstner, Dixon, and Hudson (in press) sampled a 97 km stretch of the Pecos River and found 123 adults (1.3 per km) and three juveniles (0.03 per km). Sampling in a 35 km stretch of the Devils River produced 68 adults (1.94 per km) and three juveniles (0.08 per km). For comparison, a population of *Pseudemys texana* was sampled in a 9 km stretch of the San Marcos River and found to have 9.5 adults and 2.1 juveniles per km, suggesting that the sampled populations of *P. gorzugi* had a much lower recruitment rate.

The foregoing surveys of *Pseudemys gorzugi* in the Rio Grande Basin show a relatively low population density and a distinct paucity of juveniles (Bailey, Dixon, Hudson, and Forstner, 2008). Genetic analyses of the mitochondrial ND4 gene and five microsatellite DNA markers indicate that *P. gorzugi* may be genetically relatively homogeneous throughout its geographic range. Bailey, Forstner, Dixon, and Hudson (in press) have also shown that *gorzugi* is a monophyletic group genetically distinct from other *Pseudemys*. There was no evidence of significant variation in population structure in parts of the range studied (Bailey, Dixon, Hudson, and Forstner, 2008).

CONSERVATION AND CURRENT STATUS

Low population density, low genetic diversity, declining environmental quality, and direct human depredation threaten the existence of *P. gorzugi* in Texas and adjacent Mexico.

Dixon (2000) noted that a population from San Felipe Springs and a stream in Del Rio had an abundant population of Rio Grande cooters in 1996. When the area was resampled in 1998 no turtles could be found. The premise was that the turtles were captured for sale outside the United States.

The research of Bailey, Dixon, Hudson, and Forstner (2008) shows evidence of multiple threats, including habitat degradation, introduction of fire ants, and commercial collection for human consumption outside the United States. Construction of dams, channels, flood control practices, and other water diversions have caused streams within the geographic range to become intermittent in many areas. This has probably had a direct negative impact on populations of *P. gorzugi* (Bailey, Forstner, Dixon, and Hudson, in press).

Trachemys scripta (Schoepff) 1792

Slider turtles; tortuga pinta; tortuga pintada; schmucks-childkrötte, jicotea, icotea, hicatee, tortuga negra (Iverson, 1992a; Moll and Legler, 1971). The vernacular name "slider" alludes to the characteristic habit of sliding off basking sites at the approach of danger.

ETYMOLOGY

Gr. *trachys*, rough, probably in reference to the longitudinal or radial rugosities on most adult shells; Gr. *emys*, freshwater tortoise.

L. *scriptus*, written; alluding to the inscription-like markings on the shell and linear marks on soft parts.

HISTORICAL PROLOGUE

THE GENUS *TRACHEMYS*

Use of the generic names *Chrysemys*, *Pseudemys*, and *Trachemys* has ebbed and flowed for more than a century. Agassiz (1857) placed the *scripta*-like turtles in *Trachemys*, the species *picta* in *Chrysemys*, and the remainder of the group in *Ptychemys* (= *Pseudemys*). Boulenger (1889) placed all known members of the three genera in the genus *Chrysemys*. Books of the early to mid-20th century (Stejneger and Barbour, 1923; Schmidt, 1953; Carr, 1952; Conant, 1958; Wermuth and Mertens, 1961) placed *picta* in *Chrysemys* and all the rest in *Pseudemys*, a not illogical arrangement. The work of Weaver and Rose (1967) revived the trend to lump all members of *Chrysemys*, *Pseudemys*, and *Trachemys* in the genus "*Chrysemys*" (e.g., Webb, 1970; Morafka, 1977a; Conant, 1975; McDowell, 1964; Ernst and Barbour, 1972), the last such use being by Savage (2002).

Vogt and McCoy (1980) combined morphological and biochemical data (electrophoresis of LDH and P-gen enzymes) to demonstrate convincingly that the monotypic *Chrysemys* was distinct from *Pseudemys*, and that the latter contained two distinct groups: *Trachemys* containing the mainland and Antilles *scripta*-like species; and *Pseudemys* with all the rest. Seidel and Smith (1986) redefined and confirmed the distinction of *Chrysemys* (sensu stricto, painted turtles), *Trachemys* (*scripta*-like turtles, sliders), and *Pseudemys* (all the rest, cooters). However, the use of "*Pseudemys*" for the *scripta* series persisted for at least another decade by taxonomic recalcitrants (Moll and Moll, 1990; Legler, 1990).

There was widespread use of "ornata" (Williams, 1956; Legler, 1960a; Mertens and Wermuth, 1955) as a polytypic species until Legler (1990) pointed out that its range was relatively small and confined to the Pacific Coast of Mexico.

TAXONOMY AND RELATIONSHIPS WITHIN THE GENUS *TRACHEMYS*

Trachemys remains a "difficult" group. There is substantial and erudite debate on whether the mainland populations of *Trachemys* from the Rio Grande southward should be regarded as a subspecies (Moll and Legler, 1971; Legler, 1990), species, or mixture of both (Seidel, 2002).

There are 11 named and recognizable taxa of *Trachemys* in Mexico. Seven of these occur only in Mexico. Two of the northern taxa (*gaigeae* and *elegans*) occur also in the United States, and two of the southernmost (*venusta* and *grayi*) occur in adjacent Mesoamerican countries. The Mesoamerican taxa are allopatric. Each occurs in a separate geographic range defined by drainage systems. In the rare instances that these populations have come into secondary contact, interbreeding has occurred. This has been referred to as "intergradation" by Legler (1960a, 1963, 1990) and "introgression" or "hybridization" by Seidel (2002) and Seidel, Stuart, and Degenhardt (1999). Whatever this interbreeding is called, it bespeaks the lack of isolating mechanisms (and, because of the allopatric distributions, the lack of need for such isolating mechanisms) and the relatively recent dispersal of an ancestral stock. There is little doubt expressed anywhere that these allopatric taxa are closely related, distinguishable, and of common origin.

There have been several recent major analyses of variation in the *scripta*-like *Trachemys*. Legler (1990) used a multidiscriminate analysis of 29 morphological characters in 2,308 specimens and included the entire Mesoamerican mainland complex and treated *Trachemys scripta* as a wide-ranging, polytypic species, an arrangement accepted by Seidel, Stuart, and Degenhardt (1999) and Iverson (1992a). Seidel (2002) did a cladistic analysis of 23 morphological characters with 1,200 specimens and recognized seven full species, three of them monotypic and four of them polytypic, containing eight subspecies. The two studies recognized the same number of taxa in Mexico, agreed generally on the relationships of these taxa, and differed only in hierarchical treatment and zoogeographic correlations (included by Legler). Both taxonomic arrangements may be treated as hypotheses.

Seidel's goal was to provide a species-level taxonomy for *Trachemys*. Legler's (1990) objective was to present a natural and stable classification of Mesoamerican *scripta*. As the present book nears completion, this classification is natural but has not achieved nomenclatorial stability.

There are three natural groups of *Trachemys scripta*-like turtles occupying the western hemisphere: those of the United States and extreme northern Mexico, those of Mesoamerica and South America, and a third closely related group that consists of four species occupies islands in the Antilles region (Seidel, 2002).

For the purposes of this book, it really makes no difference whether the *scripta*-like taxa in Mexico are regarded as a species or subspecies, as long as their interrelationships are adequately defined and understood. We choose to follow Legler (1990) in regarding all of them as subspecies of

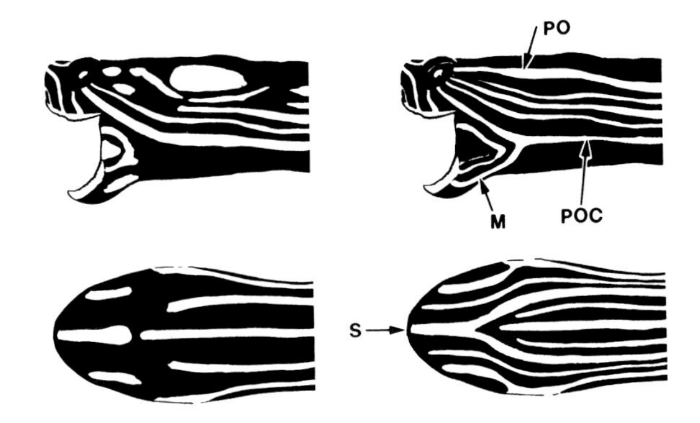

FIGURE 37.1 *Trachemys scripta*. Head striping. Hypothesized ancestral condition on right; derived conditions on left. All stripe patterns in *Trachemys* can be derived by interruptions, obliterations, and fusions of the stripes shown here. Stripes mentioned in text: PO, postorbital; POC, primary orbitocervical; M, mandibular; S, symphyseal. (Modified from Legler, 1990.) Degree of stripe continuity is defined numerically as "pattern integrity." *T. s. ornata* with few broken stripes has a high pattern integrity, whereas *T. s. callirostris* with many broken stripes has a low pattern integrity.

Trachemys scripta (Schoepff) and to regard each allopatric taxon as an incipient species.

Despite the natural relationships of the *scripta* group over a vast latitudinal range there is a major break in variation south and west of the present Rio Grande drainage. The salient features of this break and our explanations of them are discussed below. These are sympatry with other emydines, courtship patterns, sexual dimorphism, and adult size.

SYMPATRY AND HYBRIDIZATION

Trachemys scripta occurs in microsympatry with *Pseudemys gorzugi* in the Pecos River and in tributaries of the lower Rio Grande downstream from Del Rio, Texas. This is supported by specimens in collections, and Legler has personally observed the two species within 1 m of one another at several localities. *Chrysemys picta* also occurs in the Upper Pecos, but not necessarily with *Trachemys* or *Pseudemys*.

Northward and eastward in the Mississippi drainage, in the Gulf drainages, and along the eastern coastal plain of the United States, *Trachemys scripta* occurs almost always with at least one species of *Pseudemys*. *Chrysemys picta* and *Graptemys* sp. also occur in this zone of sympatry. It is therefore common in this region for several similar emydines to occur in microsympatry. All of these taxa have stripes on the head and neck. Each taxon has a unique combination of stripe pattern, sexual dimorphism, and precoital mating behavior. These differences serve as precopulatory isolating mechanisms.

South and west of the Pecos–lower Rio Grande drainage there is *no emydine sympatry* (except that which has been mentioned previously). *Trachemys scripta* is the only emydine, the only basking turtle, and the only turtle with stripes in all of Mesoamerica and South America (a few exceptions are mentioned in Legler, 1990).

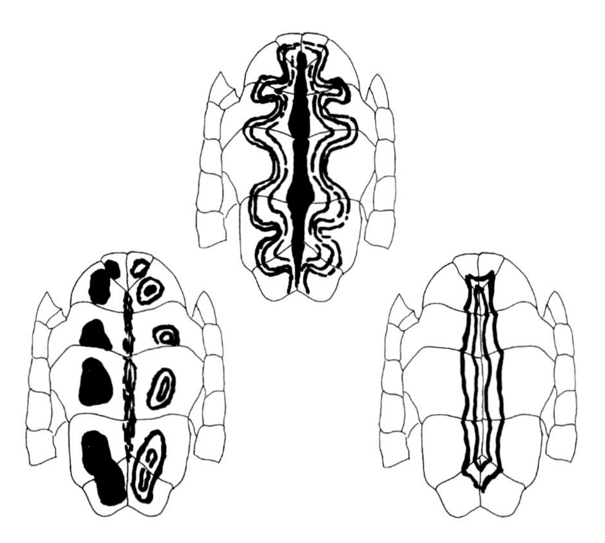

FIGURE 37.2 *Trachemys scripta*. Plastral patterns. Top illustration shows the full "*Chrysemys*" pattern. Bottom shows two derived conditions: right—reduced plastral pattern (e.g., *hartwegi*); left—a composite figure showing isolated whorls (ocelli) on one side and isolated whorls overlain by solid blotches on the other side. The latter occurs in *elegans*, *hiltoni*, and *nebulosa*; the whorls are present at hatching and the blotches develop later. (Modified from Legler, 1990.)

HEAD STRIPING

The ancestral *Trachemys scripta* had a pattern of unbroken pale stripes on the head and neck (Figure 37.1), an ocellar carapacal pattern, and a dusky, concentric plastral pattern (Figure 37.2). This pattern combination is more common in Mesoamerica than elsewhere in the broad range of *T. scripta* (Legler, 1990).

Stripes may have evolved initially as disruptive markings. Interruptions, expansions, and deletions of these stripes, and the deposition of reddish or orangish pigments in places produce the distinctive patterns of each taxon. These unique patterns may also serve as isolating mechanisms in areas of emydine sympatry. Although unique patterns persist in the *Trachemys* of Mesoamerica, they may now have no functional or adaptive significance.

Skin ground color ranges from nearly black in *venusta* to greenish olive in the young of many subspecies. Pale stripes can be distinct on an only slightly darker background. Contrast is increased by saturation of the yellow pigment in stripes and the addition of fine, black borders (e.g., *elegans*). The extremes seen in Mexico range from bright stripes on an almost black background (*venusta*) to the neutral pallidity of unbordered stripes in bare contrast to the ground color (*grayi*).

Yellows on the soft parts, on the plastron, and in the iris are often tinged with green. This greenish tint is most dramatically expressed in the iris. The iris of nonmelanistic animals is basically yellow with a greenish cast. This greenish cast reaches its epitome in populations of *T. s. ornata* along the west coast of Mexico in which the iris of both sexes may be a lime green with jewel-like transparency.

PLASTRAL PATTERN

The basic plastral pattern in *Trachemys* is a series of two to four concentric dusky lines, usually without distinct borders, occupying at least half the surface area of the plastron and extending from the gular to the anal scutes—the pattern exemplified by *Chrysemys picta belli*. The fullest version of this pattern (*taylori*, *cataspila*, and some *venusta*) has lateral whorls extending along interlaminal seams to or near the bridge and the free edges of the plastron. This pattern is modified in other Mexican *Trachemys* by isolation or loss of the lateral whorls (*elegans*, *gaigeae*, some *venusta*), foreshortening (*hartwegi*), and almost complete ontogenetic fading (*yaquia*).

General melanism, especially in adult males, may obscure or obliterate the pattern (*grayi*, *taylori*), and selective secondary deposition of melanin on the shell can create distinct patterns (*elegans*, *hiltoni*, and *nebulosa*). In these taxa, secondary deposition of melanin in the epidermis and dermis forms solid, opaque, fingerprint-like dark marks over the lateral whorls and obliterates them. These marks remain separate in *elegans* but may be interconnected to form a circumferential figure without vestige of concentricity. Shed scutes from the taxa with secondary melanism carry some of the melanin with them.

CARAPACE PATTERN

The basic carapace pattern consists of ocelli on the lateral and marginal scutes and a variable pattern on the central scutes. The pale marks of the carapace range from yellow to pinkish orange to dark orange. The lateral ocelli vary in shape but are usually vertical ellipses. Ocelli consist of two to four concentric rings that vary in their distinctness; there may be one or more linear projections in various directions from the peripheral ring. The outermost ring normally becomes the most prominent ring. The ocellar center often becomes a dark bull's-eyelike figure.

Each lateral ocellus is centered over an intercostal osseous suture, placing it usually on the posterior half of a lateral scute. There are eight costal bones; if we count the nuchal-costal suture as IC suture #1, then the ocelli are centered on even-numbered intercostal sutures. This seems to be constant. The centers of the ocelli vary in their juxtaposition to interlaminal seams because the seams vary in their juxtaposition to the sutures.

The two extremes of carapace pattern are the unmodified ocellar pattern of many Mesoamerican *scripta* (e.g., *ornata*) and the vertically barred pattern (*elegans*).

T. s. elegans usually has four vertical bars, each of which occurs near the center of a lateral scute and lies over the approximate center of an underlying costal bone (i.e., C1, 3, 5, 7). The vertical bars are modifications of the ocellar pattern, derived from the anterior parts of the ancestral ocelli. Ocelli *and* bars can be observed in many *elegans*.

The supramarginal and inframarginal ocelli are centered on the intermarginal seams. The supramarginal ocelli show more variation than the inframarginal ones. Marginal

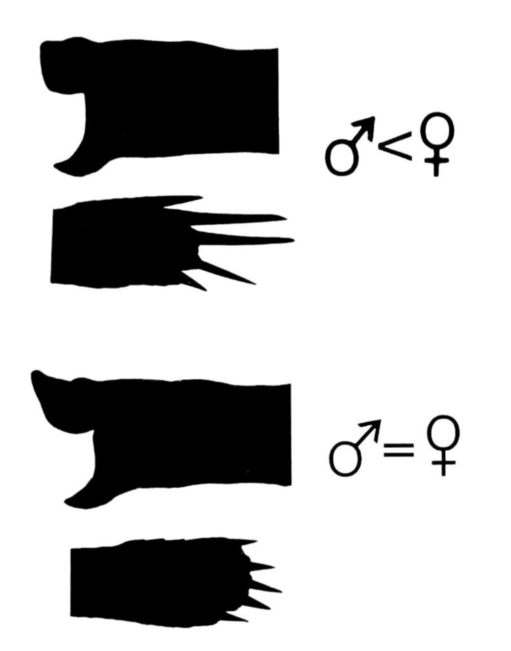

FIGURE 37.3 *Trachemys scripta*. Sexual dimorphism in basic groups of *Trachemys scripta*. In the Lower Rio Grande drainage and to the northeast: (top), males are significantly smaller than females and have elongated foreclaws, a stereotypic Leibespiel, and an unmodified snout. In populations to the south and west (bottom) there is little size dimorphism, elongated foreclaws and stereotypic Liebspiel are lacking, and mature males have a pointed, upturned snout. (Based on Legler, 1990.)

ocelli are only slightly modified in derived carapacal patterns such those of as *T. s. elegans*.

The most common feature of the central scute pattern is a pair of pale parenthetical marks, one on either side of the mid-dorsal line; these may connect in two longitudinal series or form a completely separate pair of marks on each scute. In populations with complex patterns, there is commonly an intricate pattern of triangles, rhomboids, or irregular polygons between the parenthetical marks.

Adult size in *Trachemys* seems to be correlated with latitude. There are 17 specimens of mainland *Trachemys* in Legler's database ($n = 2,071$) of 350 mm CL or larger; 13 of these are *venusta* from the Caribbean coast of Mesoamerica or *emolli* from the Lake Nicaragua drainages, all south of 16°N latitude. The two largest (424 and 414 mm CL) *venusta* are from the region of Chiriqui Lagoon, western Panama. A specimen of *T. s. grayi* (395 mm) is from Laguna Muchacha, Jutiapa, Guatemala, on the Pacific Coastal plain just west of the El Salvador–Guatemala frontier. The largest Mexican individuals for which we have data are two specimens of *T. s. ornata* (353 and 359 mm CL) from coastal lagoons near Tuxpan, Nayarit, and a specimen of *hiltoni* (351 mm) from the Rio Fuerte drainage near Alamos, Sonora. D. Moll (1994) recorded maximum adult sizes (CL) for males and females of *T. s. venusta* from Tortuguero, Costa Rica, as 350 and 440 mm, respectively, and these seem to be the record sizes for *Trachemys scripta*.

These data suggest that the largest specimens of Mesoamerican *Trachemys* may be more likely to occur in large bodies of water—large rivers, coastal lagoons, and lakes (see account of *venusta*). Many of these very large specimens are shells obtained from village middens and had been used for food.

By comparison, *Trachemys scripta* in the northern part of its range achieves a much smaller adult size. Data on 204 specimens of *T. s. elegans* from the Rio Grande and other nearby drainages of northern Mexico and adjacent United States are as follows: males, 142 mm $\pm$ 34.1 (90–214) $n = 102$; females, 163 mm $\pm$ 49.9 (90–257) $n = 102$.

SEXUAL DIMORPHISM

There is extreme sexual dimorphism in adult size in *T. s. elegans* (and the two other subspecies in the United States). Males mature at a much smaller size than females and never become as large as females; the chances of a mating pair being the same size are low. By comparison, there is much less size dimorphism in Mesoamerica; although most females are larger than most males, males often grow as large as females, and the chances of a male being as large or larger than a female are good (Moll and Legler, 1971; Legler, 1990).

The terminal (ungual) phalanges of the manus are greatly elongated in males of 12 of the 17 species in the *Chrysemys* group (not in *Malaclemys*; not in four of nine *Graptemys* species). Within *Trachemys* long foreclaws occur in the three U.S. subspecies and not in the mainland subspecies to the south. Males of all Antilles taxa have elongated foreclaws (Legler, 1990).

Mature males of most Mexican taxa attain some degree of an elongated, pointed, upturned snout. This does not occur in *elegans* or other *scripta* of the United States. The elongated snout in Mexican *Trachemys* varies from absent in *elegans* and *taylori*; to noticeable in *gaigeae*, *hartwegi*, *yaquia*, and *ornata*; to extreme in *venusta*, *cataspila*, *grayi*, *nebulosa*, and *hiltoni*. In old, large males of the "extreme" category, the snout may become bulbous and bosslike (Figure 46.6). In large adult males the anterior edges of the prefrontal bones are thickened, upturned, and inwardly beveled, and the skeletal nasal aperture (defined by the prefrontals, maxillaries, and premaxillary bones) may be huge, almost as large as the orbit (Figure 48.1).

DIAGNOSES OF *TRACHEMYS*

The following combination characterizes all U.S. *scripta* (including *elegans* in northern Mexico): (1) Male foreclaws are elongated; 2) there is a courtship Liebespiel involving the elongated foreclaws in some way; (3) males are significantly smaller than females, never attaining maximum female size and rarely an ocellar pattern on carapace; (4) there is no extreme modification of snout; (5) occurrs in microsympatry with at least one other closely related emydid having

FIGURE 37.4 *Trachemys scripta.* Stereotypic courtship behavior in *Trachemys scripta elegans.* Larger female on left. Smaller male titillates female's face and neck with rapidly vibrating, elongated foreclaws as he swims backward in front of her. (Drawing by Kerry Matz, based on photos in Cagle, 1950.)

bold stripes and a stereotypic Liebespiel involving the use of elongated male foreclaws.

The following combination characterizes all naturally occurring mainland Mesoamerican *scripta* south of the Lower Rio Grande and west of the Pecos: (1) Male foreclaws are unmodified; (2) there is no stereotyped Liebespiel involving male foreclaws (mating consists mostly of trailing, biting, and mounting—see individual accounts); (3) males attain a large average size and can become as large as females; (4) the snout of males is variously elongated; (5) never occurs in microsympatry with another emydid having bold stripes (nor with any aquatic emydine).

All Mexican *Trachemys* can be distinguished from other Mexican chelonians (including the similar *Pseudemys gorzugi*) by the following combination of characters: (1) There is a bright pattern of pale longitudinal lines on the head, neck, and limbs; 2) mandibular ramus is higher than wide in cross-section (or height and width are equal), and tomial edges slant outward; (3) median alveolar ridge of maxilla is well defined but not necessarily denticulate; (4) maxillary triturating plates are not heavily developed; (5) premaxillary notch is absent or wide and lacking lateral cusps.

All *Trachemys*, *Pseudemys*, and *Chrysemys* have a diploid chromosome number of 50 with 26 macrochromosomes and 24 microchromosomes (Killebrew, 1977a). Details may vary on the shape of the macrochromosomes, but this basic karyotype can be assumed for all species for which no specific data are available.

All other descriptive and diagnostic data appear in the accounts of subspecies.

MATING AND COURTSHIP

Trachemys s. elegans has an elaborate courtship behavior in which the male overtakes a female, positions himself in front of her, and rapidly vibrates his elongated foreclaws near her face. Rapid head bobbing is also involved (Figures 37.4 and 37.5). This Liebespiel was first noted by Taylor (1933; see account of *T. s. elegans* for detailed description). Courtship involving foreclaw titillation occurs, with unique variations, in the 12 *Chrysemys* group species that have elongated foreclaws (Carr, 1952; Ernst and Barbour, 1972; Jenkins, 1979; pers. obs.). This includes the three U.S. subspecies of *T. scripta*. There are no records of this Liebespiel in taxa that lack the elongated foreclaws; this includes all Mesoamerican and South American *Trachemys scripta* (observations have been made on *taylori*, *venusta*, *hiltoni*, *gaigeae*, *callirostris*, and *dorbigni*).

The function of elongated foreclaws is seemingly clear, but there is no demonstrable function for the elongated snout. It may permit sexual recognition by head profile in populations where males and females do not differ predictably in size. The snout could also be an erotic prod used in mating (Legler, 1990).

Lack of foreclaw titillation in Mesoamerican *Trachemys* is here regarded as a derived character (Legler, 1990; Seidel and Fritz, 1997). It also has been lost in one group of *Graptemys* (Vogt, 1980; Shealy, 1976).

FOSSILS

The genus *Trachemys* is represented in Miocene and Pliocene deposits, the earliest *T. scripta* remains being from the Pliocene Blancan of Texas. There are Pleistocene records from Sonora and Yucatán (Ernst, Lovich, and Barbour, 1994; Álvarez, 1976; Van Devender, Rea, and Smith, 1985).

Trachemys arose during or prior to the Miocene from a generalized *Chrysemys-Pseudemys*–like ancestor (similar to Pliocene *T. idahoensis* and upper Miocene *T. hilli*), which then gave rise to several Miocene and Pliocene forms with wide distribution in northern temperate America. The wide-ranging, polytypic *Trachemys idahoensis* gave rise to *T. scripta*. Fossil specimens from Florida and Texas demonstrate some of the same osteological differences seen in Recent *T. s. scripta* and *T. s. elegans*, respectively (Ernst, Lovich, and Barbour, 1994).

Fossil finds in Mexico are few. Van Devender, Rea, and Smith (1985) reported shell fragments of *Trachemys scripta* from Rancho la Brisca, north central Sonora, in Pleistocene deposits (Sangamon Interglacial., ca 150,000 YBP) with a climate similar to that of the last 4,000 years. Carapace elements represent several adults ranging up to an estimated 275 mm CL and a plastron with open fontanelles representing a juvenile and suggesting a breeding population. The site is a former marsh ca. 33 km NE of Cucurpe in the northwesternmost headwaters of the Rio Sonora drainage (Rio Santo Domingo). There is an extant marsh 5 km S of the fossil site in which adult *T. s. yaquia* now occur.

Shaw (1981, as cited by Lindsay, 1984) recorded "*Chrysemys (Pseudemys)*" from middle Pleistocene (Irvingtonian) deposits at El Golfo de Santa Clara on the northwesternmost coast of Sonora (32°41′N), near the mouth of the Colorado River. Gazin (1957) reported "*Pseudemys* sp.-turtle carapace fragments" from upper Pleistocene (Carolinian) deposits at El Hatillo near Pese,

FIGURE 37.5 Adult male of *T. s. elegans* (Illinois) induced to extend claws and begin titillation out of water. (Photo courtesy of E. O. Moll.)

Herrera Prov., western Panama. These fragments were associated with remains of giant ground sloths (*Eremotherium*), various other sloths and edentates, muscovy ducks, capybara, mastodon, horses, peccaries, and deer.

Barbour (1973, pers. comm.) collected fragmentary fossil material from a probable former effluent channel of Lake Chapala, 5 km W of Jocotepec, Jalisco. These included shell fragments of a turtle. Smith, Cavender, and Miller (1975) mentioned the "turtle" material and dated the site as late Pleistocene in age. The turtle fragments were identified as *Pseudemys* (*Trachemys*) in Smith (1980; identification verified by J. Iverson and T. Van Devender). Langebartel (1953) recorded *Trachemys* remains from caves in Yucatán dated 200–1250 AD.

The significance of the four fossil sites in Mesoamerica is discussed in the cited papers. For current purposes they constitute a useful trail of Pleistocene finds from northern Mexico to western Panama, corroborating the recency (Pleistocene) of *Trachemys* dispersal into Mesoamerica.

GEOGRAPHIC DISTRIBUTION

(See maps of subspecies.)

Trachemys scripta (sensu lato) has the greatest latitudinal range of any species of freshwater turtle, ranging (discontinuously) from southern Michigan through temperate North America, neotropical Mesoamerica, and into temperate South America for a total of 77° of latitude.

The type and type locality are unknown (Smith and Smith, 1979); the latter was restricted to Charleston, South Carolina, by Schmidt (1953). "*scripta*" has been acknowledged as the valid name for this species since the time of Boulenger (1889).

The ranges of most Mexican taxa can be defined in terms of coastal plains and perennial streams within well-defined drainage systems. The ranges of two northern subspecies, *hartwegi* and *taylori*, are circumscribed by internal drainage systems.

DISPERSAL

Trachemys scripta dispersed into Mesoamerica and the Antilles in Pleistocene times (Jackson, 1988; Seidel and Jackson, 1990). This recency of dispersal and differentiation was espoused by Legler (1990), Moll and Legler (1971), and Moll and Moll (1990) without specific dating. It is supported by Pleistocene remains (see Fossils). In the course of this dispersal, sliders rapidly exploited available aquatic niches. Whatever the steps leading to this astounding success, it was accomplished by virtue of ecological versatility and with a minimum of modification to a basic ancestral plan.

NATURAL HISTORY

Among emydids, our knowledge of the natural history of *Trachemys scripta* (sensu lato) is vast—probably rivaled only by that for *Chrysemys picta*. But most of the information comes from north of the Rio Grande and south of the Isthmus of Tehuantepec. With the exception of *venusta* in Veracruz and Chiapas, most natural history information on Mexican and South American taxa (Medem, 1962, 1975; Pritchard and Trebbau, 1984) consists of bits and pieces of information. Thorough studies have been made in Virginia (Mitchell and Pague, 1990); Illinois, Louisiana, and Tennessee (Cagle, 1950); Veracruz and Chiapas (Vogt, 1990); and Panama (Moll and Legler, 1971). Despite the gaps, existing studies provide a valuable comparison of reproductive and adult size information over a latitudinal range of ca. 32°.

Trachemys scripta is a consummate ecological generalist. Except for the timing of activity and reproductive cycles, as influenced by climate and latitude, the general life history pattern of "*scripta*" is very much the same throughout the vast range of the species. Therefore, any of the mentioned studies could be used as templates from which reasonably accurate extrapolations and interpolations could be made.

Gibbons (1970a) and Thornhill (1982) have demonstrated that populations of *Trachemys scripta* from the same geographic region (SW Illinois) grow faster, mature sooner, and have a greater reproductive potential in heated lakes than in nonheated lakes. Similarly, warmer environmental temperatures at lower latitudes permit significantly longer periods of feeding, growth, and reproduction.

In general *T. s. venusta* at low latitudes mature at larger size, ultimately attain larger size, and produce more and larger eggs than their northern temperate ancestors. Larger size is evident in coastal populations (e.g., *cataspila*, *yaquia*) immediately south of the Rio Grande but not in the isolated northern Mexican populations (*gaigeae*, *hartwegi*, and *taylori*).

HABITAT

In the United States most sliders occur in mesic regions where drainages flow annually and predictably. The same conditions occur in truly neotropical areas. Although there are vicissitudes of drought and flood, named streams do not commonly recede to pools or "permanently" become dry basins and trenches. The converse is true in northern Mexico; basins and stream disruptions caused by aridity are commonplace, and the situation is compounded by human use of water for irrigation and other essential human needs.

Optimal habitat is large, permanent lakes or slow-moving streams with associated backwaters or small tributaries and large amounts of submergent aquatic vegetation, open areas in fringing forests (as opposed to unbroken forest), and abundant basking sites (Moll and Legler 1971). However, sliders can prosper in nearly all kinds of aquatic habitats. The natural versatility of *T. scripta* is further extended by its ability to exploit anthropogenic modifications of the environment such as impoundments, ponds, clearings near villages, and golf courses.

DIET

Slider turtles are a paradigm of opportunistic omnivory wherever they occur. They can thrive on what is available, even in polluted ecosystems. Consistently available foods (typically submergent aquatic vegetation) constitute the "staple" in any given microhabitat. Moll and Legler (1971) found dietary differences in subpopulations of *venusta* in Panama only a few hundred meters apart. Animal food of some sort was present in about half of the adults studied but comprised a small percentage of total volume. The most common animal foods are insects, fish carrion, and gastropods. Juveniles consume more small invertebrates than older individuals. Windfalls are exploited by sliders of all ages; these range from fruit (e.g., figs) dropping into the water to dead animal carcasses (chiefly fish). Individuals may develop wide heads (megacephaly) after acquiring a taste for mollusks (e.g., *emolli* in Lake Nicaragua).

HABITS AND BEHAVIOR

All sliders bask when they can—whether on emergent objects or simply by floating at the surface. Elevation of body temperature seems to be the primary purpose of this behavior, but aerial basking may help to control ectoparasites and epizoic algae, and to facilitate scute shedding. Moll and Legler (1971) give a detailed analysis of basking that is applicable to all *Trachemys*.

REPRODUCTION

(See individual accounts of subspecies.)

CONSERVATION AND CURRENT STATUS

Slider turtles have always been a desideratum for hunter-gatherer societies (Tamayo and West, 1964) and are now commonly sold as food or as ornaments in markets. The same markets may also sell imported *T. s. elegans* as pets (see "Conservation" in Introduction). As common as turtles are in the markets of Veracruz, Tabasco, and Chiapas, they are a rarety in the Chiapas–Guatemala highlands. When Legler and companions made overnight roadside camps in that region, local Indians stopped on their way to market and aggressively offered to buy or barter for the large *Trachemys* shells we had laid out to dry.

DICHOTOMOUS KEY TO THE SUBSPECIES OF *TRACHEMYS SCRIPTA* IN MEXICO

The subspecies *grayi* and *cataspila* appear in more than one couplet because of ontogenetic melanism. The terminal couplet for each taxon is associated with a brief statement of geographic range.

1A. Postorbital mark a slight expansion of a continuous stripe from eye to base of neck, yellowish orange; mandibular tomium coarsely serrate (*venusta, cataspila, grayi, ornata, yaquia*) **4**

1B. Postorbital mark usually isolated from eye and/or neck stripe, color variable; if connected with another lateral stripe, contiguous stripe much narrower than 1° orbitocervical; mandibular tomium smooth or finely serrate (*taylori, elegans, gaigeae* group) **2**

2A. Postorbital mark red, connected or not to a yellowish neck stripe, connected or not to a stripe joining the eye; if isolated, the shape is a long oval (*elegans* and *taylori*) **3**

2B. Postorbital mark yellowish or orangish, not red, usually partly or completely isolated (*gaigeae* group of four taxa) **10**

3A. Markings on lateral and marginal scutes of carapace ocellar; an extensive concentric, dusky plastral figure, all parts of which are interconnected, usually partly obscured by melanism; mandibular tomium serrate; stripe on mandibular ramus isolated; mature males lack elongated foreclaws and attenuate snout. Occurs in the Basin of Cuatro Ciénegas, Coahuila. . . . . ***T. s. taylori***

3B. Markings on lateral scutes of carapace vertical bars or extremely narrow vertical ocellar figures; plastral pattern consisting of brown fingerprint-like smudges (at least anteriorly), obscured by melanism in some adult males; stripe on mandibular ramus isolated or not; mature males have elongated foreclaws. Occurs in the Lower Rio Grande and its tributaries below the Big Bend . . . . ***T. s. elegans***

4A. A clear, usually colorful pattern of ocelli on some or all of the lateral and marginal scutes of carapace; a dusky concentric plastral figure; obfuscation of pattern by melanism uncommon; postorbital mark connected to a stripe on neck, to eye, or both **5**

4B. Not as in 4a; lacking a clear, bright pattern of ocelli at some or all ontogenetic stages. **8**

5A. Ocelli centered on each lateral and occupying most of scute; orbitocervical stripe usually golden yellow and the brightest, widest, and most evident stripe in lateral aspect of head, wider than postorbital stripe or mark. Occurs in Atlantic drainages from Pta. del Moro, Veracruz, across the Yucatán Peninsula and into Mesoamerica at least to Panama .... ***T. s. venusta***

5B. Ocelli usually smaller, not well centered on laterals; 1° orbitocervical stripe paler yellow, not wider than widest part of postorbital and not the dominant stripe in lateral aspect of head (*ornata, cataspila, grayi*) **6**

6A. Ocelli on L2 and L3 (when not obscured by melanism) positioned in extreme upper posterior quadrant of scute and often dark centered; juveniles with clear ocellar pattern on carapace and dusky concentric pattern on plastron. If head striping visible, all stripes narrow, of roughly equal width, and lacking black borders. Postorbital stripe lacks evident expansion and brighter coloration. Occurs on Pacific Coastal plain from Salina Cruz, Oaxaca, to El Salvador (see also 9a) .... ***T. s. grayi***

6B. Ocelli not so positioned; stripes of varying width, postorbital stripe with at least a slight expansion of brighter color **7**

7A. Ocelli on L2 and L3 dark centered and positioned on lower posterior quadrant of scute. Plastral pattern dusky, concentric, and seam-following, without lateral whorls; clear only in juveniles, not straight-sided as in *venusta*. Occurs in Atlantic drainages of NE Mexico from the Rio San Fernando, Tamaulipas, to Tuxpan, Veracruz (see also 9a) .... ***T. s. cataspila***

7B. Ocelli centered on posterior halves of laterals and not truncated or overlapping adjacent lateral; ocelli often interconnected by longitudinal pale line extending anteriorly from the principal ring of one ocellus to the posterior edge of the next. Each ocellus has two or three pale concentric lines and a darker center that does not contrast sharply with ground color; ocellar pattern remains clear in largest adults. Plastral pattern distinct, consisting of four concentric, dusky lines on a pale yellow ground, fading somewhat with age but never completely obscured in adults. Occurs in Pacific drainages, south of the Rio Fuerte, from the region of Culiacan, Sinaloa, southward at least to the region of Cabo Corrientes, Jalisco............... ***T. s. ornata***

8A. Laterals bearing modified ocelli, each consisting of two thin peripheral rings (pale orange to yellowish) with an irregular jagged melanistic blotch at center; ocelli centered on and truncated by posterior margin of lateral or overlapping onto adjacent lateral. Carapace patterns more distinct in juveniles, obscure in adults. Dark ocellar centers in sharp contrast to ground color; entire carapace may be dark brown in old specimens. Plastral pattern indistinct (except in young), fading to uniform yellowish or obscured by brownish pigment with age. Occurs in the Sonora, Yaqui, and Mayo drainages of Sonora .... ***T. s. yaquia***

8B. Ocelli of carapace and often plastral pattern obscured by melanism; stripes indistinct.................. **9**

9A. Pale stripes on head and neck narrow, of approximately equal width and lacking sharp contrast to ground color; no evident expansion of postorbital stripe; attenuated snout of adult males produces an acute lance-shaped profile (Figure 41.4) .... ***T. s. grayi***

9B. Pale stripes on head and neck wider in general and variable in width; expansion of postorbital stripe about as wide as diameter of orbit; snout of adult males attenuate, not forming a lance-shaped profile (Figure 39.1) .... ***T. s. cataspila***

10A. Throat bearing distinct, dark-bordered, pale stripes (seven to nine at level of posterior border of tympanum, including 1° orbitocervicals); symphyseal stripe usually forming Y (*gaigeae* and *hartwegi*) **11**

10B. Throat lacking distinct, dark-bordered, pale stripes; uniformly pale or, if striped, stripes wide and indistinct (no more than five at posterior level of tympanum, including 1° orbitocervicals); symphyseal stripe seldom forming Y; mid-ventral markings on throat, if present, more often lance shaped (*hiltoni* and *nebulosa*) **12**

11A. Postorbital mark bright orange, isolated, not connected with eye or a stripe on neck; if connected with a stripe posteriorly, then that stripe much narrower than primary orbitocervical. Mandibular stripe longer than postorbital mark. Plastral pattern more or less continuous from gular to anal, consisting of two or more concentric lines (distinct at least in young), and distributed about equally on anterior and posterior halves of plastron. Occurs in the Upper Rio Grande and Rio Conchos drainages............ ***T. s. gaigeae***

11b. Postorbital mark dark yellowish orange, relatively large, usually not pointed behind, more nearly in shape of perfect oval. Mandibular stripe half or less length of postorbital mark. Plastral pattern concentrated chiefly on posterior half of plastron (behind pectoro-abdominal seam); small markings on pectorals, humerals, and gulars, if present, usually isolated from rest of pattern; pattern lacks distinct concentric arrangement. Occurs in the Rio Nazas drainage............. ***T. s. hartwegi***

12A. Throat usually uniformly pale between 1° orbitocervical stripes; mandibular stripe, if distinct, usually not contiguous with 1° orbitocervical stripe; vertical extension of 1° orbitocervical on tympanum usually contiguous with anterior postorbital stripe; plastral pattern of dark blotches narrow, not distinctly hourglass shaped, completely filled by dark pigment or extremely narrow, not enclosing wide, pale central area; gulars often lacking dark markings. Occurs in the

Rio Fuerte drainage of Sonora, Chihuahua, and Sinaloa . . . . *T. s. hiltoni*

12B. Throat uniformly pale or bearing a lance-shaped mark, expanded at level just posterior to tympana, extending from mandibular symphysis to base of neck; mandibular stripe usually distinct and contiguous with 1° orbitocervical; vertical extension of 1° orbitocervical on tympanum usually not contiguous with anterior postorbital stripe; plastral pattern of dark blotches forming hourglass figure extending from humeropectoral seam to anals, enclosing broad, pale central area (chiefly on abdominals) that may or may not include other dark marks; marks on gulars and humerals usually separated from main pattern by constriction at humeropectoral seam. Occurs in Baja California Sur . . . . *T. s. nebulosa*

Trachemys scripta elegans (Wied) 1839

Red-eared slider (Iverson, 1992a); Oreja Roja (Liner, 1994); Tortuga Japonesa (urban pet shops).

ETYMOLOGY

L. *elegans*, tasteful, choice, fine, select—probably expressing what Prince Maximilian perceived as a handsome elegance in this brightly colored turtle ("Schmuckschildkrötte" in his native language).

HISTORICAL PROLOGUE

Trachemys scripta (sensu lato) is one of the most studied turtles in the world. *T. scripta elegans* has a large natural range, is a favorite in the international pet trade, and has been introduced virtually worldwide.

The taxon was the earliest discovered in the New World to be described by the explorer himself (*Emys elegans*; Wied, 1839). Prince Maximilian A. P. zu Wied (1782–1867) and Karl Bodmer, an artist, arrived in the United States (from Germany) in July 1832 and traveled westward to explore the flora, fauna, and indigenous peoples. Illness forced him to stay the winter of 1832–1833 in New Harmony, Indiana. Thomas Say and C. A. LeSueur were there at the same time. It was at this time that the first specimen of *Emys elegans* was obtained.

Boulenger (1889) was the first to recognize *elegans* as a subspecies of *scripta* ("*Chrysemys scripta elegans*"); that subspecific epithet has been in common use since about the time of Cagle (ca. 1944), despite quarrels between 1939 and 1944 as to whether *elegans* was a subspecies of *troosti* or vice versa. Carr (1952) produced the first usable map of the U.S. subspecies of *T. scripta* as we now recognize them.

DIAGNOSIS

The following characters, in combination, distinguish *T. s. elegans* from all other members of *T. scripta* and all other emydids in Mexico.

Expansion of postorbital mark is red, black-bordered, longer than wide, often (59%) connected to the eye, and seldom (2.9%) connected to a neck stripe; mandibular stripe is often (56%) connected to a neck stripe; symphyseal stripe is connected to at least one stripe on throat (100%) and usually forms a "Y." Pattern integrity is 0.545. Markings on lateral scutes are vertical and linear, and are located on the posterior part of the scute and over the centers of odd-numbered costal plates; they are rarely ocellar. Plastron has a dusky concentric pattern on a yellowish background; lateral extensions of pattern end in whorls; in some cases only the whorls are present; ontogenetically each whorl is obscured by a melanistic fingerprint-like smudge. Males are significantly smaller than females. Males have greatly elongated foreclaws and lack an exaggerated elongation of the snout. Maxillary tomium is not serrated.

GENERAL DESCRIPTION (BASED ON SPECIMENS FROM RIO GRANDE DRAINAGE IN TEXAS AND MEXICO)

T. s. elegans ranks with the smaller sliders in Mexico. Maximum size and weight in New Mexico are given by Degenhardt, Painter, and Price (1996) as 210 mm and 1,299 g for males and 264 mm and 2,700 g for females. Tucker, Dolan, and Dustman (2006) give data on total range of sizes for 24,682 specimens in Illinois as 30–302 mm, the largest being an adult female and a record size. Such large sizes are rare, however (only three individuals >300 mm in entire sample). Maximum size (Legler database) for the Pecos and lower R. Grande drainages of Texas and Mexico is 214 for males and 248 for females. Pattern integrity is 0.386; see Figure 37.1 and legend for definition of "pattern integrity."

BASIC PROPORTIONS

(Legler Database; Abbreviations in Figure 5.1)

CW/CL: males, 0.78 ± 0.04 (0.66–0.90) $n = 75$;
females, 0.76 ± 0.033 (0.66–0.90) $n = 41$.

CH/CL: males, 0.38 ± 0.018 (0.34–0.46) $n = 75$;
females, 0.39 ± 0.021 (0.36–0.44) $n = 41$.

CH/CW: males, 0.49 ± 0.03 (0.39–0.57) $n = 74$;
females, 0.53 ± 0.04 (0.44–0.61) $n = 41$.

PWHP/PWMF: males, 0.99 ± 0.04 (0.91–1.08) $n = 72$;
females, 0.99 ± 0.03 (0.92–1.05) $n = 41$.

BR/CL: males, 0.31 ± 0.09 (0.31–0.36) $n = 60$;
females, 0.34 ± 0.06 (0.33–0.38) $n = 35$.

WH/CL: males, 0.17 ± 0.01 (0.14–0.19) $n = 73$;
females, 0.16 ± 0.011 (0.15–0.19) $n = 38$.

Modal plastral formula: $4 > 6 > 1 > 3$.

The striped pattern is bright yellow, often dark-bordered stripes on a darker background; pattern and contrast are seldom obscured by melanism in Mexican populations. Shell is relatively flat and wide, low in profile and cross-section, and squarish from above.

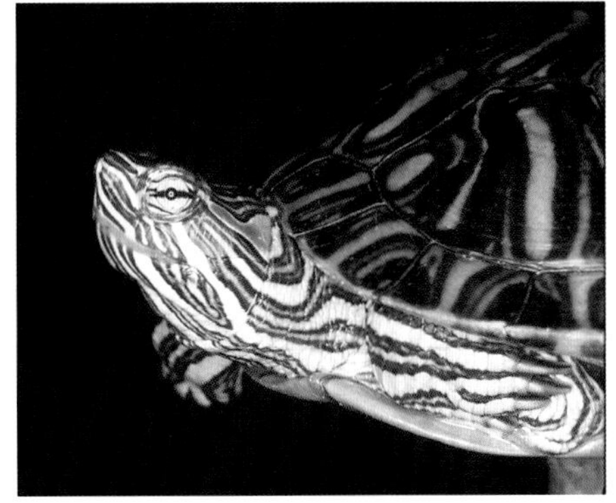

FIGURE 38.1 *Trachemys scripta elegans*. Head and anterior shell markings showing little variation over the entire range of the subspecies. Left, Illinois; right, Del Rio Texas.

Some older adult males develop melanism, which appears first along interlaminal seams and eventually obscures normal carapace and plastron patterns; no melanism is observed in females. Smudgelike marks on plastron are medium to dark brown, but never jet black. Plastral marks (or whorls) are often connected in juveniles; posterior marks are often joined in adults; at least one of the anterior two pairs is distinct in all individuals; frequently dark pigment forms an irregular line or elongate blotch along mid-plastral seams. Mid-dorsal keel is obtuse.

VARIATION

In Pecos and Lower Rio Grande populations, when the primary postorbital mark does not contact the eye, there is a small (nearly round) secondary postorbital mark that is contiguous with the orbit and narrowly separated from the 1° POM; the 2° POM is usually red but is less commonly orange. A 2° POM is characteristic of all *gaigeae* and *hartwegi*. It occurs in the following frequencies in *elegans*: Pecos, 44%; Lower Rio Grande, Langtry to Eagle Pass, 37%; Lower Rio Grande, McAllen to Matamoros, 59%. This variation is always paired with a fairly typical, slightly foreshortened red postorbital mark. A 2° POM is rare in more southern populations on both sides of Mexico. All of the specimens examined from the Rio Grande have vertical bars on the lateral scutes.

These examples of natural intermediacy (as opposed to hybridization with introductions) could be explained by limited hybridization events after the barriers between the ancient Pecos–Rio Grande and Upper Rio Grande–Conchos drainages were breached.

Seidel, Stuart, and Degenhardt (1999) judged there to be no morphometric separation among four populations of *elegans* in the Rio Grande, Pecos, Canadian, and the Colorado/Brazos drainages of northern Mexico, New Mexico, and Texas.

GEOGRAPHIC DISTRIBUTION

Map 26

Occurrence of greatest pertinence to this book is in the Pecos River in Texas and New Mexico and the Lower Rio Grande and its tributaries (principally the Rio Salado, Rio Sabinas, and Rio San Juan) from a point roughly at the intersection of 102°W, along the international border, in Coahuila, Neuvo León, and Tamaulipas, to Matamoros. Natural distribution may extend a short distance south of Matamoros in an arid zone of uncoordinated drainage ("Conjunto 1"; Tamayo, 1946); *elegans* × *cataspila* hybrids have been reported at La Laca, Tamaulipas (Legler, 1990; see account of *T. s. cataspila*).

Elsewhere there is an extensive distribution in the Mississippi and the major Gulf drainages. Intergradation with *T. s. troosti* occurs in Kentucky and Tennessee and with *T. s. scripta* in Alabama and Georgia.

The type locality was not stated in the original description (Wied, 1839) fide Iverson (1992a); Smith and Smith (1979) give it as "Fox River, near New Harmony, Posey Co., Indiana." New Harmony is on the Wabash River just north of its confluence with the Ohio River.

Introductions of *elegans* in New Mexico have been noted since the 1970s (Degenhardt and Christiansen, 1974; Stuart, 1995a,b; Degenhardt, Painter, and Price, 1996; summarized by Stuart, 2000). A breeding population of *T. s. elegans* exists in the Upper Rio Grande ("Rio Grande Valley") of New Mexico in open, natural ponds at the Rio Grande Nature Center State Park (RGNC) within the city of Albuquerque (Stuart, 2000). *Trachemys s. scripta* has been introduced in the same area and occurs, with *elegans*, in ponds and preserves along the river.

We consider the current natural range of *T. s. elegans* in New Mexico, Texas, and northern Mexico to correspond to the drainage of the ancient Lower Rio Grande (see account of *T. s. gaigeae*). *Trachemys s. elegans* occurs in the Rio Nadadores at Celemania, Coahuila, immediately outside the Basin of Cuatro Ciénegas (see account of *T. s. taylori*).

The mentioned Lower Rio Grande tributaries are the only waters in Mexico in which two species of emydids with head striping (*Pseudemys gorzugi* and *Trachemys scripta*) occur in natural microsympatry (i.e., can see each other, can be seen within a meter of one another, and are caught in the same traps).

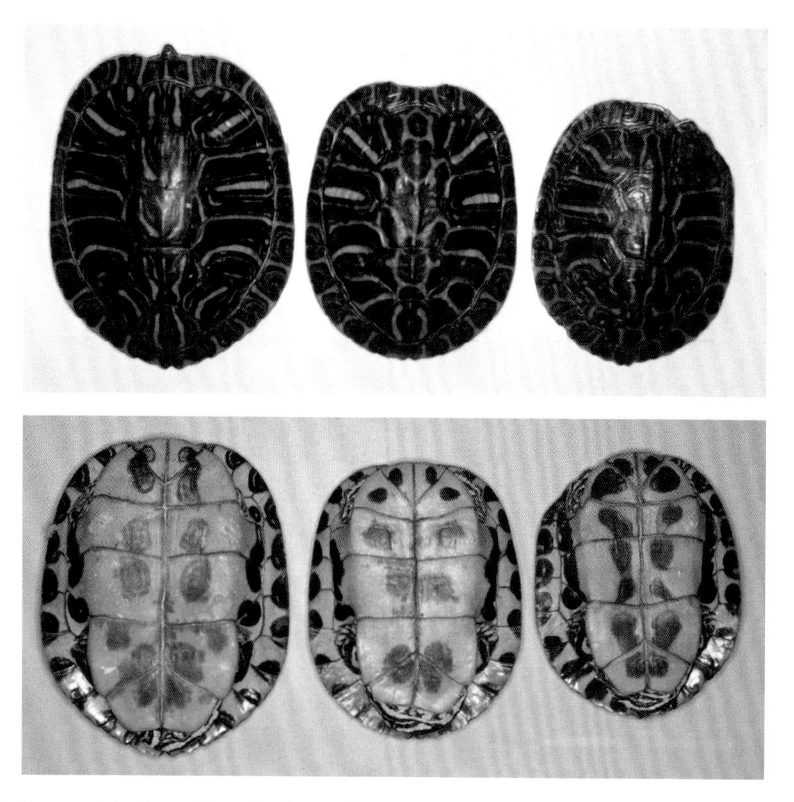

FIGURE 38.2 *Trachemys scripta elegans*. Dorsal and ventral views, three adults from Rio Grande, Del Rio, Texas (Cd. Acuña).

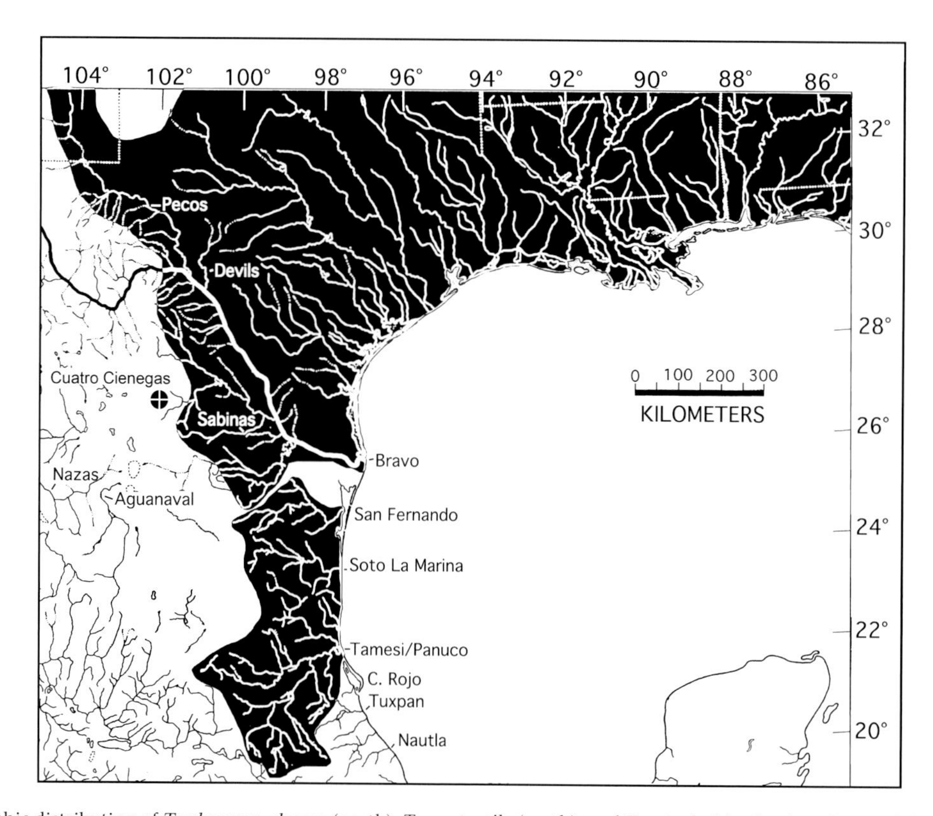

MAP 26 Geographic distribution of *Trachemys s. elegans* (north), *T. s. cataspila* (south), and *T. s. taylori* (endemic to Basin of Cuatro Ciénegas).

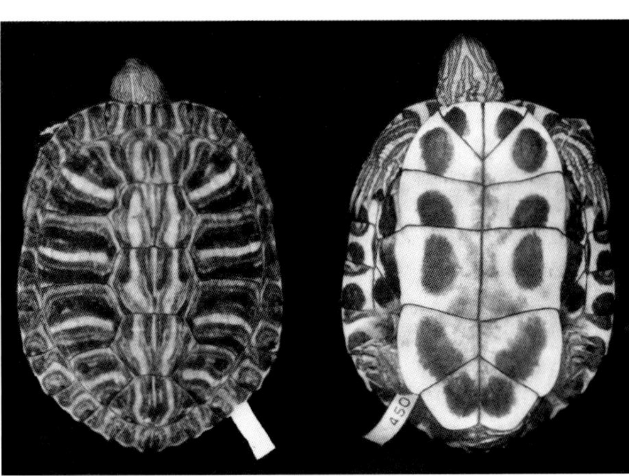

FIGURE 38.3 *Trachemys scripta elegans*. Top—first-year juvenile and hatchling (Matamoros, Tamaulipas). Note ocellar plastral markings that are obliterated by melanistic smudges early in ontogeny. Bottom—typical carapace (horizontal bars) and plastral patterns (smudges) after about 2 years of growth (Múzquiz, Coahuila).

NATURAL HISTORY

The subspecies *elegans* has been extensively studied in the United States but only scarcely within its limited range in northern Mexico.

Activity in Illinois begins when water reaches 10°C. Activity in New Mexico has been observed from early April to late October (Degenhardt, Painter, and Price, 1996; Stuart, 2000). Active sliders have been observed in winter swimming in water as cold as 2.4°C and sometimes under ice in north central Texas (Ernst, Lovich, and Barbour, 1994). Activity is said to be diurnal (Degenhardt, Painter, and Price, 1996), but they enter baited traps at night (JML, pers. obs.).

In the Upper Rio Grande of New Mexico *elegans* occurs at elevations to 1,380 m and is limited to ponds and lakes (Degenhardt, Painter, and Price, 1996). In the Lower Rio Grande they occur in and along flowing streams but prefer the slow water at bends and in beaver ponds. They are seen also in stock ponds. Smaller tributaries (e.g., Rio San Rodrigo, ca. 25 km upstream from Piedras Negras) are often crystal clear.

REPRODUCTION

Most of the data presented below are from introduced populations in the Rio Grande Valley of New Mexico (Stuart, 2000). No data are available for natural populations in the Lower Rio Grande drainage in Mexico.

Ernst, Lovich, and Barbour (1994) summarize the courtship behavior of *Trachemys scripta* ssp. Detailed descriptions of courtship and mating in *T. s. elegans* appear in Davis and Jackson (1970) and Jackson and Davis (1972). See also abbreviated account in discussion of genus *Trachemys*.

The following is from a summary of courtship in *T. s. elegans* (Jackson and Davis, 1972): (1) Male exhibits restless, "appetitive" search behavior; (2) female appears, gives no visible display, and appears oblivious to male; (3) male may begin a variable period of cloacal sniffing, trailing behind female as if following a chemical releaser; (4) female continues to move with no visible display on her part; (5) male maneuvers into frontal position facing female and extends forelimbs with palms pronated ("signaling posture"); (6) female becomes stationary; (7) male begins titillation sequence of highly variable length; (8) female closes eyes during titillation sequence, with head slightly withdrawn; (9) male moves rapidly behind female and mounts. "Titillation" is the rapid vibration of the dorsal surfaces of the male foreclaws against the sides of the female's head (see also *T. s. gaigeae* account).

After mounting, copulation is preceded by "rapid searching motion" of the tails of both individuals, which quickly become interlocked in a prehensile fashion with the male's tail hooking around the female's tail and the vents being brought into close apposition; when intromission is achieved, the male withdraws his head and forelimbs and leans back to assume an upright position at 90° to body of female; separation is achieved when the male swims rapidly to the surface, pulling the female with him until disengagement occurs. Total copulation time was about 13 minutes (Davis et al., 1970). In *T. s. scripta* the copulatory union is firm enough to hold the pair together even when the male is lifted out of the water (Lovich, pers. obs., in Ernst, Lovich, and Barbour, 1994). Jackson and Davis (1972) propose that stylized titillation has evolved "to supersede biting as a means of immobilizing the female." Stylized mating rituals

FIGURE 38.4 *Trachemys scripta elegans*. First-year juvenile from lower Pecos drainage (Eddy Co., New Mexico) with slightly different markings than individuals from Matamoros.

FIGURE 38.5 *Trachemys scripta elegans*. A typical progression of plastral melanism seen in some Mexican populations (four immature females, Rio Grande near Olmito, Texas).

(see general account of *Trachemys*) almost certainly act as precopulatory isolating mechanisms where several species of striped emydines occur in microsympatry (Legler, 1990).

At northern latitudes spermatogenesis occurs in late spring and summer and spermiogenesis in autumn. The germinal epithelium is quiescent in winter. Follicular enlargement begins in autumn (perhaps as early as August), and ovulation occurs from late April into July (Moll et al., 1971).

Size of eggs is 36.2 × 21.6 mm in Illinois and 37.7 × 22.6 mm in Louisiana (Cagle, 1950). Stuart (2000) found two gravid females in New Mexico (29 May and 19 June) and induced oviposition in both. Female size and egg data were as follows:

CL 263 mm, 2,400 g, 24 eggs: L, 34.1 ± 1.5 (31.3–36.5); W, 22.6 ± 0.6 (21.3–23.6); Wt, 10.25 g ± 0.6 (8.7–11.1). One egg was unusually large—41.8 × 26.3 mm, 17.5 g. CL 183 mm, 11 eggs: L, 31.7 ± 2.0 (29.0–35.8); W, 20.4 ± 0.8 (19.1–21.2), Wt, 7.1 g ± 0.30 (7.0–8.0).

Mean eggs per clutch is nine (4–18) in Illinois and seven (2–19) in Louisiana (Cagle, 1950) and one to three clutches are produced per season. Incubation time in the northern part of the range is 67–79 days at 24–30°C (Moll et al., 1971). In New Mexico, hatchlings emerged from 20 eggs incubated at 27–30°C after 58–62 days. Sex determination is temperature dependent: eggs incubated at 22.5, 25.0, or 27.0°C produce 100% males; at 30°C they produce only females (Ernst, Lovich, and Barbour, 1994).

GROWTH AND ONTOGENY

Data on size for 149 individuals from natural populations in the Pecos and Lower Rio Grande drainages of Texas and Mexico (JML database) are as follows: adult males, 133 ± 28.8 (90–214) n = 78; adult females, 170 ± 37.4 (107–248) n = 44; im. males, 92 ± 2.5 (90–95) n = 4; im. females, 110 ± 20.2 (84–148) n = 12; unsexed juveniles, 73 ± 13.0 (43–92) n = 11. The means and maxima fall well bellow the median in a database for 770 adults of all Mexican *Trachemys*. Most adult males are smaller than females, and sexual maturity is achieved at ca. 90 mm in males and in the range of 107–148 mm in females.

Size and weight of 20 newly emerged hatchlings in New Mexico was CL 31.5 mm ± 0.7 (30.1–32.7) and Wt 6.7 g ± 0.4 (5.6–7.4) (Stuart, 2000). Hatchling weight is 77% that of a freshly laid egg (Ewert, 1979).

CONSERVATION AND CURRENT STATUS

Ernst, Lovich, and Barbour (1994) review the role of *T. scripta* in the pet trade. In the early 1960s "turtle farms" in the southeastern United States (ca. 150) were producing 5–10 million turtles annually. This number fell to current levels (three to four million per annum) when the sale of turtles "under four inches" was banned in the United States and Canada. The reason for this was the risk of infection from *Salmonella* and related bacteria. The turtle farms were alleged to be self sustaining but were not; they removed about 9,000 adults per annum from the wild to maintain breeding stocks, seriously depleting natural populations. Juvenile sliders are still being exported from the United States to other countries (Japan, France, Italy, Hong Kong, Spain, England, Belgium, Germany, Mexico, and the Netherlands; from a summary in Ernst, Lovich, and Barbour, 1994).

T. s. elegans juveniles are the most common turtles sold in Mexican pet shops (RCV, pers. obs.). The turtles are first shipped to Japan from farms in the United States and thence to Mexico where they are referred to as "tortuga Japonesa"—all of which fuels an urban legend that they are native to Japan and will never grow to a larger size. The latter is generally true but only because most of them die (see also "Economic Use, Conservation, and the Law" in Introduction). No *T. s. elegans* are farmed in or shipped from Mexico. A few *T. s. venusta* were shipped from a turtle farm in Veracruz to Japan from 1996 to 2005 (RCV, pers. obs).

Trachemys scripta cataspila (Günther) 1885

Huastecan slider (Mast et al., 1986; Iverson, 1992a); jicotea Huasteca (Liner, 1994). The vernacular name was coined by Mast et al. (1986) "in recognition of the coincidental range of *T. s. cataspila* with the pre-Columbian Huastec Indian culture."

FIGURE 39.1 *Trachemys scripta cataspila*. Top—head of adult male, UU 9850, CL 236 mm (Arroyo Choreras, Las Norias, Tamaulipas); bottom—dorsal and ventral views of adult male, UU 9851, CL 213 mm (Arroyo Choreras, Las Norias, Tamaulipas).

ETYMOLOGY

Cataspila is a descriptive name, cleverly compounded by Günther: Gr. *kata*, down, and Gr. *spilos*, spot, speck, stain, blemish, alluding to the position of the ocelli on L2 and L3 near the lower edge of the scutes.

HISTORICAL PROLOGUE

Günther (1885) was vague about the number of types consulted in his description, but it is clear that he examined at least four specimens of 50, 127, 178, and 330 mm, the smallest being the juvenile illustrated in his Plate VI. Most of the following information was supplied by a lengthy communication from Dr. A. G. C. Grandison. The types are all in the British Museum (Natural History). Boulenger (1889) recognized seven syntypes (cotypes), three of which bore the locality "Mexico." An eighth cotype exists; it is BMNH 1947.3.4.25, a stuffed female (CL 13 inches, 330 mm) with a label identifying it as a type of *cataspila* and the data ". . . Tropical America. From the Zoological Society's Gardens." This specimen was listed by Gray (1855a,b) as *ornata* (specimen "j") and illustrated in Gray's Plate XII. Günther's reference to this plate in synonymy indicates that he probably examined the specimen and that it was the "13 inch" specimen he referred to in text. Gray's Plate XII is a mirror image. The specimen is in good condition, but the shell bears five holes that are probably spear or bullet wounds, a common circumstance in market specimens (see Legler, Smith, and Smith, 1980).

It seems logical to nominate this specimen (BMNH 1947.3.4.25) as the lectotype of *Trachemys scripta cataspila* (Günther, 1885).

Emys ventricosa Gray 1855 was placed in the synonymy of *Pseudemys scripta* by Smith and Smith (1979) based on information in Legler, Smith, and Smith (1980). The name was based on the shell of a specimen that had died in captivity and was part of a collection of reptiles emanating (in 1848) from a Mr. Warwick in Mexico, nearly all of which came from eastern Mexico. The name became a *nomen oblitum*, having been unused for more than 100 years.

RELATIONSHIPS AND TAXONOMIC STATUS

Trachemys s. cataspila is here regarded as most closely related to *T. s. taylori* and *T. s. venusta*.

DIAGNOSIS

A subspecies of *Trachemys scripta* that resembles both *taylori* and *venusta* and is distinguished from all Mexican *scripta* by the following combination of characters: (1) Postorbital

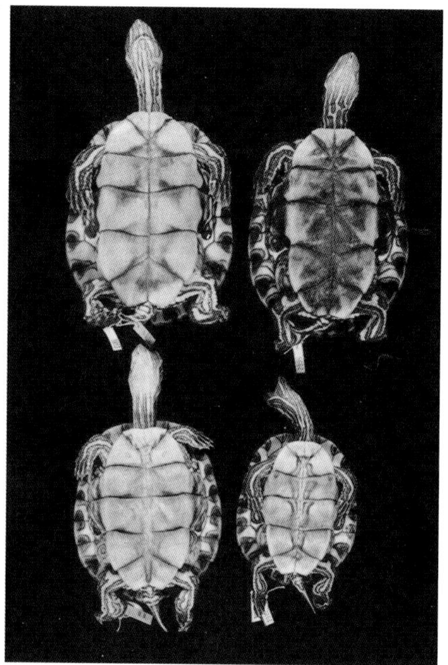

FIGURE 39.2 Hybrids (all immature) of *Trachemys scripta cataspila* × *elegans* from an intergrading population near La Laca, Tamaulipas (UU 9853-6).

stripe is expanded in temporal region and may be isolated from eye (66%) and/or neck (25%); expansion is pale orange to yellow, never bright red. (2) Primary orbitocervical is often the boldest, widest stripe in head profile (but never as evident and bold as in *venusta*). (3) Ocelli on L2 and L3 are dark centered and positioned on lower posterior quadrant of scute. (4) Carapacal pattern is commonly obscured by melanism, usually in males. (5) Plastral pattern is dusky and concentric with seam-following lateral extensions that lack terminal whorls (but seldom are straight-sided as in *venusta*). (6) Plastral pattern is clear and complete only in juveniles; it is usually obscured or obliterated by fading and then melanism in adults of both sexes.

GENERAL DESCRIPTION

Trachemys scripta cataspila ranks with *venusta* among the larger Mexican sliders, and maximum size for males (312) and females (314) is not much different.

BASIC PROPORTIONS

CW/CL: males, 0.75 ± 0.040 (0.69–0.90) $n = 44$;
females, 0.75 ± 0.033(0.71–0.83) $n = 16$.

CH/CL: males, 0.40 ± 0.019 (0.37–0.47) $n = 44$;
females, 0.43 ± 0.008 (0.38–0.48) $n = 16$.

CH/CW: males, 0.54 ± 0.03 (0.47–0.60) $n = 47$;
females, 0.56 ± 0.04 (0.50–0.65) $n = 17$.

PWHP/PWMF: males, 0.97 ± 0.04 (0.88–1.07) $n = 47$;
females, 0.96 ± 0.05 (0.85–1.05) $n = 17$.

BR/CL: males, 0.35 ± 0.009 (0.33–0.37) $n = 48$;
females, 0.36 ± 0.020 (0.33–0.38) $n = 17$.

WH/CL: males, 0.15 ± 0.013 (0.10–0.19) $n = 48$;
females, 0.15 ± 0.008 (0.14–0.17) $n = 15$.

Modal plastral formula: 4 > 6 > 3 > 1 > 5 > 2.

In the following description, colors are based on live specimens from Las Norias, Tamaulipas, and El Naranjo, San Luis Potosi.

Ground color of carapace is olive to pale or medium brown with a basically ocellar pattern. Ocelli tend to be complete, with dark centers and bordered with pale to bright orange, the orange becoming richer in older individuals. There is no yellow on carapace except for dark yellow on free margin. Carapacal pattern is often obscured by melanism in adults of both sexes but more extremely so in males. Plastral pattern is clear only in juveniles—dusky and concentric with seam-following lateral extensions that do not end in whorls; it is not straight-sided as in *venusta*.

Pattern of head and neck consists of yellow stripes, usually with fine black borders and contrasting with a dark ground. Primary postorbital stripe is sometimes simply a darker shade of yellow but usually some shade of orange, variably isolated from eye (47%) and/or neck stripe (86%) and usually expanded in temporal region; primary orbitocervical is often the boldest, widest stripe in head profile (but never as evident and bold as in *venusta*); mandibular stripe is isolated (30%) or not; pattern integrity is 0.563.

There is often a small anterior remnant of postorbital stripe in contact with the eye and isolated from principal postorbital stripe that is orange to near red in some individuals. Stripes on throat and sides of neck are generally pale yellow to cream yellow. Primary orbitocervical stripe is bright yellow to orange yellow on side of face, becoming paler posteriorly. Most stripes on dorsal field of head and neck are obscure, dark tan to brownish gold on black to blackish brown background. Outer paramedian stripes have a suggestion of very dull orange. Stripe from below nostril to mid maxilla is bright orange-yellow.

The expanded postorbital mark, whether connected to other stripes and eye or not, may not be uniformly colored. It is common to see the lower longitudinal half of the stripe as one color (usually brighter and containing orange) and the upper longitudinal portion as dull (tan, olive, etc.). When present, the isolated anterior postorbital mark tends to match the brighter part of the primary postorbital.

Iris is pale lemon yellow with very slight greenish and sometimes orange tinge. Tomial edges of mandibular sheath are serrate—coarsely so in large adults, finely in smaller individuals. There are distinct median alveolar ridges on maxilla and mandible that become more distinct with age; larger adults bear a single denticle on the anterior half of each maxillary alveolar ridge.

ONTOGENY OF PIGMENTATION

The plastron first loses the dusky juvenile pattern, becomes solidly pale with some melanistic mottling, and then undergoes melanistic darkening in adults of both sexes from posterior to anterior, a sequence similar to that in *taylori*. Slight traces of the concentric pattern may persist on the abdominals and as darkened interlaminal seams in immature individuals as large as 161 mm.

A series of large adult males from the Tamesi and Soto la Marina drainages of Tamaulipas shows a pattern of carapacal melanism that progresses with size (230–312 mm). The carapacal ocelli fade, and the ground color becomes nearly uniform pale olive; melanin is deposited along the interlaminal seams and, finally, as blotches on the scutes themselves, to produce a dark brown mottling. The same events occur in females with less intensity. Melanism is manifested also on the soft parts and eventually may obscure stripe patterns on the head and limbs.

GEOGRAPHIC DISTRIBUTION

Map 26

Trachemys s. cataspila occurs in the Gulf Coastal drainages from the Rio San Fernando of Tamaulipas (mouth at 24°45′N) southward to approximately Tuxpan, Veracruz (20°58′N) (see below). Most records are from coastal rivers and lagoons at low elevations. Observations (see Habitat) of a thriving population at El Naranjo, SLP (22°32′N, 99°14′W, ca. 250 m) demonstrate successful existence in smaller clear-water streams. Williams and Wilson (1965) record specimens from El Salto, SLP (22°31′N, 99°15′W), upstream and 14 km NW of El Naranjo, and from the Rio Verde at La Media Luna (21°55′N, 100°00′W, ca. 991 m). The Medio Luna record was confirmed by Iverson and Berry (1979) but attributed to introduction by Smith and Smith (1979) because of its high altitude. However, there is nothing unusual about sliders occurring at higher elevations. Several subspecies of Mexican *T. scripta* (*elegans*, *taylori*, *gaigeae*, *hartwegi*, and *yaquia*) occur at elevations greater than 700 m. Furthermore, El Naranjo (Rio Naranjo) and La Media Luna (Rio Verde) both drain to the Santa Maria tributary of the Rio Panuco with its mouth at Tampico. Other Panuco tributaries drain parts of eastern Queretero and Hidalgo, and the natural distribution of *cataspila* may well extend to these states.

The type locality was restricted to Tampico, Tamaulipas, Mexico, by Smith and Smith (1979).

INTERMEDIACY AND HYBRIDIZATION

Of 596 specimens of *Trachemys scripta* examined from the Gulf drainages of eastern Mexico, 82 are clearly *cataspila* and 499 are clearly *venusta*; four specimens from north of R. San Fernando were considered to be intermediate to some degree between *elegans* and *cataspila* and eight (R. Tuxpan, R. Tecolutla, and R. Nuatla) were considered to be intermediate between *cataspila* and *venusta*. These intermediates are regarded as hybrids. The natural range of *T. s. venusta* begins in the Rio Actopan (19°30′N) and Rio Antigua drainages (pers. comm., vendors and hunters in Veracruz).

The Rio Soto la Marina is the northernmost large river draining to the Gulf in northeastern Mexico. The Rio San Fernando is a modest perennial stream draining to the Laguna Madre. Northward to the mouth of the Rio Grande at Matamoros/Brownsville there are seemingly no natural, coordinated drainage systems or perennial streams. There is little suitable and no optimal habitat for sliders. This is Conjunto #1 of Tamayo (1946)—"Zona con desague deficiente."

Within this arid and relatively barren zone there are canals that run generally from NW to SE and emanate from an impoundment on the Rio Grande between Reynosa and Matamoros (Vas. Culebron, 97°50′W, based on ONC Chart H-23, 1973). The configuration of the canal system varies from map to map and probably from time to time. But, most importantly, these waterways are ultimately confluent with parts of the Rio San Fernando drainage and many of the small intermittent drainages to the Laguna Madre. This canal system and its connections could logically account for the presence of *elegans* south of its natural range and for *cataspila* north of its natural range.

Seidel, Stuart, and Degenhardt (1999) summarized the few specimens and localities for *Trachemys* that are known between the Rio San Fernando and Matamoros. All were classified as either *elegans* or *cataspila*. Their coverage of this matter purported to establish the northern limits of *cataspila* and the southern limits of *elegans* with a precision that permitted them to state that the ranges are separated by about 10 km along a line extending westward from the coast at ca. 25°24′N, near Reforma, Tamaulipas.

These specimens included the four immature females mentioned previously and by Legler (1990; UU 9853-56; CL, 98 to 133 mm; Figure 39.2). They were seined from a turbid, sterile-looking roadside pool at La Laca, 10 km NW of El Tejon, Tamaulipas (25°06′N, 98°07′W) and demonstrate a great range of variation that includes characters of both *elegans* and *cataspila*. All have an orange postorbital expansion connected posteriorly to a yellow stripe (no red in the head markings), and all have finely serrated mandibular tomium. One has a plastral pattern in which solid smudges are beginning to appear over the lateral whorls; the smallest specimen has a narrow plastral pattern more or less typical of *cataspila*, but the carapacal pattern has clear vertical bars on the laterals that are typical of *elegans*. It is concluded that this small series shows a preponderance of *cataspila* characters but a definite indication of *elegans* influence.

The eight specimens showing intermediacy between *cataspila* and *venusta* come from a 100 km stretch of coast extending from Tuxpan (20°58′N) to Nautla (20°12′N), all north of Punta del Morro, a headland extending from the Sierra Madre Oriental and interrupting the coastal plain at 19°52′N. From Nautla to Pta. del Morro (50 km) there is little typical slider habitat, and no specimens are available; the coastal plain is narrow, with the land rising abruptly to the west and dunes and a narrow beach with dunes to the east. The intermediacy observed north of Punta del Morro could be a relict of a former intergrading population or be attributed to human introductions from one population to the other. Concerning the latter, sliders are hunted and marketed both north (e.g., Tampico) and south (e.g., Cd. Veracruz) of Pta. del Morro, with the state of Veracruz producing the greatest harvest. It is quite logical that sliders collected at one venue could be sold at the other, and that escapees from vendors and buyers could occur. Also, Pta. del Morro would not preclude dispersal from one river mouth to another if turtles were washed out to sea by floods.

HABITAT

Huastecan sliders occur from sea level in swamps, estuaries, and large river habitats (e.g., Soto La Marina, Tamesi, Panuco) and in tributary streams at elevations to nearly 1,000 m in the foothills of the Sierra Madre Oriental. On the coastal road from Nautla to Pta. del Morro sliders were seen in muddy roadside ditches and in swampy habitats near mangroves. Local residents spoke of large yellow turtles in the coastal lagoons.

Most coastal habitats contained murky water. In stark contrast, habitat at El Naranjo (ca. 250 m) consisted of a number of clear, deep pools interconnected by fast, shallow riffles. Maximum depth in some pools was 5–6 m. They had a slight greenish cast but were clear enough to see bottom in the deepest places from vantage points on the high banks. Large cypresslike trees and brushy, herbaceous vegetation grew along the banks, and neighboring cultivated areas contained bananas, sugar cane, and fruit trees. Long grasses drooped into the water from overhanging banks, and there were numerous brush piles in the water. Active turtles were easily seen and very shy. Much of the river was shaded for part of the day.

DIET

Examination of eight dissected digestive tracts from the Tamesi/Panuco drainage (April 1962 and June 1965) showed a mixture of broad leaves, stems, and fruits (including figs); only one individual contained submergent vegetation (*Najas*), and none contained any trace of animal matter. These sparse data bespeak opportunistic foraging that exploits vegetation that falls into the water. At El Naranjo sliders ignored traps baited with sardines but entered those baited with fresh, prawnlike crustaceans.

REPRODUCTION

The following data suggest an annual reproductive cycle with vitellogenesis being completed in March and April, ovulation and nesting from March through June, and hatching from late June to mid October.

Females were bearing first and second clutches of oviducal eggs at El Naranjo on 4 April 1962. A gravid female (CL, 224 mm; Wt, 1,800 g) was recorded SE of Rancho Nuevo, Tampico (23°11′N, 97°47′W) on 24 June 1981 (Mast and Carr, 1986). These dates seem not significantly earlier than laying dates reported for *elegans* in Louisiana (Cagle, 1950).

Dissections of two gravid females from El Naranjo, San Luis Potosi, on 4 April 1962 revealed the following: UU 11346, CL 278 mm—10 thin-shelled oviducal eggs plus one unshelled ovum in an oviduct (a very rare phenomenon); fresh corpora lutea, 15; nine follicles, 11–25 mm; UU 11347, CL 265 mm—19 corpora lutea in two sets; four follicles, 18–20 mm; three follicles, 9–10 mm. These data suggest an annual reproductive potential of 24–26 eggs in two to three clutches.

Measurements of the 10 eggs from El Naranjo are as follows: L, 44.1 ± 1.49 (41.9–46.6); W, 28.4 ± 0.52 (27.6–29.4). Two eggs reported by Mast and Carr (1986) were as follows: 41.7 × 26.5, 16.92 g and 43.6 × 27.3, 18.85 g. These few eggs rank with the largest that have been recorded for *Trachemys scripta* anywhere.

Hatchlings with an unretracted yolk sac or a barely healed umbilicus and bearing a caruncle have been found on land from late June to mid July (J. L. Christiansen, field notes; Mast and Carr, 1986).

GROWTH AND ONTOGENY

Data on size (JML database) are as follows: adult males, 242 ± 42.9 (125–312) n = 48; adult females, 268 ± 47.6 (163–314) n = 15; im. males, 115 ± 9.2 (107–125) n = 3; im. females, 162 ± 59.3 (105–257) n = 8; juveniles, 68.0 ± 19.9

(42.0–89.0) $n = 4$. Three recently hatched individuals from different localities were all 40 mm CL.

The two largest immature females (251 and 257 mm CL, 30 km S of Cd. Mante) were ranked as subadult—they had many incipient follicles of 3–4 mm, but the bases of the ovaries contained granular follicles characteristic of juveniles; these individuals probably would have reproduced for the first time in the following year or perhaps very late in the current breeding season.

Growth rings on most specimens are too indistinct to interpret. A juvenile of 113 mm CL (UU 12737, Hacienda Tamiahua, Veracruz) is in its first or second full year of growth. Age in the following specimens from the Panuco drainage can be estimated as follows: males—UU 11351, 200 mm CL, ca. 7 years old; UU 11352, 159 mm CL, in third or fourth year; UU 6054, 175 mm CL, in fourth or fifth year; female—UU 11348, 168 mm CL, in fourth or fifth year.

The size attained by adult males is remarkable. Of six *Trachemys scripta* males larger than 300 mm (JML database), two of them (both 312 mm) are *cataspila*. The type of *Emys ventricosa* is ca. 340 mm (Legler, Smith, and Smith, 1980). Males lack elongated foreclaws. Large, old males have obvious elongated, upturned snouts that are neither sharply pointed nor bosslike; mature males smaller than 200 mm have essentially unmodified snouts, but all mature males have thickened, upturned prefrontal bones that are inwardly beveled on their anterior edges. The largest adults of both sexes come from the coastal reaches of large rivers (e.g., Soto la Marina).

FIGURE 40.1 *Trachemys scripta venusta*. Shell patterns (Alvarado, Veracruz, UU). Left, immature female; right, adult male.

CONSERVATION AND CURRENT STATUS

The paucity of published life history information for *cataspila* is surprising in light of its common occurrence along the northern Gulf Coast of Mexico, an area with roads, seaports, and railways. There are many opportunities for various local life history studies, especially where they can be done on private or protected lands. For conservation status see Introduction.

Trachemys scripta venusta (Gray) 1855

Neotropical slider (Moll and Legler, 1971); Meso-American slider (Iverson, 1992a); tortuga de Guadalupe (Liner, 1994); jicotea and tortuga pinta are local names.

ETYMOLOGY

L. *venustus*, beautiful, lovely; alluding to Venus, goddess of sexual love and physical beauty.

HISTORICAL PROLOGUE

Trachemys scripta venusta, as here defined, has the greatest geographic range and attains the largest size of any subspecies of *scripta*. It is one of the best known Mexican turtles from the standpoint of natural history. Vogt and associates conducted field studies of *venusta* in Veracruz, Chiapas, and Tabasco from 1980 to 2000. Legler and family collected and explored in Yucatán, Campeche, and Quintana Roo in 1966. Our collective database represents a total of 635 specimens, many of which were caught and then released after harvesting data. Preserved specimens are stored mainly at the Carnegie Museum of Natural History, the Chelonian Research Institute, and the University of Utah.

Moll and Legler (1971) established baseline natural history information for *venusta* in Panama (09°N), much as had been done with *T. s. elegans* at latitudes of 30–39°N in the northern temperate region (Cagle, 1950; Gibbons, 1970b; Gibbons et al., 1981; Gibbons and Semlitsch, 1982; Congdon and Gibbons, 1983). Information from intermediate tropical latitudes in southeastern Mexico results from published (Vogt, 1990) and much unpublished work by Vogt and associates, concentrated in the following areas: the Rio Lacantún and tributaries (Montes Azules Biosphere Reserve) extreme SE Chiapas (ca. 90°40′N, 16°28′W to 91°00′N, 16°15′W); the Laguna de Catazaja on the border of Tabasco and Chiapas (ca. 17°46′N, 92°02′W); various lagoons and rivers in the Biosphere Reserve of

FIGURE 40.2 *Trachemys scripta venusta*. Adult head striping (Alvarado, Veracruz, UU). Left, female; right, male.

Los Tuxtlas near Catemaco, Veracruz, and Estación de Biología Tropical "Los Tuxtlas" (ca. 18°10′N–18°45′N, and 94°42′W–95°27′W); the Rio Papaloapan drainage near Lerdo de Tejada and Alvarado, Veracruz (ca. 18°46′N, 95°45′W); and a study by Claudia Zenteno-Ruiz (1999) from 26.5 km N of Cd. Veracruz (19°10′N, 96°16′W), providing data on a population near the northern edge of the range of *venusta*.

Trachemys scripta is a neotropical organism only by virtue of occurring in the neotropics—not because of anything qualitatively distinctive about its life history. Its generalized habits and requirements evolved in the northern temperate region. As a generalist, *T. scripta* was preadapted for dispersal into the neotropics. The movement of *scripta* southward probably occurred during the Pleistocene, perhaps during periods of drought and lower sea levels when wider coastal plains provided uninterrupted dispersal corridors. Sliders probably founded aquatic niches with minimal competition from other aquatic turtles. Specific adaptations to the tropics are not identifiable and probably have not had time to evolve. Sliders can survive and prosper in a wide variety of environments including those resulting from anthropogenic works (e.g., deforestation and impoundments; Moll and Legler, 1971; Moll and Moll, 1990).

RELATIONSHIPS

Trachemys s. venusta does not occur in sympatry with any other *Trachemys*. The idea of sympatry with *T. s. grayi* was promulgated by Bogert (1961), in a review of Miguel Álvarez del Toro's "Los Reptiles de Chiapas," with a comment that Don Miguel had demonstrated sympatry in Chiapas. There is no such reference in the book, and Álvarez del Toro personally informed us that, to his knowledge, the two taxa occur in different drainages. The two subspecies are, however, kept in the same pools at the zoo in Tuxtla Gutiérrez and, fide Álvarez del Toro, do not interbreed in captivity (see also Reproduction, following).

Trachemys s. venusta displays substantial variation over its large geographic range. Much of the distribution and variation from Pta. del Morro (19°52′N) at least to the Isthmus of Panama is continuous and gradual along Atlantic coastal plains where there are few barriers. The exceptions are isolated populations in the cenotes and aguadas of the Yucatán Peninsula (see Habitat) and on islands (e.g., Cozumel).

DIAGNOSIS

A subspecies of *Trachemys scripta* most closely resembling *T. s. cataspila* and *T. s. grayi* and distinguished from all other *scripta* by the following combination of characters: (1) Postorbital stripe is not or but little expanded and is usually continuous from orbit well onto neck; there are no red markings on head. (2) Primary orbitocervical stripe is widest stripe of head and neck and dominates head profile. (3) There is a large, pale-centered ocellus near center of each lateral scute that occupies most of scute. (4) Patterns of carapace, plastron, and striping are clearly defined, bright, and colorful at all ontogenetic stages and in both sexes. (5) Plastral pattern is a dusky concentric figure varying in size from less than half to almost the full plastral surface but always straight sided and usually lacking lateral whorls (Legler, 1990: Figure 7.3c) .

GENERAL DESCRIPTION

Some of the largest Mexican *Trachemys* are from populations of *venusta*. Maximum CL (our databases): males, 302 mm; females, 311 mm.

BASIC PROPORTIONS

CW/CL: males, 0.72 ± 0.034 (0.66–0.80) $n = 99$;
females, 0.73 ± 0.027 (0.67–0.79) $n = 108$.

CL/CL: males, 0.38 ± 0.020 (0.31–0.44) $n = 99$;
females, 0.41 ± 0.010 (0.36–0.48) $n = 108$.

CL/CW: males, 0.53 ± 0.04 (0.44–0.62) $n = 72$;
females, 0.56 ± 0.04 (0.49–0.68) $n = 52$.

PWHP/PWMF: males, 0.96 ± 0.04 (0.90–1.05) $n = 89$;
females, 0.97 ± 0.04 (0.86–1.08) $n = 99$.

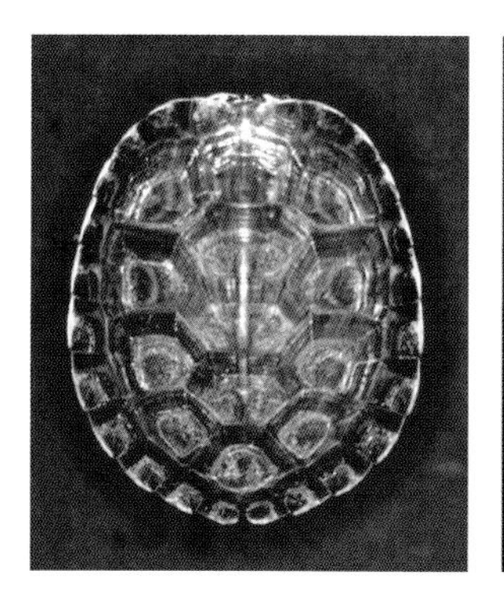

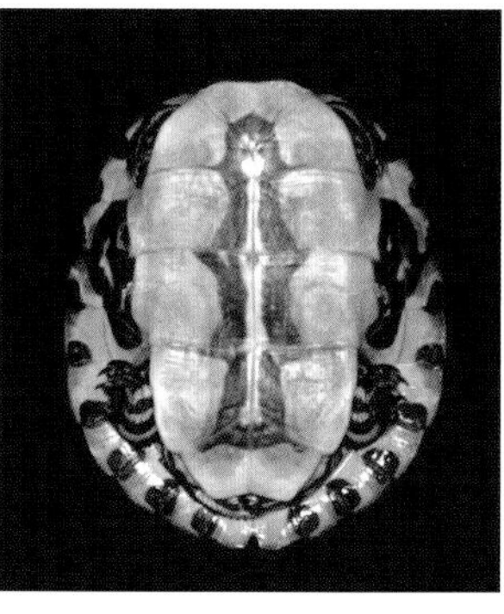

FIGURE 40.3 *Trachemys scripta venusta*. Juvenile shell (Alvarado, Veracruz, UU).

FIGURE 40.4 *Trachemys scripta venusta*. Neonate (UU, Tecolutla, Veracruz).

BR/CL: males, 0.35 ± 0.023 (0.30–0.39) $n = 89$;
females, 0.37 ± 0.01 (0.32–0.40) $n = 97$.

WH/CL: males, 0.15 ± 0.012 (0.13–0.17) $n = 80$;
females, 0.15 ± 0.010 (0.13–0.19) $n = 85$.

Modal plastral formula: $4 > 6 > 3 > 1 > 5 > 2$.

Patterns of carapace, plastron, and striping are nearly always clearly defined, bright, and colorful in adults and juveniles. Plastral pattern is usually lacking lateral whorls and secondary melanistic smudges. Pattern of head and neck consists of bright, wide, yellowish stripes contrasting with a darker ground. Symphyseal stripe often (42%) forms a Y; mandibular stripe is usually (84%) isolated. Pattern integrity is high (0.585). The primary orbitocervical stripe is wide, golden yellow, usually the widest and brightest lateral stripe, and dominates the profile of the head and neck (Figures 40.2 and 40.5). Tomial edges of jaw sheaths are serrate or dentate or, after wear, crenulate.

COLORS IN LIFE

Adults (five females, six males, from 10 km W of Alvarado, Veracruz, UU): Orbitocervical stripe is wide, bright orangish yellow; other pale marks have orangish cast; otherwise, they are bright yellow, slightly paler, and greenish on chin and throat. Iris is clear, bright, pale yellow, sometimes greenish at periphery. Carapace ground color is brownish; centers of ocelli are blackish and bordered first by pale dusky yellow and then by dusky orange (reddish orange on marginals). Ground color of plastron is rich, sometimes orangish, yellow, especially in males. Plastral figure is black and is distinct and narrow in males, but more diffuse in females.

Hatchlings (seven sibs, 2- to 3-weeks old, progeny of one of aforementioned females, UU): Ground color of plastron is lemon yellow, about same as dried shell of mother. Plastral figure is slate black, restricted to midline of plastron, interrupted along abdominofemoral seam in some specimens, extending no farther than posterior gular apex. Ground color of soft parts is slate black with exception of top of head, which is more nearly dark olive. Expanded part of postorbital stripe has egg yolk orange cast as does first part of stripe following posteriorly. Expansion is distinctly bordered above and below by dusky cream, the upper border being wider than the lower. Ground color of carapace is greenish, and each central scute has a pair of dusky orange parenthetical marks. Each lateral scute has major ocellus of dusky orange with incomplete black center. Iris is clear pale yellow.

Adult male and female (Playas de Catazaja, extreme NE Chiapas, 17°46′N, 92°02′W): Plastral ground color is dark to golden yellow, much darker than bridge or inframarginal surfaces. Plastral figure is slate, dusky, rather distinct, and narrow in male, less distinct and partly obliterated in female. Ground color of bridge and inframarginal surfaces is pale, clear, slightly greenish yellow. Inguinal pockets are cream, with a definite greenish-yellow cast. Limbs are boldly and contrastingly marked—ground color is nearly black with suggestions of brownish black in some places; there are pale stripes and other pale markings on limbs and tail about the same as ground color of bridge and inframarginal surfaces but brighter—generally pure, pale greenish-yellow.

Carapace is dark, dusky, dark brownish to brownish slate with obscure markings, especially in female. Lateral scutes of male show a black ocellus on central posterior third of scute surrounded by C-shaped dirty, pale orange ring that is bounded distinctly by black.

Postorbital stripe is expanded behind eye. Stripe from eye to expansion is dirty brownish yellow. Upper two-thirds of expansion is brownish olive, slightly paler than ground, and maintains a barely distinguishable darker border. Lower one-third of expansion is dirty greenish yellow, continuing backward into the purer greenish-yellow neck stripe. Primary orbitocervical stripe is broad and bright greenish yellow.

FIGURE 40.5 *Trachemys scripta venusta.* Sexual dimorphism in snout and head profile in adults from Dziuché, Quintana Roo, UU. Top, female; bottom, male.

ONTOGENY OF SHELL PATTERN

Melanism is less common in *venusta* than in its nearest neighbors, *cataspila* and *grayi*, and is less common in females than in males. The carapace may become darkened and its pattern less distinct, but the plastral pattern is rarely obscured by melanism. In some populations turtles become completely melanistic, particularly males (Vogt, field data). Vogt found five melanistic adult males with CL 269–295 mm and one melanistic adult female with CL 320 mm in a total of 397 adults from the Rio Lacantún and its tributaries in the Montes Azules Biosphere Reserve from 1986 to 1992.

TYPES AND TYPE LOCALITY

We restrict the type locality to Belize City, Belize (= British Honduras until 1981) in lieu of "Honduras" (Smith and Taylor, 1950a) for reasons stated in the following.

There are eight specimens in the type series, all in the British Museum (Natural History). Smith and Smith (1979) nominated BMNH 1845.8.5.26 (1947.3.4.80 current catalog number) as the lectoholotype. Gray (1855b) refers to this specimen as "e. Half-grown . . . . . Honduras. Mr. Dyson's Collection" and, more completely (1873:48) as "e. Animal. Shell, 8 1/4 in. *Emys venusta*, Gray, Cat. 24. t. 12, 8 Honduras. Dyson. 45, 8, 5, 26." Further information gathered by JML with the help of Ms. A. G. C. Grandison and Mr. A. Stimson reveals that the records for this specimen include the locality, "Belize, Honduras"; the specimen is stuffed and attached to a wooden base bearing a label, "The specimen figured." Plate XIIA (Gray, 1855b) is a mirror image of the lectoholotype, which is an immature female (CL, 210 mm).

GEOGRAPHIC DISTRIBUTION

Map 27

Atlantic drainages from the beginning of the coastal plain just south of Punta del Morro, southeastward through the lowlands of Veracruz, Oaxaca, and Tabasco, in isolated

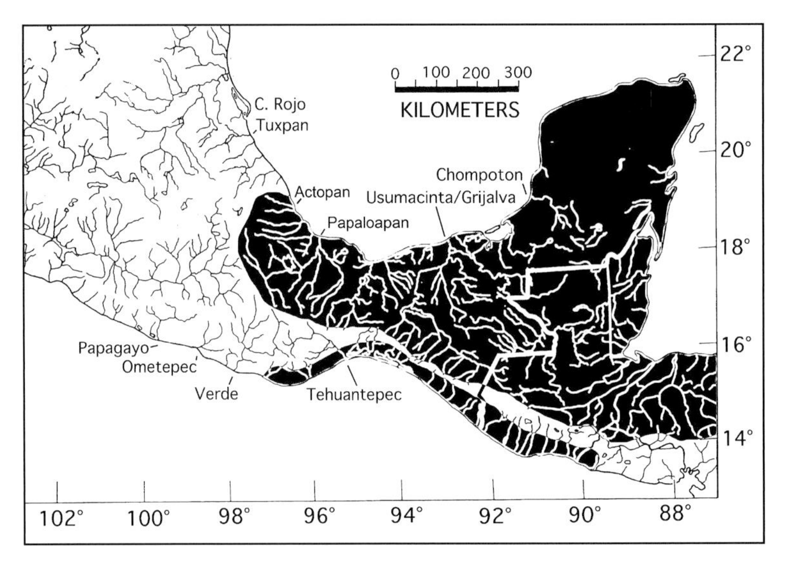

MAP 27 Geographic distribution of *Trachemys s. venusta* (Atlantic and Caribbean) and *T. s. grayi* (Pacific).

FIGURE 40.6 *Trachemys scripta venusta*. Shell patterns on the Yucatán Peninsula near Puerto Morelos, Quintana Roo. Left, immature female; right, juvenile.

aquatic habitats throughout the Yucatán Peninsula and thence eastward and southward along the Caribbean coasts of Belize, Guatemala, Honduras, Nicaragua, Costa Rica, and Panama to the Atrato drainage of Columbia. *T. s. venusta* probably crossed the Isthmus of Panama to colonize the Pacific coast from Panama northwestward to the Golfo Dulce of Costa Rica (Legler, 1990).

HABITAT

Jicoteas occur in most kinds of permanent aquatic habitats: marshes, ponds, lakes, rivers, and estuaries with detectable salinity. In rivers and large lakes they favor shoreline habitats and depths of 1–3 m. They sometimes migrate into temporary ponds during the rainy season.

Vogt (pers. obs.) found subadult *T. s. venusta* aestivating on land in ca. 10 cm of leaf litter beneath low shrubs at Lerdo de Tejada, Veracruz. The aestivation sites were between a dry marsh and the river, about 20 m from the river, and were shared by a variety of other turtles (*Kinosternon acutum*, *K. leucostomum*, *Staurotypus triporcatus*, *Claudius angustatus*, and *Chelydra rossignoni*).

T. s. venusta occurs with *Dermatemys* in large rivers and in lakes and marshes formed from oxbows. They are more likely to be found in still water, whereas *Dermatemys* prefers moving water or slow deep pools between stretches of fast water.

DIET

Jicoteas are opportunistic omnivores, consuming most of the plants and animals that occur in or fall into natural aquatic habitats. Notable are aquatic insects, crustaceans, and fish (live or as carrion). However, 80% of the diet in Veracruz and Chiapas was found to be plants: leaves, flowers, fruits, and seeds. Juveniles often contained large quantities of filamentous algae (Vogt and Guzman, 1988; Vogt, unpublished). Moll and Legler (1971) report similar results from stomach analysis in Panama, but also that dietary preferences differed between populations in the same drainage system.

HABITS AND BEHAVIOR

Jicoteas are commonly seen basking on logs or almost any other object projecting above the surface, especially in the morning. Heat gained by basking presumably optimizes metabolic functions associated with reproduction and diet. Basking may also be of value in control of epizoic algae and ectoparasites, and in the synthesis of vitamin D. Turtles of both sexes and most age groups often bask in close proximity on the same object. Communal basking may be advantageous as an early warning system for the approach of predators (see discussion in Moll and Legler, 1971).

REPRODUCTION

Data in this section, unless otherwise cited, are from Vogt (1990) or our combined unpublished data and observations.

Courtship and copulation were observed by Vogt (pers. obs.) in large artificial enclosures at Catemaco and Nanciaga, Veracruz, in January, February, and March (1981–1991). Typical behavior was for a male to swim toward a female and to press his snout on or near her cloaca. The male then swam rapidly in front of the female, came into snout-to-snout contact, and vibrated his head in a vertical plane for 1–2 minutes. When head movements were terminated, the male swam around behind the female and attempted to mount. No foreclaw titillation occurred. Seidel and Fritz (1997) also described this behavior in captives (six males and one female) from Tlacotalpan, Veracruz. They also observed foreclaw titillation behavior between males and interpreted it as aggressive. Moll and Legler (1971) observed two natural matings in Panama, each consisting of a rear approach and mounting without foreplay.

Rosado (1967) observed *venusta* and *grayi* in an artificial concrete tank at Teapa, Tabasco; the following remarks probably refer to *venusta* (sensu stricto). Mating was observed throughout the year. The male pursues the female aggressively; the female stops and the male places himself in front of her "... and sprays her with water ejected through his nostrils. Following this he caresses her neck and they finally submerge to the tank bottom where the act is consummated." "Spraying" has been reported in several

FIGURE 40.7 Aquatic habitats on the Yucatán Peninsula. Left, from top—Aguada Sayusil near Villa Union, Yucatán, July 1966: an extinct cenote that has become a lake; a small ground-level opening to an underground cenote; roadside aquatic habitat in borrow pits. Right—a small, rounded, senescent cenote near Villa Union, Yucatán; shear rock walls surround a small body of water in which *Trachemys scripta venusta* and *Kinosternon scorpioides* occurred.

other kinds of turtles (including *T. s. gaigeae*; Stuart and Miyashiro, 1998). It may simply be a way to expel water from the buccopharyngeal cavity before breathing at the surface.

The following behavior was observed by Vogt and Martha Harfush in a captive population at Centro Mexicana de la Tortuga, Mazunte, Oaxaca, in early April 2006. Adults of *T. s. grayi* and *T. s. venusta* (ca. 35 total) shared a tank of ca. 2 × 4 m and 0.5 m deep; no hybrids were evident. Male *venusta* were actively courting female *venusta* and paid no attention to either sex of *grayi*. Courtship behavior of *venusta* was as described previously. There was no mating behavior of any kind observed in *grayi*.

Vitellogenesis begins in mid September and coincides with a decrease in environmental temperature. Follicular enlargement coincides with the highest rainfall of the year (September to October). By mid October most follicles have enlarged to 7–13 mm in diameter, and some have reached 14–20 mm. Follicles are preovulatory at diameters of 20 mm or greater. The first ovulations occur in January (Vogt, 1990). Ovulation occurs at the end of the relatively cool rainy season (late January to early February) when temperatures begin to rise at the beginning of the dry season. Gravidity occurred as follows for 200 females in Chiapas (1984–1988): earliest oviducal eggs were noted 15 January; 15 January–15 February, 20% of females were gravid; 15 February–1 March, 30% were gravid; some females were gravid as late as 23 May. The incidence of gravidity in four samples from Veracruz in 1995 was as follows: 22 February, 9 of 26 (34.6%); 22 March, 20 of 59 (33.8%); 12 April, 8 of 43 (18.6%); 12 June, 2 of 37 (05.4%) (Zenteno-Ruiz, 1999).

Nesting occurs between February and June, within the dry season (late January through May), when daytime temperatures are as high as 35°C. The onset of the "summer" rainy season (late May to early June) marks the end of nesting and the initiation of a 4-month period of ovarian quiescence.

Nesting sites were similar in all areas studied in Veracruz and Chiapas. Nests along the Rio Lacantún (SE Chiapas) were on sandbars, 3–6 m from water in shaded and nonshaded locations. Nests near Lerdo de Tejada, Veracruz, were on natural mounds of sand or clay at the edges of marshes and were often overgrown by vegetation (Vogt, pers. obs.). Zenteno-Ruiz (1999) found 61 nests concentrated (1.2 nests/sq m) in a communal nesting area and 24 nests more dispersed (0.15 nests/sq m) on the shores of the same lake near Cd. Veracruz. Jicoteas are known to swim out to sea from large rivers and nest on beaches along the Caribbean coast of Mesoamerica (Moll, 1994), but this has not been documented for Mexico.

Based on ovarian examination and oviducal egg counts (23 females from the Rio Papaloapan drainage [Veracruz] and the Rio Lacantún drainage [Chiapas; Vogt, 1990]), two to four clutches are produced per year (mean, 3.5). Mean number of eggs per clutch (31 clutches) was 12.03 ± 3.75 (5–22); mean clutch mass was 143.8 g ± 69.86 (33.7–417). By rough calculation, mean annual reproductive potential would be 42 eggs per year with extremes of 10–88 eggs. There is no indication that mature females skip seasons of reproduction.

Data on size for 373 eggs from 31 clutches (southern Veracruz and Chiapas) are as follows: L, 38.1 ± 3.3 (28.0–48.8); W, 22.6 ± 3.0 (18.6–29.3); and Wt, 11.8 g ± 3.33 (6.4–22.7) (Vogt, 1990).

Egg size was not correlated with number of eggs per clutch. Female size (CL and weight) is significantly and positively correlated with egg length, width, and weight. Eggs of Mexican *venusta* are smaller than those from Panama (Moll and Legler, 1971).

Zenteno-Ruiz (1993) gathered data on eggs from nests in a captive population (a "turtle farm") at Nacajuca, Tabasco (12 km N of Villahermosa). Data are from 3,669 eggs removed from 348 nests in two complete and successive laying seasons. Number of eggs per clutch: 10 ± 2.7 (4–17) n = 285 in 1991, and 10.6 ± 3.15 (4–23) n = 421 in 1992. Nesting occurred from 28 January to 7 May in 1991 with a peak in March, and from 23 January to 8 June in 1992 with a peak in April. Mean number of eggs per clutch was eight at the beginning of the season, 10.5 at the peak, and 7.5 in the final nests of the season. Overall data on eggs were as follows: L, 40.3 ± 3.7 (30.0–49.0) n = 350; W, 23.7 ± 3.42 (19.0–27.0) n = 400; Wt, 13.28 ± 3.42 (9.64–17.20) n = 400. Clutch weight was 129.3 ± 30 (65–185) n = 50. Observed nest predators were ants, beetles, and opossums (*Didelphis marsupialis* and *Philander opossum*).

Incubation time at lab ambient temperatures (26–28°C) was 63–74 days. Laboratory incubation times at controlled temperatures were s follows: 27.5–28.5°C, 62–74 days; 28.5–29.5°C, 50–62 days; and 29.5–30.5°C, 42–57 days (Vogt, 1990). Incubation time at the same temperature is slightly shorter in southern Mexico than that for more northern populations in the United States (Vogt, 1990; Ewert, 1985).

No nest temperatures are available for tropical Mexico. Moll and Legler (1971) recorded natural nest temperatures in Panama of 22–30°C and estimated incubation time to be ca. 80 days under natural conditions. Incubation time in artificial nests constructed in shade and sun in Panama were as follows: 73 days at 29°C (23.9–38.9) and 94 days at 26.8°C (24.4–31.0), respectively.

Sex determination is temperature dependent (Vogt and Flores-Villela, 1992a). Only females are produced above 29°C; a few males (1 male:4.2 females) are produced at 28.5°C, and a 1:1 sex ratio is produced at a threshold temperature of 27.5°C. Threshold temperature is 1°C lower in neotropical Mexico than in similar sites studied by Bull, Vogt, and McCoy (1982) for *Trachemys scripta elegans* in the southern United States. Incubation times and threshold temperatures remain to be determined in other Mexican *Trachemys*. It would be interesting to study the threshold temperature of *T. scripta* populations (e.g., *taylori*, *gaigeae*, and *hartwegi*) occurring in hot desert habitats.

Eggs hatch in nests of differing ages over a period of 3 months. There is sufficient yolk in the yolk sac to sustain the hatchlings for extended periods. Emergence is synchronous and coincident with the first heavy rains in June. The emerging hatchlings (of different actual ages) then quickly make their way to water. In some situations (nests in clay soil), the hatchlings are actually trapped in the nest until rain or floodwaters moisten the soil.

GROWTH AND ONTOGENY

Mature males have an elongated, pointed, upturned snout (less pronounced than in *T. s. grayi*) and lack elongated foreclaws. The carapace is completely smooth in adults but has a mid-dorsal keel in hatchlings.

Sexual maturity occurs at carapace lengths of 110–153 mm in males and 162–206 mm in females. Turtles were sexed by gonadal examination or sexual dimorphism. Data on size (CL, mm) for 240 individuals of *venusta* from southern Mexico and Belize (JML database) are as follows: males, 188 ± 43.1 (110–302) n = 89; females, 231 ± 32.6 (162–311) n = 101; im. males, 126 ± 18.3 (113–153) n = 4; im. females, 151 ± 28.8 (106–206) n = 18; unsexed im. (probably females), 120 ± 19.4 (97–153) n = 8; unsexed juveniles, 55 ± 22.3 (34–96) n = 20.

Size at sexual maturity may vary from locality to locality and may be correlated with the size of the aquatic habitat (see "Habitat" and discussion of Yucatecan populations in the following).

Size data for 70 mature females from Laguna Oaxaca, Chiapas (16°60′N, 90°40′W), were as follows: CL 276 mm ± 34.9 (202–337); Wt, 3,185 g ± 1,235 (800–6,500). Size data for 154 hatchlings from 18 clutches (southern Mexico, laboratory incubation) were as follows: CL 31.8 mm ± 2.7 (25.0–37.8); Wt, 7.24 g ± 1.68 (3.5–10.7) (Vogt, 1990). Some of the variation in size may result from temperature during incubation. Ewert (1985) reported that smaller hatchlings were produced at higher incubation temperatures.

Size and weight of 34 adult females (Rio Lacantún) were as follows: 284 mm ± 32.3 (188–330) and 3,350 ± 974.4 g (1,100–5,050). No immature females were caught. Size and weight of 65 adult males was as follows: 241 mm ± 28.1 (187–331) and 1,896 g ± 668.4 (900–4,400); 13 subadult males were CL 162 mm ± 20.2 (128–179) and 621 g ± 223.6 (300–900). No juveniles were caught. The dearth of immature stages was attributed to crocodile predation (Vogt, 1990).

Growth is rapid under favorable conditions. Two young specimens (UU, 16 km S and 8 km W of Pto. Morelos, Quintana Roo, 29 July 1966) have carapace lengths of 58 and 60 mm; both have a clear areola and a single, broad, unbroken zone of growth and are judged to be in the season of hatching.

TRACHEMYS SCRIPTA VENUSTA ON THE YUCATÁN PENINSULA

It has been suggested that sliders on the northern half of the Yucatán Peninsula warrant taxonomic distinction (Vetter, 2005; "Yucatán form"). Much of the Yucatán Peninsula consists of irregular limestone formations in which erosion has produced fissures, sinkholes, underground streams, and caverns (cenotes). Tamayo (1949:247–248) presents a graphic sequence showing the transformation of a closed cenote to an open pit and ultimately to a lakelike aguada.

Trachemys on the Yucatán Peninsula occur in small, sometimes temporary waters—aguadas, open cenotes, roadside ditches, and borrow pits. There is little or no flowing surface water on the Yucatán Peninsula north of 18°.

Small populations of *venusta* occur in most of these situations and are isolated, either by virtue of distant proximity to other aquatic habitats or by shear rock walls that permit turtles to fall in but not to exit. These karst habitats are (or were at some time) interconnected by subterranean water systems. Marine fishes (e.g, tarpon) occur in cenotes that are distant from the ocean (Hubbs, 1936, 1938; Stuart, 1964; JML, pers. obs.).

Work on the Yucatán Peninsula by Legler and family in 1966 shed some light on differences in size and reproductive cycle.

A senescent cenote observed near Libre Union, Yucatán (Figure 40.7), consisted of a miniature lake surrounded by vertical rock walls about 14 m high. The lake had an earthen shore, contained various aquatic vegetation, and was ringed by terrestrial vegetation including mature coconut palms. One could stand on the rim and be above the crowns of the trees. Traps had to be lowered on long lines. *T. s. venusta* and *Kinosternon scorpioides* were observed and trapped in the cenote. Presumably turtles moving about on land continue to fall in.

Ingress and egress of turtles via underground connections may or may not occur, but this possibility is not fanciful. Legler observed a small cenote between Puerto Juárez and Puerto Morelos in northern Quintana Roo with a surface opening of scarcely 4 sq m (Figure 40.7). Crystal clear water welled up almost flush with ground level; a tarpon was observed deep inside and a *Trachemys* (subadult) was seen as it surfaced, breathed, and then disappeared into the depths.

Specimens from aguadas and senescent cenotes in the region of Libre Union, Yucatán (ca. 20°42'N, 88°49'W), and elsewhere on the northern Yucatán Peninsula were smaller than those from the main Mexican range of more southern *venusta* populations (e.g., Rio Usumacinta, 250 km SW in Tabasco), as follows:

Males, 153 ± 29.4 (110–204) $n = 31$.

Females, 222 ± 17.6 (187–243) $n = 10$.

Im. females, 143 ± 26.0 (116–182) $n = 8$.

The Aguada Sayusil (ca. 400 m in diameter) was the largest body of water from which specimens were obtained. Visual observations and trapping indicated large populations of *Trachemys scripta venusta* and *Kinosternon scorpioides*.

Data on size of 20 eggs from two gravid females (9 and 11 thin-shelled eggs, CL 242 and 225 mm, respectively) were as follows: L, 38.4 ± 2.36 (30.9–42.0); W, 24.8 ± 0.82 (23.5–25.9). Both females also bore 10 potentially preovulatory follicles (7–22 mm, in two groups). These were the only gravid females found by Legler, 4 July–14 August 1966.

Eight other females from the northern Yucatán Peninsula had not ovulated but bore 5–16 potentially preovulatory yolked follicles of 5–20 mm in one to three size groups. Three females (204 to 221 mm CL) bore a single yolked follicle (8–13 mm) and may have been destined to skip a breeding season. Stomachs of dissected specimens were tightly packed with vegetation (grasses, roots, and pieces of leaves from broad-leafed plants) plus a few snails, insects, and anuran eggs.

The foregoing data suggest that vitellogenesis begins long before July, and that ovulation and nesting might begin as early as July and continue at least into the months of August and September. These data constitute an incomplete view of a breeding schedule that differs significantly from populations of *venusta* in wet tropical areas, where the reproductive cycle proceeds from vitellogenesis in September to ovulation, gravidity, and nesting from January to March.

All of the above-mentioned northern Yucatecan females lacked identifiable corpora lutea. In those that had not ovulated, this could be explained logically by the regression of corpora lutea from the previous reproductive season—but not for the gravid females with thin-shelled eggs. In our experience, thin-shelled oviducal eggs are usually the result of very recent (24 hours or less) ovulations and are represented by large, open, flaccid, and often bleeding corpora lutea. We interpret this circumstance as a delay in shell formation in the earliest clutches of the year coupled with a late dry season, until such time as climatic conditions stimulated a resumption of shell formation and ultimately nesting. Regional rains began in late July in 1966.

Although the northern Yucatán region is included in the frost-free tropics of Mexico, its climate differs markedly from that over most of the range of *T. s. venusta*. Seemingly the long dry period, lasting usually until May or June (Peterson and Haug, 2005) functions, much in the manner of cold temperate zone winters, to inhibit ovulation. Only after the rains have begun in June or July do the turtles begin to ovulate.

PREDATORS, PARASITES, AND DISEASES

Crocodiles (*Crocodylus moreleti*) are probably major predators of aquatic turtles in many areas of southern Mexico. Many (ca. 25%) adult *Staurotypus triporcatus* in Laguna Oaxaca showed signs of crocodile predation (pieces of carapace missing or carapace with obvious large, round holes the size, pattern, and shape of crocodile teeth marks [Vogt, 1990]). *Trachemys* and *Dermatemys* did not have such wounds, and this suggests that when these species are attacked by crocodiles they are killed and eaten. The bottom-dwelling juveniles of *Staurotypus* were abundant, whereas juvenile *Trachemys* and *Kinosternon leucostomum* of any size, which spend more time at the surface, were rare. Crocodiles are notorious for surface feeding (Vogt, 1990).

The possibility of shark predation exists in the lower reaches of large rivers and estuaries (see remarks in Introduction).

An immature male *venusta* (Rancho Las Vegas, 11.5 km S of Puerto Juarez, QR) had a small tick attached to the neck where the skin joins the precentral scute; this uncommon occurrence might suggest that juveniles spend some of their time on land or on the floating vegetation that covers most of a local cenote (JML, pers. obs.).

POPULATIONS

Zenteno-Ruiz (1999) captured and marked 333 *T. s. venusta* in a 6.57 ha lake 26.5 km N of Cd. Veracruz from 22 February to 20 September 1995. She estimated 154 turtles/ha as follows: 97 females (58 adult) and 56 males (52 adult). Biomass for all age groups was calculated as 174 kg of females/ha and 43 kg of males/ha. The sex ratio was 37% males to 63% females.

During Vogt's 20 year tenure in Mexico (1980–1999), no attempt was made to conduct specific population studies of *Trachemys scripta venusta*, but many (>400) were caught, marked, and released in the course of work on other species in Chiapas, Tabasco, and Veracruz. All turtle communities studied in the mentioned period showed evidence of at least some human disturbance. *Trachemys s. venusta* shows remarkable versatility in habitat tolerance and preference. The following remarks demonstrate this with comparative data from various habitats. Where possible, these data are given for periods before and after major periods of harvesting by humans. The data also indicate habitat preferences for the various species mentioned.

LARGE RIVERS

Rio Tzendales: a cold, fast, clear-water tributary of the Rio Lacantún, in lowland tropical rainforest and completely protected within the Reserve of Montes Azules, Chiapas. The turtle population in 1995–1996 (n = 64) was 22% *T. s. venusta* and 78% *Dermatemys mawi*. In 2002 (n = 27) *Dermatemys* had been virtually extirpated by poachers, and relative percentages were 98% *T. s. venusta* and 2% *D. mawi*.

Rio Blanco: a large, meandering, turbid, vegetation-choked tributary of the Rio Papaloapan in Veracruz. The turtle population in 1995–1996 (n = 54) was 8% *T. s. venusta*, 77% *D. mawi*, and 15% *Staurotypus triporcatus*.

SMALL STREAMS

Rio La Margarita: a small, cold, clear-water tributary of Lake Catemaco, Veracruz: In 1995–1996 turtle populations (n = 211) were 40% *T. s. venusta*, 25% *S. triporcatus*, and 35% *Kinosternon leucostomum*. At this time a local fishing cooperative prohibited the taking of turtles. At some time prior to 2002 an intensive turtle harvesting had occurred. Samples in 2002–2003 (n = 1,065) were 7% *T. s. venusta*, 8% *S. triporcatus*, and 85% *K. leucostomum* (De la Torre Loranca, 2004). The percentage of adults decreased from 75% to 63% in *Trachemys* and from 80% to 16% in *S. triporcatus*. The population structure of *K. leucostomum* remained unchanged. There is no evidence that *Dermatemys* has ever occurred in the Catemaco drainage (Vogt, pers. obs.).

Rio Água Dulce, La Palma, Veracruz: a small isolated stream draining to the Gulf of Mexico; the part of the stream studied was forested on both sides and turtles had not been heavily harvested in 1995–1996; populations (n = 32) were 8% *T. s. venusta*, 71% *S. triporcatus*, and 21% *K. leucostomum*.

LAKES

Laguna Oaxaca: a warm, turbid, shallow, isolated oxbow lake adjacent to the Rio Lacantún, Chiapas. Access was difficult (a portage of 800 m from the river) and probably discouraged poachers. Populations in 1993 (n = 276) were 11% *T. s. venusta*, 22% *D. mawi*, and 67% *S. triporcatus*. A new road to a gravel mine subsequently provided easy access for poachers, and by 1996 turtle populations had been almost exterminated (98% calculated). Five days of trapping produced eight juvenile turtles—one *T. s. venusta*, two *D. mawi*, and five *S. triporcatus*.

MARSHES

The Pantanos de Centla near Villahermosa, Tabasco: a large marshland with clear water and abundant vegetation showed the effects of overharvesting by 1995–1996; turtle populations (n = 16) were 86% *T. s. venusta*, 7% *D. mawi*, and 7% *K. leucostomum*.

Rio San Augustin, near Lerdo de Tejada, Veracruz: a slow-moving, silty river meandering through a wetland; samples in 1995–1996 (n = 86) were 53% *T. s. venusta*, 1% *D. mawi*, 21% *S. triporcatus*, 24% *K. leucostomum*, and 1% *C. rossignoni*. Another sample (n = 284) from a marsh adjacent to the river was comprised of 3% *T. s. venusta*, 92% *Claudius*, 5% *K. leucostomum*, and 1% *C. rossignoni*. Farther inland there were scattered pools in shrubby woodland where the turtle community (n = 239) was comprised of 95% *Kinosternon acutum*, 2% *K. leucostomum*, and 3% *Claudius angustatus*.

CONSERVATION AND CURRENT STATUS

In tropical Mexico (Veracruz, Chiapas, Tabasco, and Campeche) *Trachemys s. venusta* is commonly sold in markets and are supplied by market hunters; it is a culinary item second only to *Dermatemys mawi*. Information on the decline of populations appears previously under "Populations" and under "Conservation" in the Introduction.

FIGURE 41.1 *Trachemys scripta grayi*. Left, adult male; right, adult female (Pacific coastal Oaxaca near Mazunte).

Trachemys scripta grayi (Bocourt) 1868

Tortuga negra (Álvarez del Toro, 1973); Gray's slider (Iverson, 1992a); jicotea negra (Liner, 1994).

ETYMOLOGY

A patronym honoring John Edward Gray, 1800–1875.

HISTORICAL PROLOGUE

This taxon is not well known. Sumichrast (1880 and 1882) made some observations on *grayi* that are among the earliest natural history notes on Mexican turtles. Dean (1980) provided useful observations on its natural history in an unpublished thesis devoted chiefly to *Staurotypus salvini*.

RELATIONSHIPS AND TAXONONOMIC STATUS

Trachemys s. grayi is seemingly most closely related to its nearest neighbor, *T. s. venusta* (Legler, 1990; Seidel, 2002). The adults of these taxa show a greater morphological distinction than most neighboring populations of *T. scripta*. The young do not.

DIAGNOSIS

The subspecies is distinguished from all other members of *Trachemys scripta* (but particularly from its nearest geographic neighbors, *venusta* and *ornata*) by the following combination of characters: (1) Pattern of head and neck, when visible, consists of narrow stripes of about same width and lacking dark boarders; there is slight or no postorbital expansion; no single stripe or mark dominates head pattern. (2) Ocelli on lateral scutes are always placed far posteriorly in posteromedial quadrant of scute and truncated by posterior edge of scute. (3) Plastral pattern is dusky and concentric, similar to that of *venusta* but more likely to have lateral projections along interlaminal seams. (4) Patterns of carapace, plastron, and striping commonly are clear and well defined in young but usually obscured by fading and melanism in adults. (5) Males are distinctive in having an elongated, pointed, but not upturned snout and lack elongated foreclaws. (6) Males are much lower in cross-section (CH/CW) than females (0.48 ± 0.03 vs. 0.56 ± 0.04, respectively) as opposed to the same values in *T. s. venusta* (0.53 ± 0.04 vs. 0.56 ± 0.04).

GENERAL DESCRIPTION

Trachemys s. grayi ranks among the largest of the *scripta* group. There are eight *grayi* (all females) among the largest 20 *T. scripta* in JML database. Maximum size in our database: males, 278; females 395. Sumichrast (1882) mentions individuals of 500 mm or more in length.

BASIC PROPORTIONS (LEGLER DATABASE)

CW/CL: males, 0.74 ± 0.031 (0.69–0.80) n = 11;
females, 0.74 ± 0.020 (0.70–0.81) n = 17.

CH/CL: males, 0.35 ± 0.014 (0.33–0.40) n = 11;
females, 0.42 ± 0.011 (0.38–0.47) n = 17.

CH/CW: males, 0.48 ± 0.03 (0.43–0.54) n = 12;
females, 0.56 ± 0.04 (0.48–0.63) n = 18.

PWHP/PWMF: males, 0.97 ± 0.03 (0.91–1.00) n = 12;
females, 0.98 ± 0.02 (0.93–1.02) n = 18.

BR/CL: males, 0.35 ± 0.008 (0.34–0.37) n = 12;
females, 0.38 ± 0.011 (0.34–0.40) n = 18.

WH/CL: males, 0.13 ± 0.010 (0.12–0.15) n = 9;
females, 0.16 ± 0.011 (0.14–0.19) n = 4.

Modal plastral formula: 4 > 6 > 3 > 1 > 5 > 2.

Ocelli on L2 and L3, when visible, are positioned in the extreme upper posterior quadrant of scute. Principal stripes on head and neck are narrow, are of roughly equal width, lack black borders, and are not in sharp contrast to ground color. Postorbital stripe is continuous from eye to base of neck (97%) and lacks evident expansion; mandibular stripe is isolated (91%); symphyseal stripe is connected to another stripe on one or both sides (70%). Pattern integrity is high (0.683). Tomial edges of jaws are serrated.

COLOR IN LIFE

Based on series (UU) from Mojarras, Chiapas: Male—ground color of soft parts is dusky, neutral pale brown, and is much darker above than below (in contrast to the almost black

FIGURE 41.2 *Trachemys scripta grayi*. Adult female (UU) from Mojarras, Chiapas.

FIGURE 41.3 *Trachemys scripta grayi*. Juvenile head striping (Pacific coastal Oaxaca near Mazunte). (Photo courtesy of J. R. Buskirk.)

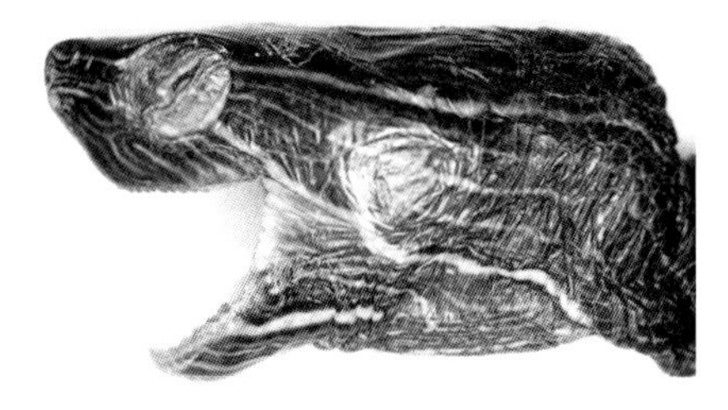

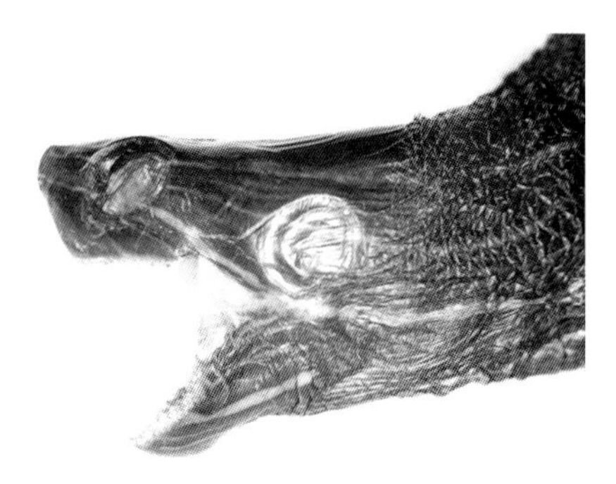

FIGURE 41.4 *Trachemys scripta grayi*. Heads of mature male (top) and female (bottom) showing distinct sexual dimorphism; profile of male is lancelike and flat from tip of snout to tip of supraoccipital process. Note also near obliteration of narrow striping. (Specimens courtesy of M. A. del Toro.)

color of Veracruz *venusta*). Head stripes are pale greenish yellow; iris is clearer and brighter yellow than head stripes and very slightly greenish at periphery; postorbital stripe is continuous from eye to axillary pocket and is interrupted by melanin behind eye in older individuals; stripe from tip of snout to eye is also interrupted; orbitocervical stripe is continuous and is bright on side of head and dimmer on neck; stripe on mandibular ramus is isolated; pale marks on limbs and tail are pale yellow like head and are nearly cream below. Markings of female are like male, but marks on side of head are darker than those in males; iris is slightly darker. Pattern of plastron is obscured by brownish melanistic blotches in all mature males and largest female in series; ground color lacks any suggestion of orange; pattern is dusky and diffuse on pale yellow ground where not obscured by melanin. Ocelli on carapace, where not obscure, have dusky central spot and pale, dusky orange border.

Patterns of carapace, plastron, and striping commonly are clear and well defined in young but usually are obscured by fading and melanism in adults. Juveniles have clear ocellar pattern on carapace and dusky concentric pattern on plastron (Figure 41.5).

GEOGRAPHIC DISTRIBUTION

Map 27

Geographic distribution corresponds to a continuous, narrow coastal plain from at least as far west as the mouth of the Rio Cozoaltepec (Oaxaca, ca. 96°45′W; M. Harfush and J. R. Buskirk, pers. comm.) eastward, along the Pacific coasts of Chiapas and Guatemala to a volcanic headland that interrupts the coast just east of Acajutla, El Salvador. The entire geographic range probably lies below 200 m elevation. The type locality is the mouth of the Rio Nagualate (= Nahualate), Suchitepequez Province, Guatemala (14°03′N, 91°32′W), as stated in Bocourt (1868).

The distribution extends various distances up short streams that drain to coastal lagoons. The old "coastal" road (#190) from Tehuantepec to Tonalá and Tapachula crosses the upper reaches of these streams; most are fast, clear, and rocky with some deeper pools. No *Trachemys* were seen in these streams in 1962 and 1964, but local residents described turtles that could have been *Trachemys scripta grayi* (JML, field notes). Don Miguel Álvarez del Toro has provided the locality "Rio Arriba, Acapetahua, Chiapas" near Tapachula at an elevation of about 150 m.

The drainages in which *T. s. venusta* and *T. s. grayi* occur are separated by a low continental divide about 65 km N of the Pacific coast. A headwater tributary of the Rio Coatzacoalcos (Atlantic drainage) is separated from a small stream flowing to the Pacific by about 12 km near Chivela (15 km NE of Ixtepec), a gap through which the transisthmian highway and railroad pass.

At the Zoological Park in Tuxtla Gutiérrez, Chiapas, captive adults of *grayi* and *venusta* are said to ignore each other and have coexisted without interbreeding for many years (Álvarez

FIGURE 41.5 *Trachemys scripta grayi*. Dorsal and ventral views of sibling neonates. Note seam-following plastral pattern (lacking whorls) and distinctly ocellate carapace pattern. (Specimens courtesy of M. A. del Toro.)

del Toro, 1973). This suggests that *venusta* and *grayi* could be sufficiently distinct to occur in natural sympatry without interbreeding and is regarded by Smith and Smith (1979) as a major reason to question the conspecificity of the two taxa.

It is possible that *grayi* and *venusta*, at the Tehuantepec Isthmus, are the termini of a single polytypic species (superspecies sensu; Mayr and Ashlock, 1991) that evolved north and west of the Isthmus (Legler, 1990).

HABITAT

P. s. grayi seems to flourish in coastal lagoons (some of which are brackish) but occurs also in a variety of lakes and slow streams in the Pacific coastal regions of the Isthmus of Tehuantepec (Sumichrast, 1882; Álvarez del Toro, 1973; JML, pers. obs.).

DIET

Gut contents of three adults (Moll and Legler, 1971) contained only vegetation (leaves, stems, roots, grass, and fruit). Sumichrast (1882) reported aquatic insect remains in gut contents that were otherwise vegetational. He also stated that turtles are known to leave the water to obtain figs falling from trees along river banks. Captives prefer meat when available (Álvarez del Toro, 1960).

HABITS AND BEHAVIOR

Dean (1980) studied a population from mid May to August 1979 at Cabeza del Toro, 2.5 km W of Puerto Arista, Chiapas. The study was done in a permanent freshwater swamp (estero) of 1.5 ha surrounded by sandy soils; there was no influx of saltwater, but the presence of mangroves indicated that it had been brackish at one time.

Slider activity began at the start of the wet season (May) and lasted until October; a period of quiescence was hypothesized for the dry season (November–April). Turtle activity was diurnal. Home range was estimated at 5,400 m^2. Basking was never observed, despite an abundance of basking sites in the estero. Cloacal temperatures were close to water temperatures (27–34°C). All of the sliders had heavy to light growths of algae on the shell.

REPRODUCTION

Sexual maturity is attained at carapace lengths of ca. 150 mm in males and slightly more than 200 mm in females.

Sumichrast (1882) reported 16–18 eggs laid in mid March; the eggs are of long oval shape and about "16 × 45 mm" (probably 26 × 45). Captives in Tuxtla Gutiérrez lay 10–20 eggs from February to April that hatch in 3 months (Álvarez del Toro, 1973).

A female of 320 mm CL (10 April) from Mojarras, Chiapas, bore 15 thin-shelled eggs, 34 corpora lutea in two or three sets, and six enlarged follicles (20–23 m). A reasonable prediction of annual reproductive potential (ARP), based on the meager data here presented, would be 40 eggs in three to four clutches.

Measurements of 12 of the eggs were as follows: L, 44.2 mm ± 1.74 (39.9–46.5); W, 28.1 mm ± 0.44 (27.3–28.7).

GROWTH AND ONTOGENY

Data on size (JML database): adult males, 207 ± 41.6 (151–285) n = 12; adult females, 297 ± 50.7 (205–395)

FIGURE 41.6 *Trachemys scripta*. Captive adults at Mazunte, Oaxaca. *T. s. grayi* (left) and *T. s. venusta* (right); *venusta* occurs naturally in Gulf drainages of the Isthmus of Tehuantepec.

n = 18; im. females, 181 ± 41.8 (133–211) *n* = 3; im. unsexed (probably females), 124 ± 28.6 (102–164) *n* = 4; juveniles unsexed, 47 ± 9.02 (38–67) *n* = 15.

Sexual maturity and age produce dimorphic changes in form that are perhaps greater than those in any other subspecies of *T. scripta*. Notable among these is the narrower head of males (WH/CL in males is 0.13 ± 0.010 vs. 0.16 ± 0.011 in females) and the unique head profile of adult males. The snout of males is elongate and pointed but not upturned, the lateral profile of head being wedge shaped and virtually flat from tip of occipital process to tip of snout (Figure 41.4). The edge of the prefrontal bone on an adult male skull is only slightly modified. There is seemingly never a bosslike enlargement of the snout (Álvarez del Toro and Smith, 1982; JML, pers. obs.). Males also have a more pronounced anal notch. Günther (1885: pls. 4–6) accurately shows a typical adult male that has most of the features here described.

The young of *grayi* and *venusta* show fewer differences than adults. Predictable ontogenetic change produces trenchant differences in adults. The plastral pattern in *grayi* is distinct in juveniles but becomes obscured by brownish melanistic blotches in all mature males and in large female adults. The ground color of the carapace becomes brownish, and the ocelli, where not obscured, have a dusky central spot and, at most, a dusky orange border.

PREDATORS, PARASITES, AND DISEASES

Dean (1980) thought it probable that shell injuries in 40% of his marked animals were inflicted by crocodilians (*Caiman crocodilus*) observed in the swamp. Deep scratchlike wounds on prepared shells are similar to known crocodilian damage of Australian chelids (JML, pers. obs.).

POPULATIONS

Based on recaptures of 47 individuals, the population of *grayi* in a 1.5 ha coastal swamp was estimated at 87.3 individuals with a density of 58.2 turtles/ha. Sex ratio was 34% males to 66% females (Dean, 1980).

CONSERVATION AND CURRENT STATUS

According to Sumichrast (1882), the flesh and eggs are good but not widely used by native people. Live specimens were on sale at the market in Tapachula, Chiapas, in April 1962 (JML, pers. obs.).

Trachemys scripta taylori (Legler) 1960

Taylor's slider; Cuatro Ciénegas slider; tortuga negra (local, but name is also used for *Terrapene coahuila*).

ETYMOLOGY

Named in honor of Professor Edward Harrison Taylor (1889–1978), pioneer in 20th-century Mexican herpetology and one of Legler's mentors.

HISTORICAL PROLOGUE

In 1958 Legler, Wendell L. Minckley, and Robert M. Wimmer visited the Basin of Cuatro Ciénegas, Coahuila, in search of an aquatic box turtle (*Terrapene coahuila*) and found that most of the aquatic organisms in the basin were endemic. Among these was a *Trachemys scripta* (later named *taylori*) that was obviously part of the *scripta* complex we knew in the United States. Making the proper comparisons just for the purpose of describing the subspecies was an ordeal because little was known about Mesoamerican *Trachemys scripta* at the time. Legler's study of sliders dates from that time (Legler, 1960a).

RELATIONSHIPS AND TAXONOMIC STATUS

We consider *taylori* to be a relict of an earlier Rio Grande stock that was isolated and differentiated in the Basin of Cuatro Ciénegas in postpluvial times. The same stock may have given rise to the *cataspila*/*venusta* series on the Gulf Coastal plain.

DIAGNOSIS

A subspecies of *Trachemys scripta* that most closely resembles *T. s. cataspila* and *T. s. elegans* and differs from all other members of the species by the following combination of characters: (1) Postorbital mark is red and usually isolated from eye and/or neck stripe; mandibular stripe is always isolated. (2) Ocelli on lateral scutes are dark-centered at all ages; dark center may eventually obliterate entire ocellus. (3) Plastral pattern covers approximately half of plastron and is dusky, concentric, and has lateral extensions ending

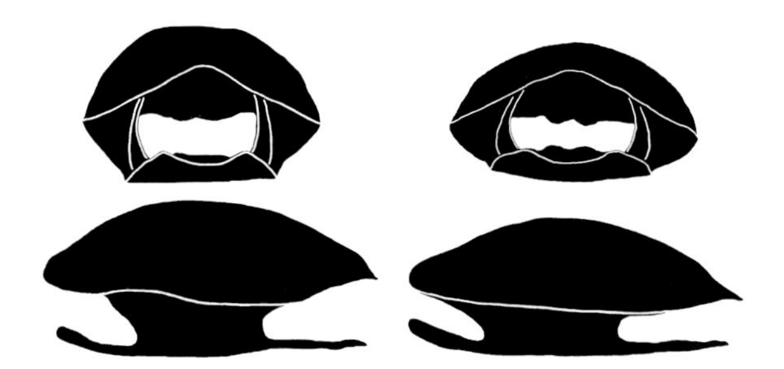

FIGURE 42.1 *Trachemys scripta taylori*. Comparison of optical cross-sections and profiles of two subspecies of Mexican *Trachemys scripta* (adult females): left, *taylori*; right, *gaigeae*. Note differences in proportions of CH and CW to CL.

FIGURE 42.2 *Trachemys scripta taylori*. Adult female, type locality.

in whorls or expansions on all or most of the interlaminal seams, and all parts are interconnected. (4) Melanism proceeding forward from rear of plastron may obscure all or part of plastral pattern ontogenetically. (5) There are no fingerprint-like, opaque smudges on plastron. (6) Mandibular tomium is coarsely serrate (Legler, 1960a).

GENERAL DESCRIPTION

Trachemys s. taylori is one of the smaller Mexican sliders (largest adult female and male, 218 mm and 179 mm, respectively). Shell is high relative to width, differing from *elegans*, *gaigeae*, and *hartwegi* and ranking with *cataspila*, *venusta*, and *grayi* in this respect. Greater height is imparted chiefly by the bridge, which (in optical cross-section) forms a distinct plane from the high carapacal rim to the plastron (Figure 42.1), rather than a continuous bulge as in *elegans*, *hartwegi*, and *gaigeae*. Height of shell margin ca. 36% of HT, 16% of CL.

BASIC PROPORTIONS

CW/CL: males, 0.75 ± 0.027 (0.66–0.79) n = 51;
females, 0.74 ± 0.027 (0.67–0.81) n = 60.

HT/CL: males, 0.39 ± 0.017 (0.36–0.45) n = 51;
females, 0.41 ± 0.020 (0.36–0.45) n = 60.

HT/CW: males, 0.52 ± 0.03 (0.46–0.58) n = 51;
females, 0.56 ± 0.04 (0.49–0.64) n = 60.

PWHP/PWMF: males, 1.05 ± 0.05 (0.93–1.19); females,
1.02 ± 0.04 (0.95–1.13) n = 60.

BR/CL: males, 0.32 ± 0.009 (0.30–0.35) n = 51;
females, 0.34 ± 0.011 (0.31–0.39) n = 61.

WH/CL: males, 0.18 ± 0.010 (0.16–0.20) n = 51;
females, 0.18 ± 0.011 (0.16–0.20) n = 58.

Modal plastral formula: 4 > 6 > 3 > 1 > 5 < = > 2; abdominal lamina longer than combined lengths of pectoral + humeral or humeral + gular.

There are one or two oval, vertically compressed ocelli on each lateral scute and one ocellus on the superior surface of each marginal scute; all principal ocelli and a variable number of supernumerary ocelli have black centers at all ontogenetic stages.

Posterior lobe of plastron is relatively narrow (narrower than forelobe) with femoral edges reflected ventrally. There is a dusky, concentric, fully interconnected, medium to full plastral figure with lateral interlaminal extensions ending in whorls on a pale yellowish ground; progressive melanism of plastron does not form fingerprint-like smudges or blotches.

Mandibular stripe is always isolated. A symphyseal "Y" is formed in 90% of specimens. Pattern integrity is 0.302.

Maxillary sheath is notched anteriorly; tomial edges are finely and unevenly serrate; maxillary triturating surfaces have distinct median alveolar ridge with fine denticulations but no large cusps. Mandibular tomium is coarsely serrate and distinctly pointed at symphysis; mandibular alveolar ridges are finely denticulate and bear a low, blunt tooth on anterior one-third.

COLORS IN LIFE

Ground color of soft parts is dark olive to slate gray or black; ground color of carapace is olive to slate gray; ground color of plastron is pale yellow; markings are blackish, tinged with brown in younger specimens and sooty black in most adults. Postorbital mark is red; other markings on soft parts are cream to yellowish with dark borders (Legler, 1960a).

GEOGRAPHIC DISTRIBUTION

Maps 26 and 28

T. s. taylori occurs only in permanent aquatic habitats of the Basin of Cuatro Ciénegas, Coahuila. The type locality was given as 16 km S of Cuatro Ciénegas (Legler, 1960a), an approximation made with a marching compass and an odometer. Based on a map in Minckley (1969) these distances are more nearly 13 km S and 1 km W of Cuatro Ciénegas.

Schmidt and Owens (1944:101) mention a large series of ". . . *troosti elegans*" and "*gaigeae*" (without reference

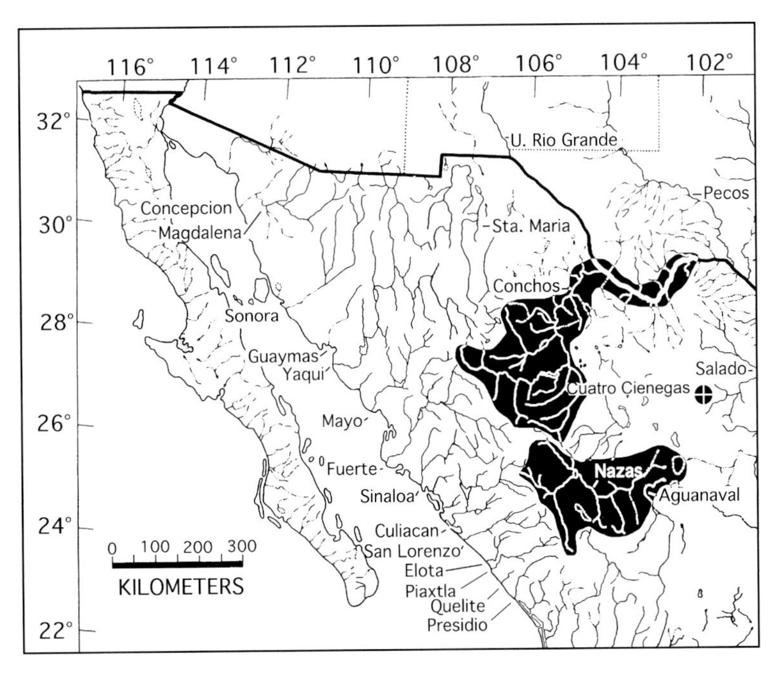

MAP 28 Geographic distribution of *Trachemys s. gaigeae* (north), *T. s. hartwegi* (south), and *T. s. taylori* (endemic to Basin of Cuatro Ciénegas).

to museum number) from various localities in Coahuila on the Lower Rio Grande drainage northeast of Cuatro Ciénegas (R. Sabinas, R. Salado, and R. San Juan). Legler has examined all of these: eight are *Pseudemys gorzugi* and the others were *T. s. elegans* (Legler, 1960a). One specimen (FMNH 55655, 22 August 1939) is *taylori* and is the earliest record of slider turtles from the Basin of Cuatro Ciénegas.

THE BASIN OF CUATRO CIÉNEGAS

The following is based chiefly on Minckley (1969) and Legler (pers. obs.). See also the accounts of *Terrapene coahuila* and *Apalone* sp. for other details.

The municipality of Cuatro Ciénegas (26°59′N, 102°04′W) lies in a small (800 km^2) intermontane basin with a floor elevation of 740 m, roughly 70 km W and 15 km N of Monclova, Coahuila, and about 270 km SSE of the Big Bend of the Rio Grande. Entrance to the basin is from the east via two successive, dramatically narrow portals (Puerto Sacramento and P. Salado) formed by the termini of two mountain ranges (Sierra de Menchaca and the Sierra San Vicente). A stream, a highway (#30), and a railway pass through these gaps, which are as narrow as 100 m in places. The effluent stream is swift and referred to locally as the Rio Nadadores. At times in the past the headwaters of the Nadadores lay in the northern part of the basin and flowed naturally to the Rio Salado. Just inside the Puerto Salado an alluvial sill currently prevents natural drainage but bears evidence of two old channels. However, the current external drainage of the basin is affected by manmade channels through the sill. This effluent channel flows northeastward, naturally, as the Rio Salado (de los Nadadores), with several tributaries en route, to the Presa de Don Martin and thence southeastward to join the Lower Rio Grande via Falcon Reservoir at the intersection of 26°50′N. There is thus evidence that the basin has been isolated from and reconnected to the Lower Rio Grande drainage at several times in the past. These isolation events are reflected in the relationships of fishes and many other freshwater organisms endemic to the basin.

INTERGRADATION AND HYBRIDIZATION

Trachemys s. taylori occurs in the Rio Nadadores where it leaves the basin, and *T. s. elegans* occurs just outside the basin. There is no evident physical barrier to dispersal of either taxon.

A few probable hybrids of *elegans* × *taylori* have been taken from a 100 m long impounded pool in an otherwise fast, narrow stretch of the Rio Nadadores 3.2 km W of Celemania, a railway station shown on few maps. It lies between the more commonly shown stations of Nadadores and La Madrid and well outside the outer portal (Sacramento) of the basin. The four specimens referred to by Legler (1960a, 1963) as "intergrades" were taken from this pool. Further observations and collecting by Legler and J. F. Berry (May 1976) show that the thriving population of *Trachemys* at this locality is *elegans* and contains a few individuals with *taylori* characters. In retrospect, the suspected hybrids showed more characters of *elegans* than *taylori* and did not represent a classic example of intergradation. They do show that *taylori* characters occur in a few *elegans* at a geographic point where the natural ranges are closely juxtaposed.

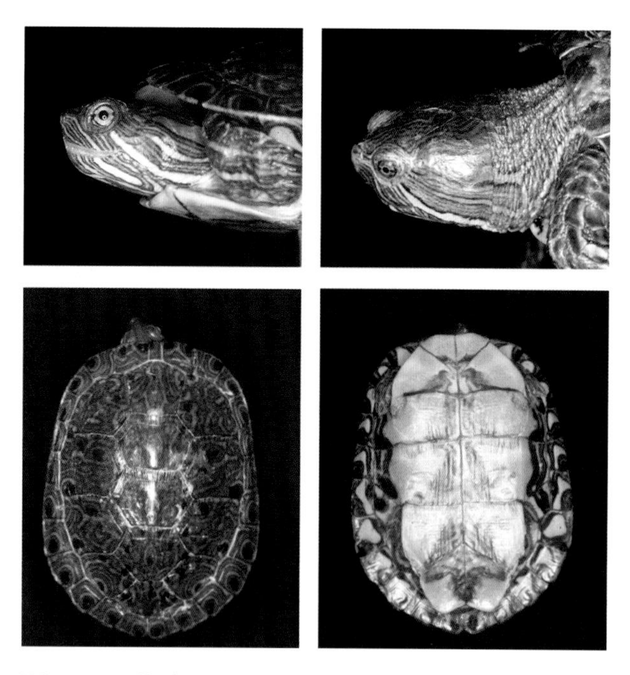

FIGURE 42.3 *Trachemys scripta taylori*. Top left, juvenile head; top right, adult female head. Bottom, dorsal and ventral views of subadult before the onset of melanism.

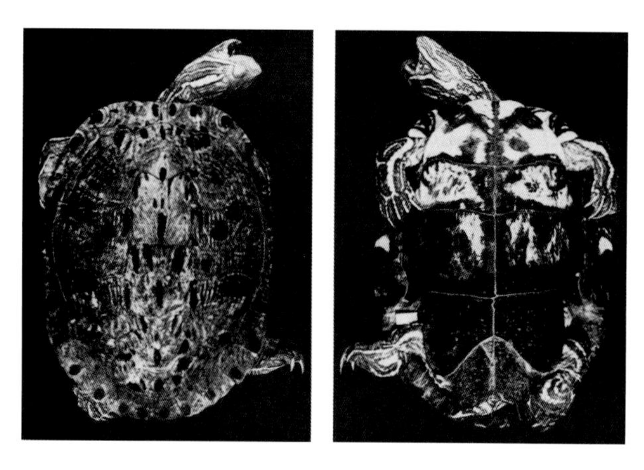

FIGURE 42.4 *Trachemys scripta taylori*. Dorsal and ventral views of holotype, adult female with substantial melanism (KU 46952, CL 202 mm). (From Legler, 1960a, with permission.)

On the other hand, there is no phenotypic expression of *elegans* characters in the populations occurring naturally within the basin (JML, pers. obs.; Susan McGaugh, pers. comm.).

HABITAT

Trachemys s. taylori flourishes in various clear, deep ponds and streams within the Basin of Cuatro Ciénegas, most of which contain abundant submergent and emergent aquatic vegetation and a soft mud bottom (see further details in account of *Apalone atra*). It is the most abundant aquatic turtle in the basin. *Apalone ater, Apalone spinifera*, and some *Terrapene coahuila* occur in the same aquatic habitats. No sliders have been seen or taken in shallow or stagnant water. Most specimens have been caught in baited hoop nets. Diving in pools is unproductive because the turtles conceal themselves under dense aquatic vegetation or in bottom mud. The Basin of Cuatro Ciénegas is one of two places in Mexico where a *Trachemys* does not occur with a *Kinosternon* (see account of *T. s. nebulosa*).

DIET

All stomachs examined (September 1960 and May 1976) contained identifiable, chopped succulent vegetable material (Legler, 1960a; and field notes).

HABITS AND BEHAVIOR

Heads of sliders are often seen in abundance in large pools in the basin, especially in the evening. Aerial basking has not been reported nor is there an abundance of good typical basking sites. The ponds contain crystal clear water with depths of up to 6 m. Water temperatures may reach 27°C in late summer.

REPRODUCTION

Davis and Jackson (1973) described the precopulatory behavior of a male *T. s. taylori* (118 mm CL) and a female *Chrysemys picta belli*. The behavior of *taylori* consisted chiefly of pursuit and vigorous biting. Bites were chiefly to the posterolateral carapace margin but could occur on any part of the body. Biting could be severe enough to damage the female's shell. As the courtship neared its end, bites were made to any body part until the female became quiescent. At this time the male bit and held onto the female's shell or skin and was dragged and shaken through the water by the female for as long as 5 minutes. Eventually the male maneuvered over the female, bit at any moving part, and mounted. The male's penis was partly everted during the biting phases of courtship. Intromission was not observed.

When a male *taylori* was approached rapidly by another turtle, a stereotyped motor pattern was evoked; forelimbs and head partially retracted, and the hind limbs rigidly extended and braced against the substrate, placing the long axis of the male at an angle of 15–20° to the substrate. This behavior could be evoked also when the male viewed its own reflection.

Legler (pers. obs.) observed a small male *taylori* in an aquarium room where several aquaria were closely spaced; the turtle was very aggressive and managed to climb from tank to tank and attempt mating with whatever (turtle) was there, some of which were as distantly related as Australian chelids.

Three females from El Mojarral (CL 145–167 mm, 25 May 1976) were judged to be subadult, based on typically juvenile oviducts and ovaries with many granular follicles, a few pale yellow follicles no larger than 2–3 mm, and no evidence that ovulation had ever taken place. These females would

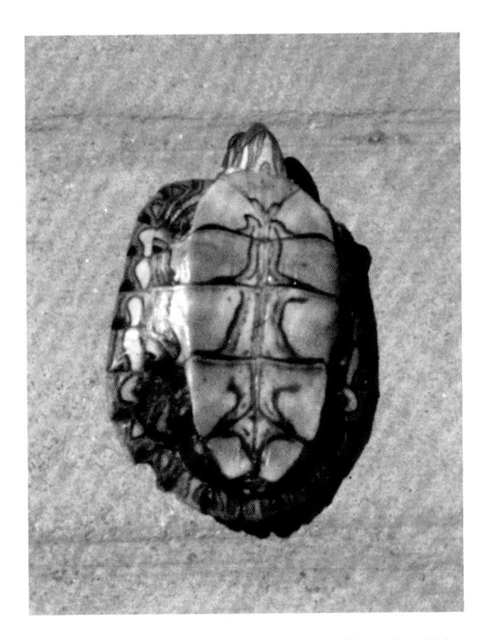

FIGURE 42.5 Trachemys scripta taylori. Juvenile male with no melanism.

probably have matured in the current season and have first ovulated no later than the following season. A female of 144 mm CL (same date and location) was definitely mature, bearing seven preovulatory follicles of 11–19 mm and three of 5–6 mm (plus minute granular follicles) with no evidence ovulation. The largest follicles could have reached ovulatory size in the month of June. This female was probably reproducing for the first time.

GROWTH AND ONTOGENY

Size data on 120 specimens (JML database) are as follows: adult males, 123 ± 21.0 (90–179) n = 51; adult females, 176 ± 23.0 (132–218) n = 54; im. males, 85; im. females, 109 ± 19.3 (88–140) n = 7; im., unknown sex, 99; juvenile females, 81 ± 2.8 (79–83) n = 2; other juveniles, 70 ± 8.1 (64–79) n = 4. The maximum size for adult females (218 mm) is the smallest among Mexican *Trachemys scripta* (Legler, 1990a). The minimum size of 93 mm recorded for adult females by Legler (1990a) is in error.

Male and female sizes are less disparate than those in *elegans*, and the chances of a mating pair being of same size are high. Males lack elongate foreclaws and an exaggerated, elongated, pointed snout. No stereotyped Liebespiel is known.

In both sexes melanism progresses with age and size and substantially alters the appearance of large adults. Melanism is especially noticeable on the plastron, beginning on the posterior half and progressing forward, virtually obliterating the original pattern. At least part of the plastron is sooty black in adults. The dark ocellar centers on the carapace enlarge and may obliterate entire ocelli. Eventually the carapace may have numerous irregularly arranged black marks on a faint reticulum of pale lines.

PREDATORS, PARASITES, AND DISEASES

Minckley (1966) saw a coyote (*Canis latrans*) observe and seemingly stalk a *T. s. taylori* in a pond 9 km SW of Cuatro Ciénegas in April 1965; the turtle was swimming 40 cm below the surface in clear water. The coyote attacked in a series of jumps and thrust its head beneath the water to capture the turtle as it tried to burrow into the bottom. The coyote carried the turtle to shore and gnawed on the front of its shell until frightened away. Remains of nine other *Trachemys* and two *Apalone* sp. were found nearby; all bore tooth marks and were crushed near the anterior end of the carapace.

CONSERVATION AND CURRENT STATUS

We know of no instance of humans taking *T. s. taylori* for food. The main threat to this species is the decline of water resources resulting from agricultural use. Ecological studies could probably be carried out in a protected, relatively pristine environment.

See account of *Apalone atra* for a brief summary of present conservation measures and agencies in the Basin of Cuatro Ciénegas.

Trachemys scripta gaigeae (Hartweg) 1939

Big Bend slider (Legler, 1960b; Iverson, 1992a; Stuart and Ernst, 2004); jicotea Big Bend (Liner, 1994); pinto (local name near Meoqui, Chihuahua).

ETYMOLOGY

A matronym honoring Helen Thompson Gaige (1890–1972), University of Michigan, and collector of the holotype. This is the only species of Mexican chelonian named for a female herpetologist.

HISTORICAL PROLOGUE

Hartweg included six paratypes from the Rio Nazas in the type description. These were included as paratypes in Legler's (1990) description of *T. s. hartwegi*. The holotype is UMMZ 66472, female, collected summer 1928 at "Boquillas, Rio Grande River, Brewster Co. Texas."

RELATIONSHIPS AND TAXONOMIC STATUS

Hartweg (1939) described *gaigeae* (sensu lato) as a subspecies of *T. scripta*. Stejneger and Barbour (1939) elevated the name to full species rank in the same year. For more than half a century the name has been used at both levels. The subspecific relationship of *gaigeae* and *hartwegi* has not been questioned, but it has become popular to regard *gaigeae* as a full species. Biological evidence for taxonomic rank is considered in Degenhardt, Painter, and Price (1996); Legler (1990); Stuart (1995a,b); Seidel, Stuart, and Degenhardt (1999); and Seidel (2002). Other mention of rank is

FIGURE 43.1 *Trachemys scripta gaigeae*. Head pattern of female (Rio Grande, Ojinaga, Chihuahau near mouth of Rio Conchos).

chiefly in surveys and reviews, the most recent of which is Stuart and Ernst (2004).

The chief reasons for considering *gaigeae* as a species are (1) its supposed sympatry with *elegans* in New Mexico (Degenhardt, Painter, and Price, 1996); (2) the lack of elongated claws in males and a courtship pattern different from that of *elegans* (Stuart and Miyashiro, 1998). However, Seidel (pers. comm.) and Stuart (1995a,b) contend that the natural ranges of these taxa do not overlap and hybridization occurs where *T. s. elegans* has been introduced in the Rio Grande of New Mexico. As to the differences in male sexual dimorphism and courtship pattern, these apply to all of the *Trachemys* occurring south and west of the old Lower Rio Grande (a group lumped by Legler [1990] as MESOSCRIP and containing all of the *Trachemys* known to occur in Mexico). The males of all taxa in this group lack the elongated foreclaws and, where mating has been observed, it lacks the stereotypic frontal approach accompanied by titillation with the foreclaws seen in U.S. *scripta*.

Although *T. s. gaigeae* is here treated as a subspecies of *T. scripta*, we recognize the arguments for its status as a full species. Our taxonomic treatment results chiefly from Legler's long-term studies of the *scripta* group, and our arguments for the status of *gaigeae* as a subspecies seem equally as cogent as those for species rank. But actually, controversy on taxonomic rank can be set aside, because it is clear what taxon is covered in this account.

DIAGNOSIS

A subspecies of *Trachemys scripta* occurring in the Upper Rio Grande of Texas and New Mexico and the Rio Conchos drainage of Chihuahua, most closely related to *T. s. hartwegi* and differing from all other members of *scripta* by the following combination of characters: (1) Postorbital mark is orange, dark bordered, and isolated from other neck stripes and is almost as high as long and shorter than mandibular stripe; if connected with a stripe posteriorly, then that stripe is much narrower than the primary orbitocervical stripe. (2) Mandibular stripe is connected to a neck stripe. (3) Symphyseal stripe is forked behind, and at least one of resulting rami (usually both) is continuous with ventral longitudinal neck stripe. (4) Mandibular tomium is smooth or finely serrate. (5) There are no well-formed, complete, or typical ocelli on lateral scutes of carapace; ocelli, if present, are vestigial and incomplete and contain a dark spot on a lateral scute. Dorsal aspect of carapace is not dominated by supramarginal pattern of black marks. (6) Abdominal is the longest scute of plastron; anal is second longest. (7) Plastral pattern is more or less continuous from gular to anal and consists of two or more concentric lines (distinct at least in young) and distributed about equally on anterior and posterior halves of plastron, obscured by secondary melanism only in some large adults.

GENERAL DESCRIPTION

The Big Bend slider ranks among the smaller subspecies of *scripta* in Mexico—*elegans* and those referred to as the "northern isolates" (*gaigeae*, *hartwegi*, and *taylori*) by Legler (1990); its small size could be an artifact of collecting techniques that do not produce large adults or to the environmental effects of living in aquatic environments surrounded by deserts. Maximum reported size is 257 mm CL (female shell, Elephant Butte Reservoir, Sierra Co., New Mexico; Stuart, Painter, and Stearns, 1993). Maximum CL in Texas and adjacent Mexico is 173 mm in males and 220 mm in females.

BASIC PROPORTIONS

CW/CL: males, 0.79 ± 0.022 (0.75–0.82) n = 16;
females, 0.79 ± 0.027 (0.73–0.84) n = 23.

HT/CL: males, 0.34 ± 0.017 (0.31–0.37) n = 16;
females, 0.37 ± 0.023 (0.31–0.40) n = 23.

HT/CW: males, 0.44 ± 0.02 (0.40–0.48) n = 15;
females, 0.46 ± 0.03 (0.41–0.52) n = 23.

PWHP/PWMF: males, 0.96 ± 0.02 (0.91–0.98) n = 16;
females, 0.94 ± 0.05 (0.87–1.05) n = 23.

BR/CL: males, 0.32 ± 0.010 (0.31–0.35) n = 16;
females, 0.34 ± 0.008 (0.32–0.36) n = 23.

WH/CL: males, 0.16 ± 0.006 (0.15–0.17) n = 13;
females, 0.16 ± 0.008 (0.15–0.17) n = 21.

Modal plastral formula: 4 > 6 > 3 > 1 = 5 > 2.

Shell is relatively low in optical cross-section and profile (Figure 42.1); mid-dorsal keels are usually indistinct or wanting. Radial corrugation on central scutes is indistinct or wanting in adults. There is a general reduction in the

FIGURE 43.2 *Trachemys scripta gaigeae*. Shell and head coloration in juvenile (lower left), male (upper left), and female (right; UU; Camargo, Chihuahua).

degree of anterior marginal/peripheral overhang (expressed by Seidel [2002] as "underlap" of precentral scute), a character shared with most of the other subspecies of *scripta* in Mexico; gular and pectoral scutes are relatively short, 12% and 16% of CL, respectively (see Figure 43.3).

Plastral pattern is continuous from gular to anal, is narrow, lacks lateral whorls and consists of two or more concentric lines (distinct at least in young), and is distributed about equally on anterior and posterior halves of plastron.

Carapacal ocelli are irregular in form and placement; at best, they are vague even on a freshly molted shell, consisting usually of a single incomplete pale border with an equally indistinct dark center that is solid or ringlike. The pale border is usually connected to the "reticular" pattern of pale lines on the carapace (Ernst, Lovich, and Barbour, 1994). Carapacal ocelli, if present, are most common on L2–4.

There are supramarginal markings on posterolateral quadrants of marginal scutes that have an incomplete pale outer ring and a solid dark center or heavy dark ring with a small pale center. Inframarginal ocelli are similarly placed and consist of a black mark with a pale center containing a black dot, resembling an eye. The marginal ocelli do not form a bold peripheral pattern that dominates either the dorsal or ventral aspect of the shell.

Mandibular stripe is longer than postorbital mark (97%), usually (55%) joining a stripe on neck to form a "Y." Symphyseal stripe joins one or two ventral neck stripes to form a complete or partial "Y." Postorbital mark is small, often teardrop shaped and pointed behind. In nearly all individuals a much smaller orange spot (a vestige of postorbital stripe) occurs immediately posterior to the eye. Pattern integrity is 0.386. Foreclaw length is not exaggerated in males.

COLOR IN LIFE

Color is accurately depicted in the excellent photographs of Ernst, Lovich, and Barbour (1994: pl. 32, immature); Degenhardt, Painter, and Price (1996: pl. 31, adult); and Stuart et al. (2004: Figures 1–3, adult male and hatchling). All colors fade in preservative. Ground color of soft skin is greenish or brownish olive; most longitudinal stripes are pale yellowish with dark borders. Some of the stripes on the antebrachium may be pale orange. Chief postorbital mark and small postorbital mark are medium orange. Ground color of plastron is yellow with a dusky median figure. (See Growth and Ontogeny for remarks on melanism.) Iris is pale yellow with a black stripe passing through pupil. Ground color of carapace is pale olive brown to tan. Pale marks on carapace are pale orange.

GEOGRAPHIC DISTRIBUTION

Map 28

The Rio Conchos and its tributaries in Chihuahua and possibly extreme northern Durango; the Rio Grande from Ojinaga at the mouth of the Rio Conchos, downstream through the Big Bend region and through a gorge to a point roughly due S of Sanderson, Terrell Co., Texas (29°45'N, 102°24'W). This downstream limit is accurately portrayed in Seidel, Stuart, and Degenhardt (1999) and Stuart and Ernst (2004). There is a cluster of substantiated records in Socorro and Sierra Counties, New Mexico, including Elephant Butte and Caballo reservoirs. From these points downstream to Ojinaga there are two records in Texas, one near El Paso and one ca. 175 km SE of El Paso (30°32'N). The gaps in distribution upstream from Ojinaga have not been explained.

Schmidt and Owens (1944) reported 10 specimens of "*gaigeae*" from Lower Rio Grande tributary waters; eight of these were *Pseudemys gorzugi* and two were *T. s. elegans* (see account of *elegans* and Legler, 1960a). Roger Conant (pers. comm.) saw a slider in a pond 24 km N of Villa Ocampo, Durango (on a small tributary of the Rio Florido, a Conchos tributary). This may be the southernmost record of *gaigeae*.

Most of the distribution in Mexico occurs below 1200 m, but it occurs as high as 1,280–1,410 m near Bosque del Apache National Wildlife Refuge, Elephant Butte, and Caballo reservoirs in New Mexico. The type locality of *gaigeae* is on the eastern side of the Big Bend (29°12'N, 102°55'W) and immediately across from Boquillas del Carmen, Coahuila. The locality could be stated equivalently as "Rio Bravo, Boquillas del Carmen, Coahuila."

The Rio Grande flows through a gorge beginning just downstream from Boquillas (29°12'N, 102°24'W) and courses generally NNE some 110 km to a point (29°40'N, 102°24'W) due south of Sanderson, Texas. After flowing out of the gorge the river makes two distinctive 90° bends and thence flows generally eastward to Langtry and to the mouth of the Pecos. The gorge contains fast-moving water interrupted by slower water at the mouths of tributary canyons and is bordered to the north by Brewster Co., Texas. *Trachemys s. gaigeae* occurs at several places in the gorge, and *T. s. elegans* has been recorded some 30 km farther

FIGURE 43.3 *Trachemys scripta gaigeae.* Top—dorsal and ventral views of juvenile, subadult female, and adult male (left to right; Camargo, Chihuahua); bottom—dorsal and ventral views of adult female, subadult female, and adult male (in order of size; Rio Grande abreast of Lajitas, Chihuahua).

downstream (ca. 102°W; Seidel, Degenhardt, and Dixon, 1997; Seidel, Stuart, and Degenhardt, 1999; Seidel, pers. comm.). As yet there is no evidence that *gaigeae* and *elegans* occur in microsympatry in the river. *Trachemys s. elegans* has also been taken or observed in ponds near the river (Seidel, pers. comm.).

Trachemys s. elegans becomes progressively more common farther downstream. At Langtry the river is about 55 m wide, shallow, and fast. Legler observed and caught one *elegans* in a side canyon with beaver ponds at Langtry, Val Verde Co., Texas, in 1963.

Stuart (2000) states that *gaigeae* is native to the Rio Grande in New Mexico. Where *gaigeae* occurs with *elegans* in New Mexico (ponds and canals near river), hybridization occurs. In these places *elegans* is almost certainly introduced (Degenhardt and Christensen,1974; Degenhardt, Painter, and Price, 1996; Seidel, Stuart, and Degenhardt, 1999).

In our opinion there is little doubt that *gaigeae* occurs naturally in the Rio Conchos drainage and the Big Bend region of the Rio Grande. But introduction of both *gaigeae* and *elegans* in New Mexico cannot be dismissed.

ZOOGEOGRAPHY: THE EVOLUTION OF THE MODERN RIO GRANDE

The literature on drainage history in northern Mexico has been reviewed by Smith and Miller (1986); Minckley, Hendrickson, and Bond (1986); and Legler (1990). These authors and others cited by them offer evidence for freshwater fishes that substantiate most of the remarks that follow.

The present Rio Grande was once two rivers. The two main tributaries of the lower Rio Grande were the Pecos River and another tributary eroding upstream through the Big Bend region. At some stage the Rio Conchos joined the Big Bend tributary; it seems clear that this confluence occurred before the Rio Conchos acquired its southern headwaters (Legler, 1990).

The upper Rio Grande originally flowed southward from Colorado to internal drainage systems in northern Chihuahua, New Mexico, and Texas. There was a large lake or series of lakes in the region in the Pleistocene. The Lower Rio Grande captured the Upper Rio Grande and parts of the internal systems in mid-Pleistocene times. The capture occurred ca. 90 km SE of El Paso (31°N), near El Porvenir, Chihuahua. Subsequently the integrated Rio Grande has flowed to the Gulf (Smith and Miller, 1986). The internal basins had also received flow from streams initially draining the east slope of the Sierra Madre Occidental but were captured partly by Pacific streams (e.g., the Rio Papagochic by the Yaqui and the Rio Tunal to the Rio Mezquital). This and the integration of the upper and lower sections of the Rio Grande reduced the flow of water into the internal basins.

This formerly well-watered area is now relatively arid and lies just south of the United States–Mexican border, and contains many basins and internal drainage systems. Most of the latter are now dry except in time of heaviest rainfall. A few (e.g., the Rio Nazas, Rio Águanaval, Rio Casas Grandes, and Rio Santa Maria) still flow in places and have fluctuating terminal lakes. All of these basins and drainages contained more water at times in the Pleistocene. The area extends southward approximately to 24°N (southern extent of Rio Águanaval), is bounded generally by the Sierra Madre Occidental and Sierra Madre Oriental, and extends northward to the Rio Grande and at least to the basins just south of Columbus, New Mexico (Pluvial Lake Palomas). This region is termed the "central Mexican interior basins" by Smith and Miller (1986).

The former confluence of basins and drainages within the region—with each other and with the Upper Rio Grande, Rio Conchos, and the Lower Rio Grande—is generally accepted. These drainages were all interconnected at some time during the Pleistocene in a manner that permitted exchange of aquatic faunas or gene flow between existing populations. These drainage connections were not all concurrent; they probably occurred in a stepwise fashion as basins overflowed and as low-gradient streams altered their courses or were captured by others (Legler, 1990).

The ancestral stock of *gaigeae*/*hartwegi* evolved in the central Mexican interior basins. Subsequent disruptions of drainage isolated the populations that became *gaigeae* in the Conchos drainage and *hartwegi* in the Nazas drainage. Accounting for the presence of the ancestral stock in the basin system is an unsolved zoogeographic problem that was briefly discussed by Legler (1990).

It seems likely (JML, pers. obs.) that the southernmost Rio Conchos tributaries were once confluent with currently internal drainages. Tributaries flowing to the interior basins are now narrowly separated from those of the Rio Florido ca. 65 km S of Jimenez, Chihuahua. The Mapimian channels to the northeast are now dry, but they are on the general drainage that terminates in lakes near Torreon. The headwaters of the Rio Nazas come into close proximity with other Rio Florido tributaries ca. 60 km upstream from the Presa El Palmito (= P. Lázaro Cardenas; Legler, 1990).

The Conchos probably became confluent with the Lower Rio Grande prior to its direct or indirect confluence with the Upper Rio Grande. Populations of *gaigeae* probably founded the slider populations in the Conchos and Big Bend regions prior to the integration of the Upper and Lower Rio Grande. *T. s. elegans* was likely already in place in the Pecos and the Lower Rio Grande.

These hypotheses are congruent with freshwater fish distributions, many of which were recognized by Seth Meek (1904) more than a century ago. Smith and Miller (1986) give the following on native freshwater fishes of the involved streams: The Rio Conchos shares 22 species with the Lower Rio Grande, 17 with the Pecos, 9 with the Upper Rio Grande, and 6 with the Nazas. The fish fauna of the Upper Rio Grande is poor in native species (16) compared to the Conchos (34) and Pecos (52).

HABITAT

In Mexico (JML, field notes) *gaigeae* occurs only in permanent bodies of water. Nearly all were taken or observed in fluviatile situations ranging from shallow, rapidly flowing parts of the Rio Grande and Conchos (671 and 732 m) to a large slow pool at the confluence of the Conchos and San Pedro (1,174 m), all of which were turbid. Typical habitat had a sandy to muddy bottom and sandy banks. Traps were set in all available aquatic situations. Only one specimen (a juvenile) was taken from an artificial pond near Ojinaga. Sliders were also observed and trapped in a narrow, clear stream forming a series of deep, bulrush-bordered pools as it descended through limestone country into Lago Toronto, 27.4 km SW of Camargo, Chihuahua.

Most of the streams in which *gaigeae* occurs have been impounded, and the sliders do not hesitate to exploit these lentic habitats.

T. s. gaigeae in Chihuahua occurs commonly with soft-shelled turtles (*Apalone spinifera emoryi*) and with one or two species of *Kinosternon* (*hirtipes* and *flavescens*). In smaller streams and most other riverine habitats *Kinosternon* vastly outnumbers (6:1) *Trachemys*; *Apalone* occurs only in larger streams and reservoirs. *Trachemys* and *Apalone* occurred in equal numbers at the mouth of the San Pedro. At Lajitas there were more *Apalone* than *Trachemys*, and *Kinosternon* were relatively rare.

The Rio Conchos near its mouth (1.6 km W of Ojinaga, Chihuahua; 732 m) has a flood plain several kilometers wide on both sides. The river is swift with intermittent pools and is nowhere deep. Streamside vegetation is chiefly tamarisk, cottonwood, and willow, giving way to creosote and mesquite desert vegetation some 2 km from the river. The Rio Grande, at Lajitas, Brewster Co., Texas (671 m), has a narrow flood plain with willows and mesquite.

Habitat in New Mexico is more often in stillwater situations near and between Elephant Butte and Caballo reservoirs. Mentioned in the literature are large impoundments, artificial ponds, flooded oxbow lakes, ditches, borrow pits, and artificial wetlands near the river (Degenhardt and Christiansen, 1974; Stuart, 1995b; Degenhardt, Painter, and Price, 1996; Stuart and Painter, 2002). Optimal habitat is reported to be ponds supporting dense growths of *Potamogeton pectinatus*, floating mats of filamentous green algae and a moderate accumulation of dead organic material (Stuart, 1995a,b; Stuart and Painter, 2002).

Other turtles associated with *gaigeae* in the area are *T. s. elegans* (introduced), *Chrysemys picta*, and *Apalone spinifera* (Degenhardt, Painter, and Price, 1996). Between Elephant Butte and Caballo reservoirs *gaigeae* was the most frequently trapped turtle in the river (Stuart, 1995a,b). The Rio Grande proper north and south of this epicenter of distribution is seemingly marginal habitat. Legler (pers. obs.) observed and caught a large series of *Chrysemys picta* (August 1960) in borrow pits paralleling the Rio Grande near Isleta, Bernalillo Co., New Mexico, ca. 25 km S of Albuquerque; no other turtles were seen or caught.

FIGURE 43.4 *Trachemys scripta gaigeae*. Top—dorsal and ventral views of juvenile (UU, confluence of Rio San Pedro and Rio Conchos, Chihuahua). Bottom—paratype, FMNH 27760, first-year juvenile (Boquillas, Rio Grande River, Brewster Co., Texas).

DIET

Gut contents from adults in Chihuahua contained only aquatic vegetation (Legler, 1960b). Food items from analyses of feces, dissection, stomach flushing, and observations of feeding turtles in New Mexico (Stuart and Painter, 2002) included vascular plants, filamentous algae of two or three kinds, a grasshopper, diatoms, protozoans, and fungi. Terrestrial grasses were eaten where turtles could enter inundated areas. An individual was observed cropping algae from submerged dead wood. Freshly captured adults defecated large quantities of loosely consolidated vegetable material. The authors made the following conclusions: Hatchlings and small juveniles up to 40 mm CL feed chiefly on small invertebrates; adults and subadults larger than 140 mm are chiefly herbivorous but opportunistically consumed small invertebrates and larger organisms such as crayfish and fish, either live or as carrion.

HABITS AND BEHAVIOR

Most authors refer to the taxa of the *scripta* group as "diurnal," probably because they are an evident part of the fauna and can be seen in the daytime. However, setting baited traps, fyke nets, and trammel nets overnight is a common practice, and every kind of aquatic turtle we have taken in Mexico can be caught at night.

Aerial basking in *gaigae* occurs only if basking sites are available (seldom in Mexico). Aquatic basking at the surface was observed at the mouth of the Rio San Pedro in morning and late afternoon.

Activity in New Mexico extends at least from April to October. A basking individual was observed on a mud bank in early February 1993. Basking and "terrestrial activity" were observed from February to November at Bosque del Apache National Wildlife Refuge (Degenhardt, Painter, and Price, 1996). Little else is known of habits except for that described under "reproduction."

REPRODUCTION

Adult females taken from Lajitas on the Rio Grande, at Ojinaga on the Rio Conchos, and at the confluence of the R. Conchos and R. San Pedro, 12–27 June 1959, contained shelled eggs or fresh corpora lutea (Legler, 1960b). These females showed evidence for a reproductive potential of more than one clutch. Size (CL) and number of oviducal eggs for four females taken on 26 June were as follows: CL 202 mm, 11 eggs; CL 186 mm, 9 eggs; CL 194 mm, 7 eggs; and CL 164 mm, 6 eggs (Legler, 1960b).

Ovaries of the largest female in JML database (220 mm CL, 27 km SW of Camargo, Chihuahua, 9 September 1960) were past the end of the reproductive season; they contained eight yolked (probably regressing) follicles of 8–10 mm and a total of 79 regressing corpora lutea in a least three groups. This represents one of the highest annual reproductive potentials seen in any Mexican nonmarine chelonian.

Stuart and Miyashiro (1998) provide an account of precopulatory courtship behavior based on observations of adults from Socorro Co., New Mexico, and Brewster Co., Texas, in an artificial outdoor pond. Typical behavior consists of a male trailing a female with "cloacal sniffing." The male then approaches the female from the front or side (rarely from above or beneath) with head and neck fully extended and pointed toward head of female. When the heads are less than 5 cm apart, the male begins rapid head bobbing (two to three bobs per second). During male head bobbing the female moves her head from side to side. Head movements of both sexes are performed with the mouths closed and without physical contact. No biting or use of foreclaws by that male was observed. Head movements ceased when a turtle surfaced. Males continued head bobbing in following a female to the surface. When the head of the male broke the surface it ejected thin jets of water from the nostrils. The authors speculated that nasal squirting was more or less continuous during courtship but was undetectable underwater. It was not observed in males surfacing for feeding or during other noncourtship activities.

Nasal squirting has been observed in other taxonomically diverse chelonians—for example, *Trachemys s. venusta* (Rosado, 1967), *Terrapene* (Brumwell, 1940), and *Emydura* (Murphy and Lamoreaux, 1978)—and its significance is not understood. It may be a simple expulsion of water from the buccopharyngeal cavity to permit normal aerial breathing. Male head bobbing or vibration is also a part of courtship in *T. s. venusta*. Nasal squirting is common in other turtles and may occur in some form in all turtles.

The only account of nesting is by Morjan and Stuart (2001) in Socorro Co., New Mexico (Bosque del Apache Wildlife Refuge), 31 May 1999. A gravid female (ca. 240 mm CL) emerged from a pond at 1610 hours, crossed a road, traveled a short distance, and spent about 5 minutes searching for a nest site. She repeatedly extended her neck upward in an attitude of circumspection and pressed her head and neck to the ground while scratching the surface with alternating strokes of the front legs. At the base of a vegetational clump she turned 180°, dug a typical nest with her hind legs, and laid 19 eggs. The nest was on level, sandy ground 25 m from water and was 19 cm deep. The nest was completed at 1740 hours after she compacted the soil with rocking movements of the plastron. Although *elegans* × *gaigeae* hybrids occur at this locality, the nesting female was clearly a *gaigeae*. This female had been captured twice previously in 1994 (gravid on 8 June, not gravid on 6 July).

Subsequent monitoring of the nest revealed hatchlings present on 20 August, 6 November, and 16 January 2000. The hatchlings were torpid in January, and they had emerged and the nest chamber had collapsed by April, after a period of substantial rainfall and warm temperatures.

Stuart and Painter (2005) provide data on number and size of eggs in the northern part of the range based on palpation of females obtained in the Upper Rio Grande Valley (southern Socorro Co., New Mexico) in a 3-year period. Females (228 to 266 mm CL) were gravid from 19 May to 11 July. Oviposition was induced with oxytocin. Mean number of eggs per clutch in 12 clutches was 15.4 ± 4.9 (6–22).

Data on egg size (n = 147) were as follows: L, 35.0 mm ± 1.3 (31.6–37.7); W, 22.5 mm ± 0.9 (20.1–24.6); mass, 10.7 g ± 1.1 (8.5–13.0) (Stuart and Painter, 2005). Female CL and pre-oviposition mass were not significantly correlated with egg size or mass. Relative clutch mass was 0.10 ± 0.02 (0.05–0.14) and EMI (mean individual egg mass × 100/spent female mass) was 0.65 ± 0.10 (0.55–0.86), a value they considered intermediate between temperate and tropical populations of *Trachemys scripta*.

Eleven clutches from New Mexico were incubated in moist vermiculite at 28–30°C. Mean incubation time was 60.8 days ± 2.4 (57–64). Data on 123 hatchlings within 1 month of hatching were as follows: CL, 29.0 mm ± 1.4 (25.2–32.7); PL, 27.4 ± 1.2 (24.1–30.6); mass, 6.0 g ± 0.8 (4.3–7.7) (Stuart and Painter, 2005).

The effect of incubation temperature on sex determination has not been reported for *gaigeae* but is known for other subspecies of *scripta* (see account of *venusta*).

Stuart and Painter (1997) discuss a record-sized clutch of 29 eggs from a dissected female (224 mm CL, Elephant Butte Reservoir, Socorro Co., New Mexico, 8 June 1988). Mean data on the eggs were as follows: L, 0.37 ± 1.6 (34.5–40.3) and W, 22.8 ± 0.6 (21.5–24.0). A small egg (13.6 × 11.1 mm) was also found in an oviduct but not counted. The oviducal eggs demonstrated some variation in degree of shelling. Enlarged ovarian follicles were in two sets (16 of ca. 15 mm and six in the range of 10–12 mm). Since five of the shelled eggs were found in the coelomic cavity (suggesting crowding in the oviduct) and the female had been in captivity for 3 weeks, it is likely that the high number of oviducal eggs represents two normal clutches and the follicles represent a potential for two more clutches. In any case this was the first evidence on which annual reproductive potential could be based—in this case, three to four clutches for a total of 51 eggs.

GROWTH AND ONTOGENY

Degenhardt, Painter, and Price (1996) gave the following mean sizes and weights for specimens from Elephant Butte Reservoir: 12 males—150 mm (119–199), 475 g (250–800); 10 females—214 mm (181–248), 1,320 g (880–2,125). Data on size for Mexican specimens (JML database) are as follows: adult males; 140 ± 18.2 (115–173) n = 16; adult females, 179 ± 25.5 (135–220) n = 23; im. males (2), 90 and 103; im. female (1), 159; all other im., 92 ± 15.2 (60–111) n = 8. Individuals much larger than any caught were seen in the Rio Conchos near Meoqui, Chihuahua (Legler, 1960b).

Difference in adult size is moderate, elongated foreclaws do not occur in males, and the snout of mature males is slightly elongated but not upturned. The transverse interlaminal seams of the plastron (less so on the interlaminal seams of the carapace) may become narrowly darkened with age, but there is no extensive melanism in either sex. In some large adults the basic concentric plastral pattern may become overlain by secondary melanistic blotches similar to those in *hiltoni* and *nebulosa*.

CONSERVATION AND CURRENT STATUS

The subspecies is classified as "vulnerable" by the IUCN (Baillie and Groombridge, 1996). Populations are stable at three locations in New Mexico (Bosque del Apache National Wildlife Refuge, and Elephant Butte and Caballo reservoirs). Much of the range in the Big Bend region of Texas is located in a national park. Local residents in northern Chihuahua say *gaigeae* is good to eat, but no one questioned had actually eaten one (JML, pers. obs.). Lowering of stream flow by drawing water for irrigation may affect suitability of habitat in all parts of the range, but the taxon will not be subject to extinction as long as impoundments of these streams are in existence.

Trachemys scripta hartwegi (Legler) 1990

Nazas slider (Legler, 1990).

ETYMOLOGY

Named in honor of Norman E. Hartweg (1904–1964), who contributed some of the earliest work on modern taxonomic studies of Mesoamerican chelonians.

HISTORICAL PROLOGUE

Legler became aware of this taxon in about 1958 and collected most of the type material in 1960. Smith and Smith (1979) used the epithet "*Pseudemys scripta hartwegi* Legler nomen nudum" in a manner that would insure its availability at a later date.

RELATIONSHIPS AND TAXONONOMIC STATUS

Trachemys scripta hartwegi is most closely related to *T. s. gaigeae*. Both are placed in a group also containing *T. s. nebulosa* and *T. s. hiltoni* (Legler, 1990; Seidel, 2002). The ancestral stock of *gaigeae* and *hartwegi* differentiated in the interior basins of northern Mexico. This stock was subsequently fragmented by increasing aridity and by changes in drainage associated with the integration of the Rio Conchos with the Rio Grande, and by stream piracy diverting some internal drainages to the Pacific slope (see account of *gaigeae*).

DIAGNOSIS

T. s. hartwegi can be distinguished from all other Mexican *Trachemys* by the following combination of characters: (1) Postorbital mark is dark yellowish orange, relatively large, rounded in front, and never contacts the orbit; it is usually (66%) not connected to neck stripe and is rounded behind. (2) Mandibular stripe is short and isolated and is half or less than half as long as postorbital mark. (3) Carapace is nearly unicolored pale brown with dark-centered ocelli on each supramarginal surface, forming a bold circumferential pattern. (4) Mandibular tomium is smooth or finely serrate. (5) Anal lamina are usually longer than abdominal. (6) Central plastral pattern is brown, relatively narrow, nonconcentric, and concentrated posterior to pectoroabdominal seam. The condition of the mandibular stripe alone is diagnostic among Mexican *Trachemys* (occurring elsewhere only in *T. s. callirostris*).

GENERAL DESCRIPTION

The Nazas slider is of medium size, ranking with *gaigeae*, *elegans*, and *taylori*. Maximum observed size (CL, Legler database): males, 149 mm; females, 278 mm.

FIGURE 44.1 *Trachemys scripta hartwegi*. Top—Holotype, UU 3802, immature, CL 95 mm; dorsal and ventral views. Bottom left, for comparison—*T. s. gaigeae*, immature female, CL 159 mm, KU 51205 (Rio Conchos, 1.6 km NW of Ojinaga, Chihuahua). Bottom right—male paratype *T. s. hartwegi*, CL 153 mm, MSU 1140.

BASIC PROPORTIONS

CW/CL: males, 0.76 ± 0.016 (0.74–0.79) n = 7;
females, 0.78 ± 0.016 (0.75–0.81) n = 14.

CH/CL: males, 0.33 ± 0.021 (0.29–0.35) n = 7;
females, 0.34 ± 0.014 (0.32–0.37) n = 14.

CH/CW: males, 0.43 ± 0.02 (0.39–0.46) n = 7;
females, 0.44 ± 0.02 (0.41–0.48) n = 14.

PWHP/PWMF: males, 1.00 ± 0.03 (0.95–1.05) n = 7;
females, 0.99 ± 0.05 (0.87–1.05) n = 13.

BR/CL: males, 0.33 ± 0.013 (0.30–0.34) n = 7;
females, 0.35 ± 0.008 (0.33–0.36) n = 14.

WH/CL: males, 0.16 ± 0.015 (0.15–0.16) n = 7;
females, 0.16 ± 0.013 (0.14–0.18) n = 13.

Modal plastral formula: $6 > 4 > 3 > 1 > 5 > 2$.

The pale brown, chiefly unmarked carapace is dominated by a bold circumferential ring of dark supramarginal ocelli (often appearing as large black dots). A mid-dorsal keel is usually distinct on posterior halves of C1–4. There are distinct radial ridges on central scutes in adults. Pattern on lateral scutes is obscure or wanting, and there are no typical ocelli. Dark spots on inframarginal surfaces bold, solid, and frequently pale bordered, contrasting sharply with ground color.

Central plastral figure is dim and does not extend onto anals or gulars, and is not at all concentric in adults. Plastral seams are heavily darkened with brownish pigment (except for gular–humeral seam). Secondary epidermal melanism may occur but not in discrete smudges. Symphyseal stripe

FIGURE 44.2 *Trachemys scripta hartwegi*. Left—adult female from type locality; Right—juvenile (Rodeo, Durango).

is not forked behind but separated by a black border and hiatus from other ventral neck stripes; posterior end is flanked, in some specimens, by a pair of spots. Pattern integrity is low (0.155). Tomial edges of jaws are smooth or finely serrated. Gular and pectoral scutes are 16% and 20% of plastral length, respectively. The anal is the longest scute of the plastron, and the abdominal is the second longest in 82% of specimens.

GEOGRAPHIC DISTRIBUTION

Map 28

The Rio Nazas drainage of Durango and Coahuila and, within historic times, at least the lower reaches of the Rio Águanaval. Both are internal drainages. The type locality is Rio Nazas, 1.2 km E of Presa Lázaro Cardenas, Durango, Mexico. This is the tailwater below and within sight of the dam itself. Other localities (Legler, 1990) are all from the Rio Nazas: Coahuila—San Pedro (de las Colonias); 21 km W of San Pedro; 8 km S of San Pedro; Durango—Rio Nazas; 1.2 km E of Lázaro Cardenas Dam; Lerdo; 24 km SW of Lerdo; 10 mi NNW of Rodeo; and 2.4 km NW of Nazas.

The Rio Nazas flows from the highlands of western Durango to a terminal basin, the Laguna de Mayran. The Rio Águanaval flows from the highlands of Zacatecas to the Laguna Viesca (25°27′N, 102°48′W), ca. 40 km S of L. Mayran and 75 km E of Torreon. Maps dated prior to 1960 show the Laguna Viesca to contain water and L. Mayran to be dry. Both rivers had ceased to flow continuously in the early 1960s because of water use for irrigation (Conant, 1963; Iverson, 1981a; Tamayo and West, 1964). However, this phenomenon was noted by Meek as early as 1904.

The Laguna Viesca was a thriving aquatic habitat in 1960 with turtles, fishes, and water snakes. There are six specimens of *Trachemys* from a small spring-fed pond south of Laguna Viesca at Baylor University. Legler has examined photographs, and they appear to be *T. s. hartwegi*. The endemic *Kinosternon hirtipes megacephalum* Iverson formerly occurred in this same pond but is now extinct. All of this indicates that *T. s. hartwegi* once occurred in the Águanaval drainage and might still occur there in permanent aquatic habitats. See account of *gaigeae* for the drainage history of the region.

FIGURE 44.3 *Trachemys scripta hartwegi*. Top—dorsal and ventral views of juvenile and immature female, type locality. Bottom—immature female (Rodeo, Durango).

HABITAT

At a low stage in September 1960 the reservoir impounded by the Lázaro Cardenas Dam was stony, with rather sterile-looking shores. A relatively small amount of water spilled from large pipes into a beautiful wide pool with clear, greenish water and a rocky to sandy bed and banks. From a vantage point on a roadway one could see catfish, carp, *Kinosternon hirtipes*, and about 30 *Trachemys*. Most of these were basking at the surface and diving sporadically; a few basked on rocks near the shore. Below the dam is a gallery forest of mature Mexican cypress (*Taxodium mucronatum*; Conant, 1963).

Elsewhere (16 km NNW of Rodeo; 1960) the Nazas is flanked by cultivated land, has fewer large trees (cottonwoods, willows, and cypress), and consists of long (ca. 400 m) stretches of slowly moving water separated by

FIGURE 44.4 *Trachemys scripta hartwegi*. Dorsal and ventral views of series from Rodeo, Durango: left, two immature females left; right, two mature males.

FIGURE 44.5 *Trachemys scripta hartwegi*. Head patterns. Left—female (Rio Nazas); right—series (Rodeo, Durango; two adult males below, two adult females above).

riffles. Sliders favor quiet pools and bends, and could be observed easily at the surface or basking on mud banks. *T. s. hartwegi* occurs with *Kinosternon hirtipes* at all localities observed by Legler.

DIET

In mid-September 1960 Legler and Raymond Lee observed literally millions of sardine-sized (ca. 100 to 120 mm long) fishes (*Astyanax fasciatus mexicanus*) below the dam at the type locality. Most were concentrated at the dam, and all were facing upstream. This was clearly an abortive upstream migration of some sort. Dead fish were floating everywhere, and both the *Kinosternon* and the *Trachemys* were feeding on them. Baited traps produced few turtles.

REPRODUCTION

A female obtained at the type locality (UU 4701, 298 mm, paratype) on 11 September 1960 bore numerous but old and uncountable corpora lutea, 18 yolked follicles of 15–18 mm, and four follicles of 8–10 mm. This is interpreted as evidence that vitellogenesis had begun in autumn, and that the maximum reproductive output could be 22 eggs in the following season.

A large female (UU 17583, 274 mm) taken 26 May 1976 at the type locality bore 41 ovarian follicles of 12–16 mm;

seven follicles of 10–12 mm, many follicles less than 5 mm, and no corpora lutea. By conservative estimate, these data show a potential for at least two and possibly three clutches laid sometime after early June. No data are available for eggs.

GROWTH AND ONTOGENY

Data on size for 34 specimens (CL, Legler database) are as follows: adult males, 139 ± 10.1 (124–149) $n = 7$; adult females, 184 ± 58.8 (120–298) $n = 14$; im. females, 130 ± 15.0 (115–145) $n = 3$; juveniles, unsexed, 61.3 ± 20.1 (36–95) $n = 10$.

Mature males have unmodified foreclaws and slightly upturned but not attenuated snouts. Mating behavior has not been observed.

Two of the smallest specimens examined (36 and 44 mm) were taken in late September and early October near the type locality; in both, the umbilicus was completely closed and the areolae showed slight growth; these were judged to be young of the year and permitted an estimation of hatchling size at ca. 32 mm CL.

Ages were estimated for two medium-sized juveniles with distinct growth zones as follows: 92 mm (June), beginning of fourth or fifth full year of growth; 95 mm (2 October), late in second or third full year.

Consideration of all meager data from ovarian examination and examination of clear growth records in young animals suggests a temperate breeding pattern in which laying of multiple clutches begins in May or June, some hatchlings emerge as early as August, and a variable amount of growth occurs in the year of hatching.

Ontogenetic changes in patterns and colors are not dramatic. Ocellar patterns on supra- and inframarginal surfaces fill with dark pigment with age but otherwise remain distinct; the plastral pattern is only vaguely concentric at hatching and also fills with dark pigment with age. Some individuals of all ages have vestiges of ocelli on the posterior lateral scutes, and the overall contrast of all shell patterns decreases with age. Hatchlings and small juveniles have a rounded mid-dorsal keel, most prominent on C1–C3, marked with variable dark dots and dashes.

CONSERVATION AND CURRENT STATUS

There is no evidence that *hartwegi* is used by humans for food anywhere in its range. Populations in the drier parts of the range (e.g., vicinity of San Pedro de las Colonias) have already disappeared, but those associated with large impoundments are likely to persist and flourish as long as the reservoirs are there.

Trachemys scripta yaquia Legler and Webb, 1970

Yaqui slider (Legler and Webb, 1970; Iverson, 1992a); jicotea yaqui (Liner, 1994).

FIGURE 45.1 *Trachemys scripta yaquia.* Top—shell of holotype (UU 6030) female. Bottom—paratype (UU 6031, Figure 1 from Legler and Webb, 1970, with permission).

ETYMOLOGY

The subspecific name refers to the land of the Yaqui Indians in northwestern Sonora.

HISTORICAL PROLOGUE

The subspecies was described in 1970 (Legler and Webb, 1970) and, although it has been mentioned in various accounts and revisions (e.g., Legler, 1990; Seidel, 2002), little has been added to our knowledge of the taxon since then.

RELATIONSHIPS AND TAXONONOMIC STATUS

Legler (1990) regarded *yaquia* to be most closely related to *T. s. ornata*. We support this hypothesis. The geographic range of *hiltoni* (R. Fuerte drainage) lies between the ranges of *ornata* and *yaquia*, presenting a complex zoogeographic problem. The distinctness of *hiltoni* and its seeming lack of intergradation may indicate an earlier colonization of Pacific drainage systems than either *ornata* or *yaquia*. Colonization of the Pacific coast by *Trachemys* is discussed in the account of *hiltoni*.

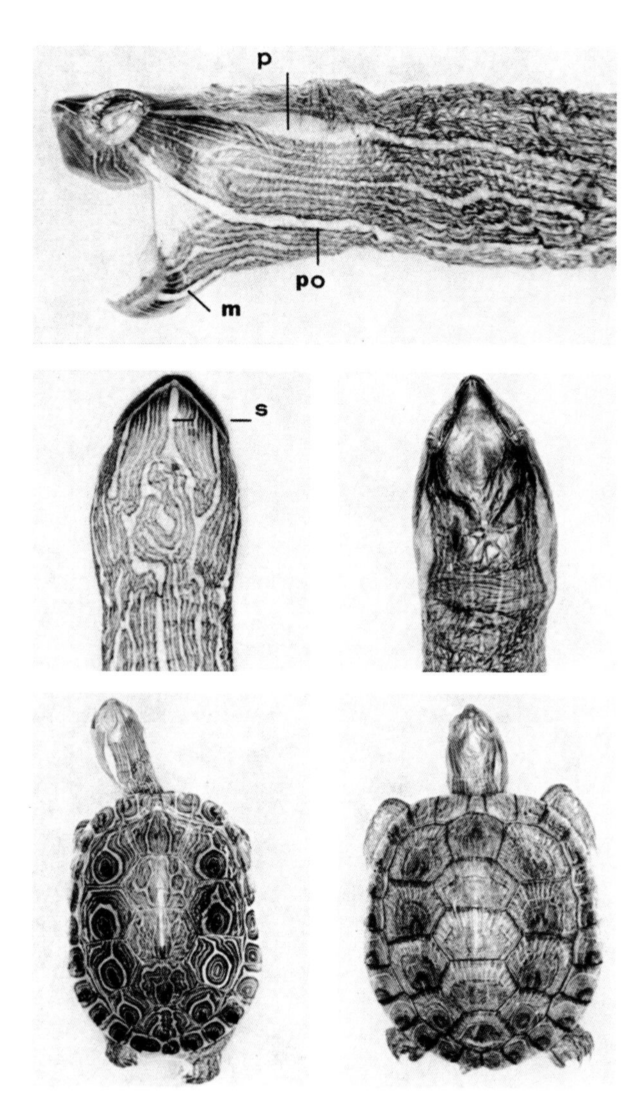

FIGURE 45.2 *Trachemys scripta yaquia*. Top and center—head of holotype showing detail of typical striping (m, mandibular; p, postorbital; po, primary orbitocervical; s, symphyseal). Bottom left—*T. s. ornata* juvenile (UU 3821; Rosario, Sinaloa); bottom right—*T. s. yaquia* juvenile (UU 12486; Nuri, Sonora).

DIAGNOSIS

A subspecies of *Trachemys scripta* most closely resembling *T. s. ornata* and distinguished from all other Mexican *Trachemys* by the following combination of characters: (1) Postorbital mark is yellowish orange, a moderate expansion of a single stripe from eye to base of neck. (2) Mandibular stripe usually (84%) is isolated and shorter than postorbital expansion. (3) Tomial edge of mandibular sheath is coarsely serrated. (4) Humeral is equal to or greater than one-half of pectoral. (5) Lateral scutes have jagged black blotches at the centers of poorly defined ocelli. (6) Mature males lack conspicuously elongated snout and foreclaws. (7) Carapacal and plastral patterns are obscure in adults of both sexes, and plastron never bears discrete, solid, dark blotches.

GENERAL DESCRIPTION

(From Legler and Webb, 1970, unless otherwise noted.)

The largest adults we have examined are as follows: males, 268 mm; females, 309 mm. Vetter (2004) gives maximum CL as 310 mm.

BASIC PROPORTIONS OF 20 ADULTS (JML DATABASE)

CW/CL: males, 0.75 ± 0.021 (0.71–0.78) n = 15;
females, 0.77 ± 0.016 (0.75–0.79) n = 5.

CH/CL: males, 0.33 ± 0.011 (0.31–0.34) n = 15;
females, 0.36 ± 0.017 (0.35–0.39) n = 5.

CH/CW: males, 0.44 ± 0.02 (0.40–0.47) n = 15;
females, 0.47 ± 0.02 (0.46–0.51) n = 5.

PWHP/PWMF: males, 1.01 ± 0.03 (0.95–1.05) n = 15;
females, 1.02 ± 0.03 (0.98–1.05) n = 5.

BR/CL: males, 0.31 ± 0.007 (0.30–0.32) n = 15;
females, 0.34 ± 0.007 (0.33–0.35) n = 5.

WH/CL: males, 0.15 ± 0.017 (0.13–0.16) n = 15;
females, 0.15 ± 0.008 (0.14–0.16) n = 5.

Modal plastral formula: 4 > 6 > 5 > 3 > 1 > 2.

Patterns of carapace and plastron are never as bright and colorful as those in *T. s. ornata*. Lateral scutes have jagged dark blotches at the centers of poorly defined ocelli. Ocelli, when visible, each consist of two thin peripheral rings (pale orange to yellowish) and a darker center with indistinct boundaries; ocelli are situated on posteromedial portions of laterals and are incomplete at posterior edges of these scutes. Ground color of carapace is pale brown; ocelli on laterals are faintly distinguishable and bear distinct dark centers in sharp contrast to ground color; dark centers are irregular and often vertically oriented and ragged; entire carapace may be dark brown in old specimens; plastral pattern is indistinct except in young, fading to uniform yellowish or obscured by brownish pigment with age.

Plastral ground color is straw yellow; darker markings are dusky blackish brown; all interlaminal seams are darkened; there is a concentric central plastral figure extending from posterior gular angle to anterior anal angle; figure is narrow except for transverse extensions along interlaminal seams, never extending to bridge or edges of plastron.

Postorbital mark is dark yellow to orangish yellow, a slight expansion of a continuous stripe from edge of eye (76%) to base of neck (100%). Symphyseal stripe usually (68%) is connected to neck stripe. Pattern integrity index is 0.650. Primary orbitocervical stripe is golden yellow; mandibular and symphyseal stripes are cream yellow. Iris is yellow to pale orange with black transverse stripe.

Tomial edges of jaw sheaths serrated, mandibular more coarsely the maxillary; sharp, pointed serrations become blunted and more nearly crenulate with age. Median alveolar ridge of maxillary denticulate.

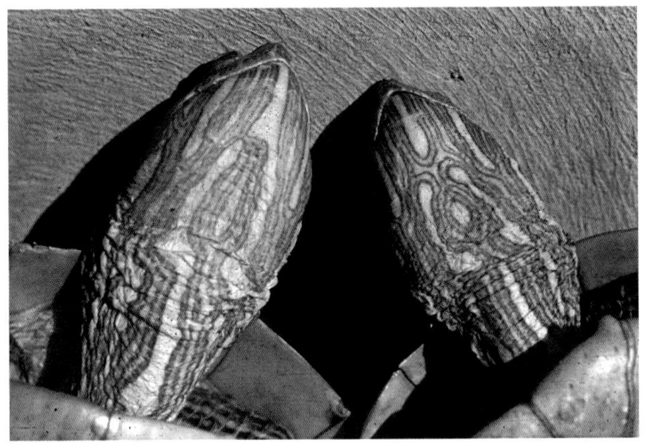

FIGURE 45.3 *Trachemys scripta yaquia*. Typical head striping in females (Rio Mayo).

FIGURE 45.4 *Trachemys scripta yaquia*. UU 6031, adult male, type locality.

GEOGRAPHIC DISTRIBUTION

Map 29

Occurrence is in the Sonora, Yaqui, and Mayo drainages of Sonora and may extend into the Yaqui and Mayo headwaters in Chihuahua. The type locality is Rio Mayo, Conicarit, Sonora, Mexico.

It has been assumed (Legler and Webb, 1970; Smith and Smith, 1979) that *T. s. yaquia* occurs only in the lower reaches of these drainages. The northernmost records are in the upper reaches of the Rio Sonora (and its tributary, Rio San Miguel) and the Rio Yaqui drainages (near the headwaters of the Rio Bavispe) at elevations of at least 1,000–1,200 m.

The ranges of *P. s. yaquia* and *P. s. hiltoni* (R. Fuerte drainage) may be separated by as little as 10–16 km in the region northeast of Alamos. No sympatry or hybridization of *yaquia* with any other *Trachemys* has been detected or reported (Legler and Webb, 1970).

Fossil remains of *Trachemys* (Pleistocene, Sangamon interglacial) have been found at Saracachi Cienega, Rancho Água Fria (ca. 30°25′N, 110°34′W), in the Magdalena drainage and an extant population of sliders (presumably *T. s. yaquia*) exists about 5 km S of the fossil site (Van Devender, Rea, and Smith, 1985). These Rio San Miguel headwaters lie scarcely 15 km E of Rio Magdalena headwater streams. Further exploration may reveal the presence of *T. s. yaquia* in the Rio Magdalena drainage if permanent aquatic habitats exist there.

Aaron Flesch (pers. comm.; and O'Brien et al., 2006) found a *Trachemys scripta* shell and observed several sliders basking on the Rio Aros (R. Yaqui tributary) near Buena Vista, Sonora (ca. 29°20′N, 109°55′W), a region in the lower western foothills of the Sierra Madre Occidental and vegetated by foothills thornscrub ("tropical-subtropical scrublands"; O'Brien et. al., 2006). Considering the remoteness of this region, we regard these observations as a record for *yaquia*. The turtles were observed in and near perennial pools. *Trachemys s. yaquia* is known more than 300 river-km farther upstream in the Bavispe drainage at Huachinera (30°12′N, 108°57′W, 1,118 m; Smith and Smith, 1979).

The Zweifel and Norris (1955) record of *T. s. hiltoni* from the Yaqui (160 km upstream from Cocorit) is applicable to *T. s. yaquia*. The specimen is MVZ 55389 (fide J. R. Buskirk, pers. comm.; not 55384 as stated by Legler and Webb, 1970).

The northernmost records for *T. s. yaquia* lie 100- to 150 km S of the international border. The Rio Yaqui drainage includes a small part of southeastern Arizona.

HABITAT

The type locality is a deep channel and several contiguous oxbow pools immediately below the Presa Mocuzari on the Rio Mayo (Figure 45.6). Many large adults were seen basking at the surface. Specimens were captured there in still pools bordered by cattails in baited hoop nets set near the bank, on a mud bottom at depths of 0.3–2.0 m. The water in the pools was clear but smelled of sulfur. There were dense beds of submergent aquatic vegetation in the main channel of the stream.

Webb caught five adults in a flood pond of the Rio Yaqui, 11 km N of Cd. Obregon, Sonora; the pond was shallow

FIGURE 45.5 *Trachemys scripta yaquia*. Group to show variation of plastral pattern; two females (largest), four males.

(average depth 0.6 m) with a soft bottom of mud and organic material up to 0.6 m deep, was fringed by cattails, and contained floating mats of water hyacinth. Specimens have also been taken in quiet pools of a small Yaqui headwater tributary (Rio Chico) 2.5 km N of Nuri, Sonora (Legler and Webb, 1970).

It is clear that Yaqui sliders survive an arid environment in pools of intermittent streams, and that they prosper in the habitats created by human hydrological works.

DIET

The digestive tracts of three specimens from the type locality contained an average of 62% (20–100) vegetation and 38% (0–80) animal matter, the latter including fish and insect parts.

REPRODUCTION

Several females collected bore neither oviducal eggs nor corpora lutea, but two of the larger females taken on

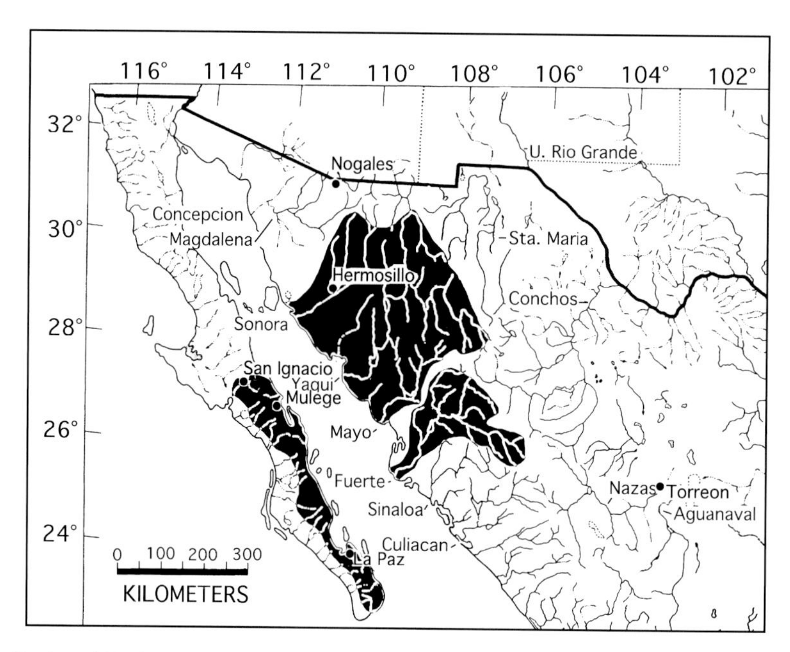

MAP 29 Geographic distribution of *Trachemys s. yaquia* (mainland, north), *T. s. hiltoni* (mainland, south), and *T. s. nebulosa* (Baja California).

FIGURE 45.6 *Trachemys scripta yaquia*. Habitat at type locality (Rio Mayo, Conicarit, Sonora). (Photo courtesy of J. L. Christianson.)

23 July 1965 bore enlarged ovarian follicles; the larger (289 mm CL) had 22 follicles of about 18 mm diameter and the smaller (241 mm) bore 14 follicles ranging from 9 to 16 mm. These data suggest a laying season in late July or August and a reproductive potential of one or two clutches. Females of *T. s. hiltoni* obtained in the Rio del Fuerte on the same date bore oviducal eggs or follicles as large as 25 mm.

GROWTH AND ONTOGENY

Mean size of adults is not greatly disparate. Data on adult CL is as follows: adult males, 223 mm ± 29.21 (162–268) n = 15; adult females, 284 mm ± 27.9 (241–309) n = 5; im. females, 196 mm ± 22.5 (173–223) n = 4. A juvenile of 57 mm was in its first full year of growth (Legler and Webb, 1970; Legler databases).

The snout is attenuated and slightly upturned in males and unmodified in females; the foreclaws of males are not elongated. The adult male carapace is darker and less distinctly patterned than that in females but the male plastron is more clearly marked. The obfuscation of ocelli by brownish, jagged patches is regarded as secondary melanism.

Trachemys scripta hiltoni (Carr) 1942

Fuerte slider (Iverson, 1992a); jicotea Fuerte (Liner, 1994); tortuga Juan (local name at type locality; Bogert and Oliver, 1945).

ETYMOLOGY

A patronym honoring John W. Hilton, collector of the type series.

HISTORICAL PROLOGUE

The types were obtained in extreme southern Sonora during the rainy season (15 June to 15 October 1941). Carr based his description on live specimens and described *hiltoni* as a subspecies of *Pseudemys scripta*, recognizing the close relationship of *hiltoni* to *nebulosa* and *gaigeae*, and initiating the recognition of Mesoamerican sliders as part of the "*scripta* series" (Carr, 1942, 1952).

RELATIONSHIPS AND TAXONONOMIC STATUS

Trachemys scripta hiltoni is clearly most closely related to *T. s. nebulosa* of Baja California. Both, in turn, are less clearly related to the subspecies *hartwegi* and *gaigeae* occurring east of the Sierra Madre Occidental. These four taxa form the "*gaigeae* group." Seidel (2002) considers *hiltoni* to be a subspecies of *nebulosa*.

DIAGNOSIS

The following combination of characters distinguishes *Trachemys scripta hiltoni* from its closest relatives and all other *Trachemys* occurring in Mexico: (1) Throat is usually uniformly pale and lacking well-defined stripes; there is no lance-shaped pale figure on throat. (2) Mandibular stripe, if distinct, often (46%) is not contiguous with another stripe. (3) Postorbital mark is isolated (18% eye, 16% neck) or clearly not a mere expansion of a stripe, almost as wide as long, and always shorter than mandibular stripe. (4) A vertical extension of primary orbitocervical on tympanum usually is contiguous with anterior postorbital stripe. (5) Plastral pattern is narrow, consists of dark blotches, is not distinctly hourglass shaped, does not enclose a wide, pale central area, and sometimes is filled by dark pigment; gular scute often lacks dark markings. Pattern integrity is low (0.324).

GENERAL DESCRIPTION

Trachemys s. hiltoni attains a much larger size than its close relatives, *gaigeae* and *hartwegi*. In a database of 1,121 Mexican *Trachemys scripta* (JML), five adult *hiltoni* (four females and one male) rank among the top 20 largest adult Mexican *Trachemys scripta*, all of which are larger the 320 mm CL. Maximum size in our series: males, 320; females, 351.

BASIC PROPORTIONS

CW/CL: males, 0.72 ± 0.022 (0.69–0.78) n = 28;
females, 0.75 ± 0.021 (0.72–0.80) n = 14.

CH/CL: males, 0.34 ± 0.014 (0.31–0.36) n = 28;
females, 0.38 ± 0.020 (0.35–0.42) n = 14.

CL/CW: males, 0.46 ± 0.02 (0.42–0.49) n = 28;
females, 0.51 ± 0.03 (0.46–0.56) n = 14.

PWHP/PWMF: males, 1.09 ± 0.04 (1.03–1.17) n = 26;
females, 1.08 ± 0.04 (1.02–1.16) n = 14.

BR/CL: males, 0.33 ± 0.007 (0.32–0.35) n = 23;
females, 0.35 ± 0.006 (0.33–0.35) n = 13.

FIGURE 46.1 *Trachemys scripta hiltoni*. Head pattern, adult female, near El Fuerte, Sinaloa.

WH/CL: males, 0.12 ± 0.012 (0.11–0.15) n = 25;
females, 0.14 ± 0.005 (0.13–0.14) n = 14.

Modal plastral formula: 4 > 6 > 5 > 3 > 1 > 2.

Carapace has brown ground color with indistinct, irregular black and orangish marks. Ocelli is distinct only on supramarginal surfaces and has black centers ringed in pale to medium orange. There are no distinct ocelli on lateral scutes. Pale reticular marks on other scutes are also dull orange. There are a few black marks on other carapacal scutes. General aspect of carapace is definitely but indistinctly tricolored—brown, black, and orange. Plastron is definitely and distinctly bicolored (yellow and black). Adult plastral pattern of epidermal melanistic blotches is narrow, not distinctly hourglass shaped, and does not enclose a wide, pale central area (completely filled by dark pigment or extremely narrow); gular scute often lacks dark markings. Stripes on throat, if present, lack dark borders. Throat usually is uniformly pale between primary orbitocervical stripes; mandibular stripe, if distinct, usually is not contiguous (46%) with primary orbitocervical; a vertical extension of primary orbitocervical usually is contiguous with anterior part of postorbital stripe. Pattern integrity is low (0.324). Jaws lack any trace of serration.

COLOR IN LIFE

(Based on numerous live specimens from region of Alamos, Sonora.)

Dorsal ground of head and neck is slate, dusky. There are scarcely any discernible stripes; those present are very obscure cream (neutral and pale). Ground of lateral and dorsolateral fields are slightly paler except that of maxillary, which is yellowish-olive, imparting a jaundiced look to lower half of snout. Anterior postorbital stripe is joined to vertical bar on tympanum. Postorbital mark is dirty pale yellow with suffusion of orange (as in *nebulosa* but not just on lower half). Stripes visible in lateral field (very few) are dirty cream to dirty pale yellow; few of these have distinct dark borders. Ventral field of head and neck, which engulfs the primary orbitocervical stripes, is almost uniformly pale—pinkish pale gray cream on chin and part of throat, pale yellow to cream where stripes have been lost. Mandibular and symphyseal stripes remain distinctly pale lemon yellow but lack dark borders. Ground of mandibular sheath is like that of maxillary sheath. Postorbital mark: pale orange, approaching flesh color in some specimens and more neutral in others; never yellowish, orange, or scarlet orange; isolated posterior part of this stripe, on neck, essentially is same color as expanded part. Limbs are slate gray with pale yellow markings. Axillary and inguinal pockets are almost uniform grayish, slightly yellowish, cream. Ground of plastron is pale yellow with darker yellow or orangish yellow anteriorly and posteriorly. Iris is pale yellow.

The ontogeny of color and pattern is much the same as that described for *T. s. nebulosa*. In the smallest specimen examined (61 mm, Guirocoba) the ground color of the carapace is olive and bears no discernible markings except for the indistinct, dark-centered supramarginal surfaces. The ground color of the plastron is pale yellow; a nebulous, narrow, concentric pattern is already overlain partially by solid brown blotches emanating from the areolae; markings of head, neck, and limbs are pale yellow to cream and are more distinct than those of adults but lack dark borders.

GEOGRAPHIC DISTRIBUTION

Map 29

The Rio Fuerte drainage of Chihuahua, Sonora, and Sinaloa. The type locality is Guirocoba, Sonora (see Habitat). A record of *hiltoni* from the Rio Yaqui (Zeifel and Norris, 1955) is surely applicable to *T. s. yaquia*. The distribution of *hiltoni* is incongruously bracketed by that of *yaquia* to the north and *ornata* to the south.

THE ORIGIN OF *HILTONI* IN THE RIO FUERTE DRAINAGE

The following hypotheses have not, as yet, been challenged or refuted. The subspecies *gaigeae*, *hartwegi*, *hiltoni*, and *nebulosa* (the "*gaigeae* group") demonstrate a sharing of characters that bespeaks common origin. An ancestral stock inhabited the central Mexican interior basins and entered a Pacific drainage stream via headwater exchange and stream capture. This has also happened in freshwater fishes.

There is no evidence of direct exchanges between the Rio Fuerte and the interior basins, but exchanges between the R. Papigochic (once part of an interior drainage system via the Rio Guzman) and the R. Yaqui occurred and are supported by freshwater fish distributions. The following hypothesis is proposed. An ancestral stock invaded the Yaqui and dispersed to and through the Mayo drainage to the Fuerte drainage. None of these drainages contained *Trachemys* at the time. The Sonoran fossils reported by Van

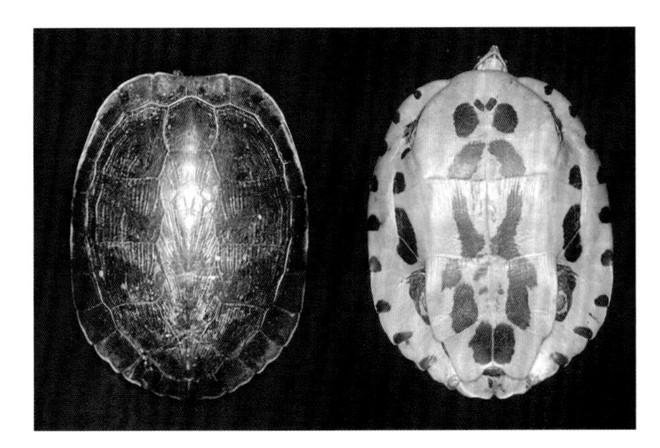

FIGURE 46.2 *Trachemys scripta hiltoni*. Dorsal and ventral views of shell pattern, same individual shown in Figure 46.1.

FIGURE 46.3 *Trachemys scripta hiltoni*. Dorsal and ventral views of juvenile in first or second year (MVZ 50911), CL 69 mm, type locality, Guirocoba, Sonora.

Devender, Rea, and Smith (1985) could represent this colonizing stock. The ancestral stock of *hiltoni* prospered in the Fuerte system but did not survive in any of the drainages to the northwest. At the time these events occurred, *T. s. ornata* had reached the Rio Culiacán via dispersal of an ancestral stock up the Pacific coast, northward from the Isthmus of Tehuantepec.

This scenario fails to account for the present distribution of *T. s. yaquia* in the Sonora, Yaqui, and Mayo drainages. John Iverson (pers. comm., 1987) suggested that *T. s. yaquia*, although it has differentiated within its current range, could have been introduced by humans in recent times from somewhere in the range of *T. s. ornata*.

HABITAT

The type locality is Guirocoba (ca. 26°53′N, 108°41′W, 453 m), the name referring to a village and a nearby ranch, some 29 km SE of Alamos, Sonora. Guirocoba lies in a small valley in the foothills of Sierra Madre Occidental. Terrestrial vegetation of the region is of mixed tropical and desert species. A small stream with deep pools runs through the grassy valley floor. In January 1959, adult *T. s. hiltoni* were seen in the stream where it parallels the road. The type specimens probably came from the stream. It joins the Rio Alamos some 25 km SW of Guirocoba near the town of Tapizuelas. Most of Legler's observations of sliders were made in the Rio Alamos near the Vado Cuchujaqui, a well-known picnic area, 5.5 km E and 8 km S of Alamos (26°57′N, 108°53′W). Under normal conditions, the river runs clear and has riffles alternating with pools as deep as 10 m with steep rock walls on one side and sandy shallows on the other (JML, pers. obs. and field notes).

DIET

Examination of three adult digestive tracts from the Rio Alamos yielded the following. Two stomachs contained parts of greenish coleopterans, a preponderance of grasses, and parts of small snails. Colon contents of the same specimens were chiefly grasslike aquatic vegetation chopped into pieces 10 to 20 mm long. Another adult had an empty stomach and a colon that was packed with brown mud containing parts of aquatic crustaceans and algal filaments.

HABITS AND BEHAVIOR

Legler (field notes, 17 June 1961) sat on a high cliff with binoculars near the Vado Cuchujaqui and watched several adults as they moved about in the water. They swam slowly beneath mats of floating or barely submergent vegetation and remained concealed for several minutes at a time; presumably they were foraging there. Trapping records suggest that the turtles also forage along the banks of the stream, some of which are solid rock. Grasses hang into the water from earthen banks in places. One individual was seen eating submergent algae in a shallow sandy part of the river. When adults moving about in shallow water (usually feeding) were startled by a thrown stone, they disappeared into dead wood snags along the bank or into deep holes below cliffs.

Encounters between two males were seen on two separate occasions. The individuals seemed to be aware of each other only when they could see each other. When they noticed each other they swam rapidly toward each other and, when they met, angled off so that they did not collide head-on (yet they must have touched their carapace margins). At this time bites were exchanged. In each instance the smaller male swam rapidly away after the encounter and the larger male remained in place. It is interesting, however, that a small male seen mating with a large female in the same area was not driven off.

REPRODUCTION

Mating was observed from the aforementioned vantage point in June. A medium-sized male was mating with a large female. He followed her about and mounted periodically.

FIGURE 46.4 *Trachemys scripta hiltoni*. Hatchling (El Atillo, Sinaloa). (Photo by J. R. Buskirk.)

FIGURE 46.5 *Trachemys scripta hiltoni*. Dorsal view of specimen shown in Figure 46.4. (Photo by J. R. Buskirk.)

The pair was observed copulating on six separate occasions within 1 hour. The female would swim from a deep hole at the base of a cliff into shallow water with a sandy bottom and the male would follow her out. The mounted male appeared sometimes to be standing on the bottom with his tail and penis extended forward for almost half the length of his plastron. No clasping was noted. The male's forefeet seemed to be resting on the lateral margins of the female's carapace. No face-to-face Liebespiel was observed.

The ovaries of three large females (UU 3840, 41, 43) taken in the Rio Alamos on 18 June 1961 demonstrate high fecundity and provide data on the timing of reproduction. None had ovulated, but all had follicles of ovulatory size (20–23 mm) plus multiple sets of enlarged (9–19 mm), yolked follicles representing potential future clutches. Size and annual reproductive potential for these females was as follows: 351 mm, 83 eggs, three to six clutches; 330 mm, 67 eggs, three to four clutches; 323 mm, 94 eggs, four to six clutches.

Two females taken 22 July 1965 from the Rio Fuerte at the Highway 15 crossing bore thin-shelled oviducal eggs or fresh corpora lutea representing a first clutch and contained enlarged, yolked follicles for a second clutch (CL and ARP: 281 mm, 29 eggs; 269 mm, 22 eggs). Dimensions of 12 eggs from the larger female were as follows: L, 42.8 ± 1.48 (41.0–46.3); W, 27.6 ± 0.81 (26.4–29.3). Carr (1942) depicts an egg from a paratype of 43 × 27 mm; it was laid in New York in mid November after the types had been held live for several weeks. These data suggest an annual cycle that could be a slight modification of a northern temperate pattern (Moll and Legler, 1971).

Buskirk (1983) found a recently hatched juvenile (CL 37 mm) on 10 July 1983 in the Rio Fuerte and another (CL 48 mm) with a single growth zone in a roadside pool on 17 November 1981.

GROWTH AND ONTOGENY

Size data for specimens examined are as follows: adult males, 244 ± 39.13 (158–320) n = 28; adult females, 284 ± 42.5 (207–351) n = 14; im. males, 122 and 152; im. females, 137 ± 14.91 (126–169) n = 7; one juvenile, 61 mm. The smallest juvenile (61 mm) was taken 15 August 1950 at the type locality. The scutes have an areola plus two zones of growth, and the animal is judged to be well into its first full year of growth.

Large females from Alamos have a flat to slightly concave plastron. In larger adult males the elongated snout develops a bosslike terminal enlargement that contributes to a very distinctive profile. This sexually dimorphic character is shared with *T. s. nebulosa* but not with other long-snouted subspecies (see general account of *Trachemys scripta*).

POPULATIONS, CONSERVATION, AND CURRENT STATUS

In January 1959 and in June 1961 adult *T. s. hiltoni* were abundant and could be easily observed in the Rio Alamos and near Guirocoba. All of the specimens obtained were from baited traps, and no live specimens or shells were brought to us by local residents. This combination of circumstances suggests that sliders were not an important source of food for humans at that time. During cursory exploration of the Rio Alamos in June 1978, no sliders were seen. At that time the lovely streamside area at the Vado Cuchujaqui had been severely littered and perturbed by human use. Population declines may have been well under way at that time.

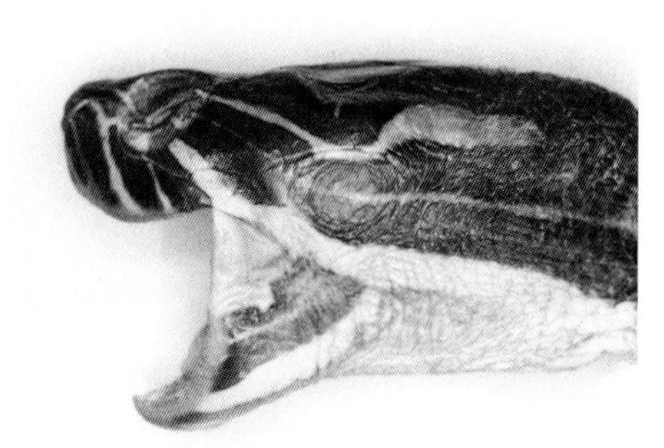

FIGURE 46.6 *Trachemys scripta hiltoni*. Sexual dimorphism in snout development, near Alamos, Sonora: top, adult female (UU 3844); bottom, large adult male (UU 3827). The enlargement of proboscis-like snout progresses with age in mature males.

Trachemys scripta nebulosa (Van Denburgh) 1895

Baja California slider (Iverson, 1992a) ; tortuguita del Rio (Grismer, 2002); caguama de Baja California (Liner, 1994); lower California turtle (Van Denburgh, 1922).

ETYMOLOGY

L. *nebulosus*, cloudy (misty, dark, indefinite). Since the holotype is a juvenile, the etymon could allude to the indistinct plastral pattern. However, the allusion could as well be to the indistinct pattern of head striping at all ontogenetic stages.

HISTORICAL PROLOGUE

Van Denburgh (1895) described *Chrysemys nebulosa* on the basis of four specimens from Baja California Sur. The holotype (CAS 2244, juvenile, 80 mm CL) exists and has been examined (JML notes). Three plates (IV–VI) in Van Denburgh (1895) accurately show most details of shell, head, and limb markings. Plate VI is in error in showing the postorbital mark connected to a neck stripe on both sides. No connection exists (JML, pers. obs.). The artist worked from a preserved specimen with partially retracted head and (forgivably) guessed at this.

Three paratypes of adult size (194–283 mm, no sexes given) were listed; one was from the type locality and the others from "Água Caliente" and San Jose del Cabo. They were destroyed in the San Francisco earthquake and fire of 1906. However, Van Denburgh mentions all of them in his book of 1922.

RELATIONSHIPS AND TAXONOMIC STATUS

Based on external characters, *nebulosa* is most closely related to, and in fact almost identical to, *T. s. hiltoni*, its mainland counterpart. Both taxa are clearly more closely related to *gaigeae* and *hartwegi* than to their closest geographic neighbors on the mainland (*yaquia* and *ornata*). Carr (1942, 1952) considered *nebulosa* as a subspecies of *scripta* and recognized its close relationship to *hiltoni* and *gaigeae* (based on head markings). Seidel (2002) regards *nebulosa* as a full species comprised of the subspecies *nebulosa* and *hiltoni*. Grismer (2002) considers it a full species but does not discuss relationships.

DIAGNOSIS

The boldly marked plastron of adults and the webbed (nonelephantine, nonpaddle-like) feet easily diagnose *nebulosa* among the few chelonians of Baja California (*Actinemys marmorata*, *Gopherus agassizi*, and marine species).

The following combination of characters distinguishes *nebulosa* from all other Mexican *Trachemys* and pertains particularly to its distinction from *hiltoni*: (1) Throat is uniformly pale or bears a pale lance-shaped mark, with its point at mandibular symphysis, expanding just posterior to tympanum and extending to base of neck. (2) Mandibular stripe is usually distinct and contiguous with primary orbitocervical. (3) Postorbital mark is isolated or clearly not a mere expansion of a stripe, is almost as wide as long, and is always shorter than mandibular stripe. (4) Vertical extension of primary orbitocervical stripe on tympanum is usually not contiguous dorsally with another stripe. (5) Adults have a plastral pattern of dark blotches that forms an hourglass or flask-shaped figure and extends from humeropectoral seam to anals and encloses a broad, pale central area (chiefly on abdominal) that may or may not include other dark marks; marks on gulars and humerals are usually serrated from main pattern by constriction at humoropectoral seam.

GENERAL DESCRIPTION

Trachemys s. nebulosa ranks, with *T. s. hiltoni*, among the largest *T. scripta*. Maximum recorded size is a female of 370 mm (Grismer, 2002; Carr, 1952).

BASIC PROPORTIONS

(Legler Database)

CW/CL: males, 0.70 ± 0.029 (0.65–0.73) n = 7;
females, 0.71 ± 0.009 (0.70–0.72) n = 5.

CH/CL: males, 0.34 ± 0.013 (0.32–0.36) n = 7;
females, 0.37 ± 0.012 (0.35–0.38) n = 5.

CH/CW: males, 0.48 ± 0.03 (0.44–0.54) n = 7;
females, 0.52 ± 0.02 (0.49–0.54) n = 5.

PWHP/PWMF: males, 1.06 ± 0.03 (1.03–1.11) n = 7;
females, 1.06 ± 0.05 (1.0–1.1) n = 5.

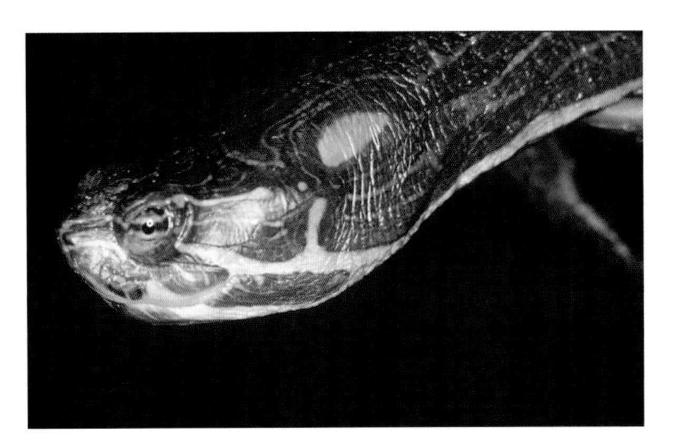

FIGURE 47.1 *Trachemys scripta nebulosa*. Head and neck of adult female, (La Purisima, Baja California Sur).

BR/CL: males, 0.34 ± 0.013 (0.31–0.35) $n = 7$;
females, 0.36 ± 0.000 (0.36–0.36) $n = 5$.

WH/CL: males, 0.14 ± 0.012 (0.12–0.19) $n = 7$;
females, 0.13 ± 0.005 (0.13–0.14) $n = 5$.

Modal plastral formula: $4 > 6 > 5 > 3 > 1 > 2$.

Dark-centered ocelli on the supramarginal surfaces form a peripheral pattern that dominates the dorsal aspect of the carapace; distinct, dark-centered ocelli never form on the lateral scutes. Basic plastral pattern is similar to *hiltoni* but smudges are narrower and more linearly arranged. Plastral figure is constricted at humeropectoral seam and is nearly twice as wide at abdominofemoral seam. Widest dark smudge is less than half the width of widest portion of central pale area. Stripes on throat are ill defined (or absent) and lack dark borders. Throat usually is uniformly pale or bears a lance-shaped mark, is expanded at level just posterior to tympana, and extends from mandibular symphysis to base of neck; mandibular stripe usually is distinct and contiguous with primary orbitocervical; vertical extension of primary orbitocervical on tympanum usually is not contiguous with anterior part of postorbital stripe. Postorbital mark is isolated or clearly not a mere expansion of a stripe, is almost as wide as long, and is always shorter than mandibular stripe. Pattern integrity is low (0.245). Jaws are not serrated or serrations are fine.

COLORS IN LIFE

Color in life (based on two adult females, 279 and 285 mm [UAZ]: Rio Purissima, 18.2 km NE of La Purissima; Baja California Sur). Dorsal field of head and neck has ground of dark, neutral brownish slate. Pale markings over skull roof are irregular. Primary orbitocervical stripe and vertical tympanic bar are pale greenish yellow with some darker areas of lemon yellow. Mandibular stripe and a maxillary stripe are bright lemon yellow and the boldest, brightest stripes in lateral aspect. Symphyseal stripe is same color. Other pale marks are cream to pale grayish yellow. Ground of throat is pale grayish brown; chin is much paler. Ground of limbs is black to brownish slate; pale stripes show same range of pale yellows as for neck; there are many yellow marks with at least a suggestion of green. Inguinal pockets are grayish cream and mottled with pale brownish slate. Ground of plastron is pale yellow to slightly orangish-yellow to darker brownish color. Plastral figure is dark brown to black. Iris is pale yellow with suggestion of green to darker bluish-green. Grismer (2002) states that overall coloration is darker in areas with dark substrate than in areas of pale substrate.

ONTOGENY OF SHELL PATTERN

The markings of juveniles are significantly different than those of adults. Plastral pattern is dusky on a pale ground; pattern on each scute terminates abruptly at interlaminal seam; pattern covers half or less of plastron and is widest on femorals; lateral extensions, where present (chiefly on femorals), are solid, not concentric; there are traces of the concentric pattern along the plastral midline on most of the scutes; carapace pattern in dorsal aspect is dominated by a ring of complete or incomplete dark-centered marginal ocelli; pattern on laterals is more or less reticular, showing only vague indications of ocelli.

Major ontogenetic changes occur with maturity; the ground color of the plastron becomes more yellowish and dark, and opaque patches of secondary melanism form over the individual lateral extensions of the plastral pattern; these patches may remain isolated or be connected to form a pale centered figure that is wider posteriorly than anteriorly and is usually constricted at the humeropectoral seam and at the middle of femoral scute, the entire figure suggesting the outline of a flask or hourglass. Plastral pattern changes occur at sizes of 130 to 150 mm CL, before sexual maturity is attained (JML, pers. obs.). The carapace ground darkens to a brownish color with few paler markings except for the orange peripheral rings of the marginal ocelli.

GEOGRAPHIC DISTRIBUTION

Map 29

The lower half of the peninsula of Baja California, in Baja California Sur from San Ignacio (27°48′N, 112°56′W) to San Jose del Cabo on the southern tip of the peninsula. Most of the specimens and records are from ponds and streams in the western foothills of the Sierra de la Giganta. The type locality is Los Dolores, Baja California Sur (Vandenburgh, 1895). Los Dolores (25°05′N, 110°52′W) is situated west of the northern tip of San Jose Island and is shown on some maps as "Bahia Dolores." Grismer (2002) gives a very clear map of distribution. Distribution is disjunct and limited to isolated permanent bodies of fresh water, including nearly all major drainage systems in southern Baja California.

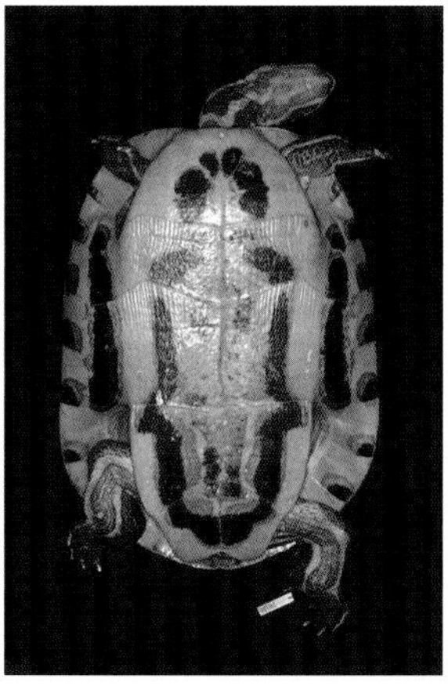

FIGURE 47.2 *Trachemys scripta nebulosa.* Dorsal and ventral views of adult female *T. s. nebulosa.* (La Purisima, Baja California Sur).

Grismer (2002) attributes much of the distribution north of the Cape Region to human introduction.

THE ORIGIN OF *T. S. NEBULOSA* IN BAJA CALIFORNIA

T. s. hiltoni and *T. s. nebulosa* are so similar that it would be imprudent to separate them taxonomically were they not separated by the Gulf of California. The virtually certain relationship of *T. s. hiltoni* to ancestors in the interior basins (regardless of its possible route to the Rio Fuerte) qualifies it as the logical ancestor of *nebulosa.* Their similarity bespeaks a more recent origin of *nebulosa* than other subspecies of *scripta.*

Grismer (2002) concluded that *nebulosa* had a natural origin in Baja California that was related to the formation of the peninsula and was probably thereafter transported throughout southern Baja California as a food source in more recent times. It has been hypothesized (Conant, 1969) that *hiltoni* and *nebulosa* are relicts of an older distribution around the entire Gulf of California. *Trachemys* fossils have been found at the mouth of the Colorado River in Irvingtonian Pleistocene beds at El Golfo (Shaw, 1981) and in northern central Sonora (Van Devender, Rea, and Smith, 1985). Scenarios invoking isolation on a Miocene plate (Murphy, 1983) or even Pleistocene vicariance events seem to be too old to account for the slight differences between the two taxa.

A crossing of the Gulf in recent times, by rafting or human introduction, is an attractively simple working hypothesis. We know of no instances of large mats of flotsam emanating from the mouths of the Rio Fuerte, but larger floods may have occurred prior to the advent of hydrologic works.

The possibility of human introduction has been alluded to by Conant (1963), Smith and Smith (1979), Murphy (1983), and Legler (1990). Grismer (2002) questions the idea of introduction because culinary use of nonmarine turtles is not mentioned in the "Jesuit literature." Actually, Jesuit writings provide some useful clues as to how sliders might have been introduced into the southern half of the peninsula by mainland aboriginals, long before the period of Jesuit hegemony (ca. 1697–1767).

The following provide evidence that there were freshwater turtles of some kind in Baja California in the mid-18th century. Clavigero (1789:66) stated, "Among the turtles, besides the common land variety and the freshwater ones, there are two other species of large sea turtles." Miguel del Barco (1757, 1980 translation) made the most specific remark of any of the Jesuit writings: "In some pools of arroyos are found small freshwater turtles."

There is also written evidence of a mainland native technology that permitted crossing of the Gulf in dugout canoes that were powered by sails and paddles. Baegert himself (1772) crossed the Gulf of California in a dugout canoe (a "hollow tree") of ca. 11 m in May 1751. The trip took about 2.5 days, during two of which they were out of sight of land. The course was SW from "the twenty-eighth degree of northern latitude" (near Guaymas) to Loreto (ca. 200 km). No further details of the trip are given, but Baegert must have been accompanied and must have carried food and water. Miguel del Barco (1980) made several references to native use of "canoes" for pearl diving near San Jose Island and that many of these canoes came from "outfitters" in Sonora.

These writings demonstrate that the Gulf could be crossed in dugouts and bespeaks at least elementary navigation at least 250 years ago (Baegert reached Loreto, which was where he wished to go). This aboriginal technology could have dated back many hundreds of years.

If humans could traverse the Gulf they had to carry food and water. Live turtles and iguanas are often trussed up and kept for long periods as a supply of fresh meat (witness any tropical market). Turtles could have been carried in both directions across the Gulf.

There is a troubling element in any of the forgoing scenarios—they do not include a species of *Kinosternon.* Some species of *Kinosternon* could prosper in the dry environment of Baja California. The only other place in Mexico where *Trachemys* does not occur with a species of *Kinosternon* is Cuatro Ciénegas, Coahuila (*T. s. taylori*).

HABITAT

Large bodies of water with muddy bottoms are preferred, but high densities of turtles have been seen in shrinking ponds that form in the bends of drying streams. In the region of San Ignacio turtles occur in lagoons along the course of the river and migrate from one pool or lagoon to another only after rare heavy rains. The water is clear and slightly brackish due to sandstone outcrops in the river bed. These lagoons lie in small artesian oases lined with date palms, fig trees, and other fruit trees. Turtles are easy to observe from an overlook along the edge of the Rio

Mulege (ca. 26°53′N, 110°52′W), just behind the Mission (Grismer, 2002; Carr, 1952).

HABITS AND BEHAVIOR

Aerial basking occurs on floating logs and shoreside rocks; as many as 20 turtles can be seen basking at once. Population estimates of nearly 100 individuals from a single pond have been made (Roberts, 1982). The turtles are wary and difficult to approach. In winter they bury themselves in mud bottoms and occasionally come out during sunny weather to bask. Most activity occurs between mid March and late October (Grismer, 2002).

REPRODUCTION

Eric Waering (Carr, 1952) observed a female laying eggs in a nest 11 m from the shore of a lagoon near San Ignacio in June 1943. The nest was the size and depth of the female's carapace, in the shade of a small tree, and was full of liquid, "presumably urine." Hatchlings have been reported from July through October in the La Presa region, suggesting spring and summer nesting (Grismer, 2002). No data exist on eggs, gonadal examination, or mating behavior.

GROWTH AND ONTOGENY

Data on size (CL, Legler database): Adult males, 227 ± 52.9 (176–324) *n* = 7; adult females, 266 ± 19.9 (235–285) *n* = 5; im. males, 154; im. females, 149; im. specimens of indeterminate sex, 126 ± 19.8 (105–144) *n* = 3; juveniles, 80.0 and 42.0. There is only slight dimorphism in size, and males lack elongated foreclaws. Males have spectacularly developed, elongated snouts that develop a prominent bosslike enlargement with age (as in *T. s. hiltoni*). The holotype is a juvenile (80 mm CL) in its second or third full year of growth. The smallest specimen examined by us (San Ignacio) has a 42 mm CL, shows very slight growth, and has a large, distinct umbilical scar; it is judged to be near hatchling size.

CONSERVATION AND CURRENT STATUS

Trachemys scripta nebulosa is a source of food for humans in remote areas in southern Baja California (e.g., La Presa region and ranches SW of San Ignacio), as judged by shells in kitchen middens. Turtles are collected by hand, with nets, or by diving and are held for future consumption. Fried or boiled meat of the limbs is said to be "gamey" (Grismer, 2002).

Trachemys scripta ornata (Gray) 1831

Jicotea, local name; ornate slider (Liner, 1994; Iverson, 1992a).

ETYMOLOGY

L. *ornatus*. Handsome, bedecked, splendid, ornata in allusion to its generally attractive pattern.

HISTORICAL PROLOGUE

The name "ornata" was formerly in wide use for many or all populations of Mexican and Mesoamerican *Trachemys* (see general account of *T. scripta*). *Trachemys scripta ornata* is here regarded as the most generalized of the Mexican subspecies—a phenotype from which all other character states and their combinations could logically be derived. Reference is made chiefly to the little-modified pattern of head stripes, the complete ocelli on the lateral scutes, a minimum of sexual dimorphism, and a nearly full (albeit ontogenetically faded) concentric plastral pattern. The most closely related taxa in Mexico are *T. s. grayi* to the south and *T. s. yaquia* to the north (but see comments in the account of *hiltoni*).

DIAGNOSIS

The following characters, in combination, will distinguish *T. s. ornata* from all other *Trachemys* in Mexico: (1) Postorbital mark is orange, an expansion of a yellow postorbital stripe from eye to base of neck, and always connected to orbit and to a neck stripe on one or both sides; mandibular stripe usually (86%) is isolated; symphyseal Y usually (79%) is complete on one or both sides. (2) Mandibular tomium (and usually maxillary tomium in large individuals) is coarsely serrated in adults. (3) Distinct ocelli are centered on posterior halves of lateral scutes and not truncated by overlapping adjacent laterals. (4) Plastral pattern is distinct and consists of four concentric, dusky lines on a pale yellow ground that fade somewhat with age but are never completely obscured in adults.

GENERAL DESCRIPTION

Trachemys scripta ornata ranks, with *hiltoni* and *grayi*, among the largest sliders in Mexico, as judged by representation in JML database. Maximum sizes for adult males and females are 359 and 353 mm, respectively.

BASIC PROPORTIONS

CW/CL: males, 0.76 ± 0.031 (0.70–0.82) *n* = 16;
females, 0.77 ± 0.031 (0.72–0.82) *n* = 16.

CH/CL: males, 0.37 ± 0.017 (0.33–0.39) *n* = 16;
females, 0.39 ± 0.023 (0.37–0.42) *n* = 15.

CH/CW: males, 0.48 ± 0.03 (0.42–0.53) *n* = 15;
females, 0.50 ± 0.03 (0.46–0.58) *n* = 13.

PWHP/PWMF: males, 1.01 ± 0.02 (0.98–1.05) *n* = 17;
females, 1.00 ± 0.03 (0.93–1.05) *n* = 19.

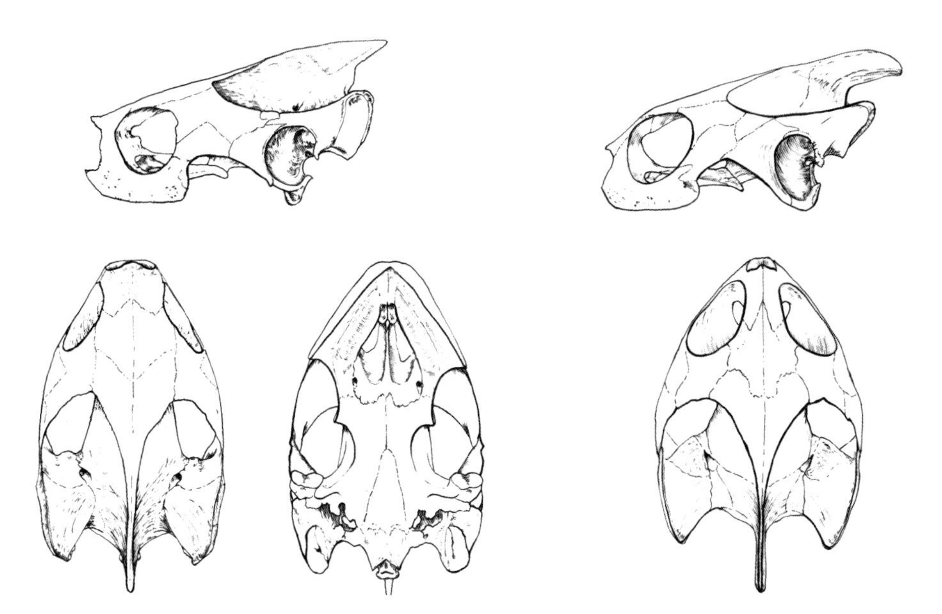

FIGURE 48.1 Skulls of two subspecies of *Trachemys scripta*. Left—*T. s. hiltoni*, adult male, 15 km ESE of Alamos, Sonora (UU 3832, CBL 48.4, Greatest width 30 mm; right—*T. s. ornata*, adult male, Rio Baluarte, Rosario, Sinaloa (UU 3809, CBL 31.1, GB 24.7). Skulls shown at same size for comparison. Note profiles of snout region. (Drawings by Elizabeth Lane.)

BR/CL: males, 0.33 ± 0.013 (0.31–0.37) n = 17;
females, 0.35 ± 0.016 (0.34–0.39) n = 19.

WH/CL: males, 0.15 ± 0.010 (0.14–0.17) n = 16;
females, 0.14 ± 0.014 (0.13–0.18) n = 17.

Modal plastral formula: 4 > 6 > (5 < = > 3) > 1 > 2.

Ground color of carapace is dark brown and distinctly ocellate; each major ocellus has two or three pale concentric lines and a darker center that does not contrast sharply with ground color. Ocelli often are interconnected by a longitudinal pale line that extends anteriorly from the anterior edge of one ocellus to the posterior edge of the next. Ocellar pattern is never lost ontogenetically; dark ocellar centers do not contrast sharply with ground color. There is a full plastral figure that fades ontogenetically.

In general, there is a high degree of pattern integrity (0.701). Markings on head and neck are generally distinct, linear, and unmodified by melanism or fading. There is a postocular stripe beginning at posterior edge of orbit and continuing unbroken (both sides) to base of neck; it expands to a width of 6–7 mm, beginning at level of anterior edge of tympanum and continuing posteriorly for distance equal to twice the diameter of tympanum. There is an orbitocervical stripe beginning at lower posterior border of orbit and continuing unbroken to base of neck. Markings on throat are distinct, not clouded; pale marks here and on other soft parts are yellowish cream in alcohol (probably bright yellow in life). Mandibular stripe is long but isolated. Symphyseal stripe is variable, isolated or not. Upper and lower tomia are serrated, the lower always heavily so, the upper at least moderately; serration is not correlated with size and age.

COLOR IN LIFE

(Based on specimens from 2 km N of El Dorado, Sinaloa.)

Plastron is pale yellow with distinct, slightly narrowed, dusky concentric figure. Ground color of soft parts is dark brownish olive. Postorbital mark is dark orangish-yellow to dirty (not bright) orange, darker above, paler below. Head stripes generally are dark to golden yellow, paler on throat. Iris of most specimens is a beautiful pale green. Largest female has an iris with more yellow but still distinctly greenish. Shell patterns usually are bright and distinct throughout life. Plastral pattern fades somewhat with age but is never completely obscured in adults.

TYPES AND TYPE DESCRIPTION

The following information on the two cotypes in the British Museum (Natural History) was provided by Ms. A. G. C. Grandison (pers. comm., 1965).

Boulenger (1889) refers to two young preserved specimens as the "types" of *Emys ornata* (Gray, 1831). They are BMNH 1946.1.22.40 (CL 43.7 mm) and BMNH 1946.1.22.41 (CL 40.5 mm) and were obtained in Mazatlan, Sinaloa, by Alexander Collie, British Royal Navy, in the course of an expedition from 1825 to 1831 (received 1831 or earlier). The larger specimen (BMNH 1946.1.22.40) is in better condition and is here nominated as lectotype on the

FIGURE 48.2 *Trachemys s. ornata.* Habitat (Laguna Rio Viejo, 2 km N of El Dorado, Sinaloa).

recommendation of Ms. A. G. C. Grandison. The type locality was originally stated as "Mazetland" or "Mazet Land" and corrected to Mazatlan (Sinaloa) by Gray (1855a,b). Günther (1885) examined the types and stated that one of them had been "coarsely" illustrated by Gray in Beechey (1839).

The two beautifully illustrated specimens in Günther's account (1885, Plate 1) are not the types; they are specimens referred to as 2 and 5 inches in length collected by Forrer at Presidio (10 km SW of Villa Union, Sinaloa). The legend states that the specimens are drawn at "nat. size"; in this case the carapace lengths of these turtles, as measured from a facsimile reprint of Günther (1885), were 38 and 87 mm. Specimens listed by Smith and Smith (1979) from Presidio, Rio del Presidio, and Villa Union can be regarded as topotypic in provenance.

GEOGRAPHIC DISTRIBUTION

Map 30

Occurrence is continuous in Pacific drainages below 300 m, from the region of Culiacán, Sinaloa southward at least to the region of Cabo Corrientes in Jalisco, and thence, discontinuously to the region of Acapulco, Guerrero. No records for *ornata* from the Rio Sinaloa exist, but all our experience suggests it could occur in still-water habitats that parallel the lower course of the river.

At Cabo Corrientes, Jalisco (20°25′N), a major headland extending to the sea precludes simple coastwise dispersal but not mouth-to-mouth dispersal resulting from floods. To the southeast there are smaller narrow patches of coastal plain (e.g., Manzanillo and Playa Azul) separated by minor headlands, all of which contain coastal lagoons and might support sliders. Acapulco lies in the center of the largest "patch." Coastal lagoons occur for 100 km to the NW and SE. Sliders could occur in all of these lagoons. Sliders have been rather vaguely reported from this region (Duges, 1890; Velasco, 1892, 1896; Casas Andreu, 1967). Specimens exist from two lagoons.

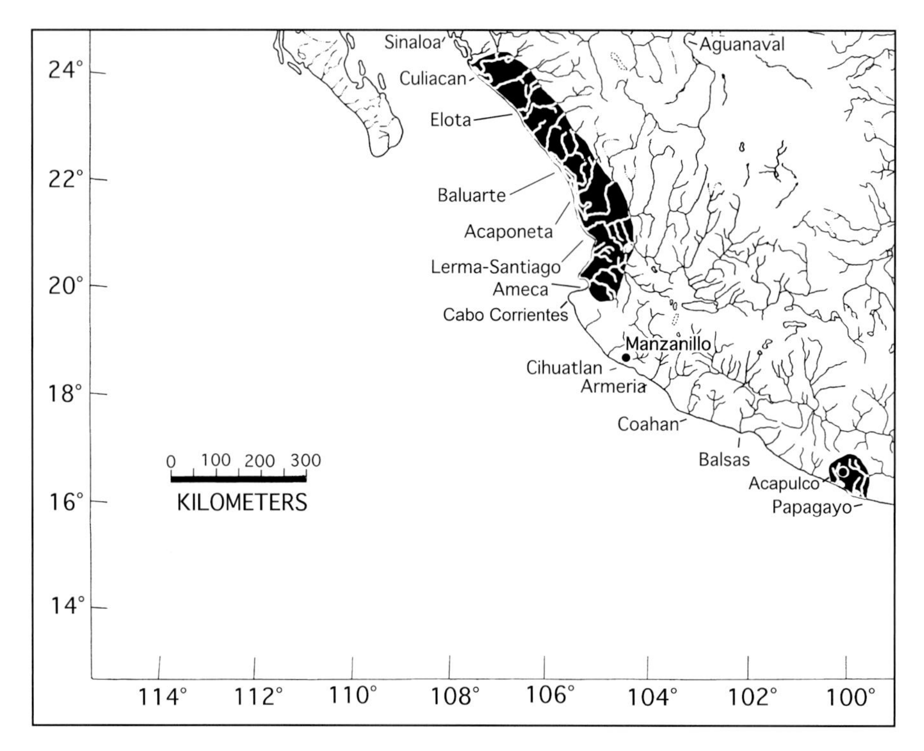

MAP 30 Geographic distribution of *Trachemys scripta ornata*. Isolated populations may occur between Acapulco and Cabo Corrientes.

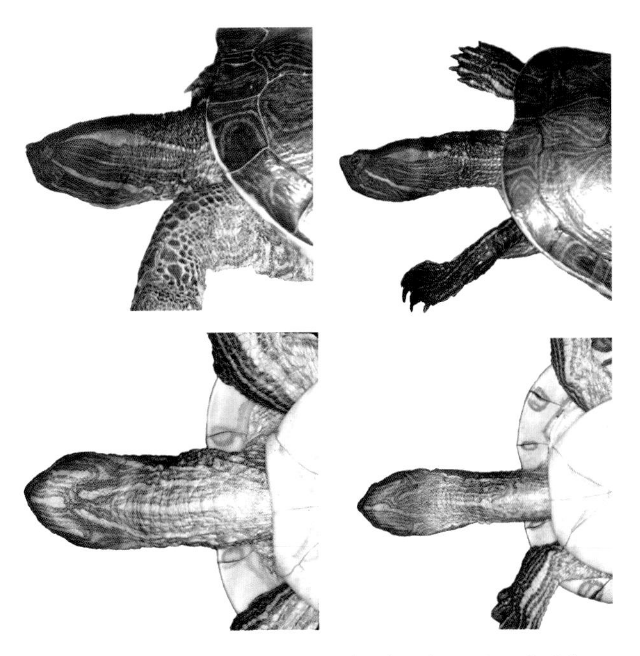

FIGURE 48.3 *Trachemys s. ornata*. Head and neck striping of adult *T. s. ornata* (Laguna Rio Viejo): left, female; right, male (UU).

FIGURE 48.4 *Trachemys s. ornata*. Dorsal and ventral views of adult *T. s. ornata* (Laguna Rio Viejo): left, female; right, male (UU).

James L. Christiansen (field notes, July 1965) visited a seaside fishing village and lagoon (16 km N of Coyuca, Guerrero) and learned that sliders and *Kinosternon integrum* occurred there. Market hunting had reduced their numbers. One hunter had obtained only two sliders in the past year. A shell of *T. s. ornata* (UU 6048, 326 mm) was obtained. A denser population occurred in a small lake, Pie de la Questa, WNW of Acapulco near the southeastern tip of Laguna Coyuca. Adult sliders from Pie de la Questa (UU 6044-47,48) were purchased in the Acapulco market. Trapping and observation of another lagoon 24 km SE of Acapulco (near airport, slightly saline, with hyacinth and cattails) produced only *Kinosternon integrum*. Flowing streams in the region (e.g., Rio Omitlan) are fast and unsuitable habitat for sliders.

Legler (1990) hypothesized that a *venusta*-like ancestral stock crossed the Isthmus of Tehuantepec and gave rise to *grayi*, *ornata*, and *yaquia* by dispersing along the Pacific coast during a period of lower ocean levels. When ocean levels rose to their current height, the mentioned headlands isolated the separate populations.

HABITAT

T. s. ornata thrives in slow-water habitats, natural or those resulting from anthropogenic works—impoundments, irrigation systems, bridge construction, and ponds of various kinds. Usually it does not occur in nearby fast-flowing waters, where flood debris may show flooding 3 m above normal level. In all instances known to us, *ornata* occurs with *Kinosternon integrum*. In some cases (e.g., Ticha, Nayarit) they occur, at least seasonally, in coastal marismas that are at least somewhat saline. But in many coastal areas fishermen say they do not catch sliders.

At Culiacán, Sinaloa, inquiries produced no evidence that sliders occurred at Culiacán or downstream to the mouth of the river. However, fishermen told JML that "large turtles with stripes" occurred in slow water upstream at the confluence of the R. Hamaya. These turtles had been caught and removed with decreasing success over the years.

Trachemys scripta ornata is common in freshwater marshes and ponds near Teacapan in extreme southern coastal Sinaloa (22°33′N, 105°45′W) and can be seen basking on banks or overhanging branches in spring and summer (Scott, 1962).

Sliders were common in an isolated oxbow of the Rio San Lorenzo ("Laguna Rio Viejo") 2 km N of El Dorado, Sinaloa, a water body ca. 400 × 100 m. Tules border the lake on floating mats, and the water is dark and murky, 3 to 5 m deep, with a fine mud bottom and a few rooted tree stumps. Hyacinth covered 10% of the surface. Jacanas were observed. *Trachemys* heads were nearly always in evidence. Baited traps (sardine bait, 764 trap hours) produced 12 *T. s. ornata* and 155 *K. integrum* (101 saved). One *ornata* was caught by hand while sleeping at night on the surface. Otters were present and also entered the traps.

FIGURE 48.5 *Trachemys s. ornata*. Large adult female (AMNH 76128; Tuxpan, Nayarit) showing heavy serration of jaws (one row maxilla, two rows mandible).

FIGURE 48.6 *Trachemys s. ornata*. Adult female, *T. s. ornata* (Laguna Pie de La Questa, WNW of Acapulco. Guerrero). (Photo courtesy of J. L Christianson.)

Numerous sliders were observed in isolated ponds from the highway bridge over the Rio Presidio in Sinaloa. An irrigation canal off the Rio Baluarte at Rosario, Sinaloa, with deep (1 to 2 m) pools connected by slow water supported a good population from which a series of 16 *ornata* was taken on 22 June 1961 (UU 3807-22; JML, field notes).

DIET

Dissections of an adult female and an adult male from 2 km E of El Dorado revealed roots, grass stems, and pieces of broad leaves in all parts of the gut. No animal matter was present (E. O. Moll, field notes).

REPRODUCTION

Scott (1962) presented evidence of enlarged ovarian follicles in late April and of nesting in early May near Teacapan, Sinaloa. A gravid female (338 mm CL, 5.9 kilos), taken 31 March 1964, 2 km N of El Dorado, Sinaloa, contained 20 thin-shelled oviducal eggs and a corresponding number of fresh corpora lutea for a first clutch of the season. She bore a total of 51 yolked ovarian follicles (9–25 mm) that could be potentially ovulated in the current season for an ARP of 71 eggs in three to four clutches. By contrast, a smaller female (260 mm) taken in the same trap had ovaries with only four enlarged follicles (6–9 mm) and no evidence of recent ovulation. Dimensions of eight eggs from a specimen purchased in Acapulco (UU 6049, CL 270 mm, Laguna Pie de la Questa) were as follows: L, 40.3 ± 0.91 (38.2–41.0); W, 25.3 ± 1.4 (23.3–27.6).

GROWTH AND ONTOGENY

Sizes (CL) of specimens in the Legler database are as follows: adult males, 201 mm ± 57.1 (126–359) n = 16; adult females, 249 mm ± 72.7 (153–353) n = 15; im. females, 115 mm ± 8.40 (101–126) n = 8; juveniles, 63 mm ± 25.2 (34–80) n = 3. Ewert (1979) gives the CL for one hatchling as 36.5 mm.

There seems to be minimal sexual difference in adult size. Males lack elongated foreclaws and have a slightly attenuated, narrowed, upturned snout but lack the extreme elongation of *grayi* and the knoblike enlargement of *hiltoni*.

PREDATORS, PARASITES, AND DISEASES

Scott (1962) reports a breeding population of *Crocodilus acutus* near Teacapan, Sinaloa. It would be unusual, if the turtles occurred in the same waters as the crocodiles, that the crocodiles did not prey on turtles of all sizes.

POPULATIONS

A series of 16 specimens (UU) obtained in baited traps in the Rio Baluarte near Rosario, Sinaloa, is comprised of five males, five females, three immature females, and three juveniles. The same traps produced a similarly balanced series of 16 *Kinosternon integrum* in the same period. Another series of 13 *ornata* from 2 km N of El Dorado, Sinaloa, contained eight adult males and five adult females.

CONSERVATION AND CURRENT STATUS

Fishermen near Acapulco, Guerrero, and Culiacán, Sinaloa, indicated a decline in their capture of sliders from modest to virtually nil over several years (field notes, James L. Christiansen, 1965; JML, 1964).

FIGURE 48.7 *Trachemys s. ornata*. Plastron of individual in Figure 48.6. (Photo courtesy of J. L. Christianson.)

FIGURE 48.8 *Trachemys s. ornata*. Laguna Pie de La Questa, WNW of Acapulco, Guerrero. (Photo courtesy of J. L. Christianson.)

Genus *Terrapene* Merrem, 1820

Box turtles; tortugas de caja (Liner, 1994).

ETYMOLOGY

The name is said to be derived from the English vernacular "terrapin," which in turn is derived from American Indian (Algonquin) *torope*, referring presumably to diamondback terrapins of the genus *Malaclemys* (Smith and Smith, 1979; several dictionaries). However, it is possible that Merrem compounded the name with a scholarly view to the Latin and French roots *terrestris* and *terre*, respectively, alluding to the well-known terrestrial habits of box turtles.

DIAGNOSIS OF GENUS

Unless otherwise cited, based on Boulenger, 1889; McDowell, 1964; Smith and Smith, 1979; Milstead, 1969; and Legler, 1960.

A genus in the Family Emydidae of small adult size and chiefly terrestrial habits. Carapace is highly arched; CH/CL, 0.50–0.60; CW/CL, 0.80–0.85; maximum width is at posterior one-third of carapace; eighth neural bone is absent, permitting midline articulation of eighth and sometimes seventh costals. Plastron has a freely kinetic hinge between hyoplastron and hypoplastron (along pectoro-abdominal seams), and hinge is rudimentary or absent in young; buttresses are reduced to low processes on hyo- and hypoplastron and are movably articulated to carapace; combined plastral lobes are about as long or slightly longer than carapace; plastral lobes are capable of complete or nearly complete closure of shell. A mid-dorsal keel is present in juveniles, but is reduced or absent in adults. Entoplastron is traversed by humeropectoral seam. Inguinal and usually axillary scutes are absent. Scapular rod (blade) is at right angle to acromion process; a suprascapula and an episcapula form a series of three synovial joints between scapula and carapace. Phalangeal formula is variable but usually is reduced (23332 – 22222 for manus and 23332 – 23331 for pes). Frontal bone enters orbit; maxilla reaches infratemporal fossa. Temporal roofing of skull is reduced, usually extremely so; postorbital bar usually is reduced; zygomatic arch is complete or incomplete; jugal tapers to a point ventrally, not in contact with pterygoid, not excluding maxilla from border of inferior temporal fossa. Triturating surface of maxilla is without ridging. Upper jaw sheath is hooked, notched, or flat. Head lacks discrete scales. Cloacal bursae are usually absent or minute. Diploid chromosome number is 50: 26 macrochromosomes, 24 microchromosomes (Killebrew, 1977a; Dodd, 2001).

GENERAL DESCRIPTION

The genus is a diverse assemblage adapted to a wide variety of habitats (Minx, 1996). Box turtles are the smallest emydids in Mexico (adult size ca. 90–216 mm; maximum CL and Wt, 235 mm and 2.1 kg; *T. carolina major*; Jackson and Brechtel, 2006) and are habitually terrestrial and omnivorous. The shell is variably arched, the digits are shortened, and the interdigital webbing is vestigial or lacking. A hinged kinetic plastron develops early in life. The plastral lobes can close the shell completely. The only other North American emydid with a hinged plastron is the monotypic (aquatic) genus *Emydoidea*, which does not occur in Mexico.

Waagen (1972) found one or two axillary musk glands in *Terrapene ornata* and *T. nelsoni*; the glands were lacking in 78% of the *carolina* examined. Musk is most often secreted by turtles (especially juveniles) in response to trauma or to being handled and becomes less common as individuals

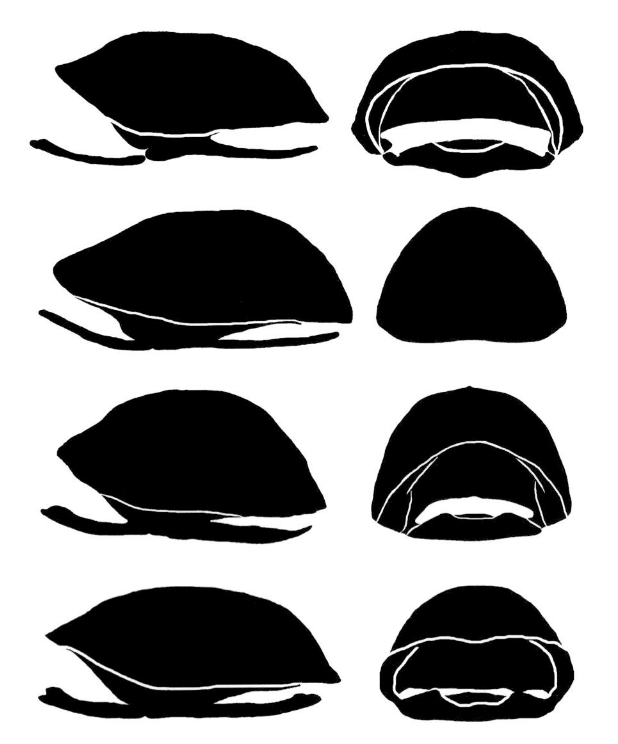

FIGURE 49.1 Typical shell profiles (left) and optical cross sections (right) of four species of *Terrapene*. Top to bottom—*T. o. ornata,* male, CL 127 mm, UU 11579, Kansas; *T. n. nelsoni,* male, holotype, USNM 46252, CL 129 mm; *T. yucatana,* female, CL 146 mm, UU 9804, Libre Union, Yucatán; *T. coahuila,* male, CL 131 mm, UU 17565, Basin of Cuatro Ciénegas, Coahuila.

become inured to handling (Niell, 1948; Legler, 1960c). Bird dogs that detect unseen box turtles (Carr, 1952; Wied, 1865; Devaux and Buskirk, 2001) may be smelling musk. Although musk has a strong odor, it seemingly does not deter predators.

Relationships within the genus are complex, and variation within *Terrapene carolina* is especially problematic. The monophyly of the genus has not been questioned (but see account of *T. coahuila*). Most *Terrapene* are as terrestrial as *Gopherus*. *Terrapene ornata* is the most highly modified for terrestrial existence. At the other extreme is the aquatic *Terrapene coahuila*. Minx (1996) is the most recent revision of the genus and recognizes four extant species (*carolina, coahuila, ornata,* and *nelsoni*) comprising 11 named taxa. Mexican taxa are poorly known. Milstead (in Milstead and Tinkle, 1967) defined two groups within *Terrapene*, which were accepted and refined by Minx (1996). The *ornata* group includes *T. ornata* and *T. nelsoni* and is characterized by a relatively low, flat-topped carapace, extreme reduction of temporal roofing, and males with a uniquely modified first toe that acts as a clasper (Figure 50.1). The *carolina* group includes the seven "subspecies" of *T. carolina* and is characterized by a high, rounded cross-section, less reduced temporal roofing, and males with an unmodified first toe. *Terrapene coahuila* is usually included in this group but could easily stand by itself. Details can be found in the original papers by Minx (1992, 1996) and Milstead (1969) or in Dodd (2001). The fossils *Terrapene putnami* and *T. longinsulae* are placed in the *carolina* and *ornata* groups, respectively (Milstead and Tinkle, 1967).

Of the 11 recognized taxa, six or seven occur in Mexico. Various works have recognized these Mexican taxa as full species (Legler, 1960c) or as subspecies (Milstead and Tinkle, 1967; Minx, 1996). The genus *Terrapene* is well enough understood morphometrically to be ripe for molecular analysis.

DISCUSSION OF CHARACTERS USED IN THE CLASSIFICATION OF *TERRAPENE*

The conformation of the maxillary beak (e.g., notched, hooked, or flat) is often used to characterize species of *Terrapene*. The beak is a keratinized epidermal structure and is subject to wear. Its anterior portion is thicker laterally than mesially. Wear increases the depth of the notch. An unworn sheath is more likely to be flat or slightly hooked than notched (see remarks by Legler, 1960c, for *Terrapene ornata*).

The temporal roof of most emydid turtles is greatly reduced (as, say, compared to that of a sea turtle). Dorsal (posterior) emargination and ventral emargination of the temporal roof combine to form two distinct bars of bone. The vertical postorbital arch defines the posterior edge of the orbit, containing the postorbital bone and often the jugal bone. The horizontal zygomatic arch spans the gap between the postorbital arch and the quadrate bone, containing, variably, the quadratojugal and part of the jugal bone (Figure 49.2).

There has been error and confusion in the literature on these structures in the past century. The postorbital bone is referred as the "postfrontal" by Taylor (1895) and many subsequent authors (e.g., Hay, 1908b; Cahn, 1937; Carr, 1952; Dodd, 2001). Milstead and Tinkle (1967) incorrectly include the squamosal bone in the zygomatic arch, and Milstead (1969) attributes cartilaginous stages to the purely dermal bones of the temporal roofing. The terminology for arches and bones used here follows Minx (1996), Romer (1956), and Zangerl (1969).

Reduction of temporal roofing reaches its extreme in *Terrapene*. Its remnants are represented by one or two relatively thin arches of bone—the postorbital arch and the zygomatic arch. The zygomatic arch is progressively reduced or absent. This involves the reduction in all three of its elements but chiefly in the jugal and quadratojugal bones. The zygomatic arch is widest and most substantial in *Terrapene coahuila* and present but reduced *Terrapene c. major*. In the *ornata* group the arch is always incomplete and the quadratojugal is vestigial or absent (Figure 49.2).

In *Terrapene* the kinetic plastron is not fully developed in juveniles, structurally or functionally (Legler, 1960c, *T. ornata*; Carr, 1952; and Minton, 1972, *T. carolina*). Legler (1960c) gives a detailed account of the changes in skeleton

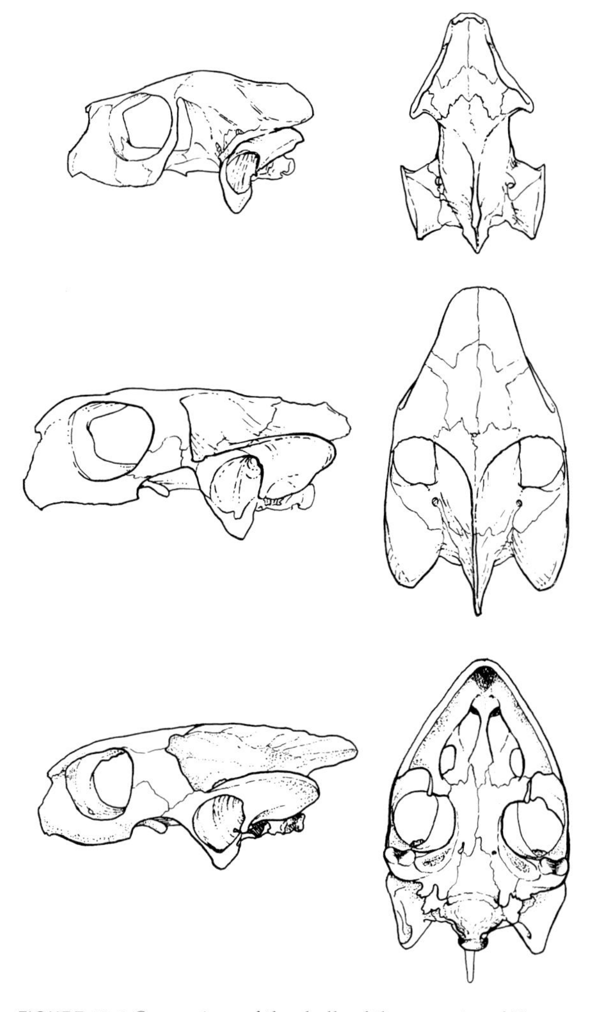

FIGURE 49.2 Comparison of the skulls of three species of *Terrapene*. Top to bottom: *T. o. ornata*, male, CBL 30, KU 1172, Kansas; *T. carolina* ssp., adult, CBL 38, KU 39742, northern Florida; *T. coahuila*, female, greatest width 30 mm, FMNH 47372, paratype, Basin of Cuatro Ciénegas, Coahuila. Note especially the differences in lateral temporal ("zygomatic") arch. (Top and middle images from Legler, 1960, with permission, drawn by Lucy Remple; bottom image drawn by Connie Spitz.)

and overlying scutes that must occur ontogenetically in the first 4 years of life before the shell can be completely closed (Figures 2.4 and 3.4).

There is a great range of pattern and color in the genus. The carapace tends to have a pale pattern of some sort on a darker ground color, and the plastron is usually unicolored. Much of the variation in patterned species can be understood in terms of ontogenetic change. Hatchlings often have a rather simple pattern of single pale dots or splotches on the areola of each carapacal scute; as growth occurs, the neonatal spot develops into some kind of radial pattern. This can produce sharply defined radians. The radians fuse variably to form splotches and rosettes, or it can break up into smaller dots. The process of aging after sexual maturity is accompanied by melanism and/or loss of pattern associated with wear to produce a uniformly pale (straw- or horn-colored) shell. This process occurs when growth has slowed or stopped. Melanin deposits along the interlaminal seams, especially of the carapace, can produce striking dark borders on pale laminae. Melanin may then spread to form blotches and even to darken the entire shell. Carapace pattern may be cryptic (Legler, 1960c, and pers. obs.; Dodd, 2001; Ernst, Lovich, and Barbour, 1994; and Carr, 1952).

FOSSILS

Thus far no *Terrapene* fossils have been reported for Mexico. Remains of *T. yucatana* have been found at archeological sites in Yucatán (see account of *T. yucatana*). The fossil history of *Terrapene* is reviewed by Holman and Corner (1985) and Holman and Fritz (2005). Presence of fossils resembling *T. ornata* from the Barstovian Middle Miocene of south-central Nebraska suggest that the *ornata* and *carolina* groups were established by then (ca. 11.5–13.0 MYBP) and that the genus *Terrapene* evolved well prior to the middle Miocene. A new species, *Terrapene corneri* (Holman and Fritz, 2005) is described from the Middle Miocene (a complete anterior plastral lobe). It shares characters with both *ornata* and *carolina* "groups" and shares unique characters with extant *T. coahuila*. Fossils of *T. carolina* are known from later in the Miocene (ca. 8.5–10 MYBP). Geographic locations of all fossils of *T. ornata* and *T. carolina* are shown by Ward (1978) and Ernst and McBreen (1991), respectively.

GEOGRAPHIC DISTRIBUTION

The genus occurs from southern Michigan, southern Wisconsin and southern Maine, east of the Rocky Mountains southward to Arizona, Texas, and Florida and thence southward in Mexico to the Yucatán Peninsula in the east and, from Sonora, at least to northwestern Nayarit (22°30'N) west of the Sierra Madre Occidental. This constitutes a total latitudinal range of ca. 29°. This distribution is virtually continuous east of the Rocky Mountains in the United States, but is spotty and discontinuous in Mexico. At northern latitudes box turtles occur at elevations of 150–215 m in New England, to 1,300 m in North Carolina (Dodd, 2001) and up to 1,500 m in western Mexico. The ranges of *Terrapene ornata* and *Terrapene carolina* overlap in the central United States, and there are reports of hybridization. Most alleged natural hybrid individuals have been examined and have proved to be one species or the other (Legler, 1960c; Milstead and Tinkle, 1967).

NATURAL HISTORY

Habitat ranges from aquatic to terrestrial to harsh desert. Box turtles produce few eggs per clutch. Opportunistic omnivory is common (Dodd, 2001). The natural history of *Terrapene* has been well studied in the United States (Stickel, 1949; Schwartz and Schwartz, 1974, 1991; Blair,

1976; Legler, 1960c; Metcalf and Metcalf, 1985; Dodd, 2001), but, with the outstanding exception of *T. coahuila* (Brown, 1974), there have been no major studies in Mexico. Ages of at least 70 years are known for the genus (Dodd, 2001).

A KEY TO THE SPECIES AND SUBSPECIES OF *TERRAPENE* IN MEXICO

(Based on Smith and Smith, 1979; Minx, 1996; and Legler notes.)

1A. Plastron with a pattern of radiating yellow lines on a darker background or plastron uniformly yellowish; if latter, then highest part of carapace at or anterior to plastral hinge or first central sloping at 45° or less, or mid-dorsal keel absent or weak and restricted to posterior part of carapace. . . . . ***Terrapene ornata*, 2**

1B. Plastron without radiating pale lines on darker ground; if uniformly pale, then highest part of carapace posterior to plastral hinge or first central lamina sloping 50° or more, or median mid-dorsal keel prominent anteriorly as well as posteriorly 3

2A. Second lateral scute bearing five to eight wide, pale radiations on a darker background; radial pattern retained throughout life; may occur in NE Mexico . . . . ***Terrapene o. ornata***

2B. Second lateral scute bearing 11–14 fine, pale radiations on a darker background; carapace becomes uniform yellowish or straw colored with age; occurs in Chihuahuan and Sonoran deserts. . . . . ***Terrapene o. luteola***

3A. Small, round, pale dots on dark brown areas of carapace and plastron; large individuals with unicolor yellowish carapace but plastron extensively dark with or without pale dots. . . . . ***Terrapene nelsoni*, 4**

3B. Neither carapace nor plastron with pale dots. 5

4A. Ground color of carapace tan to dark brown and profusely marked with pale yellowish spots; rarely unicolored; spots also on soft parts; slope of anterior shell profile steep, mean 38°; flaring of anterior shell margin not exaggerated; occurs from northern central Sonora and southwestern Chihuahua to northern Sinaloa . . . . ***T. n. klauberi***

4B. Ground color of carapace pale brown to pale olive with fewer spots, often unicolored; slope of anterior shell profile gradual, mean 30°; anterior shell margin widely flared; currently known from northern Nayarit and northwestern Jalisco . . . . ***T. n. nelsoni***

5A. Carapace relatively long, low and narrow; CL/CL .40–.46; intergular seam usually more than 45% length of anterior plastral lobe; carapace unicolor, dark slaty brown or olive with pale patches on worn areas; plastron usually uniformly pale straw to grayish olive with darkened interlaminal seams; occurs only in the basin of Cuatro Ciénegas, Coahuila . . . . ***Terrapene coahuila***

5B. Carapace relatively high and wide; HT/CL .47–.55; intergular seam usually less than 45% of anterior plastral lobe; carapace pattern ranges from unicolor (straw) to extremely variable arrays of dots, radians and blotches 6

6A. Interfemoral seam relatively long, mean 21% length of posterior plastral lobe; interhumeral seam relatively long, mean 33% length of anterior plastral lobe; usually (92%) four claws on hind foot; occurs on the Yucatán Peninsula . . . . ***T. yucatana***

6B. Interfemoral seam relatively short, mean 15% length of posterior plastral lobe; interhumeral seam relatively short, mean 23% length of anterior plastral lobe; usually (93%) three claws on hind foot; occurs in NE Mexico N of 24° . . . . ***T. mexicana***

The *Terrapene Ornata* Group

Diagnosis and Definition

(Based on Milstead and Tinkle, 1967; Milstead, 1969; and JML database.)

Adult size can be as large as ca. 160 mm in extant forms. Anterior profile of carapace is inclined at angle of less than 45°, often imparting a flattened or "scooped" look to the anterior one-third of profile (Figure 49.1). Marginal flaring is common anteriorly and posteriorly. Basic proportions are relatively wide, low, and at least somewhat flat topped in profile and cross-section. Mid-dorsal keel is absent or weak in adults; highest point of shell profile is anterior to plastral hinge. Plastral hinge usually is at level of M5–6 or M6. C1 is nearly always straight sided in adults; M1 usually is irregularly oval or triangular. Interhumeral seam is relatively short, 11–19% of anterior lobe; interfemoral seam is relatively long, 16–23% of posterior lobe. Plastral edge usually is lacking an indentation at level of femoro-anal seam. Pes always has four clawed digits. Osseous zygomatic arch is always incomplete, without posterior jugal projections; quadratojugal is absent. Posterior plastral lobe is flat or only slightly concave in adult males; first toe of adult male is capable of medial rotation and is used as a clasping organ (Figure 50.1). Cloacal bursae are absent. There are one or two anterior musk glands with orifice(s) at M4 or at M4 and M5. Shell usually is marked with pale radial lines or pale spots on a darker background, but old adults may have unicolored shells. Membership: *Terrapene ornata* and *T. nelsoni*.

Evolution and Phylogeny of the *Terrapene Ornata* Group

The following is a summary by Milstead (1969) based on Auffenberg and Milstead (1965), Milstead (1967), and Milstead and Tinkle (1967). The hypothetical ancestral stock of the *ornata* group was similar to the fossil *T. longinsulae* and was adapted to arid grasslands on the Great Plains. Subsequent evolution was correlated with alternately mesic

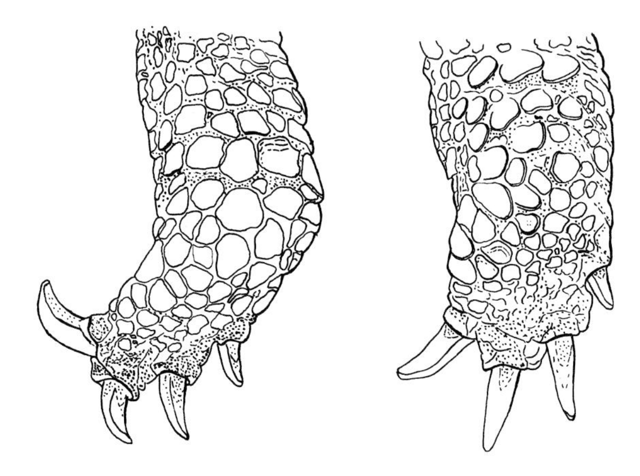

FIGURE 50.1 Plantar views of right hind foot in a male (left) and female (right), *Terrapene o. ornata*. The widened, thickened, and inturned distal phalanx of the first toe is used to clasp the female during mating and characterizes the *T. ornata* group. (Drawing by Lucy Remple, from Legler, 1960, with permission.)

and arid climates on the Great Plains in the Pliocene and Pleistocene. During wetter periods, forests extended into the Great Plains from both east and west, forcing *ornata*-like box turtles southward and westward. *T. o. ornata* could have arisen by later reinvasion from the south and west or from a stock that was isolated to the north or east of the main population, in a relict area of prairie like the prairie peninsula of the midwest. In either case this stock later underwent a more extensive distribution on the modern Great Plains. The stock to the southwest became *T. o. luteola* and remained in what is now the Chihuahuan Desert.

While the foregoing hypothesis of *T. ornata* evolution is plausible, the derivation of *T. nelsoni* from a southwestern extension of the *ornata* stock may not be. Minx (1996) agrees that *ornata* and *nelsoni* are sister taxa but demonstrates that *nelsoni* retains too many primitive characters and *ornata* possesses too many unique derived characters to support Milstead's contention (1969) that *nelsoni* was derived from *ornata*.

Terrapene ornata (Agassiz) 1857

Ornate box turtles (Iverson, 1992a); tortuga de adornos (Liner, 1994).

ETYMOLOGY

L. *ornatus*, decorated, adorned; in reference to the usual bright, pale, radial lines of the shell on a darker background.

HISTORICAL PROLOGUE

The species is one of the best studied of the Mexican box turtles, but no studies have been conducted on Mexican populations. *Terrapene ornata* has almost always been recognized as a full species without question. Say (1825) seems to have made the first published reference to the species but failed to describe it. There are two extant subspecies (*ornata* and *luteola*). They are well documented and generally accepted. Subspecific designation was first attempted by Cragin (1894) in his description of *Terrapene ornata cimaronensis* (an invalid name now in the synonymy of *T. ornate*; Smith and Smith, 1979). Milstead (1969) suggested that *luteola* was virtually indistinguishable from the fossil *longinsulae*.

DIAGNOSIS

Carapace is relatively low and wide, flat topped in transverse section and profile, and rounded in dorsal view (HT/CL ca. 50%; CW/CL ca. 85%). Profile of C1 slopes flatly and continuously to anterior edge of carapace at angle of less than 45°; marginal flare (overhang) is minimal anteriorly and posteriorly; highest part of carapace is at or anterior to hinge; dorsal keel is absent or obscure; lateral margins of plastron are deficient, leaving gaps between carapace and plastron when plastral lobes are closed; plastral hinge usually is opposite M6, or M5–M6 (Figure 49.1). Interlaminal seams are expressed as a percentage of hindlobe: abdominal, 32% or less; femoral, 18% or more; anal 47% or more. Zygomatic arch is absent without vestige of quadratojugal bone (Legler, 1960c: Figure 2, 534). Phalangeal formula is 22222 for manus and 23331 for pes; there are four clawed digits on pes. Soft parts are dark brown, irregularly blotched with yellow; scales on forelegs are pale centered; jaws are yellow; head has irregular pale markings (diagnosis based on Milstead, 1969; Smith and Smith, 1979; Legler, 1960c; Minx, 1992, 1996; Iverson, 1992a; and pers. obs.).

GENERAL DESCRIPTION

Terrapene ornata is the smallest and most specialized member of the genus, having a reduced number of phalanges; a relatively thin, lightweight, and loosely articulated shell; a reduced plastron; and a skull without vestige of a zygomatic arch. Maximum carapace length is 149–154 mm; usual adult size is 100–136 mm.

Most individuals are easily distinguishable by a pattern of yellow radiating lines on a darker ground on the carapace and plastron (see remarks on unicolored shell in old *luteola*).

Two subspecies are recognized: *T. o. ornata* and *T. o. luteola* (Figure 50.2). Unless otherwise noted, the following remarks pertain to *luteola*.

BASIC PROPORTIONS FOR SUBSPECIES OF *TERRAPENE ORNATA*

Terrapene o. ornata
Kansas (Legler database and Legler, 1960c)

CW/CL: males, 0.86 ± 0.03 (0.78–0.93)
n = 60; females, 0.85 ± 0.033 (0.77–0.95)
n = 135; im. males, 0.85 ± 0.02 (0.83–0.87)
n = 4; im. females, 0.85 ± 0.028 (0.80–0.89)
n = 6; im. unsexed, 0.88 ± 0.038 (0.81–1.00)
n = 34.

FIGURE 50.2 Left—*Terrapene o. ornata*. Adult male, eastern New Mexico. The narrow carapace stripes distinguish it from the subspecies *luteola*. Right—*T. o. luteola*, male, Carizozo, New Mexico.

CH/CL: males, 0.48 ± 0.023 (0.43–0.53) *n* = 57; females, 0.50 ± 0.032 (0.41–0.60) *n* = 135; im. males, 0.52 ± 0.031 (0.50–0.56) *n* = 4; im. females, 0.52 ± 0.031 (0.47–0.55) *n* = 6; im. unsexed, 0.51 ± 0.027 (0.46–0.58) *n* = 33.

CH/CW: males, 0.54 ± 0.106 (0.59–0.64) *n* = 59; females, 0.60 ± 0.031 (0.53–0.67) *n* = 135; im. males, 0.62 ± 0.021 (0.60–0.64) *n* = 4; im. unsexed, 0.58 ± 0.034 (0.50–0.65) *n* = 33.

Terrapene o. luteola
Southwestern United States and Northwestern Mexico (Legler database)

CW/CL: males, 0.79 ± 0.063 (0.70–0.90) *n* = 28; females, 0.79 ± 0.057 (0.68–0.90) *n* = 42; unsexed adults, 0.76 ± 0.024 (0.73–0.79) *n* = 4; im. females, 0.81 ± 0.027 (0.79–0.85) *n* = 4; juveniles, 0.88 ± (0.85–0.91) *n* = 2.

CH/CL: males, 0.48 ± 0.033 (0.42–0.54) *n* = 28; females, 0.79 ± 0.057 (0.68–0.90) *n* = 42; unsexed adults, 0.52 ± 0.024 (0.49–0.54) *n* = 4; im. females, 0.53 ± 0.048 (0.50–0.60) *n* = 4; juveniles, 0.52 ± 0.025 (0.50–0.53) *n* = 2.

CH/CW: males, 0.61 ± 0.068 (0.49–0.71) *n* = 28; females, 0.62 ± 0.051 (0.54–0.74) *n* = 41; unsexed adults, 0.68 ± 0.038 (0.62–0.71) *n* = 4; im. females, 0.65 ± 0.038 (0.63–0.71) *n* = 4; juveniles, 0.59 ± 0.001 (0.59–0.59) *n* = 2.

Modal plastral formulae: 6 always longest, 2 always shortest; modal 6 > 1 > 4 > 5 > 3 > 2 (55%), 6 > 1 > 4 > 5 > 3 > 2 (28%).

Terrapene ornata luteola Smith and Ramsey, 1952

Desert box turtle (Iverson, 1992a); Western box turtle (Collins, 1997; Nieuwolt, 1997); tortuga ornada del desierto (Liner, 1994).

ETYMOLOGY

L. *luteolus*, a diminutive of L. *luteus*, yellow; in allusion to the almost uniformly yellowish straw color of older adults (as seen in the type specimens).

GEOGRAPHIC DISTRIBUTION

Map 31

Terrapene ornata luteola occurs in the northern portions of the Chihuahuan and Sonoran deserts from southeastern Arizona, southwestern New Mexico, and Texas (west of the Pecos River) southward into northeastern Sonora and northern Chihuahua. The type locality of *luteola* is 27 km S of Van Horn, Culberson Co., Texas. Mexican records for *luteola* describe an inverted triangle with its apex at General Trias (45 km SW of Cd. Chihuahua) and its broad base extending from the Nogales–Imuris region along the U.S.–Mexican border to Ojinaga at the mouth of the Rio Conchos. Records are concentrated along roads, especially the main highway and railroad from Cd. Juárez to Cd. Chihuahua. The subspecies has not been taken below 800 m. A record for Guaymas, Sonora (Legler, 1960c), is justly rejected (Smith and Smith, 1979; Milstead, 1969; Ward, 1978).

The two subspecies of *Terrapene ornata* intergrade in southeastern New Mexico and trans-Pecos, Texas. The presence of *T. o. ornata* in Mexico is not yet supported by specimens, but occurrence is likely southeast of the Big Bend region, in the northernmost parts of Coahuila, Nuevo León, and Tamaulipas.

NATURAL HISTORY

Terrapene ornata has not yet been studied in Mexico. However, the species is well studied in the United States from Wisconsin, Nebraska, and Kansas to extreme SE Arizona. Most of the natural history information cited in the following comes from the fairly recent studies of Nieuwolt (1996, 1997), Socorro Co., New Mexico, northern Chihuahuan Desert; Plummer (2003, 2004) and Plummer,

FIGURE 50.3 *Terrapene ornata luteola*. Adult head profiles (Chihuahua). Left, old male with advanced melanism; right, adult female with retained head pattern and showing partial loss of shell pattern.

Williams, Skiver, and Carlyle (2003), Sulphur Springs Valley, Cochise Co., Arizona; and a recently begun study by Dr. Dawn Wilson (2006). All of these findings, near the southwestern limit of the range, would be logically applicable to Mexican populations only a few kilometers to the southwest.

HABITAT

Terrapene ornata luteola is adapted to the open, arid, and rigorous environment of the southwest where activities are limited by winter cold and summer aridity. In Socorro Co., New Mexico, Western box turtles occur at altitudes of 1,300–1,600 m on level semiarid grassland and gravelly foothill slopes where soil types (sand, red clay, silts, and fine gravels) were suitable for burrowing and nesting. Microhabitats in Arizona range from open bare soil and low grass to high grass (20–50 cm) and shrubs (Plummer, 2003). Annual rainfall is about 200 mm, and most of it occurs in July and August. Temperature range in the season of activity is about 21–43°C.

Mammal burrows, chiefly those of kangaroo rats (*Dipodomys spectabilis*), are used as refugia wherever *luteola* has been studied. Several box turtles may be found in the burrows of a large *Dipodomys* mound. The turtles also make body forms in soft soil, some of which are deep enough to cover the turtle. Low vegetation (e.g., creosote bush) is used for shade (Norris and Zweifel, 1950; Nieuwolt, 1996, 1997). Box turtles are easiest to find along roads in wet weather, but it seems clear that the numbers of individuals seen in desert habitat are never as high as reported for *T. o. ornata* in Kansas (Legler, 1960c).

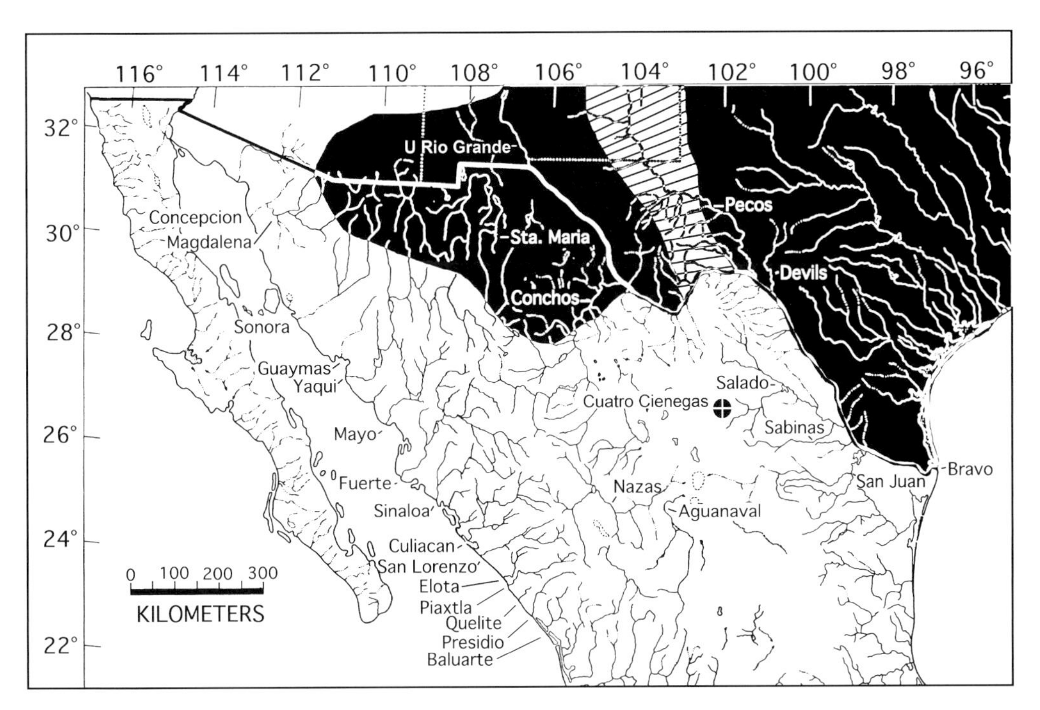

MAP 31 Geographic distribution of *Terrapene ornata*. *T. o. ornata* (east) and *T. o. luteola* (west). A known range of intergradation is marked with diagonal lines.

FIGURE 50.4 *Terrapene ornata luteola*. Adult females (Chihuahua), showing progressive loss of shell pattern, dorsal and ventral views.

DIET

There are few dietary records for *luteola*. Plummer, Williams, Skiver, and Carlyle (2003) reported the eating of arthropods in burrows. Foods eaten by captives include carrion (dead box turtles and collared lizards mentioned), buffalo gourds (*Cubcurbita foetidissima*), and tadpoles of spadefoot toads (*Scaphiopus hammondi*). Metamorphosed toadlets were rejected, and the box turtles who had tried to eat them made wiping movements of the mouth with the forelimbs (Norris and Zweifel, 1950).

HABITS AND BEHAVIOR

Annual period of activity is approximately May to October, varying substantially from year to year but about 153 days if half of May and half of October are included. Periods of inactivity are spent chiefly in enlarged kangaroo rat burrows.

Activity is correlated with moisture; turtles are easier to find and observe during and after rain. Under optimum conditions activity is confined to daylight hours, but in hot, dry weather turtles may be active before sunrise and near sunset (Degenhardt, Painter, and Price, 1996). Turtles are active at ambient temperatures of 13–24°C when vapor pressure deficit is below 20 unibars and during solar radiation of 200 joules/cm^2/h or less (Nieuwolt, 1996).

On days when activity occurs, the activity budget is as follows: resting, 80–90%; walking, 10–20%; breeding and social behavior, less than 4%. Overall, 36% of resting time was spent in burrows or clumps of vegetation. Females rested less and walked more than males and juveniles, but males moved faster than females.

The following is from Plummer (2003, 2004) and Plummer, Williams, Skiver, and Carlyle (2003). Most of the annual cycle is spent in underground refugia—burrows in which the turtles are 10 to 150 cm below the surface. This is an 8-month period of subterranean life (November through June) comprised of a 5-month winter dormancy followed by a 3-month summer dormancy. Entrance into winter dormancy varied from 18 October to 6 December in different years, depending on temperature and rainfall. Annual activity occurs in a concentrated 90-day period from July through September and is coincident with monsoon rains. In that period box turtles are active for an average of about 5 hours per day. Therefore, only about 5% of a desert box turtle's life is spent on the surface, and as much as 95% is spent in burrows. What happens in the burrows remains to be studied, but basking at burrow mouths and feeding, mating, and nesting (see following text) in burrows have been observed.

Turtles in burrows may attain body temperature levels suitable for activity many weeks before they actually emerge. Turtles emerging at the beginning of the monsoon are severely dehydrated. Body temperatures at the beginning of winter domancy were 14–18°C; from December through February body temperatures averaged 9°C and were near that of soil temperature. Body temperatures of actively moving turtles in burrows in April approached those for surface activity. Minimum temperatures for moving turtles was 18°C. Mean body temperatures of actively moving turtles were 29.6°C ± 3.88 (18.8–36.5) and 29.9°C ± 2.38 (20.3–34.4) for turtles in burrows.

Desert box turtles congregate around temporary rain pools (Norris and Zweifel, 1950; Gehlbach, 1956); in one case at least 10 individuals were seen in one large pool. The turtles were actually diving and swimming underwater before surfacing, in the manner of aquatic emydids and with far greater dexterity than reported by Legler (1960c) for *T. o. ornata* in Kansas. This may bespeak efficient ballasting after imbibing water.

Home ranges vary greatly in size. Overall mean size is 1.6 ha ± 1.31 (extremes, 0.03–4.14) when calculated by the minimum polygon method and 276 m ± 178 (32–526) by greatest diameter method. Overlap of home range occurs within and between sexes, and home ranges of individuals do not change from year to year or within a given year. Males have larger home ranges than females when calculated by minimum polygon and smaller home ranges than females when calculated by the maximum diameter

FIGURE 50.5 *Terrapene ornata luteola*. Adults (Chihuahua), showing advanced ontogenetic color changes. Male (left) with black head and female (right) with nearly complete loss of pattern and some darkening of interlaminal seams.

method. The use of a road affected the major axes and probably the sizes of home ranges in New Mexico (Nieuwolt, 1996). Germano and Nieuwolt (1999) recorded the homing of an adult female over a distance of 9.15 km. Each of the 11 individuals marked by Plummer (2004) had a home burrow. Home ranges were smaller in a dry year than in a wet year (1.1 ha vs. 1.7 ha) and each included the same water tank.

REPRODUCTION

Females reach sexual maturity at 107 to 108 mm CL and weights of 327–349 g (Nieuwolt, 1997). Overall mean size of 137 adult females in New Mexico was as follows: CL, 119 mm ± 5.8 (105–139); Wt, 430 g ± 57 (272–590) (Nieuwolt, 1997). Mean CL for 16 gravid females from Cochise Co., Arizona, was as follows: 126 mm ± 4.8 (120–135) (D. Wilson, pers. comm.).

An encounter between two males lasting 8 minutes is reported in detail by Nieuwolt (1996). Male A approaches male B; A lifts front legs over B periodically and bumps B's shell; B turns and moves away; A follows and continues to bump B from behind; A and B are then positioned at right angles and attempt to bite each other; B moves away and ends the encounter. There was a female 114 m away.

Norris and Zweifel (1950) describe a captive mating. The male had achieved intromission; his tail was curled under that of the female, and his hind legs grasped her beneath the posterior marginals just in front of her hind legs. Copulatory movements of the male's tail and legs were observed for 2 hours, during most of which the male was on his back and being dragged about by the female. This behavior is characteristic of other observed *Terrapene* copulations and bespeaks the firm union of the engorged penis and the grasping efficiency of the hind feet and legs. There is variation in the juxtaposition of the hind legs in copulating *T. ornata*; Legler (1960c) observed the male's legs to be posterior to those of the female and pinned forcibly by flexion of the female's legs.

In a 3-year study by Nieuwolt (1997), 77 of the 137 females caught and X-rayed were gravid. Gravidity occurred from mid May to mid July and rarely in early August. Number of gravid females was greatest in years with high spring rainfall and least in years with drier spring conditions. Females are gravid at the end of winter dormancy. Eggs may be laid in a brief period between emergence and summer dormancy or be retained until the end of that period. The total range of gavidity suggests that at least some multiple clutches were produced.

The only data on nests of *luteola* are from Cochise Co., Arizona (D. Wilson, pers. comm.). Females make a typical nest in a burrow at night. They leave the burrow the next morning. The nests observed were up to 220 cm below the ground surface. Nesting in burrows, if it occurs in the northern range of *T. ornata*, could explain the dearth of information on nests and hatchlings.

Reported number of eggs per clutch ranges from one to four, with a mean of 2.68 ± 0.82 (n = 77) (Nieuwolt, 1997) and 3.06 ± 0.57 (n = 16) by D. Wilson (2006), and does not vary significantly from year to year (Nieuwolt, 1997). In New Mexico females commonly skip reproductive seasons. As few as 57% of the females lay eggs in a given year. Some females skipped two successive seasons. Plummer (2004) mentions drought as a cause of skipped reproductive seasons. Nieuwolt (1997) found no evidence for multiple clutches, but her conclusions were based solely on radiographs taken in June. Without ovarian examination for follicles and corpora lutea, multiple clutches cannot be ruled out. Harvesting of road kills could help resolve such questions.

Mean width of 203 eggs was 26.6 mm ± 0.11 (23.8–29.0) as estimated from X-ray examination. Eggs were retained (from one X-ray to another) for periods of 8–50 days (Nieuwolt, 1997). Wilson (2006) measured egg width by the same technique (65 eggs, 16 females) as 26.9 mm ± 1.3 (25–32). The eggs of *luteola* are much wider than those reported for *T. o. ornata* (20.0–23.3 mm; Legler, 1960c). Freshly laid eggs have finely granulated, hard shells.

Natural incubation time for *T. o. ornata* in Kansas was estimated at 65 days (Legler, 1960c). Hatchlings of *luteola* are of about 30 mm CL. Nieuwolt (1993) recorded a juvenile of 34.6 mm CL and a weight of 20 g at the beginning of its first full year of growth.

In *Terrapene ornata* sex is determined by incubation temperature, and only males result from incubation temperatures of 25°C and lower. Ewert and Nelson (1991) classify the *Terrapene ornata* incubation pattern as "Ia": a species with females larger than males and in which the smaller gender predominates in eggs incubated at lower temperatures.

GROWTH AND ONTOGENY

Size data for *Terrapene o. ornata* (CL, Legler database): males, 116 ± 6.5 (101–129) n = 60; females, 117 ± 7.6 (100–136) n = 136; im. males, 101 ± 5.9 (92–105) n = 4; im. females, 101 ± 6.7 (94–112) n = 6; im. unsexed, 78.2 ± 18.8 (35.7–107) n = 34.

Size data for *Terrapene o. luteola* (CL, Legler database): males, 135 ± 13.2 (109–157) n = 28; females, 136 ± 14.5 (108–170) n = 43; unsexed adults, 146 ± 5.7 (139–152) n = 4; im. females, 109 ± 8.7 (101–121) n = 4; im. unsexed, 98.5 ± 3.5 (96–101) n = 2.

Females of *T. o. ornata* attain larger size than males (Legler, 1960c; Degenhardt, Painter, and Price, 1996; Dodd, 2001), and the same probably occurs in *luteola*. Males have a distinctive inturned first toe, mentioned elsewhere (Figure 50.1). The posterior lobe of the plastron in adult males is only shallowly concave.

Markings on the head and neck of most adults are obscure. In adult males the top and sides of the head anterior to the tympanum become uniformly grayish green or bluish green. Adult males have a bright orange or red iris.

Mature adults of both sexes gradually lose the radial pattern of the shell and become unicolored. This may be partly a function of wear on the shell. Males of *T. o. luteola* may also become melanistic as they age (males of *T. o. ornata* do not); the soft skin darkens to grayish, bluish, and nearly black, obliterating any brighter coloration. The iris darkens but retains a reddish coloration (Legler, 1960c). This combination of colors in old males suggests a military fatigue uniform (see Dodd, 2001: pl. 30).

In the period of regular annual growth, number of growth rings is commensurate with age (see discussion in Legler, 1960c).

PREDATORS AND DEFENSE

Turtles void odoriferous fluid from the cloaca when handled, and Norris and Zweifel (1950) thought this could be a predator deterrent. If so, one wonders at the trade-off involved in a desert organism voiding its stored water. There is also speculation as to whether the odoriferous secretions of the musk glands act to deter predators. Given the number of shells found with mammalian tooth marks on them (in Kansas; Legler, 1960c) this seems unlikely. Kool (1981) was unable to demonstrate that the musk of *Chelodina longicollis* was a deterrent to various native predators in Australia.

There are few firsthand accounts of predators, but all larger mammals and birds are suspect. Germano (1999) recorded a turkey vulture (*Carthartes aura*) perched atop a box turtle on a dirt road in Socorro Co., New Mexico, pecking at its closed shell and exposing an area of bone about the area of a lateral scute; the shell was not penetrated and the wound healed. The closable shell armor, which is ipso facto protective, is clearly the best defense a box turtle has and will usually prevent fatal wounds by all but carnivores with a large enough gape to crush the entire turtle or the patience to wait for the protrusion of a head or limb. Legler (1960c) recorded chigger mites (*Trombicula*) and bot fly larvae (*Sarcophaga cistudinis*) as common albeit chiefly innocuous ectoparasites of *T. ornata* in Kansas.

POPULATIONS

Of 261 individuals studied by Nieuwolt (1996), the observed population structure was as follows: 27.2% males, 52.9% females, and 18.2% immature. Norris and Zweifel (1950) recorded only one juvenile among the 70 animals they observed.

Nieuwolt's (1996) study strongly suggests that box turtles in desert habitat have larger home ranges and lower population densities than do populations of *T. o. ornata* in the more mesic environments of the Great Plains (see Legler, 1960;c and Metcalf and Metcalf, 1970).

CONSERVATION AND CURRENT STATUS

Plummer (2003) regards *T. o. luteola* to be dependent on the burrows of *Dipodomys spectabilis* and infers that the success of the box turtles is linked to that of the rodents. Kangaroo rats have declined dramatically in some areas since the mid 1980s. Rosen, Sartorious, Schwalbe, Holm, and Lowe (1996) consider *luteola* to be a "sensitive species" in Arizona.

Ornate box turtles are "protected" in a perfunctory legal sense wherever they occur. *Terrapene ornata* is the state reptile of Kansas (as of 1986). Box turtles are seldom used for food by humans but are popular as pets and are often gathered by commercial collectors for this purpose. By crossing roads they subject themselves to vehicular hazard, especially by some drivers who try to hit them. We visited the site of Legler's (1960c) study of *T. o. ornata* in early July 2002 and found no box turtles in 2 hours of walking the areas of former high density and saw no road kills between Lawrence and the study area. Current residents of the Damm Farm could recall only occasional recent sightings of box turtles.

Terrapene nelsoni Stejneger, 1925

Spotted box turtle; Nelson's box turtle (Iverson, 1992a; tortuga manchas (Liner, 1994); Sierra box turtle (Milstead and Tinkle, 1967).

ETYMOLOGY

A patronym honoring Edward William Nelson (1855–1934), former chief biologist of the U.S. Biological Survey.

FIGURE 51.1 *Terrapene n. nelsoni.* Holotype, USNM 46252.

HISTORICAL PROLOGUE

The species was briefly described from one specimen (USNM 46252; Pedro Pablo, Nayarit) collected by E. A. Goldman and E. W. Nelson in August 1897. Actually the first record of *Terrapene nelsoni* is a specimen recorded by Mocquard (1899) as *Cistudo carolina* and taken by León Digeut at Sierra de Nayarit, in 1896 or 1897 (Smith and Smith, 1979). This is a stuffed specimen in the Museum d'Histoire Naturelle de Paris (see photo in Devaux and Buskirk, 2001). For nearly 30 years *T. nelsoni* remained almost unknown. Bogert described *Terrapene klauberi* in 1943. Since the type of *T. nelsoni* was in special storage during WW II, Bogert relied on existing published photos of *T. nelsoni* in Ditmars (1934). Alas, the legends for photos of *T. nelsoni* and *T. goldmani* (a synonym of *Terrapene mexicana*) had been transposed and Bogert unwittingly compared *T. klauberi* to a member of the *carolina* group. Bogert, in fact, later stated that he would not have described *klauberi* had he been able to see the type of *nelsoni* (Shaw, 1952; Milstead and Tinkle, 1967). No other specimens were found until 1963 when Milstead and Tinkle (1967) conducted fieldwork near Pedro Pablo. They obtained only one specimen. People of the village collected 36 adult specimens in the following month. All of the 37 new specimens were collected in June and July.

The close relationship of *nelsoni* and *klauberi* has been apparent for some time, but their taxonomic status remains uncertain. Mertens and Wermuth (1955) treated them as subspecies; Wermuth and Mertens (1961) and Shaw (1952) treated them as full species. Milstead and Tinkle (1967) and Milstead (1969) actually studied most of the material available and concluded that they were subspecies.

DIAGNOSIS

The diagnosis and descriptions of *Terrapene nelsoni* and its subspecies are based on Milstead (1969), Smith and Smith (1979), Ernst and McBreen (1991), Minx (1992 and 1996), Dodd (2001), and Devaux and Buskirk (2001). The following two characters are seemingly unique:

1. Margins of shell are flared to create a distinctive shell shape; in profile the anterior part of carapace slopes gradually (30–38°) from posterior edge of first central scute to the free anterior edge of carapace or to a point where C1 abuts on the enlarged anterior marginals; the flared marginals may lie on the same plane as C1 or project more nearly horizontally. In either case there is an anterior shelf beneath which the head can be pulled without closing the plastron (Figures 51.1 and 51.2); the entire anterior profile of the carapace may be shallowly concave ("scooped out," fide Milstead and Tinkle, 1967) (see Figure 49.1 and illustrations in Dodd, 2001; Devaux and Buskirk (2001); and Milstead and Tinkle, 1967). The condition is exaggerated in adult males, is more extreme in *nelsoni* than *klauberi*, and does not occur in *ornata*. It is suggestive of the anterior shell profile and marginal overhang of some tortoises (e.g., *Kinixys* sp.).
2. Anterior surface of forelimb bears enlarged osteodermal scales in males of both subspecies at junction of brachium and antebrachium (two to three large hemispherical scales with secondary lumplike projections, surrounded by four to five smaller osteoderms). These are the "bulbous," "convex," or "hypertrophied" scales mentioned by Minx (1996) and Devaux and Buskirk (2001).

The following nonunique characters are diagnostic in combination: Shape of shell is domed but at least slightly flat topped in profile and transverse section; dorsal view is an elongated oval (much less often round). A weak middorsal keel is present in 66% of specimens. Plastral hinge is opposite M6 or M5–M6.

GENERAL DESCRIPTION

Terrapene nelsoni is the larger species of the *T. ornata* group with average carapace length of 134 mm, maximum about 160 mm. Interhumeral seam is 16–19% of anterior plastral lobe (11–16% in *ornata*); interabdominal seam is 38–39% of posterior plastral lobe (27–32% in *ornata*); anterior plastral lobe is relatively short, 65–66% of length of posterior lobe (66–72% in *ornata*); axillary scute usually is absent.

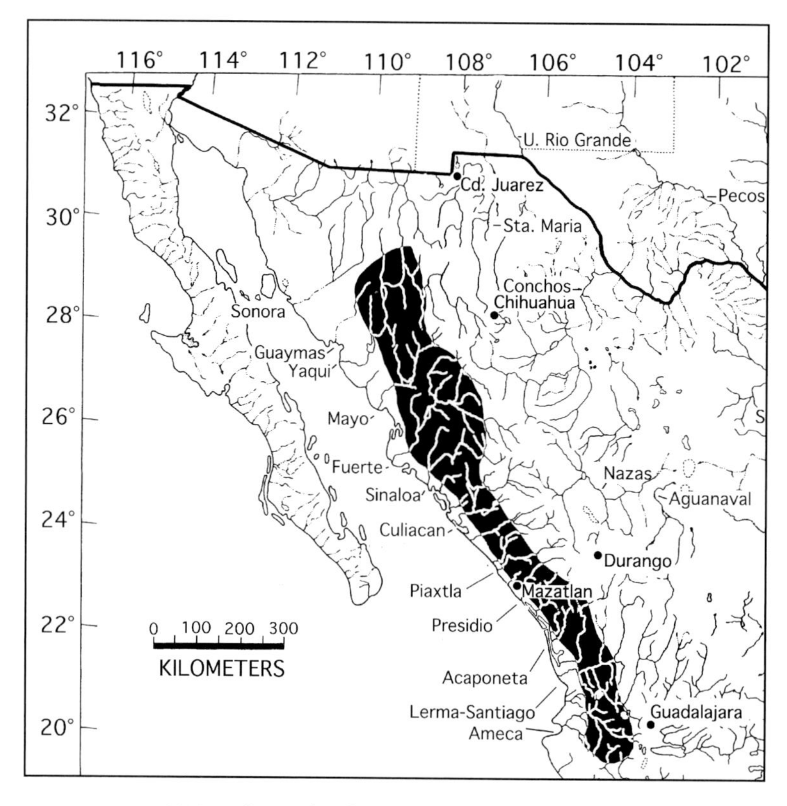

MAP 32 Geographic distribution of *Terrapene nelsoni*.

There are four clawed digits on pes; phalangeal formula is 22222 for manus and 23331 or 23332 for pes; postorbital bar is extremely thin; zygomatic arch is absent, without vestige of quadratojugal bone; head is large; maxillary jaw sheath is hooked anteriorly, notched or not.

Ground color of carapace ranges from tannish to dark brown and usually has a pattern of pale yellowish spots; plastron is dark brown, bordered by yellow, and bears pale spots or streaks or is unmarked; ground color of head is pale to dark brown, marked or not with pale spots; carapace of large individuals may be uniformly straw or horn colored, but plastron is extensively dark with or without pale marks; there are no pale radial markings anywhere on shell.

DISCUSSION OF CHARACTERS

The spots of *Terrapene nelsoni* suggest a radial arrangement and may be derived from an ancestral pattern of pale, radial markings like those in *T. ornata* (Legler, 1960c; Milstead, 1969).

The dorsal anterior carapace in old males of *T. n. nelsoni* is buttressed by three vaguely defined parallel ridges of thickened bone—the mid-dorsal keel on C1 and two lateral ridges running anteriorly from the areolar area of L1 to the edge of the shell between M1 and M2 (Figure 51.1). This sculpturing is reminiscent of a well-made shovel (Legler, pers. obs.). The functional significance of this anatomy is unknown. P. C. H. Pritchard (pers. comm.) suggests it could be used in burrowing. On the other hand, since it is pronounced in males and males are known to engage in combat (Devaux and Buskirk, 2001), it might function as a battering ram or protection (Legler, pers. obs.). Similarly the distinctive osteodermal scales on the forelimbs could act as armor in combat (J. R. Buskirk, pers. comm.).

GEOGRAPHIC DISTRIBUTION

Map 32

Intermediate elevations (400–1,520 m) on western slopes of the Sierra Madre Occidental from central Sonora, extreme southwestern Chihuahua, and northern Sinaloa to northwestern Jalisco, discontinuously, between the latitudes 29°42′N and 20°32′N. There are wide gaps in Sinaloa and between the type locality in Nayarit and the record for Jalisco. *Terrapene nelsoni* (sensu lato) is the only box turtle in Mexico with a pattern of numerous pale dots on a darker (brownish) background. Any naturally occurring *Terrapene* found on the Pacific slopes of the Sierra Madre Occidental south of about 29° may be regarded as *Terrapene nelsoni* (see accounts of subspecies).

EVOLUTION

The following account is based on Milstead (1965, 1969) and Milstead and Tinkle (1967). Fluctuating humidity in Pliocene and Pleistocene times opened and closed a Rocky

FIGURE 51.2 *Terrapene n. nelsoni.* Adult female from type locality in Nayarit. (Photo by P. Minx, used with permission.)

Mountain corridor across the continental divide in southern New Mexico and Arizona (Auffenberg and Milstead, 1965). In pluvial periods, humid conditions permitted expansion of woodlands on the Great Plains and the extension of open, grassy forests over the Rocky Mountain corridor. It is over this route that an ancestral stock of *T. ornata* (*longinsulae* or *luteola*) migrated to the western foothills of the Sierra Madre Occidental. Decreasing humidity during arid periods closed the corridor. The southwestern population, isolated by the closing of the corridor, ultimately became *T. nelsoni*. The oak savanna association where *nelsoni* now occurs is at approximately the same altitude as the old Rocky Mountain corridor. The current distribution of *T. o. luteola* overlies the old corridor and marks the route westward.

SEXUAL DIMORPHISM

Shaw (1952) discussed sexual dimorphism and commented that it could explain some of the original distinctions between *nelsoni* and *klauberi*. Males have shallowly concave plastra (which are convex or flat in females) and are capable of turning (adducting and medially rotating) the first toe inward at right angles to the main axis of the pes where it serves as a clasping organ in mating (Figure 50.1). The tail is thicker and longer in males than in females. General spotting on the shell and limbs is less distinct in males than in females. Devaux and Buskirk (2001) summarize other sexual dimorphism as follows: head is larger in males; head tends to be unspotted in males; there are orange to reddish marks on the male chin, contrasting with a generally darker coloration of the jaws to form a mustache-like figure as compared to the unmarked jaw sheaths in females; male beak is more likely to be notched than that of the female; iris in males red and is brown in females. The large head, red eyes, notched beak, and bright head markings give males a "malevolent appearance" that is said to frighten some people.

CONSERVATION AND CURRENT STATUS

The species is referred to as rare (Smith and Smith, 1979; Devaux and Buskirk, 2001) on the basis of number of specimens seen or placed in museums and may be unrelated to actual population levels. There are known to be collectors offering substantial rewards for the species near Alamos, Sonora. Since box turtles of this species (and, indeed, all Mexican box turtles) are notoriously hard to find, are not large enough to be hunted for food, and occur in rugged and remote areas with relatively few roads, little agriculture, and relatively few people, conservation, at this moment, does not seem to be a pressing issue. Collecting for the pet trade may constitute the greatest current threat to this species.

SUBSPECIES

See key in account of genus *Terrapene*.

Although we are cognizant of the distinction of the taxa *klauberi* and *nelsoni*, we regard their rank as subspecies versus species as moot; neither has been demonstrated unequivocally. We therefore diagnose these taxa separately.

The southernmost record of *klauberi* (6.5 km W of Terreros = 71 km N of Culiacán, Sinaloa; Smith and Smith, 1979) is about 350 km NW and 2.5° of latitude north of Pedro Pablo. Three of the dots shown in Sinaloa by Iverson (1992a) are hearsay accounts unsubstantiated by specimens and occur (e.g., Los Mochis and Mazatlan) at low elevations. Only one of these (PCH Pritchard collection, 1842; adult female skeleton, 50 km NW of Mazatlan, Sinaloa, on Hwy. 15) is supported by a specimen; it is rejected because it was obtained on a major highway near sea level. It remains to be determined whether present gaps in distribution are the result of incomplete exploration or bespeak isolated clumps of natural distribution.

Very little of the natural history of either subspecies is known except for the contributions of Milstead and Tinkle (1967) on habitat and reproduction in *T. n. nelsoni*, the ongoing studies of Montes-Ontiveros and colleagues in Jalisco (Montes-Ontiveros, 2000), and personal communications on *klauberi* in Sonora from George M. Ferguson and Aaron Flesch.

Terrapene nelsoni nelsoni (Stejneger) 1925

Southern spotted box turtle (Dodd, 2001); tortuga manchada meridional (Liner, 1994); Sierra box turtle (Milstead and Tinkle, 1967); Nelson's box turtle (Iverson, 1992a).

DIAGNOSIS AND GENERAL DESCRIPTION

Slope of anterior shell profile is gradual ("low," mean 30°), exaggerated, and slightly concave ("scoop shaped") in old males; marginals are widely flared anteriorly. Carapace is an elongated oval in dorsal view and is relatively flat topped in profile; HT/CL is 45%. A "lateral keel" may be present in both sexes. Interhumeral seam is relatively short (mean 16% of anterior lobe); interpectoral seam is relatively long (mean 35%). Average CL is 134 mm; maximum is 149 mm. Carapace typically is pale brown to pale olive

FIGURE 51.3 *Terrapene n. nelsoni*. Adults from type locality. Top—old female; middle left—old melanistic male; middle right—adult female with full pattern of spots (all photos); bottom left (Nayarit)—female (left) and male meeting head to head; bottom right—immature individual with full pattern, unaltered by ontogenetic changes. (Images in top and middle rows by J. B. Iverson, used with permission; bottom row by P. Minx, used with permission.)

brown and bears pale yellowish spots (spots are less numerous and larger than those in *klauberi*); carapace may lose spots and become uniformly straw or horn colored in older adults (presumably the culmination of an ontogenetic process); plastron is extensively but irregularly dark with contrasting pale yellow areas at periphery, usually unspotted; pale spots on soft parts are reduced or absent. Upper jaw is notched or hooked. Phalangeal formula is 22222 for manus and 23331 for pes.

BASIC PROPORTIONS

(Minx and Legler Databases)

CW/CL: males, 0.68 ± 0.027 (0.65–0.73) n = 6; females, 0.67 ± 0.025 (0.62–0.72) n = 13.

CH/CL: males, 0.39 ± 0.021 (0.38–0.44) n = 7; females, 0.41 ± 0.022 (0.38–0.45) n = 13.

CH/CW: males, 0.58 ± 0.018 (0.56–0.60) n = 6; females, 0.62 ± 0.030 (0.57–0.67) n = 13.

Data on adult size (CL, Minx and Legler databases): males, 139 mm ± 5.8 (130–147) n = 7; females, 137 mm ± 6.04 (123–145) n = 13.

GEOGRAPHIC DISTRIBUTION

Map 32

T. n. nelsoni was formerly known only from the environs of the type locality, Pedro Pablo, Nayarit. Montes-Ontiveros and Ponce-Campos (2006) established its presence in northwestern Jalisco as follows: ca. 30 km WSW of Ameca; 1,250 m; 20°32′N, 104°22′W (ca. 107 km WSW of Guadalajara). This extends the known range 237 km SSE of Pedro Paulo. Several individual turtles have been observed, and the population is under study. It seems likely that careful searching in rugged, dry topical forest and oak woodland at elevations above the coastal plain and in suitable habitat could extend the range much farther southeastward into Michoacán and Guerrero.

Although scientists (Minx, 1996; Milstead and Tinkle, 1967) have been to a village named Pedro Pablo, that name appears only on the fold-out map of Goldman (1951). The locality is cited as 22 miles (35.4 km) E of Acaponeta, 3,500 ft (1068 m) (Goldman, 1951, followed by Milstead and Tinkle, 1967). However, Goldman's map clearly shows Pedro Pablo to be 34 km *NE* of Acaponeta. Using this bearing, the calculated coordinates of the village are as follows: 22°35′N, 105°03′W, 8 km N and 33 km E of Acaponeta and less than 10 km W of the border of extreme southwestern Durango.

None of the entries for "Pedro Pablo" in the NIS Gazetteer of Mexico (1956) or the GNS (GEOnet Names Server) for "Pedro Pablo" approximate either of these coordinates. The GNS lists "Pedro Pablo" as a "short form" for "Mesa de Pedro Paulo" at 22°26′00″N, 105°10′27″W. We have found the name "Mesa de Pedro y Paulo" at these coordinates only on one map of Nayarit (Maps of Mexico.com), and this is assumed to be the locality visited by scientists since the time of Goldman. This locality lies about 21 km SW of the locality on Goldman's map and is 21 km SE of Acaponeta (20 km E and 8 km S). Whether Goldman's map was in error or there are two localities, *Terrapene nelsoni* occurs at both. The point where Digeut obtained the first specimen of *nelsoni* is probably not far east of "Pedro Pablo."

HABITAT

The following description of habitat is based largely on the observations of Nelson and Goldman in August 1897 (Goldman, 1951) and confirmed by Milstead and Tinkle in 1967. The village of Pedro Pablo sits at an altitude of 1,068 m. Acaponeta is in the Nayarit-Guerrero biotic province of Goldman and Moore (1945) and Goldman (1951). The locality is in a region of scrub forest with many thorny bushes, cacti, and low trees, 7 to 9 m high; the foothills and low mountain chains are in an ecotonal area between the Nayarit-Guerrero, Sinaloan, and Sierra Madre Occidental biotic provinces. There is a 150 m deep valley to one side of the village. Milstead and Tinkle (1967) spent most of their time collecting on the luxuriant valley floor where they found only *Kinosternon integrum*. Residents of Pedro Pablo say that box turtles live on a small hill with open savanna and many small oak trees above the town. The authors obtained their only specimen of *nelsoni* on a rocky trail at the foot of a bluff near the summit of the hill. It seemed that *T. n. nelsoni* in that region was limited to the oak savanna association (which Goldman [1951] records as occurring above 1,200 m). Residents of a nearby village (El Oro) at 450 m had no knowledge of box turtles.

DIET, HABITS, AND BEHAVIOR

The following are preliminary findings from a natural population near Ameca, Jalisco, by Omar Montes-Ontiveros and colleagues (pers. comm.). Data were gathered during all seasons of the year. A total of 14 *T. n. nelsoni* were captured and marked (four males, six females, four immature). Nests were found and opened in November and December (dates of laying unknown); each nest contained three eggs. Diet consisted of adult and larval insects and worms.

REPRODUCTION

Milstead and Tinkle (1967) examined ovarian follicles, corpora lutea, and oviducal eggs in 19 dissected females collected near Pedro Pablo in the months of June and July 1963. All were at approximately the same reproductive stage—they had just laid eggs; they contained oviducal eggs; or they contained preovulatory follicles. This suggests that laying occurs early in the wet season (June through September) with hatching predicted by the end of the wet season. There was no indication of more than one clutch per season. Only small yolked follicles in the 5 to 10 mm range remained after this single clutch was produced (and were interpreted as potential for the next year's clutch). Follicular atresia was observed and was considered the fate preovulatory-sized follicles remaining in the presence of corpora lutea.

Mean number of eggs per clutch was three (one to four), whether calculated from corpora lutea or preovulatory follicles, or from actual counts of oviducal eggs. There was no significant relationship between female size and number or size of eggs. Mean dimensions of 17 eggs were 47 × 27 mm (no extremes given).

M. Ewert (pers. comm.) provided the following data based on gravid females obtained by P. Minx near Pedro Pablo in July 1987: the weight of a fresh egg (45.0 × 27.1 mm) was 19.6 g; incubation time at 25°C was estimated at 80–82 days in an abortive clutch. One slightly deformed hatchling had a CL of 37.7 mm. The carapace had a pale brown ground color and a few (ca. 65) large, evenly distributed yellowish spots on the larger scutes of the carapace. The plastron is uniform tan with an indistinct paler margin (measurements and photos courtesy P. C. H. Pritchard).

The above data indicate that *Terrapene nelsoni* produces the largest eggs of any member of the *ornata* group and ranks with extreme northern populations (Doroff and Keith, 1990) and extreme southern populations (Nieuwolt, 1997) of *Terrapene ornata* in average number of eggs per clutch and annual reproductive potential.

Terrapene nelsoni klauberi (Bogert) 1943

Northern spotted box turtle (Dodd, 2001); tortuga manchada septentrional (Liner, 1994); Klauber's box turtle; vernacular name in Sonora, tortuga de chispitas (turtle with sparks; Devaux and Buskirk, 2001); vernacular name near Guirocoba, Sonora, tortuga del monte (Bogert and Oliver, 1945).

ETYMOLOGY

A patronym honoring Laurence Monroe Klauber (1883–1968).

FIGURE 51.4 *Terrapene nelsoni klauberi*. Top and middle—adult female, 22 km W of Yecora, Sonora, 1,490 m; bottom—adult female, 15 km E of Nacori Chico, Sonora, near northern edge of range. (Courtesy G. M. Ferguson, photos by S. Jacobs and A. Flesch.)

DESCRIPTION AND DIAGNOSIS

BASIC PROPORTIONS

(Minx and Legler Databases)

CW/CL: males, 0.73 ± 0.023 (0.69–0.75) n = 8;
females, 0.73 ± 0.034 (0.68–0.78) n = 14;
im. unsexed, 0.75.

CH/CL: males, 0.43 ± 0.031 (0.36–0.47) n = 8;
females, 0.45 ± 0.025 (0.41–0.50) n = 14;
im. unsexed, 0.46.

CH/CW: males, 0.59 ± 0.040 (0.53–0.66) n = 8;
females, 0.62 ± 0.036 (0.54–0.68) n = 14.

Modal plastral formula: 6 > 4 > 1 > 3 > 5 > 2.

Slope of anterior shell profile is steep (mean 38°; steeper than *nelsoni* and more gradual than *T. ornata*); marginals are less flared than in *nelsoni*. Carapace shape in dorsal view is variable, round or oval (*ornata*-like) to elongate (*nelsoni*-like) and flat topped in profile; a "lateral keel" may be present in both sexes; interhumeral seam is relatively long (mean 19% of anterior lobe); interpectoral seam is relatively short (mean 33%). CL average is 131 mm; maximum is 159 mm. Carapace typically is tan to dark brown with pale spots; spots are smaller and more numerous than in *nelsoni*; spots may be dim but are rarely absent; uniform carapace coloration is less common than in *nelsoni*; plastron is dark colored with contrasting peripheral pale areas, and the dark area usually bears spots; pale spots are profuse on soft parts. Upper jaw is notched. Phalangeal formula is 22222 for manus and 23332 or 23331 for pes.

Data on adult size (CL, Minx and Legler databases): males, 137 mm ± 9.4 (125–150) n = 8; females, 135 mm ± 12.5 (108–149) n = 14.

GEOGRAPHIC DISTRIBUTION

Map 32

Eastern, central, and extreme southeastern Sonora and adjacent extreme southwestern Chihuahua to northern Sinaloa. Iverson (1992a) summarized known localities as of 1992.

The type locality, Rancho Guirocoba (ca. 26°53′N, 108°41′W; 453 m; near Alamos; type series of four specimens), lies in southern Sonora near the borders of Sinaloa and Chihuahua) (see account of *Trachemys scripta hiltoni*). The taxon is also known from nearby Alamos (Bogert and Oliver, 1945), where specimens are known to have been sold. Myers (1945) extended the range 280 km NNW to Matape (Pesqueiro), central Sonora (29°08′N, 109°58′W), a locality just west of Presa Novillo. A specimen from near Terreros, Sinaloa (25°12′N, 107°52′W), extended the range ca. 200 km southward from of Guirocoba (Hardy and McDiarmid, 1969). Lemos-Espinal, Smith, and Chiszar (2001) provided the first record for Chihuahua (Batopilas, 27°01′N, 107°44′W), 96 km ENE of Guirocoba.

From 1986 to 2005 George M. Ferguson (pers. comm.) and colleagues (in particular, Aaron Flesch—see O'Brien et al., 2006) gathered seven new records of distribution in the course of biological survey work in eastern Sonora. They have generously shared this information with us as follows: two records near type locality, 42 km NE and 32 km NE of Alamos, 1,400 and 950 m; a coastal record from San Javier, 26 km W of the Rio Yaqui bridge (28°33′N, 109°45′W), 520 m; a cluster of three records near Yecora (28°22′N, 109°04′W), 850–1,520 m, ca. 160 km N of type locality and 125 km NNW of Matape); and, most

FIGURE 51.5 *Terrapene nelsoni klauberi*. Dorsal and ventral views of old, melanistic male (Alamos, Sonora; KU 51430).

importantly, the northernmost record of *klauberi*; two specimens just east of Nacori Chico (29°42′N, 108°53′W) 1,000 m, in small side canyons of a Rio Yaqui tributary (R. El Riito). This northernmost record is slightly more that 300 km almost due north of the type locality and ca. 200 km SSE of Douglas, Arizona. There is an excellent chance that the species could occur another 100 km northward, in a narrowing tongue of thorn scrub forest, to the region north of Presa Angostura. In any case, the foregoing records (substantiated by photos and specimens) bring the range of *klauberi* into close geographic proximity to that of *Terrapene ornata luteola* and, indeed, to the United States. The range of *ornata* probably begins in semidesert and desert grassland habitats at the northeastern edge of the foothills thornscrub.

A specimen from Chihuahua (Smith and Smith, 1979; USNM 104627, Rio Santa Maria, near Progreso) has been examined and is *Terrapene ornata luteola* (JML, pers. obs.).

HABITAT

Typical habitat is woodland in the western foothills of the Sierra Madre Occidental at known elevations of 400 to at least 1,520 m. In general the habitat is rugged, remote from major roads, and difficult of access. Bogert and Oliver (1945) described habitat at the type locality (ca. 450 m) as "densely vegetated hills."

The following observations on habitat and vegetation in relation to altitude and latitude are from George M. Ferguson (pers. comm.) unless otherwise credited:

520 m, 28°30′N (San Javier), borderline tropical deciduous forest, predominantly foothills thornscrub, with some Sonoran desert elements.

600 m, 25°12′N (Terreros), a steep hillside on an isolated rocky peak in low canopy forest; numerous burrows nearby, some of which were attributed to *Terrapene* (Hardy and McDiarmid, 1969).

850 m, 28°28′N (San Nicolas Yecora), tropical deciduous forest at lower edge of oak woodland.

950 m, 27°06′N (near type locality) in oak woodland at upper edge of tropical deciduous forest.

1,000 m, 29°42′N (Nacori Chico), a dry arroyo with foothill thornscrub (maximum height, 7 m), near the lowest margin of the oaks (Aaron Flesch, pers.comm.).

1,200 m, 29°08′N (Matape), rolling granitic hills with volcanic intrusions, covered with desert scrub (Myers, 1945).

1,400 m, 27°19′N (near type locality), pine-oak woodland.

1,490–1,520 m, 28°22′N–28°23′N (Yecora and Maicoba), pine-oak woodland.

NATURAL HISTORY

Activity is coincident with the wet, misty, rainy weather of the southwestern summer monsoon period, July to September (Bogert and Oliver, 1945; G. M. Ferguson and Aaron Flesch, pers. comm.; see also account of *T. ornata luteola*).

A female paratype (CL 133 mm) contained two ovarian follicles of 15 mm diameter; presumably the specimen was preserved between June 15 and October 15, during the rainy season (Bogert, 1943).

This taxon ranks with *Terrapene mexicana* in paucity of life history information. A female from the Nacori Chico locality had crimson stains on the jaw sheaths, suggesting ingestion of organ-pipe cactus fruit. Mating behavior is similar to that of other box turtles, and captive males may fight viciously with one another. Captives have survived for 6–12 years in Tucson, Arizona, and San Francisco, California (Devaux and Buskirk, 2001; J. R. Buskirk, pers. comm.).

The *Terrapene Carolina* Group in Mexico

Terrapene carolina (Linnaeus, 1758) is a polytypic species that occurs widely and contiguously in the eastern United States (from northern Michigan and Canada to southeastern Texas) and contains the subspecies *T. c. carolina*,

T. c. major, T. c. bauri, and *T. c. triunguis* (Ernst and McBreen, 1991). None of these taxa occur in Mexico. Since the revision of Milstead (1969) it has been customary to treat the two Mexican representatives of this group (*mexicana* and *yucatana*) as subspecies of *T. carolina*. We treat these taxa as full species and do not include *Terrapene coahuila* in the *carolina* group.

DIAGNOSIS AND DEFINITION OF THE *CAROLINA* GROUP

Adult size is larger than 300 mm in extinct forms and up to 216 mm in extant taxa. Anterior profile of carapace slopes steeply (>50°) to free anterior edge; marginals are not or are weakly flared (Figure 49.1). Basic proportions are relatively narrow (CW/CL, 0.75 ± 0.04 (0.68–0.87) and highly arched (CH/CW, 0.68 ± 0.04 [0.59–0.79]; HT/CL, 0.51 ± 0.03 [0.45–0.60]); curve of carapace is continuous in cross-sections and profile and is not flat topped (Figure 49.1); carapace usually is elongate in dorsal view. Mid-dorsal keel is prominent. Highest point of shell profile is posterior to plastral hinge. Plastral hinge usually is at level of M5. C1 is always urn or wedge shaped, never straight sided in adults; M1 usually is rectangular. Interhumeral seam is relatively long, 18–33% of anterior lobe; interfemoral seam is relatively short, 10–21% of posterior lobe. Plastral edge usually has an indentation at junction of femoro-anal seam. Three or four clawed digits on Pes. Osseous zygomatic arch usually is incomplete, with or without posterior jugal projections and/or vestigial quadratojugal. Posterior plastral lobe is flat to deeply concave in adult males; first toe of adult males is incapable of medial rotation. Cloacal bursae are absent. Musk glands usually are absent; if present, there is one anterior gland with orifice at M4. Coloration of shell is highly variable—uniformly straw colored to a full radial pattern in Mexican taxa; carapace commonly is unicolored with darkened interlaminal seams. Membership: *T. mexicana*, and *T. yucatana* (based on Milstead and Tinkle, 1967; Milstead, 1969; and JML database).

DISCUSSION OF CHARACTERS

The shape of C1 in adults is the result of ontogenetic change; the scute is straight sided in immature stages (see Dodd, 2001: Figures 16, 19, 24). The character should be regarded as useful only for large adults.

The carapace may be raised slightly at the areola of any one of the central scutes but especially at C3. This can produce (in *T. c. triunguis, T. mexicana*, and *T. yucatana*) a lumpy (rather than evenly curved) shell profile, but the condition is variable and seldom reaches the extremes depicted by Milstead (1969: Figure 2). It is of doubtful value for identification or characterization.

Müller (1936) referred to webbing as distinct in *yucatana* but absent in *mexicana*. Interdigital webbing in all *Terrapene* is best considered as absent or vestigial and is not useful for identifying or characterizing taxa (JML, pers. obs.).

EVOLUTION OF *TERRAPENE* IN EASTERN MEXICO

There are two distinct, widely separated taxa of the *T. carolina* group in Mexico—*T. mexicana* and *T. yucatana*. These taxa are clearly of common origin and most closely related to *T. c. triunguis*. *Terrapene mexicana*, on the coastal plain of northeastern Mexico, is separated from *yucatana* on the Yucatán Peninsula by at least 700 coastwise kilometers in which no other box turtles occur. The range of *T. mexicana* is likewise separated from that of *T. c. triunguis* by about 260 km (Ernst and McBreen, 1991; Map 33).

The current ranges of *yucatana* and *mexicana* can be explained by the disjunction of a formerly continuous ancestral population along the Gulf Coast of Mexico during Pleistocene times (Milstead, 1969; Lee, 1980). Lee (1980) further attributes this disjunction to human influence in pre-Columbian times (without explanation). Milstead (1969) suggests the following: During a glacial stage when the Pleistocene sea levels were low and a coastal plain existed around the Gulf Coast from Florida to Yucatán, an ancestral *Terrapene* (probably close to the fossil *putnami*) ranged throughout the available habitat. Rising sea levels later inundated the coastal plain in southeastern Mexico, isolating the population that became *yucatana* on the Yucatán Peninsula. The northern stock was fragmented by arid conditions, with *triunguis* retreating to the north and *mexicana* differentiating in northeastern Mexico. The greater degree of difference between *yucatana* and *mexicana* than between *mexicana* and *triunguis* supports this chronology to some extent. *Terrapene c. triunguis* now ranges no farther south or west than eastern Texas and is separated from *mexicana* by the arid Tamaulipan biotic province (Dice, 1943; Goldman and Moore, 1945; Blair, 1950; Goldman, 1951). The geographic ranges of *mexicana* and *yucatana* lie in easily accessible terrain and are well documented (relative to *nelsoni*) and not artifacts of incomplete collecting. Further collecting is unlikely to alter their overall configuration.

Terrapene mexicana Gray, 1849

Mexican box turtle (Iverson, 1992a; Ditmars, 1934); tortuga Mexicana (Liner, 1994).

ETYMOLOGY

The specific name is a toponym implying residence or occurrence Mexico.

DIAGNOSIS

The following combination of characters is diagnostic within the *carolina* group of *Terrapene*: Length of anterior plastral lobe is 72% of posterior lobe; humeral scute is shorter than pectoral (means 23% and 36%, respectively); anal scute is at least three times the length of femoral (means 52% and 15% of posterior lobe, respectively); usually (94%) three clawed toes on pes; phalangeal formula 23322 for manus and 23331 (89%) or 23330 (11%) for pes. A "lateral keel" sometimes is

FIGURE 52.1 *Terrapene mexicana*. Adult male (Soto la Marina, Tamaulipas).

present above bridge area (Milstead, 1969). The intergular and interhumeral seam ratios of *mexicana* are intermediate between those of *Terrapene carolina triunguis* and *T. yucatana*.

GENERAL DESCRIPTION

Terrapene mexicana is the largest *Terrapene* in Mexico (mean adult CL 145 mm; maximum 195 mm, adult female; Müller, 1936).

BASIC PROPORTIONS

CW/CL: males, 0.74 ± 0.017 (0.71–0.76) n = 11; females, 0.76 ± 0.035 (0.69–0.81) n = 13; unsexed adults, 0.74 ± 0.038 (0.68–0.87) n = 31; im. females, 0.81 ± 0.032 (0.79–0.85) n = 3.

CH/CL: males, 0.51 ± 0.021 (0.48–0.54) n = 11; females, 0.51 ± 0.027 (0.47–0.55) n = 13; unsexed adults, 0.51 ± 0.033 (0.45–0.60) n = 31; im. females, 0.54 ± 0.055 (0.50–0.60) n = 3.

CH/CW: males, 0.68 ± 0.024 (0.63–0.71) n = 11; females, 0.68 ± 0.048 (0.62–0.75) n = 12; unsexed adults, 0.69 ± 0.050 (0.59–0.79) n = 31; im. females, 0.66 ± 0.042 (0.63–0.71) n = 3.

Modal plastral formulae: 6 > 4 > 1 > 3 > 2 > 5 or 6 > 4 > 1 > 3 > 5 > 2.

The pattern of the adult carapace is more variable than that of any other Mexican box turtle. This high variability is seen also in the subspecies of *carolina* occurring on or near the Gulf Coastal plain in the United States. Milstead (1969) and Smith and Smith (1979) have attempted to categorize this variation. Two common basic patterns seem to exist, and each is modified by melanism of some kind. The most common pattern consists of unicolored pale laminae (horn or straw colored) with narrowly darkened interlaminal seams (see Dodd, 2001: Figure 23); this is modified by widening of the dark interlaminal seams and the appearance of irregular blotches of melanin of varying size that may ultimately darken the entire carapace. These changes are ontogenetic.

The other common pattern is a dark background with pale radiating lines. These radians may be bold or fine (Gray, 1849; Ditmars, 1934: Figure 28). Anastomoses of the radians produces blotches, and interruptions produce spots or dashes. See also Müller (1936: Figures 1–4).

The plastron is usually dull yellowish with darkened interlaminal seams. There are no bright colors anywhere on the shell. Color of the soft skin (other than the head) in general is dull yellowish, tannish, or brownish with some contrast of pale and dark coloring on the enlarged antebrachial scales. Milstead (1969) alluded to occasional specimens with white heads. Müller described 29 specimens from the region of Tampico and found white heads to be common. At least one of these was a female. He also found colors ranging from grayish blue to ultramarine associated with the white head in 25% of the specimens. The white head is known also in *T. c. major* (Milstead, 1969). P. Minx (pers. comm.) has seen a pale unicolored head in *T. n. nelsoni*. Illustrations of white-headed *mexicana* appear in Ernst and McBreen (1991: Figure 4). It seems likely that the white, blue, and other pale colors of the head have the same ontogenetic progression in all of the *Terrapene* that display them (see account of *T. yucatana*).

GEOGRAPHIC DISTRIBUTION

Map 33

The following is based chiefly on Milstead (1967), Smith and Smith (1979), and Iverson (1992a).

Below 500 m on the coastal plain of eastern San Luis Potosi, Tamaulipas, and in northern Veracruz, southward to ca. 20°N where mountains interrupt the coastal plain at Punta del Morro. Distribution is limited by mountains to the west and south, by coastal marshes to the east, and by aridity to the north. Occurrence in lowland areas of extreme eastern, south-central Nuevo León at lower elevations (<400 m) near Linares (Rio Purificación drainage) is likely. The northernmost locality is Padilla, Tamaulipas (24°01′N, 98°47′W) and the southernmost locality is Casitas, Veracruz (20°15′N, 96°47′W). The type locality is "Mexico," later restricted to "Vicinity of Tampico," Tamaulipas, by Müller (1936).

HABITAT

Habitat is "austroriparian-like" and ecotonal between the Tamaulipan, Veracruz, and Sierra Madre Occidental biotic provinces (Milstead, 1967)—dry lowland habitats, including tropical deciduous forest, thorn forest, and possibly thorn scrub. Humid tropical habitats (including

FIGURE 52.2 *Terrapene mexicana*. Variation in adult shell pattern in two females (top) and four males (middle and bottom), all from Tamaulipas. (Courtesy of P. Scanlan, Gladys Porter Zoo, and R. M. Winokur.)

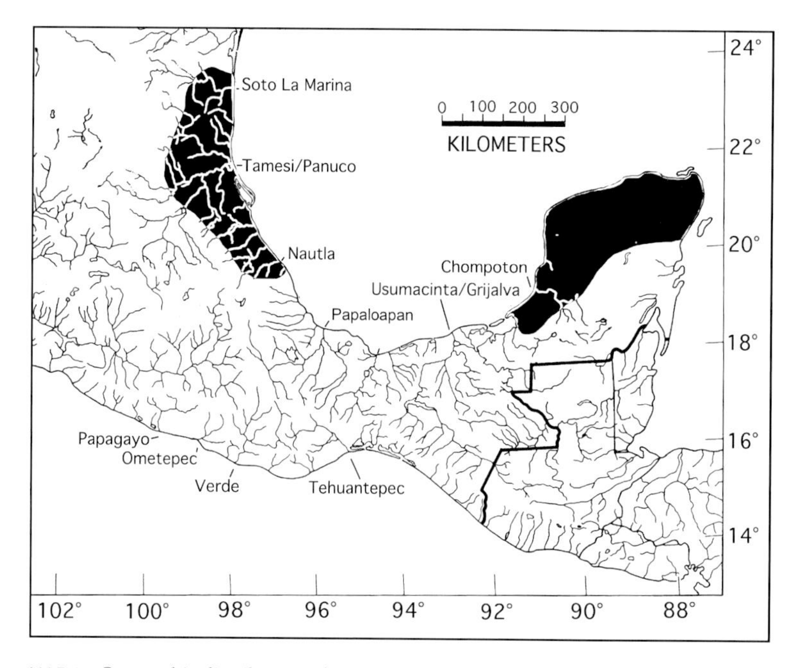

MAP 33 Geographic distribution of *Terrepene mexicana* (west) and *T. yucatana* (east).

FIGURE 52.3 *Terrapene mexicana*. Variation in head color and pattern in four males (top and middle) and two females (bottom), all from Tamaulipas. (Numbered left to right and top to bottom: 1, 4, 6, J. R. Buskirk; 2, 3, R. M. Winokur; 5, P. Scanlan.)

FIGURE 52.4 *Terrapene mexicana*. Carapace pattern in three adult females from Tamaulipas. (Courtesy of P. Scanlan, Gladys Porter Zoo.)

FIGURE 52.5 *Terrapene mexicana*. Neonate showing slight growth, caruncle, and distinct mid-dorsal keel. (Courtesy of P. Scanlan, Gladys Porter Zoo.)

rain forest) in southern Veracruz probably isolated populations of *Terrapene* on the Yucatán Peninsula from those on the coastal plain. This distribution pattern is paralleled by other reptilian species (Martin, 1958).

GROWTH AND ONTOGENY

Data on size (CL): males, 148 mm ± 6.2 (136–157) n = 11; females, 155 mm ± 13.58 (138–195) n = 14; unsexed adults, 148 mm ± 11.06 (118–168) n = 31; im. females, 112 mm ± 8.08 (105–121) n = 3.

Müller (1936) assumed that the most common color patterns were derived from a radially patterned condition. This was logical in that it is precisely how the uniform coloration of *Terrapene ornata luteola* develops. However, Dodd (2001: Figure 24) shows a picture of a juvenile *mexicana* with a uniformly tan carapace marked only by narrowly darkened interlaminal seams. Milstead (1969) was therefore correct in hypothesizing that the most common pattern begins in young turtles with nearly uniform coloring and slightly darkened seams.

NATURAL HISTORY

No information has been recorded on this taxon. Omnivory is assumed. Carr (1952) recorded egg size in the closely related *Terrapene carolina triunguis* as 35 × 23 mm and 38.5 × 20 mm. Most specimens have been found, dead or alive, along roads. It would be productive to harvest road kills for reproductive and dietary information.

CONSERVATION AND CURRENT STATUS

The entire geographic range is traversed by motor roadways, subjecting these animals to road hazards and illegal collecting, probably to the same extent as *Terrapene ornata* on the Great Plains. Mittermeier (1971) did not find *Terrapene* sp. in Mexican markets or curio shops.

Terrapene yucatana (Boulenger) 1895

Yucatán box turtle (Iverson, 1992a; Ditmars, 1934); Yucatecan box turtle; tortuga Yucateca (Liner, 1994); "Ac" or "Coc ac," in Mayan (Buskirk, 1993); tortuga de caja (Lee, 1996).

ETYMOLOGY

The specific name is a toponym implying residence or occurrence in Yucatán.

DIAGNOSIS

The following combination of characters is diagnostic within the *carolina* group of *Terrapene*: Anterior plastral lobe is 64% of posterior lobe; humeral and pectoral scutes are of subequal length (means 33% and 33%, respectively); anal scute is about twice as long as femoral (means 47% and 21%, respectively); there are usually (92%) four clawed toes on pes; phalangeal formula is 23332 for manus and 23332 (89%) or 23331 (11%) for pes; there is no "lateral keel" above bridge area (Milstead, 1969; Minx, 1996).

GENERAL DESCRIPTION

Terrapene yucatana is nearly as large as *mexicana* (mean adult CL 145 mm, maximum 155 mm; Milstead, 1969).

BASIC PROPORTIONS

(Legler Database)

CW/CL: males, 0.75 ± 0.011 (0.74–0.76) n = 2;
females, 0.73 ± 0.018 (0.71–0.74) n = 2;
unsexed adult, 0.76.

CH/CL: males, 0.52 ± 0.003 (0.52–0.52) n = 2;
females, 0.52 ± 0.027 (0.49–0.55) n = 3;
unsexed adult, 0.53.

CH/CW: males, 0.70 ± 0.014 (0.69–0.71) n = 2;
females, 0.73 ± 0.008 (0.73–0.74) n = 2;
unsexed adult, 0.71.

The small series examined presents difficulty in stating "modal" plastral formula. Plastral scutes 6 or 4 are always longest; 1, 2, or 3 is shortest.

Color photos of the same adult appear in Lee (1996: Figures 283–284) and Dodd (2001: pl. 25). We found no illustrations of hatchlings or juveniles. The following remarks on color and sexual dimorphism are based on the descriptions of Buskirk (1993) and Milstead (1969) and briefer remarks by Duellman (1965) and Lee (1996).

Adults exhibit two basic carapacal patterns: (1) uniformly pale straw or horn color with darkened interlaminal seams of varying width; (2) a similar unicolored pattern upon which a pattern of dark radial lines is superimposed. The plastron ranges from uniform black to uniform yellowish with many intermediate grades of variation (J. R. Buskirk,

FIGURE 53.1 *Terrapene yucatana*. Adults (top, female; bottom, male) near Villa Union, Yucatán; boldly colored head patterns develop in males.

pers. comm.). Both of the carapacal patterns are modified (presumably with age) by irregular melanistic blotching in the originally unicolored areas. Adults with completely dark carapaces are common. The ontogeny of these color patterns is not clear. Buskirk (1993) comments that many female *yucatana* could be mistaken for the least colorful *T. c. triunguis*.

In large adults of both sexes there is a reddish coloration to the inframarginal surfaces, chiefly posteriorly. The reddish coloration is most intense where soft skin joins the marginal scutes (J. R. Buskirk, pers. comm.).

GEOGRAPHIC DISTRIBUTION

Map 33

The northwestern two-thirds of the Yucatán Peninsula, in the states of Campeche, Yucatán, and Quintana Roo. Neill (1960) listed its occurrence in the dry northern part of Belize, but this seems never to have been substantiated (Smith and Smith, 1979). Habitat requirements exclude *yucatana* from the wetter parts of the peninsula (including Guatemala and Belize). *Terrapene yucatana* is the southernmost *Terrapene* and is more distantly isolated from its closest relatives than any other *Terrapene*. The indefinite type locality, "North Yucatán" (Boulenger, 1895) has been restricted to Chichen Itza, Yucatán (Smith and Taylor, 1950b). The three syntypes in the British Museum (Natural History; 94.3.23.2–4) are females and were collected by a Mr. Gaumer. Distribution is based on Lee (1996), Buskirk (1993), Smith and Smith (1979), and UU collections.

HABITAT

Terrapene yucatana occurs principally in tropical deciduous scrub forest (Buskirk 1993; Legler, pers. obs.) in the northern and central parts of the Yucatán Peninsula. It inhabits both forested and open situations, although it appears to prefer the former. Individuals have been found in marshy areas, pastures, thorn forests, and tropical evergreen forests, but not in the tall mesic forests at the base of the peninsula. The turtles are terrestrial but may occasionally enter shallow water (Lee, 1996). J. R. Buskirk (pers. comm.) observed a fuzzy algal growth on the posterior carapace of a specimen from near Libre Union, Yucatán, and Legler (pers. obs.) has seen an example of the reddish-brown water deposit frequently found on aquatic turtles (e.g., *Kinosternon*) on the shell of a specimen from the same locality (see family Kinosternidae).

Langebartel (1953) recorded a fragment (peripheral bone) of *Terrapene* from Actun Coyok Cave (at Oxkutzcab, 16 km southeast of Ticul, southwestern Yucatán) and Mercer (1896) recorded *"Cistudo"* (= *Terrapene*) in Loltun Cave. These finds were made in strata showing signs of human use and could have been dated anywhere between 200 and 1250 AD. *Kinosternon creaseri* and *Trachemys* sp. were also recorded from these and other caves. Langebartel (1953) thought the aquatic species were brought to the caves for human consumption but that the *Terrapene* could simply have fallen in.

DIET

There are no reports of natural diet, but omnivory can be assumed. Captives observed by Buskirk (1993) did not eat raw meat but were fond of fresh orange sections.

HABITS AND BEHAVIOR

Activity is chiefly during the rainy season, from June to early November (Buskirk, 1993). Box turtles are sometimes found after the slashing and burning of agricultural clearing, a practice that has occurred for "thousands of years" on the Yucatán Peninsula (Brainerd, 1953). Shells often bear evidence of healed burns and lacerations. Burned turtles are sometimes eaten by local people. Box turtles are popular as pets with Mayans and are kept in large, walled enclosures for known periods of 10 years or more. In one instance two males were kept in the same enclosure with females; the males were viciously territorial (the dominant male inflicting wounds that bled); when one of the males was removed, matings began to occur. Juveniles sometimes appear in these enclosures, but hatchlings have never been observed (Buskirk, 1993).

REPRODUCTION

Nothing on reproduction has been recorded in the literature. Dissections made of three females by Legler (pers. obs.) from the vicinity of Libre Union and Merida, Yucatán, 9–17 July 1966 revealed the following:

CL, 146 mm; 3 shelled oviducal eggs; 3 fresh corpora lutea, no sign of other ovulations; 2 follicles, 15 mm; 1 follicle, 9 mm; 14 follicles, 4–7 mm.

CL, 142 mm; 2 shelled oviducal eggs; 2 fresh corpora lutea, no sign of other ovulations; 2 follicles, 20 mm; 1 follicle, 12 mm; 2 follicles, 5 mm.

CL, 151 mm (obviously an old female); ovary short, baglike, and seemingly inactive, containing a few follicles of 4 mm and many smaller; seemingly some vitellogenesis in larger follicles; no sign of recent ovulation but presumed to have potential to lay in following season.

Mean dimensions of five eggs were as follows: L, 47.2 mm ± 2.53 (44.2–50.4); W, 27.5 mm ± 0.63 (26.7–28.1).

From these meager data the following is inferred: clutches are small (two to three eggs); a first clutch is laid in July; follicular potential for a second clutch exists; mature females may skip seasons of reproduction. The eggs of *Terrapene n. nelsoni* and *Terrapene yucatana* are of similar size and are the largest recorded for any *Terrapene*. Each of these taxa is the southernmost representative of its species group, is of moderate adult size, and produces a few large eggs per clutch (see Dodd [2001] and Ewert [1979] for comparative data).

GROWTH AND ONTOGENY

Data on size (CL): males, 145 mm ± 3.53 (143–148) n = 2; females, 152 mm ± 2.0 (150–154) n = 3; adult unsexed, 144, n = 1.

Males are seemingly larger than females (mean CL of eight females, 147.4 mm, and for five males, 154.5 mm) and have more massive heads (Buskirk, 1993). Males have a flat or only slightly concave plastron and a thicker, longer tail with the vent exposed beyond the posterior margin of the carapace. The mid-dorsal keel is more evident in females than in males (Buskirk, 1993).

Adult males have the darkest carapaces; the dark radial pattern is characteristic of the oldest females, and younger females having only narrowly dark-edged laminae. The darkest plastra occur in young adult females (Buskirk, 1993).

There is a striking sexual dichromatism of neck, head, and iris that is reviewed by Buskirk (1993) and mentioned by Duellman (1965) and Lee (1996). The soft parts of females are dull—pale brownish limbs and heads that may grade into yellow and tan. Mature males have unicolored pale, usually white (but often bluish) heads. This male head coloration may contain blue or pink flecking, especially on the eyelids and the throat. This is similar to the description of pale greenish-blue head coloration described by Legler (1960c) for *Terrapene ornata*. The blue coloration seems to be an intermediate stage in the loss of melanin (rather than an actual blue pigment). Moll, Matson, and Krehbiel (1981) review seasonal and sexual dichromatism in chelonians. Moll actually sectioned the skin of a white-headed adult male *yucatana* (UU 9800) and found an epidermis of normal thickness in which there were few or no melanophores. Thus, it appears that the white head is a permanent feature of males, but it remains to be demonstrated whether seasonal changes are superimposed upon this pattern.

Buskirk (1993) reports that iris color of five males varied from off-white to the same yellowish tint exhibited by nine females. Legler (unpublished data) found the iris in eight females to range from golden (most common) to olive gray or brown with gold flecks. One male had a pale-orange iris that contrasted sharply with the white head and was quite evident at a distance of several paces. Since Legler's series was taken during the breeding season in early July (two of three females gravid) and Buskirk's series was observed in November, it is possible that some seasonal change in iris color may occur. It has been assumed that sexual dichromatism of the head and eyes functions in gender identification (Legler, 1960c).

CONSERVATION AND CURRENT STATUS

Many authors have commented on the rarity of box turtles in Mexico, and this seems to be based largely on the paucity of specimens in collections. It is clear that *Terrapene* populations everywhere may be declining as the result of human encroachment on natural habitats and the building of roads along which box turtles and herpetologists seem to travel. In rural Mexico there are relatively few roads to search. Even people who have spent time looking for these somewhat cryptic creatures (and know how to look for them) have difficulty finding them. John Iverson is cited as having never found a box turtle in Yucatán (Buskirk, 1993). Legler and family lived on the Yucatán Peninsula for a month (June to July 1966) and did a lot of exploring in good box turtle habitat. This included low forest in the immediate vicinity of an aguada (Águada Sayusil). We found no box turtles, but local residents said they were common and brought several to us. Buskirk (1993) comments that *yucatana* is not abundant, that rural residents think its numbers are decreasing, and that human population growth and habitat loss are the causes of this decline.

FIGURE 54.1 *Terrapene coahuila*. Top—adult female typifying the general appearance of the species (from Howeth and Brown, 2011; photo by Dean Hendrickson, with permission). Bottom—head of adult male (left) and female (right).

Terrapene coahuila Schmidt and Owens, 1944

Coahuilan box turtle (Iverson, 1992a; Dodd, 2001); tortuga negra (Webb, Minckley, and Craddock, 1963); tortuga de Cuatro Ciénegas (Liner, 1994).

ETYMOLOGY

A toponym referring to occurrence in the state of Coahuila.

HISTORICAL PROLOGUE

The original description of this species is generally uninformative. The types were collected in the Basin of Cuatro Ciénegas in 1937 by E. G. Marsh Jr. (Marsh, 1937). Although the species is aquatic (and known as such to residents), the authors (Schmidt and Owens, 1944) were seemingly unaware of this. Either Marsh did not report this fact or he did not personally collect the specimens. Little information was gathered in the period between 1944 and 1958, and it was inexplicable why more specimens had not appeared. In August 1956 Legler examined the type series in the Field Museum of Natural History. One of the paratopotypes (FMNH 47372, female adult) had a growth of epizoic algae on the posterior margin of the carapace. This sparked new interest in an unusual *Terrapene* and precipitated an expedition to the Basin of Cuatro Ciénegas in September 1958 (J. M. Legler, W. L. Minckley, and R. Wimmer). Submerged, baited traps caught *T. coahuila* almost immediately, and the first adequate series was established, all from aquatic habitats. Simultaneously, the type series of *Trachemys scripta taylori* and *Apalone atra* were obtained, and it became clear that most of the aquatic organisms in the basin were distinctive in some way. Minckley went on to make long-term studies of the biology of the basin, especially the fishes.

William S. Brown (1971) published a morphometric study of *T. coahuila* and later (Brown, 1974) a landmark 3-year autecological study of the species. These two works comprise most of the knowledge of *T. coahuila* and make it one of the best known Mexican chelonians.

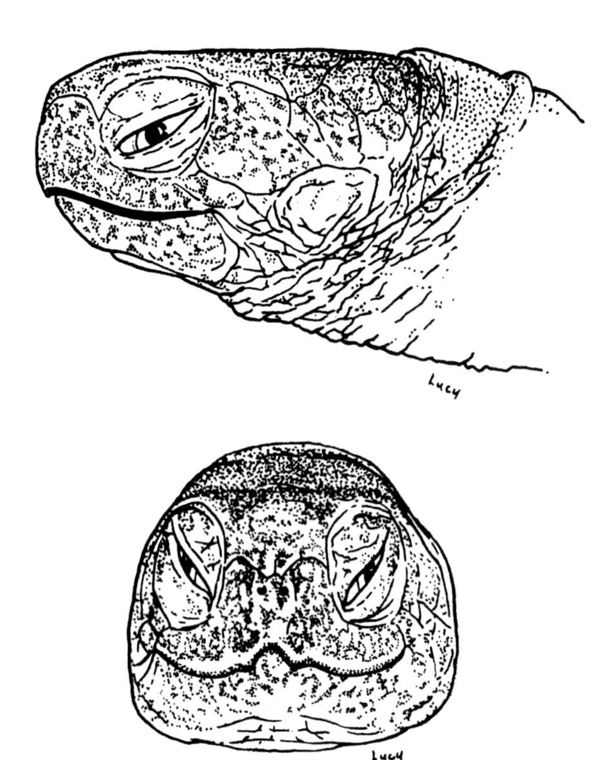

FIGURE 54.2 *Terrapene coahuila*. Frontal and lateral views of head, adult female, paratype, CNMH 47372. Aquatic algae were first found on this specimen and precipitated early exploration of the Cuatro Ciénegas Basin. (Drawing by Lucy Remple.)

See accounts of *Trachemys scripta taylori* and *Apalone atra* for discussion of isolation events and their relationship to chelonian speciation in the Basin of Cuatro Ciénegas.

DIAGNOSIS

Terrapene coahuila is a unique box turtle and a unique emydid. The species is distinguished from other members of the genus by the following combination of characters. It is longer and narrower than any other *Terrapene;* cross-section is a slightly flattened dome; highest point of carapace is posterior to hinge. There is little flaring and no serration of posterior carapace margin; hatchlings have a prominent median keel that is scarcely evident in adults. Anterior plastral lobe is short, 63% of posterior lobe; plastral hinge is opposite M5. Plastral scutes as percentages of plastral lobes are as follows: anterior lobe—intergular 49%, interhumeral 20%, interpectoral 30%; posterior lobe—interabdominal 35%, interfemoral 11%, interanal 54%. Axillary scute usually is present. Temporal roofing is more extensive than in any other *Terrapene*; postorbital bar is broad and heavily developed; zygomatic arch is complete and broad; postorbital is in broad contact with jugal and quadratojugal (Figure 49.2, skulls). Phalangeal formula is 23332 for manus and pes; there are four clawed toes on pes. Cloacal bursae are present and larger than in any other *Terrapene* (10 mm long and 3 mm at orifice, Williams, Smith, and Chrapliwy, 1960). Plicae internae of penis are reduced (Zug, 1966).

GENERAL DESCRIPTION

Mean adult carapace length (in the population studied by Brown [1974]) is 133 mm, maximum 168 mm. The species may be considered aquatic. Its habits more nearly resemble those of aquatic emydids than members of the genus *Terrapene* (Brown, 1974). Its morphology and some of its behavior have caused some experienced herpetologists, at first glance, to think it a kinosternid. The shell, although arched, is relatively long and narrow, and the entire animal is dull colored, usually lacking any sort of distinct pattern or bright coloration.

BASIC PROPORTIONS

(From the population studied by W. S. Brown [1974].)

CW/CL: males, 0.68 ± 0.021 (0.63–0.72) n = 81;
females, 0.69 ± 0.019 (0.64–0.75) n = 94.

CH/CL: males, 0.43 ± 0.017 (0.38–0.47) n = 82;
females, 0.46 ± 0.015 (0.42–0.49) n = 94.

CH/CW: males, 0.64 ± 0.025 (0.59–0.69) n = 78;
females, 0.67 ± 0.027 (0.60–0.73) n = 94.

There is significant variation in size structure (carapace length) across the small geographic range of *T. coahuila* (J. Howeth, unpublished data). This geographic variation in body size and potentially age structure may be a consequence of site-specific differences in resource availability and predation pressure, thereby affecting growth and mortality rates. Although there is sexual size dimorphism in the species, the geographic differences in size structure cannot be attributed to spatially or temporally skewed sex ratios (Howeth and Brown, 2011).

Modal plastral formula: 6 > 4 > 1 > 3 > 2 > 5.

The carapace is a dark, dull combination of slaty olives, browns, and blacks with patches as pale as straw where wear has occurred. In an adult shell this combination produces and irregular series of large blending patches and is effective camouflage. The plastron and inframarginal surfaces are distinctly paler, grayish to pale olive to straw, and without intrinsic pattern. All interlaminal seams are at least narrowly darkened, and those of the plastron are in sharp contrast to the ground color. In juveniles the carapace has a subtle pattern of fine yellowish reticulations, and there is a pale postorbital stripe. The stripe soon disappears, and the adult head is pale to dark slate, often with fine darker marbling. The carapace pattern is retained to varying degrees in older turtles; it may appear

FIGURE 54.3 *Terrapene coahuila*. Top—lateral view of dry shell, adult male, UU; bottom—plastron, adult male, UU.

in patches on turtles of any age. Close scrutiny of pattern on a clean shell suggests a field of short, fat worms of alternating pale and dark color. The markings radiate out from the areolae in a manner that suggests the pattern of *Emys orbicularis*. Description and diagnosis are based on Smith and Smith (1979), Milstead (1969), Minx (1992), Brown (1971), and Webb, Minckley, and Craddock (1963), unless otherwise cited.

RELATIONSHIPS

The phylogenetic position of *Terrapene coahuila* is still moot. Auffenberg (1958) and Legler (1960c) treated *T. coahuila* and its aquatic habits as primitive. Holman (1987) regards *T. coahuila* as the least derived *Terrapene* by comparison to the oldest known fossil (*Terrapene corneri*; Holman and Fritz, 2005; middle Miocene; Nebraska). Based on studies of plastral hinge mechanisms and limb musculature, Bramble (1974) referred to *T. coahuila* as "... the pliesiomorphic facies of the genus." The presence of cloacal bursae in *T. coahuila* and their absence in most other members of the genus would argue against the recent acquisition of aquatic habits.

Based on morphology Brown (1974) and Milstead (1969) favor a fairly recent origin of *coahuila* from an isolated stock of *T. carolina* in subrecent times and its aquatic habits as recently acquired. Various cladistic analyses (reviewed by Feldman and Parham, 2002) have placed *T. coahuila* either as the sister group of "*carolina*" (Feldman and Parham, 2002; Bickham, Lamb, Minx, and Patton, 1996) or as the sister group to all other *Terrapene*.

Terrapene coahuila is, in fact, an exception to nearly everything in a logical taxonomic arrangement of the genus, and its inclusion in the *carolina* group (and, indeed, in the genus *Terrapene*) may be more a matter of convenience than of relatedness. We do not recognize it as a member of either of the main groups of *Terrapene*.

GEOGRAPHIC DISTRIBUTION

Map 31

The species occurs in disjunct populations only within the Basin of Cuatro Ciénegas in central Coahuila. This isolated desert basin of some 800 km^2 contains the few marshes and other aquatic habitats that collectively constitute a minuscule geographic range that is only a fraction of total surface area of the basin floor. As such, the natural geographic ranges of *T. coahuila* and two other endemic chelonians in the basin (*Apalone atra* and *Trachemys scripta taylori*) are among the smallest known for chelonians.

HABITAT

All remarks on natural history are based on Brown (1974) or personal observations by Legler unless otherwise noted.

The Cuatro Ciénegas Basin floor lies at 740 m and is ringed by mountains 1,500 to 3,000 m high. Habitat consists chiefly of north-south trending spring-fed marshes with mud bottoms, shallow water, and dense submergent and emergent aquatic vegetation. Marsh habitats are fed by warm springs; water temperatures are as warm as 35–39°C in summer and 7.2–16.5°C in winter. The bottom mud is at least 5°C cooler than the water in summer. The mud may not be covered by water on a hot day, but the water is replenished at night. Marshes are isolated by much greater areas of intervening desert grassland and shrub, but turtles are common only in the marshes. Horses and mules do some damage to marshes by eating *Eleocharis*. Goats don't stay in the marshes long enough to do damage. *Terrapene coahuila* inhabits both permanent and seasonal wetlands that are widely distributed across the valley floor.

DIET

Terrapene coahuila is an opportunistic omnivore, concentrating on aquatic vegetation and aquatic insect larvae in equal amounts. Diet is more nearly like that of aquatic emydids (e.g., *Chrysemys picta* and *Glyptemys muhlenbergi*) than other species of *Terrapene* (Brown, 1974).

Foraging usually occurs on vegetational mats in shallow water (2–6 cm) with head in the water and part of carapace above the surface and dry. The limbs are used to move vegetational obstructions as turtles forage; some turtles were observed to pull a mass of vegetation out of the water and pick it apart. The turtles were able to scan objects

FIGURE 54.4 *Terrapene coahuila*. Top—dorsal and ventral views of hatchling (from Howeth and Brown, 2011). Middle—first-year juvenile showing early form of head pattern. Bottom—first-year juvenile swimming at surface. (All photos by P. Scanlan, Gladys Porter Zoo.)

underwater with enough precision to be selective in their choice of insect larvae. The head is frequently raised in circumspection during foraging.

The turtles are able actually to capture live fish at various depths (Brown, 1974; Williams, Smith, and Chrapliwy, 1960), but this is uncommon. Usually the turtles ignore the small fishes that seek the particulate debris of foraging.

Dietary "bingeing" occurs with any plentiful animal or plant food (e.g., *Eleocharis* seeds, amphipods). Overall, the principal contents of 45 stomachs (by percentage of total volume) were as follows: plant material, 45.6%; Diptera, 19.6%; Coleoptera, 11.7%; and Odonata, 11.3%.

HABITS AND BEHAVIOR

There is no period of hibernation; the turtles remain active all year except for short periods of quiescence during environmental extremes. In the hottest time of the year, activity tends to be in early morning, late afternoon, and night. Turtles found on land in December with body temperatures higher than the air were assumed to have been basking.

Daily activity is crepuscular and is more closely linked to water temperature than to air temperature. Movement through a marsh is typically by wading in a sinuous path over vegetational mats (*Chara*) and around tussocks of sedge with only part of the body submerged. Despite Milstead's (1967) allusion to awkward swimming (of captives), Coahuilan box turtles can swim well and rapidly. They seem to be best at rapid swimming near the bottom (as with kinosternids), but can also cruise at various depths in open water with agility and at least some grace (Legler, pers. obs., 1978). Williams and Han (1964) found *T. coahuila* to be denser (ca. 0.95 g/cm^3) than *T. carolina* by about 23%, suggesting that this might influence the ability to stay underwater with ease.

Stephens and Wiens (2008) measured the swimming speed and endurance of *T. coahuila*. Speed was equivalent to that of *Emydoidea* and *Glyptemys* but slower than fully aquatic emydids (e.g., *Graptemys*, *Trachemys*). Endurance, measured after 20 minutes of continuous locomotion, was the second highest of the 16 emydids tested. The relative terrestrial speed of *T. coahuila* was slower than that of *T. carolina* and *T. ornata*.

Coahuilan box turtles make frequent, short movements in a local wetland and infrequent, long-range movements among wetland habitats (Howeth, McGaugh, and Hendrickson, 2008). Brown (1974) found that about half of the turtles he marked from 1964 to 1966 were recaptured in a different wetland than where originally marked. Mean diameter of home range was 25.6 m. There were 133–156 adult turtles per marsh hectare. Movements within marshes averaged 13 m between captures. Most turtles remained within a given marsh, but 20% had moved to another marsh and had presumably done so by overland locomotion (despite the existence of underground waterways). Turtles are commonly seen on roads during rains (Legler, pers. obs., 1958).

Terrapene coahuila is socially tolerant under natural conditions. No territoriality or aggression was noted in the wild among individuals as close as 3 m to one another. Savage fights have been recorded among captives (Webb, Minckley, and Craddock, 1963).

There is wide seasonal and daily variation in body temperature. Thermoregulation within the limited habitat and small home ranges of *T. coahuila* seems to be a matter chiefly of seeking comfort. Minimum air temperatures may be near but seldom below freezing (and there is no record of ice on the marshes). The turtles rarely leave the marshes and may never venture into the harsh desert environments surrounding them. When in water, turtles characteristically keep the upper half of the shell exposed and dry. Of

254 cloacal temperatures taken, most approximated that of the water with a tendency to be slightly lower than water temperature in morning and higher late in the day. The extremes of recorded body temperature for active turtles (14.8° in January and 33.5° in April) were considered to represent the voluntary minimum and maximum tolerances. Turtles in water of 7.2–16.5°C in December had cloacal temperatures of 7.2–12.4°C and were lethargic. Escape from dangerously high air or water temperatures can be accomplished by burrowing into the mud. It is evident that aerial basking is used to elevate body temperature when air and/or water temperatures are suboptimal. On a clear, sunny day in December, air temperatures ranged from 17°C to 19°C, and turtles on land maintained average body temperatures of 29.2°C. Turtles may spend several days of quiescence in mud in the hottest and coldest parts of the year.

No optimum temperature has been predicted or estimated. For turtles placed experimentally in crushed ice at 0.0°C and then allowed to warm, first coordinated movements usually occurred from 10.2°C to 14.4°C (extremes, 6.0–16.4°C). Body temperatures near 40°C are assumed to be lethal. A tethered adult female in direct sunlight showed distress at 35°C after 7 minutes (gaping and frothing at mouth); at 22 minutes and 37.5°C the turtle was unable to close the shell.

Foraging turtles, when disturbed, raise their heads and then pull quietly into the shell and/or dive into the soft bottom substrate. When handled they invariably pull into the shell and tightly close both plastral lobes until left undisturbed for several minutes.

REPRODUCTION

Nearly all the following data on reproduction in nature is from Brown (1974) unless otherwise cited. We also present pertinent data resulting from captive colonies.

Sexual maturity is attained at carapace lengths of about 90 mm. The smallest mature male and female were of 93.1 and 90.7 mm CL, respectively, and the largest immature specimens 85.1 and 89.2 mm CL. Data on mean adult CL and statistically predicted extremes are as follows: males, 108.9 mm (93–114) n = 70; females, 100.9 mm (95–105) n = 94.

Complete captive matings last up to 2 hours and have been observed in all months except October, January, and February (Tempe, Arizona; W. L. Minckley, pers. comm.). The male follows the female, pushing at her shell with his extended head. This is followed by neck retraction and butting or bumping her shell with his, using enough force to be heard by a human standing nearby. After mounting he grips her with his rear claws on the skin of the gluteal region and makes biting movements at her anterior soft parts. He then is in a rather precarious vertical position during which intromission probably occurs. He then falls backward and experiences the remainder of the sexual bout on his back, sometimes being dragged about by the female. It is not clear whether the firm connection to the female is the engorged penis, a locking of the males legs by the females legs and plastron, or both. Disturbed copulating pairs disengage and seek cover. Most of these elements of mating behavior occur also in other *Terrapene* (Evans,1953, 1968; Brumwell, 1940; Legler, 1960c; Nieuwolt, 1997; Brown, 1974).

Matings of free-living *T. coahuila* have been witnessed in April, November, and December. Brown (1974) suggested that mating could occur year round. Mating may occur out of water but is more common in water. Matings in water sometimes occur at depths that submerge both turtles but most often in depths that leave both sexes free to breathe. Mating in an aquarium with straight glass sides resulted in the drowning of a female (Legler, pers. obs.)

The male gonadal cycle is typical for northern temperate turtles: spermatogenesis occurs in summer; sperm overwinter in the epididimydes and are used to fertilize ova in the following season. Testes are small in July and August, become enlarged in late August, and are largest in April when sperm are most abundant. Neotropical emydids are known to extend the spermatogenetic cycle into the "winter" (Moll and Legler, 1971), and Brown (1974) suggests this may be happening to some extent in the relatively warm environments of Cuatro Ciénegas at 27°N.

Enlargement of ovarian follicles begins in late August and continues at least until April. Ovulation begins in early April, is concentrated in late April, and extends into late August and perhaps early September. Preovulatory follicles are 15–17 mm in diameter.

Oviposition probably occurs from late April or early May to early September. Natural nests and nesting have not been described, but it is logical that nesting occurs on the drier margins of the marshes or in the soil of sedge tussocks. Captive females typically lay clutches in loose, slightly moist substrate (e.g., gravel, sand) near vegetation, late in the evening (Cerda and Waugh, 1992; Bauer and Jasser-Häger, 2006).

In the context of the thermal environment of Cuatro Ciénegas, Brown predicted a natural incubation period of 70 days with hatchlings appearing from mid September through early November. A juvenile no older than 3 months was taken on the study area in mid October. There is no evidence that "overwintering" of either hatchlings or eggs occurs (nor any reason to think it advantageous in the mild climate of Cuatro Ciénegas).

Measurements of nine eggs in three clutches were as follows: L, 33.2 mm ± 0.67 (30.5–36.3); W, 16.9 mm ± 0.28 (15.8–18.2); Wt, 5.66 g ± 0.30 (4.44–6.81). These eggs rank among the smallest recorded for *Terrapene*. Brown (1974) estimated PL at hatching to be in the range of 26–29 mm, and Ewert (1979) gave hatchling CL as 30.4 mm.

Mean incubation period per egg was 46.3 days in an incubator. Temperature-dependent sex determination occurs: a temperature of 28°C produces females and temperatures of 26–27°C produce males (Bauer and Jasser-Häger, 2006). Newly hatched *T. coahuila* in a captive colony weighed 4.4 g on average (Cerda and Waugh, 1992).

Other breeders report an average of 10–12 g (Bauer and Jasser-Häger, 2006).

Data on reproductive potential from all sources but based mainly on ovarian examination are as follows: overall mean number of eggs per clutch is 2.3 (2–4) with clutches of two or three eggs occurring most commonly; about half of the adult females produce a second clutch averaging 2.4 eggs (1–4); and about one-third produce a third clutch averaging 1.7 eggs (1–3). A female producing three clutches in 1 year would have a mean annual reproductive potential of 6.8 eggs (maximum 11 eggs). Number of eggs per clutch is positively correlated with size of female. There is no evidence that females skip seasons of reproduction, and follicular atresia is minimal.

Combining census and reproductive data, Brown (1974) hypothesized that 90 females could have an annual reproductive potential of 409 eggs if: (1) each female produced one clutch of 2.7 eggs (243 eggs); (2) 47 females (52%) produced a second clutch of 2.4 eggs (113 eggs); and (3) 31 females (34%) produced a third clutch of 1.7 eggs (53 eggs). Considering probable low adult mortality Brown thought this potential of some 400 eggs could withstand "high losses" from oviposition to puberty.

GROWTH AND ONTOGENY

Data on size (CL) in the population studied by Brown (1974) in 11 marshes near the northern tip of the Sierra de San Marcos were as follows: males, 114 mm ± 15.7 (82–158) n = 82; females, 101 mm ± 5.6 (89–114) n = 94.

Adults were larger in marshes peripheral to Brown's study area (Jennifer Howeth, pers. comm.): males, 145 mm ± 23.6 (97–230) n = 312; females, 128 mm ± 11.5 (100–199) n = 343; unsexed adults, 126 mm ± 7.16 (118–132) n = 3; im. males, 108 mm ± 8.2 (92–125) n = 24; im. females, 109 mm ± 7.35 (95–119) n = 18; juvenile, 65.7 mm, n = 1.

Males are slightly larger than females and have a distinctly concave posterior plastral lobe and longer and thicker tail. The carapace is lower in males (CH/CL, 0.43 [0.38–0.47] n = 70) than in females (0.46 [0.42–0.49] n = 94). The iris of males is yellowish and flecked with brown.

A range-wide survey of seven additional subpopulations by Howeth in 2002 and 2003 (sites detailed in Howeth, McGaugh, and Hendrickson, 2008) reinforces the findings of sexual size dimorphism by Brown (1971), where males are significantly larger than females (mean straight-line carapace length: male—143.8 mm, n = 363; female—126.5 mm, n = 381; J. Howeth, unpublished). Interestingly, the mean size of both sexes in this range-wide study was greater than that found by Brown (1971) in a single subpopulation and suggests geographic variation in age structure or individual growth rates.

According to Bauer and Jasser-Häger (2006) sex can be reliably determined at a minimum mass of 100 g, which corresponds to 3 years of age for captive individuals, and females of ca. 100 mm CL and 4 years of age are sexually mature.

Little has been recorded on ontogenetic change. No scute shedding has been observed, but the shell becomes completely smooth at carapace lengths of 90 mm or less, and growth rings are rarely distinct enough to estimate growth or age past 2 or 3 years. Co-osification of dermal shell elements is progressive with age but is never as incomplete as in *Terrapene ornata* and probably less extreme than in some *T. carolina*.

Brown (1974) was able to determine the following growth increments in a few preserved juveniles: 17%, 25%, and 28% (of starting size) for three individuals in the season of hatching and 20%, 28%, and 49% in the first full year of growth. Longevity is unknown and may have to rely on techniques other than growth rings (e.g., Zug, 1990, 1991).

Miscellanious collecting and surveys of the basin have produced only one long term recapture—an adult female marked in 1965 (at CL 103 mm) and recaptured in 1974 had grown only to 105 mm CL in 1980 (Howeth and Brown, 2011).

PREDATORS, PARASITES, AND DISEASES

There seems to be no interest in *Terrapene coahuila* for human food.

Brown (1974) found nematodes in the gut and small, unidentified leeches attached to the soft skin and ventral margin of the carapace. Epizoic blue and green algae, including *Basicladia*, grow on the shell. Minckley (1966) reported predation by a coyote on *Trachemys scripta* and *Apalone* sp. in the basin, but no box turtle remains were found in the midden created by the predator. One injured *Terrapene* was found (in December) on its back with its carapace and plastron severely scratched but alive; there was abundant coyote sign in the immediate vicinity. Paucity of *T. coahuila* remains could bespeak either the ease with which this small turtle could be carried off by a coyote or the greater protection afforded by a tightly closed shell.

Of 218 specimens examined, 24 (11%) bore the marks of injury; 3% had burn scars; 3% were missing limbs; and 3% had other scars on the shell. Burn scars can be attributed to regular grass burning in the basin. The worst survived burn injury had damaged most of the carapace, leaving exposed dead bone (comparable to that shown in Legler, 1960c, Pl. 29:2).

POPULATIONS

Brown (1974) marked a total of 169 individuals and recaptured them a total of 271 times; 164 of these were adults of known sex, and only three were juveniles. Of the adults, 70 (43%) were males and 94 (57%) were females.

Brown (1974) estimated a population density of 148 individuals (122–189) per hectare from mark to recapture. Based on this estimate, associated *T. coahuila* biomass (32.3 kg/ha) is higher than the average for semiaquatic turtle species (19.2 kg/ha; Iverson, 1982). Population density in *T. coahuila* evaluated ca. 40 years after the study by Brown (1974), which addressed a different region of the species' range and assessed number of individuals per capture effort (not population size), suggests lower absolute densities that vary greatly by wetland habitat area. Howeth, McGaugh, and Hendrickson (2008) found that *T. coahuila* density ranged from 0.60 to 66.7 individuals per hectare per capture effort, where larger wetland habitats supported significantly higher turtle densities.

CONSERVATION AND CURRENT STATUS

Terrapene coahuila is *ipso facto* endangered because of a diminutive geographic range that is shared by humans. The ultimate fate of the Basin of Cuatro Ciénegas was not an issue with the two young biologists (Legler and Minckley) who pondered its endemism in 1958. But the consequences of continuous and catastrophic aquatic habitat loss in Cuatro Ciénegas were foreshadowed by several early biologists. In 1963, Webb, Minckley, and Craddock suggested that "prolonged and intensive collecting, as well as deterioration of habitat through the construction of irrigation canals and agrarian development, could endanger the species because of the isolation of some habitats." Subsequently Smith and Smith (1979) also noted cause for concern on lowering of the water table. In the years since these statements, insufficient progress has been made toward habitat preservation for *T. coahuila* and the other endemic aquatic species of Cuatro Ciénegas.

Various factors and events portend the demise of aquatic habitats and their endemic faunae in the basin. Many former habitats are now completely dry and devoid of aquatic life. Recruitment in *T. coahuila* has seemingly declined: only two hatchlings were seen in the period 2002–2007, and its actual total range is now calculated at 360 km^2. The completely aquatic organisms (e.g., fishes, snails, aquatic plants, aquatic turtles) will go first and *Terrapene coahuila* will go last. These taxa will not become extinct provided captive colonies are kept by responsible zoos and breeders. All these factors are related to water use and management and are detailed by Howeth and Brown (2011).

Family Geoemydidae

Family Geoemydidae Theobald 1868—The familial name "Geoemydidae" has been in proper but unpopular use since about 1994 (David, 1994; Spinks, Shaffer, Iverson, and McCord, 2004). Membership of the family is equivalent to the formerly used "Bataguridae" (Gaffney and Meylan, 1988) or "Batgurinae" (McDowell, 1964). The geoemydids are the chiefly Asiatic representatives of the once broadly inclusive Family Emydidae. Bour and Dubois (1986) showed that the name, Geoemydidae, has priority over Bataguridae or Batagurinae.

Geoemydids share many characters with the familiy Emydidae (see account of Emydidae), most of which distinguish the two families from the Testudinidae. Bickham et al. (2007) recognize 19 genera, 64 species, and 91 terminal taxa.

The family can be distinguished from the emydids as follows: There is a single articular socket between the fifth and sixth cervical centra. Angular bone is excluded from contact with Meckel's cartilage by a longitudinal process of the articular. There is a strong lateral tuberosity on the basioccipital bone (the "batagurine process" of McDowell, 1964).

Geographic distribution of the family is Europe, North Africa, Asia, and the East Indies in the Old World and from northern Mexico (ca. 27°N) to Brazil (ca. 02°S, 55°W) and Ecuador (05°S, 80°W) in the New World, where *Rhinoclemmys* is the only genus.

Genus *Rhinoclemmys* Fitzinger, 1835

Rhinoclemmys is comprised of seven or eight species occurring in North America, Central America, and South America. In this book we cover the three species and seven terminal taxa that occur in Mexico.

DIAGNOSIS

Rhinoclemmys can be distinguished by the following combination of characters: It retains deep growth rings with inflated-looking growth zones and a rounded mid-dorsal keel well into adulthood. Plastral buttresses are well defined; at least the anterior half of plastron is suturally and immovably attached to carapace; there are no true plastral hinges. Interdigital webbing is lacking in Mexican species. Rostral pores usually are present (Winokur and Legler, 1974), and mental glands are lacking in Mexican species (Winokur and Legler, 1975).

Eggs are large with thick, brittle shells and are usually too large to pass through the posterior orifice of shell without the aid of plastral kinesis (see "Egg Size and Plastral Kinesis" in Introduction).

Diploid chromosome number is 52 (Killebrew, 1977a; Stock, 1972). The three Mexican species have karyotypes that are nearly identical, but the nucleolus organizer regions (NOR) are in a slightly different position than in Asiatic geoemydids with 52 chromosomes (Bickham and Baker, 1976a; Carr and Bickham, 1986). The karyotype of *Rhinoclemmys pulcherrima* is as follows: $2N = 52$ with six pairs of Group A macrochromosomes, five pairs of Group B macrochromosomes, and 15 pairs Group C microchromosomes (Bickham and Baker, 1976a).

The genus contains two groups that are most easily defined by the presence or absence of interdigital webbing (referred to as "aquatic" and "terrestrial," respectively, by Iverson, 1992a).

Among testudinoids *Rhinoclemmys*, in general, bridges a morphological and ecological gap between highly adapted terrestrial (e.g., *Gopherus* and *Terrapene*) and completely aquatic forms (e.g., *Trachemys* and *Pseudemys*).

The three species occurring in Mexico have virtually no webbing and seem usually to be terrestrial to semiaquatic with habits somewhat like box turtles. They share these characters with one other species (*R. annulata*) that occurs from Ecuador to Nicaragua.

DICHOTOMOUS KEY TO THE MEXICAN SPECIES AND SUBSPECIES OF *RHINOCLEMMYS*

1A. Apex of maxillary tomium hooked, not notched; top of head bears a pale horseshoe-shaped figure. . . . . **(*Rhinoclemmys rubida*) 2**

1B. Apex of maxillary tomium notched, flanked or not by cusps; no horseshoe-shaped figure on top of head. . . **3**

2A. Carapace pale brownish, unmarked, and with unflared margins; gular scute twice length of humeral; pale postorbital mark elongate . . . . ***Rhinoclemmys rubida rubida***

2B. Margins of carapace flared and marginal scutes distinctly paler than lateral; gular and humeral scutes subequal; pale postorbital mark oval . . . . ***Rhinoclemmys rubida perixantha***

3A. Ground color of dorsal head dark, chiefly devoid of paler stripes; bridge entirely pale, lacking dark marks or fields; carapace relatively high; a small dark-bordered yellow spot on each lateral scute. . . . . ***Rhinoclemmys areolata***

3B. Several reddish transverse bars or stripes across front of rostrum; often a median dorsal reddish line on dorsal head that intersects one or more stripes on rostrum forming an arrowlike figure; bridge dark or with dark marks; carapace flattened to high domed; each lateral scute bears a central bright orange or yellow spot . . . . **(*Rhinoclemmys pulcherrima*) 4**

4A. Plastron with a narrow dark central figure, much less than half width of plastron, wider at ends and usually divided (forked) anteriorly; a narrow, dark longitudinal stripe on bridge (between axillary and inguinal notches) separates pale lateral field of plastron from a narrow, parallel pale stripe on bridge that in turn borders the ornate inframarginal surfaces (Figure 56.5); inferior surface of each marginal with two vertical, dark-bordered pale streaks (that may be connected ventrally to form a "U"); a tiny dark-edged ocellus with a pale center on the areolar area of each lateral scute; carapace low. Coastal Guerrero and Oaxaca . . . . ***Rhinoclemmys pulcherrima pulcherrima***

4B. Central plastral figure unbroken and wide: bridge unmarked, completely **5**

5A. Central plastral figure less than half plastral width; inferior surface of each marginal with one vertical pale mark; a large, pale, dark-edged ocellus on each lateral lamina. Carapace high (somewhat domed). Eastern Oaxaca to El Salvador . . . . ***Rhinoclemmys pulcherrima incisa***

5B. Central plastral figure unbroken and wide, more than half width of plastron; dorsal carapace essentially unmarked, solidly brownish; inframarginal surfaces; carapace relatively low and wide; each inframarginal surface with one vertical mark; Southern Sonora to Colima . . . . ***Rhinoclemmys pulcherrima rogerbarbouri***

Rhinoclemmys areolata (Duméril and Bibron) 1851

Mojina (Liner, 1994); furrowed wood turtle (Iverson, 1992a).

ETYMOLOGY

L. *areolatus*, with small spaces, alluding to a conspicuous pale spot on each lateral scute in juveniles.

HISTORICAL PROLOGUE

The taxonomy of this species has been stable, and no synonyms have been acquired. There have been comments on geographic variation in head color and pattern (Ernst, 1978; Pérez-Higareda and Smith, 1987), but no subspecies have been described. This species and its Mexican congeners would benefit from a good biosystematic study combining morphologic and molecular data.

DIAGNOSIS

Diagnosis and description are based broadly on Smith and Smith (1979), Ernst (1978), and Iverson (1992a).

Apex of maxillary tomium is notched and may be flanked by cusps; there is no distinctive U-shaped mark on head. Ground color of dorsal head is dark and chiefly devoid of paler stripes. Bridge is entirely pale, lacking dark marks or fields. Carapace is relatively high. There is a small, dark-bordered yellow spot on each lateral scute.

GENERAL DESCRIPTION

Maximum measurements for males are CL 188 mm and PL 178 mm and for females are CL 207 mm and PL 202 mm (Platt, Finger, Rainwater, and Woodke, 2004).

BASIC PROPORTIONS

(RCV Database)

CW/CL: males, 0.70 ± 0.02 (0.69–0.72) n = 5; females, 0.67 ± 0.02 (0.64–0.71) n = 13; im. males, 0.81 ± 0.10 (0.74–0.96) n = 4; im. females, 0.76 ± 0.06 (0.68–0.92) n = 12.

CH/CL: males, 0.40 ± 0.01 (0.39–0.42) n = 5; females, 0.40 ± 0.02 (0.37–0.43) n = 13; im. males, 0.38 ± 0.03 (0.35–0.42) n = 4; im. females, 0.40 ± 0.04 (0.30–0.43) n = 12.

CH/CW: males, 0.57 ± 0.02 (0.54–0.60) n = 5; females, 0.59 ± 0.03 (0.54–0.66) n = 13; im. males, 0.48 ± 0.09 (0.37–0.56) n = 4; im. females, 0.53 ± 0.08 (0.32–0.61) n = 12.

BASIC PROPORTIONS

(Ernst, 1978)

HT/CL: 0.42 (0.36–0.49).

CW/CL: 0.72 (0.65–0.83).

PL/CL: 0.95 (0.88–1.04).

BR/PL: 0.42 (0.35–0.48).

Modal plastral formula: 4 > 3 > 5 > 6 > 1 > 2.

The carapace becomes smooth in old adults but is rugose with distinct growth rings in juveniles and subadults. Central scutes are usually wider than long, each bearing a low, blunt median keel. Precentral scute is often bifurcated posteriorly. Sides of carapace usually are flattened but slightly concave at level of bridge. M8–12 are often flared or upturned; posterior marginals form a weakly serrated border. There is an obtuse plastral notch.

Head is small and narrow, and snout slightly protrudes. Maxillary tomium is weakly serrated. Dentary symphysis is not heavily developed, and its length is much less than orbital diameter. Anterior antebrachia bear 7–15 transverse rows of enlarged scales.

COLORATION

The carapace is brown with dark seams and yellow mottling, forming a "lichenlike" pattern (Ernst, 1980). The lateral scutes have a red or yellow dark-bordered central pale spot. These spots fade with age in some populations and appear to be lacking altogether in Belize populations (Vogt, Platt, and Rainwater, 2009).

Plastron is yellow with a dark central figure and often darkened seams; bridge is entirely yellow with dark seams. There is a small, pale (yellowish) spot on areolar region of each lateral scute (fades with age). Inframarginal surfaces are yellow with olive free edges. Ground color of head is reddish brown with some dark mottling dorsally; a yellow or red supratemporal stripe extends posteriorly from orbit downward to side of neck where it joins another neck stripe. There are two elongated red or yellow spots on nape, another between orbit and tympanum, and a pale bar on each eyelid (Günther, 1885: plate 8b).

FIGURE 55.1 *Rhinoclemmys areolata*. Head and neck of adult (Yucatán Peninsula).

Iris is dark. Jaws and chin are yellow, with the former bearing small black spots or ocelli. Ground color of neck is olive to brown with stripes dorsally and dark spots ventrally.

GEOGRAPHIC DISTRIBUTION

Map 34

Occurrence is from the Gulf Coastal plain just south of Veracruz eastward, through the entire Yucatán Peninsula to the Caribbean drainages of Belize, Guatemala, and extreme northwestern Honduras (near Tela), at generally low altitudes. The species also occurs on Isla Cozumel and the Turneffe Atoll (Platt, Karesh, Thorbjarnarson, and Rainwater, 1999; Platt, Meerman, and Rainwater, 1999). There is no evidence that *R. areolata* overlaps the range of any other *Rhinoclemmys*. The type locality was originally given as "Province du Peten (amer. centr.)." The most recent restriction is that of Dunn and Stuart (1951)—"Florres" (El Peten, Guatemala).

GEOGRAPHIC VARIATION

Ernst (1978) describes differences in populations over the entire range of the species but recognizes no subspecies. The population on Isla Cozumel has a mean lower shell height (CH/CL, 0.39 ± 0.02) than mainland populations.

HABITAT

Mojinas inhabit savannas, grasslands, thorn scrub woodlands, and marshes. Some individuals spend more time than others in aquatic habitats. Often they can be found foraging in ephemeral pools and creeks.

Artner (2009) includes recent photos of natural habitat on the Yucatán Peninsula.

HABITS

These turtles use the burrows of armadillos as refugia and do not make their own burrows. Six adults tracked in Belize

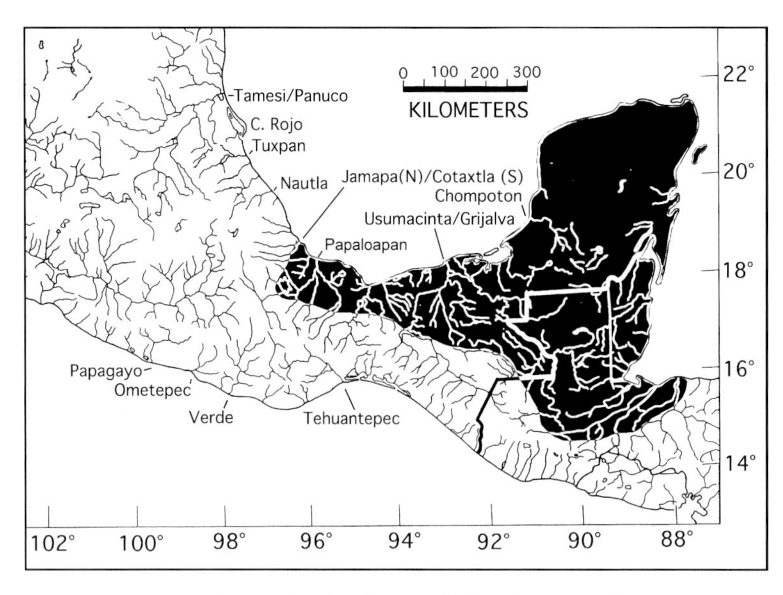

MAP 34 Geographic distribution of *Rhinoclemmys areolata*.

with thread spool devices remained in armadillo burrows for up to 48 hours. Another was seen entering a burrow, and turtle tracks were found at burrow entrances (Platt, Finger, Rainwater, and Woodke, 2004). Artner (2009) mentions the occasional turtle falling into open (senescent) cenotes.

DIET

Mojinas are omnivores, with a large part of their diet being plant material, including herbs, legumes, fruits, and fungi. The diet of *R. areolata* varies seasonally in Belize (Vogt, Platt, and Rainwater, 2009). During the dry season (January to mid June) *Sauvagesia erecta*, an understory herb, is the primary food. After the wet season begins in mid to late June the diet is diversified with the addition of herbaceous legumes and fruits such as *Eugenia* and *Miconia*. Craboo seeds (*Byrsonima crassifolia*) require 2–4 days to pass through the gut. It is probable that *R. areolata* is an important seed disperser for craboo and perhaps other woody shrubs. Craboo cease fruiting in late September, at which time herbaceous vegetation again becomes the dietary mainstay. In Veracruz it is also probable that *R. areolata* is a seed disperser since they swallow many seeds whole and these seeds can germinate after defecation (Lopéz-León and Vogt, 2000).

The small intestine was found to be densely packed with nematodes. This association is characteristic of herbivory in turtles (Bjorndal and Bolten, 1990) and may aid in the digestive process. Coleopteran exoskeletons, eggshells, crayfish, crustaceans, snake skin, mammal hair and bones, and bird feathers have been found in scats (Platt, 1993). One turtle was found consuming gray fox scat (*Urocyon cinereoargenteus*) containing rodent hair and partially digested *Eugenia* fruits, suggesting some mammal remains may be secondarily ingested through coprophagy (Vogt, Platt, and Rainwater, 2009). However, the observation of three adult *R. areolata* feeding on a nine-banded armadillo killed by a jaguar indicates that at least some vertebrate remains found in scats are due to direct consumption. The scat of many larger turtles contained numerous small stones, which possibly aid in digesting fibrous plant material (Sokol, 1971) that comprises much of the diet.

REPRODUCTION

Data on sexual maturity for 38 individuals from Cozumel (Vogt, pers. obs.) were as follows: smallest adult male 158 mm, largest immature male 144 mm; smallest adult female 146 mm, largest immature female 137 mm. These data plus growth-ring analysis suggest that sexual maturity occurs at about 145–150 mm at ages of 9–10 years in both sexes.

Copulation has been observed under natural conditions (in Belize) in May, July, August, and September. Males appear to track females using scent trails (Vogt, Platt, and Rainwater, 2009). Courtship and copulation were observed in March, April, and May in Tabasco (Pérez-Higareda and Smith, 1988); they describe the courtship and copulatory behavior as taking place in the water both in nature and under captive conditions. Copulation lasted up to 45 minutes. Olfaction appears to be important since cloacal sniffing is followed by vigorous head rubbing by the male prior to mating. Artner (2009) observed numerous copulations in a pair of captives; the male and female were kept in separate enclosures and copulation began immediately after they were placed together. A direct rear approach is made by the male before mounting (head rubbing is mentioned), and the event lasts 15 minutes.

Based on ovarian examination (Vogt, Platt, and Rainwater, 2009), *Rhinoclemmys areolata* appears to be laying

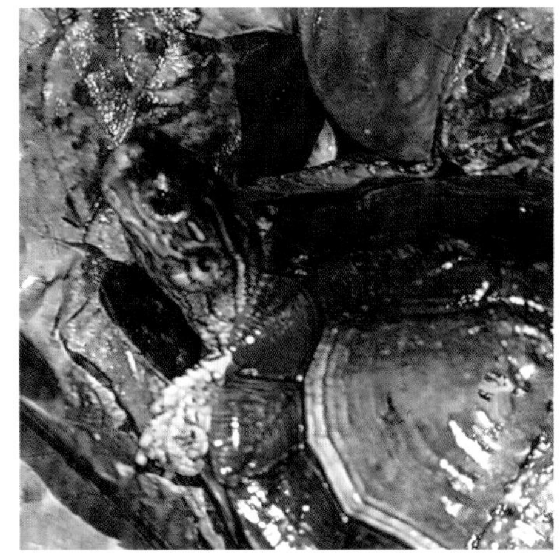

FIGURE 55.2 *Rhinoclemmys areolata*. Variation in head pattern.

up to five one-egg clutches per year in Quintana Roo and Belize. In Cozumel one female had three corpora lutea of different diameters, and two atretic follicles. Another had three corpora lutea and one enlarged follicle at the end of the reproductive season on 14 July (Vogt, unpublished). Artner (2009) reports one-egg clutches in captives. There is also evidence for four clutches in Belize (Vogt, Platt, and Rainwater, 2009).

Nesting has been observed in May and June in Quintana Roo; in June, July, and October in Veracruz; and in May, June, and July in Belize. Each laying typically consists of one large egg.

The skeletal posterior orifice formed by the carapace and plastron expands during oviposition; kinesis occurs between the inguinal buttresses and the carapace (Ernst and Barbour, 1989). This phenomenon occurs also in other *Rhinoclemmys* (Joel Friedman, pers. comm.).

S. G. Platt (Vogt, Platt, and Rainwater, 2009) found two nests in Belize; one egg was only partially buried beneath a sedge clump and was exposed following a fire. The other was in a shallow hole about 5 cm deep in sandy, well-drained soil. Nesting was observed in July 1985 at the edge of a marsh near Villahermosa, Tabasco (Pérez-Higareda and Smith, 1988). The female was covering a hole containing one egg. Subsequently (in captivity) she made three one-egg nests at 5-day intervals. S. G. Platt (Vogt, Platt, and Rainwater, 2009) found one female in Belize laying a clutch of two eggs.

Data on size and weight of 15 eggs from captives at Mazunte, Oaxaca (Martha Harfush, pers. comm.) were as follows: L, 59.0 ± 5.1 (51.6–71); W, 32.9 ± 3.96 (20.5–38.6); Wt, 38.8 ± 6.9 (28.6–50).

The eggs are brittle and smooth shelled with pores. Shell thickness is 0.31–0.43 mm, of which the fibrous layer comprises 19% and the mineral layer 81% (Ewert, 1979).

Ewert, Firth, and Nelson (1984) reported three eggs that had been retained in the oviducts of one individual; these eggs acquired additional egg membranes and shells during the period of retention, seemingly as subsequent ovulations occurred. Legler (pers. obs.) has observed multiple shell deposition in long-retained eggs for *R. funerea* in Central America.

Mean incubation periods and incubation temperatures are as follows: 120 days at 25–25.5°C, 103 days at 26–27°C, and 66.3 days at 29.5–30°C (Ewert, 1979). Eggs hatch in 89–90 days under natural conditions in Belize (Vogt, Platt, and Rainwater, 2009). Sex is determined by incubation temperature, males being produced at 25°C and females at 30°C (Ewert and Nelson, 1991).

GROWTH AND ONTOGENY

Size and weight in a series of 38 individuals from Isla Cozumel (Vogt, 11 July 1982) were as follows: males—CL, 140 mm ± 16.2 (114–160); Wt, 356 g ± 97 (211–475) n = 9; females—CL, 173 mm ± 11 (158–195); Wt, 643 g ± 113 (441–857) n = 13; immature males—CL, 91 mm ± 26 (54–109); Wt, 131.5 g ± 82 (25–201) n = 4; immature females—CL, 112 mm ± 29.5 (61–180); Wt, 244 g ± 192 (28–788) n = 12.

Females grow slightly larger than males. Males have a slightly concave plastron. Tails in males are elongated and thicker than those of females.

Hatchlings range from 44 to 48 mm CL and weigh 18–22 g. Ewert (1979) reported captive hatchlings of 55 mm CL and 34 g. Size and weight of two hatchlings from Mazunte, Oaxaca, were as follows: CL, 48.0 mm ± 3.25 (45.7–53.3); CW, 38.7 ± 2.97 (36.6–40.8); Wt, 22.7 ± 0.42 (22.4–23). A 5-week-old juvenile had 70 mm CL and 58 mm CW and weighed 63 g (Martha Harfush, pers. comm.).

Artner's captive study (2009) contains numerous photos of ontogenetic stages from hatching to adulthood.

PREDATORS, PARASITES, AND DISEASES

Dyer and Carr (1990) found helminth parasites (*Nematophila grandis*) in two of seven *R. areolata* examined from Tabasco and Chiapas (see also Diet).

FIGURE 55.3 *Rhinoclemmys areolata*. Dorsal and ventral views of adult female from captive colony in Mazunte, Oaxaca.

POPULATIONS

During 20 years of fieldwork with freshwater turtles in Veracruz, Tabasco, Chiapas, Yucatán, and Quintana Roo Vogt found this species only sporadically. When specifically looking for mojinas they could be easily found in Tabasco and were often seen crossing roads in Quintana Roo, particularly after rain showers. The largest and densest population known is on Isla de Cozumel. At 0930 hours in July 1982 RCV collected 38 adults in less than an hour after a thundershower. The series consisted of 14 mature females, 11 immature females, nine mature males, and four immature males. Most were located by the sound of them voraciously eating plant material. Perhaps due to the paucity of predators the population grew unchecked. In January 1993 the population was still dense, but foxes had appeared. The island is designated as a reserve so this population is protected by law.

Surveys conducted from 1992 to 1999 show that *R. areolata* is widespread and abundant in the lowland pine habitats of northern Belize (Platt, Meerman and Rainwater, 1999; Platt, Karesh, Thorbjarnarson and Rainwater, 1999). Densities of five to six turtles per hectare were estimated from a sample of 141 adults (61 males and 80 females; Vogt, Platt, and Rainwater, 2009).

CONSERVATION

Some local populations are impacted by road mortality and collection for food, and for the pet and curio trades. *Rhinoclemmys areolata* is occasionally consumed by Mexican compesinos and has been sold, stuffed, and dried in curio stands (see Introduction). Stuart (1935) noted that Maya Indians in Guatemala hunted *R. areolata* for food. Rural dwellers in Belize occasionally take *R. areolata* for food.

The species is currently propagated in captivity at the Nacajuca turtle facility in Tabasco, where copulation, nesting, hatching, and rapid growth of neonates occurs under seminatural conditions. Adult turtles are maintained in a dirt-floored enclosure with access to a cement pond and feed readily on native and cultivated fruits (jobo, nanche, banana, tomato). Hatchlings and subadults are housed in an aquatic environment and grow rapidly on a diet of fresh fish and banana. The indigenous people of Tabasco, Yucatán, and Quintana Roo, Mexico, often keep large numbers (20–40) of *R. areolata* for long periods in small corrals near their kitchens along with ducks, chickens, and pigs. These turtles are fed table scraps and whatever fruit falls in naturally. Because most households lack refrigeration, these turtles serve as an important alternative protein source when other meat is unavailable. Vogt has never found the species offered for sale in restaurants in Mexico, nor did Legler see them in the market in Belize City. Neither of us has eaten the flesh. The specimens that we dissected did not appear to have enough meat to make commercial production for food an economically feasible endeavor.

The origin and significance of the name "mojina" is uncertain. Dr. K. H. Berendt lived and studied languages in Tabasco and furnished Cope (1865) with natural history information. Natives of Tabasco use the name "mojina." He reported that mojinas were common pets near dwellings and that they utter ". . . a soft melancholy piping, which is rather touching when they are killed."

Mr. J. R. Buskirk (pers. comm.) was told by several sources that these sounds, plus a shedding of tears during butchering, were the reasons they were not commonly used as food in the parts of Tabasco he visited and were not sold in the markets there (1966–1968). Nothing pertinent can be found for "mojina" in dictionaries. Some dictionaries define "mohino, na" as "sad" or "mournful" and this could be the origin of the vernacular name.

Rhinoclemmys pulcherrima (Gray) 1855

Sabanera (local use in Mexico); painted wood turtle (Iverson, 1992a).

ETYMOLOGY

L. *pulcher*, beautiful; pulcherrima is a superlative meaning the most beautiful, alluding to the brightly colored shell and head; L. *incisus*, *incisa*; to cut into, in reference to the

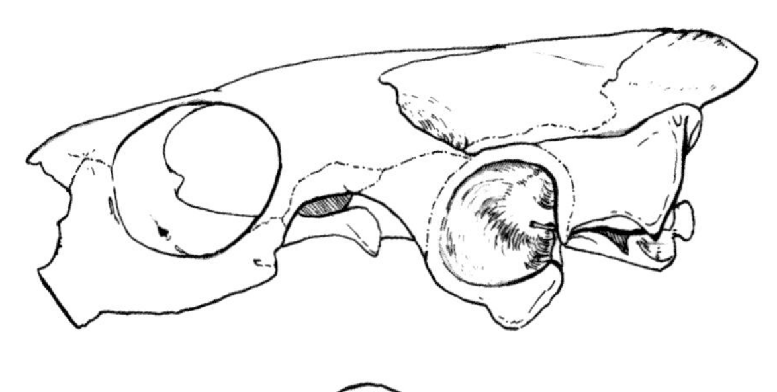

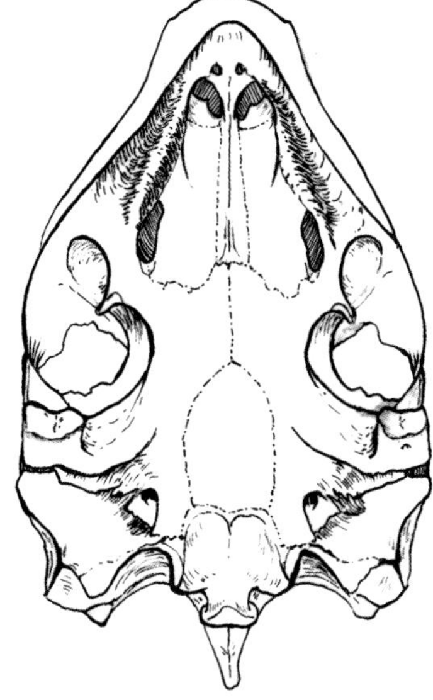

FIGURE 56.1 *Rhinoclemmys pulcherrima incisa*. Lateral and ventral views of skull (UU 12741, immature female, El Salvador; condylobasilar length ca. 29 mm). Batagurine processes not clearly shown. (Drawing by Elizabeth Lane.)

notched, cusped maxillary tomium. *Rhinoclemmys pulcherrima rogerbarbouri* is a patronym honoring the late Roger W. Barbour, University of Kentucky.

The diagnosis and descriptions are based broadly on Smith and Smith (1979), Ernst (1978, 1981), Iverson (1992a), and photographs supplied by Martha Harfush, Mazunte, Oaxaca.

DIAGNOSIS

Apex of maxillary tomium is notched and may be flanked by cusps; there is no distinctive U-shaped mark on head. There are several reddish transverse bars or stripes across front of rostrum; often a median dorsal line on head intersects one or more stripes on rostrum to form an arrowlike figure. Bridge is dark or with dark marks. Carapace proportions are variable, flattened to high domed. There are bright red or yellow markings on lateral scutes.

GENERAL DESCRIPTION

Maximum known CL is 181 mm in males and 206 mm in females (Ernst, 1978, 1981a). Maximum for "adults" is given as 230 mm by Ernst (1981). These maxima and minima vary among the subspecies.

BASIC PROPORTIONS

(M. Harfush Database)

CW/CL: males, 0.69 ± 0.07 (0.59–0.75) n = 4; females, 0.74 ± 0.03 (0.72–0.78) n = 4; im., 0.84 ± 0.04 (0.78–0.87) n = 4.

CH/CL: males, no data; females, 0.35 ± 0.05 (0.30–0.40) n = 3; im., 0.39 ± 0.03 (0.35–0.41) n = 3.

CH/CW: males, no data; females, 0.47 ± 0.08 (0.41–0.56) n = 3.

BASIC PROPORTIONS

(Ernst, 1978)

HT/CL: 0.42 (0.31–0.53).

CW/CL: 0.78 (0.65–0.84).

PL/CL: 0.95 (0.88–1.06).

BR/PL: 0.44 (0.39–0.49).

Modal plastral formula: 4 > 3 > 5 > 6 > 1 > 2.

Texture of scutes is rough because of distinct growth rings. Central scutes are more broad than long with a blunt median keel. Precentral scute is narrow, rectangular. and bifurcated posteriorly. Carapace is notched posteriorly. Sides of carapace are flat to slightly convex. Posterior marginals form a slightly serrated border and are sometimes turned upward. Growth rings are prominent except in old individuals in whom they are worn smooth. There is a wide anal notch.

Forelegs are covered with large red or yellow scales, each scale having two black spots; the spots are aligned to form longitudinal rows. Hind limbs have smaller scales that are olive to brown posteriorly; yellow or red scales have small black spots anteriorly. The tail is red or yellow with dark dorsal lines. Head is small with only slightly projecting snout. Maxillary tomium is slightly serrated. Length of dentary symphysis is less than diameter of orbit.

COLORATION

Rhinoclemmys pulcherrima is the most colorful member of the genus in Mexico, as reflected by its name (see Günther [1885], plates 7 and 8a).

Patterns on lateral scutes range from unicolor, to a single dark-bordered red or yellow areolar spot, to bright yellow or red lines or ocelli. Central scutes are unicolor or dark flecked, or have evident red or yellow lines. Supramarginal surfaces show same range of color and markings. Inframarginal surfaces are brown with one to three yellow bars.

There are as many as six red to orange stripes (continuous or broken) beginning on the side of the head or neck and passing around the front of the face (rostrum and jaw sheaths); at least two of these pass across the top of the snout; the longest of these stripes (often continuous) passes

FIGURE 56.2 *Rhinoclemmys p. pulcherrima* (Mazunte, Oaxaca). Top—adult female (left) and male (right); bottom left—immature individual in second or third year of fast growth typified by inflated-looking growth zones; bottom right—young adult female.

from the side of the neck over the tympanum and orbit and traces an indented outline of the dorsal aspect of the head and neck. Also, there is usually a mid-dorsal stripe that intersects one of the stripes passing over the top of the snout, forming a figure that suggests an anteriorly pointing arrow. The stripes are orange-red to red on a much darker olive to dark slate ground color.

Any one of the mentioned stripes may be broken; the entire pattern may be fragmented; or certain stripes may be missing entirely, but the remaining bits and pieces suggest the paths described previously (Figures 58.2–58.4 this work; Günther,1885; Vetter, 2005).

GEOGRAPHIC DISTRIBUTION

Maps 35 and 36

Overall: Guirocoba, southern Sonora (ca. 26°53′N, 108°41′W), thence southward along the Pacific slope and coastal plain of western Mexico, western Guatemala, El Salvador, western Honduras, and the Pacific coastal area of Nicaragua and Costa Rica to just south of San Jose. Distributions of subspecies appear in the following.

The restricted type locality, as corrected by Ernst (1978), is "vicinity of San Marcos, Guerrero" (Iverson, 1992a).

There is a shell of *R. pulcherrima* (USNM 104626) that was collected by Hobart M. Smith "near Progreso, Rio Santa Maria, Chihuahua" 11 October 1938. We discount this as an error. It probably was procured by one of his collectors ". . . in the mountains to the west" (HMS, pers. comm.). See also the account of *Terrapene nelsoni* for a related error involving Progreso.

GEOGRAPHIC VARIATION

Four subspecies have been described; three of them occur in Mexico. The overall known range in Mexico is fragmented, and subspecies are distinctly separated. Body form and color vary over this wide range. All patterns and colors become more pronounced and bright to the south and east. The most recent taxonomic revision of the genus and species is that of Ernst (1978). Neither cladistic analyses nor modern molecular phylogenetic studies have been undertaken.

Rhinoclemmys pulcherrima pulcherrima (Gray) 1855

Coastal Guerrero and Oaxaca southward to ca. 98°40′W.

Plastron has a narrow, dark central figure, much less than half the width of plastron, wider at ends, and usually divided (forked) anteriorly; a narrow, dark longitudinal stripe on bridge (between axillary and inguinal notches) separates the pale lateral field of plastron from a narrow parallel pale stripe on bridge that in turn borders the ornate inframarginal surfaces (Figure 58.4); inferior surface of each marginal has two vertical, dark-bordered, pale streaks (that may be connected ventrally to form a "U"); there is a tiny dark-edged ocellus with a pale center on the areolar area of each lateral scute. Carapace is low.

Rhinoclemmys pulcherrima incisa (Bocourt) 1868

Pacific coastal regions from extreme eastern and central Oaxaca (ca. 96°W), thence southward through Pacific regions of Guatemala, El Salvador, and Honduras to western

FIGURE 56.3 *Rhinoclemmys p. pulcherrima*. Top row and bottom left—variation in head pattern near Mazunte, Oaxaca; bottom right—head pattern in female *R. p. incisa* (El Salvador). (Photo courtesy of Kerry Matz.)

FIGURE 56.4 *Rhinoclemmys p. pulcherrima*. Plastral patterns (Mazunte, Oaxaca). Top—immature male (left) and adult male (right); bottom—immature female (left) and adult female (right).

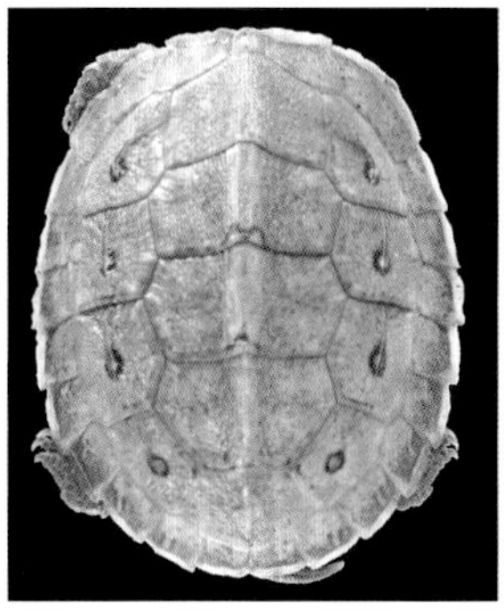

FIGURE 56.5 *Rhinoclemmys p. pulcherrima*. Dorsal and ventral views of neonate showing slight growth (Mazunte, Oaxaca). Plastral patterns are variable in the hundreds of photos provided by M. Harfush and B. Horne from the colony at Mazunte. Note projections at anterior tip of plastron.

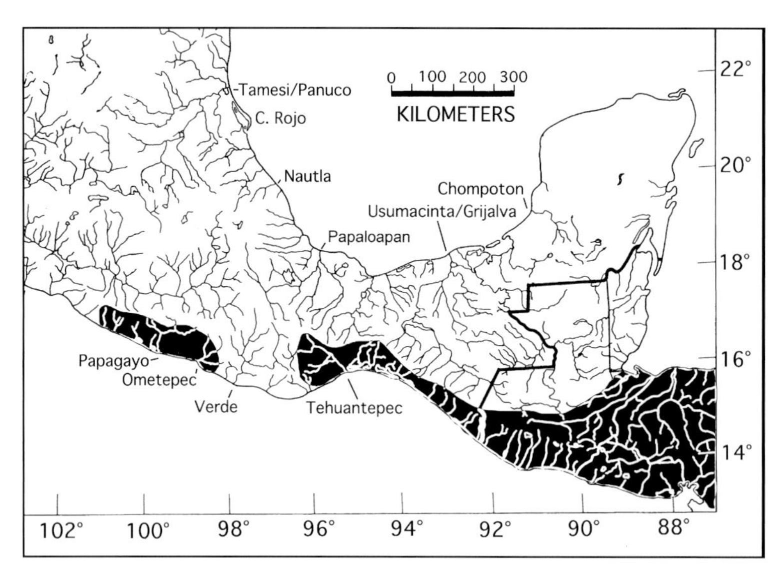

MAP 35 Geographic distribution of *Rhinoclemmys p. pulcherrima* (west) and *R. p. incisa* (east).

Nicaragua where it probably intergrades with *R. p. manni* in the region of the Nicaraguan lakes. The type locality is in El Salvador (see Iverson, 1992a:161 for details).

Central plastral figure is unbroken and wider than that of *R. p. pulcherrima* but is less than half the plastral width; bridge is solidly dark and lacks parallel dark and pale lines; inferior surface of each marginal has one vertical pale mark; there is a larger, pale, dark-edged ocellus on each lateral lamina. Carapace is high (and somewhat domed).

Rhinoclemmys pulcherrima rogerbarbouri Ernst, 1978

Southern Sonora (Guirocoba, 27°N) to Colima (ca. 104°W). The type locality is "Guirocoba, Sonora, Mexico" (see account of *Trachemys scripta hiltoni* for description of area).

Central plastral figure is unbroken and wide, and is more than half the width of the plastron; bridge is unmarked and completely dark; carapace is essentially unmarked and solidly brownish; lateral scutes are unmarked; each inframarginal surface has one vertical mark. Carapace is relatively low and wide (Figure 56.6).

RELATIONSHIPS

Rhinoclemmys pulcherrima is most closely related to *R. rubida* according to Ernst (1978), based chiefly on color patterns and external shell shape.

HABITAT

The little natural history information for Mexico is from the subspecies *pulcherrima* and *incisa*. We cover it under *Rhinoclemmys pulcherrima* sensu lato. Buskirk (2001) has provided an excellent overview of natural history information emanating from captive colonies.

Sabaneras thrive in neotropical deciduous woodlands from the coastal plain to the steep hillsides, usually near intermittent or permanent streams. Most of the coastal plain forests have now been destroyed, and the species exists mainly in the mountainous regions where local people refer to it as the "montera" (Buskirk, 2001). The northern subspecies, *rogerbarbouri*, is sometimes found in the same xeric habitats as *Gopherus agassizi*. In the dry season sabaneras favor muddy pools in shaded arroyos, often in steep-sided gullies. They are more nearly terrestrial during the rainy season and can be found in lush vegetation far from water.

DIET

Sabaneras are omnivorous and feed on a wide variety of succulent terrestrial plants, grasses, fruits, and flowers as well as invertebrates (e.g., snails) that are available to them in the natural enclosure at Mazunte, Oaxaca. No specific dietary studies have been conducted in the wild. Captives prefer fruits if there is a choice (e.g., choosing apples in lieu of carrots) and do not eat the nopal cactus favored by *R. rubida* (M. Harfush, pers. comm.).

HABITS AND BEHAVIOR

Routine activity occurs during early morning, and the turtles rest in various refugia (e.g., the shade of small trees) for the remainder of the day.

REPRODUCTION

We have no data on size at sexual maturity but estimate it at ca. 120 mm in both sexes.

Mating was most often observed at Mazunte, Oaxaca, in May and June. Vogt observed a mating in April.

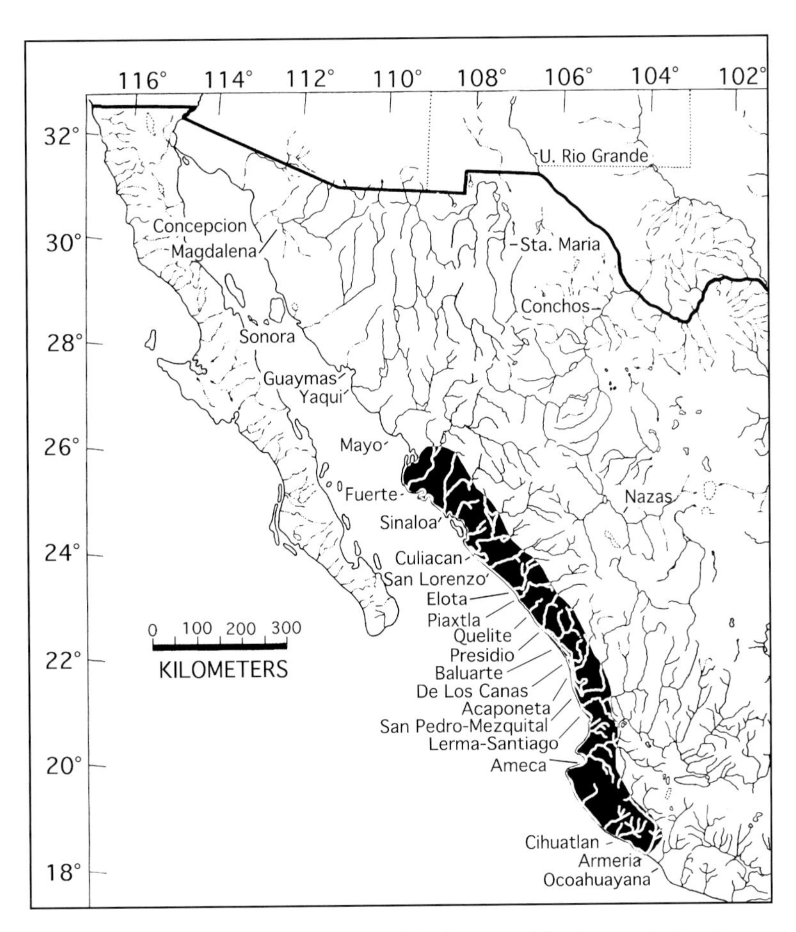

MAP 36 Geographic distribution of *Rhinoclemmys pulcherrima rogerbarbouri*.

Nesting usually occurs at night and less often in the early morning.

One of the most thorough studies of courtship behavior in any species of turtle was conducted on *R. p. incisa* by Hugo Hidalgo in El Salvador (1982), from which the following is extracted.

Courtship and mating occur during the rainy season (May to October), usually after or during rainfall. He observed 80 incomplete and 57 complete courtship sequences in an outdoor enclosure, as well as others in nature. Courtship behavior was divided into three phases: (1) male actively searches, sniffs, and trails the female and vibrates his head and neck; (2) female actively engages in nose-to-nose contact with the male and displays biting behavior; (3) male mounts female followed by coition. He found 12 cues and responses to these cues associated with courtship and mating. Both visual and chemical cues (e.g., sniffing ground while following a female) were used in specific and sex recognition.

Some of the male behavioral patterns were direct and evident (e.g., males approaching females followed by cloacal sniffing). Trailing of females sometimes lasted days or weeks (mean 19 days, range 5–27 days). When the trailing male finally finds the female stationary he places himself in front of her at a right angle and vibrates his head and neck in the vertical plane. This elicits biting by the female. After the male has been bitten repeatedly he moves quickly behind her and mounts. Tail grasping involves the male probing for the female's tail while performing lateral oscillations on stiffened hind legs, gaping, snapping, and salivating profusely. Salivation produces air bubbles on the sides of the mouth. A "meowing-like" sound is produced with an open mouth. After tail grasping the cloacae are juxtaposed and intromission occurs. While the pair is in copula, the male falls to the side of the female and begins pumping actions that last 10 minutes to 2 hours. These pumping actions may be required to keep the penis engorged and prevent disengagement.

Redmer (1987) noted that captives mated in water or on land, and that the male usually regurgitates its stomach contents while mounted.

Martha Harfush of the Centro Mexicano de La Tortuga in Mazunte, Oaxaca, has been collecting data from a well-managed captive population of *R. p. pulcherrima* since 2001. Captives are maintained in outside enclosures and are subject to natural weather conditions. The following information on reproduction is nearly all from her. The eggs may be covered with dead leaves or laid in an excavation. They are carefully positioned with a hind limb as they are laid. In the

FIGURE 56.6 *Rhinoclemmys pulcherrima rogerbarbouri* (Alamos, Sonora). Top, immature female; bottom, plastra of adult male (left) and female (right). (Photos by R. M. Winokur.)

mentioned 5-year period 102 nests were observed. Nesting usually occurs at night. The nesting season extends from the beginning of the rainy season (mid July) to late February.

Mean clutch size was 1.5 ± 0.57 (1–3) n = 102 clutches.

Overall data for size and weight of 148 eggs from 96 clutches were as follows: L, 52.5 ± 4.3 (41.6–61.7); W, 31.1 ± 2.3 (21.3–39.8); Wt, 30.6 g ± 5.95 (17.2–51.7). Among the 96 clutches, 52 had one egg, 41 had two eggs, and four had three eggs. Ewert (1979) reported *R. pulcherrima* eggs of 52 × 31 mm.

Hatching occurred from late January to late July. Mean days from laying to emergence (at environmental ambient temperatures) was 214 ± 53.6 (121–346) n = 78 clutches. These long incubation periods suggest that both embryonic diapause and aestivation occur (see general account of reproduction under "Natural History"). Ewert (1985) found an incubation period of 186–246 days under experimental conditions.

McCormick (1998) reported a captive female to have laid two eggs in June and three in September. Mean egg size was 51 × 28 mm.

Christensen (1975a,b) reported two clutches of five eggs each in September and October and two clutches of three eggs each in December and January in captives of *R. p. incisa* from Chiapas. Size range of eggs was as follows: L, 37–52 mm × W, 24–32 mm. The first hatchling emerged after 120 days of incubation at ambient temperatures of 16–29°C. Hatchlings were nearly circular in shape; CL of three hatchlings was 32, 33, and 43 mm (Figure 56.5).

GROWTH AND ONTOGENY

Data on size for *R. pulcherrima* (Harfush database; Buskirk, 2001): males, 136.75 ± 26.4 (100–160) n = 4; females, 136.70 ± 17.24 (126–162) n = 4; im. female, 145; unsexed adults, 162 ± 41.6 (105–213) n = 5; im., 92 ± 6.17 (84–97) n = 4; juveniles in first 2 months of life, 55.8 ± 10.23 (46–68) n = 5.

Ernst (1978) stated that sexual dimorphism was minimal. Adult females grow larger than males. The plastron is concave in males and slightly upturned anteriorly in females.

In some hatchlings the anterolateral angles of the first plastral scute each bear a hornlike projection (some are said to bear a third projection but this has not been observed in hundreds of photos of the Mazunte population of *R. p. pulcherrima*; Figure 56.5).

Growth is rapid in the first years of life. Buskirk (2001) reported an average increment of 31% in CL in the first 2 months after hatching for three well-fed *R. p. incisa* captives. McCormick (1998) reported longevity of 27 years in captivity.

Data on size and weight of 90 hatchlings from eggs incubated at Mazunte are as follows: CL, 49 mm ± 3.3 (41.6–55.8); CW, 44 ± 3.7 (29.5–50.7); Wt, 21.9 ± 3.2 (12.3–31.2). The relationship of hatchling weight, expressed as a percentage of egg weight, at Mazunte was as follows: 0.71 ± 0.134 (0.47–1.00) n = 57. Hatchlings reported by Ewert (1979) had a mean size of 51 mm CL and 45 mm PL.

PREDATORS, PARASITES, AND DISEASES

Indigenous people in Guerrero are known to eat this species, and the turtles are often kept as pets (Buskirk, 2001). *Rhinoclemmys p. incisa* was usually on sale at markets in El Salvador in the 1960s (JML, pers. obs.). Janzen (1980) suggested that the bright red and yellow colors on the ventral marginal surfaces perhaps mimic the color of a coral snake (*Micrurus*) and might dissuade some predators.

POPULATIONS

No life history studies have been conducted on natural populations of this species, but Buskirk (2001) reports them to be coexisting with humans in areas where traditional agriculture is practiced.

CONSERVATION AND CURRENT STATUS

See general account under Economic Uses and Conservation.

Rhinoclemmys rubida (Cope) 1869

Cabeza amarilla (local name, Mazunte, Oaxaca); Mexican spotted wood turtle (Iverson, 1992a).

FIGURE 57.1 *Rhinoclemmys r. rubida*. Adult female from vicinity of Mazunte, Oaxaca.

FIGURE 57.2 *Rhinoclemmys r. rubida*. Neonate from captive colony at Mazunte, Oaxaca. Note crease at mid body, which "unfolds" soon after hatching. (Photo by M. Harfush.)

ETYMOLOGY

L. *rubidus*, reddish; referring to the bright coloration on the head and limbs.

Gr. *peri*, around; Gr. *xanthos*, yellow; refers to the distinctively paler tan coloration of the marginals.

HISTORICAL PROLOGUE

Rhinoclemmys rubida was first collected by F. Sumichrast near Juchitan, Oaxaca, and described by Cope in 1869. Mosimann and Rabb published a description of the subspecies *Rhinoclemmys rubida perixantha* in 1953. Otherwise little is known about this turtle. It occurs within the Chamela Biological Station of UNAM and could easily be studied there. However, Vogt was unable to find the subspecies there in about 20 days of exploration over 10 years. All of the following data are for *R. r. rubida*.

DIAGNOSIS

Apex of maxillary tomium is hooked, not notched; head is marked with a broad red to yellow horseshoe-shaped mark with its blunt apex on the snout and open end posterior; there are few markings on lateral scutes; carapace is relatively low.

GENERAL DESCRIPTION

Maximum CL: males, 230 mm; females, 179 mm (Ernst, 1978).

BASIC PROPORTIONS

(M. Harfush Database)

CW/CL: males, 0.69 ± 0.07 (0.59–0.75) $n = 4$;
females, 0.74 ± 0.03 (0.72–0.78) $n = 4$;
im., 0.84 ± 0.04 (0.78–0.87) $n = 4$.

CH/CL: males, no data; females, 0.35 ± 0.05 (0.30–0.40) $n = 3$; im., 0.39 ± 0.03 (0.35–0.41) $n = 3$.

CH/CW: males, no data; females, 0.47 ± 0.08 (0.41–0.56) $n = 3$; im., 0.47 ± 0.05 (0.41–0.51) $n = 3$.

BASIC PROPORTIONS

(Ernst, 1978)

CW/CL: 0.76 (0.68–0.83).

HT/CL: 0.40 (0.31–0.50).

PL/CL: 0.95 (0.86–1.06).

BR/PL: 0.44 (0.40–0.49).

Modal plastral formula: 4 > 3 > 5 > 6 > 1 > 2.

Carapace is oval and flattened to slightly domed, widest at marginals 6–7, highest at seam between centrals 2–3, and slightly notched posteriorly. Sides of carapace are often concave at level of bridge. Posterior marginals form a serrated posterior carapacal edge. Growth rings are distinct into adulthood. Central scutes usually are wider than long, each bearing a low, blunt median keel. Precentral scute is often bifurcated posteriorly. Marginals are variously flared. There is a wide anal notch. Head is large with a well-developed snout. Maxillary tomium is weakly serrated. Length of dentary symphysis is equal to diameter of orbit. Anterior antebrachia bear 8–13 rows of enlarged scales.

COLORATION

Ground color of carapace is yellow to brown with dark seams and dark mottling on each scute; yellow spots may occur on areolae of central and lateral scutes. Supramarginal surfaces usually bear a rectangular yellow blotch; inframarginal surfaces are brown with yellow centers or yellow mottling. Plastron is yellow with a dark central figure wider than one-half the width of the plastron; bridge is solidly dark.

Color of head markings is highly variable—olive to red; there are several pale bars across snout, another between orbit and tympanum, and another from corner of mouth to tympanum; tympanum is pale spotted or mottled (Günther, 1885: plate 8c).

FIGURE 57.3 Mating pair of *Rhinoclemmys rubida rubida*. Male is "wiping" his rostrum on the chin and throat of larger female. (From a digital movie by M. Harfush, Mazunte, Oaxaca.)

FIGURE 57.4 *Rhinoclemmys rubida perixantha*. Dorsal and ventral views of adults (male on left, female on right) near Chamela, central coastal Jalisco, July. (Photos by E. Goode and M. Rodrigues, used with permission.)

Iris is yellow to red; jaws and chin are yellow with small dark markings. Color of forelimbs is yellow to red with dark spotting and vermiculation. Hind limbs are yellow on posterior surface and olive to gray anteriorly. Tail is yellow with fine dark dorsal stripes.

GEOGRAPHIC VARIATION AND DISTRIBUTION

Map 37

Two subspecies are recognized.

Rhinoclemmys rubida rubida (Cope) 1869

Semi-mesic Pacific lowlands from central Oaxaca (Totolapan) to extreme southeastern Chiapas at elevations up to 1,200 m.

Carapace is more or less uniform pale brownish above; sides of carapace are a little flared; there is a pale postorbital mark that is elongated. Carapacal scutes are uniformly pale brown; gular scute is about twice as long as humeral; there is an elongated temporal spot.

Rhinoclemmys rubida perixantha Mosimann and Rabb 1953

Central coastal Jalisco (Chamela) through Colima to central Michoacan (perhaps into Nayarit and western Oaxaca) at elevations up to 300–400 m and perhaps higher.

Periphery of carapace is distinctly paler than sides; marginals are distinctly flared; interhumeral seam is nearly as long as intergular; there is a pale postorbital mark that is oval; lateral scutes are darker than centrals or marginals; there is a short, oval temporal spot.

The geographic distance and rugged terrain that separate the two subspecies is such that current gene flow between them is unlikely.

HABITAT

Hartweg and Oliver (1940) found this species to be common on the rocky hillsides and mountains near Tehuantepec, in low areas of erosion above an alluvial plain. Schmidt and Shannon (1947) found them near a well-shaded, small stream in lowland scrub forest near Apatzingan, Michoacán. In Colima, Oliver (1937) reported the species in wooded areas near the coast. In coastal Michoacán and Jalisco they inhabit thorn scrublands and can be found in the more humid areas along arroyos. Vogt spent several days in July 1982 attempting to find them at Chamela without success.

DIET

Hartweg and Oliver (1940) found one individual (of 31 captured) to be feeding on a caterpillar. In captivity they feed on nopales (*Opuntia*; Martha Harfush, pers. comm.) and squash.

Three wild individuals (96 to 130 mm CL) in the Cuixmal-Chamela Biosphere Reserve on the Jalisco coast were found near or feeding on ripe papaya (*Carica mexicana*) lying on the forest floor in August 2001. Captive turtles from the same area fed on a variety of fruits including papaya and

FIGURE 57.5 *Rhinoclemmys rubida perixantha.* Top, head of adult male; bottom, copulating pair (near Chamela, central coastal Jalisco, July). (Photos by E. Goode and M. Rodrigues, used with permission.)

lettuce, snails, and earthworms (Alvarado-Diaz, Estrada-Virgen, Garcia-Parra, and Suazo-Ortuno, 2003). No stomachs have been flushed.

HABITS AND BEHAVIOR

Activity is mainly in early morning at Chamela (Alvaro Miranda, pers. comm.).

REPRODUCTION

The species occurs naturally in the region of Mazunte, Oaxaca, and a sizable captive population is maintained at the Centro Mexicano de la Tortuga.

Martha Harfush (pers. comm. ex Mazunte, Oaxaca) has provided us with several short sequences ("movies") made with a digital camera. These depict a pair of *R. rubida* displaying unmistakable courtship behavior in the captive compound at Mazunte. The movies showed several different pairs at various stages of mating. The following is an articulation and free interpretation of seven such sequences (by JML).

The male chases the larger female, ultimately overtakes her, and blocks her progress. He then circles the retracted female, attempting to prise up the margin of her shell with forefoot and snout. At times the female seems to be near the tipping point. Some biting at the anterior end of the female may occur. In the course of the circling, the female may be pushed as much as a meter across a flat substrate.

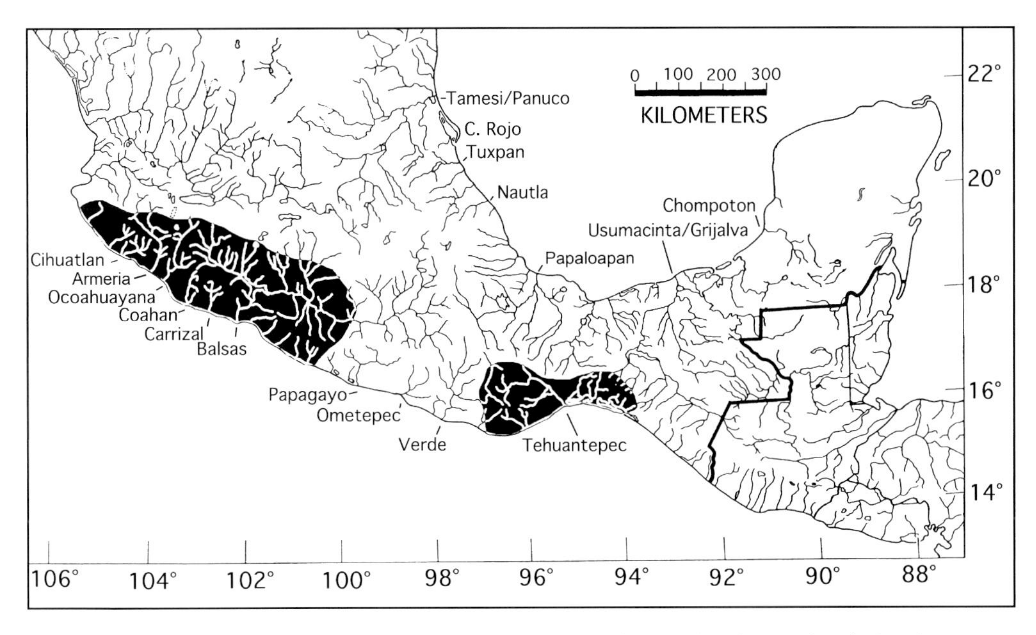

MAP 37 Geographic distribution of *Rhinoclemmys rubida rubida* (east) and *R. r. perixantha* (west).

Eventually the pair lies side by side; the male forces himself under the edge of the female's plastron and tilts it at an angle of almost 45°; the heads are in close proximity, and the head of the male is clearly positioned below that of the female. He repeatedly (15 times in 33 seconds) "wipes" the top of his rostrum across the chin and throat of the female; there is no question about the movements being deliberate (Figure 57.3). Between each "wipe" he rapidly bobs his head vertically and the sequence is then repeated. In the course of this head-to-head contact, the female repeatedly faces the male and gapes widely several times but does not bite him. At one point the female is tipped (is lying on her back) and the male is displaying interest in her cloacal region. He nuzzles and bites the rear margin of her carapace.

Ewert (1979) gave mean egg size as 62 × 25 mm and hatchling size as 52 mm CL (49 mm PL). Size and weight of two eggs from Mazunte, Oaxaca (Martha Harfush, pers. comm.) were as follows: L, 59.0 ± 10.1 (51.8–66.1); W, 28.9 ± 1.27 (28–29.8); Wt, 29.6 ± 4.1 (26.7–37.5).

Size and weight for four hatchlings from Mazunte were as follows: CL, 47.7 mm ± 2.62 (45–50.3); CW, 39.5 ± 6.7 (32.4–47.5); Wt, 22.1 ± 0.42 (45–50.3).

GROWTH AND ONTOGENY

Sizes for a small series of *R. r. rubida* (Martha Harfush database): males, 137 ± 26.4 (100–160) $n = 4$; females, 129 ± 23.1 (97–162) $n = 5$; im. females, 90 ± 6.0 (84–96) $n = 3$.

The plastron is concave in males and slightly upturned anteriorly in both sexes. Males have longer, thicker tails with the cloaca extending posterior to the carapace margin.

CONSERVATION AND CURRENT STATUS

There is a small remaining colony of reproducing adults at the Centro Mexicano de la Tortuga (Mazunte, Oaxaca) that could be used for reintroduction to protected areas if the origin of the adults is known. There is a viable natural population in the Cuixmal-Chamela Biosphere Reserve on the Jalisco coast (Álvarez-Diaz, Estrada-Virgen, Garcia-Parra, and Suazo-Ortuno, 2003). This species is not sought for human food. However, they are sought for the pet trade; 90% of the previous captive colony at Mazunte was stolen in 2005, presumably for the pet trade.

SUPERFAMILY CHELYDROIDEA

Snapping Turtles

Family Chelydridae

Family Chelydridae Gray, 1831—The relationship of American snapping turtles (chelydrids) to Asiatic big-headed turtles (*Platysternon megacephalum*) has been debated for more than a century. Most recent taxonomic works place the genera *Chelydra* and *Macrochelys* in the family Chelydridae and *Platysternon* in the Family Platisternidae (see Smith and Smith, 1979, p. 341, for discussion of family name authorship).

Agassiz (1857) placed *Chelydra, Macrochelys* ("*Gypochelys*"), and *Platysternon* in the same family on the basis of sheer external morphological similarity. This premise has been fortified in bits and pieces since that time. The only major exception (based on actual, erudite study) was that of Williams (1950), who ranked chelydrids close to kinosternids on the basis of cervical vertebral morphology. This classification was followed by some workers until the late 20th century (e.g., Pritchard, 1979). Iverson (1992a) clearly recognizes the close relationship of chelydrids to *Platysternon* but places the latter in the monotypic Family Platysternidae, contrary to the evidence presented by Whetstone (1977) and Frair (1982).

Shaffer, Meylan, and McKnight (1997) finally assembled evidence from various and extensive morphological and molecular studies to corroborate the natural relationships of platysternids and chelydrids, and suggested both be placed in the Family Chelydridae (as well as clarifying the relationships of many other groups of turtles).

After due consideration we fully recognize the close relationship but divergence of chelydrids and platysternids, leave them in their respective families, and present the following diagnosis of the Chelydridae as equivalent to that of Gaffney's (1975b) subfamily Chelydrinae.

The Family Chelydridae contains the genera *Chelydra* (snapping turtles, four terminal taxa) and *Macrochelys temmincki* (alligator snapping turtles). Only *Chelydra rossignoni* and (probably) *Chelydra serpentina* (sensu stricto) occur in Mexico.

Genus *Chelydra* Schweigger, 1812

Snapping turtles (Iverson, 1992a).

ETYMOLOGY

Gr. *chelydros*, a water serpent.

DIAGNOSIS OF GENUS *CHELYDRA*

Chelydrid turtles have the following combination of characters:

Posterior margin of carapace is serrated. Plastron is cruciform and connected ligamentously to carapace by narrow bridge; there are no plastral buttresses; five pairs of plastral scutes meet on midline; fourth plastral scutes are displaced laterally from midline and lie on the bridges; entoplastron is T-shaped. Plastron has median fontanels that may persist in adults. Adult tail is as long as or longer than plastron at all ontogenetic stages and bears a mid-dorsal serrated crest of enlarged scales.

Proximal rib ends are bowed ventrally, creating a commodious epaxial space; nuchal costiform processes are well developed, long, and slender, extending to P3 and reaching or underlapping the second trunk rib (or nearly so). There are two suprapygals; inframarginal are scutes present; supramarginal scutes are absent; there are 11 peripheral bones, 12 marginal scutes, and 8 neurals, quadrilateral to octagonal. Phalangeal formula is 2 – 3 – 3 – 3 – 3.

There is a hooked premaxillary beak; tomial edges of jaws form efficient shearing mechanism (Figure 2.7); alveolar surfaces are narrow and lack ridges. Head is wide. Temporal and postorbital emargination is relatively shallow; there is virtually no secondary palate; quadrate completely encloses stapes; there is a relatively small jugal; frontals usually do not enter orbit (but see species account); there is no parietosquamosal contact; maxilla does not contact quadratojugal; pterygoid lacks a ventral, posteriorly directed ridge; vomer has median ridge; orbits are dorsolaterally oriented; centrum of eighth cervical vertebra is biconvex.

Top of head is covered with well-defined, juxtaposed scales; scales on limbs are not enlarged or osteodermal; limbs are heavily developed: feet are webbed and heavily clawed. There is one or more pairs of gular barbels; body skin (especially head and neck) is tuberculate to nearly villose, never smooth. Cloacal bursae are present. There is a linear series of three musk glands, two inframarginal and one axillary, each gland with one duct and one orifice; inframarginal ducts are short and pass through fibrous connective tissue in carapacoplastral articulation.

Diploid chromosome number is 52 with 28 microchromosomes and 24 macrochromosomes.

DIAGNOSIS

(Based on Smith and Smith, 1979; Gaffney, 1975b; Gaffney and Meylan, 1988; Ernst and Barbour, 1972; Ernst, Gibbons, and Novak, 1988; Ernst, Lovich, and Barbour, 1994; Feuer, 1966; Waagen, 1972; Williams, 1950; Stock, 1972; Bickham and Baker, 1976a; Killebrew, 1977b; Haiduk and Bickham, 1982.)

The practical identification of *Chelydra* is simple, absolute, and not at all confusing. It is the only taxon in Mexico that has a cruciform plastron and a prehensile tail that is at least as long as the plastron at all ages, and bears a mid dorsal series of large keel-like scales. Adults are usually large and muscular, and can reach weights of at least 20 kg.

GENERAL DESCRIPTION

The head-to-tail profile of a *Chelydra* on land and at rest is an impressive sight (if one is lucky enough to see one at rest), dominated by the very long tail and a relatively scanty shell coverage that reveals muscles of the shoulder and hip not evident in chelonians with extensive marginal overhang—suggestive of a weight lifter in a singlet (Figure 58.1). Superficial similarities to *Proganochelys* (Figure 1.1) are impressive.

There are four distinguishable taxa within the genus *Chelydra*. They have chiefly allopatric distribution from Canada to Ecuador and are variously referred to as subspecies (Feuer, 1966; Gibbons, Novak, and Ernst, 1988) or species (Babcock, 1932; Cope, 1872; Stejneger, 1918). Subspecies rank is based chiefly on intermediacy and probable intergradation between the taxa *osceola* and *serpentina* in northern Florida (Carr, 1952; Feuer, 1971). *Chelydra serpentina* (sensu lato) is separated by a wide gap (which includes most of Mexico) from the two Mesoamerican species.

KEY TO THE TAXA OF *CHELYDRA*

Based on Smith and Smith (1979) and Feuer (1966).

1A. Length of anterior plastral lobe (hyo-hypoplastral suture to anterior tip) greater than 40% of CL (0.42 ± 0.005); width of bridge greater than 8% of CL (0.082 ± 0.003); angle formed by gulars is greater than 130° of (140 ± 2.9) . . . . *Chelydra acutirostris* and *Chelydra rossignoni*, 2

1B. Length of anterior plastral lobe less than 40% of CL (0.38 ± 0.002); width of bridge less than 8% of CL (0.075–0.002); angle of gulars is less than 130° (except in some juveniles) (120° ± 1.5) . . . . *C. osceola* and *C. serpentina*, 3

2A. Width of third central scute greater than 25% maximum CW; neck villose in appearance, ornamented with long, flat, pointed cutaneous appendages; occurs from Veracruz to northern Honduras . . . . *C. rossignoni*

2B. Width of third central scute less than 25% maximum CW; neck tubercular or rough in appearance, ornamented with low, rounded warts; occurs from Honduras to Ecuador. . . . . *C. acutirostris*

3A. Width of third central scute greater than 31% maximum CW; neck villose in appearance in most (80%) specimens; ornamented with long, flat, pointed cutaneous appendages; more than 10 (13.9 ± 1.2) scales between postocular scales and soft skin of neck; occurs in southern Florida . . . . *C. osceola*

3B. Width of third central scute less than 31% maximum CW; neck tubercular or rough in appearance, ornamented with low, round warts; 10 or fewer scales (7.4 ± 1.7) between postocular scales and soft skin of neck; occurs in Canada and the United States . . . . *C. serpentina*

Chelydra rossignoni (Bocourt) 1868

Servengue (Veracruz); chiquiguao (Tabasco and Chiapas); Central American snapping turtle (Iverson, 1992a).

ETYMOLOGY

The specific name is a patronym honoring Jule Rossignon, who was a correspondent of the French Commission Scientifique au Mexique, a classic publication on the reptiles

FIGURE 58.1 A rare view of a relaxed snapping turtle (adult *C. serpentina*) stretched full length, emphasizing characters mentioned in description and the etymology of the specific epithet. (From personal papers of Dr. L. H. Stejneger, with permission.)

and amphibians of Mexico and Central America, published by Duméril, Bocourt, Mocquard, and Brocchi (1870–1890) was based on these collections.

HISTORICAL PROLOGUE

Testudo serpentina was described by Linnaeus in 1758. Schweigger (1812) was the first to use the combination *Chelydra serpentina*. Bocourt described *Emysaurus rossignoni* in 1868.

FIGURE 58.2 A large (CL 344 mm) adult male *Chelydra rossignoni* held by Austin (left) and Edward Legler (UU 9451, near Alvarado, Veracruz). The "shaggy" epidermal texture of the head and neck and the long barbels can be seen.

Chelydra is one of the most widely distributed freshwater turtle genera, ranging from Canada to Ecuador. Until recently, the genus was considered monotypic—*Chelydra serpentina* with four subspecies (Feuer, 1966; Iverson, 1992a). Phillips et al. (1996) presented convincing biochemical evidence that the two allopatric Mesoamerican taxa (*acutirostris* and *rossignoni*) were genetically distinct enough to be regarded as species.

In the present work, we consider the four recognized taxa of *Chelydra* to represent three species, as follows: *Chelydra serpentina* (to include *osceola*), *C. acutirostris*, and *C. rossignoni*. This classification is espoused by Phillips, Dimmick, and Carr (1996); Bickham et al. (2007); and Fritz and Havas (2007). Earlier authors have treated these taxa as subspecies of *C. serpentina* (Smith and Smith, 1979; Iverson, 1992a; Feuer, 1966) or considered them barely worthy of recognition (Carr, 1952; Medem, 1977; Frair, 1982).

DIAGNOSIS AND GENERAL DESCRIPTION

Since both species of Mesoamerican *Chelydra* are poorly known, our illustrations include a comparison of their diagnostic characters.

Chelydra rossignoni rivals *Dermatemys mawi* as the largest freshwater turtle in Mexico, reaching a CL of at least 470 mm and a weight of 20 kg (Lee, 1996). The combination of a cruciform plastron and a tail as long as or longer than the plastron distinguishes *Chelydra* from any other turtle in Mexico. *Chelydra rossignoni* can be distinguished from both *C. acutirostris* and *C. s. serpentina* by the integumentary appendages on its head and neck. In general these appendages are short, blunt, and conical or cylindrical in *acutirostris* and *s. serpentina*. In *rossignoni* they are longer, flatter, and flaplike. The following description is based on specimens from Veracruz.

Rims of eyelids are adorned with flattened, rounded, scalloplike tubercular scales. There are one to three pairs of gular barbels of varying length. The longest barbels usually are equal to the diameter of the orbit, flattened, and more nearly straplike than cylindrical or conical.

Neck is ornamented with distinctive, flattened, triangular, flaplike appendages (the "tubercles" of other authors). Each flap bears one or two slender, long, barbel-like extensions at its distal apex; flaps are arranged in irregular longitudinal rows; in a large adult (UU 9451, CL 345 mm),

FIGURE 58.3 Dorsal and ventral views of a hatchling *C. rossignoni* (MCZ-R-29098, Tela, Honduras). (From personal papers of Dr. L. H. Stejneger, with permission.)

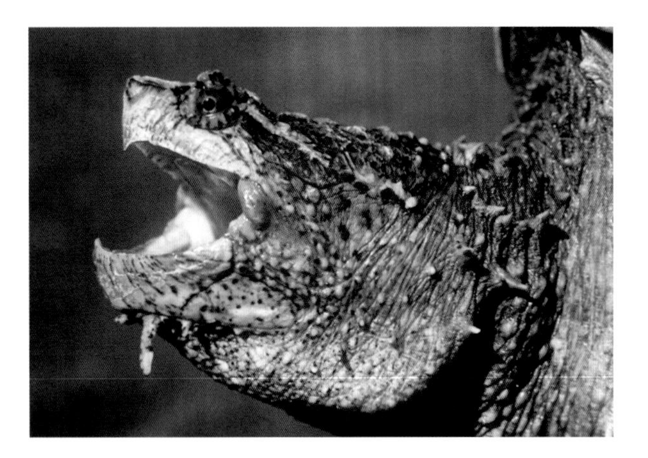

FIGURE 58.4 Immature *C. rossignoni* showing head pattern and long barbels.

a large single flap is an equilateral triangle of 13 mm per side and 1 to 2 mm thick; the flaps are thin enough to be moved by slight fluid currents; their function is unknown. Appendages on the ventral surfaces of limbs may also be flaplike and triangular but are smaller than those on head and neck. The total effect of the integumentary appendages, at least on the neck, creates the "villous," "papillose," or "fuzzy" image alluded to by other authors.

By contrast, *C. acutirostris* and *C. s. serpentina* lack the specialized eyelid scales and have barbels in the form of narrow, elongated cones (shorter than orbital diameter) and neck ornamentation consisting of smaller conical tubercles. Triangular appendages, if present, are much smaller, less membranous, and tipped by shorter tubercles, creating an overall visage that is not "papillose."

Skulls of the three species are shown in Figure 58.8. Most of the characters discussed by Feuer (1966) are evident. Feuer found only one character that was unique to *rossignoni*: the frontal bones enter the orbital rim on one or both sides.

In profile, the dorsal surface of the skull is depressed in the frontoparietal region, accentuating the supraorbital ridges; the temporal roof is moderately eroded ventrally; and the supraoccipital process is shallow (*rossignoni* and *serpentina*). *Acutirostris* has a virtually planar dorsal surface in profile, a deep supraoccipital process, and little ventral erosion of the temporal roofing. *Chelydra rossignoni* has the most dorsally placed orbits.

BASIC PROPORTIONS FOR ADULTS LARGER THAN 200 MM

Used with permission from R. C. Feuer database.

CW/CL: *serpentina* 0.92 ± 0.07 (0.81–1.03) *n* = 18;
rossignoni 0.85 ± 0.02 (0.81–0.89) *n* = 12;
acutirostris 0.83 ± 0.04 (0.75–0.89) *n* = 19.

HT/CL: *serpentina* 0.51 ± 0.02 (0.48–0.55) *n* = 5;
rossignoni 0.42 ± 0.04 (0.37–0.49) *n* = 9;
acutirostris 0.37 ± 0.02 (0.34–0.41) *n* = 8.

HT/CW: *serpentina* 0.51 ± 0.02 (0.48–0.55) *n* = 5;
rossignoni 0.49 ± 0.05 (0.44–0.57) *n* = 9;
acutirostris 0.44 ± 0.04 (0.39–0.51) *n* = 11.

PL/CL: *serpentina* 0.74 ± 0.03 (0.69–0.78) *n* = 6;
rossignoni 0.78 ± 0.02 (0.75–0.83) *n* = 20;
acutirostris 0.77 ± 0.04 (0.70–0.87) *n* = 31.

WH/CL: *serpentina* 0.24 ± 0.01 (0.22–0.26) *n* = 16;
rossignoni 0.26 ± 0.02 (0.23–0.29) *n* = 11;
acutirostris 0.25 ± 0.02 (0.20–0.28) *n* = 17.

WH/CW: *serpentina* 0.26 ± 0.03 (0.22–0.31) *n* = 16;
rossignoni 0.30 ± 0.02 (0.28–0.33) *n* = 10;
acutirostris 0.30 ± 0.02 (0.26–0.33) *n* = 17.

These small samples show *rossignoni* to be intermediate between *serpentina* and *acutirostris* in shell proportions. It is evident to us (but not so far statistically supported) that *rossignoni* has a relatively wider head than either of the other species. The iris is tan to pale orange and has four equally spaced black blotches creating a stellate pattern; the blotches are similar to others on the eyelids and other circumorbital tissue, and these patterns tend to camouflage the opened eye. The carapace is dark brown, black, or some nuance of a neutral grayish color and unmarked; the plastron is also unmarked and varies from straw to horn color. The head and limbs are brownish gray above and paler (cream to yellow) below. The hatchling carapace is a uniform dark slate, but the plastron and inframarginal surfaces are jet black with irregular stark-white speckles. The jaws are usually streaked with black, and various arrangements of dark marks and pale ground colors may form irregular patterns.

Most *Chelydra* have a pale, poorly defined stripe on the side of the head, passing from side of rostrum through eye to temporal region. The limbs are about the same color as the plastron below and the carapace above. Most patterns and nuances of color are lost in the largest individuals, which may become solidly dark. In an anterior view of the

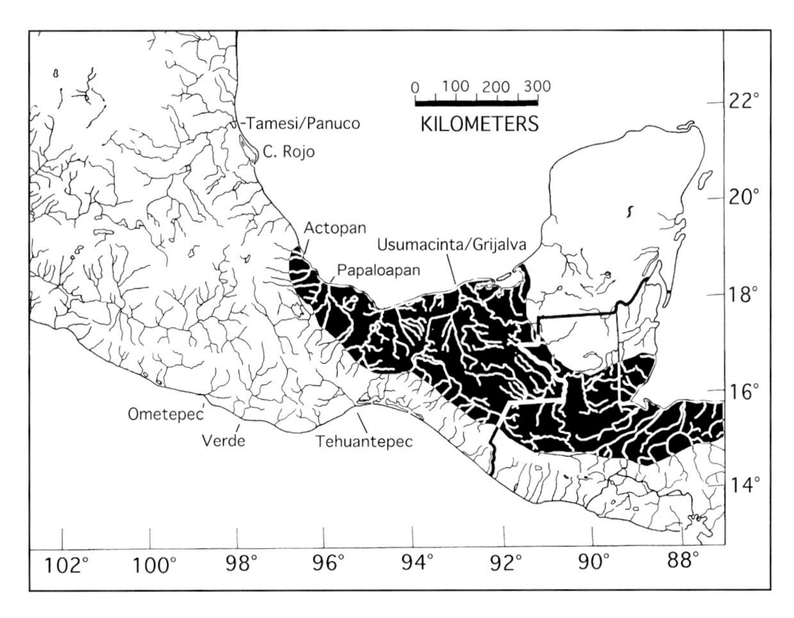

MAP 38 Geographic distribution of *Chelydra rossignoni*.

head, the combination of supraorbital ridges, the recessed frontoparietal region, and the protruding eyelid scales create the appearance of a small, slightly concave deck over the orbits.

In the following accounts of life history data, it is clear that *Chelydra rossignoni* is not a common animal in the neotropics (and that we do not fully understand this circumstance).

FOSSILS

No fossils have been reported from Mexico. The oldest fossils of *Chelydra serpentina* (sensu lato) are from the Pliocene Hemphillian and Blancan of Kansas and the Blancan of Nebraska. Pleistocene records are from Irvingtonian deposits in Kansas and Maryland, and Rancholabrean sites in many states from Nevada to Pennsylvania (reviewed by Ernst, Lovich, and Barbour, 1994).

GEOGRAPHIC DISTRIBUTION

Map 38

The Gulf Coastal plain from central Veracruz (ca. 19°32′N) and adjacent Oaxaca southeastward across the base of the Yucatán Peninsula in Tabasco, Campeche, and Chiapas to northern Guatemala, southern Belize, and extreme northwestern Honduras (Tela, 87°27′W; Lee, 1996; Smith and Smith, 1979; Iverson, 1992a). The species is absent from the internal drainages of the Yucatán Peninsula. Few specimens exist in collections.

There are two or three syntypes in the National Museum of Natural History, Paris (MNHN; Bocourt, 1868; Medem, 1977). Stuart (1963) nominated MNHN 2130 as the lectotype and restricted the stated type locality ("des marais de Pansos, pres le Rio Polochic (Guatemala)") to "Panzos, near the Rio Polochic, Guatemala" (Iverson, 1992a).

Of the many localities listed for Veracruz and Oaxaca by Pérez-Higareda (1978), most lack authentication.

PHYLOGENY AND ZOOGEOGRAPHY

Feuer (1966) studied morphological variation in *Chelydra serpentina* (sensu lato) that considered nearly all available museum specimens (1398), including 19 *C. rossignoni* and 20 *C. acutirostris*, which were identified to sex but, though they ranged from hatchlings to large adults, were all considered in the same analyses without regard to ontogenetic allometry.

Feuer recognized the usual four terminal taxa (*serpentina*, *osceola*, *rossignoni*, and *acutirostris*). He based his conclusions on relatedness chiefly on the morphology of integumentary appendages of the head and neck (e.g., barbels and tubercles). He ranked *serpentina/acutirostris* and *rossignoni/osceola* as closely related pairs, despite their distant geographic proximity.

Chelydra is an "old northern" faunal element that reached South America from North America by migrating through Mesoamerica in Tertiary times. Ancestral *Chelydra* sp. was widely distributed east of the Rocky Mountains by at least Pleistocene times, essentially within most of its present range.

The following hypothetical scenario is adapted in part from Feuer (1966). The ancestral stock of *Chelydra* existed in North America in late Miocene times. Its distribution, at a time of low ocean levels, included a broad, continuous Gulf Coastal plain that reached the lowlands at the base of the Yucatán Peninsula. These were contiguous with those of the (less extensive) lowlands of Belize, Guatemala, and Honduras. This ancestral stock crossed the Nicaraguan portal in its late archipelagic stages (see Introduction for zoogeographic details).

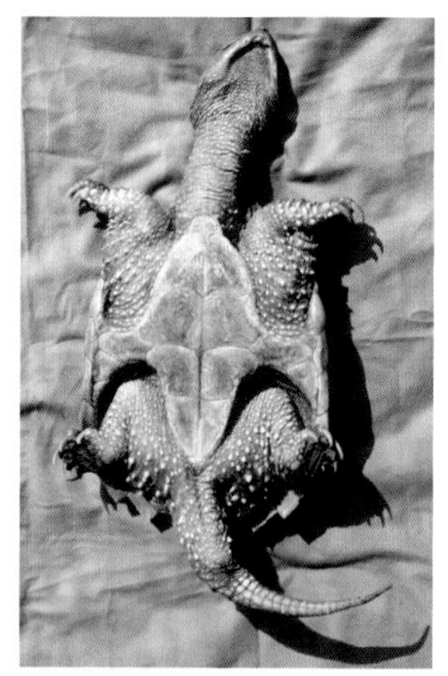

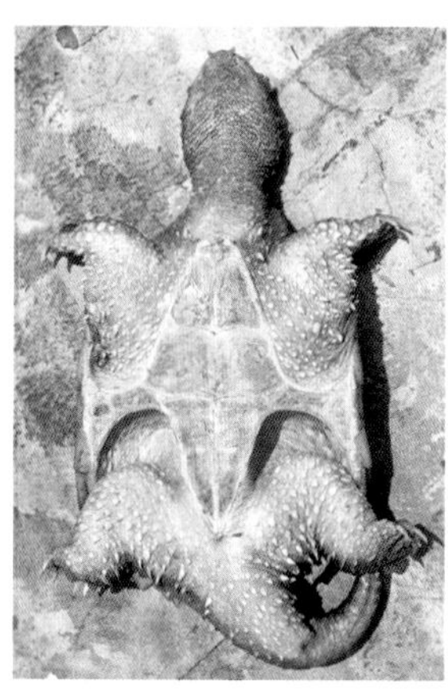

FIGURE 58.5 Dorsal and ventral views of adult male neotropical *Chelydra*. Top—*C. acutirostris*, UU 5055, Puerto Cabezas, Nicaragua; bottom—*C. rossignoni*, CL 344 mm, UU 9451, Alvarado, Veracruz.

The stock on the southeastern side of the portal became *C. acutirostris* and reached Colombia and Ecuador with the close of the Panamanian and Colombian portals.

The stock to the north of the Nicaraguan portal became *C. rossignoni* and was further isolated by the Tehuantepec portal. When the Tehuantepec portal closed, *C. rossignoni* extended its range northeastward along the Gulf Coast of Mexico approximately to Punta del Moro, Veracruz, achieving its present range. *Chelydra* seems never to have reinvaded the coastal plain north of Pta. del Moro.

Chelydra s. osceola in lower peninsular Florida could be a terminal relict of the old *rossignoni* stock (Feuer, 1966), or it could have evolved from a northern population of *C. serpentina* when the lower end of the peninsula was isolated during a Pleistocene interglacial period (Auffenberg, 1958; see Feuer [1966] for details). *C. s. serpentina* intergrades with *C. s. osceola* in central Florida (Feuer, 1971).

The extensive distribution of *Chelydra s. serpentina* may be approximately the same as it was in late Miocene times, except for its seeming absence on the Gulf Coast (from just north of Brownsville to Pta. del Moro).

Feuer (1966) discussed possibly ancestral characters and found them to exist in *C. s. serpentina* and *C. acutirostris*.

Mitochondrial DNA analysis (Phillips, Dimmick, and Carr, 1996) suggests an *acutirostris/rossignoni* pair and an *osceola/serpentina* pair, the pairs substantially distinct from one another. Based on an mtDNA divergence time scale of 0.2–0.4% per million years (Bowen et al., 1979; Avise, Bowen, Lamb, Meylan, and Bermingham, 1992), the time since *acutirostris* and *rossignoni* diverged would be 4.25–8.5 million years and ca. 11–22 million years for the North American and Mesoamerican pairs.

The current ranges of *rossignoni* and *acutirostris* roughly bracket the old Nicaraguan portal. The southernmost records of *C. rossignoni* are from narrow, discontinuous coastal plains near the mouth of the Rio Ulua, Honduras. One of the specimens (UU) is from a freshwater estuary, partially covered with hyacinth, its mouth blocked by a sandbar (Estero Prieta, Masca, Cortes Prov. Honduras; JML, pers. obs.).

Eastward from Tela (the southeasternmost record of *C. rossignoni*) the coast is punctuated by headlands. The few records from farther east in Honduras and in southern Nicaragua are from mountainous regions where large rivers (e.g., Coco, Paulaya, and Patuca) have cut back 150–300 km from the coast. The coastal plain in eastern Nicaragua is virtually continuous into Panama. The gap between the ranges of *rossignoni* and *acutirostris* remains undocumented.

HABITAT

The abundance of *Chelydra serpentina* in the north (United States) contrasts sharply with the paucity of neotropical *C. rossignoni*. Vogt found only 12 *rossignoni* in 30 years of work in neotropical Mexico, and there are only six specimens in the UU collections (vs. 34 specimens of *C. acutirostris*).

In Veracruz and Chiapas snapping turtles occur mainly in the marshy borders of shallow lakes and associated slow, meandering, muddy streams choked with water hyacinth (e.g., Laguna Oaxaca, the marshlands of Lerdo de Tejada and Tlacotalpan, and marshy lakes near Cosamaloapan). They seem not to utilize the deep portions of lakes or inhabit large rivers. Nowhere has the species been found in abundance (based on observations, trapping, and information from fishermen). Records from the volcanic Lago de Catemaco (Pérez-Higareda, 1978) are unsubstantiated by specimens (and unknown to fishermen).

The species was once common in the marshes near Villahermosa, Minatitlan, and Coatzacoalcos—wetlands that are now virtually devoid of wildlife following petrochemical pollution. A viable population remains in a reserve in the Rio Playa near Comalcalco, Tabasco.

Many of the *Chelydra* obtained by Legler in Mesoamerica (chiefly *acutirostris*) were taken at night in standard baited hoop nets from small, clear streams in forested habitat where no turtles were visible, and there seemed little hope of catching large snappers. It is therefore evident that the turtles were able to conceal themselves in some manner during the day and then forage at night.

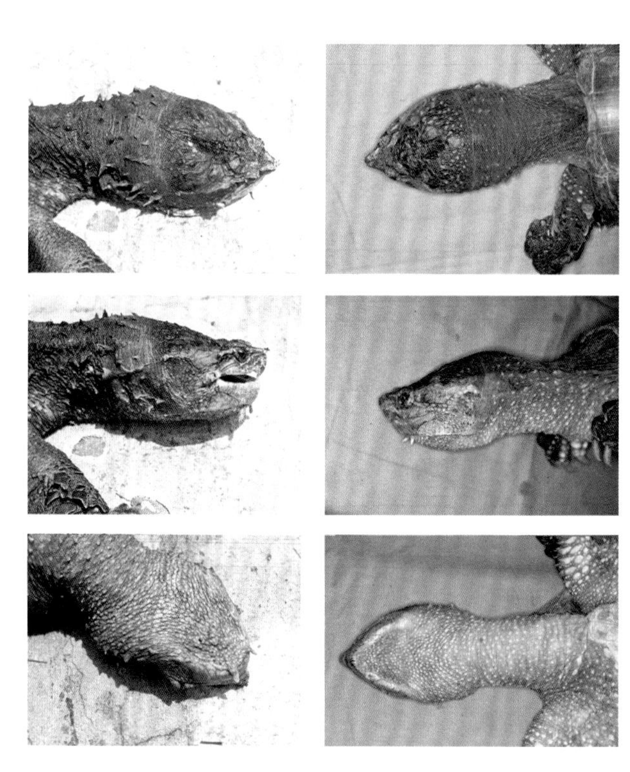

FIGURE 58.6 Dorsal, lateral, and ventral views of head and neck in adult male neotropical *Chelydra*. Left—*C. rossignoni*, CL 344 mm, UU 9451, Alvarado, Veracruz; right—*C. acutirostris*, UU 5055, Puerto Cabezas, Nicaragua. Differences in epidermal specializations of head and neck are clearly shown.

DIET

For snapping turtles generally, the shearing jaw armature (Figure 2.7), the rapid and accurate strike, and general pugnacity suggest a carnivorous diet. It was thus an interesting paradox when Lagler (1943) actually examined gut contents and reported that *C. s. serpentina* was actually an omnivore with gut contents showing about equal amounts of plant and animal matter (by volume) and a preponderance of plant material by frequency. Feuer (1966) examined gut contents of *acutirostris* and *rossignoni* obtained by Legler in Panama, Costa Rica, and Honduras; they contained 85–100% fruit and other plant material.

Stomach contents of *rossignoni* in Veracruz suggest an opportunistic diet including small invertebrates (with a preponderance of crustaceans), fruits, seeds, and aquatic vegetation such as duckweed (*Lemna*). Álvarez del Toro (1974) reported feeding on hatchling crocodiles in Chiapas; younger individuals feed on duckweed, aquatic invertebrates, and terrestrial invertebrates that fall into the water, as well as small fish and carrion. Carnivorous feeding was observed in captives at Los Tuxtlas by RCV. A turtle usually sits and waits for prey to move within range. As a prey organism approaches, the turtle stalks it with outstretched neck; when in striking range, it forms a stable three-point stance with the hind limbs and base of tail (Figure 58.7), retracts the head, and strikes accurately. Small prey is ingested in the gape and suck manner (see Introduction). Larger prey is caught in the jaws and torn apart with the forelimbs. The maxillary beak is an effective holdfast structure.

HABITS AND BEHAVIOR

Aerial basking is uncommon. Ruthven (1912) collected a hatchling basking on a rock in the center of the Rio Hueyapan, near San Juan, Veracruz. The species is secretive and nocturnal and rarely leaves the water except to nest or aestivate. RCV found a young individual (CL 200 mm) aestivating in leaf litter near Lerdo de Tejada in May. The aestivation site was shared with other immature turtles (*Staurotypus triporcatus*, *Claudius angustatus*, and *Trachemys scripta venusta*). Open water was available about 4 m away (the Rio San Augustin, a permanent stream). The aestivating turtle had seemingly migrated from an adjacent dry marsh. Captive individuals (Nacajuca, Tabasco) aestivated during the dry season (May to July) by burying themselves in the mud bottoms of their enclosures until summer rains raised the water levels and reduced water temperature in July and August.

Chelydra rossignoni is known for its aggressive, pugnacious temperament. When confronted they will strike out rapidly and bite anything in close proximity. Even hatchlings will attempt to bite.

All species of *Chelydra* have a reputation for being vicious, irascible, and even dangerous. These notions come mostly from newspapers and popular books and, if firsthand, are based entirely on snappers encountered out of water. Under these conditions they behave defensively with widely gaping maw, aimed at and tracking the source of annoyance, followed by repeated strikes. They hold on tenaciously to anything they can bite. Whether or not this behavior serves to intimidate predators is moot. Certainly they could inflict serious lacerations, but they do not sever broomsticks and canoe paddles or amputate toes. We know of no instances in which humans have been bitten under the water by snappers. RCV, at the age of 12, was standing on what felt like a rock in the muddy bottom of the Yahara River in Madison, Wisconsin, and when he reached down to feel the rock he found it to be a 10 kg snapping turtle; it did not attempt to bite underwater. Legler's limited diving experience with *Chelydra* sp. suggests that snappers behave underwater like most other turtles and avoid humans by moving off at speed. The safe way to hold any snapper is by the tail with the dorsal surface facing away and well away from one's body.

REPRODUCTION

The sparse reproductive data presented here were collected over a 20-year period from females (principally from southern Veracruz) that ranged in size from 260 to 330 mm CL and 2,545–7,500 g in weight. Slightly more is known about *acutirostris* (Acuña Mesén, 1993), and there is a vast

FIGURE 58.7 Adult male *Chelydra rossignoni* poised for strike in "three point stance." The same stance is used in underwater feeding.

amount of information on *C. s. serpentina* in the northern temperate region (summarized by Ernst, Lovich, and Barbour, 1994). In the following account, information from these closely related species is used for comparison, but mostly to guess, by extrapolation, at what we do not know about *rossignoni*.

Mating has not been studied in *rossignoni*. Courtship and mating of captive *C. serpentina* in Kansas was described by Taylor (1933) and Legler (1955) and reviewed by Ernst, Lovich, and Barbour (1994).

Vitellogenesis has not been studied. Examination of gravid *rossignoni* indicates that vitellogenesis is finished by March. Our limited observations and dissections suggest that *rossignoni* produces a single clutch of many eggs each year. Size and weight of eight adult females was 304 mm ± 25.6 (260–330) and 5,356.1 g ± 1,774.2 (2,545–7,500). Six of these were gravid, obtained at the Lerdo de Tejada site in March 1984 and March 1985. Ovulation was induced with oxytocin. All bore one set of corpora lutea and no enlarged follicles, suggesting that they lay only one clutch per season.

Reproductive data on three additional females were not highly different, except that one of the captives at Los Tuxtlas laid 52 eggs on 22 December 1982. This early date was perhaps due to captive conditions. The eggs were infertile.

Another female from Lerdo de Tejada was induced (with oxytocin) to lay 15 eggs on 7 April 1988 and then laid another 15 eggs on 26 April. Her ovaries bore two sets of corpora lutea but no enlarged follicles, suggesting that these were two distinct clutches. This is the only evidence that might suggest the production of multiple clutches.

Mean number of eggs per clutch observed by Vogt in Veracruz was 38 ± 10.8 (26–55) n = 6 clutches. Álvarez del Toro (1982) reported 20–30 eggs from April to July in Chiapas. The eggs are hard shelled and spherical, not unlike a ping-pong ball in size and texture. Size and weight for 227 eggs are as follows: diameter, 29.7 ± 1.02 (28.4–30.9); Wt, 14.4 g ± 1.68 (12.1–16.6). These females were fairly small, and larger females might be expected to produce more eggs per clutch. Vogt (1981a) has reported 10–96 eggs per clutch for *Chelydra serpentina* in Wisconsin.

The nesting season in nature is February through May. An attempted nesting was observed on the banks of the Rio Lacantún at Playa de Galaxia on 22 February 1992 at 0900 hours (Vogt and R. Mittermeier, pers. obs.). In Costa Rica *Chelydra acutirostris* nests from February to July, but nesting peaks in June (Acuña Mesén, 1993).

Incubation time at the same temperatures is longer in *rossignoni* than for *serpentina* in Wisconsin: 103 days at 25°C and 75–84 days at 29°C versus 60 days at 25°C and 45 days at 29°C, respectively.

Development is direct, lacking embryonic diapause or aestivation within the egg. Sex determination is temperature controlled in *rossignoni*. Mostly males are produced at 25°C and mostly females above 28°C. Threshold temperature is 26.5°C, two degrees lower than that for *C. serpentina* (Vogt and Flores-Villela, 1992a). Shorter incubation times in the north would favor hatching prior to winter freezing.

GROWTH AND ONTOGENY

Males of *C. serpentina* grow larger than females and have a distinctly longer pre-anal portion of the tail. Nuances of sexual difference exist in various proportions, but there seem to be no other trenchant sexually dimorphic characters (Ernst, Lovich, and Barbour, 1994).

Overall, in the northern temperate region, minimum size at sexual maturity for *serpentina* ranges from 186 to 208 mm CL (Vogt, 1981; Hammer, 1996; Galbraith, Brooks, and Obbard, 1989). In Michigan sexual maturity is usually reached in 11–12 years but ranges from 6 to 17 years (Congdon, Breitenbach, Van Loben Sels, and Tinkle, 1987).

Size and weight of 62 hatchlings from Veracruz (incubated at 25–29°C) are as follows: CL, 31.8 mm ± 3.04 (26.0–36.6); Wt, 31.8 g ± 1.35 (7.2–15.2). Ewert (1979) gives the mean CL of four hatchlings as 37.7 mm.

The adult shell coloration begins to form in *rossignoni* at 2–3 years. The carapace of hatchlings has three weak keels that are lost with growth. The tail is much longer than the carapace in hatchlings (Figure 58.3).

Longevity of 23 years in captivity has been reported for *serpentina* (Bowler, 1977).

For a closely related group ranging over ca. 51° of latitude, one might expect reproductive differences and trends in *Chelydra* such as those reported by Moll and Legler (1971) for *Trachemys scripta*.

PREDATORS, PARASITES, AND DISEASES

Small individuals are known to be eaten by *Staurotypus triporcatus*, *Crocodylus moreleti*, otters (*Lutra canadensis*),

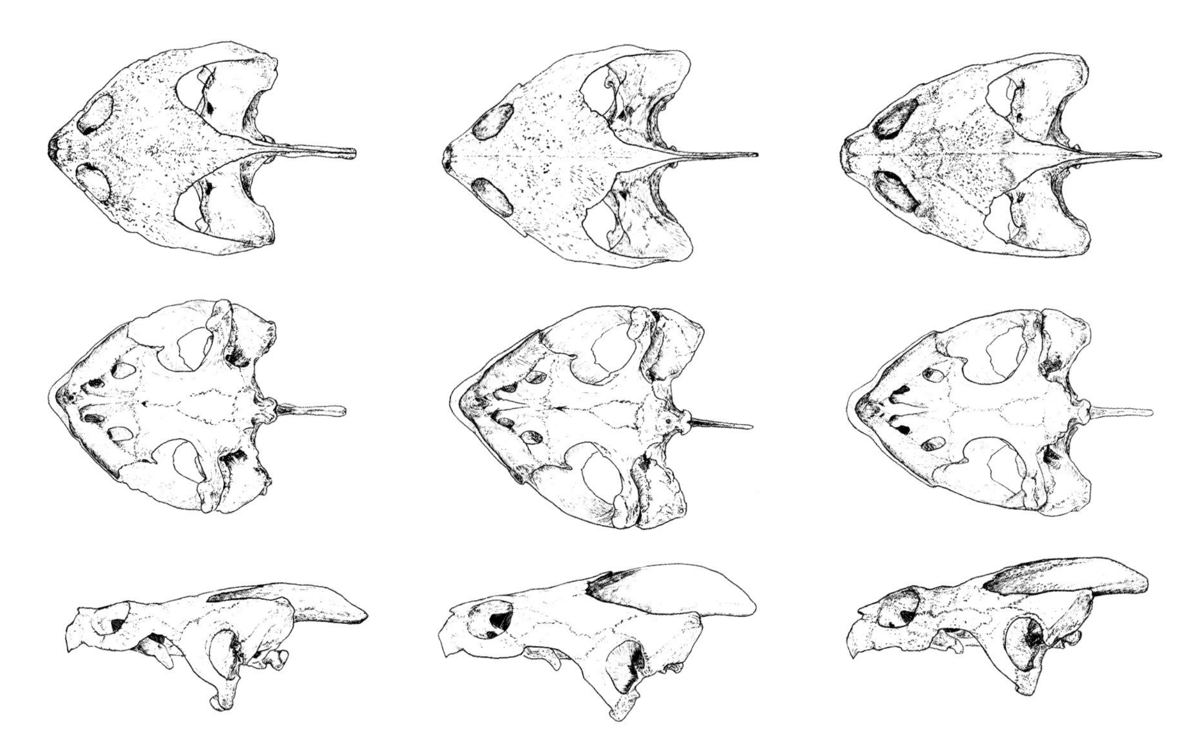

FIGURE 58.8 Comparison of adult *Chelydra* skulls. From left to right by column: *C. rossignoni*, CNHM 5322, CBL 86, sex unknown, San Pedro, Guatemala; *C. acutirostris*, UU 3948, CBL 86, female, Almirante, Panama; *C. serpentina*, UU 3935, CBL 83, female, New York. Skulls presented at same size for comparison. (Drawings by Elizabeth Lane.)

and herons (Álvarez del Toro, 1974, 1982). Ruthven (1912) reported snakes (*Drymarchon*) feeding on hatchlings. Adults are probably preyed upon only by crocodiles and humans. *C. rossignoni* is not commonly sold in markets and seems never to have been sold as mummified curios.

POPULATIONS

The rarity of the species precludes conclusions of any kind on populations. Feuer (1966) thought populations of Mesoamerican *Chelydra* were less dense than those in the United States and considered that this could be a characteristic of a recently established population. Smith and Smith (1979) also comment on the rarity of *rossignoni* based on the few documented localities.

CONSERVATION AND CURRENT STATUS

The following remarks are based on the sum total of trapping and observations by Vogt from 1968 to 2000 and on museum specimens. *Chelydra rossignoni* is unquestionably "rare." Nowhere has it been found to be abundant. It is seldom seen or caught even in the places it is known to occur; relatively few specimens exist in collections; it has not commonly been sold as food or souvenirs in the last 30 years; and we have seen little indication that it is being eaten by local peoples.

Explanations of this rarity could logically be: (1) there is something we don't know about its life history that makes it difficult to see or catch; (2) individuals have a long life span and can recruit at low population densities; (3) populations were once higher and existing populations were decimated by hunting. It remains logical that no subsistence hunter would pass up the amount of good meat yielded by a snapping turtle, but *rossignoni* is not currently being sent to market.

The two Mesoamerican species are never abundant and are never the dominant turtle in an ecosystem (as was *Chelydra serpentina* in Wisconsin when RCV worked there in the 1970s).

In long-term (1984–1996), intensive trapping at Laguna Oaxaca, Chiapas, only three of some 2,000 turtles captured were *C. rossignoni*. Legler (pers. obs.) found *Chelydra acutirostris* to be only slightly more abundant from Honduras to Isthmian Panama.

Chelydra serpentina (Linnaeus) 1758

Common snapping turtle (Iverson) 1992a

The presence of this species has never been authenticated in Mexico, but it may occur there. Carr (1952) shows the distribution extending into the northern parts of Coahuila, Nuevo León, and Tamaulipas, and on both sides of the Rio Grande from Trans Pecos, Texas, to Brownsville.

Feuer (1966) shows dots in Texas north of the Rio Grande but includes both sides of the Rio Grande in the mapped hypothesized range.

Smith and Smith (1979) conclude that there are no positive records for *C. s. serpentina* in Mexico but include a discussion of where it might (and probably does) occur. We concur with this speculation and would urge others to search the northward-flowing Lower Rio Grande tributaries south of Del Rio, where *Pseudemys gorzugi* is known to occur—the Salado, Sabinas, and San Juan drainages in northern Coahuila, Nuevo León, and Tamaulipas (see account of *Trachemys scripta taylori*). *Chelydra s. serpentina* occurs in the canal systems on both sides of the Rio Grande near Brownsville but is known to have been introduced there (A. Rentfro and M. Forstner, pers. comm.).

LITERATURE CITED

Abel, J. H., and R. A. Ellis. 1966. Histochemical and electron microscopic observations on the salt secreting lacrymal glands of marine turtles. Am. J. Anat. 118:337–358.

Acuña Mesén, R. A. 1990. Impact of fire and drought on the population structure of *Kinosternon scorpioides* at Palo Verde, Guanacaste, Costa Rica. Brenesia 33:85–97.

Acuña Mesén, R. A. 1993. Las Tortugas Continentales de Costa Rica. San José, Costa Rica: Editorial ICER.

Adest, G. A., and G. Aguirre-León. 1995. Natural and life history of the Bolsón tortoise, *Gopherus flavomarginatus*. Publicaciones de La Sociedad Herpetologica Mexicana 2:1–5.

Adest, G. A., G. Aguirre-León, D. J. Morafka, and J. V. Jarchow. 1989a. Bolsón tortoise *(Gopherus flavomarginatus)* conservation: 1. Life history. Vida Silvestre Neotropical 2(1):7–13.

Adest, G. A., G. Aguirre-León, D. J. Morafka, and J. V. Jarchow. 1989b. Bolsón tortoise *(Gopherus flavomarginatus)* conservation: 2. Husbandry and reintroduction. Vida Silvestre Neotropical 2(1):14–20.

Adest, G. A., J. Jarchow, and B. Brydolf. 1988. A method for manual ventilation of tranquilized tortoises. Herp. Rev. 19(4):80.

Adest, G.A., M. A. Recht, G. Aguirre-León, and D. J. Morafka. 1988. Nocturnal activity in the bolsón tortoise *(Gopherus flavomarginatus)*. Herp. Rev. 19(4):75–76.

Advise, J. D., and C. G. Jackson, Jr. 1973. Notes on the courtship of a captive male *Chrysemys scripta taylori*. Herpetologica 29(1):62–64.

Agassiz, L. 1857a. Contribution to the Natural History of the United States of America. First Monograph. Vol. I, Pt. I, Essay on Classification; Pt. II, North American Testudinata. Boston: Little, Brown and Co.

Agassiz, L. 1857b. Contribution to the Natural History of the United States of America. First Monograph. Vol. II, Pt. III, Embryology of the Turtle. Boston: Little, Brown and Co.

Aguirre-León, G. 1995. Conservation of the Bólson tortoise, *Gopherus flavomarginatus*. Publicaciones de La Sociedad Herpetologica Mexicana 2:6–9.

Aguirre-León, G., D. J. Morafka, and G. A. Adest. 1997. Conservation strategies for the Bolsón tortoise, *Gopherus flavomarginatus,* in the Chihuahuan Desert. pp. 333–338. In: J. Van Abbema (Ed.). Proceedings: Conservation, Restoration, and Management of Tortoises and Turtles—An International Conference. SUNY Purchase, New York. New York Turtle and Tortoise Society and WCS.

Aguirre-León, G., E. Cazares, and B. Sanchez. 2002. Conservacion y aprovechamiento del Chopontil *(Claudius angustatus)*. Instituto de Ecologia, Xalapa, Veracruz, México.

Aguirre-León, G., G. A. Adest, and D. J. Morafka. 1984. Home range and movement patterns of the Bolsóln tortoise, *Gopherus flavomarginatus*. Acta Zoologica Mexicana Nueva Serie No. 1 1984:1–28.

Alberts, A. C., D. C. Rostal, and V. A. Lance. 1994. Studies on the chemistry and social significance of chin gland secretions in the desert tortoise, *Gopherus agassizii*. Herp. Monogr. 8:116–124.

Alford, R. A. 1980. Population structure of Gopherus polyphemus in northern Florida. J. Herpetol. 14:177–182.

Allen, E. R. 1950. Sounds produced by Suwannee terrapin. Copeia 1950:62.

Alvarado-Diaz, J., A. Estrada-Virgen, D. Garcia-Parra, and I. Suazo-Ortuno. 2003. *Rhinoclemmys rubida* (Mexican spotted wood turtle). Diet. Herp. Rev. 34(4):363.

Álvarez, S. T. 1976. Restos oseos de las excavaciónes de Tlatilco, Edo. de México. Restos oseos rescatudos del cenote sagrado de Chichén Itzá, Yucatán. Apuntes Arqueológia, México 15:1–39.

Álvarez del Toro, M. 1960. Los Réptiles de Chiapas. Instituto Zoológico del Estado, Tuxtla Gutierrez, Chiapas, México.

Álvarez del Toro, M. 1973. Los Réptiles de Chiapas. 2nd ed. Instituto Zoológico del Estado, Tuxtla Gutierrez, Chiapas, México.

Álvarez del Toro, M. 1974. Los crocodylia de México (estudio comparativo). México, D. F. Instituto de Recursos Naturales Renovables.

Álvarez del Toro, M. 1982. Los Réptiles de Chiapas. 3rd ed. Instituto Zoológico del Estado, Tuxtla Gutierrez, Chiapas, México.
Álvarez del Toro, M., and H. M. Smith. 1982. Comparisons of pattern and color in life of the subspecies of the turtle *Pseudemys scripta* in Chiapas, Mexico. Bull. Maryland Herp. Soc. 18(4):194–195.
Álvarez del Toro, M., R. A. Mittermeier, and J. B. Iverson. 1979. River turtle in danger. Oryx 15(2):170–173.
Anon. 1987. Carta de México: topográfica 1:250 000. Publisher México, D.F.: Instituto Nacional de Estadística, Geografía e Informática, 1982 (1987 printing).
Appleton, A. B. 1980. Bolsón hatchlings at the Research Ranch, Elgin, Arizona. Desert Tortoise Council Proceedings of Symposium 5 1980:147–150.
Appleton, A. B. 1986. Further observations of bolsón tortoises *(Gopherus flavomarginatus)* at the Appleton-Whittell Research Ranch. Desert Tortoise Council Proceedings of Symposium No. 8 1986:108–114.
Aquino Cruz, O. 2003. Habitos alimentarios de la Tortuga *Kinosternon herrerai* Stejneger 1925, en arroyos del sureste del municipio de Xalapa, Veracruz, Mexico. Unpublished MS thesis, Univ. Veracruzana, Fac. de Biologia-Xalapa, Mexico.
Artner, H. 2009. Field observations and captive breeding of the furrowed wood turtle *Rhinoclemmys areolata* (Duméril and Bibron, 1851). Emys, 16(3):4–19.
Ashton, R. and P. Ashton. 2008. The Natural History and Management of the Gopher Tortoise, *Gopherus polyphemus* (Daudin). Kreiger Press, Malabar, Fl.
Auffenberg, W. 1958. Fossil turtles of the genus *Terrapene* in Florida. Bull. Florida State Mus. 3:53–92.
Auffenberg, W. 1964. Notes on the courtship of the land tortoise *Geochelone travancorica* (Boulenger). J. Bombay Nat. Hist. Soc. 61:247–253.
Auffenberg, W. 1966. The carpus of land tortoises (Testudininae). Bull. Florida State Mus. Biol. Sci. 10:159–192.
Auffenberg, W. 1969. Tortoise behavior and survival. Rand McNally and Co., Chicago.
Auffenberg, W. 1974. Checklist of fossil land tortoises (Testudinidae). Bull. Florida State Mus. Biol. Sci. 18:121–251.
Auffenberg, W. 1976. The genus *Gopherus* (Testudinidae): Pt. I. Osteology and relationships of extant species. Bull. Florida State Mus. Biol. Sci. 20:47–110.
Auffenberg, W. 1978. Courtship and breeding behavior in *Geochelone radiata*. Herpetologica 34(3):277–287.
Auffenberg, W., and R. Franz. 1978a. *Gopherus*. Catalog. Amer. Amphib. Rept. 211:1–2.
Auffenberg, W., and R. Franz. 1978b. *Gopherus agassizii*. Catalog. Amer. Amphib. Rept. 212:1–2.
Auffenberg, W., and R. Franz. 1978c. *Gopherus berlandieri*. Catalog Amer. Amphib. Rept. 213:1–2.
Auffenberg, W., and R. Franz. 1978d. *Gopherus flavomarginatus* Legler. Bolson tortoise. Catalog. Amer. Amphib. Rept. 214:1–2.
Auffenberg, W., and W. G. Weaver Jr. 1969. *Gopherus berlandieri* in southeastern Texas. Bull. Florida State Mus. Biol. Sci. 13:141–203.
Auffenberg, W., and W. W. Milstead. 1965. Reptiles in the Quaternary of North America. In: H. E. Wright, Jr., and D. G. Frey (Eds.). The Quaternary of the United States. Princeton Univ. Press, Princeton, N. J. pp. 557–568.
Avise, J. C. 2004. Molecular Markers, Natural History, and Evolution. 2nd ed. Sinauer Associates, Sunderland, Massachusetts.
Avise, J. C., B. W. Bowen, T. Lamb, A. B. Meylan, and E. Bermingham. 1992. Mitochondrial DNA evolution at a turtle's pace: evidence for low genetic variability and reduced microevolutionary rate in the testudines. Mol. Biol. Evol. 9:457–473.
Ayala Guerrero, F., A. Calderón, and M. C. Peréz. 1988. Sleep patterns in a chelonian reptile *(Gopherus flavomarginatus)*. Physiology and Behavior 44(3):333–337.
Babcock, H. L. 1932. The American snapping turtles of the genus *Chelydra* in the collection of the Museum of Comparative Zoology, Cambridge, Massachusetts, U.S.A. Proc. Zool. Soc. London 4:873–874.
Baegert, J. J. 1772. Nachrichten von der amerikanischen Halbinsel Californien: mit einem zweyfachen Anhang falscher Nachrichten. Geschrieben von einem Priester der Gesellschaft Jesu, welcher lang darinn diese letztere Jahr gelebt hat. Translated as "Observations in Lower California" by M. M. Brandenburg and Carl L. Baumann; Berkeley, Univ. of California Press
Bagatto, B., C. Guyer, B. Hauge, and R. P. Henry. 1997. Bimodal respiration in two species of Central American turtles. Copeia 1997(4): 834–839.
Bailey, L. A., J. R. Dixon, R. Hudson, and M. R. J. Forstner. 2008. Minimal genetic structure in the Rio Grande Cooter *(Pseudemys gorzugi)*. Southwest. Natur. 53(3):406–411.
Bailey, L. A., M. R. J. Forstner, J. R. Dixon, and R. Hudson. In press. Contemporary status of the Rio Grande Cooter (Testudines; Emydidae; *Pseudemys gorzugi*) in Texas; phylogenetic, ecological and conservation considerations.
Baillie, J. and B. Groombridge. 1996. IUCN Red List of Threatened Animals. Gland, Switzerland: IUCN.
Baird, I. L. 1970. The anatomy of the reptilian ear. In: C. Gans and T. S. Parsons (Eds.). Biology of the Reptilia, Morphology B, Vol. 2, pp. 193–275. New York: Academic Press.
Baird, S. F., and C. Girard. 1852. Descriptions of new species of reptiles collected by the U.S. Exploring Expedition under the command of Capt. Charles Wilkes, U.S.N. Proc. Acad. Natur. Sci., Philadelphia 6:174–177.
Barbour, C. D. 1973. A biogeographical history of *Chirostoma* (Pisces: Atherinidae): a species flock from the Mexican Plateau. Copeia 1973 (3): 533–556.
Barco, M. del. 1980. The Natural History of Baja California. Historia natural y coronica de la antigua California. English translation by Froylan Tiscareno. Originally published Madrid, 1757 Dawson's Book Shop, Los Angeles.
Barrett, S. L. 1990. Home range and habitat of the desert tortoise *(Xerobates agassizi)* in the Picacho Mountains of Arizona. Herpetologica 46:202–206.
Barrett, S. L., and J.A. Humphrey. 1986. Agonistic interactions between *Gopherus agassizii* (Testudinidae) and *Heloderma suspectum* (Helodermatidae). Southwest. Natur. 31:261–263.
Barrow, J. 1979. Aspects of ecology of the desert tortoise *Gopherus agassizi*, in Joshua Tree National Monument, Pinto Basin, Riverside County, California. Proc. Symp. Desert Tortoise Council 1979(1):105–131.
Bartlett, R. D., and P. Bartlett. 2003. Reptiles and Amphibians of the Amazon. Florida Univ. Press Gainesville.

Bartram, W. 1794. Travels through North and South Carolina, Georgia, East and West Florida. 2nd ed. London.

Batsch, August J. G. C. 1788. Versuch einer Anleitung zur Kenntniss und Geschichte der Thiere und Mineralien. Erster Theil. Allgemeine Geschichte der Natur; besondre der Säugthiere, Vögel, Amphibien und Fische. Jena: Akademischen Buchandlung, Vol. 1, pp. 528.

Bauer, M., and I. Jasser-Häger. 2006. The Mexican Aquatic Box Turtle, *Terrapene coahuila* Schmidt and Owens, 1944. Husbandry and breeding to the F2 generation. Radiata 15:40–49.

Baur, G. 1887. Ueber den ursprung der extremitäten der Ichthyopterygia. Jahrbr. Mitt. Oberrhein Geol. Verein 20:17–20.

Baur, G. 1891. Notes on some little known American fossil tortoises. Proc. Acad. Nat. Sci. Philadelphia, 1891:411–430.

Begon, M. 1979. Investigating Animal Abundance: Capture Recapture for Biologists. Edward Arnold, London.

Belkin, D. A. 1963. Anoxia: tolerance in reptiles. Science 139:492–493.

Bell, T. 1831. A Monograph of the Testudinata. London.

Bellairs, A. 1969. The Life of Reptiles. Weidenfeld and Nicolson, London.

Berry, J. F. 1978. Variation and systematics in the Kinosternon Scorpioides and K. leucostomum complexes (Reptilia: Testudines: Kinosternidae) of Mexico and Central America. Ph.D. dissertation, Univ. of Utah. {See Dissertation Abstracts 39(7), No. 7824683}.

Berry, J. F., and C. M. Berry. 1984. A reanalysis of geographic variation and systematics in the yellow mud turtle, *Kinosternon flavescens* (Agassiz). Ann. Carn. Mus. Pittsburgh 53(7):185–206.

Berry, J. F., and J. B. Iverson. 1980. *Kinosternon herrerai*. Cat. Amer. Amph. Rept. 239.1–239.2.

Berry, J. F. and J. B. Iverson. 2001a. *Kinosternon leucostomum*. Cat. Amer. Amph. and Rept. 724:1–8.

Berry, J. F., and J. B. Iverson. 2001b. *Kinosternon scorpioides*. Cat. Amer. Amph and Rept. 725:1–11.

Berry, J. F., and J. B. Iverson. 1980. A new species of mud turtle, genus, *Kinosternon,* from Oaxaca, Mexico. J. Herp. 14(4):313–320.

Berry, J. F., and J. M. Legler. 1980. A new turtle (genus *Kinosternon*) from northwestern Mexico. Contr. Sci Nat. Hist. Mus. Los Angeles Co. (325):1–12.

Berry, J. F., M. E. Seidel, and J. B. Iverson. 1997. A new species of mud turtle (genus *Kinosternon*) from Jalisco and Colima, Mexico, with notes on its natural history. Chel. Cons. Biol. 2 (3):329–337.

Berry, J. F., and R. Shine. 1980. Sexual size dimorphism and sexual selection in turtles (order Testudines). Oecologia (Berlin) 44:185–191.

Berry, K. H. 1976. A comparison of size classes and sex ratios in four populations of the desert tortoise. Proc. Symp. Desert Tortoise Council 1976:38–50.

Berry, K. H. 1978. Livestock grazing and the desert tortoise. Trans. 43rd N.A. Wildl. Natur. Res. Conf. pp. 505–519.

Berry, K. H. 1986. Desert tortoise *(Gopherus agassizii)* relocation: implications of social behavior and movements. Herpetologica 42:113–125.

Berry, K. H. 1989. *Gopherus agassizii,* desert tortoise. In: I. R. Swingland and M. W. Klemmens, eds. The conservation biology of tortoises. Occasional Papers no. 5.IUCN Species Survival Committee, Gland Switzerland, pp. 5–7.

Berry, K. H. 1991. The status of the desert tortoise in 1990: current population issues in California. In, pp. 80–82. Beaman, K. R., F. Caporaso, S. McKeown, and M. D. Graff (eds.). Proceedings of the First International Symposium on turtles and tortoises: conservation and captive husbandry. Chapman Univ., Orange, California.

Berry, K. H. 2002. Using growth ring counts to age juvenile desert tortoises *(Gopherus agassizii)* in the wild. Chel. Cons. Biol. 4(2):416–424.

Berry, K. H., and F. B. Turner. 1984. Notes on the behavior and habitat preferences of juvenile desert tortoises *(Gopherus agassizii)* in California. Proc. Symp. Desert Tortoise Council 1984:111–130.

Bickett, J. E. 1980. A behavioral study of captive Bolson tortoises, *Gopherus flavomarginatus,* at the Research Ranch, Elgin, Arizona. Desert Tortoise Council Proceedings of Symposium 5 1980:143–145.

Bickham, J. W. 1975. A cytosystematic study of turtles in the genera *Clemmys, Mauremys,* and *Sacalia*. Herpetologica 31:198–304.

Bickham, J. W. 1981. Two-hundred million year old chromosomes: deceleration of the rate of karyotypic evolution in turtles. Science 212:1291–1293.

Bickham, J. W., J. B. Iverson, J. F. Parham, H. Philippen, A. G. J. Rhodin, B. Shaffer, P. Q. Spinks, and P. P. Van Dijk. 2007. An annotated list of modern turtle taxa with comments on areas of taxonomic instability and recent change. In: B. Shaffer, N. FitzSimmons, A. Georges, and A. G. J. Rhodin (Eds.). Defining Turtle Diversity: Proceedings of a Workshop on Genetics, Ethics, and Taxonomy of Tortoises and Freshwater Turtles. Chelonian Research Monographs 4:168–194.

Bickham, J. W., and J. L. Carr. 1983. Taxonomy and phylogeny of the higher categories of cryptodiran turtles based on a cladistic analysis of chromosomal data. Copeia 1983:918–932.

Bickham, J. W., M. D. Springer, and B. J. Galloway. 1984. Distributional survey of the yellow mud turtle *(Kinosternon flavescens)* in Iowa, Illinois, and Missouri: a proposed endangered species. Southwest. Natur. 29:123–132.

Bickham, J. W., and R. J. Baker. 1976a. Karyotypes of some neotropical turtles. Copeia 1976:703–708.

Bickham, J. W., and R. J. Baker. 1976b. Chromosome homology and evolution of Emydid turtles. Chromosoma (Berlin) 54:201–219.

Bickham, J. W., T. Lamb, P. Minx, and J. C. Patton. 1996. Molecular systematics of the genus *Clemmys* and the intergeneric relationships of emydid turtles. Herpetologica 52:89–97.

Bienz, A. 1896. *Dermatemys mavii* Gray, eine osteologische Studie mit Beitragen zur Kenntnis der Schildkröten. Rev. Suisse Zool. 3:61–135.

Bishop, S. C., and F. J. W. Schmidt. 1931. The painted turtles of the genus *Chrysemys*. Field Mus. Nat. Hist., Zool. Ser. Publ. No. 293. 18:123–139.

Bjorndal, K. A., and A. B. Bolten. 1990. Digestive processing in a herbivorous freshwater turtle: consequences of small intestine fermentation. Physiol. Zool. 63:1232–1247.

Blair, W. F. 1950. The biotic provinces of Texas. Texas J. Sci. 2(1):93–117.

Blair, W. F. 1976. Some aspects of the biology of the ornate box turtle, *Terrapene ornata*. Southwest. Natur. 21:89–104.

Bleakney, J. S. 1958. Postglacial dispersal of the turtle *Chrysemys picta*. Herpetologica 14:101–104.

Boarman, W. I. 1993. Predation on turtles and tortoises by "subsidized predators." Conservation, Restoration, and Management of Tortoises and Turtles: An International Conference. Abstracts, p. 21.

Bocourt, F. 1868. Description de queilques Cheloniens nouveaux appartenant a la fauna Mexicaine. Ann. Sci. Natur. Zool., Paris (5)10:121–122.

Bogert, C. M. 1943. A new box turtle from southeastern Sonora, Mexico. Amer. Mus. Nov. 1226:1–7.

Bogert, C. M. 1960. Influence of sound on amphibians and reptiles. In: W. E. Lanyon and W. N. Tavolga (Eds.). Animal Sounds and Communication, pp. 137–220. American Inst. Biol. Sci., Washington, DC.

Bogert, C. M. 1961. Los reptiles de Chiapas by M. Álvarez del Toro. (Book Review). Copeia 1961(4):506–507.

Bogert, C. M., and J. A. Oliver. 1945. A preliminary analysis of the herpetofauna of Sonora. Bull. Amer. Mus. Nat. Hist. 83(6):301–425.

Boulenger, E. G. 1914. Reptiles and Batrachians. Dent and Sons Ltd., London.

Boulenger, G. A. 1889. Catalogue of the Chelonians, Rhynchocephalians, and Crocodiles in the British Museum (Natural History). Taylor and Francis, London.

Boulenger, G. A. 1895. On the American box tortoises. Ann. Mag. Natur. Hist. 15:330–331.

Boulenger, G. A. 1913. On a collection of batrachians and reptiles made by Dr. H. G. F. Spurrell, F.Z.S., in the Choco, Colombia. Proc. Zool. Soc. London 1913(4):1019–1038.

Bour, R., and A. Dubois. 1984. *Xerobates* Agassiz, 1857, synonyme plus ancien de *Scaptochelys* Bramble, 1982 (Reptilia, Chelonii, Testudinidae). Bull. Soc. Linn. Lyon. 53:30–32.

Bour, R., and A. Dubois. 1986. Nomenclature ordinale et familieale des tortues (Reptilia). Bulletin Mensuel del la Societe Linneenne de Lyon. 55:87–90.

Bowen, W. M., M. George, Jr., and A. C. Wilson. 1979. Rapid evolution of animal mitochondrial DNA. Proc. Nat. Acad. Sciences (USA) 76:1967–1971.

Bowler, J. K. 1977. Longevity of reptiles and amphibians in North American collections. SSAR Herp. Circ 6:1–32.

Brainerd, G. W. 1953. Archeological findings. pp. 107–119 in: R. T. Hatt, Faunal and archeological researches in Yucatan caves. Cranbrook Inst. Sci. Bull. 33:1–119.

Bramble, D. M. 1973. Media dependent feeding in turtles. Amer. Zool. 13:1342.

Bramble, D. M. 1974. Emydid shell kinesis: biomechanics and evolution. Copeia 1974:707–727.

Bramble, D. M. 1982. *Scaptochelys:* generic revision and evolution of gopher tortoises. Copeia 1982:852–867.

Bramble, D. M., and D. Wake. 1985. Feeding mechanisms of lower tetrapods. In: M. Hildebrand, D. M. Bramble, K. F. Liem, and D. B. Wake (Eds.). Functional Vertebrate Morphology. Ch. 13, pp. 230–261, 13 illus. The Belknap Press of Harvard Univ. Press, Cambridge.

Bramble, D. M., and J. H. Hutchison. 1981. A reevaluation of plastral kinesis in African turtles of the genus *Pelusios*. Herpetologica 37:205–212.

Bramble, D. M., J. H. Hutchison, J. L. Carr, and J. M. Legler. 2009. Homology and functional evolution of the plastron in soft-shelled turtles (Trionychidae). Abstract, Gaffney Turtle Symposium: October 17–18; Royal Tyrrell Museum, Drumheller, Alberta, Canada.

Bramble, D. M., and J. L. Carr. (Unpublished manuscript) Homology, mechanics and evolution of the trionychid plastron.

Brattstrom, B. H. 1965. Body temperatures of reptiles. Amer. Midland Natur. 73(2):376–422.

Bravo-Hollis, M. 1944. Trematodo parasito del intestino de *Kinosternon integrum*. An. Inst. Biol. Univ. Nac. Aut. México 16:41–45.

Bravo-Hollis, M., and J. Caballero Deloya. 1973. Catalogo de la Colección Helmintologíca del Instituto de Biología. Publ. Esp. Inst. Biol. Univ. Nac. Aut. México(2):1–138.

Breckenridge, W. J. 1955. Observations on the life history of the soft-shelled turtle *Trionyx ferox,* with especial reference to growth. Copeia 1955(1):5–9.

Brown, B. C. 1950. An Annotated Checklist of the Reptiles and Amphibians of Texas. Baylor Univ. Studies, Baylor Univ. Press, Waco, TX.

Brown, W. R. 1954. Composition of Scientific Words. Rev. Ed. Smithsonian Institution Press, Washington D.C.

Brown, W. S. 1971. Morphometrics of *Terrapene coahuila* (Chelonia, Emydidae), with comments on its evolutionary status. Sorthwest. Natur. 16 (2):171–184.

Brown, W. S. 1974. Ecology of the aquatic box turtle, *Terrapene coahuila* (Chelonia, Emydidae) in northern Mexico. Bull. Florida State. Mus. Biol. Sci., 19 (1):1–67.

Brumwell, M. J. 1940. Notes on the courtship of the turtle, *Terrapene ornata*. Trans. Kansas Acad. Sci. 43:391–392.

Bruno, M.R. 1983. Nictitating membranes and associated ocular glands in recent turtles. Abstract. SS A R., Herp. League Joint Meeting, Univ. of Utah, Aug. 1983.

Bull, J. J. 2004. Prevalance of TSD in vertebrates. Perspectives on sex determination: Past and future. *In* Valenzuela, N. & V. Lance 2004. Temperature-dependent sex determination in vertebrates. Smithsonian Books, Washington, D.C.

Bull, J. J., and R. C. Vogt. 1979. Temperature-dependent sex determination in turtles. Science 206:1186–1188.

Bull, J. J., and R. C. Vogt. 1981. Temperature-sensitive periods of sex determination in emydid turtles. J. Exp. Zool. 218:435–440.

Bull, J. J., R. C. Vogt, and C. J. McCoy. 1982. Sex determining temperatures in turtles: a geographic comparison. Evolution 36(2):326–332.

Bull, J. J., R. G. Moon, and J. M. Legler. 1974. Male heterogamety in kinosternid turtle (genus *Staurotypus*). Cytogenet. Cell Genet. 13(5):419–425.

Burge, B. L., and W. G. Bradley. 1976. Population density structure and feeding habits of the desert tortoise, *Gopherus agassizi* in a low desert study area in southern Nevada. Proc. Symp. Desert Tortoise Council 1976:51–76.

Burke, A. C. 1989a. Critical Features in Chelonian Development: The Ontogeny and Phylogeny of a Unique

Tetrapod Bauplan. Unpublished PhD dissertation, Harvard Univ., Cambridge, Mass.
Burke, A. C. 1989b. Development of the turtle carapace; Implications for the evolution of a novel bauplan. J. Morphol. 199:363–378.
Burke, A. C. 1989c. Epithelial-mesenchymal interactions in the development of the chelonian bauplan. In H. Splechtna and J. Hilgers (Eds.). Trends in vertebrate morphology. Prog. Zoology 35:206–209, Fisher Verlag, Stuttgart.
Bury, R. B. 1970. *Clemmys marmorata*. Cat. Amer. Amphib. Rept. 100:1–3.
Bury, R. B. 1972. Habits and Home Range of the Pacific Pond Turtle, *Clemmys marmorata,* in a Stream Community. PhD dissertation. Univ. California, Berkeley.
Bury, R. B. 1989. Turtle of the month—*Clemmys marmorata*—a true western turtle (Pacific pond). Tortuga Gazette 25(2):3–4.
Bury, R. B. and C. H. Ernst. 1977. *Clemmys*. Catalog. Amer. Amphib. Rept. 203:1–2.
Bury, R. B. and D. J. Germano. 1994. Biology of North American tortoises: introduction. Fish and Wildlife Research 13:1–5.
Bury, R. B., D. J. Germano, T. R. Van Devender, and B. E. Martin. 2002. The desert tortoise in Mexico: distribution, ecology, and conservation. pp. 86–108 in: T. R. Van Devender (Ed.), The Sonoran Desert Tortoise: natural history, biology, and conservation. Tucson: Univ. AZ Press
Bury, R. B., D. J. Morafka, and C. J. McCoy. 1988. The ecogeography of the Mexican Bolsón tortoise *(Gopherus flavomarginatus):* derivation of its endangered status and recommendations for its conservation. Part 1. Distribution, abundance and status of the Bolson tortoise. Ann. Carnegie Mus. Nat. Hist. 57(1)1988:5–30.
Bury, R. B., and E. L. Smith. 1986. Aspects of the ecology and management of the tortoise *Gopherus berlandieri* at Laguna Atascosa, Texas. Southwest. Natur. 31:387–394.
Bury, R. B., R. A. Luckenbach, and L. R. Munoz. 1978. Observations on Gopherus agassizii from Isla Tiburon, Sonora, Mexico. Proc. Symp. The Desert Tortoise Council 1978:69–72.
Bury, R. B., T. C. Esque, L. A. DeFalco, and P. A. Medica. 1994. Distribution, habitat use, and protection of the desert tortoise in the eastern Mojave Desert. Fish and Wildlife Research 13:57–72.
Buskirk, J. R. 1983. Life history notes: *Pseudemys (Chrysemys) scripta hiltoni*. Reproduction. Newsl. Chel. Doc. Cent. 2(2–4):33.
Buskirk, J. R. 1990. An overview of the western pond turtle, *Clemmys marmorata*. In: K. R. Beaman, F. Caporaso, S. McKeown, and M. D. Graff (Eds.). pp. 16–23. Proceedings of the First International Symposium on turtles and tortoises: conservation and captive husbandry. Chapman Univ., Orange, California.
Buskirk, J. R. 1993. Yucatan box turtle *Terrapene carolina yucatana*. Tortuga Gazette 29(5):1–4.
Buskirk, J. R. 2001. The Mexican painted wood turtle, *Rhinoclemmys pulcherrima*. Emys 8(2):20–28.
Buskirk, J. R. 2002. The western pond turtle, *Emys marmorata*. Radiata, 11(3):3–30.
Caballero y Caballero, E. 1938. Algunos trematodos de reptiles de México An. Inst. Biol. Univ. Nac. Aut. México 9:103–120.
Caballero y Caballero, E. 1939. Nematodes de los reptiles de México V. An. Inst. Biol. Univ. Nac. Auton. México 10:275–282.
Caballero y Caballero, E. 1940. Trematodos de las tortugas de México. An. Inst. Biol. Univ. Nac. Aut. México 11:559–572.
Caballero y Caballero, E. 1942. Trematodos de las tortugas de México. II. Descripción de un nuevo genero de la familia Pronocephalidae Looss, 1902 y descripción de una nueva espécie del genero *Octangioides* Price, 1937. An. Inst. Biol. Univ. Nac. Mex. 13(1):81–90.
Caballero y Caballero, E. 1961. Trematodos de las tortugas de México. VIII. Descripción de un nuevo genero que parásita a tortugas de agua dulce. Ciencia 21(2):61–64.
Caballero y Caballero, E., and D. Sokoloff. 1935. A new tematode (*Schizamphistomoides tabascensis* n. sp.) from the intestine of a fresh water turtle, *Dermatemys mawii* Gray. Trans. Am. Microsc. Soc. 54:135–137.
Caballero y Caballero, E. and E. Herrara-Rosales. 1947. Trematodos de las tortugas de México. V. An. Inst. Biol. Univ. Nac. Aut. México 18:159–164
Caballero y Caballero, E. and G. Caballero. 1967. Trematodos de la tortugas de México. XI. *Acanthostomum nuevoleonensis* n. sp. parásito de Trionyx spinifer emoryi (Agassiz, 1857). An. Esc. Nac. Cienc. Biol., Mex. 13(1/4):83–90.
Caballero y Caballero, E., and G. Rodriguez. 1960. Estudio de trematodos digeneos de algunas tortugas comestibles de México. Ph.D. Thesis, Univ. Nac. Auton. México.
Caballero y Caballero, E., and M. C. Cerecero. 1943. Nematodes de los reptiles de México VIII. Descripción de tres nuevas espécies. An. Inst. Biol. Univ. Nac. Auton. México 14(2):527–539.
Cagle, F. R. 1944a. Sexual maturity in the female of the turtle *Pseudemys scripta* elegans. Copeia 1944(3):149–152.
Cagle, F. R. 1944b. Home range, homing behavior and migration in turtles. Misc. Publ. Mus. Zool. Univ. Mich. 61:1–34.
Cagle, F. R. 1950. The life history of the slider turtle, *Pseudemys scripta troostii* (Holbrook). Ecol. Monogr. 20:31–54.
Cagle, F. R., and A. H. Chaney. 1950. Turtle populations in Louisiana. Amer. Midl. Nat. 43:383–388.
Cahn, A. R. 1937. The turtles of Illinois. Ill. Biol. Monogr. 16(1–2):1–218.
Campbell, H. W., and W. E. Evans. 1967. Sound production in two species of tortoises. Herpetologica 23:204–209.
Campbell, H. W., and W. E. Evans. 1972. Observations on the vocal behavior of chelonians. Herpetologica 28(3):277–280.
Campbell, J. A. 1998. Amphibians and Reptiles of Northern Guatemala, the Yucatan, and Belize. Univ. Oklahoma Press, Norman.
Cann, J., and J. M. Legler. 1994. The Mary River Tortoise: a new genus and species of short-necked chelid from Queensland, Australia (Testudines: Pleurodira). Chel. Cons. Biol. 1(2):81–96.
Carpenter, C. C. 1957. Hibernation, hibernacula and associated behavior of the three-toed box turtle *(Terrapene carolina triunguis)*. Copeia 1957(4):278–282.
Carr, A. F., Jr. 1942. A new *Pseudemys* from Sonora, Mexico. Amer. Mus. Novitates 1181:1–4.
Carr, A. F. 1952. Handbook of Turtles: The Turtles of the United States, Canada, and Baja California. Cornell Univ. Press, Ithaca, New York.

Carr, A. F., L. Ogren, and C. McVea. 1980. Apparent hibernation by the Atlantic Loggerhead Turtles *Caretta caretta* off Cape Canaveral, Florida. Biol. Conserv. 19:7–14.

Carr, J., and R. Mast. 1988. Natural history observations of *Kinosternon herrerai* (Testudines: Kinosternidae). Trianea (Act. Ceint. Yecn. Inderena) 1:87– 97.

Carr, J. L., and A. Almendáriz. 1990. Contribución al conocimiento de la distribución geográfica de los quelonios del Ecuador occidental. Politecnica 14:75–103.

Carr, J. L. 1993. *Kinosternon oaxcacae*. Range extension. Herp. Review 24(3):108.

Carr, J. L., J. W. Bickham, and R. H. Dean. 1981. The karyotype and chromosomal banding patterns of the Central American river turtle *Dermatemys mawii*. Herpetologica 37(2):92–95.

Carr, J. L., and J. W. Bickham. 1986. Phylogenetic Implications of karyotypic variation in the batagurinae (Testudines: emydidae). Genetica 70:89–106.

Carroll, R. L. 1988. Vertebrate Paleontology and Evolution. New York: W.H. Freeman and Company.

Casas Andreu, G. 1967. Contribución al conocimiento de las tortugas dulceacuicolas de México. Tesis Profesional, Mexico, D. F.

Castillo-Centeno, O. 1986. Factores ecológicos y de mercado de la reproducción de *Rhinoclemmys pulcherrima* y *Kinosternon scorpioides* (Testudines: Emydidae y Kinosternidae) en Costa Rica. Unpublished Thesis. Ciudad Universitária "Rodrigo Facio," San José, Costa Rica.

Castro-Aguire, J. L. 1978. Catalogo sistematico de los peces marinos que penetran a las agus continentales de México con aspectos zoogeograficos y ecologicos. Dir. Inst. Nac. Pesca, Ser. Sci., Mex. 19:I–xi,1–298.

Cazares-Hernández, E. 2004. Aspectos ecológicos de la tortuga *Kinosternon herrerai,* Stejneger, 1925 (Reptilia: Testudines: Kinosternidae) el arroyo "La Bomba", Municipio de Xalapa, Veracruz. Tesis Licenciado en Biologia, Facultad de Biologia Zona Xalapa , Universidad Veracruzana.

Cerda-Ardura, A., and D. Waugh. 1992. Status and management of the Mexican box terrapin *Terrapene coahuila* at the Jersey Wildlife Preservation Trust. Dodo 28:126–142.

Cerda-Ardura, A., F. Soberón-Mobarak, S. E. McGaugh, and R. C. Vogt. 2008. *Apalone spinifera atra* (Webb and Legler 1960) black spiny softshell turtle, Cuatrocienegas softshell, tortuga concha blanda, tortuga negra de Cuatro Cienegas. In: A. G. J. Rhodin, P. C. H. Pritchard, P. P. van Dijk, R. A. Saumure, K. A. Buhlmann, and J. B. Iverson (Eds.). Conservation Biology of Freshwater Turtles and Tortoises; A compilation Project of the IUCN/SSC Tortoise and Freshwater Turtle Specialist Group. Chelonian Research Monographs No. 5, pp. 021.1-021.4, doi;10.3854/crm.5.021.atra.vl.2008, http://www.iucn-tftsg.org/cbftt/.

Christensen, R. H. 1975a. Breeding Central American wood turtles. Chelonia 2(2):8–10.

Christensen, R. H. 1975b. An update to "Breeding Central American wood turtles." Chelonia 2(3):20.

Christiansen, J. L., and A. E. Dunham. 1972. Reproduction of the yellow mud turtle *(Kinosternon flavescens flavescens)* in New Mexico. Herpetologica 28(2):130–137.

Christiansen, J. L., and B. J. Gallaway. 1984. Raccoon removal, nesting success, and hatchling emergence in Iowa turtles with special reference to *Kinosternon flavescens* (Kinosternidae). Southwest. Natur. 29(3):343–348.

Christiansen, J. L., B. J. Gallaway, and J. W. Bickham. 1990. Population estimates and geographic distribution of the yellow mud turtle *(Kinosternon flavescens)* in Iowa. J. Iowa Acad. Sci. 97(3):105–108.

Christiansen, J. L., and E. O. Moll. 1973. Latitudinal reproductive variation within a single subspecies of painted turtle, *Chrysemys picta belli*. Herpetologica 29:152–163.

Christiansen, J. L., J. A. Cooper, J. W. A. Bickham, B. J. Gallaway, and M. D. Springer. 1985. Aspects of the natural history of the yellow mud turtle *Kinosternon flavescens* (Kinosternidae) in Iowa: a proposed endangered species. Southwest. Nat. 30(3):413–425.

Clark, P. J., M. A. Ewert, and C. E. Nelson 2001. Physical apurtures as constraints on egg size and shape in the common musk turtle, Sternotherus odoratus, Functioinal Ecology, 15(1):70–71.

Clavigero, F. J. 1789. Storia della California. Venice. Translated and annotated as The history of [lower] California, by S. E. Lake and A. A. Gray, 1937. Stanford: Stanford Univ. Press.

Cochran, P. A., and D. R. McConville. 1983. Feeding by *Trionyx spiniferus* in backwaters of the upper Mississippi. J. Herpetol. 17:82–86.

Collins, J. T. 1997. Standard common and current scientific names for North American amphibians and reptiles. 4th ed. SSAR Herp. Circ. 25:1–44.

Compagno, L. J. V. 1984. FAO species catalogue. Vol. 4. Sharks of the World. An annotated and illustrated catalogue of shark species known to date. Pt. 2. Carcharhiniformes. FAO Fish. Synop. (125)2:251–655.

Conant, R. 1958. A Field Guide to Reptiles and Amphibians of the United States and Canada East of the 100th Meridian. Houghton Mifflin Co., Boston.

Conant, R. 1963. Semi aquatic snakes, of the Genus *Thamnophis* from the isolated drainage system of the Rio Nazas and adjacent areas in Mexico. Copeia 1963 (3):473–499.

Conant, R. 1969. A review of the water snakes of the genus *Natrix* in Mexico. Bull. American Mus. Nat. Hist., 142(1):1–140.

Conant, R. 1975. A Field Guide to Reptiles and Amphibians of Eastern and Central North America, 2nd Ed. Houghton Mifflin Co., Boston.

Conant, R., and J. F. Berry. 1978. Turtles of the Family Kinosternidae in the Southwestern United States and adjacent Mexico: identification and distribution. American Museum Novitates 2642:1–18.

Congdon, J. D., and D. W. Tinkle. 1982. Reproductive energetics of the painted turtle *(Chrysemys picta)*. Herpetologica 38:228–237.

Congdon, J. D., and J. W. Gibbons. 1983. Relationships of reproductive characteristics to body size in *Pseudemys scripta*. Herpetologica 39:147–151.

Congdon, J. D., G. L. Breitenbach, R. C. Van Loben Sels, and D. W. Tinkle. 1987. Reproduction and nesting ecology of snapping turtle *(Chelydra serpentina)* in southeastern Michigan. Herpetologica 43(1):39–54.

Congdon, J. D., R. D. Nagle, O. M. Kinney, and R. C. Van Loben Sels. 2001. Hypotheses of aging in a long-lived

vertebrate, Blanding's turtle (Emydoidea blandingii). Experimental Gerontology 36(2001):813–827.
Cooper, J. G. 1863. Description of *Xerobates agassizii*. Proc. California Acad. Sci. 2:118–123.
Cope, E. D. 1865. Third contribution to the herpetology of tropical America. Proc. Acad. Nat. Sci. Philadelphia 17:185–198.
Cope, E. D. 1869. Seventh contribution to the herpetology of tropical America. Proc. Amer. Philosoph. Soc. 11:147–169.
Cope, E. D. 1872. Synopsis of the species of Chelydrinae. Proc. Acad. Nat. Sci. Philadelphia 1872:22–29.
Cragin, F. W. 1894. Herpetological notes from Kansas and Texas. Colo. Coll. Studies 5:37–39.
Crumly, C. R. 1984. The genus name for North American gopher tortoises. Proc. Symp. Desert Tortoise Council 1984:147–148.
Crumly, C. R. 1994. Phylogenetic systematics of North American tortoises (genus *Gopherus*): evidence for their classification. Fish and Wildlife Research 13:7–32.
Crumly, C. R., and L. L. Grismer. 1994. Validity of the tortoise *Xerobates lepidocephalus* Ottley and Velazques in Baja California. Fish and Wildlife Research 13:33–37.
Dalrymple, G. H. 1979. Packaging problems of head retraction in trionychid turtles. Copeia 1979 (4):655–660.
Dantzler, W. H. 1976. Renal function (with special emphasis on nitrogen excretion). In: C. Gans and W. R. Dawson, Biology of the Reptilia, 5: Physiology A, 447–504.
Dantzler, W. H., and B. Schmidt-Nielsen. 1966. Excretion in fresh-water turtle *(Pseudemys scripta)* and desert tortoise *(Gopherus agassizii)*. Am. J. Physiol. 210:198–210.
David, P. 1994. Liste des reptiles actuels du monde. I. Chelonii. Dumerilia 1:7–127.
Davis, J. D., and C. G. Jackson, Jr. 1970. Copulatory behavior in the red-eared turtle, *Pseudemys scripta elegans* (Wied). Herpetologica 26:238–240.
Davis, J. D., and C. G. Jackson, Jr. 1973. Notes on the courtship of a captive male *Chrysemys scripta taylori*. Herpetologica 29(1):62–64.
Davis, W. B., and J. R. Dixon. 1961. Reptiles (exclusive of snakes) from the Chilpancingo region, Mexico. Proc. Biol. Soc. Washington 74:37–56.
De Beer, G. R. 1949. Caruncles and egg teeth: some aspects of the concept of homology. Proc. Linnean Soc. London 161(2):218–224.
De la Torre Loranca, M. A. 2004. Propuesto de manejo de las poblaciónes de tortugas (*Kinosternon leucostomum* y *Staurotypus triporcatus* en el Ejido " La Margarita," Catemaco, Veracruz, México. MS thesis, Instituto de Ecologia, A. C. Jalapa, Veracruz, Mexico.
Dean, R. H. 1980. Selected Aspects of the Ecology of the Central American Mud Turtle, *Staurotypus salvinii*. Unpublished MS thesis, Texas A&M Univ., College Station.
deBraga, M., and O. Rieppel. 1997. Reptile phylogeny and the interrelationships of turtles. Zool. J. Linn. Soc. 120:281–354.
Degenhardt, W. G., C. W. Painter, and A. H. Price. 1996. Amphibians and Reptiles of New Mexico. Univ. New Mexico Press, Albuquerque.
Degenhardt, W. G., and J. L. Christiansen. 1974. Distribution and habitats of turtles in New Mexico. Southwest. Natur. 19:21–46.
Deraniyagala, P. E. P. 1939. The Tetrapod Reptiles of Ceylon. Vol. 1. Testudinates and Crocodilians. Dulau and Co., Ltd., London.
DeSola, R. 1931. Sex determination in a species of the Kinosternidae, with notes on sound production in reptiles. Copeia 1931: (3)124–125.
Dessauer, H. C. 1970. Blood chemistry of reptiles: physiological and evolutionary aspects. In: C. Gans and T. S. Parsons, Biology of the Reptilia, 3: Morphology C, 1–72.
Dessauer, H. C., and W. Fox. 1956. Characteristic electrophoretic patterns of plasma proteins of orders of Amphibia and Reptilia. Science 124:225–226.
Devaux, B., and J. Buskirk. 2001. Les Terrapene. La Tortue 55:14–31.
Dice, L. R. 1943. The Biotic Provinces of North America. Ann Arbor: Univ. Michigan Press.
Diemer, J. E. 1989. *Gopherus polyphemus,* Gopher tortoise. In: I. R. Swingland, and M. W. Klemens (Eds.). The Conservation Biology of Tortoises. Occas. Pap. IUCN Species Survival Commission No. 5, pp. 14–16.
Ditmars, R. L. 1934. A review of the box turtles. Zoologica 17–18:1–44.
Dixon, J. R. 2000. Amphibians and Reptiles of Texas with Keys, Taxonomic Synopses, Bibliography, and Distribution Maps. 2nd ed. Texas A and M Univ. Press, College Station, Texas.
Dodd, C. K., Jr. 1978. A note on the defensive posturing of turtles from Belize, Central America. Herp. Review 9(1):11.
Dodd, C. K., Jr. 1982. A controversy surrounding an endangered species listing: the case of the Illinois mud turtle. Smithson. Herp. Inform. Serv. 55:1–22.
Dodd, C. K., Jr. 1983. A review of the status of the Illinois mud turtle *Kinosternon flavescens spooneri* Smith. Biol. Conserv. 27:141–156.
Dodd, C. K., Jr. 2001. North American Box Turtles. A Natural History. Univ. Oklahoma Press, Norman.
Doroff, A. M., and L. B. Keith. 1990. Demography and ecology of an ornate box turtle *(Terrapene ornata)* population in south-central Wisconsin. Copeia 1990:387–399.
Dorrian, C., and H. Ehmann. 1988. Aquatic defenses of the Eastern snake-necked turtle, *Chelodina longicollis*. Abstract. Australian Bicentennial Herpetological Conference, Queensland Museum, Brisbane, 17–22 August 1988.
Dowler, R. C., and J. W. Bickham. 1982. Chromosomal relationships of the tortoises (family Testudinidae). Genetica 58:189–197.
Duellman, W. E. 1958a. A monographic study of the colubrid snake genus *Leptodeira*. Bull. Am. Mus. Nat. Hist. 94:567–594.
Duellman, W. E. 1958b. A preliminary analysis of the herpetofauna of Colima, Mexico. Occ. Pap. Mus. Zool. Univ. Mich. 589:1–22.
Duellman, W. E. 1961. The amphibians and reptiles of Michoacan, Mexico. Univ. Kansas Publ. Mus. Nat Hist. 15:1–148.
Duellman, W. E. 1963. Amphibians and reptiles of the rainforests of southern El Petén, Guatemala. U. Kans. Publ. Mus. Nat. Hist. 15(5):205–249.
Duellman, W. E. 1965. Amphibians and reptiles from the Yucatan Peninsula, Mexico. Univ. Kansas Publ. Mus. Nat. Hist. 15:577–614.

Duméril, A. H. A. 1870. Etude sur les reptiles. Mission scientifique au Mexique et dan l'Amerique Centrale. Recherches zoologiques, 3rd part; Paris, Imprimerie Imperial, (first section. Livr. 1), 32p., pls. 1-7, 9. 11, 12.

Duméril, A. H. A., M. F. Bocourt, and F. Mocquard. 1870–1890. Mission scíentifique au Mexique et dans l'Amerique Centrale. Recherches zoologiques. Troisieme Partie. Paris: Imprimerie Imperiale.

Duméril, A. M. C., and A. H. A. Duméril. 1851. In: Catalogue methodique de la collection des reptiles du Museum d'Histoire Naturelle. pp. i–iv, 1–224. Gide and Boudry, Paris.

Dunn, E. R. 1931. The herpetological fauna of the Americas. Copeia 1931(3):106–119.

Dunn, E. R. 1940. Some aspects of herpetology in lower Central America. Trans, New York Acad. Science, vol. 2, pp. 156–158.

Dunn, E. R., and L. C. Stuart. 1951. Comments on some recent restrictions of type-localities of certain South and Central American amphibians and reptiles. Copeia 1951(1):55–61.

Dunson, W. A. 1960. Aquatic respiration in *Trionyx spinifer asper*. Herpetologica 16:277–283.

Dunson, W. A. 1976. Salt glands in reptiles. In C. Gans and W. R. Dawson, Biology of the Reptilia, 5: Physiology A, 413–446.

Dyer, W. G., and J. L. Carr. 1990. Some digeneans of the Neotropical turtle genus Rhinoclemmys in Mexico and South America. J. Helminthol. Soc. Wash. 57:12–14.

Edmund, A. G. 1969. Dentition. In C. Gans, A. d'A. Ballairs, and T. S. Parsons, Biology of the Reptilia, 1: Morphology A, 117–200.

Edwards, T., R. Murphy, K. Berry, C. Melendez Torres, M. Villa Andrade, M. Mendez de la Cruz, and M. Vaughn. 2009a. Unraveling the genetic history of desert tortoises *(Gopherus agassizii)* in Sonora, Mexico. Abstracts: pp. 7–8, The Desert Tortoise Council. 34th annual symposium, Henderson, Nevada, 20–22 Feb. 2009.

Edwards, T., V. Buzzard, R. Murphy, K. Bonine, J. Phillips, and J. Truet. 2009b. Genetic assessment of a captive population of bolsón tortoises (Gopherus flavomarginatus) in New Mexico. Abstracts: pp. 8–9, The Desert Tortoise Council. 34th annual symposium, Henderson, Nevada, 20–22 Feb. 2009.

Ehrenfeld, J. G., and D. W. Ehrenfeld. 1973. Externally secreting glands of freshwater and sea turtles. Copeia 1973:305–314.

Eisner, T., T. H. Jones, J. Meinwald and J. M. Legler. 1978. Chemical composition of the odorous secretion of the Australian turtle, *Chelodina longicollis*. Copeia 1978(4):714–715.

Engstrom, T. N., H. B. Shaffer, and W. P. McCord. 2004. Multiple data sets, high homoplasy, and the phylogeny of softshell turtles (Testudines: Trionychidae). Systematic Biology 53:693–710.

Ernst, C. H. 1971. Observations of the painted turtle, *Chrysemys picta*. J. Herpetol. 5:216–220.

Ernst, C. H. 1978. A revision of the neotropical turtle genus *Callopsis* (Testudines: Emydidae: Batagurinae). Herpetologica 34(2):113–134.

Ernst, C. H. 1980. *Rhinoclemmys areolata*. Cat. Amer. Amph. Rept 251.1–251.2.

Ernst, C. H. 1981. *Rhinoclemmys pulcherrima*. Cat. Amer. Amph. Rept. 275.1–275.2.

Ernst, C. H. 1990. *Pseudemys gorzugi*. Cat. Amer. Amphib. Rept. 461.1–2.

Ernst, C. H., J. E. Lovich, and R. W. Barbour. 1994. Turtles of the United States and Canada. Washington: Smithsonian Institution Press.

Ernst, C. H., and J. F. McBreen. 1991. *Terrapene carolina*. Catalog. Amer. Amphib. Rept. 512:1–13.

Ernst, C. H., J. W. Gibbons, and S. S. Novak. 1988. *Chelydra*. Cat. Amer. Amphib. Rept. 419.1–4.

Ernst, C. H., and R. W. Barbour. 1972. Turtles of the United States. Univ. Press, Kentucky, Lexington.

Ernst, C. H., and R. W. Barbour. 1989. Turtles of the World. Smithsonian Institution Press.

Ernst, E. M., and C. H. Ernst. 1977. Synopsis of helminthes endoparasitic in native turtles of the United States . Bull. Maryland Herpetol. Soc. 13(1)1–75.

Espejel, V. G. 2004. Aspectos biologícos del manejo del chopontil *Claudius angustatus* (Testudines: Staurotypidae). MS thesis, Manejo de Fauna Silvestre, Instituto de Ecológia, A. C., Xalapa Veracruz.

Espejel, V. G., R. C. Vogt, and M. A. López-Luna. 1998. Aspectos de ecologia del chopontil *Claudius angustatus* (Testudines; Staurotypidae), en el sureste de Veracruz. Resumenes V. Reunion Nacional de Herpetologia. Sociedad Herpetologos Méxicanoa, Xalapa, Veracruz.

Esque, T. C., and E. L. Peters. 1994. Ingestion of bones, stones, and soil by desert tortoises. Fish and Wildlife Research 13:105–111.

Etchberger, C. R., M. A. Ewert, J. B. Phillips, C. E. Nelson, and H. D. Prange. 1992. Physiological responses to carbon dioxide in embryonic red-eared slider turtles, *Trachemys scripta*. J. Exp. Zool. 264:1–10.

Evans, L. T. 1953. The courtship pattern of the box turtle, *Terrapene c. carolina*. Herpetologica 9:189–192.

Evans, L. T. 1961. Structure as related to behavior in the organization of populations of reptiles. In: F. Blair (Ed.). Vertebrate speciation. Austin: Univ. of Texas Press, pp. 148–178.

Evans, L. T. 1968. The evolution of courtship in the turtle species, Terrapene carolina. Amer. Zool. 8:695–696.

Ewert, M. A. 1979. The embryo and its egg: development and natural history. In: M. Harless and H. Morlock (Eds.). Turtles: Perspectives and Research. New York: John Wiley and Sons, pp. 333–413.

Ewert, M. A. 1981. Embryonic diapause and embryonic aestivation in turtle eggs: definition and two environmental correlates. Bull. Ecol. Soc. America 62(2):156.

Ewert, M. A. 1985. Embryology of turtles. In: C. Gans, F. Billett, and P. F. A. Maderson (Eds.). Biology of the Reptilia, Vol. 14, Development A. New York: John Wiley and Sons, pp. 75–267.

Ewert, M. A. 1991. Cold torpor, diapause, delayed hatching and aestivation in reptiles and birds. In: D. C. Deeming and M. W. J. Ferguson (Eds.) Egg Incubation: Its Effects on Embryonic Development in Birds and Reptiles. Cambridge, England, pp. 173–191.

Ewert, M. A., and C. E. Nelson. 1991. Sex determination in turtles: diverse patterns and some possible adaptive values. Copeia 1991(1):50–69.

Ewert, M. A., and D. S. Wilson. 1996. Seasonal variation of embryonic diapause in the striped mud turtle *(Kinosternon baurii)* and general considerations for conservation planning. Chelonian Conservation Biology 2(1):43–54.

Ewert, M. A., D. R. Jackson, and C. E. Nelson. 1994. Patterns of temperature-dependent sex determination in turtles. J. Exper. Zool. 270:3–15.

Ewert, M. A., and J. M. Legler. 1978. Hormonal induction of oviposition in turtles. Herpetologica 34:314–318.

Ewert, M. A., S. J. Firth, and C. E. Nelson. 1984. Normal and multiple egg shells in batagurine turtles and their implications for dinosaurs and other reptiles. Can. J. Zool. 62:1834–1841.

Ewing, H. E. 1943. Continued fertility in female box turtles following mating. Copeia 1943:112–114.

Feldman, C. R., and J. F. Parham. 2002. Molecular phylogenetics of emydine turtles: taxonomic revision and the evolution of shell kinesis. Molecular Phylogenetics and Evol. 22 (3):388–398.

Feldman, M. 1982. Notes on reproduction in *Clemmys marmorata*. Herp. Review 13:10–11.

Felger, R. S., K. Clifton, and P. J. Regal. 1976. Winter dormancy in sea turtles: independent discovery and exploitation in the Gulf of California by two local cultures. Science 191:283–285.

Feuer, R. C. 1966. Variation in Snapping Turtles, *Chelydra serpentina* (Linnaeus): A Study in Quantitative Systematics. PhD dissertation. Univ. Utah, Salt Lake City.

Feuer, R. C. 1971. Intergradation of the snapping turtles *Chelydra serpentina serpentina* (Linnaeus, 1758) and *Chelydra serpentina osceola* Stejneger, 1918. Herpetologica 27:379–384.

Fichter, L. S. 1969. Geographical distribution and osteological variation in fossil and recent specimens of two species of *Kinosternon* (Testudines). J. Herpet. 3(3–4):113–119.

Finneran, L. C. 1948. Reptiles at Branford, Connecticut. Herpetologica 4:123–126.

Fioroni, P. 1962. Der Eizahn und die Eischwiele der reptilien, Acta Anat. 49:328–366.

Fitch, H. S. 1982. Reproductive cycles in tropical reptiles. Occ. Pap. Mus. Natur. Hist. Univ. Kansas. 96:1–53.

Fitch, H. S. 1985. Variation in clutch and litter size in New World reptiles. Univ. Kansas Mus. Natur. Hist. Misc. Publ. 76:1–76.

Fitzinger, L. J. 1826. Neue Classification der Reptilien, nach ihren Natürlichen Verwandtschaften nebst einer Verwandtschafts-Tafel und einem Verzeichnisse der Reptilien-Sammlung des k.k. Zoologischen Museum zu Wien. Wien: J.G. Hübner Verlagen.

Fitzinger, L. J. 1835. Entwurf einer systematischen Anordnung der Schildkröten nach den Grundsatzen der naturlichen Methode. Ann. Mus. Naturgesch. Wien 1:105–128.

Flannery, K. V. 1967. Vertebrate fauna and hunting patterns. Chapter 8, pp. 132–177 *In* Byers, Douglas S. 1967. The prehistory of the Tehuacan Valley. Vol.1 Environment and Subsistence. Univ. Texas Press, Austin and London.

Flickinger, E. L., and B. M. Mulhern. 1980. Aldrin persists in yellow mud turtle. Herp. Rev. 11(2):29–30.

Flores-Villela, O. A. 1980. Reptiles de importancia economíca en México, ´Tesis de licenciatura em Biología UNAM, 250 p.

Flores-Villela, O. A. 1986. Ciclo reproductivo de la hembra de Claudius *angustatus* en el sur de Veracruz. Abst. Simp. Tortugas Agua Dulce Neotropicales.

Flores-Villela, O. A., E. H. García, and A. N. Montes Oca. 1991. Catalogo de Anfibios y Reptiles del Museo de Zoologia, Facultad de Ciéncias, Universidad Nacional Autonoma de México, México D. F.

Flores-Villela, O. A., and G. R. Zug. 1994. Reproductive biology of the chopontil, *Claudius angustatus* (Testudines: Kinosternidae), in southern Veracruz, México. Chelonian Conservation and Biology 1(3):181–186.

Flores-Villela, O. A., H. M. Smith, and D. Chiszar. 2004. The history of herpetological exploration in Mexico. Bonner zoologishe Beitrage, 52:3/4, 311–335.

Flores-Villela, O. A., and R. C. Vogt. 1984. Food habits and morphometrics of the chopontil *Claudius angustatus* in Veracruz, México. Abstr. ASIH/SSAR/HL Norman Mtgs.

Fox, H. 1977. The urogenital system of reptiles. In C. Gans and T. S. Parsons, , Biology of the Reptilia, 6: Morphology E,. 1–158. Academic Press, New York.

Frair, W. 1964. Turtle family relationships as determined by serological tests. In: C. A. Leone (Ed.) Taxonomic Biochemistry and Serology, pp. 535–544. The Ronald Press Company, New York.

Frair, W. 1982. Serological studies of *Emys,* Emydoidea, and some other testudinid turtles. Copeia 1982(4):976–978.

Frair, W., R. G. Ackman, and N. Mrosovsky. 1972. Body temperature of *Dermochelys coriacea:* warm turtle from cold water. Science 177:791–793.

Fritts, T. H., and R. D. Jennings. 1994. Distribution, habitat use, and status of the desert tortoise in Mexico. Fish and Wildlife Research 13:49–56.

Fritz, U., and P. Havas. 2007. Checklist of Chelonians of the World. Published online at http://www.Vertebrate-Zoology.

Gadow, H. 1909. Amphibia and Reptiles. MacMillan and Co., London.

Gaffney, E. S. 1975a. A phylogeny and classification of the higher categories of turtles. Bull. Amer. Mus. Natur. Hist. 155:387–436.

Gaffney, E. S. 1975b. Phylogeny of the chelydrid turtles: a study of shared derived characters in the skull. Fieldiana: Geol. 33:157–178.

Gaffney, E. S. 1979. Comparative cranial morphology of recent and fossil turtles. Bull. Amer. Mus. Natur. Hist. 164:65–376.

Gaffney, E. S. 1990. The comparative osteology of the triassic turtle *Proganochelys*. Bul. Amer. Mus. Nat. Hist. 194:1–263.

Gaffney, E. S., J. C. Balouet, and F. de Broin. 1984. New occurrences of extinct meiolaniid turtles in New Caledonia. Am. Mus. Novitates 2800:1–6

Gaffney, E.S., and P. A. Meylan. 1988. A phylogeny of turtles. In M. J. Benton, The Phylogeny and Classification of the Tetrapods. Vol. I, Amphibians, Reptiles, Birds. Sys. Assoc., spec. vol., 35A, pp. 157–219. Clarendon Press, Oxford.

Gage, S. H. 1884. Pharyngeal respiration in the soft-shelled turtle *(Aspidonectes spinifer)*. Proc. Amer. Assoc. Adv. Sci. 32:316.

Gage, S. H., and S. P. Gage. 1886. Aquatic respiration in soft-shelled turtles: a contribution to the physiology of respiration in vertebrates. Amer. Natur. 20:233–236.

Galbraith, D. A. 1993. Multiple paternity and sperm storage in turtles. Herpetol. J. 3(4):117–123.

Galbraith, D. A., R. J. Brooks, and M. E. Obbard. 1989. The influence of growth rate on age and body size at maturity in female snapping turtles *(Chelydra serpentina).* Copeia 1989(4):896–904.

Galeotti, P., R. Sacchi, D. P. Rosa, and M. Fasola. 2005b. Female preference for fast-rate, high-pitched calls in Hermann's tortoises—*Testudo hermanni*. Behav. Ecol. 16:301–308.

Galeotti, P., R. Sacchi, M. Fasola, D. Pellitteri, D. P. Rosa, M. Marchesi, and D. Ballasina. 2005a. Courtship displays and mounting calls are honest condition-dependent signals that influence mounting success in Hermann's tortoises. Can. J. Zool. 83:1306–1313.

Gans, C., and G. M. Hughes. 1967. The mechanism of lung ventilation in the tortoise *Testudo graeca* Linné. J. Exp. Biol. 47:1–20.

García Aguayo, A., and G. Ceballos. 1994. Guía de campo de los reptiles y anfibios de la costa de Jalisco. Field Guide to the reptiles and amphibians of the Coast of Jalisco. Fundación Ecologica de Cuixmala, A. C., Instituto de Biología, UNAM, México D. F.

Gardner, J. D., A. P. Russell, and D. B. Brinkman. 1995. Systematics and taxonomy of soft-shelled turtles (family Trionychidae) from the Judith River Group (mid-Campanian) of North America. Can. J. Earth Sci. 32:631–643.

Gaunt, A. S., and C. Gans. 1969. Mechanics ofrespiration in the snapping turtle, *Chelydra serpentina* (Linné). J. Morph. 128(2):195–227.

Gazin, C. L. 1957. Exploration for the remains of giant ground sloths in Panama. Smithsonian Institution Annual Report 4279:341–354.

Gehlbach, F. 1956. Annotated records of southwestern amphibians and reptiles. Trans. Kansas Acad., Sci. 59(3):364–372.

Gehlbach, F. H. 1965. Amphibians and reptiles from the Pliocene and Pleistocene of North America: A chronological summary and selected bibliography. Texas J. Sci. 17(1):56–70.

George, J. C., and R. V. Shah. 1954. The occurrence of a striated outer muscular sheath in the lungs of *Lissemys punctata granosa* Schoepff. Jour. Anim. Morph. and Phys. 1:13–16.

Georges, A., and M. Adams. 1992. A phylogeny for Australian chelid turtles based on allozyme electrophoresis. Aust. J. Zool. 40:453–476.

Germano, D. J. 1992. Longevity and age size relationships of populations of desert tortoises. Copeia 1992:367–374.

Germano, D. J. 1993. Shell morphology of North American tortoises. Amer. Midl. Nat. 129(2):319–335

Germano, D. J. 1994a. Comparative life histories of North American tortoises. In R. B. Bury and D. J.Germano (Eds.). Biology of North AmericanTortoises, pp. 174–185. Fish and Wildlife Research 13, Technical Report Series, U.S. Department of the Interior, National Biological Survey, Washington, D.C.

Germano, D. J. 1994b. Growth and age at maturity of North American tortoises in relation to regional climates. Can. J. Zool. 72(5):918–931.

Germano, D. J. 1999. *Terrapene ornata luteola* (desert box turtle). Attempted predation. Herp. Rev. 30(1): 40–41.

Germano, D. J., and M. A. Joyner. 1988. Changes in a desert tortoise *(Gopherus agassizii)* population after a period of high mortality. In: R. C. Szaro, K. E. Severson, and D. R. Patton (Eds.). Management of Amphibians, Reptiles, and Small Mammals in North America. pp. 190–198. USDA Tech Serv. Gen. Tech. Rep. RM-166.

Germano, D. J., and P. M. Nieuwolt-Dacanay. 1999. *Terrapene ornata luteola* (desert box turtle). Homing behavior. Herp. Rev. 30(2):96.

Germano, D. J., T. H. Fritts, and P. A. Medica. 1994. Range and habitats of the desert tortoise. Fish and Wildlife Research 13:73–84.

Gibbons, J. W. 1968. Population structure and survivorship in the painted turtle, *Chrysemys picta*. Copeia 1968:260–268.

Gibbons, J. W. 1970a. Reproductive dynamics of a turtle *(Pseudemys scripta)* population in a reservoir receiving heated effluent from a nuclear reactor. Can. J. Zool. 48:881–885.

Gibbons, J. W. 1970b. Sex ratios in turtles. Res. Popul. Ecol. 12:252–254.

Gibbons, J. W. 1976. Aging phenomena in reptiles. In: M. F. Elias, B. E. Eleftheriou, and P. K. Elias (Eds.). Special Review of Experimental Aging Research, pp. 453–475. EAR, Bar Harbor, Maine.

Gibbons, J. W. 1987. Why do turtles live so long? Bioscience 37:262–269.

Gibbons, J. W., and J. L. Greene. 1979. X-ray photography: a technique to determine reproductive patterns of freshwater turtles. Herpetologica 35:86–89.

Gibbons, J. W., R. D. Semlitsch, J. L. Greene and J. P. Schubauer. 1981. Variation in age and size at maturity of the slider turtle *(Pseudemys scripta).* Amer. Natur. 117:841–845.

Gibbons, J. W., and R. D. Semlitsch. 1982. Survivorship and longevity of a long-lived vertebrate species: how long do turtles live? J. Anim. Ecol. 51:523–527.

Gibbons, J. W., S. S. Novak, and C. H. Ernst. 1988. *Chelydra serpentina*. Cat. Amer. Amphib. Rept. 420:1–4.

Giles, J. C. 2005. The underwater acoustic repertoire of the long-necked, freshwater turtle *Chelodina oblonga*. PhD dissertation. Murdoch Univ., Western Australia.

Giles, J. C., J. A. Davis, R. D. McCauley, and G. Kuchling. 2009. Voice of the turtle: the underwater acoustic repertoire of the long-necked freshwater turtle, *Chelodina oblonga*. Acoustical Society of America (DOI:10.1121/1.3148209).

Gilmore, C. W. 1922. A new fossil turtle, *Kinosternon arizonense,* from Arizona. Proc. U.S. Nat. Mus. 62:1–8.

Gist, D. H., and J. D. Congdon. 1998. Oviductal sperm storage as a reproductive tactic of turtles. J. Exp. Zoolog. 282 (4–5):526–534, illustr.

Gist, D. H., and J. M. Jones. 1989. Sperm storage within the oviduct of turtles. J. Morphol. 199:379–384.

Glass, B. and N. Hartweg. 1951. *Kinosternon murrayi,* a new musk turtle of the *hirtipes* group from Texas. Copeia 1951(1):50–52.

Glenn, J. L. 1983. A note on the longevity of a captive desert tortoise *(Gopherus agassizi)*. Proc. Symp. Desert Tortoise Council 1983:131–132.

Glenn, J. L., R. C. Straight, and J. W. Sites, Jr. 1990. A plasma protein marker for population genetic studies of the desert tortoise *(Xerobates agassizi)*. Great Basin Natur. 50:1–8.

Goldman, E. A. 1951. Biological investigations in Mexico. Smith. Misc. Coll. 155:1–476.

Goldman, E. A., and R. T. Moore. 1945. The biotic provinces of Mexico. J. Mammal. 26(4):347–360.

González-Trapaga, R. G. 1995. Reproduction of the Bolsón tortoise, *Gopherus flavomarginatus*, Legler 1959. Publicaciones De La Sociedad Herpetologica Mexicana 2:32–36.

González-Trapaga, R. G., G. Aguirre, and G. A. Adest. 2000. Sex-steroids associated with the reproductive cycle in male and female bolsón tortoise, *Gopherus flavomarginatus*. Acta Zoologica Mexicana (n.s.) 80:101–117.

Goode, J. M. 1991. Breeding semi-aquatic and aquatic turtles at the Columbus Zoo. In: K. R. Beaman, F. Caporaso, S. McKeown, and M. D. Graff (Eds.). Proceedings of the First International Symposium on Turtles and Tortoises: Conservation and Captive Husbandry, 1990. pp. 66–76, Los Angeles: Circle Printing Co.

Goode, J. M. 1994. Reproduction in captive neotropical musk and mud turtles (*Staurotypus triporcatus, S. salvinii*, and *Kinosternon scorpioides*) In J. B. Murphy, K. Adler, and J. T. Colins (Eds.). Captive Managements and Conservation of Amphibians and Reptiles. SSAR, Ithaca, (N.Y.). Contributions to Herpetology 11:275–295.

Goodman, R. H., Jr. 1997. Occurrence of double clutching in the southwestern pond turtle, *Clemmys marmorata pallida*, in the Los Angeles Basin. Chel. Cons. Biol. 2(3):419–420.

Graham, T. E., and V. H. Hutchison. 1969. Centenarian box turtles. Int. Turtle Tortoise Soc. J. 3:25–29.

Granda, A. M., and D. F. Sisson. 1992. Retinal function in turtles. In C. Gans and P. S. Ulinski, , Biology of the Reptilia, 17: Neurology C, pp. 136–174. Academic Press, New York.

Grant, C. 1960a. Differentiation of the southwestern tortoises (genus *Gopherus*) with notes on their habits. Trans. San Diego Soc. Natur. Hist. 12:441–448.

Grant. C. 1960b. *Gopherus*. Herpetologica 16:29–31.

Gräper, L. 1931. Zur vergleichenden Anatomie der Schildkrötenlunge. Gegenbaur's Morphologisches Jahrbuch. Abteilung I, Morphologie und Mikroskopische Anatomie, 78: 22–374.

Gray, J. E. 1831. Synopsis Reptilium or Short Descriptions of the Species of Reptiles. Part 1. Cataphracta, Tortoises, Crocodiles, and Enaliosaurians. London.

Gray, J. E. 1847. Description of a new genus of Emydae. Proc. Zool. Soc. London 1847 (6):55–56.

Gray, J. E. 1849. Description of a new species of box tortoise from Mexico. Proc. Zool. Soc. London 17:16–17.

Gray, J. E. 1855a. Description of a new genus and some new species of tortoises. Ann. Mag. Nat. Hist. (2)15:67–69.

Gray, J. E. 1855b. Catalogue of the Shield Reptiles in the Collection of the British Museum. Part 1. Testudinata (Tortoises). Taylor and Francis, London.

Gray, J. E. 1864. Description of a new species of *Staurotypus* (*S. salvinii*) from Guatemala. Proc. Zool. Soc. London 1864:127–128.

Gray, J. E. 1869. Notes on the families and genera of tortoises (Testudinata) and on the classification afforded by the study of their skulls. Proc. Zool. Soc. London 1869:165–225.

Gray, J. E. 1870. Supplement to the Catalogue of Shield Reptiles in the Collection of the British Museum. British Mus., London.

Gray, J. E. 1873. Hand list of the specimens of shield reptiles in the British Museum. London p.1–124.

Greenbaum, E., and J. L. Carr. 1996. Sexual differentiation in spiny softshell turtles *(Apalone spinifera)*. Proposal to Chelonian Research Foundation Linnaeus Fund.

Greenbaum, E., and J. L. Carr. 2001. Sexual differentiation in the spiny softshell turtle *(Apalone spinifera)*, a species with genetic sex determination. J. Exp. Zool. 290:190–200.

Greer, A. E., J. D. Lazell, Jr., and R. M. Wright. 1973. Anatomical evidence for a counter-current heat exchanges in the leatherback turtle *(Dermochelys coriacea)*. Nature 244:181.

Grismer, L. L. 2002. Amphibians and Reptiles of Baja California Including Its Pacific islands and the Islands in the Sea of Cortes. Univ. of California Press, Berkeley and Los Angeles.

Grismer, L. L., and J. A. McGuire. 1993. The oases of central Baja California, Mexico. Part I. A preliminary account of the relict mesophilic herpetofauna and the status of the oases. Bull. Southern California Acad. Sci. 92:2–24.

Günther, A. C. L. G. 1885. Biología Centrali-Americana. Reptilia and Batrachia. In: O. Salvin and F. D. Godman (Eds.). Biologíca Centrali-Americana. Porter, London.

Gutzke, W. H. N., and J. J. Bull. 1986. Steroid hormones reverse sex in turtles. Gen. Comp. Endocrinol. 64:368–372.

Haiduk, M. W., and J. W. Bickham. 1982. Chromosomal homologies and evolution of testudinoid turtles with emphasis on the systematic placement of *Platysternon*. Copeia 1982(1):60–66.

Halk, J. H. 1986. Life history notes. *Trionyx spiniferus* (spiny softshell turtle). Size. Herp. Rev. 17(4):91.

Hammer, D. A. 1969. Parameters of a marsh snapping turtle population, Lacreek Refuge, South Dakota. J. Wildl. Manag. 33:995–1005.

Harding, J. H., and S. K. Davis. 1999. Natural history notes. *Clemmys insculpta* (wood turtle) and *Emydoidea blandingii* (Blanding's turtle). Hybridization. Herp. Rev. 30(4): 225–226.

Hardy, L. M., and R. W. McDiarmid. 1969. The amphibians and reptiles of Sinaloa, Mexico. Univ. Kansas Publ. Mus. Nat Hist. 18:39–252.

Harper, F. 1940. Some works of Bartram, Daudin, Latreille and Sonnini and their bearing upon North American herpetological nomenclature. Amer. Midl. Nat. 23: 692–723.

Harry, J. L., and D. A. Briscoe. 1988. Multiple paternity in the loggerhead turtle *(Caretta caretta)*. J. Heredity 79:96–99.

Hartweg, N. 1934. Description of a new kinosternid from Yucatan. Occ. Pap. Mus. Zool. U. of Michigan 277:1–2.

Hartweg, N. 1938. *Kinosternon flavescens stejnegeri*, a new turtle from northern Mexico. Occ. Papers Mus. Zool. Univ. Mich. 371:1–5.

Hartweg, N. 1939. A new American *Pseudemys*. Occ. Pap. Mus. Zool. Univ. Michigan 397:1–4.

Hartweg, N., and J. A. Oliver. 1940. A contribution to the herpetology of the Isthmus of Tehuantepec. IV. Misc. Publ. Univ. of Michigan 47:1–31.
Hausmann, P. 1968. Claudius angustatus. Int. Turt. Tort. Soc. Jour. 2(3):14–15.
Hay, O. P. 1908a. Descriptions of five species of North American fossil turtles, four of which are new. Proc. US National Mus., 35 (1640).
Hay, O. P. 1908b. Fossil turtles of North America. Carnegie Inst., Washington, Publ. 75, iv + 568 pp., 113 pls., 704 figs.
Head, J. J., O. A. Aguilera, and M. R. Sanchez-Villagra. 2006. Past colonization of South America by Trionychid turtles: fossil evidence from the Neogene of Margarita Island, Venezuela. J. Herp. 40:(3)378–381.
Hedges, S. B., and L. L. Poling. 1999. A molecular phylogeny of reptiles. Science 283:998–1001.
Hellgren, E, C., R. T. Kazmaier, D. C. Ruthven III, and D. R. Synatzske. 2000. Variation in tortoise life history: demography of *Gopherus berlandieri*. Ecology 81:1297–1310.
Hennig, W. 1966. Phylogenetic Systematics. Univ. Chicago Press, Chicago.
Heringhi, H. L. 1969. An ecological survey of the herpetofauna of Alamos, Sonora, Mexico. Master's thesis, Arizona State Univ.
Hidalgo, H. 1982. Courtship and mating behavior in Rhinoclemmys pulcherrima incisa (Testudines: Emydidae: Batagurinae). Trans. Kansas Acad. Sci. 85(2)82–95.
Himmelstein, J. 1980. Observations and distributions of amphibians and reptiles in the state of Quintana Roo, México. HERP. Bulletin of the New York Herpetological Society. 16:18–34.
Hirayama, R. 1984. Cladistic analysis of batagurine turtles (Batagurinae: Emydidae: Testudinoidea): A preliminary result. In: "Studia Palaeocheloniologica" (F. DeBroin and E. Jimenez-Fruentes Eds.) Stud. Geol. Univ. Salamanca 1:141–157.
Hisaw, F. 1926. Experimental relaxation of the pubic ligament of the guineae pig. Proc. Soc. Exp. Biol. Med. 23: 661–663.
Hofmeyr, M. D., B. T. Henen, and V. J. T. Loehr 2005. Overcoming environmental and morphological constraints: egg size and pelvic kinesis in the smallest tortoise, Homopus signatus. Can. J. Zool. 83: 1343–1352.
Holbrook, J. E. 1842. North American Herpetology; or a Description of the Reptiles Inhabiting the United States. Ed. 2. Philadelphia, J. Dobson. [First edition, vol. 1, 1836]
Holland, D. C. 1988. *Clemmys marmorata* (Western pond turtle). Behavior. Herp Review 19:87–88.
Holland, D.C. 1994. The Western Pond Turtle: Habitat and History. Final report. Report to U.S. Dept. Energy, Bonneville Power Admin., Project No. 92-068.
Holman, J. A. 1963. Observations on dermatemyid and staurotypine turtles from Veracruz, Mexico. Herpetologica 19(4):277–279.
Holman, J. A. 1966. Some Pleistocene turtles from Illinois. Trans. Illinois Acad. Sci. 59(3):214–216.
Holman, J. A. 1969. The Pleistocene amphibians and reptiles of Texas. Publ. Mus. Michigan State Univ. Biol. Ser. 4: 163–192.
Holman, J. A. 1986. Butler Springs herpetofauna of Kansas (Pleistocene, Illinoian) and its climatic significance. J. Herpetol. 20:568–570.
Holman, J. A. 1987. Herpetofauna of the Egelhoff Site (Miocene: Barstovian) of north-central Nebraska. J. Vert. Paleontol. 7:109–120.
Holman, J. A. 2002. Additional specimens of the Miocene turtle Emydoidea hutchisoni Holman 1995-new temporal occurrences, taxonomic characters, and phylogenetic inferences. Journal of Herpetology 36(3):436–446.
Holman, J. A., and A. J. Winkler. 1987. A mid-Pleistocene (Ivingtonian) herpetofauna from a cave in south central Texas. Texas Mem. Mus., Univ. Texas Pearce- Sellards Ser. (44):1–17.
Holman, J. A., and M. E. Schloeder. 1991. Fossil herpetofauna of the Lisco C Quarries (Pliocene: early Blancan) of Nebraska. Trans. Nebraska Acad. Sci. 18:19–29.
Holman, J. A., and R. G. Corner. 1985. A Miocene Terrapene (Testudines: Emydidae) and other Barstovian turtles from south-central Nebraska. Herpetologica 41:88–93.
Holman, J. A., and U. Fritz. 2001. A new emydine species from the middle Miocene (Barstovian) of Nebraska, USA with a new generic arrangement for the species of *Clemmys* sensu McDowell (1964) (Reptilia: Testudines: Emydidae). Zool. Abh. Mus. Teirkde. Dresden 51 (20): 331–354.
Holman, J. A., and U. Fritz. 2005. The box turtle genus Terrapene (Testudines: Emydidae) in the Miocene of the USA. Herpetological Journal, 15:81–90.
Holroyd, P., and J. F. Parham. 2003. The antiquity of African tortoises. Journ. Vert. Paleontology 23(3):688–690.
Holte, D. L. 1998. Nest Site Characteristics of the Western Pond Turtle, *Clemmys marmorata,* at Fern Ridge Reservoir in West Central Oregon. MS dissertation in Wildlife Sciences, Oregon State Univ., Corvallis.
Houseal, T. W., J. W. Bickham, and M. D. Springer. 1982. Geographic variation in the yellow mud turtle, *Kinosternon flavescens.* Copeia 1982(3):567–580.
Howeth, J. G., S. E. McGaugh, and D. A. Hendrickson. 2008. Contrasting demographic and genetic estimates of dispersal in the endangered Coahuilan box turtle: a contemporary approach to conservation. Molecular Ecology 17: 4209–4221.
Howeth, J. G., and W. S. Brown. 2011. Terrapene coahuila Schmidt and Owens 1944—Coahuilan Box Turtle. Chelonian Research Monographs No. 5, Chelonian Research Foundation.
Hubbs, C. L. 1936. Fishes of the Yucatan peninsula. Carnegie Inst. Wash. Publ. 457(17):157–287.
Hubbs, C. L. 1938. Fishes from the caves of Yucatan Carnegie Inst. Wash. Publ. 491(21):261–295.
Hubbs, C. L., and V. G. Springer. 1957. A review of the Gambusia nobilis species group, with descriptions of three new species, and note on their variation, ecology, and evolution. Texas J. Sci. 9:279–327.
Hughes, R. C., J. W. Higgenbotham, and J. W. Clary. 1941. The trematodes of reptiles. Part II. Host Catalogue. Proc. Oklahoma Acad. Sci. 21(1):35–43.
Hughes, R. C., J. W. Higgenbotham, and J. W. Clary. 1942. The trematodes of reptiles. Part I. Systematic section. Amer. Midland Natur. 27(1):109–134.
Hulse, A. C. 1974. Food habits and feeding behavior in Kinosternon sonoriense (Chelonia: Kinosternidae). J. Herp 8:195–199.

Hulse, A. C. 1976a. Carapacial and plastral flora and fauna of the Sonora mud turtle, *Kinosternon sonoriense* LeConte (Reptilia, Testudines, Kinosternidae). J. Herp. 10:45–48.

Hulse, A. C. 1976b. Growth and morphometrics of *Kinosternon sonoriense* (Reptilia, Testudines, Kinosternidae). J. Herpetol. 10:341–348.

Hulse, A. C. 1982. Reproduction and population structure in the turtle, *Kinosternon sonoriense*. Southwest Nat. 27: 447–456.

Hutchison, J. H. 1991. Early Kinosterninae (Reptilia: Testudines) and their phylogenetic significance. J. Vert. Paleo. 11:145–167.

Hutchison, J. H., and D. M. Bramble. 1981. Homology of the plastral scales of the Kinosternidae and related turtles. Herpetologica 37:73–85.

Hutchison, J. H., and J. D. Archibald. 1986. Diversity of turtles across the Cretaceous/Tertiary boundary of northeastern Montana. Palaeogeography, Palaeoclimatology, Palaeoecology 55:1–22.

Hutchison, V. H., A. Vinegar, and R. Kosh. 1966. Critical thermal maxima in turtles. Herpetologica 22 (1):32–41.

International Commission on Zoological Nomenclature. 1985. Opinion 1343. Kinosternon alamosae Berry and Legler, 1980 and *Kinosternon oaxacae* Berry and Iverson, 1980 (Reptilia, Testudines): conserved. Bull. Zool. Nomencl. 42:266–268.

Iverson, J. B. 1975. Notes on Nebraska reptiles. Trans. Kansas Acad. Sci. 78:51–62.

Iverson, J. B. 1976. *Kinosternon sonoriense*, Sonora Mud Turtle. Cat. Amer. Amph. Rept. 176.1–2

Iverson, J. B. 1978. Distributional problems of the genus *Kinosternon* in the American Southwest and adjacent Mexico. Copeia 1978:476–479.

Iverson, J. B. 1979a. On the validity of *Kinosternon arizonense* Gilmore. Copeia 1979 (1):175–177.

Iverson, J. B. 1979b. A taxonomic reappraisal of the yellow mud turtle, *Kinosternon flavescens* (Testudines: Kinosternidae). Copeia 1979(2):212–225.

Iverson, J. B. 1980. *Kinosternon acutum*. Cat. Amer. Amph. and Rept. 261:1–2

Iverson, J. B. 1981a. Biosystematics of the *Kinosternon hirtipes* species group (Testudines, Kinosternidae). Tulane Stud. Zool. Bot. 23:1–74.

Iverson, J. B. 1981b. Geographic variation in sexual dimorphism in the Mud turtle, *Kinosternon hirtipes*. Abstr. ASIH Corvallis Mtgs. p. 47.

Iverson, J. B. 1982. Biomass in turtle populations: a neglected subject. Oecologia (Berlin). 55:69–76.

Iverson, J. B. 1983a. *Kinosternon creaseri*. Catalog. Amer. Amph. Rept. 312.1

Iverson, J. B. 1983b. *Kinosternon oaxacae*. Cat. Amer. Amph. Rept. 338.1–2

Iverson, J. B. 1985. Geographic variation in sexual dimorphism in the mud turtle Kinosternon hirtipes. Copeia 1985 (2):388–393.

Iverson, J. B. 1986a. A checklist with distribution maps of the turtles of world. 1st ed., 283 pp., Privately Printed, Richmond, Indiana [ISBN 0-9617431- 0-7]

Iverson, J. B. 1986b. Notes on the natural history of the Oaxaca Mud turtle, *Kinosternon oaxacae*. J. Herp. 20(1): 119–123.

Iverson, J. B. 1988a. Distribution and status of Creaser's mud turtle, *Kinosternon creaseri*. Herpetol. Jour. 1:285–291.

Iverson, J. B. 1988b. Neural bone patterns and the phylogeny of the turtles of the subfamily Kinosterninae. Contrib. Biol. Geol. Milwaukee Publ. Mus. 75:1–12.

Iverson, J. B. 1989a. Natural history of the Alamos mud turtle, *Kinosternon alamosae* (Kinosternidae). Southwest. Nat. 34:134–142.

Iverson, J. B. 1989b. The Arizona mud turtle, *Kinosternon flavescens arizonense* (Kinosternidae), in Arizona and Sonora. Southwest. Nat. 34:356–368.

Iverson, J. B. 1990a. *Kinosternon alamosae*. Cat. Amer. Amph. Rept. 460:1–2.

Iverson, J. B. 1990b. Nesting and parental care in the mud turtle, *Kinosternon flavescens*. Can. J. Zool. 68(2):230–233.

Iverson, J. B. 1991a. Patterns of survivorship in turtles (order Testudines). Can. J. Zool. 69:385–391.

Iverson, J. B. 1991b. Phylogenetic hypotheses for the evolution of modern kinosternine turtles. Herp. Monogr. 5:1–27.

Iverson, J. B. 1991c. Life history and demography of the yellow mud turtle, Kinosternon flavescens. Herpetologica 47:373–395.

Iverson, J. B. 1992a. A Revised Checklist with Distribution Maps of the Turtles of the World. Published by the author, Richmond, Indiana.

Iverson, J. B. 1992b. Correlates of reproductive output in turtles (Order Testudines). Herpetol. Monogr. 5:25–42.

Iverson, J. B. 1998. Molecules, morphology, and mud turtle phylogenetics (family Kinosternidae). Chelonian Conservation and Biology 3(1):113–117.

Iverson, J. B. 1999. Reproduction in the Mexican mud turtle Kinosternon integrum. J. Herpet. 33:144–148.

Iverson, J. B. 2010. Reproduction in the Red-Cheeked Mud Turtle *(Kinosternon scorpioides cruentatum)* in Southeastern Mexico and Belize, with comparisons across the species range. Chel. Cons. Biol. 9(2):250–261.

Iverson, J. B, C. A. Young, and J. F. Berry. 1998. *Kinosternon integrum* Leconte, Mexican Mud Turtle. Cat. Amer. Amph. Rept. 652.1–6.

Iverson, J. B., C. P. Balgooyen, K. K. Byrd, and K. K. Lyddan. 1993. Latitudinal variation in egg and clutch size in turtles. Can. J. Zool. 71:2448–2461.

Iverson, J. B., E. L Barthelmess, G. R. Smith, and C. E. de Rivera. 1991. Growth and reproduction in the mud turtle *Kinosternon hirtipes* in Chihuahua. Mexico. J. Herp. 25(1) 64–72.

Iverson, J. B., and G. R. Smith. 1993. Reproductive ecology of the painted turtle (Chrysemys picta) in the Nebraska sandhills and across its range. Copeia 1993(1):1–21.

Iverson, J. B., and J. F. Berry. 1979. The mud turtle genus *Kinosternon* in northeastern Mexico. Herpetologica 35: 318–324.

Iverson, J. B., and J. F. Berry. 1980. *Claudius, C. angustatus*. Cat. Amer. Amph. Rept. 236.1–2.

Iverson, J. B., and R. A. Mittermeier. 1980. Dermatemydidae, *Dermatemys*. Cat. Amer. Amph. Rept. 237:1–4.

Jackson, C. G., Jr., and F. T. Awbrey. 1972. Mating bellows of the Galapagos tortoise, *Geochelone elephantopus*. Herpetologica 34(2):134–136.

Jackson, C. G., Jr., and J. D. Davis. 1972. A quantitative study of the courtship display of the red-eared turtle, *Chrysemys scripta elegans* (Wied). Herpetologica 28:58–64.

Jackson, C. G., Jr., T. H. Trotter, J. A. Trotter, and M. W. Trotter. 1978. Further observations of growth and sexual maturity in captive desert tortoises (Reptilia: Testudines). Herpetologica 34:225–227.

Jackson, D. R. 1988. A re-examination of fossil turtles of the genus *Trachemys* (Testudines: Emydidae). Herpetologica 44:317–325.

Jackson, D. R., and B. Brechtel. 2006. *Terrapene, T. carolina, T. c. bauri, T. c.* major (American Box Turtle, Florida Box Turtle, Gulf Coast Box Turtle). Maximum Size. Herp. Rev. 37(3):342–343.

Jackson, O. F., and J. E. Cooper. 1981. Nutritional diseases. In: J. E. Cooper and O. F. Jackson (Eds.). Diseases of the Reptilia, Vol. 2, Chapter 12, pp. 409–428. Academic Press, Inc. (London) Ltd., London.

Jacobshagen, E. 1920. Zur Morpholgie des Oberflachenreliefs der Rumpfdarmschleimhaut der Reptilien. Jena. Z. Naturw. 56:361–430.

Jacobshagen, E. 1937. IV. Mittel-und enddarm. In: Handbuch der Vergleichenden Anatomie der Virbeltiere. Vol. 3. pp. 563–724. Urban und Schwarzenberg, Berlin.

Jacobson, E. R., J. M. Gaskin, M. B. Brown, R. K. Harris, J. L. Gardiner, H. P. LaPointe, H. P. Adams, and C. Reggiardo. 1991. Chronic upper respiratory tract disease of free-ranging desert tortoises *(Xerobates agassizii).* J. Wildl. Dis. 27:296–316.

Jaeger, E. C. 1955. A Source Book of Biological Terms and Names. 3rd ed. C. C. Thomas, Springfield, IL.

Jaekel O. 1914. Ueber die Wirbeltierfunde in der oberen Trias von Halberstadt. Palaeontol. Z. 1:155–215.

Jaekel, O. 1918. Die Wirbeltierfunde aus dem Keuper von Halberstadt. Serie II. Testudinata. Teil 1 Stegochelys dux, n.g., n.sp. Palaeontol. Z. 2:88–214.

Janzen, D. H. 1980. Two potential coral snake mimics in the tropical deciduous forest. Biotropica 12(1)77–78.

Janzen, F. J., S. L. Hoover, and H. B. Shaffer. 1997. Molecular phylogeography of the western pond turtle *(Clemmys marmorata):* preliminary results. Linnaeus Fund Research Report. Chel. Cons. Biol. 2(4):623–626.

Jenkins, J. D. 1979. Notes on the courtship on the map turtle *Graptemys pseudogeographica* (Gray) (Reptilia, Testudines, Emydidae). J. Herp. 13(1):129–131.

Jennings, G. 1980. Aztec. New York: Athenian Press.

Jennings, M. R. 1981. *Gopherus agassizi* (Desert tortoise). Longevity. Herp. Review 12:81–82.

Jennings, M. R., and M. P. Hayes. 1994. Amphibian and reptile species of special concern in California. File Report, Calif. Dept. Fish and Game, Contract No. 8023, Rancho Cordova.

Jones, M. T. 2010. *Glyptemys insculpta* (wood turtle) maximum adult size. Herp. Rev. 41(1):71.

Judd, F. W., and F. L. Rose. 1977. Aspects of the thermal biology of the Texas tortoise, *Gopherus berlandieri* (Reptilia, Testudines, Testudinidae). J. Herp. 11(1):147–153.

Judd, F. W., and F. L. Rose. 1983. Population structure, density and movements of the Texas tortoise *Gopherus berlandieri.* Southwest. Natur. 28:387–398.

Judd, F. W., and F. L. Rose. 1989. Egg production by the Texas tortoise, *Gopherus berlandieri,* in southern Texas. Copeia 1989:588–596.

Judd, F. W., and F. L. Rose. 2000. Conservation status of the Texas tortoise *Gopherus.* Occ. Pap. Mus. Texas Tech Univ. 106:1–11.

Judd, F. W., and J. C. McQueen. 1980. Incubation, hatching, and growth of the tortoise, *Gopherus berlandieri.* J. Herp.14:377–380.

Judd, F. W., and J. C. McQueen. 1982. Notes on longevity of *Gopherus berlandieri* (Testudinidae). Southwest. Natur. 27:230–232.

Kandel, E. R., J. H. Schwartz, and T. M. Jessell. 2000. Principles of Neural Science. 4th ed. McGraw-Hill , New York.

Kangas, D. A. 1986. Population size and some statistical predictors of abundance of *Kinosternon flavescens* in north Missouri. Trans. Missouri Acad. Sci. 20:98.

Kaufmann, J. H. 1992. The social behavior of wood turtles, *Clemmys insculpta,* in central Pennsylvania. Herpetol. Monogr. 6:1–25.

Kazmaier, R. T., E. C. Hellgren, and D. C. Ruthven III. 2002. Home range and dispersal of Texas tortoises, *Gopherus berlandieri,* in a managed thornscrub ecosystem. Chel. Cons. Biol. 4(2):488–496.

Kiernan, J. A. 2004. Barr's The Human Nervous System. 8th ed. J. B. Lippincott Company. Philadelphia.

Killebrew, F. C. 1977a. Mitotic chromosomes of turtles IV. The Emydidae. Texas J. Sci. 29:245–253.

Killebrew, F. C. 1977b. Mitotic chromosomes of turtles. V. The Chelydridae. Southwest. Natur. 21(4):547–548.

Klauber, L. M. 1941. Four papers on the application of statistical methods to herpetological problems. Bull. Zool. Soc. San Diego 17:1–95.

Koob, T. J. 1998. Relaxin, nonmammalian. In: E. Knobil and J. D. Neill (Eds.). Encyclopedia of Reproduction. Vol. 4, pp. 223–231. Acad. Press. San Diego.

Kool, K. 1981. Is the musk of the Long-necked Turtle, *Chelodina longicollis,* a deterrent to predators? Aust. J. Herp. 1:45–54.

Krenz, J. G., G. J. P. Naylor, H. B. Shaffer, and F. J. Janzen. 2005. Molecular phylogenetics and evolution of turtles. Molec. Phylogen. Evol. 37:178–191.

Küchling, G. 1999. The Reproductive Biology of the Chelonia. Berlin: Springer, Zoophysiology 28.

Kuhn, O. 1964. Testudines. Fossilim Catalogues I: Animalia, Pars 107:1–299.

Lagler, K. F. 1943. Food habits and economic relations of the turtles of Michigan with special reference to fish management. Amer. Midl. Natur. 29:257–312.

Lamb, T., and C. Lydeard 1994. A molecular phylogeny of the gopher tortoises, with comments on familial relationships within the Testudinoidea. Molec. Phylogen. Evol. 3(4): 283–291.

Lamb, T., J. C. Avise, and J. W. Gibbons. 1989. Phylogeographic patterns in mitochondrial DNA of the desert tortoise *(Xerobates agassizi),* and evolutionary relationships among the North American gopher tortoises. Evolution 43:76–87.

Lamothe-Argumedo, R. 1972. Monogeneos de reptiles: I. Redescripcion del cuatro especies de monogenea (Polstomatidae) parasitos de la vejiga urinaria de tortugas de México. An. Inst. Biol. Univ. Nac. Auton. Mexico 43(1):1–16.

Lane, H. H. 1909. A paired entoplastron in *Trionyx* and its significance. Proc. Indiana Acad. Sci. 1909:345–350.

Langebartel, D. A. 1953. The reptiles and amphibians. pp. 91–108 in: R. T. Hatt, Faunal and archeological researches in Yucatan caves. Cranbrook Inst. Sci. Bull. 33:1–119.

Lardie, R. L. 1975a. Courtship and mating behavior in the yellow mud turtle, *Kinosternon flavescens flavescens*. J. Herp. 9:223–227.
Lardie, R. L. 1975b. Notes on the eggs and young of *Clemmys marmorata* marmorata (Baird and Girard). Occ. Pap. Mus. Nat. Hist. Univ. Puget Sound. 47:654.
Lardie, R. L. 1975c. Observations on reproduction in *Kinosternon*. J. Herpl. 9:260–264.
Lardie, R. L. 1978. Additional observation on courtship and mating in the plains yellow mud turtle, *Kinosternon flavescens flavescens*. Bull. Oklahoma Herpetol. Soc. 3:70–72.
Lardie, R. L. 1983. Aggressive interactions and territoriality in the yellow mud turtle. Bull. Oklahoma Herpetol. Soc. 8:68–83.
Laurito, M. C., A. L. Valerio Z., L. D. Gómez, J. I. Mead, E. A. Pérez G., and L. G. Pérez. 2005. A Trionychidae (Reptilia: Testudines, Cryptodira) from the Pliocene of Costa Rica, Southern Centralamerica. Revista Geológica de América Central, 32: 7–11, 2005 ISSN: 0256-7024.
LeConte, J. 1830. Description of the species of North American tortoises. Ann. Lyceum Natur. Hist. New York. 3:91–131.
LeConte, J. 1854. Description of four new species of *Kinosternum*. Proc. Acad. Natur. Sci. Philadelphia 7:180–190.
Ledig, J. 1988. Haltung und Zucht von *Claudius angustatus*. Elaphe (2):24.
Lee, J. C. 1980. An ecogeographic analysis of the herpetofauna of the Yucatan Peninsula. Univ. of Kans. Mus. Nat. Hist. Misc. Publ. 67:1–75.
Lee, J. C. 1996. The Amphibians and Reptiles of the Yucatan Peninsula. Comstock Publishing Associates.
Lee, M. S. Y. 1993b. The origin of the turtle body plan: bridging a famous morphological gap. Science 261:1716–1720.
Lee, M. S. Y. 1993a. The origin of turtles reconsidered (again). Second World Congress of Herpetology, Abstracts.
Lee, M. S. Y. 1994. The turtle's long-lost relatives. Nat. Hist. 103(6):63–65.
Lee, M. S. Y. 1996. Correlated progression and the origin of turtles. Nature. 379: 812–814.
Lee, M. S. Y. 1997. Pareiasaur phylogeny and the origin of turtles. Zool. J. Linn. Soc, 120: 197–280.
Lee, R. C. 1969. Observing the Tortuga Blanca. Int. Turt. Tort. Soc. Jour. 3(3):32–34.
Legler, J. M. 1955. Observations on the sexual behavior of captive turtles. Lloydia 18: 95–99.
Legler, J. M. 1958. The Texas slider *(Pseudemys floridana texana)* in New Mexico. Southwest Natur. 19583(1–4): 230–231.
Legler, J. M. 1959. A new tortoise, genus *Gopherus,* from north-central Mexico. Univ. Kans. Publ. Mus. Nat. Hist. 11(5):335–343.
Legler, J. M. 1960a. A new subspecies of slider turtle *(Pseudemys scripta)* from Coahuila, Mexico. Univ. Kansas Publ. Mus. Natur. Hist. 13(3):78–84.
Legler, J. M. 1960b. Remarks on the natural history of the Big Bend slider, *Pseudemys scripta gaigeae* Hartweg. Herpetologica 16:139–140.
Legler, J. M. 1960c. Natural history of the ornate box turtle, *Terrapene ornata ornata* Agassiz. Univ. Kansas Publ., Mus. Nat. Hist. 11(10):527–669.
Legler, J. M. 1960d. A simple and inexpensive device for trapping aquatic turtles. Proc. Utah. Acad. Sci. Arts. Lett. 37:63–66.
Legler, J. M. 1963. Further evidence for intergradation of two Mexican slider turtles *(Pseudemys scripta)*. Herpetologica 19(2):142–143.
Legler, J. M. 1965. A new species of turtle, genus *Kinosternon,* from Central America. Univ. Kans. Publ., Mus. Nat. Hist. 15(13):615–625.
Legler, J. M. 1977. Stomach flushing: A technique for chelonian dietary studies. Herpetologica 33:281–84.
Legler, J. M. 1978. Observations on behavior and ecology in *Chelodina expansa* (Chelonia: Chelidae). Canadian J. Zool. 56(11):2449–2453.
Legler, J. M. 1985. Australian chelid turtles: Reproductive patterns in wide-ranging taxa. In: . G, Grigg, R. Shine, and H. Ehmann (Eds.). Australasian Frogs and Reptiles, pp. 117–123. Royal Zool. Soc. New South Wales.
Legler, J. M. 1989. Diet and head size in Australian chelid turtles, genus *Emydura*. Annals. Soc. R. Zool. Belg., 119, suppl. 1, Brussels [abstract].
Legler, J. M. 1990. The genus *Pseudemys* in Mesoamerica: taxonomy, distribution, and origins. In: J. W. Gibbons (Ed.). Life History and Ecology of the Slider Turtle. pp. 82–105. Smithsonian Institution Press, Washington, D.C.
Legler, J. M. 1993a. General description and definition of the order Chelonia. In: C. J. Glasby, G. J. B. Ross, P. L. Beesley (Eds.). Fauna of Australia. Vol. 2A. Amphibia and Reptilia, pp. 104–107. Canberra: Australian Government Publishing Service.
Legler, J. M. 1993b. Morphology and physiology of the Chelonia. In: C. J. Glasby, G. J. B. Ross, and P. L. Beesley (Eds.). Fauna of Australia. Vol. 2A. Amphibia and Reptilia, pp. 108–119. Canberra: Australian Government Publishing Service.
Legler, J. M., and A. Georges. 1993. Family Chelidae. *In:* Glasby, Christopher J.; Ross, Graham J. B.; Beesley, Pamela L. (Eds.). Fauna of Australia. Vol. 2A. Amphibia and Reptilia. Canberra: Australian Government Publishing Service, pp. 142–152.
Legler, J. M., and A. Georges. 1993a. Biogeography and phylogeny of the Chelonia. In: C. J. Glasby, G. J. B. Ross, and P. L. Beesley (Eds.). Fauna of Australia. Vol. 2A, pp. 129–132. Amphibia and Reptilia. Canberra: Australian Government Publishing Service.
Legler, J. M., and E. O. Moll. (Unpublished manuscript). Biosystematic studies of the genus *Staurotypus* (Chelonia: Kinosternidae).
Legler, J. M., H. M. Smith, and R. B. Smith. 1980. *Testudo scripta* Schoepff, 1792: *Emys cataspila* Günther 1885; Proposed Conservation (Reptilia, Testudines). Z. N. (S.) 2315. Bull. Zool. nomencl. 37(4):240–246.
Legler, J. M., and J. Cann. 1980. A new genus and species of chelid turtle from Queensland, Australia. Contrib. Sci. Nat. Hist. Mus. Los Angeles Co. 324:1–18.
Legler, J. M., and L. J. Sullivan. 1979. The application of stomach flushing to lizards and anurans. Herpetologica. 35(2):107–110.
Legler, J. M., and R. G. Webb. 1961. Remarks on a collection of Bolson Tortoises, *Gopherus flavomarginatus*. Herpetologica 17(2):26–37.
Legler, J. M., and R. G. Webb. 1970. A new slider turtle *(Pseudemys scripta)* from Sonora, Mexico. Herpetologica 26(2):157–168.

Legler, J. M., and R. M. Winokur. 1979. Unusual neck tubercles in an Australian chelid turtle, *Elseya latisternum* (Testudines: Chelidae). Herpetologica 35(4):325–329.

Lemos-Espinal, J. A., H. M. Smith, and D. Chiszar. 2001. *Terrapene nelsoni klauberi* (northern spotted box turtle). Herp. Rev. 32(4): 274.

Lemos-Espinal, J. A., H. M. Smth, and D. Chiszar. 2004. Introducion a los anfibios y reptiles del estado de Chihuahua. Univ. Nac. Aut. Mexico.

Lenk, P., U. Fritz, U. Joger, and M. Wink. 1999. Mitochondrial phylogeography of the European pond turtles, *Emys orbicularis* (Linnaeus 1758). Mol. Ecol. 8: 1911–1922.

Leopold, A. S. 1959. Wildlife of Mexico. The Game Birds and Mammals. Berkeley, Univ. California Press.

Li, C., X-C. Wu, O. Rieppel, L-T. Wang, and L-J. Zhao. 2008. An ancestral turtle from the late Triassic of southwestern China. Nature 456:497–501.

Lieberman, S. S., and D. J. Morafka. 1988. The ecogeography of the Mexican Bolson tortoise *(Gopherus flavomarginatus)*: derivation of its endangered status and recommendations for its conservation. Part 2. Ecological distribution of the Bolson tortoise. Ann.Carnegie Mus. Nat. Hist. 57(1):31–46.

Ligon, D. B., and P. A. Stone. 2003a. Radiotelemetry reveals terrestrial estivation in Sonoran mud turtles *(Kinosternon sonoriense)*. J. Herp. 37(4):750–754.

Ligon, D. B., and P. A Stone. 2003b. *Kinosternon sonoriense* (Sonoran mud turtle) and *Bufo punctatus* (red-spotted toad). Predator-prey. Herp. Rev. 34(3):241–242.

Lindquist, K. L., and A. B. Appleton. 1982. Some observations on activity patterns of captive Bolson tortoises *(Gopherus flavomarginatus)*. Desert Tortoise Council Proceedings of Symposium 7:162–172.

Lindsay, E. H. 1984. Late Cenozoic mammals from northwestern Mexico. J. Vert. Paleo. 4:208–215.

Liner, E. A. 1994. Scientific and common names for the amphibians and reptiles of Mexico in English and Spanish. SSAR Herp. Circular 23.

Linnaeus, C. 1758. Systema Naturae. 10th ed. Vol. 1. Holmiae, Sweden. 1:1–824.

Linnaeus, C. 1766. Systema Naturae. 12th ed. Halae Magdeborgicae, Sweden. 1:1–532.

Long, D. R. 1972. Clutch formation in the turtle *Kinosternon flavescens* (Testudinidae: Kinosternidae) Southwest Naturalist 31:1–8.

Long, D. R. 1985. Lipid utilization during reproduction in female *Kinosternon flavescens*. Herpetologica 41(1):58–65.

Long, D. R. 1986. Lipid content and delayed emergence of hatchling yellow mud turtles. Southwest. Natur. 31: 244–246.

López León, N. P. 2001. Eficiencia digestiva en tres especies de tortugas de agua dulce: *Rhinoclemmys areolata, Trachemys scripta*, y *Dermatemys mawi*. Tesis de Licenciado en Biologia. Universidad de Ciencias y Artes del Estado de Chiapas, Tuxtla Gutierrez Chipas.

López-León, N. P. 2008. Diseño de una propuesta de manejo de tres especies de tortugas dulceacuicolas (*Kinosternon scorpioides cruentatum, Staurotypus salvinii*, y *Trachemys venusta grayi*) en dos localidades de La Reseva De La Biosfera La Encrucijada, Chiapas. MS thesis, Manejo de Fauna Silvestre, Instituto de Ecologia, A. C. Jalapa, Veracruz.

López-León, N. P., and R. C. Vogt. 2000. (abstract) Digestive efficiency in three species of freshwater turtles: *Dermatemys mawi, Rhinoclemmys areolata*, and *Trachemys scripta venusta*. Annual Meeting American Society of Icthyologists and Herpetologists, Universidad Autonoma de Baja California Sur, 2000. La Paz, B.C.S.

López-Luna, M. A., R. C. Vogt, and V. G. Espejel. 1998. Habitos alimenticios de dos especies simpatricas de tortugas dulceacuicolas en el sur de Veracruz. In: V. Reunion Nacional de Herpetologia, Jalapa.

Loveridge, A., and E. E. Williams. 1957. Revision of the African tortoises and turtles of the suborder Cryptodira. Bull. Mus. Comp. Zool. Harvard 115:163–557.

Lovich, J. E., and K. Meyer. 2002. The western pond turtle *(Clemmys)* in the Mojave River, California, USA: highly adapted survivor or tenuous relict? J. Zool. Lond. 256: 537–545.

Lovich, R., C. Mahrdt, and B. Downer. 2005. Geographic distribution. *Actinemys marmorata*. Herp. Rev. 36(2): 200–201.

Lovich, R., T. Akre, J. Blackburn, T. Robison, and C. Mahrdt. 2007. Geographic distribution. *Actinemys marmorata*. Herp. Rev. 38(2):216–217.

Luckenbach, R. A. 1982. Ecology and management of the desert tortoise *(Gopherus agassizii)* in California. In: R. B. Bury (Ed.). North American Tortoises: Conservation and Ecology, pp. 1–39. U.S. Fish and Wildlife Service Wildlife Research Report (12).

Lydekker, R. 1889. Catalogue of the Fossil Reptilia and Amphibia in the British Museum. Part III. Chelonia. London: British Museum of Natural History.

Macdonald, D. W., F. G. Ball, and N. G. Hough. 1980. The evaluation of home range size and configuration using radiotracking data. In: J. Amlaner, Jr., and D. W. Macdonald (Eds.). A Handbook of Biotelemetry and Radiotracking, pp. 405–424. Pergamon, Oxford.

Macip-Ríos, R. 2005. Ecología poblacional e historia de vida de la tortuga *Kinosternon integrum* en la localidad de Tonatico, Estado de México. MS thesis, Biologia Ambiental, Instituto de Biologia, Universidad Nacional Autonoma de Mexico.

Mahmoud, I. Y. 1967. Courtship behavior and sexual maturity in four species of kinosternid turtles. Copeia 1967:314–319.

Mahmoud, I. Y. 1968. Feeding behavior in kinosternid turtles. Herpetologica 24(4):300–305.

Mahmoud, I. Y. 1969. Comparative ecology of the kinosternid turtles of Oklahoma. Southwest Natur. 14:31–66.

Malkin, B. 1958. Cora ethnozoology, herpetological knowledge: a bio-ecological and cross cultural approach. Anthrop. Quart. 31:73–90.

Mallouf, R. J. 1986. Prehistoric cultures of the northern Chihuahuan Desert. In: J. C. Barlow, A. M. Powell, and B. Timmermann (Eds.). Second Symposium on the Resources of the Chihuahuan Desert Region, pp. 69–78. Chihuahuan Desert Research Institute, Alpine, Texas.

Marchand, L. J. 1942. A Contribution to a Knowledge of the Natural History of Certain Freshwater Turtles. Master's thesis, Univ. of Florida, Gainsville.

Marchand, L. J. 1944. Notes on the courtship of a Florida terrapin. Copeia 1944(3):191–192.

Marchand, L. J. 1945. Water goggling: a new method for the study of turtles. Copeia 1945(1):37–40.

Mares, M. A. 1971. Coprophagy in the Texas tortoise, *Gopherus berlandieri*. Texas J. Sci. 23:300–301.

Marlow, R. W., and K. Tollestrup. 1982. Mining and exploration of natural mineral deposits by the desert tortoise, *Gopherus agassizii*. Anim. Behav. 30:475–478.

Marsh, E. G., Jr. 1937. Biological Survey of the Santa Rosa and Del Carmen Mountains of Northern Coahuila, Mexico. U.S. Nat. Park Service, mimeo rep.

Martin, P. S. 1958. A biogeography of reptiles and amphibians in the Gomez Farias region, Tamaulipas, Mexico. Misc. Publ. Mus. Zool. Univ. Michigan 101:1–102.

Mason, M. 1995. Tortoises as seed dispersers in semi-arid rangelands (valley bushveld). Gonfaron, France: International Congress of Chelonian Conservation, Abstracts.

Mast, R. B., and J. L. Carr. 1986. Life history notes. Testudines. *Trachemys scripta cataspila* (Haustecan slider). Herp. Rev. 17(1):25.

Mata-Silva, V., A. Ramirez-Bautista, M. Paredes-Flores, and M. Espino-Ocampo. 2002. Kinosternon herrerai (Herrera's Mud Turtle). Mexico: Mexico. Herp. Rev. 33(2):147.

Mayr, E., and P. Ashlock. 1991. Principles of systematic zoology. McGraw Hill, New York.

McCormick, B. 1998. The Mexican wood turtle, *Rhinoclemmys pulcherrima*. Tortuga Gazette 34(1):1–2.

McDowell, S. B. 1964. Partition of the genus *Clemmys* and related problems in the taxonomy of the aquatic Testudinidae. Proc. Zool. Soc. London. 143:239–279.

McGaugh, S. E. 2008. Color variation correlated with habitat type and background coloration in *Apalone spinifera* in Cuatro Ciénegas, Coahuila, Mexico. Journal of Herpetology 42:347–353.

McGaugh, S. E., C. M. Eckerman, and F. J. Janzen. 2008. Molecular phylogeography of *Apalone spinifera* (Testudines: Trionychidae). Zoologica Scripta 37:289–304.

McGaugh, S. E., and F. J. Janzen. 2008. The status of *Apalone ater* populations in Cuatro Ciénegas, Coahuila, Mexico, preliminary data. Chelonian Conservation and Biology 7(1):88–95.

McGinnis, S. M., and W. G. Voigt. 1971. Thermoregulation in the desert tortoise, *Gopherus agassizii*. Comp. Biochem. Physiol. 40A:119–126.

McKeown, S., and R. G. Webb. 1982. Softshell turtles in Hawaii. J. Herp. 16(2):107–111.

Medem, F. 1962. La distribución geográphica y ecológia de los Crocodylia y Testudinata en el departmento del Chóco. Rev. Acad. Colombiana Cien. Exactas, Fis. y Natur. 11 (44):279–303.

Medem, F. 1977. Contribucion al conocimiento sobre la taxonomia, distribución geografica y ecologia de la tortuga "bache" *(Chelydra serpentina acutirostris)*. Caldasia 12:41–98.

Medem, F. 1975. La reproducción de la Icotea *(Pseudemys scripta callirostris)*, (Testudines, Emydidae). Caldasia 11(53):83–106.

Medica, P. A., C. L. Lyons, and F. B. Turner. 1982. A comparison of 1981 populations of desert tortoises *(Gopherus agassizi)* in grazed and ungrazed areas in Ivanpah Valley, California. Proc. Symp. Desert Tortoise Council 1982:99–124.

Medica, P., C. Lyons, and F. Turner. 1986. "Tapping": a technique for capturing tortoises. Herp Rev. 17(1):15–16.

Medica, P. A., R. B. Bury, and R. A. Luckenbach. 1980. Drinking and construction of water catchments of the desert tortoise, *Gopherus agassizi*, in the Mojave Desert. Herpetologica 36:301–304.

Meek, S. E. 1904. The fresh-water fishes of Mexico north of the Isthmus of Tehuantepec. Zool. Ser., Field Columbian Mus., vol. 5.

Melville, R. V. 1983. Comments on the proposed suppression of *Kinosternon alamose* and *K. oaxacae* Pritchard, 1979. Bulletin Zool. Nom. 40(2): 71.

Mercer, H. C. 1896. The hill-caves of Yucatan. Philadelphia: J, B. Lippincott.

Merrem, B. 1820. Tentamen systematis amphibiorum. Marburg.

Mertens, R. 1956. Uber reptilienbastarde, II. Senck. Biol. 37:383–394.

Mertens, R., and A. Zilch. 1952. Die Amphibien und Reptilien von El Salvador. Abh. Senck. Naturf. Gesell., Frankfurt (487):1–83.

Mertens, R., and H. Wermuth. 1955. Die rezenten Schildkröten, Krokodile und Bruckenechsen. Eine kritische Liste der heute lebenden Arten und Rassen. Zool. Jahrb. 83:323–440.

Metcalf, A. L. and E. Metcalf. 1985. Longevity in some ornate box turtles *(Terrapene ornata ornata)*. J. Herpetol. 19:157–158.

Metcalf, E. L., and A. L. Metcalf. 1970. Observations on ornate box turtles (*Terrapene ornata ornata* Agassiz). Trans. Kansas Acad. Sci. 73:96–117.

Meyer, J. R., and L. D. Wilson. 1973. A distributional checklist of the turtles, crocodilians, and lizards of Honduras. Nat. Hist. Mus. Los Ang. Co. Contrib. Sci. 224:1–1039.

Meylan, P. A. 1984. Evolutionary relationships of Recent trionychid turtles: evidence from shell morphology. Studia Geologica Salmanticensia (Studia Palaeochelonologica). 1 (special):169–188.

Meylan, P. A. 1987. The phylogenetic relationships of soft-shelled turtles (family Trionychidae). Bull. Amer. Mus. Natur. Hist. 186:1–101.

Meylan, P. A., R. Schuler, and P. Moler. 2002. Spermatogenic cycle of the Florida Softshell Turtle, *Apalone ferox* Copeia 2002(3): 779–786.

Miller, L. 1932. Notes on the desert tortoise *(Testudo agassizii)*. Trans. San Diego Soc. Natur. Hist. 7:187–208.

Miller, R. R. 1946. The probable origin of the soft-shelled turtle in the Colorado River basin. Copeia 1946(1):46.

Miller, R. R. 1961. Man and the changing fish fauna of the American Southwest. Mich. Acad. Sci., Arts, Lett., Pap. 46: 365–404.

Miller, R. R. 1968. A drainage map of Mexico. Systematic Zool. 17(2):174–175.

Miller, R. R. 1975. Five new species of Mexican poeciliid fishes of the genera Poecilia, Gambuusia, and Poeciliopsis. Occ. Pa. Mus. Zool. Univ. of Michigan 672:1–44

Miller, R. R. 1983. Checklist and key to the mollies of Mexico (Pisces: Poeciliidae: *Poeciliia*, subgenus *Mollienesia*). Copeia 1983:817–822.

Milstead, W. M. 1967. Fossil box turtles *(Terrapene)* from central North America and box turtles of eastern Mexico. Copeia 1967(1):168–179.

Milstead, W. M., and D. W. Tinkle. 1967. *Terrapene* of western Mexico, with comments on the species groups in the genus. Copeia 1967 (1):180–187.

Milstead, W. W. 1965. Notes on the identities of some poorly known fossils of box turtles (Terrapene). Copeia 4:513–514.

Milstead, W. W. 1967. Fossil box turtles *(Terrapene)* from central North America and box turtles of eastern Mexico. Copeia 1967:168–179.

Milstead, W. W. 1969. Studies on the evolution of box turtles (genus *Terrapene*). Bull Florida State Mus. Biol. Sci. 14:1–108.

Minckley, W. L. 1966. Coyote predation on aquatic turtles. J. Mammal. 47:137.

Minckley, W. L. 1969. Environments of the bolson of Cuatro Cienegas, Coahuila, Mexico Univ. Texas El Paso Sci. Ser. 2: 1–65.

Minckley, W. L., D. A. Hendrickson, and C. E. Bond. 1986. Geography of western North American freshwater fishes: Description and relationships to intracontinental tectonism. In: C. H. Hocutt and E. O. Wiley (Eds). Zoogeography of North American Freshwater Fishes, pp. 519–614. John Wiley and Sons, Inc. New York.

Minckley, W. L., and R. K. Koehn. 1965. Re-discovery of the fish fauna of the Sauz basin, northern Chihuahua, Mexico. Southwest. Natur. 10(4):313–315.

Minton, S. A., Jr. 1972. Amphibians and reptiles of Indiana. Indiana Acad. Sci. Monogr. 3:1–346.

Minx, P. 1992. Variation in the phalangeal formulae in the turtle genus *Terrapene*. J. Herpetol. 26:234–238.

Minx, P. 1996. Phylogenetic relationships among box turtles, genus *Terrapene*. Herpetologica 52:584–597.

Mitchell, J. C., and C. A. Pague. 1990. Body size, reproductive variation, and growth in the slider turtle at the northeastern edge of its range. In: J. W. Gibbons (Ed.). Life History and Ecology of the Slider Turtle, pp. 146–151. Smithsonian Institution Press, Washington, D.C.

Mittermeier, R. A. 1971. Status—the market in So. E. Mexico. Int. Turt. Tort. Soc. Jour. 5(3):36–38.

Mlynarski, M. 1976. Testudines. In: O. Kuhn (Ed.). Encyclopedia of Paleoherpetology. Part 7, pp. 1–30. Gustav Fischer Verlag, Stuttgart, New York.

Mocquard, M. F. 1899. Contribution a la faune herpetologique de la Basse-Californie. Nouv. Arch. Mus. d'Hist. Nat. Paris, Ser. 4. 1:297–344.

Moll, D. 1990. Population sizes and foraging ecology in a tropical freshwater stream turtle community. J. Herp. 24(1):48–53.

Moll, D. L. 1979. Subterranean feeding by the Illinois mud turtle, *Kinosternon flavescens spooneri* (Reptilia, Testudines, Kinosternidae). J. Herpetol. 13:371–373.

Moll, D. L. 1986. The distribution, status and level of exploitation of the freshwater turtle *Dermatemys mawi* in Belize, Central America. Biol. Conserv. 35:87–96.

Moll, D. L. 1989. Food and feeding behavior of the turtle, *Dermatemys mawei,* in Belize. J. Herp. 23(4):445–447.

Moll, D. L. 1990. Population sizes and foraging ecology in a tropical freshwater stream turtle community. J. Herpet. 24:48–53.

Moll, D. L. 1994. The ecology of sea beach nesting in slider turtles *(Trachemys scripta venusta)* from Caribbean Costa Rica. Chel. Cons. Biol. 1(2):107–116.

Moll, D. L., and E. O. Moll. 1990. The slider turtle in the neotropics: adaptation of a temperate species to a tropical environment. In: J. W. Gibbons (Ed.). Life History and Ecology of the Slider Turtle, pp. 152–161. Smithsonian Institution Press, Washington, D.C.

Moll, E. O, and J. M. Legler. 1971. The life history of a neotropical slider turtle, *Pseudemys scripta* (Schoepff), in Panama. Bull. Los Angeles Co. Mus. Nat. Hist. 11:1–102.

Moll, E. O., K. E. Matson, and E. B. Krehbiel. 1981. Sexual and seasonal dichromatism in the Asian river turtle *Callagur borneoensis*. Herpetologica 37(4):179–193.

Moll, E. O., and K. L. Williams. 1963. The musk turtle *Sternotherus odoratus* from Mexico. Copeia. (1):157.

Montes-Ontiveros, O. E. 2000. Ecología y biología de la tortuga pericota *Terrapene nelsoni*. Sexta Reunion Nacional de Herpetologia. Tuxtla Gutierrez Chiapas, Mexico. (abstract).

Montes-Ontiveros, O. and P. Ponce-Campos. 2006. *Terrapene nelsoni nelsoni*. (Southern Spotted Box Turtle). Herp. Rev. 37(2):239.

Moodie, K. B., and T. R. Vandevender. 1974. Pleistocene turtles from the Whetlock oil well locality, Graham County, Arizona. (Abstract). J. Ariz. Acad. Sci. 9:35 (Proceedings supplement).

Moon, R. G. 1974. Heteromorphism in a kinosternid turtle. Mammal. Chrom. News. 15:10–11.

Mooser, O. 1980. Pleistocene fossil turtles from Aguascalientes, state of Aguascalientes. Rev. Inst. Geol. Univ. Auton. México. 4:63–66.

Morafka, D. J. 1977a. The status of the Mexican Bolson tortoise Gopherus flavomarginatus. Desert Tortoise Council Proceedings of Symposium 2 1977:167–168.

Morafka, D. J. 1977b. A biogeographical analysis of the Chihuahuan desert through its herpetofauna. Biogeographica 9:1–313.

Morafka, D. J. 1982. The status and distribution of the Bolson tortoise *(Gopherus flavomarginatus)*. U S Fish and Wildlife Service Wildlife Research Report 12:71–94.

Morafka, D. J. 1988. The ecogeography of the Mexican Bolson tortoise (*Gopherus* flavomarginatus): derivation of its endangered status and recommendations for its conservation. Part 3. Historical biogeography of the Bolson tortoise. Annals of the Carnegie Museum 57(1):47–72.

Morafka, D. J., and C. J. McCoy (Eds.). 1988. The ecogeography of the Mexican Bolson tortoise *(Gopherus flavomarginatus):* derivation of its endangered status and recommendations for its conservation. Ann. Carnegie Mus. Nat. Hist. 57(1):1–3.

Morafka, D. J., G. A. Adest, G. Aguirre, and M. Recht. 1981. The ecology of the Bolson tortoise *Gopherus flavomarginatus*. Publicaciones Instituto De Ecologia Mexico. 8:35–78.

Morafka, D. J., K. H. Berry, and K. E. Spangenberg. 1997. Predator-proof field enclosures for enhancing hatching success and survivorship of juvenile tortoises: a critical evaluation. J. Van Abbema (Ed.). Proceedings: Conservation, Restoration, and Management of Tortoises and Turtles—An International Conference, 11–16 July 1993, State Univ. of New York, Purchase, New York. New York Turtle and Tortoise Society and WCS.

Morafka, D. J., L. G. Aguirre, and R. W. Murphy. 1994. Allozyme differentiation among gopher tortoises *(Gopherus):* conservation genetics and phylogenetic and taxonomic implications. Can. J. Zool. 72(9):1665–1671.

Morafka, D. J., M. G. Aguirre, M. Recht, and G. Adest. 1980. Activity population structure and thermoregulation of Bolson tortoises. Desert Tortoise Council Proceedings Of Symposium 5 1980:141–142. Abstract.

Morafka, D. J., R. A. Yates, J. Jarchow, J. Rosskopf, Jr., G. A. Adest, and G. Aguirre. 1986. Preliminary results of microbial and physiological monitoring of the bolson tortoise, *Gopherus flavomarginatus*. In: Z. Rock (Ed.). Studies in herpetology. Proceedings of the European Herpetological Meeting (3rd Ordinary General Meeting of the Societas Europaea Herpetologica) 1985, pp. 657–662. Charles Univ., Prague.

Morales-Verdeja, S. A. 2000. Estudio histologíco del testículo y epidídimo de Kinosternon leucostumum (Chelonia: Kinosternidae) en un ciclo anual. Tesis de Maestro en Ciencias Facultad de Ciencias, UNAM, México D. F.

Morales-Verdeja, S. A., and R. C. Vogt. 1997a. *Kinosternon leucostomum*. In: E. Gonzáles, R. Diozo, and R. C. Vogt (Eds.). Historia Natural de Los Tuxtlas, pp. 488–490. U.N.A.M., Mexico, D.F.

Morales-Verdeja, S. A., and R. C. Vogt. 1997b. Terrestrial movements in relation to aestivation and the annual reproductive cycle of Kinosternon leucostomum. Copeia 1997(1):123–130.

Morjan, C. L. and J. N. Stuart. 2001. Nesting record of a Big Bend Slider Turtle *(Trachemys gaigeae)* in New Mexico, and overwintering of hatchlings in the nest. Southwest. Natur. 46(2):230–234.

Mosimann, J. E. 1956. A morphometric analysis of allometry in shells of the turtles: *Graptemys geographica, Chrysemys picta,* and *Sternotherus odoratus*. Ph.D. U. of Michigan, Ann Arbor. Diss. Ab. Int. 17/06:1420 Order No:00-21341.

Mosimann J. E. and G. B. Rabb. 1953. A new subspecies of the turtle *Geoemyda rubida* (Cope) from western Mexico. Occas. Pap. Mus. Zool. Univ. of Mich. 548:1–7.

Mrosovsky, N. 1972. Spectrographs of the sounds of leatherback turtles. Herpetologica 28(3):256–258.

Mookergee, I., N. Solly, S. Royce, G. Tregear, C. Samuel, and M. Tang. 2006. Endogenous relaxin regulates collagen deposition in an animal model of allergic airway disease. Endocrinology 147(2):754–61.

Müller, L. 1936. Beitrage zur Kenntnis der Schildkrötenfauna von Mexico. Zool. Anz. 113(5/6):97–114.

Murphy, J. B., and W. E. Lamoreaux. 1978. Mating behavior in three Australian chelid turtles (Testudines: Pleurodira: Chelidae). Herpetologica 34:398–405.

Murphy, R. W. 1983. Paleobiogeography and genetic differentiation of the Baja California Herpetofauna. Occ. Pap. California Acad. Sci. 137:1–48.

Myers, G. S. 1945. A third record of the Sonoran box turtle. Copeia 1945 (3):172.

Nagle, R. D., V. J. Burke, and J. D. Congdon. 1998. Egg components and hatchling lipid reserves: parental investment in kinosternid turtles from the southeastern United States. Comp. Biochem. Physiol. 120:145–152.

Nagy, K. A. 1988. Seasonal patterns of water and energy balance in desert vertebrates. J. Arid Environm. 14:201–210.

Nagy, K. A., and P. A. Medica. 1977. Seasonal water and energy relations of free-living desert tortoises in Nevada: a preliminary report. Proc. Symp. Desert Tortoise Council 1977:152–157.

Nagy, K. A., and P. A. Medica. 1986. Physiological ecology of desert tortoises in southern Nevada. Herpetologica 42: 73–92.

Neill, W. T. 1960. Nature and men in British Honduras. Maryland Naturalist, 30(1/4):2–14.

Neill, W. T., and E. R. Allen. 1959. Studies on the amphibians and reptiles of British Honduras. Publ. Res. Div. Ross Allen Rept. Inst. 2(1):1–76.

Nichols, U. G. 1953. Habits of the desert tortoise, *Gopherus agassizii*. Herpetologica 9:65–69.

Nielsen, M. T., and J. M. Legler. Cloacal Bursae in Recent Turtles. Unpublished manuscript.

Nieuwolt, M. C. 1993. The Ecology of Movement and Reproduction in the Western Box Turtle in Central New Mexico. Unpublished PhD dissertation, Univ. New Mexico, Albuquerque.

Nieuwolt, P. M. 1996. Movement, activity, and microhabitat selection in the western box turtle, *Terrapene ornata luteola,* in New Mexico. Herpetologica 52(4):487–495.

Nieuwolt, P. M. 1997. Reproduction in the western box turtle, *Terrapene ornata luteola*. Copeia 997(4):819–826.

Noble, G. A. and E. R. Noble. 1940. A Brief Anatomy of the Turtle. Stanford Univ. Press, California.

Norris, K. S., and R. G. Zweifel. 1950. Observations on the habits of the ornate box turtle, *Terrapene ornata* (Agassiz). Natur. Hist. Misc. (5)8:1–4.

Nussbaum, R. A., E. D. Brodie, Jr., and R. M. Storm. 1983. Amphibians and Reptiles of the Pacific Northwest. Univ. Press Idaho, Moscow.

O'Brien, C., A. D. Flesch, E. Wallace, M. Bogan, S. E. Carrillo-Percastegui, S. Jacobs, and C. van Riper III. 2006. Biological Inventory of the Rio Aros, Sonora, Mexico: A River Unknown. Final report to T & E, Inc., Tucson, AZ.

Ogushi, K. 1913. Anatomische Studien an der japanischen dreikralligen Lippenschildkröte *(Trionyx japonicus)*. II. Mitteilung. Morphol. Jahrb. 46:299–562.

Oliver, J. A. 1937. Notes on a collection of amphibians and reptiles from the State of Colima, Mexico . Occ. Papers. l Mus. Zool. Univ. of Mich. 360:1–28.

Oliver, J. A. 1955. The natural history of North American amphibians and reptiles. Van Ostrand, Princeton, NJ.

Ottley, J. R., and V. M. Velazques Solis. 1989. An extant indigenous tortoise population in Baja California Sur, Mexico, with the description of a new species of *Xerobates* (Testudines: Testudinidae).Great Basin Natur. 49:496–502.

Packard, G. C., T. L. Taigen, M. J. Packard, and R. D. Shuman. 1979. Water-vapor conductance of testudinian and crocodilian eggs (class Reptilia). Respir. Physiol. 38:1–10.

Packard, M. J., and G. C. Packard. 1979. Structure of the shell and tertiary membranes of eggs of softshell turtles *(Trionyx spiniferus)*. J. Morphol. 159:131–144.

Packard, M. J., G. C. Packard, and T. J. Boardman. 1982. Structure of eggshells and water relations of reptilian eggs. Herpetologica 38:136–155.

Packard, M. J., K. F. Hirsch, and J. B. Iverson. 1984. Structure of shells from eggs of Kinosternid turtles. J. Morph. 181:9–20.

Paladino, F. V., M. P. O'Connor, and J. R. Spotila. 1990. Metabolism of leatherback turtles, gigantothermy, and thermoregulation of dinosaurs. Nature 344:858–860.

Parmley, D. 1990. A late Holocene herpetofauna from Montague County, Texas. Tex. J. Sci. 42:412–415.

Parmley, D. 1992. Turtles from the Late Hemphillian (Latest Miocene) of Know County, Nebraska. Tex. J. Sci. 44(3): 339–348.

Parsons, T. S. 1970. The nose and Jacobson's organ. In C. Gans and T. S. Parsons, Biology of the Reptilia, 2: Morphology B, 99–192. Academic Press, New York.

Parsons, T. S., and J. E. Cameron. 1977. Internal relief of the digestive tract. In C. Gans and T. S. Parsons, Biology of the Reptilia, 6: Morphology E. 159–224. Academic Press, New York.

Patterson, R. 1976. Vocalization in the desert tortoise. Proc. Symp. Desert Tortoise Council 1976:77–83.

Patterson, R. 1982. The distribution of the desert tortoise *(Gopherus agassizii).* In: R. B. Bury (Ed.). North American tortoises: conservation and ecology. U. S. Fish Wildl. Serv. Wildl. Res. Rep. (12), pp. 51–55.

Pauler, I. 1981. Zur Pflege und Zucht von *Claudius angustatus.* Herpetofauna (13):6–8.

Pawley, R. 1968. The hidden tortoise of Torreon. Int. Turtle Tortoise Soc. J.2(6):20–23,36.

Pawley, R. 1969. and at the zoo. Int. Turtle Tortoise Soc. J. 3(6):18–19, 33–37.

Pawley, R. 1975. Man and tortoise. Field Mus. Nat. Hist. Bull. 46(10):13–18.

Pearse, D. E., R. B. Dastrup, O. Hernandez, and J. W. Sites, Jr. 2006. Paternity in an Orinoco population of endangered Arrau River turtles, *Podocnemis expansa* (Pleurodira; Podocnemididae), from Venezuela. Chel. Cons. Biol. 5: 232–238.

Peréz-Higareda, G. 1978. Check list of freshwater turtles of Veracruz, Mexico. I. Southeastern portion of the state (Testudines: Cryptodira). Bull. MD. Herp. Soc. 14:215–222.

Peréz-Higareda, G. 1980. Check list of freshwater turtles of Veracruz, Mexico. II. Central portion of the state (Testudines: Cryptodira). Bull. MD. Herp. Soc. 16(1):27–34.

Peréz-Higareda, G., and H. M. Smith. 1988. Courtship behavior in *Rhinoclemmys areolata* from western Tabasco, México (Testudines: Emydidae). Great Basin Nat. 48: 263–266.

Peréz-Higareda, G., and H. M. Smith. 1987. Comments on geographic variation in *Rhinoclemmys areolata* (Testudines). Bull. Maryland Herpetol. Soc. 23(3):113–118.

Peréz Reyes, R. 1964. Estudios sobre protozoarios intestinales. I. Los flagelados del genero *Trimitus* Alexeieff, 1910. An. Esc. Nac. Ceinc. Biol. Mexico 13:59–66.

Peterson, E. H. 1992. Retinal structure. In C. Gans and P. S. Ulinski. Biology of the Reptilia 17: Neurology C, 1–135. Academic Press, New York.

Peterson, L., and G. Haug. 2005. Climate and the collapse of Maya Civilization. American Scientist. 93(4):332/

Phillips, C. A., W. W. Dimmick, and J. L. Carr. 1996. Conservation genetics of the common snapping turtle *(Chelydra serpentina).* Conservation Biology 10:397–405.

Pieau, C. 1976. Données recentes sur la différenciation sexuelle en fonction de la température chez les embryons d'Emys orbicularis L. (Chélonien). Bull. Soc. Zool. Fr. (Suppl. 4) 101:46–53.

Platt, S. G. 1993. Life history notes. *Rhinoclemmys areolata* (Furrowed wood turtle). Diet. Herp. Rev. 24(1):32.

Platt, S. G., A. G. Finger, T. R. Rainwater, and H. Woodke. 2004. *Rhinoclemmys areolata* (Furrowed wood turtle). Maximum size. Herp. Rev. 35:383.

Platt, S. G., J. C. Meerman, and T. R. Rainwater. 1999. Diversity, observations, and conservation of the herpetofauna of Turneffe, Lighthouse, and Glovers Atolls, Belize. British Herpetological Soc. Bull. 66(2):1–13.

Platt, S. G., T. R. Rainwater, and S. B. Brewer. 2004. Aspects of the burrow ecology of nine banded armadillos in northern Belize. Mammalian Biology 69:217–224.

Platt, S. G., W. B. Karesh, J. B. Thorbjarnarson, and T. R. Rainwater. 1999. Occurrence of the furrowed wood turtle *(Rhinoclemmys areolata)* on Turneffe Atoll, Belize. Chel. Cons. Biol. 3:490–491.

Plummer, M. V. 2003. Activity and thermal ecology of the Box Turtle, *Terrapene ornata,* at its southwestern range limit in Arizona. Chel. Cons. Biol. 4(3):1–9.

Plummer, M. V. 2004. Seasonal inactivity of the Desert Box Turtle, *Terrapene ornata luteola,* at the species' southwestern range limit in Arizona. J. Herp. 38(4):589–593.

Plummer, M. V., B. K. Williams, M. M. Skiver, and J. C. Carlyle. 2003. Effects of dehydration on the critical thermal maximum of the Desert Box Turtle *(Terrapene ornata luteola).* J. Herp. 37(4):747–750.

Polisar, J. 1992a. Hickatee: Summary of the Reproductive Biology, Exploitation, and Conservation of *Dermatemys mawi* in Belize. Manuscript.

Polisar, J. 1992b. Reproductive Biology, Exploitation, and Conservation of *Dermatemys mawi* in Northern Belize. Summary of Follow-Through Trip. Manuscript.

Polisar, J. 1996. Reproductive biology of a flood-season nesting freshwater turtle of the northern neotropics: *Dermatemys mawi* in Belize. Chelonian Conservation and Biology 2(1):13–25.

Pope, C. H. 1939. Turtles of the United States and Canada. Alfred A. Knopf Inc, New York.

Pregill, G., and D. W. Steadman. 2004. South Pacific iguanas: human impacts and a new species. J. Herp. 38(1):15–21.

Pritchard, P. C. H. 1967. Living Turtles of the World. T. F. H. Publ., Inc., Jersey City, N.J.

Pritchard, P. C. H. 1971. Numerical reduction of bony plastral elements in the kinosternid turtle *Claudius angustatus.* Copeia 1971(1):151–152.

Pritchard, P. C. H. 1979. Encyclopedia of Turtles. T. F. H. Publications, Inc. Ltd., Neptune, New Jersey.

Pritchard, P. C. H. 1984. Evolution and zoogeography of South American turtles. Stud. Geol. Salmanticensia. Vol. Espl (Stud.Paleochel. 1:225–233.

Pritchard, P. C. H., and N. Pronek. 1982. Request for suppression of *Kinosternon alamose* and *K. oaxacae* Pritchard, 1979 (Reptilia, Testudines). Bull. Zool. Nom. 39(3):212–213.

Pritchard, P. C. H., and P. Trebbau. 1984. The turtles of Venezuela. SSAR. Contrib. Herpetol. 2:1–403.

Proctor, V. M. 1958. The growth of *Basicladia* on turtles. Ecology 39:634–645.

Propst, D. L., and J. A. Stefferud. 1994. Distribution and status of the Chihuahua chub (Teliostei: Cyprinidae: *Gila nigrescens*), with notes on its ecology and associated species. Southwest. Natur. 39(3):224–234.

Punzo, F. 1974. A qualitative and quantitative study of food items of the yellow mud turtle, *Kinosternon flavescens* (Agassiz). J. Herp. 8(3):269–271.

Rafinesque, C. S. 1815. Analyse de la Nature ou Tableau de l'Univers et des Corps Organisés. Palermo.

Rafinesque, C. S. 1832. Description of two new genera of soft shell turtles of North America. Atlantic Jour. and Friend of Knowledge, Philadelphia 1:64–65.

Rainboth, W. J., D. C. Buth, and F. B. Turner. 1989. Allozyme variation in Mojave populations of the desert tortoise, *Gopherus agassizi*. Copeia 1989:115–123.

Raisz, Erwin. 1959. Landforms of Mexico. 1:1 Million Topographic Map. Prepared for Geography Branch, Office of Naval Research. Cambridge, Mass.

Redmer, M. 1987. Notes on the courtship and mating behaviour of the central American wood turtle, *Rhinoclemmys pulcherrima incisa*. Bull. Chic. Herpetol. Soc. 22(6–7):117–118.

Reese, R. W. 1971. Notes on a small herpetological collection from northeastern Mexico. J. Herp. 5:67–69.

Refsider, J. M. 2009. High frequency of multiple paternity in Blanding's turtle *(Emys blandingii)*. J. Herp.43(1):74–81.

Reisz, R. R., and J. J. Head. 2008. Turtle origins out to sea. Nature 456:27.

Reisz, R. R., and M. Laurin. 1991. Owenetta and the origin of turtles. Nature, 349:324–326.

Reyes Osorio, S. R., and R. B. Bury. 1982. Ecology and status of the desert tortoise *(Gopherus agassizii)* on Tiburon Island, Sonora. In: R. B. Bury (Ed.). North American tortoises: conservation and ecology, pp. 39–49. U. S. Fish Wildl. Serv. Wildl. Res. Rep. (12).

Reynolds, S. L., and M. E. Seidel. 1982. Sternotherus odoratus. Catalog. Amer. Amphib. Rept. 287:1–4.

Reynolds, S. L., and M. E. Seidel. 1983. Morphological homogeneity in the turtle *Sternotherus odoratus* (Kinosternidae) throughout its range. J. Herp. 17(2): 113–120.

Reynoso, V., and M. Montellano-Ballesteros. 2004. A new giant turtle of the genus *Gopherus* (Chelonia: Testudinidae) from the Pleistocene of Tamaulipas, Mexico, and a review of the phylogeny and biogeography of gopher tortoises. Journal of Vertebrate Paleontology 24(4):822–837.

Rhodin, A. G. J. (Ed.) 2002. The gopherine tortoises, *Gopherus* spp. Chel. Cons. Biol., Special Focus Issue. 4 (2): 247–515.

Rieppel, O., and M. deBraga. 1996. Turtles as diapsid reptiles. Nature 384:453–455.

Risley, P. L. 1933. Observations on the natural history of the common musk turtle, *Sternotherus odoratus* (Latreille). Pap. Michigan Acad. Sci. Arts Lett. 17:685–711.

Ritgen, F. A. 1828. Versuch einer natürlichen eintheilung der amphibien. Nova Acta Physico-Med. Acad. Caes. Leopold.-Carol. Nat. Curio. 14: 246.

Roberts, N. 1982. A preliminary report on the status of Chelydridae, Trionychidae, and Testudinidae in the region of Baja California, Mexico. Proc. Desert Tortoise Council 1982:154–161.

Robinson, K. M., and G. G. Murphy. 1978. The reproductive cycle of the eastern spiny softshell turtle *(Trionyx spiniferus spiniferus)*. Herpetologica 34(2):137–140.

Rogers, K. L. 1976. Herpetofauna of the Beck Ranch local fauna (Upper Pliocene: Blancan) of Texas. Publ. Mus. Mich. State. Univ., Paleontol. Ser. 1:163–200.

Rogner, M. 1996. Schildkröten 2. Hürtgenwald: Heidi-Rogner-Verlag.

Romer, A. S. 1956. The osteology of the reptiles. Univ. Chicago Press.

Romer, A. S., and T. S. Parsons. 1977. The Vertebrate Body. 5th ed. W. B. Saunders Company, Philadelphia, London, Toronto.

Rorabaugh, J. C, E., S. Montoya, and M. M. Gomez-Sapiens. 2008. *Apalone spinfera* (Spiny Softshell. Mexico Sonora. Herp. Rev. 39(3):365.

Rosado, R. D. 1967. La "Jicotea." J. Int. Turtle and tortoise Soc. 1 (3):16–19, 42.

Rose, F. L. 1970. Tortoise chin gland fatty acid composition: behavioral significance. Comp. Biochem. Physiol. 32: 577–580.

Rose, F. L. 1983. Aspects of the thermal biology of the bolson tortoise, *Gopherus flavomarginatus*. Occas. Pap. Mus.Texas Tech Univ. 89:1–8.

Rose, F. L., and F. W. Judd. 1975. Activity and home range size of the Texas tortoise, *Gopherus berlandieri*, in south Texas. Herpetologica 31:448–456.

Rose, F. L., and F. W. Judd. 1982. The biology and status of Berlandier's tortoise *(Gopherus berlandieri)*. In: R. B. Bury (Ed.). North American Tortoises: Conservation and Ecology, pp. 57–70. U. S. Fish Wildl. Serv. Wildl. Res. Rep. (12).

Rose, F. L., and F. W. Judd. 1989. *Gopherus berlandieri*, Berlandier's tortoise, Texas tortoise. In: I. R. Swingland and M. W. Klemens (Eds.). The Conservation Biology of Tortoises, pp. 8–9. Occ. Pap. IUCN Spec. Surv. Comm.

Rose, F. L., and F. W. Judd. 1991. Egg size versus carapace-xiphiplastron aperture size in *Gopherus berlandieri*. J. Herp. 25:248–250.

Rose, F. L., J. Koke, R. Koehn, and D. Smith. 2001. Identification of the etiological agent for necrotizing scute disease in the Texas tortoise. Journal of Wildlife Diseases 37(2):223–228.

Rosen, P. C. 1987. Variation of Female Reproduction Among Populations of Sonoran Mud Turtles *(Kinosternon sonoriense)*. MS thesis, Arizona State Univ., Tempe.

Rosen, P. C., C. R. Schwalbe, D. A. Rarazek Jr., P. A. Holm, and C. H. Lowe. 1994. Introduced aquatic vertebrates in the Chiricahua region: effect on declining native ranid frogs. In: L. F. Debano, P. F. Folliott, A. Ortega-Rubio, R. H. Hamre, and C. B. Edminster (Eds.). Biodiversity and Management of the Madrean Archipelago: The Sky Islands of the Southwestern United States and Northwestern Mexico, pp. 262–266. Genera Tech. Rpt. RM-GTR-264, USDA Forest Service, Fort Collins, CO.

Rosen, P. C., S. S. Sartorious, C. R. Schwalbe, P. A. Holm, and C. H. Lowe. 1996. Annotated Checklist of the Amphibians and Reptiles of the Sulphur Springs Valley, Cochise County, Arizona. Final Report, Part I. Heritage Program, Arizona Game Fish Dept.

Rostal, D. C., T. W. Wibbels, J. S. Grumbles, V. A. Lance, and J. R. Spotila. 2002. Chronology of sex determination in the desert tortoise *(Gopherus agassizii)*. Chel. Cons. Biol. 4(2):313–318.

Rostal, D. C., V. A. Lance, J. S. Grumbles, and A. C. Alberts. 1994. Seasonal reproductive cycle of the desert tortoise *(Gopherus agassizii)* in the eastern Mojave desert. Herpetological Monographs 8:72–82.

Rudloff, H. W. 1986. Schlammschildkröten-Terrarientiere der Zukunft. Aquar. Terrar. 33:166–169.
Ruthven, A. G. 1912. The amphibians and reptiles collected by the University of Michigan–Walker expedition in southern Veracruz, Mexico. Zool. Jahrb. 32:295–332.
Sacchi, R., P. Galeotti, and M. Fasola. 2003. Vocalizations and courtship intensity correlate with mounting success in marginated tortoises *Testudo marginata,* Behav. Ecol. Sociobiol. 55:95–102.
Sacchi, R., P. Galeotti, M. Fasola, and G. Gerzeli. 2004. Larynx morphology and sound production in three species of Testudinae. J. Morphol. 261:175–183.
Sachsee, W., and A. A. Schmidt. 1976. Naachzuct in der zweiter generation von Staurotypus salvini mit weiteren beobachtungen zum fortpflanzungsverhalten (Testudines, Kinosternidae). Salamandra. 12(1):5–16.
Sanchez-Montereo, P., C. G. Romero, R. C. Vogt, N. P. Lopez-León, and A. A. Dadda. 2000. Habitos alimenticios de *Kinosternon scorpioides abaxillare* en Piedra Parada, Chiapas, Mexico. Abstract. Sexta Reunion Nacional de Herpetologia Mexicana, Tuxtla Gutierez, Chiapas.
Sanchez-Villagra, M. R., H. Muller, C. H. Sheil, T. M. Scheyer, H. Nagashima, and S. Kuratani. 2009. Skeletal development in the Chinese soft-shelled turtle *Pelodiscus sinensis* (Testudines: Trionychidae) Journ. Morph. 270:1381–1399.
Savage, J. M. 1966. The origins and history of the Central American herpetofauna. Copeia 1966 (4):719–766.
Savage, J. M. 2002. The amphibians and reptiles of Costa Rica: a herpetofauna between two continents, between two seas. Univ. Chicago Press, Chicago and London.
Say, T. 1825. On the fresh water and land tortoises of the United States. Jour. Acad. Nat. Sci. Philadelphia 4:203–219.
Schamberger, M. L., and F. B. Turner. 1986. The application of habitat modeling to the desert tortoise *(Gopherus agassizii).* Herpetologica 42:134–138.
Schmidt, A. A. 1970. Zur fortpflanzung der kreuzbrustschildkröte *(Staurotypus salvini)* in gefangenschaft. Salamandra 6(1):10–31.
Schmidt, K. P. 1941. The amphibians and reptiles of British Honduras. Zool. Ser. Field Mus. Nat. Hist. 22(8):473–510.
Schmidt, K. P. 1953. A Checklist of North American Amphibians and Reptiles, 6th ed. Amer. Soc. Ichthyologists and Herpetologists. Univ. Chicago Press, Chicago.
Schmidt, K. P., and D. W. Owens. 1944. Amphibians and reptiles of northern Coahuila, Mexico. Zool. Ser., Field Mus. Nat. Hist. 29(6):97–115.
Schmidt, K. P., and F. A. Shannon. 1947. Notes on amphibians and reptiles of Michoacán, Mexico. Fieldiana Zool. 31(9):63–85.
Schneider, J. G. 1783. Allgemeine Naturgeschichte der Schildkröten, nebst einem System. Verseichnisse der einzeinen Arten, Leipzig.
Schneider, J. S., and G. D. Everson. 1989. The desert tortoise *(Xerobates agassizii)* in the prehistory of the southwestern Great Basin and adjacent areas. J. California Great Basin Anthropol. 11:175–202.
Schoepff, J. D. 1792–1801. Historia Testudinum iconibus illustrata. J. J. Palm, Erlangen. [Latin version of Naturgeschichte der Schildkrîten mit Abbildungen erlautert.].
Schoepff, J. David. 1792a. Historia testudinium. Erlangae Io. Iac. Palmii.
Schoepff, J. D. 1792b. Naturgeschichte der Schildkroten. Erlangen, Johann Jacob Palm.
Schwartz, C. W., and E. R. Schwartz. 1974. The three-toed box turtle in central Missouri: its population, home range, and movements. Missouri Dept. Conserv. Terr. Ser. (5):1–28.
Schwartz, E. R., and C. W. Schwartz. 1991. A quarter-century study of survivorship in a population of three-toed box turtles in Missouri. Copeia 1991:1120–1123.
Schweigger, F. 1812. Monographiae Cheloniorum. Naturwiss. Math. 1:271–368, 406–458.
Scott, N. J. 1962. The Reptiles of Southern Sinaloa: An Ecological and Taxonomic Study. M.Sci. Thesis, Humbolt St. Col. Arcata, Cal.
Seeliger, L. M. 1945. Variation in the Pacific mud turtle. Copeia 1945:150–159.
Seidel, M. E. 1978. *Kinosternon flavescens.* Cat. Amer. Amph. Rept. 216.1–4.
Seidel, M. E. 2002. Taxonomic observations on extant species and subspecies of slider turtles, genus *Trachemys.* J. Herp. 36(2):285–292.
Seidel, M. E., and D. R. Jackson. 1990. Evolution and fossil relationships of slider turtles. In: J. W. Gibbons (Ed.). Life History and Ecology of the Slider Turtle, pp. 68–73. Smithsonian Institution Press, Washington, D.C.
Seidel, M. E., and H. M. Smith. 1986. *Chrysemys, Pseudemys, Trachemys* (Testudines: Emydidae): Did Agassiz have it right? Herpetologica 42(2):242–248.
Seidel, M. E., J. B. Iverson, and M. D. Adkins. 1986. Biochemical comparisons and phylogenetic relationships in the family Kinosternidae (Testudines). Copeia 1986: 285–294.
Seidel, M. E., J. N. Stuart, and W. G. Degenhardt. 1999. Variation and species status of the slider turtles (Emydidae: *Trachemys*) In the southwestern United States and adjacent Mexico. Herpetologica 55(4):470–487.
Seidel, M. E., and R. V. Lucchino. 1981. Allozymic and morphological variation among the musk turtles *Sternotherus carinatus, S. depressus* and *S. minor* (Kinosternidae). Copeia 1981(1):119–128.
Seidel, M. E., and S. L. Reynolds. 1980. Aspects of evaporative water loss in the mud turtles *Kinosternon hirtipes* and *Kinosternon flavescens.* Comp. Biochem. Physiol. 67A: 593–598.
Seidel, M. E., S. L. Reynolds, and R. V. Lucchino. 1981. Phylogenetic relationships among musk turtles (genus *Sternotherus*) and genetic variation in *Sternotherus odoratus.* Herpetologica 37:161–165.
Seidel, M. E., and U. Fritz. 1997. Courtship behavior provides additional evidence for a monophyletic *Pseudemys,* and comments on Mesoamerican *Trachemys* (Testudines: Emydidae). Herp. Rev. 28(2):70–72.
Seidel, M. E., W. G. Degenhardt, and J. R. Dixon. 1997. Geographic distribution: *Trachemys gaigeae.* Herp. Rev. 28(3):157.
Seifert, W. E., S.W. Gotte, T. L. Leto, and P. J. Weldon. 1994. Lipids and proteins in the Rathke's gland secretions of the North American mud turtle *(Kinosternon subrubrum).* Comp. Biochem. Physiol. 109B(2/3):459–463.
Serb, J. M., C. A. Phillips, and J. B. Iverson. 2001. Molecular phylogeny and biogeography of *Kinosternon flavescens* based

on complete mitochondrial control region sequences. Molec. Phylogen. Evol. 18:149–162.
Sexton, O. J. 1960. Notas sobre la reproducción de una tortuga Venezolana la *Kinosternon scorpioides*. Mem. Soc. Cien. Natur. 20 (57):189–197.
Shaffer, H. B., P. A. Meylan, and M. L. McKnight. 1997. Tests of turtle phylogeny: Molecular, morphological, and paleontological approaches. Syst. Biol. 46: 235–268.
Shaw, C. A. 1981. The Middle Pleistocene El Golfo Local Fauna from Northwestern Sonora, Mexico. Unpublished MA thesis, Cal. St. Univ., Long Beach.
Shaw, C. E. 1952. Sexual dimorphism in *Terrapene klauberi* and the relationship of *T. nelsoni*. Herpetologica 8:39–41.
Shealy, R. M. 1976. The natural history of the Alabama map turtle, *Graptemys pulchra* Baur, in Alabama. Bull. Florida St. Mus. Biol. Sci. 21:47–111.
Sheil, C.A. 2003. Osteology and skeletal development of *Apalone spinifera* (Reptilia: Chelonii: Trionychidae). J. Morphol. 256(1):42–78.
Shine, R. 1994. Allometric patterns in the ecology of Australian snakes. Copeia 1994: 851–867.
Siebenrock, F. 1906. Schildkröten aus Sudmexico. Zool. Anz. 30:3–4.
Siebenrock, F. 1907. Die Schildkröten familie Cinosternidae. Sit zungsber. K.K. Akad. Wiss. (math.nat.Cl.) 116(1): 527–599.
Simmons, J. L. 2002. Herpetological collecting and collections management. SSAR Herp. Circular 31.
Sites, J. W. Jr., J. W. Bickham, and, M. W. Haiduk. 1979. Derived X chromosome in the turtle genus *Staurotypus*. Science (Washington, D.C.) 206(4425):1410–1412.
Smith, H. M. 1939. Notes on Mexican amphibians and reptiles. Field Mus. Nat. Hist. Zool. Ser. 24:15–35.
Smith, H. M., and E. H. Taylor. 1950a. An annotated checklist and key to the reptiles of Mexico exclusive of the snakes. Smithsonian Inst. U.S. Nat. Mus., Bull. 199:1–253.
Smith, H. M., and E. H. Taylor. 1950b. Type localities of Mexican reptiles and amphibians. Kansas Univ. Sci. Bull. 33(2):313–380.
Smith, H. M., and L. W. Ramsey. 1952. A new turtle from Texas. Wasmann J. Biol. 10:45–54.
Smith, H. M., and R. A. Brandon. 1968. Data nova herpetologica mexicana. Trans. Kansas Acad. Sci. 71:49–61.
Smith, H. M., and R. B. Smith. 1979. Synopsis of the Herpetofauna of Mexico. Vol. VI. Guide to Mexican turtles. Bibliographic addendum III. John Johnson, North Bennington, Vermont.
Smith, M. L. 1980. The Evolutionary and Ecological History of the Fish Fauna of the Rio Lerma Basin, Mexico. PhD Thesis, Univ. Michigan, Ann Arbor.
Smith, M. L., and R. R. Miller. 1986. The evolution of the Rio Grande Basin as inferred from its fish fauna. In: C. H. Hocutt and E. O. Wiley (Eds.). Zoogeography of North American Freshwater Fishes, pp. 457–486. John Wiley and Sons, Inc., New York.
Smith, M. L., T. M. Cavender, and R. R. Miller. 1975. Climatic and biogeographic significance of a fish fauna from the late Pliocene-early Pleistocene of the Lake Chapala Basin (Jalisco, Mexico). In: Studies on Cenozoic Paleontology and Statigraphy in Honor of Claude W. Hibbard. Univ. Michigan Pap. Paleontology, 12:29–38.
Smith, P. W. 1951. A new frog and a new turtle from the western Illinois sand prairies. Bull. Chicago Acad. Sci. 9:189–199.
Smith, P. W., and M. M. Hensley.1957. The mud turtle *Kinosternon flavescens stejnegeri* Hartweg, in the United States. Proc. Bio. Soc. Wash. 70:201–204.
Snider, Andrew T., and J. K. Bowler. 1992. Longevity of reptiles and amphibians in North American collections. Second Edition. SSAR. Herp. Circular 21:1–40.
Sokol, O. 1971. Lithophagy and geophagy in reptiles. J. Herp. 5:67–71.
Sonnini de Manoncourt, C. S., and P. A. Latreille. 1801. Histoire naturelle des reptiles avec figures dessinees d'apres nature. I. Deterville, Paris.
Sowerby, J., and E. Lear. 1872. Tortoises, terrapins and turtles. London.
Spinks, P. Q., H. B. Shaffer, J. B. Iverson, and W. P. McCord. 2004. Phylogenetic hypotheses for the turtle family Geoemydidae. Molec. Phylogen. Evol. , 32:164–182.
Spix, J. B. 1824. Animalia Nova sive Species Novae Testudinum et Ranarum. Monachii.
Spotila, J. R., and E. A. Standora. 1985. Environmental constraints on the thermal energetics of sea turtles. Copeia 1985(3):694–702.
Spotila, J. R., L. C. Zimmerman, C. A. Binckley, J. S. Grumbles, D. C. Rostal, A. J. List, E. C. Beyer, K. M. Phillips, and S. J. Kemp. 1994. Effects of incubation conditions on sex determination, hatching success, and growth of hatchling desert tortoises, *Gopherus agassizii*. Herpet. Monogr. 8: 103–116.
St. Amant, J. A. 1976. State report—California Proc. Symp. The Desert Tortoise Council 1976:5–7.
Standora, E. A., J. R. Spotila, J. A. Keinath, and C. R. Shoop. 1984. Body temperatures, diving cycles, and movement of a subadult Leatherback Turtle, *Dermochelys coriacea*. Herpetologica 40(2):169–176.
Stebbins, R. C. 1954. Amphibians and Reptiles of Western North America. McGraw Hill Book Co., Inc. New York.
Stebbins, R. C. 1985. A Field Guide to Western Reptiles and Amphibians. 2nd ed. Houghton Mifflin Co., Boston.
Stejneger, L. H. 1918. Description of a new lizard and a new snapping turtle from Florida. Proc. Biol. Soc. Washington 31:89–92.
Stejneger, L. H. 1925. New species and subspecies of American turtles. J. Washington Acad. Sci. 15:462–463.
Stejneger, L. H., and T. Barbour. 1917. A check list of North American amphibians and reptiles. Harvard Univ. Press. Cambridge, Mass.
Stejneger, L. H., and T. Barbour. 1923. A check list of North American amphibians and reptiles. 2nd ed. Cambridge: Harvard Univ. Press.
Stejneger, L. H., and T. Barbour. 1939. A Check List of North American Amphibians and Reptiles. 4th ed. Cambridge, Mass., Harvard Univ. Press.
Stephens, P. R., and J. J. Wiens. 2008. Testing for evolutionary trade-offs in a phylogenetic context: ecological diversification and evolution of locomotor performance in emydid turtles. J. Evol. Biol. 21:77–87.
Stickel, E. L. 1949. Population and home range relationships of the box turtle, *Terrapene carolina* (Linnaeus).Microfilm Abstr. 9:195–196.

Stock, A. D. 1972. Karyological relationships in turtles (Reptilia:Chelonia). Can. J. Genet. Cytol. 14:859–868.
Stone, P. A. 2001. Movements and demography of the Sonoran mud turtle, *Kinosternon sonoriense*. SW. Nat. 46(1):41–53
Stone, P. A. 2006. Terrestrial Activity in Sonoran Mud Turtles, *Kinosternon sonoriense*. Abstract, Powdermill freshwater turtle symposium, Portal Arizona, August 2006.
Stone, P. A., M. E. Babb, B. D. Stanila, G. W. Kersey, and Z. S. Stone. 2005. *Kinosternon sonoriense* (Sonoran mud turtle). Diet. Herpetolog. Rev. 36(2):167–168.
Storer, T. I. 1930. Notes on the range and life history of the Pacific fresh water turtle, *Clemmys marmorata*. Univ. California Publ. Zool. 32:429–441.
Strain, W. S. 1966. Blancan mammalian fauna and Pleistocene formations, Hudspeth County, Texas. Bull. Texas Mem. Jus., 10:1–31.
Strauch, A. 1890. Bemerkungen uber die Schildkrötensammlung. Mem. Acad. Sci. St. Pete. 38(): 1–127.
Stromsten, F. A. 1917. The development of the musk glands in the loggerhead turtle. Proc. Iowa Acad. Sci. 24:311–313.
Stuart, J. N. 1995a. Geographic distribution: *Trachemys scripta scripta*. Herp. Rev. 26(2):107.
Stuart, J. N. 1995b. Notes on aquatic turtles of the Rio Grande drainage, New Mexico. Bull. Maryland Herpetol. Soc. 31(3):147–157.
Stuart, J. N. 2000. Additional notes on native and non-native turtles of the Rio Grande Drainage basin, New Mexico. Bull. Chicago Herp. Soc. 35(10):229–235.
Stuart, J. N., and C. H. Ernst. 2004. *Trachemys gaigeae*. Cat. Amer. Amphib. Rept. 787:1–6.
Stuart, J. N., and C. W. Painter. 1997. *Trachemys gaigeae*. Reproduction. Herp. Rev. 28(3):149–150.
Stuart, J. N., and C. W. Painter. 2002. Observations on the diet of *Trachemys gaigeae* (Testudines: Emydidae). Bull. Maryland Herp. Soc. 38(1):15–22.
Stuart, J. N., and C. W. Painter. 2005. *Trachemys gaigeae* (Mexican Plateau Slider). Reproductive Characteristics. Herp. Rev.: Natural History Note.
Stuart, J. N., C. W. Painter, and B. C. Stearns. 1993. Life history notes. *Trachemys gaigeae* (Big Bend slider). Maximum size. Herp. Rev. 24(1):32–33.
Stuart, J. N., and J. B. Miyashiro. 1998. *Trachemys gaigeae* courtship behavior. Herp. Rev. 29:235–236.
Stuart, L. C. 1935. A contribution to a knowledge of the herpetofauna of a portion of the savanna region of central Peten, Guatemala. Misc. Publ. Mus. Zool., Univ. Michigan 29:1–56.
Stuart, L. C. 1950. A geographic study of the herpetofauna of Alta Verapaz, Guatemala. Contr. Lab. Vert. Biol. 45:1–77.
Stuart, L. C. 1951. The herpetofauna of the Guatemalan Plateau, with special reference to its distribution on the southwestern highlands. Contrib. Lab. Vert. Biol. Univ. Mich. (49):1–71.
Stuart, L. C. 1963. A checklist of the herpetofauna of Guatemala. Misc. Publ. Mus. Zool. Univ. Mich. 122:1–150.
Stuart, L. C. 1964. Fauna of Middle America. In: R. Wauchope and R. C. West (Eds.). Handbook of Middle American Indians, Vol. 1: Natural Environment and Early cultures, pp. 316–62. Austin: Univ. of Texas Press.
Stuart, L. C. 1966. The environment of the Central American cold-blooded vertebrate fauna. Copeia 4:684–699.
Sumichrast, F. 1880. Contribution a l'histoire naturelle du Mexique. 1. Notes sur un collection des reptiles et des batraciens de le partie occidentale de l'Istme de Tehuantepec. Bull. Zool. Soc. France 5:162–190.
Sumichrast, F. 1882. Contribución a la historia natural de México. I. Notas acerca de una colección de reptiles y baraciaos de la parte occidental del Istmo de Tehuantepec. Naturaleza 5: 268–293.
Tamayo, J. L. 1946. Datos para la Hidrología de la Republica Mexicana. Inst. Panamer. d. Geogr. e. Hist., 84:1–448.
Tamayo, J. L. 1949. Geografia General de Mexico. 2 Vols. Institutuo Mexicano de Investigaciones Economicas, Mexico.
Tamayo, J. L. 1962. Geografia General de Mexico. 3 vols. 2nd ed. Institutuo Mexicano de Investigaciones Economicas, Mexico.
Tamayo, J. L., and R. C. West. 1964. The hydrography of Middle America. In: R. Wauchope and R. C. West (Eds.). Handbook of Middle American Indians, Vol. 1, pp. 84–121. Univ. TX. Press, Austin, TX.
Taylor, E. H. 1933. Observations on the courtship of turtles. Univ. Kansas Sci. Bull. 21:269–271.
Taylor, W. E. 1895. The box turtles of North America. Proc. U.S. Natl. Mus. 17:573–588.
Thatcher, V. E. 1963. Trematodes of turtles from Tabasco, Mexico, with a description of a new species of *Dadytrema* (Trematoda: Paramphjistomidae). Amer. Midland Natur. 70(2)347–355.
Theobald, W. 1868. Catalogue of the reptiles of British Burma, embracing the provinces of Pegu, Martaban and Tenasserim, with descriptions of new or little known species. Jour. Linn. Soc. London 10:4–68.
Thompson, M. B. 1985. Functional significance of the opaque white patch in eggs of *Emydura macquarii*. In: G. Griff, R. Shine, and H. Ehmann (Eds.). Biology of Australasian Frogs and Reptiles, pp. 387–395. Royal Zoological Society of New South Wales.
Thornhill, G. M. 1982. Comparative reproduction of the turtle, *Chrysemys scripta elegans*, in heated and natural lakes. J. Herpet. 16:347–353.
Tom, J. 1994. Microhabitats and use of burrows of bolson tortoise hatchlings. U.S. Fish and Wildlife Service, Fish and Wildlife Research 13 :138–146.
Tomko, D. S. 1972. Autumn breeding of the desert tortoise. Copeia 1972:895.
Trevino, E., D. J. Morafka, and G. Aguirre. 1997. A second reserve for the Bolson tortoise, *Gopherus flavomarginatus*, at Rancho Sombreretillo, Chihuahua, Mexico. In: J. Abbema (Ed.). Proceedings: Conservation, Restoration, and Management of Tortoises and Turtles—An International Conference, 11–16 July 1993, State Univ. of New York, Purchase, New York. New York Turtle and Tortoise Society and WCS.
Trevino, E., D. J. Morafka, and G. Aguirre-León. 1995. Morphological distinctiveness of the northern population of the Bolsón tortoise, *Gopherus flavomarginatus*. Publicaciones De La Sociedad Herpetologica Mexicana 2, 1995.

Troyer, K. 1982. Transfer of fermentation microbes between generations in a herbivorous lizard. Science 216:540–542.
Tucker, J. K., C. R. Dolan, and E. A. Dustman. 2006. Chelonian species record carapace lengths for Illinois. Herp. Rev. 37(4):453.
Turner, F. B., P. A. Medica, and C. L. Lyons. 1984. Reproduction and survival of the desert tortoise *(Scaptochelys agassizii)* in Ivanpah Valley, California. Copeia 1984:811–820.
Turner, F. B., P. Hayden, B. L. Burge, and J. B. Roberson. 1986. Egg production by the desert tortoise *(Gopherus agassizii)* in California. Herpetologica 42:93–104.
Tuttle, S. E., and D. M. Carroll. 2005. Movements and behavior of hatchling wood turtles *(Glyptemys insculpta)*. Northeastern Naturalist 12(3):331–348.
Underwood, G. 1970. The eye. In: C. Gans. (Ed.). Biology of the Reptilia, Vol. 2. Morphology B, pp. 1–93. Academic Press, New York.
Valenzuela, N., and V. Lance. 2004. Temperature-dependent sex determination in vertebrates, Smithsonian Books, Washington, D.C.
Vallen, E. 1944. Über die Entwicklung der Moschusdrüsen bei einigen Schildkröten. Acta Zool 25:193–249.
Van Denburgh, J. 1895. A review of the herpetology of Lower California. Part I. Reptiles. Proc. California Acad. Sci., Ser. 2, 5:77–162.
Van Denburgh, J. 1922. The reptiles of western North America. Vol. II. Snakes and turtles. Occ. Papers Calif. Acad. Sci. 10:623–1028.
Van Devender, T. R. (Ed.) 2002. The Sonoran Desert Tortoise: Natural History, Biology, and Conservation. Tucson, AZ: Univ. of Arizona Press.
Van Devender, T. R., A. M. Rea, and M. L. Smith. 1985. The Sangamon interglacial vertebrate fauna from Rancho la Brisca, Sonora Mexico. Trans. San Diego Soc. Natur. Hist. 21:23–55.
Van Devender, T. R., and W. Van Devender. 1975. Ecological notes on two Mexican skinks. Southwest Natur. 20 (2):279–282.
Van Loben Sels, R. C., J. D. Congdon, J. Austin, and V. Austin. 2006. Life history and ecology of the aquatic sonoran mud turtle *(Kinosternon sonoriense)*: living in uncertain environments (1990–2006). Abstract, Powdermill freshwater turtle symposium, Portal Arizona, August 2006.
Van Loben Sels, R. C., J. D. Congdon, and J. T. Austin. 1997. Life history and ecology of the Sonoran mud turtle *(Kinosternon sonoriense)* in Southeastern Arizona: a preliminary report. Chelon. Cons. Biol. 2:338–344.
Van Loben Sels, R. C., J. D. Congdon, W. P. Hollett, M. Cameron, and N. A. Dickson. 2008. Nesting movements and activity, nest site characteristics, and embryo development in the Sonoran Mud Turtle *(Kinosternon sonoriense)*. Abstract, "Current Research on Herpetofauna of the Sonoran Desert," 11–13 April 2008, Tucson Herpetological Society.
Velasco, A. L. 1892. Geografia y estadistica del estado de Guerrero. Geografia y estadistica de la Republica Mexicana. Vol. 10, Reptiles and Amphibians, pp. 76–76. Mexico, D. F. Secr. Formento.
Velasco, A. L. 1896. Geografia y estadistica del estado de Colima. Geografia y Estadistica de la Republica Mexicana. Vol. 18. Secr. Fomento. Mexico, D. F.
Vermersch, T. G. 1992. Lizards and Turtles of South-Central Texas. Eakin Press, Austin, Texas.
Vetter, H. 2004. Turtles of the World. Vol 2. North America. Chimaira Buchhandelgeselschaft mbH, Frankfurt.
Vetter, H. 2005. Turtles of the World. Vol 3. Central and South America. Chimaira Buchhandelgeselschaft mbH, Frankfurt.
Vogt, R. C. 1980. Natural history of the map turtles *Graptemys pseudogeographica* and *G. ouachitensis* in Wisconsin. Tulane Stud. Zool. Bot. 22:17–48.
Vogt, R. C. 1981b. Demografia de poblaciones de *Kinosternon leucostomum* Duméril y Bibron, en la Estacion de Biologia Tropical "Los Tuxtlas," Veracruz (Informe).
Vogt, R. C. 1981a. Natural History of Amphibians and Reptiles in Wisconsin. Milwaukee Publ. Mus., Milwaukee, Wisconsin.
Vogt, R. C. 1986. Life history notes. *Claudius angustatus* (Chopontil). Coloration. Herpetological Rev. 17(3): 64.
Vogt, R. C. 1987. Herbivory in tortuga blanca *Dermatemys mawi*. [Abstract 67th Reunion Annual de la American Society of Ichthyologists and Herpetologists. Albany, New York.
Vogt, R. C. 1988. Ecologia y status de la tortuga blanca *Dermatemys mawei*. Reporte Tenico Final de Proyecto, CONACyT. Mexico.
Vogt, R. C. 1990. Reproductive parameters of *Trachemys scripta venusta* in southern Mexico. In: J. W. Gibbons (Ed.). Life History and Ecology of the Slider Turtle, pp. 162–168. Smithsonian Institution Press, Washington, D. C.
Vogt, R. C. 1991. Effect of Estradiol on Sex Determination in Turtles. Abstract. Joint Annual Meeting of American Society of Ichthyologists and Herpetologists, SSAR and the Herpetologists League. State College, Pennsylvania.
Vogt, R. C. 1993. Systematics of the false map turtles (*Graptemys pseudogeographica* complex: Reptilia, Testudines, Emydidae). Annals Carnegie Museum 62 (1):1–46.
Vogt, R. C. 1997c. Ecologia de las comunidades y status de las poblaciones de tortugas dulceaculcolas del sureste de Mexico. Reporte de proyecto 96-06-040-V CONACYT-SIGOLFO, Mexico.
Vogt, R. C. 1997a. *Claudius angustatus* (Chopontil, Taiman, Joloque). In: E. González, R. Dirzo, and R. C. Vogt (Ed.). Historia natural de los tuxtlas, pp. 480–481. UNAM, Mexico, D. F.
Vogt, R. C. 1997b. *Staurotypus triporcatus*. In E. González, E., R. Dirzo, and R. C. Vogt (Eds.). Historia natural de los tuxtlas, pp. 494–495. UNAM, Mexico, D. F.
Vogt, R. C. 1998. Ecologia de las comunidades y estatus de las poblaciones de tortugas dulceacuicolas del sureste de Mexico. Proyecto SIGOLFO 96-06-040-v. Jalapa, Veracruz.
Vogt, R. C. 2008. Amazon Turtles. Biblios, Lima, Peru.
Vogt, R. C., and C. J. McCoy. 1980. Status of the emydine turtle genera *Chrysemys* and *Pseudemys*. Ann. Carnegie Mus. Nat. Hist. 49:93–102.
Vogt, R. C., D. E. Sever, and G. R. Moreira. 1998. Esophageal papillae in Pelomedusid turtles. J. Herp. 32:279–282.
Vogt, R. C., and J. B. Iverson. 2011. *Kinosternon acutum*. In: P. C. H. Pritchard and A. G. J. Rhodin (Eds.). The

Conservation Biology of Freshwater. Turtles 1993. IUCN/ SSC Tortoise and Freshwater Turtle Specialists Group.
Vogt, R. C., and J. J. Bull. 1982. Genetic sex determination in the spiny softshell, *Trionyx spiniferus* (Testudines: Trionychidae).Copeia 1982(3):699–700.
Vogt, R. C., J. J. Bull, C. J. McCoy, and T. W. Houseal. 1982. Incubation temperature influences sex determination in kinosternid turtles. Copeia 1982; (2):480–482.
Vogt, R. C., M. Dath, V. G. Espejel, and N. López-Luna. 2000. Demography of *Kinosternon acutum* in Veracruz, Mexico. Abstract, Joint meeting of Ichthyologists and Herpetologists. La Paz, Baja California Sur, Mexico.
Vogt, R. C., and O. Flores-Villela. 1986. Determinación del sexo en tortugas por la temperatura de incubación de los huevos. Ciencia 37:21–32.
Vogt, R. C., and O. Flores-Villela. 1992a. Aspectos de la ecología de la tortuga blanca *Dermatemys mawii* en La Reserva de Montes Azules, Chiapas. Reserva de la Biosfera Montes Azules, Chiapas. Reserva de la Biosfera Montes Azules, Selva Lacandona: Investigación para su Conservación. Publ. Esp. Ecosfera 1:221–132.
Vogt, R. C., and O. Flores-Villela. 1992b. Effects of incubation temperature on sex determination in a community of neotropical freshwater turtles in southern Mexico. Herpetologica 48(3): 265–270.
Vogt, R. C., and S. G. Guzman. 1988. Food partitioning in a Neotropical freshwater turtle community. Copeia 1988(1):37–47.
Vogt, R. C., S. G. Platt, and T. R. Rainwater. 2009. *Rhinoclemmys areolata* (Duméril and Bibron 1851)—furrowed wood turtle, black-bellied turtle, mojina. In: A. G. J. Rhodin, P. C. H. Pritchard, P. P. van Dijk, R. A. Saumure, K. A. Buhlmann, J. B. Iverson, and R. A. Mittermeier (Eds.). Chelonian Research Monographs No. 5, pp. 022.1–022.7, doi:10.3854/crm.5.022.areolata.v1.2009, http://www.iucn-tftsg.org/cbftt/. Conservation Biology of Freshwater Turtles and Tortoises: A Compilation Project of the IUCN/SSC Tortoise and Freshwater Turtle Specialist Group.
Voigt, W. G., and C. R. Johnson. 1976. Aestivation and thermoregulation in the Texas tortoise, *Gopherus berlandieri*. Comp. Biochem. Physiol. 53A:41–44.
Voigt, W. G., and C. R. Johnson. 1977. Physiological control of heat exchange rates in the Texas tortoise, *Gopherus berlandieri*. Comp. Biochem. Physiol. 56A:495–498.
Waagen, G. N. 1972. Musk Glands in Recent Turtles. MS thesis, Univ. of Utah, Salt Lake City.
Wagler, J. G. 1830. Natürliches System der Amphibien, mit vorangehender Classification der Saugethiere und Vogel. Ein Beitrag zur vergleichender Zoologie. Munchen, Stuttgart and Tubigen.
Walbaum, J. J. 1782. Chelonographia oder beschreibung einiger Schildkröten nach natürlichen Urbildern. Lubeck and Leipzig, Johann Friedrich Gledrisch.
Walker, W. F., Jr. 1973. The locomotor apparatus of Testudines. In: C. Gans and T. S. Parsons (Eds.). , Biology of the Reptilia, 4: Morphology D, 1–100. Academic Press, New York.
Walker, W. F., Jr. 1979. Locomotion. In: M. Harless and H. Morlock (Eds.). Turtles: Perspectives and Research, pp. 435–454. John Wiley and Sons, Inc., New York.
Walls, G. L. 1942. The vertebrate eye and its adaptive radiation. Bull. Cranbrook Inst. Sci. 19., Bloomfield Hills, Michigan.
Ward, J. P. 1970. Accessory Respiratory Mechanisms Developed in Response to Hypoxia in Three Species of Chelonia. MA thesis, Univ. of Missouri, Columbia.
Ward, J. P. 1978. *Terrapene ornata* (Agassiz). Ornate box turtle. Catalog Amer. Amphib. Rept. 217:1–4.
Ward, J. P. 1984. Relationships of chrysemyd turtles of North America. Spec. Publ. Mus. Texas Tech Univ. 21:1–50.
Ward, W. C. 1985. Quaternary geology of northeastern Yucatan Peninsula. In: W. C. Ward, A. E. Weidie, and W. Back (Eds.). Geology and Hydrogeology of the Yucatan and Quaternary Geology of Northeastern Yucatan Peninsula, Part 2, pp. 23–53. New Orleans Geological Soc. New Orleans, La.
Watson, D. M. S. 1914. Eunotosaurus africanus Seeley, and the ancestors of Chelonia. Proc. Zool. Soc. London 2:1011–1020.
Weaver, W. G. and F. L. Rose. 1967. Systematics, fossil history, and evolution of the genus *Chrysemys*. Tulane Stud. Zool. 14:63–73.
Weaver, W. G., Jr. 1970. Courtship and combat behavior in *Gopherus berlandieri*. Bull. Florida St. Mus. Biol. Sci. 15:1–43.
Webb, R. G. 1956. Size at sexual maturity in the male softshell turtle, *Trionyx ferox* emoryi. Copeia 1956(2):121–122.
Webb, R. G. 1962. North America recent soft-shelled turtles (Family: Trionychidae). Univ. Kansas Publ., Mus. Nat. Hist., 13(10):429–611.
Webb, R. G. 1970. Reptiles of Oklahoma. U. Oklahoma Press, Norman.
Webb, R. G. 1973. *Trionyx ater* Webb and Legler: Black softshell turtle. Cat. Amer. Amph. Rept. 137.1.
Webb, R. G. 1973. *Trionyx spiniferus*. Cat. Amer. Amph. Rept. 140.1–140.4
Webb, R. G. 1980. The trionychid turtle *Trionyx steindachneri* introduced in Hawaii? J. Herpet. 14(2):206–207.
Webb, R. G. 1984. Herpetogeography in the Mazatlan-Durango region of the Sierra Madre Occidental, Mexico, In: R. A. Seigel, L. E. Hunt, J. L. Knight, L. Malaret, and N. L. Zuschlag (Eds.). Vertebrate Ecology and Systematics. A Tribute to Henry S. Fitch, pp. 217–241. Univ. Kansas Mus. Nat. Hist. Spec. Publ. No. 10 Lawrence, Kansas.
Webb, R. G. 1990. *Trionyx*. Catalog. Amer. Amphib. Rept. 487:1–7.
Webb, R. G., and J. M. Legler. 1960. A new softshell turtle (Genus *Trionyx*) from Coahuila, Mexico. Univ. of Kansas Science Bulletin, 40:21–30.
Webb, R. G., and M. Hensley. 1959. Notes on reptiles from the Mexican state of Durango. Publ. Mus. Mich. Biol. Ser. 1(6):249–258.
Webb, R. G., W. L. Minckley, and J. E. Craddock. 1963. Remarks on the Coahuilan box turtle *Terrapene coahuila* (Testudine: Emydidea). Southwestern Natur. 8(2):89–99.
Weinstein, M. N., and K. H. Berry. 1987. Morphometric analysis of desert tortoise populations. Bur. Land Mgmt. Report, CA950-CT7-003.
Weisrock, D. W., and F. J. Janzen. 2000. Comparative molecular phylogeography of North American softshell turtles *(Apalone):* implications for regional and wide-scale

historical evolutionary forces. Molecular Phylogenetics Evolution 14:152–164.
Welsh, H. H., Jr. 1988. An ecogeographic analysis of the herpetofauna of the Sierra San Pedro Martir region, Baja California, with a contribution to the biogeography of the Baja California herpetofauna. Proc. California Acad. Sci., 4th ser., 46:1–72.
Wermuth, H., and R. Mertens. 1961. Schildkröten *Krokodile* Bruckenechsen. VEB Gustav Fischer Verlag. Jena, Germany.
West, G. R., and J. P. Augelli. 1976. Middle America, Its Land and Peoples. 2nd ed. Prentice-Hall, Inc. Englewood Cliffs, N. J.
Whetstone, K. N. 1997. *Platysternon* and the evolution of chelydrid turtles. Herp. Rev. 8(3):20.
White, F. N. 1976. Circulation. In: C. Gans and W. R. Dawson (Eds.). Biology of the Reptilia, 5: Physiology A, pp. 275–334.
Wied, M. A. zu. 1839. Reise in das inneere Nord-America in den Jahren 1832 bis 1834. J. Hoelscher, Coblenz.
Wied, M. A. zu. 1865. Verzeichnis der Reptilien welche auf einer Reise im nordlichen America beobachtet wurden. Nova. Act. Acad. Leopold Carol. Nat. Curios.
Wiegmann, A. F. A. 1828. Beiträge zur amphibienkunde. Isis von Oken 21:364–383.
Wiewandt, T. A., C. H. Lowe, and M. W. Larson. 1972. Occurrence of *Hypopachus variolosus* (Cope) in the short-tree forest of southern Sonora Mexico. Herpetologica 28: 162–164.
Wiley, E. O. 1981. Phylogenetics: The Theory and Practice of Phylogenetic Systematics. John Wiley and Sons, New York.
Wiley, E. O., D. Siegel-Causey, D. R. Brooks, and V. A. Funk. 1991. The complete cladist. A primer of phylogenetic procedures. Univ. Kansas Mus. Nat. Hist., Spec. Publ. 19:1–158.
Willey, G. R., G. F. Ekholm, and R. F. Millon. 1964. The patterns of farming life and civilization. In: Wauchoe, R. and West, R. C. (Eds.) Handbook of middle American Indians, Vol. 1: Natural Environment and Early Cultures. Austin: Univ. Texas Press.
Williams, E. E. 1950. Variation and selection in the cervical central articulations of living turtles. Bull. Amer. Mus. Nat. Hist. 94 (9):505–562.
Williams, E. E. 1956. *Pseudemys scripta callirostris* from Venezuela with a general survey of the scripta series. Bull. Mus. Comp. Zool., 115(5):145–160.
Williams, E. E., and S. B. McDowell, Jr. 1952. The plastron of soft-shelled turtles (Testudinata, Trionychidae): a new interpretation. J. Morph. 90:263–280.
Williams, K. L., H. M. Smith, and P. S. Chrapliwy. 1960. Turtles and lizards from northern Mexico. Trans. Illinois St. Acad. Sci. 53:36–45.
Williams, K. L., and L. D. Wilson. 1965. Noteworthy Mexican reptiles in the Louisiana State Univ. Museum of Zoology. Proc. Louisiana Acad. Sci, 28:127–130.
Williams, K. L., and P. Han. 1964. A comparison of the density of Terrapene coahuila and *T. carolina*. J. Ohio Herp. Soc. 4(4):105.
Williams, T. A., and J. L. Christiansen. 1981. The niches of two sympatric softshell turtles, *Trionyx muticus* and *Trionyx spiniferus*. Iowa. J. Herpetol. 15(3):303–308.
Wilson, D. S. 2006. The Desert Box Turtle, *Terrapene ornata luteola:* What We Know and Where Do We Go. Abstract and presentation, Powermill Turtle Symposium, SW Res. Sta., Portal AZ, 1–3 Sept. 2006.
Winokur, R. M. 1981. Erectile tissue and smooth muscles in the snouts of turtles. [Abstract] Am. Zool. 21(4):959.
Winokur, R. M. 1968. The Morphology and Relationships of the Soft-Shelled turtles of the Cuatro Cienegas Basin, Coahuila, Mexico. MS dissertation, Arizona St. Univ., Tempe.
Winokur, R. M. 1982a. Integumentary appendages of chelonians. J. Morph. 172:59–74.
Winokur, R. M. 1982b. Erectile tissue and smooth muscle in snouts of *Carettochelys insculpta,* trionichids and other Chelonia. Zoomorphology, 101:83–93.
Winokur, R. M. 1988. The buccopharyngeal mucosa of the turtles (Testudines). J. Morph. 196:33–52.
Winokur, R. M. 1988. The buccopharyngeal mucosa of the turtles (Testudines). J. Morph. 196: 33–52.
Winokur, R. M., and J. M. Legler. 1974. Rostral pores in turtles. J. Morph. 143:107–120.
Winokur, R. M., and J. M. Legler. 1975. Chelonian mental glands. J. Morph. 147(3):275–292.
Wolf, S. 1933. Zur Kenntnis von Bau und Funktion der Reptilien Lunge. Zool. Jahrb. Abt. Anat. Ont. 57: 139–190.
Woodbury, A. M. 1952. Hybrids of *Gopherus berlandieri* and *G. agassizii*. Herpetologica 8(1):33–36.
Woodbury, A. M., and R. Hardy. 1948. Studies of the desert tortoise, *Gopherus agassizii*. Ecol. Monogr. 18:145–200.
Yamaguti, S. 1958. Systema helminthum. Volume I. The digenetic trematodes of Vertebrates. Part 1. Interscience Pub. New York.
Yntema, C. L. 1968. A series of stages in the embryonic development of *Chelydra serpentina*. J. Morph. 125: 219–252.
Yntema, C. L. 1976. Effects of incubation temperatures on sex differentiation in the turtle, *Chelydra serpentina*. J. Morph. 150:453–462.
Zangerl, R. 1969. The turtle shell. In: C. Gans, A d'A. Bellairs, and T. S. Parsons (Eds.). Biology of the Reptilia, Vol. 1, Morphology A, pp. 311–339. Academic Press, New York.
Zardoya, R., and A. Meyer. 1998. Complete mitochondrial genome suggests diapsid affinities of turtles. Proceedings of the National Academy of Sciences USA. 95: 14226–14231.
Zenteno-Ruiz, C. E. 1993. Estudo de la reproducción de tres especies de tortugas de agua dulce en el estado de Tabasco. Tesis para Licenciado en Biología. Universidad Juárez Autonoma de Tabasco. Division Academica de Ciencias Biologicas.
Zenteno-Ruiz, C. E. 1999. Caracterización demografica de la tortuga pinta *(Trachemys scripta venusta)* y sus potencialidades de aprovechamiento en la Laguna Experimental del Campus Veracruz. Tesis de Maestro en Ciencias. en Agrosistemas Tropicales, Colegio de Postgraduados Instituto de Recursos Naturales, Campus Veracruz.
Zug, G. R. 1966. The penial morphology and the relationships of cryptodiran turtles. Occ. Pap. Mus. Zool. Univ. Michigan. 647:1–24.

Zug, G. R. 1971. Buoyancy, locomotion, morphology of the pelvic girdle and hindlimb, and systematics of cryptodiran turtles. Misc. Publ. Mus. Zool. Univ. Michigan 142:1–98.

Zug, G. R. 1986. *Sternotherus* Gray. Cat. Amer. Amph. Rept. 397:1–3.

Zug, G. R. 1990. Age determination of long-lived reptiles: some techniques for seaturtles. Ann. Sci. Natur. Zool. Paris. Ser. 13(11):219–222.

Zug, G. R. 1991. Age determination in turtles. Soc. Stud. Amphib. Rept. Herpetol. Circ. (20):1–28.

Zug, G. R., L. J. Vitt, and J. P. Caldwell. 2001. Herpetology. An introductory Biology of Amphibians and Reptiles. 2nd ed. Academic Press, San Diego.

Zweifel, R. G., and K. S. Norris. 1955. Contributions to the herpetology of Sonora, Mexico: descriptions of new subspecies of snakes (*Micruroides euryzxanthus* and *Lampropeltis getulus*) and miscellaneous collecting notes. Amer. Midland Natur. 54(1):230–249.

INDEX

Note: Turtle families are indicated with **boldface** type.